Springer Proceedings in Mathematics & Statistics

Volume 110

Springer Proceedings in Mathematics & Statistics

This book series features volumes composed of selected contributions from workshops and conferences in all areas of current research in mathematics and statistics, including operation research and optimization. In addition to an overall evaluation of the interest, scientific quality, and timeliness of each proposal at the hands of the publisher, individual contributions are all refereed to the high quality standards of leading journals in the field. Thus, this series provides the research community with well-edited, authoritative reports on developments in the most exciting areas of mathematical and statistical research today.

More information about this series at http://www.springer.com/series/10533

Peter K. Friz · Jim Gatheral
Archil Gulisashvili · Antoine Jacquier
Josef Teichmann
Editors

Large Deviations and Asymptotic Methods in Finance

Springer

Editors
Peter K. Friz
Institut für Mathematik
Technische Universität Berlin
Berlin
Germany

and

Weierstraß-Institut für Angewandte Analysis und Stochastik
Berlin
Germany

Jim Gatheral
Department of Mathematics
City University of New York Baruch College
New York, NY
USA

Archil Gulisashvili
Department of Mathematics
Ohio University
Athens, OH
USA

Antoine Jacquier
Department of Mathematics
Imperial College London
London
UK

Josef Teichmann
Department of Mathematics
ETH Zürich
Zürich
Switzerland

ISSN 2194-1009 ISSN 2194-1017 (electronic)
Springer Proceedings in Mathematics & Statistics
ISBN 978-3-319-11604-4 ISBN 978-3-319-11605-1 (eBook)
DOI 10.1007/978-3-319-11605-1

Library of Congress Control Number: 2015935733

Mathematics Subject Classification (2010): 91G80, 60H30, 60F10, 91G20

Springer Cham Heidelberg New York Dordrecht London

Printed on acid-free paper

Springer International Publishing AG Switzerland is part of Springer Science+Business Media (www.springer.com)

Contents

Introduction

In a sense, this book is a celebration of the Black-Scholes model. Widely criticized for its shortcomings, ever since the dramatic Long Term Capital Management meltdown in 1998, this 'first generation' model is still the first benchmark in financial modelling. May it be the Heston, Stein-Stein, Bergomi, local volatility, local-stochastic volatility, Lévy, uncertain volatility or fancy hybrid model: they are all perturbations of the Black-Scholes model, typically either making volatility stochastic or introducing jumps. If Black-Scholes assumes, say 20–40 % volatility, all the above extensions more or less agree with this order of magnitude for overall volatility, and indeed the around-the-money volatility smile is plainly a perturbation of the flat implied volatility corresponding to Black-Scholes.

The aforementioned extensions do not come with tractable option price formulae, since most diffusion processes do not admit closed-form transition densities. The second best situation is a closed-form Fourier Transform of the transition density, and many of the aforementioned extensions share this property, additionally to explaining some stylized facts from the dynamics of the implied volatility surface.

An alternative to the Fourier approach is given by asymptotic expansions of transition densities of stochastic processes, say in the short-time (or more generally small-noise) limit. Such investigations go back to S.R. Srinivasa Varadhan in the late 1960s and are intimately connected to his theory of large deviations: it is pretty unlikely for a particle starting at some position to diffuse to some other position if there is almost no time to do so (or if the driving noise is switched off). The beauty of large deviations is to explicitly identify a precise scale rough enough to be computable (or at least to be characterized in terms of some variational problem), and fine enough to capture the most important leading-order behaviour of the system. In the context of transition densities, or heat-kernels in PDE terminology, complete expansions have been derived in the 1970s and 1980s, with a bulk of geometric information hidden in the coefficients. The Russian school has also been fundamental in the development of (sample path) large deviations for stochastic processes, in particular through the works of Mark Freidlin and Alexander Wentzell in the 1970s. On a historical note, it is interesting to remember that large deviations theory was originally developed (in the finite-dimensional case) by Harald Cramér in the 1930s for actuarial mathematics.

A widely circulated preprint by Patrick Hagan et al. (following the famous SABR paper), first presented in 2001 by Andrew Lesniewski at the Courant Finance Seminar, intensified the connection between heat-kernels, geometry and finance. The resulting SABR formula has become industry standard in fixed income modelling (and presumably a long-time headache for quants tortured by risk management). The topic was further explored by a number of people including Marco Avellaneda, Christian Bayer, Gérard Ben Arous, Jérôme Busca, Jean-Dominique Deuschel, Martin Forde, Pierre Henry-Labordère, Elton Hsu, Peter Laurence, Cheng Ouyang and many others (including, unsurprisingly, all the editors of this volume).

Despite the undisputed mathematical depth of this development, the agenda has been largely initiated by people in or near the industry, a quick publication not always being their first priority. This, at least, is our only explanation for the fact that some key papers have remained preprints ever since, though widely circulating and used for years. We also note that the derivation of closed-form approximation formulae in various non-tractable models remains a constant topic in major academic and industry meetings alike, not to mention some specialist meetings (Vienna 2009, Berlin 2011, London 2013) organized by factions of the present group of editors in different constellations.

The present proceedings grew on this fertile ground. Contributions include some unpublished classics (in brushed-up versions), notably the aforementioned preprint by Patrick Hagan et al. as well as recent works touching the theme of large deviations and/or asymptotic expansions in mathematical finance.

The editors have known each other for a long time. The idea for this book project was born in July 2013, but the first step towards realization was overshadowed by a sad event: we are still shocked that our esteemed colleague and friend, whom we had invited to co-edit this volume, has never received his invitation: Peter Laurence passed away unexpectedly in August 2013.

Peter Laurence was born in New York, NY, on 27 March 1952. After undergraduate courses at the Wharton School of Finance and Commerce at the University of Pennsylvania, he obtained a Bachelor of Science in Mathematics and Philosophy degree in 1973. He also obtained a Master of Science degree (1977) and a Ph.D. degree (1981) from the University of Wisconsin Madison. From 1974–1991, Peter was a faculty member at the University of Wisconsin, the Courant Institute of Mathematical Sciences at New York University, Worcester Polytechnic Institute, Pennsylvania State University, and the University of Milano, Italy. From 1991 till his untimely death in 2013, Peter was a professor at Sapienza Università di Roma and a visiting scholar at the Courant Institute. Peter published more than 60 research papers, co-authored a book "Quantitative Methods of Derivative Securities: From Theory to Practice" with Marco Avellaneda, and was one of the editors of the volume "Quantitative Energy Finance: Modeling, Pricing, and Hedging in Energy and Commodity Markets". His long-term friend Marco Avellaneda remembers: *Peter had an infinite joie de vivre… This involved a lot of research in Math Physics, one of his passions. I enjoyed discussing Math Physics with him. We also began our interest in finance in the 90s and co-authored [our] book. He had a kind heart. May he rest in peace and live in us who are still here.* Or as Bruno Dupire articulates it: *He was a gentleman and will be missed.*

Each of us has stories to tell about Peter and his inexhaustible passion for mathematics and its impact on finance. Instead of trying to fit them in this introduction we rather let him speak through mathematics: some of Peter Laurence's final contributions to mathematical finance do appear in these proceedings, with the kind agreement of the respective co-authors.

We are indebted to all the reviewers who helped us achieving this work. It is also our pleasure to thank Magdalena Mueller-Laurence, as well as the Springer Proceedings team, without whom this book would never have appeared.

January 2015

Peter K. Friz
Jim Gatheral
Archil Gulisashvili
Antoine Jacquier
Josef Teichmann

Probability Distribution in the SABR Model of Stochastic Volatility

Patrick Hagan, Andrew Lesniewski and Diana Woodward

Abstract We study the SABR model of stochastic volatility (Wilmott Mag, 2003 [10]). This model is essentially an extension of the local volatility model (Risk 7(1):18–20 [4], Risk 7(2):32–39, 1994 [6]), in which a suitable volatility parameter is assumed to be stochastic. The SABR model admits a large variety of shapes of volatility smiles, and it performs remarkably well in the swaptions and caps/floors markets. We refine the results of (Wilmott Mag, 2003 [10]) by constructing an accurate and efficient asymptotic form of the probability distribution of forwards. Furthermore, we discuss the impact of boundary conditions at zero forward on the volatility smile. Our analysis is based on a WKB type expansion for the heat kernel of a perturbed Laplace-Beltrami operator on a suitable hyperbolic Riemannian manifold.

Keywords SABR · Heat kernel expansion · WKB expansion · Implied volatility · Asymptotic smile formula

1 Introduction

The SABR model [10] of stochastic volatility attempts to capture the dynamics of smile in the interest rate derivatives markets which are dominated by caps/floors and swaptions. It provides a parsimonious, accurate, intuitive, and easy to implement framework for pricing, position management, and relative value in those markets. The model describes the dynamics of a single forward (swap or LIBOR) rate with stochastic volatility. The dynamics of the model is characterized by a function $C(f)$ of the forward rate f which determines the general shape of the volatility skew, a

P. Hagan (✉) · D. Woodward
Gorilla Science, 7700 NE Palm Way, Boca Raton, Fl 33487, USA
e-mail: pathagan1954@yahoo.com

A. Lesniewski
Department of Mathematics, Baruch College, CUNY 1 Bernard Baruch Way, New York, NY 10010, USA
e-mail: andrew.lesniewski@baruch.cuny.edu

P.K. Friz et al. (eds.), *Large Deviations and Asymptotic Methods in Finance*, Springer Proceedings in Mathematics & Statistics 110,
DOI 10.1007/978-3-319-11605-1_1

parameter v which controls the level of the volatility of volatility, and a parameter ρ which governs the correlation between the changes in the underlying forward rate and its volatility. It is an extension of Black's model: choosing $v = 0$ and $C(f) = f$ reduces SABR to the lognormal Black model, while $v = 0$ and $C(f) = 1$ reduces it to the normal Black model.

The main reason why the SABR model has proven effective in the industrial setting is that, even though it is too complex to allow for a closed form solution, it has an accurate asymptotic solution. This solution, as well as its implications for pricing and risk management of interest derivatives, has been described in [10].

In this paper we further refine the results presented in [10]. Our developments go in two directions. Fist, we present a more systematic framework for generating an accurate, asymptotic form of the probability distribution in the SABR model. Secondly, we address the issue of low strikes, or the behavior of the model as the forward rate approaches zero.

Our way of thinking has been strongly influenced by the asymptotic techniques which go by the names of the *geometric optics* or the *WKB method*, and, most importantly, by the classical results of Varadhan [19, 20] (see also [13, 18] for more recent presentations and refinements). These techniques allow one to relate the short time asymptotics of the fundamental solution (or the *Green's function*) of Kolmogorov's equation to the differential geometry of the state space. From the probabilistic point of view, the Green's function represents the transition probability of the diffusion, and it thus carries all the information about the process.

Specifically, let $\mathcal{U}$ denote the state space of an n-dimensional diffusion process with no drift, and let $G_X(s, x)$, $x, X \in \mathcal{U}$, denote the Green's function. We also assume that the process is time homogeneous, meaning that the diffusion matrix is independent of s. Then, Varadhan's theorem states that

$$\lim_{s \to 0} s \log G_X(s, x) = -\frac{d(x, X)^2}{2}.$$

Here $d(x, X)$ is the geodesic distance on $\mathcal{U}$ with respect to a Riemannian metric which is determined by the coefficients of the Kolmogorov equation. This gives us the leading order behavior of the Green's function. To extract usable asymptotic information about the transition probability, more accurate analysis is necessary, but the choice of the Riemannian structure on $\mathcal{U}$ dictated by Varadhan's theorem turns out to be key. Indeed, that Riemannian geometry becomes an important book keeping tool in carrying out the calculations, rather than merely fancy language. Technically speaking, we are led to studying the asymptotic properties of the perturbed Laplace-Beltrami operator on a Riemannian manifold.

In order to explain the results of this paper we define a universal function $D(\zeta)$:

$$D(\zeta) = \log \frac{\sqrt{\zeta^2 - 2\rho\zeta + 1} + \zeta - \rho}{1 - \rho},$$

where ζ is the following combination of today's forward rate f, strike F, and a volatility parameter σ (which is calibrated so that the at the money options prices match the market prices):

$$\zeta = \frac{v}{\sigma} \int_F^f \frac{du}{C(u)}.$$

The function $D(\zeta)$ represents a certain metric whose precise meaning is explained in the body of the paper. The key object from the point of view of option pricing is the probability distribution of forwards $P_F(\tau, f)$. Our main result in this paper is the explicit asymptotic formula:

$$P_F(\tau, f) = \frac{\exp\left\{-D(\zeta)^2 / 2\tau v^2\right\}}{\sqrt{2\pi\tau}\, \sigma C(F) \left(\cosh D(\zeta) - \rho \sinh D(\zeta)\right)^{3/2}} \left(1 + \cdots\right).$$

In order not to burden the notation, we have written down the leading term only; the complete formula is stated in Sect. 5. To leading order, the probability distribution of forwards in the SABR model is Gaussian with the metric $D(\zeta)$ replacing the usual distance.

From this probability distribution, we can deduce explicit expressions for implied volatility. The normal volatility is given by:

$$\sigma_{\mathrm{n}} = \sigma C(F) \left(\cosh D(\zeta) - \rho \sinh D(\zeta)\right) \left(1 + \cdots\right).$$

Precise formulas, including the subleading terms and the impact of boundary conditions at zero forward, are stated in Sect. 5. To calculate the corresponding lognormal volatility one can use the results of [11].

We would like to mention that other stochastic volatility models have been extensively studied in the literature (notably among them the Heston model [12]). Useful presentations of these models are contained in [5, 17].

A comment on our style of exposition in this paper. We chose to present the arguments in an informal manner. In order to make the presentation self-contained, we present all the details of calculations, and do not rely on general theorems of differential geometry, stochastic calculus, or the theory of partial differential equations. And while we believe that all the results of this paper could be stated and proved rigorously as theorems, little would be gained and clarity might easily get lost in the course of doing so.

The paper is organized as follows. In Sect. 2 we review the model and formulate the basic partial differential equation, the backward Kolmogorov equation. We also introduce the Green's and discuss various boundary conditions at zero. Section 3 is devoted to the description of the differential geometry underlying the SABR model. We show that the stochastic dynamics defining the model can be viewed as a perturbation of the Brownian motion on a deformed Poincare plane. The elliptic operator in the Kolmogorov equation turns out to be a perturbed Laplace-Beltrami operator. This differential geometric setup is key to our asymptotic analysis of the model which is carried through in Sect. 4. In Sect. 5 we derive the explicit formulas for the probability distribution and implied volatility which we have discussed above. In Appendix A we review the derivation of the fundamental solution of the heat equation on the Poincare plane. This solution is the starting point of our perturbation expansion. Finally, Appendix B contains some useful asymptotic expansions.

2 SABR Model

In this section we describe the SABR model of stochastic volatility [10]. It is a two factor model with the dynamics given by a system of two stochastic differential equations. The state variables of the model can be thought of as the forward price of an asset, and a volatility parameter. In order to derive explicit expressions for the associated probability distribution and the implied volatility, we study the Green's function of the backward Kolmogorov operator.

2.1 Underlying Process

We consider a European option on a forward asset expiring T years from today. The forward asset that we have in mind can be for instance a forward LIBOR rate, a forward swap rate, or the forward yield on a bond. The dynamics of the forward in the SABR model is given by[1]:

$$\begin{aligned} dF_t &= \Sigma_t C\left(F_t\right) dW_t, \\ d\Sigma_t &= v\Sigma_t dZ_t. \end{aligned} \tag{1}$$

Here F_t is the forward rate process, and W_t and Z_t are Brownian motions with

$$\mathsf{E}\left[dW_t dZ_t\right] = \rho dt, \tag{2}$$

where the correlation ρ is assumed constant. We supplement the dynamics (1) with the initial condition

$$\begin{aligned} F_0 &= f, \\ \Sigma_0 &= \sigma. \end{aligned} \tag{3}$$

[1]Note that our notation departs somewhat from the notation used in [10]: we use Σ_t instead of α_t and v_t instead of ν_t. The name SABR is an acronym for "Stochastic Alpha Beta Rho" which was the name of the model originally used at Paribas.

Note that we assume that a suitable numeraire has been chosen so that F_t is a martingale. The process Σ_t is the stochastic component of the volatility of F_t, and v is the volatility of Σ_t (the "volvol") which is also assumed to be constant.

The function $C(x)$ is defined for $x > 0$, and is assumed to be positive, smooth, and integrable around 0;

$$\int_0^K \frac{du}{C(u)} < \infty, \text{ for all } K > 0. \tag{4}$$

Two examples of C, which are particularly popular among financial practitioners, are functions of the form:

$$C(x) = x^\beta, \text{ where } 0 \leq \beta < 1 \tag{5}$$

(stochastic CEV model), or

$$C(x) = x + a, \text{ where } a > 0 \tag{6}$$

(stochastic shifted lognormal model).

Our analysis uses an asymptotic expansion in the parameter $v^2 T$, and we thus require that $v^2 T$ be small. In practice, this is an excellent assumption for medium and longer dated options. Typical for shorter dated options are significant, discontinuous movements in implied volatility. The SABR model should presumably be extended to include such jump behavior of short dated options.

The process Σ_t is purely lognormal and thus $\Sigma_t > 0$ almost surely. Since, depending on the choice of $C(x)$, F_t can reach zero with non-zero probability, we should take into account the boundary behavior of the process (1), as F_t approaches 0. This can easily be done in the case of zero correlation between W_t and Z_t, $\rho = 0$. We extend the function $C(x)$ to all values of x by setting

$$C(-x) = C(x), \text{ for } x < 0. \tag{7}$$

The so extended $C(x)$ is an even function, $C(-x) = C(x)$, for all values of x, and thus the process (1) is invariant under the reflection $F_t \to -F_t$. The state space of the extended process is thus the upper half plane. Later on in this paper we shall discuss the Dirichlet and Neumann boundary conditions for the SABR model.

A special case of (1) which will play an important role in our analysis is the case of $C(x) = 1$, and $\rho = 0$. In this situation, the basic equations of motion have a particularly simple form:

$$\begin{aligned} dF_t &= \Sigma_t dW_t, \\ d\Sigma_t &= v\Sigma_t dZ_t, \end{aligned} \tag{8}$$

with $\mathsf{E}[dW_t dZ_t] = 0$. We shall refer to this model as the *normal SABR model*.

Local volatility [4, 6], is defined as the conditional expectation value

$$\sigma_K(T, f, \sigma)^2\, dT = \mathbb{E}\left[(dF_t)^2 \mid F(0) = f,\ F_t = K,\ \Sigma(0) = \sigma\right], \tag{9}$$

or, explicitly,

$$\sigma_K(T, f, \sigma)^2 = C(K)^2\, \mathbb{E}\left[(\Sigma_t)^2 \mid F(0) = f,\ F_t = K,\ \Sigma(0) = \sigma\right]. \tag{10}$$

Our analysis in the following sections enables us, in particular, to derive an explicit expression for σ_K.

2.2 Green's Function

Green's functions arise in finance as the prices of Arrow-Debreu securities. Equations (1)–(3) correspond to the Arrow-Debreu security whose payoff at time T is given by Dirac's delta function $\delta(F_T - F, \sigma_T - \Sigma)$. The time $t < T$ price $G = G_{T,F,\Sigma}(t, f, \sigma)$ of this security is the solution to the following parabolic partial differential equation:

$$\frac{\partial G}{\partial t} + \frac{1}{2}\sigma^2\left(C(f)^2\frac{\partial^2 G}{\partial f^2} + 2v\rho C(f)\frac{\partial^2 G}{\partial f\partial\sigma} + v^2\frac{\partial^2 G}{\partial\sigma^2}\right) = 0, \tag{11}$$

with the terminal condition:

$$G_{T,F,\Sigma}(t, f, \sigma) = \delta(f - F, \sigma - \Sigma), \text{ at } t = T. \tag{12}$$

This equation should also be supplemented by a boundary condition at infinity such that G is financially meaningful. Since the payoff takes place only if the forward has a predetermined value in a finite amount of time, the value of the Arrow-Debreu security has to tend to zero as F and Σ become large:

$$G_{T,F,\Sigma}(t, f, \sigma) \to 0, \qquad \text{as } F,\ \Sigma \to \infty. \tag{13}$$

Thus $G_{T,F,\Sigma}(t, f, \sigma)$ is a Green's function for (11). Once we have constructed it, we can price any European option. For example, the price $C_{T,K}(t, f, \sigma)$ of a European call option struck at K and expiring at time T can be written in terms of $G_{T,F,\Sigma}(t, f, \sigma)$ as

$$C_{T,K}(t, f, \sigma) = \int (F - K)^+ G_{T,F,\Sigma}(t, f, \sigma)\, dF d\Sigma, \tag{14}$$

where, as usual, $(F - K)^+ = \max(F - K, 0)$, and where the integration extends over the upper half plane $\{(F, \Sigma) \in \mathbb{R}^2 : \Sigma > 0\}$.

Note that the process (1) is time homogeneous, and thus $G_{T,F,\Sigma}(t, f, \sigma)$ is a function of the time to expiry $\tau = T - t$ only. Denoting

$$G_{F,\Sigma}(\tau, f, \sigma) \equiv G_{T,F,\Sigma}(t, f, \sigma),$$

and

$$C_K(\tau, f, \sigma) \equiv C_{T,K}(t, f, \sigma),$$

we can reformulate (11)–(12) as the initial value problem:

$$\frac{\partial G}{\partial \tau} = \frac{1}{2}\sigma^2\left(C(f)^2 \frac{\partial^2 G}{\partial f^2} + 2v\rho C(f) \frac{\partial^2 G}{\partial f \partial \sigma} + v^2 \frac{\partial^2 G}{\partial \sigma^2}\right), \tag{15}$$

and

$$G_{F,\Sigma}(\tau, f, \sigma) = \delta(f - F, \sigma - \Sigma), \text{ at } \tau = 0. \tag{16}$$

Introducing the marginal probability distribution

$$P_F(\tau, f, \sigma) = \int_0^\infty G_{F,\Sigma}(\tau, f, \sigma)\, d\Sigma, \tag{17}$$

we can express the call price (14) as

$$C_K(\tau, f, \sigma) = \int_{-\infty}^\infty (F - K)^+ P_F(\tau, f, \sigma)\, dF. \tag{18}$$

This formula has familiar structure, and one of our main goals will be to derive a useful expression for $P_F(\tau, f)$.

It is also easy to express the local volatility in terms of the Green's function. Indeed,

$$\sigma_K(\tau, f, \sigma)^2 = \frac{C(K)^2 \int_0^\infty \Sigma^2 G_{K,\Sigma}(\tau, f, \sigma)\, d\Sigma}{\int_0^\infty G_{K,\Sigma}(\tau, f, \sigma)\, d\Sigma}, \tag{19}$$

or

$$\sigma_K(\tau, f, \sigma) = C(K)\sqrt{\frac{M_K^2(\tau, f, \sigma)}{P_K(\tau, f, \sigma)}}, \tag{20}$$

where

$$M_K^2(\tau, f, \sigma) = \int_0^\infty \Sigma^2 G_{K,\Sigma}(\tau, f, \sigma)\, d\Sigma \tag{21}$$

is the conditional second moment.

We will solve (15)–(17) by means of asymptotic techniques. In order to set up the expansion, it is convenient to introduce the following variables:

$$s = \frac{\tau}{T}, \; x = f, \; X = F, \; y = \frac{\sigma}{v}, \; Y = \frac{\Sigma}{v},$$

and the rescaled Green's function:

$$K_{X,Y}(s, x, y) = vTG_{X,vY}(Ts, x, vy).$$

In terms of these variables, the initial value problem (15) and (16) can be recast as:

$$\begin{aligned} \frac{\partial K}{\partial s} &= \frac{1}{2}\,\varepsilon y^2 \left(C(x)^2 \frac{\partial^2 K}{\partial x^2} + 2\rho C(x) \frac{\partial^2 K}{\partial x \partial y} + \frac{\partial^2 K}{\partial y^2} \right), \\ K(0, x, y) &= \delta(x - X, \; y - Y), \end{aligned} \tag{22}$$

where $K = K_{X,Y}$, and

$$\varepsilon = v^2 T. \tag{23}$$

It will be assumed that ε is small and it will serve as the parameter of our expansion. The heuristic picture behind this idea is that the volatility varies slower than the forward, and the rates of variability of f and σ/v are similar. The time T defines the time scale of the problem, and thus s is a natural dimensionless time variable. Expressed in terms of the new variables, our problem has a natural differential geometric content which is key to its solution.

Finally, let us write down the equations above for the normal SABR model:

$$\begin{aligned} \frac{\partial K}{\partial s} &= \frac{1}{2}\,\varepsilon y^2 \left(\frac{\partial^2 K}{\partial x^2} + \frac{\partial^2 K}{\partial y^2} \right), \\ K(0, x, y) &= \delta(x - X, \; y - Y). \end{aligned} \tag{24}$$

We will show later that this initial value problem has a closed form solution.

2.3 Boundary Conditions at Zero Forward

The problem as we have formulated it so far is not complete. Since the value of the forward rate should be positive,[2] we have to specify a boundary condition for the Green's function at $x = 0$. Three commonly used boundary conditions are [9]:

---

[2]Recent history shows that this is not always necessarily the case, but we regard such occurances as anomalous.

- *Dirichlet (or absorbing) boundary condition.* We assume that the Green's function, denoted by $K^D_{X,Y}(s, x, y)$, vanishes at $x = 0$,

$$K^D_{X,Y}(s, 0, y) = 0. \tag{25}$$

- *Neumann (or reflecting) boundary condition.* We assume that the derivative of the Green's function at $x = 0$, normal to the boundary (and pointing outward), vanishes. Let $K^N_{X,Y}(s, x, y)$ denote this Green's function; then

$$\frac{\partial}{\partial x} K^N_{X,Y}(s, 0, y) = 0. \tag{26}$$

- *Robin (or mixed) boundary condition.* The Green's function, which we shall denote by $K^R_{X,Y}(s, x, y)$, satisfies the following condition. Given $\eta > 0$,

$$\Big(-\frac{\partial}{\partial x} + \eta\Big) K^R_{X,Y}(s, 0, y) = 0. \tag{27}$$

From the financial point of view, the relevant boundary conditions are the Dirichlet and Neumann conditions. It is well known that the Green's functions corresponding to these different boundary conditions obey the following *conditioning inequalities*:

$$K^D \leq K \leq K^N. \tag{28}$$

Since the Dirichlet boundary condition corresponds to the stochastic process being killed at the boundary, the total mass of the Green's function is less than one:

$$\int K^D_{X,Y}(s, x, y)\, dx\, dy < 1. \tag{29}$$

The remaining probability is a Dirac's delta function at $x = 0$. On the other hand, for the free and Neumann boundary conditions,

$$\int K_{X,Y}(s, x, y)\, dx\, dy = \int K^N_{X,Y}(s, x, y)\, dx\, dy = 1, \tag{30}$$

and so they are *bona fide* probability distributions.

Our method allows for deriving explicit expressions for the Green's functions in the case of zero correlation. In this case, the differential operator in (22) is invariant under a $\mathbb{Z}_2$ group action given by the reflection $x \to -x$ of the upper half plane. This allows us to construct the desired Green's functions by means of the method of images. Namely, let $K_{X,Y}(s, x, y)$ denote now the solution to (22) with $C(x)$

extended to the entire upper half plane, as explained in Sect. 2.1.[3] Then, one verifies readily that

$$K^D_{X,Y}(s,x,y) = K_{X,Y}(s,x,y) - K_{X,Y}(s,-x,y), \tag{31}$$

and

$$K^N_{X,Y}(s,x,y) = K_{X,Y}(s,x,y) + K_{X,Y}(s,-x,y) \tag{32}$$

are the solutions to the Dirichlet and Neumann problem, respectively.

2.4 Solving the Initial Value Problem

It is easy to write down a formal solution to the initial value problem (22). Let L denote the partial differential operator

$$L = \frac{1}{2}\, y^2 \left(C(x)^2 \frac{\partial^2}{\partial x^2} + 2\rho C(x) \frac{\partial^2}{\partial x \partial y} + \frac{\partial^2}{\partial y^2} \right) \tag{33}$$

supplemented by a suitable boundary condition at $x = 0$. Consider the one-parameter semigroup of operators

$$U(s) = \exp(s\varepsilon L). \tag{34}$$

Then U solves the following initial value problem:

$$\begin{aligned} \frac{\partial U}{\partial s} &= \varepsilon L U, \\ U(0) &= I, \end{aligned}$$

and thus the Green's function $K_{X,Y}(s,x,y)$ is the integral kernel of $U(s)$:

$$K_{X,Y}(s,x,y) = U(s)(x,y;X,Y). \tag{35}$$

In order to solve the problem (22) it is thus sufficient to construct the semigroup $U(s)$ and find its integral kernel. Keeping in mind that our goal is to find an explicit formula for $K_{X,Y}(s,x,y)$, the strategy will be to represent L as the sum

$$L = L_0 + V, \tag{36}$$

where L_0 is a second order differential operator with the property that

$$U_0(s) = \exp(s\varepsilon L_0) \tag{37}$$

[3] This solution ignores any boundary condition at $x = 0$ and is sometimes referred to as the Green's function with a *free boundary condition*.

can be represented in closed form. Specifically, we will proceed in several steps. We start with the normal SABR model defined in Sect. 2.1, and notice that the corresponding operator L is a well known object, namely the generator of the Brownian motion on the upper half-plane. The integral kernel of the semigroup $U(s)$ generated by this operator can be represented as an explicit integral over the real axis. Next we observe that the general SABR model can naturally be mapped on the normal SABR model by means of a suitable diffeomorphism ϕ. We find that, under this mapping, the operator L is the sum of two parts: (i) the pullback of the generator of the Brownian motion on the upper half-plane, denoted by L_0, and (ii) a perturbation V. The kernel of the semigroup generated by L_0 has an explicit integral representation. The operator V turns out to be a differential operator of first order, and we will treat it as a small perturbation of the operator L_0.

The semigroup $U(s)$ can now be expressed in terms of $U_0(s)$ and V as

$$U(s) = Q(s)\, U_0(s)\,. \tag{38}$$

Here, the operator $Q(s)$ is given by the well known regular perturbation expansion:

$$Q(s) = I + \sum_{1\leq n<\infty} \int_{0\leq s_1\leq \ldots s_n\leq s\varepsilon} e^{s_1 \mathrm{ad}_{L_0}}(V)\ldots e^{s_n \mathrm{ad}_{L_0}}(V)\, ds_1 \ldots ds_n, \tag{39}$$

where ad_{L_0} is the commutator with L_0:

$$\mathrm{ad}_{L_0}(V) = L_0 V - V L_0. \tag{40}$$

We will use the first few terms in the expansion above in order to construct an accurate approximation to the Green's function $K_{X,Y}(s, x, y)$:

$$Q(s) = I + s\varepsilon V + \frac{1}{2}(s\varepsilon)^2 \left(\mathrm{ad}_{L_0}(V) + V^2\right) + O\left((s\varepsilon)^3\right). \tag{41}$$

We shall disregard the convergence issues associated with this series, and use it solely as a tool to generate an asymptotic expansion.

3 Stochastic Geometry of the State Space

In solving our model we find that the normal SABR model represents Brownian motion on the Poincare plane. Generally, when $\rho \neq 0$, or $C(x) \neq 1$, the model amounts to Brownian motion on a two dimensional manifold, the *SABR plane*, perturbed by a drift term. In this section we summarize a number of basic facts about the differential geometry of the state space of the SABR model. The fundamental geometric structure is that of the Poincare plane. We will show that the state space of the SABR model can be viewed as a suitable deformation of the Poincare geometry.

3.1 SABR Plane

We begin by reviewing the Poincare geometry of the upper half plane which will serve as the standard state space of our model. For a full (and very readable) account of the theory the reader is referred to e.g. [1].

The *Poincare plane* (also known as the hyperbolic or Lobachevski plane) is the upper half plane $\mathbb{H}^2 = \{(x, y) : \ y > 0\}$ equipped with the Poincare line element

$$ds^2 = \frac{dx^2 + dy^2}{y^2}. \tag{42}$$

This line element comes from the metric tensor given by

$$h = \frac{1}{y^2} \begin{pmatrix} 1 & 0 \\ 0 & 1 \end{pmatrix}. \tag{43}$$

The Poincare plane admits a large group of symmetries. We introduce complex coordinates on $\mathbb{H}^2$, $z = x + iy$ (the defining condition then reads Imz > 0), and consider a Moebius transformation

$$z' = \frac{az + b}{cz + d}, \tag{44}$$

where a, b, c, d are real numbers with $ad - bc = 1$. We verify easily the following two facts.

- Transformation (44) is a biholomorphic map of $\mathbb{H}^2$ onto itself.
- The Poincare metric is invariant under (44).

As a consequence, the Lie group

$$SL\left(2, \mathbb{R}\right) = \left\{ \begin{pmatrix} a & b \\ c & d \end{pmatrix} : \ a, b, c, d \in \mathbb{R}, \ ad - bc = 1 \right\} \tag{45}$$

acts holomorphically and isometrically on $\mathbb{H}^2$. This symmetry group plays very much the same role in the hyperbolic geometry as the Euclidean group in the usual Euclidean geometry of the plane $\mathbb{R}^2$.

In order to study the SABR model with the Dirichlet or Neumann boundary conditions at zero forward, we define the following reflection $\theta : \mathbb{H}^2 \to \mathbb{H}^2$:

$$\theta\left(x, y\right) = \left(-x, y\right) \tag{46}$$

(clearly, this is a reflection with respect to the y-axis). The key fact about θ is that it is an involution, i.e.

$$\theta \circ \theta\left(z\right) = z. \tag{47}$$

One can also write θ as $\theta(z) = -\bar{z}$, which shows that it is an anti-holomorphic map of $\mathbb{H}^2$ into itself. It is easy to find the set of fixed points of θ, namely the points on the Poincare plane which are left invariant by θ:

$$\theta(x, y) = (x, y) \Leftrightarrow x = 0. \tag{48}$$

i.e. it is the positive y-axis.

Let $d(z, Z)$ denote the geodesic distance between two points $z, Z \in \mathbb{H}^2$, $z = x + iy$, $Z = X + iY$, i.e. the length of the shortest path connecting z and Z. There is an explicit expression for $d(z, Z)$:

$$\cosh d(z, Z) = 1 + \frac{|z - Z|^2}{2yY}, \tag{49}$$

where $|z - Z|$ denotes the Euclidean distance between z and Z. In particular, if $x = X$, then $d(z, Z) = |\log(y/Y)|$. We also note that the reflection θ is an isometry with respect to this metric, $d(\theta(z), \theta(Z)) = d(z, Z)$.

We also note that since $\det(h) = y^{-4}$, the invariant volume element on $\mathbb{H}^2$ is given by

$$\begin{aligned} d\mu_h(z) &= \sqrt{\det(h)}\, dx\, dy \\ &= \frac{dx\, dy}{y^2}. \end{aligned} \tag{50}$$

The state space associated with the general SABR model has a somewhat more complicated geometry. Let $\mathbb{S}^2$ denote the upper half plane $\{(x, y) : y > 0\}$, equipped with the following metric g:

$$g = \frac{1}{(1 - \rho^2) y^2 C(x)^2} \begin{pmatrix} 1 & -\rho C(x) \\ -\rho C(x) & C(x)^2 \end{pmatrix}. \tag{51}$$

This metric is a generalization of the Poincare metric: the case of $\rho = 0$ and $C(x) = 1$ reduces to the Poincare metric. In fact, the metric g is the pullback of the Poincare metric under a suitable diffeomorphism. To see this, we define a map $\phi : \mathbb{S}^2 \to \mathbb{H}^2$ by

$$\phi(z) = \left(\frac{1}{\sqrt{1 - \rho^2}} \left(\int_0^x \frac{du}{C(u)} - \rho y \right), \ y \right), \tag{52}$$

where $z = (x, y)$. The Jacobian $\nabla\phi$ of ϕ is

$$\nabla\phi(z) = \begin{pmatrix} \frac{1}{\sqrt{1-\rho^2}\, C(x)} & -\frac{\rho}{\sqrt{1-\rho^2}} \\ 0 & 1 \end{pmatrix}, \tag{53}$$

and so $\phi^* h = g$, where ϕ^* denotes the pullback of ϕ. The manifold $\mathbb{S}^2$ is thus isometrically diffeomorphic with the Poincare plane. A consequence of this fact is that we have an explicit formula for the geodesic distance $\delta(z, Z)$ on $\mathbb{S}^2$:

$$\begin{aligned}\cosh \delta(z, Z) &= \cosh d(\phi(z), \phi(Z)) \\ &= 1 + \frac{\left(\int_X^x \frac{du}{C(u)}\right)^2 - 2\rho(y - Y)\int_X^x \frac{du}{C(u)} + (y - Y)^2}{2(1 - \rho^2)\, yY},\end{aligned} \tag{54}$$

where $z = (x, y)$ and $Z = (X, Y)$ are two points on $\mathbb{S}^2$. Since $\det(g) = y^{-4} C(x)^{-2}$, the invariant volume element on $\mathbb{S}^2$ is given by

$$\begin{aligned}d\mu_g(z) &= \sqrt{\det(g)}\, dx\, dy \\ &= \frac{dx\, dy}{C(x) y^2}.\end{aligned} \tag{55}$$

In the case of $\rho = 0$, the manifold $\mathbb{S}^2$ carries an isometric reflection θ which commutes with (52):

$$\theta \circ \phi(z) = \phi \circ \theta(z), \tag{56}$$

i.e. θ is inherited from the corresponding reflection θ of the Poincare plane. Explicitly, $\theta(x, y) = (-x, y)$. Strictly speaking, this holds only holds if $x \neq 0$, as the metric (51) explodes at the boundary $x = 0$.

3.2 Brownian Motion on the SABR Plane

It is no coincidence that the SABR model leads to the Poincare geometry. Indeed, the dynamics of the normal SABR model is given by the Brownian motion on the Poincare plane. In this section we shall establish this relationship, and use it in Sect. 3.3 in order to find an explicit representation of the integral kernel of (37).

Recall [13] that the Brownian motion on the Poincare plane is described by the following system of stochastic differential equations:

$$\begin{aligned}dX_t &= Y_t dW_t, \\ dY_t &= Y_t dZ_t,\end{aligned} \tag{57}$$

with the two Wiener processes W_t and Z_t satisfying

$$\mathsf{E}[dW_t dZ_t] = 0. \tag{58}$$

Comparing this with the special case of the normal SABR model (8), we see that (8) reduces to (57) once we have made the following identifications:

$$\begin{aligned} X_t &= F_{v^2 t}, \\ Y_t &= \frac{1}{v}\,\Sigma_{v^2 t}, \end{aligned} \tag{59}$$

and used the scaling properties of a Wiener process:

$$\begin{aligned} dW_{v^2 t} &= v\,dW_t, \\ dZ_{v^2 t} &= v\,dZ_t. \end{aligned}$$

Note that the system (57) can easily be solved in closed form: its solution is given by

$$\begin{aligned} X_t &= X_0 + Y_0 \int_0^t \exp\left(Z(s) - \frac{s^2}{2}\right) dW(s), \\ Y_t &= Y_0 \exp\left(Z_t - \frac{t^2}{2}\right). \end{aligned} \tag{60}$$

Let us now compare the SABR dynamics with that of the diffusion on the SABR plane. In order to find the dynamics of Brownian motion on the SABR plane we use the fact that there is a mapping (namely, (52)) of $\mathbb{S}^2$ into $\mathbb{H}^2$. Using this mapping and Ito's lemma yields the following system

$$\begin{aligned} dX_t &= \frac{1}{2}\,Y_t^2 C(X_t)\,C'(X_t)\,dt + Y_t C(X_t)\,dW_t, \\ dY_t &= Y_t dZ_t, \end{aligned} \tag{61}$$

with the two Wiener processes W_t and Z_t satisfying

$$\mathsf{E}[dW_t dZ_t] = \rho dt. \tag{62}$$

Note that this is not exactly the SABR model dynamics. Indeed, one can regard the SABR model as the perturbation of the Brownian motion on the SABR plane by the drift term $-\frac{1}{2}\,Y_t^2 C(X_t)\,C'(X_t)\,dt$.

As in the case of the Poincare plane, it is possible to represent the solution to the system (61) explicitly:

$$\begin{aligned} \int_{X_0}^{X_t} \frac{du}{C(u)} &= Y_0 \int_0^t \exp\left(Z(s) - \frac{s^2}{2}\right) dW(s), \\ Y_t &= Y_0 \exp\left(Z_t - \frac{t^2}{2}\right). \end{aligned} \tag{63}$$

Parenthetically, we note that, within Stratonovich's calculus, (61) can be written as

$$\begin{aligned} dX_t &= Y_t C\,(X_t) \circ dW_t, \\ dY_t &= Y_t \circ dZ_t. \end{aligned}$$

Therefore, the stochastic differential equations of the SABR model, if interpreted according to Stratonovich, describe the dynamics of Brownian motion on the SABR plane.

3.3 Laplace-Beltrami Operator on the SABR Plane

It will be convenient to use invariant notation. Let $z^1 = x$, $z^2 = y$, and let $\partial_\mu = \partial/\partial z^\mu$, $\mu = 1, 2$, denote the corresponding partial derivatives. We denote the components of g^{-1} by $g^{\mu\nu}$, and use g^{-1} and g to raise and lower the indices: $z_\mu = g_{\mu\nu} z^\nu$, $\partial^\mu = g^{\mu\nu}\partial_\nu = \partial/\partial z_\mu$, where we sum over the repeated indices. Explicitly,

$$\begin{aligned} \partial^1 &= y^2 \left(C(x)^2 \partial_1 + \rho C(x) \partial_2 \right), \\ \partial^2 &= y^2 \left(\rho C(x) \partial_1 + \partial_2 \right). \end{aligned}$$

Consequently, the initial value problem (22) can be written in the following geometric form:

$$\begin{aligned} \frac{\partial}{\partial s} K_Z\,(s, z) &= \frac{1}{2}\,\varepsilon\,\partial^\mu \partial_\mu K_Z\,(s, z), \\ K_Z\,(0, z) &= \delta\,(z - Z)\,, \end{aligned} \tag{64}$$

where $\delta(z - Z) = \delta(x - X, y - Y)$ denotes the two-dimensional Dirac's delta function.

Recall that the Laplace-Beltrami operator Δ_g on a Riemannian manifold $\mathcal{M}$ with metric tensor g is defined by

$$\Delta_g f = \frac{1}{\sqrt{\det g}} \frac{\partial}{\partial x^\mu} \left(\sqrt{\det g}\; g^{\mu\nu} \frac{\partial f}{\partial x^\nu} \right), \tag{65}$$

where f is a smooth function on $\mathcal{M}$. It is a natural generalization of the familiar Laplace operator to spaces with non-Euclidean geometry. Its importance for probability theory comes from the fact that it serves as the infinitesimal generator of Brownian motion on such spaces (see e.g. [7, 8, 13]).

In the case of the Poincare plane, the Laplace-Beltrami operator has the form:

$$\Delta_h = y^2 \left(\frac{\partial^2}{\partial x^2} + \frac{\partial^2}{\partial y^2} \right). \tag{66}$$

As anticipated by our discussion in Sect. 3.2, this operator is closely related to the operator L in the normal SABR model. In fact, in this case,

$$L = \frac{1}{2}\, \Delta_h, \tag{67}$$

and thus the problem (24) turns out to be the initial value problem the heat equation on $\mathbb{H}^2$:

$$\begin{aligned} \frac{\partial K_Z}{\partial s} &= \frac{1}{2}\, \varepsilon \Delta_h\, K_Z, \\ K_Z\,(0, z) &= \delta\,(z - Z)\,. \end{aligned} \tag{68}$$

The key fact is that the Green's function for this equation can be represented in closed form,

$$K_Z^h\,(s, z) = \frac{e^{-s\varepsilon/8}\sqrt{2}}{(2\pi s\varepsilon)^{3/2}\, Y^2} \int_{d(z,Z)}^{\infty} \frac{u e^{-u^2/2s\varepsilon}}{\sqrt{\cosh u - \cosh d\,(z, Z)}}\, du. \tag{69}$$

This formula was originally derived by McKean [16] (see also [13] and references therein). We have added the superscript h to indicate that this Green's function is associated with the Poincare metric. In Appendix A we outline an elementary derivation of this fact.

Let us now extend the discussion above to the general case. We note first that, except for the case of $C(x) = 1$, the operator $\partial^\mu \partial_\mu$ does not coincide with the Laplace-Beltrami operator Δ_g on $\mathbb{S}^2$ associated with the metric (51). It is, however, easy to verify that

$$\begin{aligned} \partial^\mu \partial_\mu f &= \Delta_g f - \frac{1}{\sqrt{\det g}}\, \frac{\partial}{\partial x^\nu} \left(\sqrt{\det g}\; g^{\mu\nu} \right) \frac{\partial f}{\partial x^\mu} \\ &= \Delta_g f - \frac{1}{\sqrt{1 - \rho^2}}\; y^2 C C'\, \frac{\partial f}{\partial x}, \end{aligned}$$

and thus

$$\begin{aligned} L &= \frac{1}{2}\, \Delta_g - \frac{1}{2\sqrt{1 - \rho^2}}\; y^2 C C' \frac{\partial}{\partial x} \\ &= L_0 + V, \end{aligned}$$

where L_0 is essentially the Laplace-Beltrami operator:

$$L_0 = \frac{1}{2}\Delta_g\,, \tag{70}$$

and $V(x)$ is lower order:

$$V = -\frac{1}{2\sqrt{1-\rho^2}}\, y^2 C(x) C'(x)\, \frac{\partial}{\partial x}. \tag{71}$$

Let us first focus on the Laplace-Beltrami operator Δ_g. The key property of the Laplace-Beltrami operator is that it commutes with isometries of Riemannian manifolds. In particular, this implies that

$$\phi \circ \Delta_g = \Delta_h \circ \phi, \tag{72}$$

and, thus the Laplace-Beltrami operator Δ_g is the pullback of Δ_h under ϕ. As a consequence, the heat equation

$$\frac{\partial K}{\partial s} = \frac{1}{2}\,\varepsilon\,\Delta_g K$$

on $\mathbb{S}^2$ can be solved in closed form! The Green's function $K_Z^g(s,z)$ of this equation is related to (69) by

$$K_Z^g(s,z) = \det(\nabla\phi(Z))\, K_{\phi(Z)}^h(s,\phi(z))\,. \tag{73}$$

Explicitly,

$$K_Z^g(s,z) = \frac{e^{-s\varepsilon/8}\sqrt{2}}{(2\pi s\varepsilon)^{3/2}\sqrt{1-\rho^2}\,Y^2 C(X)} \int_\delta^\infty \frac{u e^{-u^2/2s\varepsilon}}{\sqrt{\cosh u - \cosh\delta}}\, du\,, \tag{74}$$

where $\delta = \delta(z, Z)$ is the geodesic distance (54) on $\mathbb{S}^2$. This is the explicit representation of the integral kernel of the operator $U_0(s)$.

4 Asymptotic Expansion

In principle, we have now completed our task of solving the initial value problem (24). Indeed, its solution is given by

$$K_Z(s,z) = Q(s)\, K_Z^g(s,z)\,, \tag{75}$$

where $Q(s)$ is the perturbation expansion given by (39). In order to produce clear results that can readily be used in practice we perform now a perturbation expansion on the expression above. Our method allows one to calculate the Green's function of the model to the desired order of accuracy.

Let us start with the Green's function $K_Z^h(s,z)$ which is defined on the Poincare plane. In Appendix B we derived an asymptotic expansion (117) for the heat kernel on the Poincare plane. After rescaling as in (106), we arrive at

$$K_Z^h(s,z) = \frac{1}{2\pi\lambda Y^2} \exp\left(-\frac{d^2}{2\lambda}\right) \times \sqrt{\frac{d}{\sinh d}} \left(1 - \frac{1}{8}\left(\frac{d\coth d - 1}{d^2} + 1\right)\lambda + O\left(\lambda^2\right)\right),$$

where we have introduced a new variable,

$$\lambda = s\varepsilon. \tag{76}$$

We can now extend the expression to the general Green's function $K_Z^g(s,z)$. Using (73) or (74) we find that $K_Z^g(s,z)$ has the following asymptotic expansion:

$$K_Z^g(s,z) = \frac{1}{2\pi\lambda\sqrt{1-\rho^2}Y^2C(X)} \exp\left(-\frac{\delta^2}{2\lambda}\right) \times \sqrt{\frac{\delta}{\sinh\delta}} \left(1 - \frac{1}{8}\left(\frac{\delta\coth\delta - 1}{\delta^2} + 1\right)\lambda + O\left(\lambda^2\right)\right).$$

To complete the calculation in the case of general $C(x)$ we need to take into account the contribution to the Green's function coming from perturbation V defined in (71). Let us define the function:

$$\begin{aligned} q(z,Z) &= \sinh\delta(z,Z)\, V\delta(z,Z) \\ &= -\frac{yC'(x)}{2\left(1-\rho^2\right)^{3/2}Y}\left(\int_X^x \frac{du}{C(u)} - \rho(y-Y)\right). \end{aligned} \tag{77}$$

From (117) and (118),

$$\begin{aligned} K_Z(s,z) &= (I + \lambda V)\, K_Z^g(s,z) \\ &= \frac{1}{\sqrt{1-\rho^2}\, Y^2 C(X)}\left(K_Z(s,z) + \lambda\, \frac{q}{\sinh\delta}\frac{\partial}{\partial\delta} K_Z(s,z)\right), \end{aligned} \tag{78}$$

which yields the following asymptotic formula for the Green's function:

$$K_Z(s,z) = \frac{1}{2\pi\lambda\sqrt{1-\rho^2}\, Y^2 C(X)} \exp\left(-\frac{\delta^2}{2\lambda}\right) \times \sqrt{\frac{\delta}{\sinh\delta}} \left[1 - \frac{\delta}{\sinh\delta} q - \left(\frac{1}{8} + \frac{\delta\coth\delta - 1}{8\delta^2} - \frac{3(1-\delta\coth\delta) + \delta^2}{8\delta\sinh\delta} q\right)\lambda + O\left(\lambda^2\right)\right]. \quad (79)$$

In a way, this is the central result of this paper. It gives us a precise asymptotic behavior of the Green's function of the SABR model, as $\lambda \to 0$.

5 Volatility Smile

We are now ready to complete our analysis. Given the explicit form of the approximate Green's function, we can calculate (via another asymptotic expansion) the marginal probability distribution. Comparing the result with the normal probability distribution allows us to find the implied normal and lognormal volatilities, as functions of the model parameters. We conclude this section by deriving explicit formulas for the case of the CEV model $C(x) = x^\beta$ and the shifted lognormal model $C(x) = x + a$.

5.1 Marginal Transition Probability

First, we integrate the asymptotic joint density over the terminal volatility variable Y to find the marginal density for the forward x. To within $O\left(\lambda^2\right)$,

$$P_X(s,x,y) = \int_0^\infty K_Z(s,z)\, dY = \frac{1}{2\pi\lambda\sqrt{1-\rho^2}\, C(X)} \int_0^\infty e^{-\delta^2/2\lambda} \sqrt{\frac{\delta}{\sinh\delta}} \left[1 - \frac{\delta}{\sinh\delta} q - \frac{1}{8}\lambda\left(1 + \frac{\delta\coth\delta - 1}{\delta^2} - \frac{3(1-\delta\coth\delta) + \delta^2}{\delta\sinh\delta} q\right)\right] \frac{dY}{Y^2}. \quad (80)$$

Here the metric $\delta(z, Z)$ is defined implicitly by (54). We evaluate this integral asymptotically by using Laplace's method (steepest descent). This analysis is carried out in Appendix B.2. The key step is to analyze the argument Y of the exponent

$$\phi(Y) = \frac{1}{2}\delta(z, Z)^2, \quad (81)$$

in order to find the point Y_0 where this function is at a minumum. Let us introduce the notation:

$$\zeta = \frac{1}{y}\int_X^x \frac{du}{C(u)}.$$

Since $yC(u)$ is basically the rescaled volatility at forward u, $1/\zeta$ represents the average volatility between today's forward x and at option's strike X. In other words, ζ represents how "easy" it is to reach the strike X. Some algebra shows that the minimum of (81) occurs at $Y_0 = Y_0(\zeta, y)$, where

$$Y_0 = y\sqrt{\zeta^2 - 2\rho\zeta + 1}. \tag{82}$$

The meaning of Y_0 is clear: it is the "most likely value" of Y, and thus $Y_0 C(X)$ (when expressed in the original units) should be the leading contribution to the observed implied volatility. Also, let $D(\zeta)$ denote the value of $\delta(z, Z)$ with $Y = Y_0$. Explicitly,

$$D(\zeta) = \log\frac{\sqrt{\zeta^2 - 2\rho\zeta + 1} + \zeta - \rho}{1-\rho}. \tag{83}$$

The analysis in Appendix B.3 shows that the probability distribution for x is Gaussian in this minimum distance, at least to leading order. Specifically, it is shown there that to within $O\left(\lambda^2\right)$,

$$\begin{aligned} P_X(s,x,y) = {} & \frac{1}{\sqrt{2\pi\lambda}}\,\frac{1}{yC(X)\,I^{3/2}}\exp\left\{-\frac{D^2}{2\lambda}\right\}\left\{1 + \frac{yC'(x)\,D}{2\sqrt{1-\rho^2}\,I}\right. \\ & -\frac{1}{8}\lambda\left[1 + \frac{yC'(x)\,D}{2\sqrt{1-\rho^2}\,I} + \frac{6\rho yC'(x)}{\sqrt{1-\rho^2}\,I^2}\cosh(D)\right. \\ & \left.\left. -\left(\frac{3\left(1-\rho^2\right)}{I} + \frac{3yC'(x)\left(5-\rho^2\right)D}{2\sqrt{1-\rho^2}\,I^2}\right)\frac{\sinh(D)}{D}\right] + \cdots\right\}, \end{aligned} \tag{84}$$

where

$$\begin{aligned} I(\zeta) &= \sqrt{\zeta^2 - 2\rho\zeta + 1} \\ &= \cosh D(\zeta) - \rho\sinh D(\zeta). \end{aligned} \tag{85}$$

As this expression may be useful on its own, we rewrite it in terms of the original variables:

$$P_F(\tau, f, \sigma) = \frac{1}{\sqrt{2\pi\tau}} \frac{1}{\sigma C(F) I^{3/2}} \exp\left\{-\frac{D^2}{2\tau v^2}\right\} \left\{1 + \frac{\sigma C'(f) D}{2v\sqrt{1-\rho^2} I} - \frac{1}{8}\tau v^2 \left[1 + \frac{\sigma C'(f) D}{2v\sqrt{1-\rho^2} I} + \frac{6\rho\sigma C'(f)}{v\sqrt{1-\rho^2} I^2} \cosh(D) - \left(\frac{3(1-\rho^2)}{I} + \frac{3\sigma C'(f)(5-\rho^2) D}{2v\sqrt{1-\rho^2} I^2}\right) \frac{\sinh(D)}{D}\right] + \cdots\right\}, \quad (86)$$

where we have slightly abused the notation. This is the desired asymptotic form of the marginal probability distribution.

5.2 Implied Volatility

The normal implied volatility is given by Sect. 2.2, and we are thus left with the task of calculating the conditional second moment. Explicitly,

$$M_X^2(s, x, y) = \int_0^\infty Y^2 K_Z(s, z)\, dY = \frac{1}{2\pi\lambda\sqrt{1-\rho^2} C(X)} \int_0^\infty e^{-\delta^2/2\lambda} \sqrt{\frac{\delta}{\sinh\delta}} \left[1 - \frac{\delta}{\sinh\delta} q - \frac{1}{8}\lambda\left(1 + \frac{\delta\coth\delta - 1}{\delta^2} - \frac{3(1-\delta\coth\delta) + \delta^2}{\delta\sinh\delta} q\right)\right] dY. \quad (87)$$

In Appendix B.3 we show that

$$M_X^2(s, x, y) = \frac{1}{\sqrt{2\pi\lambda}} \frac{\sqrt{I}}{yC(X)} \exp\left\{-\frac{D^2}{2\lambda}\right\} \left\{1 + \frac{yC'(x) D}{2\sqrt{1-\rho^2} I} + \frac{1}{8}\lambda \left[1 - \frac{yC'(x) D}{2\sqrt{1-\rho^2} I} + \frac{2\rho y C'(x)}{\sqrt{1-\rho^2} I^2} \cosh(D) + \left(\frac{3(1-\rho^2)}{I} + \frac{2yC'(x)(3\rho^2 - 4) D}{\sqrt{1-\rho^2} I^2}\right) \frac{\sinh(D)}{D}\right] + \cdots\right\}.$$

Despite their complicated appearances, the two expressions have a lot in common, and their ratio has a rather simple form. After the dust settles, we find that

$$\sigma_K(\tau, f, \sigma)^2 = \sigma^2 C(f)^2 I(\zeta) \times \left[1 + \frac{2\sigma C'(f)(\rho \cosh(D) - \sinh(D))}{\sigma C'(f) DI + 2\sqrt{1-\rho^2}\, I^2 v} \tau v^2 + \cdots \right], \tag{88}$$

or

$$\sigma_K(\tau, f, \sigma) = \sigma C(f) I(\zeta) \times \left[1 + \frac{\sigma C'(f)(\rho \cosh(D) - \sinh(D))}{\sigma C'(f) DI + 2\sqrt{1-\rho^2}\, I^2 v} \tau v^2 + \cdots \right]. \tag{89}$$

This is a refinement of the original asymptotic expression for implied volatility in the SABR model.

It is easy to apply this formula to the specific choice of the function $C(f)$. In case of the stochastic CEV model, $C(f) = f^\beta$, with $0 < \beta \leq 1$. If $\beta = 1$, then

$$\zeta = \frac{v}{\sigma} \log\left(\frac{f}{F}\right). \tag{90}$$

For $0 < \beta < 1$,

$$\zeta = \frac{v}{\sigma} \frac{f^{1-\beta} - F^{1-\beta}}{1-\beta}. \tag{91}$$

In the shifted lognormal model, $C(f) = f + a$, where $a > 0$. Consequently,

$$\zeta = \frac{v}{\sigma} \log\left(\frac{f+a}{F+a}\right). \tag{92}$$

5.3 *Implied Volatility at Low Strikes*

Our analysis so far has been base on the assumption that we were boundary conditions at zero forward. In the case of $\rho = 0$, we can tackle the Dirichlet and Neumann boundary conditions explicitly.

As explained in Sect. 2.3, the Green's functions corresponding to the Dirichlet and Neumann boundary conditions at zero forward can easily be calculated, using the method of images, in terms of the Green's function with free boundary conditions.

This, in turn, allows us to express the marginal probability distributions in terms of (86):

$$
\begin{aligned}
P_F^{\text{Dirichlet}}(\tau, f, \sigma) &= P_F(\tau, f, \sigma) - P_F(\tau, -f, \sigma), \\
P_F^{\text{Neumann}}(\tau, f, \sigma) &= P_F(\tau, f, \sigma) + P_F(\tau, -f, \sigma).
\end{aligned}
\tag{93}
$$

Analogous formulas hold for the conditional second moments. We can now easily find asymptotic expressions for the implied volatilities corresponding to these boundary conditions.

In order to keep the appearance of the otherwise unwieldy formulas reasonable, we shall introduce some additional notation. Let

$$
I^\theta = I\left(\zeta^\theta\right), \tag{94}
$$

where

$$
\zeta^\theta = \frac{1}{y} \int_X^{\theta(x)} \frac{du}{C(u)}. \tag{95}
$$

Furthermore, let us define the ratio

$$
\gamma = \sqrt{\frac{I}{I^\theta}}. \tag{96}
$$

and note that $\gamma < 1$. Finally, we set:

$$
\eta = \begin{cases} -1, & \text{for the Dirichlet boundary condition,} \\ 0, & \text{for the free boundary condition.} \\ 1, & \text{for the Neumann boundary condition.} \end{cases} \tag{97}
$$

It is now easy to see that:

$$
\begin{aligned}
\sigma_K^\eta(\tau, f, \sigma) &= \sigma C(K) I \frac{1}{\sqrt{1 - \eta\gamma + \eta^2\gamma^2}} \\
&\times \left[1 + \frac{\sigma C'(f)(\rho \cosh(D) - \sinh(D))}{\sigma C'(f) D I + 2\sqrt{1-\rho^2}\, I^2 v} \tau v^2 + \cdots\right].
\end{aligned}
\tag{98}
$$

It is worthwhile to note that for large strikes all three of these quantities are practically equal, and one might as well work with the free boundary condition expression. Indeed, in this case, $\gamma \approx 0$, and so $\sqrt{1 - \eta\gamma + \eta^2\gamma^2} \approx 1$. Also, we see from this expression that, at least asymptotically,

$$
\sigma_K^{\text{Dirichlet}}(\tau, f, \sigma) < \sigma_K^{\text{free}}(\tau, f, \sigma) < \sigma_K^{\text{Neumann}}(\tau, f, \sigma). \tag{99}
$$

This result is intuitively clear, and (98) quantifies it in a way that can be used for position management purposes. The decision which boundary condition to adopt should be made based on specific market conditions.

Appendix A Heat Equation on the Poincare Plane

In this appendix we present an elementary derivation of the explicit representation of the Green's function for the heat equation on $\mathbb{H}^2$. This explicit formula has been known for a long time (see e.g. [16]), and we include its construction here in order to make our calculations self-contained.

A.1 Lower Bound on the Laplace-Beltrami Operator

We shall first establish a lower bound on the spectrum of the Laplace-Beltrami operator on the Poincare plane. Let $\mathcal{H} = L^2\left(\mathbb{H}^2,\, d\mu_h\right)$ denote the Hilbert space of complex functions on $\mathbb{H}^2$ which are square integrable with respect to the measure (50). The inner product on this space is thus given by:

$$(\Phi|\Psi) = \int_{\mathbb{H}^2} \overline{\Phi\,(z)}\Psi\,(z)\,\frac{dx\,dy}{y^2}. \tag{100}$$

It is easy to verify that the Laplace-Beltrami operator Δ_h is self-adjoint with respect to this inner product.

Consider now the first order differential operator Q on $\mathcal{H}$ defined by

$$Q = i\left(y\frac{\partial}{\partial y} - \frac{1}{2}\right) + y\frac{\partial}{\partial x}. \tag{101}$$

Its hermitian adjoint with respect to (100) is

$$Q^\dagger = i\left(y\frac{\partial}{\partial y} - \frac{1}{2}\right) - y\frac{\partial}{\partial x}, \tag{102}$$

and we verify readily that

$$\frac{1}{2}\left(QQ^\dagger + Q^\dagger Q\right) = -\Delta_h - \frac{1}{4}. \tag{103}$$

This implies that

$$\begin{aligned}(\Phi| - \Delta_h \Phi) &= \frac{1}{2}\left(\Phi|QQ^\dagger\Phi\right) + \frac{1}{2}\left(\Phi|Q^\dagger Q\Phi\right) + \frac{1}{4}(\Phi|\Phi) \\ &= \frac{1}{2}\left(Q^\dagger\Phi|Q^\dagger\Phi\right) + \frac{1}{2}(Q\Phi|Q\Phi) + \frac{1}{4}(\Phi|\Phi) \\ &\geq \frac{1}{4}(\Phi|\Phi),\end{aligned}$$

where we have used the fact that $(\Psi|\Psi) \geq 0$, for all functions $\Psi \in \mathcal{H}$. As a consequence, we have established that the spectrum of the operator $-\Delta_h$ is bounded from below by $\frac{1}{4}$! This fact was first proved in [16].

A.2 Construction of the Green's Function

Let us now consider the the following initial value problem:

$$\begin{aligned}\frac{\partial}{\partial s} G_Z(s,z) &= \Delta_h G_Z(s,z), \\ G_Z(0,z) &= Y^2 \delta(z - Z),\end{aligned} \tag{104}$$

where $z, Z \in \mathbb{H}^2$. In addition, we require that

$$G_Z(s,z) \to 0, \quad \text{as } d(z,Z) \to \infty. \tag{105}$$

Note that, up to the factor of Y^2 in front of the delta function and a trivial time rescaling, this is exactly the initial value problem (68):

$$G_Z(s,z) = Y^2 K_Z(2s/\varepsilon, z). \tag{106}$$

The Green's function $G_Z(s,z)$ is also referred to as the heat kernel[4] on $\mathbb{H}^2$. The reason for inserting the factor of Y^2 in front of $\delta(z - Z)$ is that the distribution $Y^2\delta(z - Z)$ is invariant under the action (44) of the Lie group $SL(2,\mathbb{R})$. In fact, we verify readily that

$$Y^2\delta(z - Z) = \frac{1}{\pi}\,\delta(\cosh d(z,Z) - 1).$$

[4]It is the integral kernel of the semigroup of operators generated by the heat equation.

Now, since the initial value problem (105) is invariant under $SO(2, \mathbb{R})$, its solution must be invariant and thus a function of $d(z, Z)$ only. Let $r = \cosh d(z, Z)$, and write $G_Z(s, z) = \varphi(s, r)$. Then the heat equation in (105) takes the form

$$\frac{\partial}{\partial s}\varphi(s, r) = \left(r^2 - 1\right)\frac{\partial^2}{\partial r^2}\varphi(s, r) + 2r\frac{\partial}{\partial r}\varphi(s, r). \tag{107}$$

We have established above that the operator $-\Delta_h$ is self-adjoint on the Hilbert space $\mathcal{H}$, and its spectrum is bounded from below by $\frac{1}{4}$. Therefore, we shall seek the solution as the Laplace transform

$$\varphi(s, r) = \int_{1/4}^{\infty} e^{-s\lambda} L(\lambda, r)\, d\lambda \tag{108}$$

which yields the following ordinary differential equation:

$$\left(1 - r^2\right)\frac{d^2}{dr^2}L(\lambda, r) - 2r\frac{d}{dr}L(\lambda, r) - \lambda L(\lambda, r) = 0. \tag{109}$$

We write

$$\lambda = -\nu(\nu + 1),$$

where

$$\begin{aligned}\nu &= -\frac{1}{2} \pm i\sqrt{\lambda - \frac{1}{4}}\\ &= -\frac{1}{2} \pm i\omega,\end{aligned}$$

and recognize in (109) the Legendre equation. Note that, as a consequence of the inequality $\lambda \geq \frac{1}{4}$, ω is real and $\operatorname{Re}\nu = -\frac{1}{2}$.

In the remainder of this appendix, we will use the well known properties of the solutions to the Legendre equation, and follow Chaps. 7 and 8 of Lebedev's book on special functions [15]. The general solution to (109) is a linear combination of the Legendre functions of the first and second kinds, $P_{-1/2+i\omega}(r)$ and $Q_{-1/2+i\omega}(r)$, respectively:

$$L\left(\frac{1}{4} + \omega^2,\ r\right) = A_\omega P_{-1/2+i\omega}(r) + B_\omega Q_{-1/2+i\omega}(r). \tag{110}$$

As $d \to 0$ (which is equivalent to $r \to 1$),

$$Q_{-1/2+i\omega}(\cosh d) \sim const\ \log d, \tag{111}$$

which would imply that $\varphi(s, \cosh d)$ is singular at $d = 0$, for all values of $s > 0$. Since this is impossible, we conclude that $B_\omega = 0$. Note that, on the other hand,

$$P_{-1/2+i\omega}(1) = 1, \tag{112}$$

i.e. $P_{-1/2+i\omega}(\cosh d)$ is non-singular at $d = 0$. We will now invoke the Mehler-Fock transformation of a function[5]:

$$\widetilde{f}(\omega) = \int_1^\infty f(r)\, P_{-1/2+i\omega}(r)\, dr\,, \tag{113}$$

$$f(r) = \int_0^\infty \widetilde{f}(\omega)\, P_{-1/2+i\omega}(r)\, \omega \tanh(\pi\omega)\, d\omega. \tag{114}$$

In particular, (112) implies that the Mehler-Fock transform of $\delta(r-1)$ is 1, and thus (remember that we need to divide $\delta(r-1)$ by π):

$$A_\omega = \frac{1}{2\pi} \tanh(\pi\omega)\,.$$

Note that this relation can be viewed as a spectral representation for the unbounded self-adjoint Laplace-Beltrami operator on the Poincare plane.

Now, the Legendre function of the first kind $P_{-1/2+i\omega}(r)$ has the following integral representation:

$$P_{-1/2+i\omega}(\cosh d) = \frac{\sqrt{2}}{\pi} \coth(\pi\omega) \int_d^\infty \frac{\sin(\omega u)}{\sqrt{\cosh u - \cosh d}}\, du, \tag{115}$$

which is valid for all real ω. Therefore

$$L\left(\frac{1}{4} + \omega^2,\ \cosh d\right) = \frac{1}{\sqrt{2}\,\pi^2} \int_d^\infty \frac{\sin(\omega u)}{\sqrt{\cosh u - \cosh d}}\, du\,,$$

and we can easily carry out the integration in (108) to obtain

$$G_Z(s, z) = \frac{e^{-s/4}\sqrt{2}}{(4\pi s)^{3/2}} \int_{d(z,Z)}^\infty \frac{u e^{-u^2/4s}}{\sqrt{\cosh u - \cosh d(z, Z)}}\, du. \tag{116}$$

This is McKean's closed form representation of the Green's function of the heat equation on the Poincare plane [16].

Going back to the original normalization conventions of (68) yields formula (69).

[5]Strictly speaking, we will deal with distributions rather than functions. A rigor oriented reader can easily recast the following calculations into respectable mathematics.

Appendix B Some Asymptotic Expansions

In this appendix we collect a number of asymptotic expansions used in this paper.

B.1 Asymptotics of the McKean Kernel

We shall first establish a short time asymptotic expansion of McKean's kernel. This expansion plays a key role in the analysis of the Green's function of the SABR model.

In the right hand side of (116) we substitute $u = \sqrt{4sw + d^2}$:

$$G_Z(s, z) = \frac{e^{-s/4}\sqrt{2}}{4\pi^{3/2}\sqrt{s}}\, e^{-d^2/4s} \int_0^\infty \frac{e^{-w}\, dw}{\sqrt{\cosh\sqrt{4sw + d^2} - \cosh d}}.$$

Expanding the integrand in powers of s yields

$$\frac{1}{\sqrt{\cosh\sqrt{4sw + d^2} - \cosh d}} = \sqrt{\frac{d}{\sinh d}} \times \left(\frac{1}{\sqrt{2sw}} - \frac{d\coth d - 1}{4d^2}\sqrt{2sw} + O\left(s^{3/2}\right)\right).$$

Integrating term by term over w we find that

$$G_Z(s, z) = \frac{e^{-s/4}}{4\pi s} \exp\left(-\frac{d^2}{4s}\right) \times \sqrt{\frac{d}{\sinh d}}\left(1 - \frac{1}{4}\,\frac{d\coth d - 1}{d^2}\, s + O\left(s^2\right)\right),$$

and we thus obtain the following asymptotic expansion of the McKean kernel:

$$G_Z(s, z) = \frac{1}{4\pi s} \exp\left(-\frac{d^2}{4s}\right) \times \sqrt{\frac{d}{\sinh d}}\left[1 - \frac{1}{4}\left(\frac{d\coth d - 1}{d^2} + 1\right) s + O\left(s^2\right)\right], \tag{117}$$

Taking the derivative of $G_Z(s, z)$ with respect of $d(z, Z)$ in the expansion above, we find that

$$\frac{\partial}{\partial d} G_Z(s, z) = \frac{1}{4\pi s} \exp\left(-\frac{d^2}{4s}\right) \times \sqrt{\frac{d}{\sinh d}} \left[-\frac{d}{2s} + \frac{d}{8}\left(1 + 3\frac{1 - d\coth d}{d^2}\right) + O(s)\right]. \tag{118}$$

B.2 Laplace's Method

Next we review the Laplace method (see e.g. [2, 3]) which allows one to evaluate approximately integrals of the form:

$$\int_0^\infty f(u) e^{-\phi(u)/\epsilon} du. \tag{119}$$

We use this method in order to evaluate the marginal probability distribution for the Green's function.

In the integral (119), ϵ is a small parameter, and $f(u)$ and $\phi(u)$ are smooth functions on the interval $[0, \infty)$.[6] We also assume that $\phi(u)$ has a unique minimum u_0 inside the interval with $\phi''(u_0) > 0$. The idea is that, as $\epsilon \to 0$, the value of the integral is dominated by the quadratic approximation to $\phi(u)$ around u_0.

More precisely, we have the following asymptotic expansion. As $\epsilon \to 0$,

$$\int_0^\infty f(u) e^{-\phi(u)/\epsilon} du = \sqrt{\frac{2\pi\epsilon}{\phi''(u_0)}} e^{-\phi(u_0)/\epsilon} \times \left\{ f(u_0) + \epsilon \left[\frac{f''(u_0)}{2\phi''(u_0)} - \frac{\phi^{(4)}(u_0) f(u_0)}{8\phi''(u_0)^2} - \frac{f'(u_0)\phi^{(3)}(u_0)}{2\phi''(u_0)^2} + \frac{5\phi^{(3)}(u_0)^2 f(u_0)}{24\phi''(u_0)^3} \right] + O\left(\epsilon^2\right) \right\}. \tag{120}$$

To generate this expansion, we first expand $f(u)$ and $\phi(u)$ in Taylor series around u_0 to orders 2 and 4, respectively (keep in mind that the first order term in the expansion of $\phi(u)$ is zero). Then, expanding the regular terms in the exponential, we organize the integrand as $e^{-\phi''(u_0)(u-u_0)^2/2\epsilon}$ times a polynomial in ϵ. In the limit $\epsilon \to 0$, the integral reduces to calculating moments of the Gaussian measure; the result is (120). It is straightforward to compute terms of order higher than 1 in ϵ, even though the calculations become increasingly complex as the order increases.

[6] It can be an arbitrary interval.

Finally, let us state a slight generalization of (120), which we use below. In the integral (119), we replace $f(u)$ by $f(u) + \epsilon g(u)$. Then, as $\epsilon \to 0$,

$$\int_0^\infty [f(u) + \epsilon g(u)] e^{-\phi(u)/\epsilon} du = \sqrt{\frac{2\pi\epsilon}{\phi''(u_0)}} e^{-\phi(u_0)/\epsilon} \times \left\{ f(u_0) + \epsilon \left[g(u_0) + \frac{f''(u_0)}{2\phi''(u_0)} - \frac{\phi^{(4)}(u_0) f(u_0)}{8\phi''(u_0)^2} - \frac{f'(u_0)\phi^{(3)}(u_0)}{2\phi''(u_0)^2} + \frac{5\phi^{(3)}(u_0)^2 f(u_0)}{24\phi''(u_0)^3} \right] + O\left(\epsilon^2\right) \right\}. \tag{121}$$

This formula follows immediately form (120).

B.3 Application of Laplace's Method

We shall now apply formula (121) to evaluate the integrals (80) and (87). Each of these integrals is of the form given by the right hand side of (121). We find easily that the minimum Y_0 of the function

$$\phi(Y) = \frac{1}{2} \delta(z, Z)^2$$

is given by

$$Y_0 = y\sqrt{\zeta^2 - 2\rho\zeta + 1},$$

where

$$\zeta = \frac{1}{y} \int_X^x \frac{du}{C(u)}.$$

Also, we let $D(\zeta)$ denote the value of $\delta(z, Z)$ with $Y = Y_0$:

$$D(\zeta) = \log \frac{\sqrt{\zeta^2 - 2\rho\zeta + 1} + \zeta - \rho}{1 - \rho},$$

and

$$I(\zeta) = \sqrt{\zeta^2 - 2\rho\zeta + 1}.$$

Finally, we note that the second derivative $\phi''(Y_0)$ of $\phi(Y)$ with respect to Y is

$$\phi''(Y_0) = \frac{D}{\left(1-\rho^2\right) y^2 I \sinh D},$$

where we have suppressed the argument ζ in $D(\zeta)$ and $I(\zeta)$. Likewise,

$$\phi^{(3)}(Y_0) = -\frac{3D}{\left(1-\rho^2\right) y^3 I^2 \sinh D},$$

and

$$\phi^{(4)}(Y_0) = \frac{3\left(1-D\coth D\right)}{\left(1-\rho^2\right)^2 y^4 I^2 \sinh^2 D} + \frac{12D}{\left(1-\rho^2\right) y^4 I^3 \sinh D}.$$

It is actually easier to begin the calculation with (87). In order to evaluate the various terms on the right hand side of (121), let us define

$$f(Y) = \sqrt{\frac{\delta}{\sinh\delta}}\left(1-\frac{\delta}{\sinh\delta}q\right),$$

and

$$g(Y) = -\sqrt{\frac{\delta}{\sinh\delta}}\left(\frac{1}{8}+\frac{\delta\coth\delta-1}{8\delta^2}-\frac{3\left(1-\delta\coth\delta\right)+\delta^2}{8\delta\sinh\delta}q\right).$$

Then, after some manipulations we find that:

$$f(Y_0) = \sqrt{\frac{D}{\sinh D}}\left(1+\frac{yC'(x)D}{2\sqrt{1-\rho^2}\,I}\right),$$

$$f'(Y_0) = -\left(\frac{D}{\sinh D}\right)^{3/2}\frac{C'(x)\left(\sinh(D)-\rho\cosh(D)\right)}{2\left(1-\rho^2\right)^{3/2} I^2},$$

$$f''(Y_0) = \sqrt{\frac{D}{\sinh D}}\,\frac{1-D\coth D}{2\left(1-\rho^2\right) y^2 I D\sinh D}\left(1+\frac{3yC'(x)D}{2\sqrt{1-\rho^2}\,I}\right)$$
$$+\left(\frac{D}{\sinh D}\right)^{3/2}\frac{C'(x)\left(\sinh(D)-\rho\cosh(D)\right)}{\left(1-\rho^2\right)^{3/2} y I^3},$$

and

$$g(Y_0) = -\frac{1}{8}\sqrt{\frac{D}{\sinh D}}\left[1 - \frac{1 - D\coth D}{D^2} + yC'(x)\,\frac{3(1 - D\coth D) + D^2}{2\sqrt{1-\rho^2}\,ID}\right].$$

Putting all these together we find that

$$\begin{aligned} M_X^2(s, x, y) = \frac{1}{\sqrt{2\pi\lambda}}\, yC(X)\sqrt{I}\,\exp\left\{-\frac{D^2}{2\lambda}\right\}\Bigg\{1 + \frac{yC'(x)\,D}{2\sqrt{1-\rho^2}\,I} \\ + \frac{1}{8}\,\lambda\Bigg[1 - \frac{yC'(x)\,D}{2\sqrt{1-\rho^2}\,I} + \frac{2\rho yC'(x)}{\sqrt{1-\rho^2}\,I^2}\cosh(D) \\ + \left(\frac{3\left(1-\rho^2\right)}{I} + \frac{2yC'(x)\left(3\rho^2-4\right)D}{\sqrt{1-\rho^2}\,I^2}\right)\frac{\sinh(D)}{D}\Bigg] + O\left(\lambda^2\right)\Bigg\}, \end{aligned}$$

as claimed in Sect. 5.

Let us now compute (80). We note that the functions f and g in (121) occurring in this integral are obtained from the corresponding functions in (80) by dividing them by Y^2. We thus define

$$\widetilde{f}(Y) = \frac{f(Y)}{Y^2},$$

and

$$\widetilde{g}(Y) = \frac{g(Y)}{Y^2}.$$

Then,

$$\begin{aligned} \widetilde{f}(Y_0) &= \frac{f(Y_0)}{y^2 I^2}, \\ \widetilde{f}'(Y_0) &= -\frac{2f(Y_0)}{y^3 I^3} + \frac{f'(Y_0)}{y^2 I^2}, \\ \widetilde{f}''(Y) &= \frac{6f(Y_0)}{y^4 I^4} - \frac{4f'(Y_0)}{y^3 I^3} + \frac{f''(Y_0)}{y^2 I^2}, \end{aligned}$$

and

$$\widetilde{g}(Y_0) = \frac{g(Y_0)}{y^2 I^2}.$$

Combining all the terms we find that

$$P_X(s,x,y) = \frac{1}{\sqrt{2\pi\lambda}} \frac{1}{yC(X)\, I^{3/2}} \exp\left\{-\frac{D^2}{2\lambda}\right\} \left\{1 + \frac{yC'(x)\, D}{2\sqrt{1-\rho^2}\, I} \right.$$
$$-\frac{1}{8}\lambda\left[1 + \frac{yC'(x)\, D}{2\sqrt{1-\rho^2}\, I} + \frac{6\rho y C'(x)}{\sqrt{1-\rho^2}\, I^2}\cosh(D)\right.$$
$$\left.\left. -\left(\frac{3\left(1-\rho^2\right)}{I} + \frac{3yC'(x)\left(5-\rho^2\right) D}{2\sqrt{1-\rho^2}\, I^2}\right)\frac{\sinh(D)}{D}\right] + O\left(\lambda^2\right)\right\},$$

as stated in Sect. 5.

References

1. Beardon, A.F.: The Geometry of Discrete Groups. Springer, New York (1983)
2. Bender, C.M., Orszag, S.A.: Advanced Mathematical Methods for Scientists and Engineers. Springer, New York (1999)
3. Bleistein, N., Handelsman, R.A.: Asymptotic Expansions of Integrals. Dover Publications, New York (1986)
4. Derman, E., Kani, I.: Riding on a smile. Risk **7**(2), 32–39 (1994)
5. Duffie, D., Pan, J., Singleton, K.: Transform analysis and asset pricing for affine jump diffusions. Econometrica **68**, 1343–1377 (2000)
6. Dupire, B.: Pricing with a smile. Risk **7**(1), 18–20 (1994)
7. Elworthy, K.D.: Geometric aspects of diffusions on manifolds. Ecole d'Ete de Probabilites de Saint Flour, vol. XVII. Springer, New York (1987)
8. Emery, M.: Stochastic Calculus in Manifolds. Springer, Berlin (1989)
9. Guenther, R.B., Lee, J.W.: Partial Differential Equations of Mathematical Physics and Integral Equations. Prentice Hall, Englewood Cliffs (1988)
10. Hagan, P.S., Kumar, D., Lesniewski, A., Woodward D.E.: Managing smile risk, Wilmott Mag. (2003)
11. Hagan, P.S., Woodward, D.E.: Equivalent black volatilities. Appl. Math. Financ. **6**(3), 147–157 (1999)
12. Heston, S.: A closed form solution for options with stochastic volatility with applications to bond and currency options. Rev. Financ. Stud. **6**, 327–343 (1993)
13. Hsu, E.P.: Stochastic Analysis on Manifolds. American Mathematical Society, Providence (2002)
14. Kevorkian, J., Cole, J.D.: Perturbation Methods in Applied Mathematics. Springer, Berlin (1985)
15. Lebedev, N.N.: Special Functions and their Applications. Dover Publications, New York (1972)
16. McKean, H.P.: An upper bound to the spectrum of Δ on a manifold of negative curvature. J. Differ. Geom. **4**, 359–366 (1970)
17. Lewis, A.L.: Option Valuation Under Stochastic Volatility. Finance Press, Newport Beach (2000)

18. Molchanov, S.A.: Diffusion processes and Riemannian geometry. Russ. Math. Surv. **30**, 1–63 (1975)
19. Varadhan, S.R.S.: On the behavior of the fundamental solution of the heat equation with variable coefficients. Commun. Pure Appl. Math. **20**, 431–455 (1967)
20. Varadhan, S.R.S.: Diffusion processes in a small time interval. Commun. Pure Appl. Math. **20**, 659–685 (1967)

Asymptotic Implied Volatility at the Second Order with Application to the SABR Model

Louis Paulot

Abstract We provide a general method to compute a Taylor expansion in time of implied volatility for stochastic volatility models, using a heat kernel expansion. Beyond the order 0 implied volatility which is already known, we compute the first order correction exactly at all strikes from the scalar coefficient of the heat kernel expansion. Furthermore, the first correction in the heat kernel expansion gives the second order correction for implied volatility, which we also give exactly at all strikes. As an application, we compute this asymptotic expansion at order 2 for the SABR model and compare it to the original formula.

Keywords Stochastic volatility · Asymptotic expansion · Implied volatility · Heat kernel · SABR

1 Introduction

The most known model for pricing derivatives is the Black-Scholes-Merton model, where the underlying is supposed to follow a geometric Brownian motion. Popular extensions include local volatility models and stochastic volatility models. As an example the SABR model [6] combines the local volatility of the CEV model [4] and a lognormal volatility process. Closed formulas for European options can be obtained for a few models; it is the case of the CEV model or for a stochastic volatility example the Heston model [10]. These are however special cases and there are generally no closed form formulas. Finite difference methods or Monte-Carlo simulations can be used to price derivatives. Approximations have also been computed to achieve faster pricing, especially for calibration processes.

For short maturities, Hagan, Kumar, Lesniewski and Woodward provide an approximation for the implied volatility of the SABR model they introduce [6]. Berestycki, Busca and Florent [2, 3] and Henry-Labordère [8] give general methods

L. Paulot (✉)
Misys, 42 Rue Washington, 75008 Paris, France
e-mail: louis.paulot@misys.com

P.K. Friz et al. (eds.), *Large Deviations and Asymptotic Methods in Finance*, Springer Proceedings in Mathematics & Statistics 110,
DOI 10.1007/978-3-319-11605-1_2

to compute short maturity asymptotics of stochastic volatility models. These expansions give the implied volatility at first order in maturity. In addition some quantities are approximated by their value at the money, which can produce errors in the wings of the distributions.

In this paper, we leverage on the heat kernel methods introduced for the study of stochastic processes by Varadhan [11, 12] and used for the SABR model in [7, 8]. Using the heat kernel expansion of DeWitt [5], we provide a method to compute exactly a Taylor expansion of the implied volatility at all strikes. The stochastic volatility diffusion is formulated as a diffusion on a Riemannian manifold. The geodesic distance gives the implied volatility at null maturity. The multiplicative factor of the heat kernel expansion provides the first order (in time) correction to implied volatility. The first corrective term of the heat kernel is translated into the second order correction to implied volatility and similarly for higher order corrections. We perform a detailed computation up to order 2 of the Taylor expansion in time of implied volatility, without other approximations.

More generally, our method can be used to approximate a stochastic volatility model by an other model for which a closed form solution exists, with an implied parameter computed as a Taylor expansion.

As an application, we compute the asymptotic SABR volatility at order 2 and compare it to finite difference method results and to the original SABR expansion.

Our results can be useful for pricing short maturities options or even long maturities options with low volatility of volatility. When the approximation is not valid, a numerical method such as a finite difference method (FDM) has to be used. When our approximation is valid, it gives much faster results. At very short maturities, the prices are even more precise. Calibration at short maturities appears to be more stable using this approximation.

In Sect. 2 we recast the financial model in physical and geometric terms and fix our conventions. In Sect. 3 we use a heat kernel expansion to compute a short maturity expansion of Black or more generally CEV implied volatility. Finally in Sect. 4 we apply the method to the SABR model and compare the results to FDM and to the original formula.

2 Diffusion Equation in Covariant Form

A stochastic volatility model for some asset with pure diffusion (no jumps) is described by two risk-neutral processes: the asset price S and a variable V which describes the stochastic part of volatility. In the Heston model V would be the variance whereas in the SABR model it is a factor of volatility. The diffusion is given by the stochastic differential equations

$$\begin{aligned} \mathrm{d}S &= \mu_S(S)\mathrm{d}t + \sigma_S(S, V)\mathrm{d}W_1 \\ \mathrm{d}V &= \mu_V(V)\mathrm{d}t + \sigma_V(V)\mathrm{d}W_2 \end{aligned} \tag{1}$$

where dW_1 and dW_2 are two standard Brownian processes with correlation ρ. The dependence of parameters in variables S and V we have written is the more common, it may be more general with all parameters depending on both variables.

Stochastic volatility models can be seen as diffusions on a Riemann surface. More precisely, prices of securities are sections of a line bundle over this Riemann surface which are solutions of a diffusion (or heat) equation.

A introduction to this subject and its applications to finance can be found in [9]. We present here the formalism and define all quantities we use in order to set our conventions.

2.1 Diffusion Equation

Let us consider a general model with n state variables $X^i(t)$ (which will be the spot and the volatility) which follow a pure diffusion process, without jumps. For simplicity we consider a European payoff of some maturity T. The price $P(X(t), t)$ of such a payoff is the solution of a diffusion equation

$$-\partial_t P = \mu^i \partial_i P + \frac{1}{2}\Sigma^{ij}\partial_i\partial_j P - rP \tag{2}$$

where Σ^{ij} is the covariance matrix, μ_i the drifts and r the numéraire rate. All coefficients can depend on state variables $X^i(t)$ and time t. Unless explicitly staten, we adopt Einstein sum convention: repeated indices are summed. The price of the European option is given by the solution of this equation with terminal boundary condition at maturity T given by the payoff.

The covariance matrix Σ^{ij} can be seen geometrically as the inverse $g^{ij} = \Sigma^{ij}$ of a metric g_{ij} on the space of variables. The diffusion equation describes the diffusion over a Riemannian manifold: the state of variables endowed with the metric $g_{ij} = (\Sigma^{-1})_{ij}$.

Examples

1. The Black-Scholes equation in the monetary account numéraire with volatility σ, risk free rate r and dividend yield q reads

$$-\partial_t P = (r-q)S\partial_S P + \frac{1}{2}\sigma^2 S^2 \partial_S\partial_S P - rP$$

   This is Eq. (2) with $\mu^S = (r-q)S$, $\Sigma^{SS} = \sigma^2 S^2$ and $r = r$.
2. For the stochastic volatility model described by Eq. (1), μ^i is a two-dimensional vector

$$\mu = \begin{pmatrix} \mu_S \\ \mu_V \end{pmatrix}.$$

The covariance matrix Σ^{ij} is

$$\Sigma = \begin{pmatrix} \sigma_S{}^2 & \rho\sigma_S\sigma_V \\ \rho\sigma_S\sigma_V & \sigma_V{}^2 \end{pmatrix}.$$

In what follows we restrict ourselves to the case of time-homogeneous models: there is no explicit time-dependence in parameters. The generalization to time-dependent cases is not difficult.

2.2 Gauge Structure

There are several *gauge* transformations which are natural for such systems:

1. Change of numéraire:
$$P(X,t) \longrightarrow \frac{P(X,t)}{\Phi(X,t)}$$
where $\Phi(X,t)$ is the price of a security which is always nonzero. Mathematically it is a real function which is positive everywhere, that we denote thus by $\Phi(X,t) = e^{-\phi(X,t)}$.
2. Change of variables
$$X \longrightarrow X'(X).$$

The natural way to handle a system with gauge freedom is to introduce covariant derivatives. The coordinate freedom is handled through the *Levi-Civita* connection which acts respectively on scalars, vectors and 1-forms as

$$\begin{aligned} D_i f &= \partial_i f \\ D_i f^j &= \partial_i f^j + \Gamma^j_{ik} f^k \\ D_i f_j &= \partial_i f_j - \Gamma^k_{ij} f_k \end{aligned}$$

where Γ^j_{ik} are the *Christoffel symbols*. The action on tensors with more indices is obtained by acting on all indices with the Christoffel symbols. Christoffel symbols can be computed from the metric as

$$\Gamma^k_{ij} = \frac{1}{2} g^{kl} \left(\partial_i g_{lj} + \partial_j g_{il} - \partial_l g_{ij} \right).$$

A fundamental property of the Levi-Civita connection is the covariance of the metric:

$$D_i g_{jk} = 0.$$

The metric is used to transforms vectors into 1-forms and conversely, i.e. lowering or raising indices:

$$A^i = g^{ij} A_j$$
$$A_i = g_{ij} A^j.$$

The numéraire gauge freedom is handled through a line bundle $\mathcal{L}$ (i.e. with sections in $\mathbb{R}$). Geometrically, P is a section of $\mathcal{L}$. A $\mathbb{R}$-valued connection[1] is defined with spatial and time components given by a 1-form A_i and a scalar[2] Q:

$$\nabla_i P = (D_i - A_i)P$$
$$\nabla_t P = (\partial_t - Q)P.$$

Under the change of numéraire

$$P \longrightarrow e^{\phi(X,t)} P$$

these operators are covariant,

$$\nabla_i P \longrightarrow e^{\phi(X,t)} \nabla_i P$$
$$\nabla_t P \longrightarrow e^{\phi(X,t)} \nabla_t P,$$

provided that A_i and Q are shifted as

$$A_i \longrightarrow A_i - \partial_i \phi$$
$$Q \longrightarrow Q - \partial_t \phi.$$

Using these connections, the diffusion equation (2) can be rewritten as

$$-\nabla_t P = \frac{1}{2} \nabla^i \nabla_i P. \qquad (3)$$

Identifying terms between Eqs. (2) and (3), the $\mathbb{R}$ connection must be

$$A_i = g_{ij} \left(-\frac{1}{2} \Gamma^j_{kl} g^{kl} - \mu^j \right) \qquad (4)$$
$$Q = -\frac{1}{2} g^{ij} \left(\partial_i A_j - A_i A_j - \Gamma^k_{ij} A_k \right) + r. \qquad (5)$$

---

[1] This connection is similar to the connection which described the electromagnetic potential, except that the fibre of the gauge bundle is $\mathbb{R}$ instead of $U(1)$. This causes a difference of a factor i in equations.

[2] There is a breaking of symmetry between time and spatial directions. The diffusion equation can be seen as a non-relativistic limit of a pure wave equation in imaginary time.

In addition with

$$g^{ij} = \Sigma^{ij}$$

this translates the set of financial parameters into geometrical quantities.

2.3 Kolmogorov Forward Equation

The Kolmogorov backward Eq. (3) leads to a dual Kolmogorov forward equation.

We suppose that all prices are expressed with respect to a numéraire which is a traded asset that does not pay any coupon or dividend. The price of the numéraire security itself is identically 1; this reads mathematically

$$-\nabla_t 1 = \frac{1}{2}\nabla^i \nabla_i 1$$

If $p(X, t)$ is a risk-neutral probability density to get in state X at time t starting from state X_0 at time 0, then the price of a European payoff of maturity $T \geq t$ can be written as

$$P(X_0, 0) = \int \mathrm{d}X p(X, t) P(X, t) \, .$$

As t does not appear on the left-hand side, the derivative of the integral with respect to t must vanish.

$$\int \mathrm{d}X \partial_t (p(X, t) P(X, t)) = 0.$$

We define an action of the gauge group on p with a plus sign instead of a minus sign when acting on P:

$$\nabla_i p = (D_i + A_i) p$$
$$\nabla_t p = (\partial_t + Q) p.$$

This means that they p and P have opposite charges under the numéraire $\mathbb{R}$ gauge group, such that pP is neutral and $\nabla_t(pP) = \partial_t(pP)$. We have thus

$$\int \mathrm{d}X \left(\nabla_t p(X, t) P(X, t) + p(X, t) \nabla_t P(X, t) \right) = 0 \, .$$

Using Eq. (3) for $\nabla_t P(X, t)$ and integrating by part on the spatial directions, this equation becomes

$$\int \mathrm{d}X \left(\nabla_t p(X, t) - \frac{1}{2} \nabla^i \nabla_i p(X, t) \right) P(X, t) = 0 \, . \qquad (6)$$

This equation will be automatically satisfied if

$$\nabla_t p = \frac{1}{2} \nabla^i \nabla_i p. \tag{7}$$

Moreover, if the market is complete Eq. (6) must be true for all functions $P(\cdot, t)$ which imposes Eq. (7). This is the Kolmogorov forward equation, written in a covariant way.

It should be noted that $p(X, t)$ is a density, which means that the Levi-Civita connection does not reduce to a partial derivative as would be the case for a scalar. More precisely, the transition probability $p(X_0, 0; X, t)$ has value in $\mathcal{L} \boxtimes \mathcal{L}^* \otimes \wedge^d (T^* M)$. Numéraire gauge tranformations associated with the line bundle $\mathcal{L}$ acting on p gives the well-known change of measure which are usually obtained from the Girsanov formula.

3 Asymptotic Implied Volatility

We consider a stochastic volatility model where the variable is a forward price or rate F with a volatility variable V:

$$\begin{aligned} \mathrm{d}F &= \sigma_F(F, V)\mathrm{d}W_1 \\ \mathrm{d}V &= \mu_V(V)\mathrm{d}t + \sigma_V(V)\mathrm{d}W_2 \end{aligned}$$

with $\langle \mathrm{d}W_1 \mathrm{d}W_2 \rangle = \rho \mathrm{d}t$.

Our computation of an asymptotic expansion at short time of implied volatility at strike K involves four steps:

1. Compute an asymptotic value of the transition probability from initial state F_0, V_0 at time 0 to K, V at time t using a heat kernel expansion;
2. Compute $\mathbb{E}\big[\sigma_F^2(K, V)\delta(F - K)\big]$ using a saddle point method;
3. Integrate over time to compute the time value;
4. Compare to the same formula for the Black-Scholes model to extract the implied volatility.

3.1 *Heat Kernel Expansion*

In order to keep exposition as simpler and clear as possible, we will skip here technical details and refer the reader to [5, 13] for mathematically precise statements.[3]

[3]In finance we will usually consider noncompact manifolds, possibly with boundaries as in the SABR model for $0 < \beta < 1$.

At short time the solution of Eq. (7) with initial condition $p(X, 0) = \delta(X - X_0)$ is asymptotically given by a heat kernel expansion[4]

$$p(X,t) = \frac{\sqrt{g(X)}}{(2\pi t)^{n/2}} \sqrt{\Delta(X_0, X)} \mathcal{P}(X_0, X) e^{-\frac{d^2(X_0, X)}{2t}} \sum_{k \geq 0} a_k(X_0, X) t^k. \quad (8)$$

$g(X)$ is the determinant of the metric at point X:

$$g = \mathrm{Det}(g_{ij}).$$

$d(X_0, X)$ is the geodesic distance between the starting point X_0 and the end point X, this is the minimal distance between X_0 and X. It can also be written as

$$\frac{d^2(X_0, X)}{t} = \min_{X(s)} \int_0^t \mathrm{d}t \, g_{ij} \dot{X}^i \dot{X}^j$$

where the minimum is taken on all paths going from $X(0) = X_0$ to $X(t) = X$. (This is independent of t.) We denote by $\mathcal{C}$ this geodesic path. $\Delta(X_0, X)$ is the Van Vleck–Morette determinant

$$\Delta(X_0, X) = \frac{\mathrm{Det}\left(-\frac{1}{2} \frac{\partial^2 d^2(X_0, X)}{\partial X^i \partial X_0^j} \right)}{\sqrt{g(X_0) g(X)}}.$$

$\mathcal{P}(X_0, X)$ is the parallel transport along the geodesic with respect to the $\mathbb{R}$ connection. It is such that its covariant derivative along the geodesic path is null:

$$\mathcal{P}(X_0, X) = e^{-\int_{\mathcal{C}} A_i \dot{X}^i \mathrm{d}t} = e^{-\int_{\mathcal{C}} A_i \mathrm{d}X^i}$$

where the integral is computed on the geodesic path $\mathcal{C}$. Finally, $a_i(X_0, X)$ are functions which are defined recursively with

$$a_0 = 1$$

[4]Using Feynman path integral, the solution to Eq. (7) can be written up to some normalization factor as

$$p(X,t) \propto \int [DX] e^{-\int_0^t \mathrm{d}t \left(\frac{1}{2} g_{ij} \dot{X}^i \dot{X}^j + A_i \dot{X}^i + Q \right)}$$

where $[DX]$ means integrating over all path $X(s)$ going from $X(0) = X_0$ to $X(t) = X$. The normalization factor is the inverse of the same quantity with the integral computed over all paths with starting point X_0, so that the total probability is 1. It is generally not possible to compute this integral exactly. However it gives some hints on the asymptotic solution at short time: the solution will be dominated by the path corresponding to the minimal value of the integrand inside the exponential, which will be close to the geodesic path.

and a_i's satisfy the differential equations

$$(k + (\nabla^i d) d \nabla_i) a_k = \mathcal{P}^{-1} \Delta^{-1/2} \left(\frac{1}{2} \nabla^i \nabla_i - Q \right) \mathcal{P} \Delta^{1/2} a_{k-1}.$$

Along a given geodesic curve parameterized by its geodesic distance from X_0, this equation reads

$$(k + d \partial_d) a_k = \mathcal{P}^{-1} \Delta^{-1/2} \left(\frac{1}{2} \nabla^i \nabla_i - Q \right) \mathcal{P} \Delta^{1/2} a_{k-1}$$

which can be integrated as

$$a_k = \frac{1}{d^k} \int_0^d ds \, s^{k-1} \mathcal{P}^{-1} \Delta^{-1/2} \left(\frac{1}{2} \nabla^i \nabla_i - Q \right) \mathcal{P} \Delta^{1/2} a_{k-1}.$$

Functions a_k are sections of a $\mathcal{L} \boxtimes \mathcal{L}^*$ bundle. The parallel transport $\mathcal{P}$ and the connexion with respect to the numéraire gauge group act on the second factor of this external product. (The first factor is related to the numéraire at $t = 0$.) Also note that $\frac{p(X,t)}{\sqrt{g(X)}}$ is a scalar with respect to the Levi-Civita connection.

In order to produce a first order expansion of the implied volatility, only the common multiplicative factor of expansion (8) is needed. In order to compute a second order term for the implied volatility, we will also make use of the first corrective term $a_1 t$ with

$$a_1 = \frac{1}{d} \int_0^d ds \, \mathcal{P}^{-1} \Delta^{-1/2} \left(\frac{1}{2} \nabla^i \nabla_i - Q \right) \mathcal{P} \Delta^{1/2}. \tag{9}$$

3.2 Expected Variance

We now compute $\mathbb{E}\left[\sigma_F^2(F_t, V_t)\delta(F_t - K)\right]$. This quantity can be written as an integral over the terminal volatility variable V:

$$\mathbb{E}\left[\sigma_F^2(F_t, V_t)\delta(F_t - K)\right] = \int \mathrm{d}V \, \sigma_F^2(K, V) p(K, V; t)$$

where $p(F, V; t)$ is given by the heat kernel expansion (8) with $X = \begin{pmatrix} F \\ V \end{pmatrix}$ and $n = 2$. The integrand can be written as

$$\sigma_F^2(K, V) p(K, V; t) = \frac{1}{2\pi t} e^{-\frac{B}{t} - C - Dt + o(t)} \tag{10}$$

with

$$B = \frac{1}{2} d(F_0, V_0; K, V)^2 \tag{11}$$

$$C = -2\ln(\sigma_F(K, V)) - \frac{1}{2}\left[\ln(g(K, V)) + \ln(\Delta(F_0, V_0; K, V))\right] + \mathcal{M}(K, V) \tag{12}$$

$$D = -a_1(K, V) \tag{13}$$

where $\mathcal{M}$ is the integral of the $\mathbb{R}$ connection

$$\mathcal{M}(K, V) = -\ln(\mathcal{P}(K, V)) = \int_{\mathcal{C}} A_i \mathrm{d}X^i$$

on $\mathcal{C}$, the geodesic curve joining (F_0, V_0) to (K, V), and a_1 is given in Eq. (9) as an integral over the geodesic path.

The integral over (10) will be dominated at short time by the B term. More precisely, it will be dominated by the volatility $V_{\min}$ which minimizes $B(K, V) = \frac{1}{2} d(F_0, V_0; K, V)^2$. This is the final volatility which minimizes the distance between the initial conditions and the strike K. Expanding all functions in the neighborhood of $V_{\min}$, where $B'(V_{\min}) = 0$, the integrand is

$$\sigma_F^2(K, V) p(K, V; t) = \frac{1}{2\pi t} e^{-\frac{B}{t} - C - Dt} e^{-\frac{B''}{2t}\delta V^2}$$
$$\left[1 - \frac{1}{2}\left(C'' - C'^2\right)\delta V^2 - \left(\frac{1}{24}B^{(4)} - \frac{1}{6}B^{(3)}C'\right)\frac{\delta V^4}{t} + \frac{1}{72}B^{(3)2}\frac{\delta V^6}{t^2} + o(t) + \text{odd terms}\right]$$

where derivatives are with respect to V, all functions B, C, D and their derivatives are taken at $(K, V_{\min})$ and $\delta V = V - V_{\min}$. When writing $o(t)$, we have anticipated that after integration $\delta V^2 \sim \frac{1}{t}$. We have also anticipated that odd terms in δV will not give contributions to the integral.

Integrating over δV, and using that the first even moments of the standard normal distribution are $M_2 = 1$, $M_4 = 3$ and $M_6 = 15$, we get for the integral

$$\mathbb{E}\left[\sigma_F^2(F_t, V_t)\delta(F_t - K)\right] = \frac{1}{\sqrt{2\pi t B''}} e^{-\frac{B}{t} - C - Dt}$$
$$\left[1 - \frac{1}{2}\left(C'' - C'^2\right)\frac{t}{B''} - \left(\frac{1}{24}B^{(4)} - \frac{1}{6}B^{(3)}C'\right)\frac{3t}{B''^2} + \frac{1}{72}B^{(3)2}\frac{15t}{B''^3} + o(t)\right]$$

This can be rewritten as

$$\mathbb{E}\Big[\sigma_F^2(F_t, V_t)\delta(F_t - K)\Big] = \frac{1}{\sqrt{2\pi t}} e^{-\frac{B}{t} - \widetilde{C} - \widetilde{D}t + o(t)} \tag{14}$$

with

$$\widetilde{C} = C + \frac{1}{2}\ln(B'') \tag{15}$$

$$\widetilde{D} = D + \frac{1}{2B''}\left[C'' - C'^2 + \frac{1}{4}\frac{B^{(4)}}{B''} - \frac{B^{(3)}}{B''}C' - \frac{5}{12}\left(\frac{B^{(3)}}{B''}\right)^2\right]$$

$$= D + \frac{1}{2B''}\left[\widetilde{C}'' - \widetilde{C}'^2 - \frac{1}{4}\frac{B^{(4)}}{B''} + \frac{1}{3}\left(\frac{B^{(3)}}{B''}\right)^2\right] \tag{16}$$

where all derivatives are with respect to V and all functions and their derivatives are taken at $(K, V_{\min})$.

3.3 Time Value

The price of a Call of maturity T and strike K can be written as the payoff integrated against the risk-neutral distribution:

$$\mathrm{Call}(K, T) = e^{-rT}\int \mathrm{d}F\,(F - K)^+ p(F, T)$$

where $p(F, t)$ is the marginal probability density of F_t. This can be written also as a double integral over forward and time as

$$\mathrm{Call}(K, T) = e^{-rT}\left[(F_0 - K)^+ + \int \mathrm{d}F \int_0^T \mathrm{d}t\,(F - K)^+\partial_t p(F, t)\right]. \tag{17}$$

As F_t is a forward, it is a driftless process and the Kolmogorov forward equation reduces to

$$\partial_t p(F, t) = \frac{1}{2}\partial_F^2\left(\sigma_{loc}^2(F, t)p(F, t)\right) \tag{18}$$

where $\sigma_{loc}^2(F, t)$ is the local (normal) volatility

$$\sigma_{loc}^2(F, t) = \mathbb{E}\Big[\sigma_F^2(F_t, V_t) \mid F_t = F\Big] = \frac{\mathbb{E}\big[\sigma_F^2(F_t, V_t)\delta(F_t - F)\big]}{p(F, t)}.$$

Plugging the Kolmogorov equation (18) in Eq. (17) and integrating twice by part on the F variable, the Call price is finally obtained as an integral over time at strike K:

$$\mathrm{Call}(K,T)=e^{-rT}\left[(F_0-K)^+ +\frac{1}{2}\int_0^T \mathrm{d}t\,\mathbb{E}\Big[\sigma_F^2(F_t,V_t)\delta(F_t-K)\Big]\right]. \quad (19)$$

Using expression (14) for the integrand, the integral over time can be computed:

$$\frac{1}{2}\int_0^T \mathrm{d}t\,\mathbb{E}\Big[\sigma_F^2(F_t,V_t)\delta(F_t-K)\Big]=$$

$$\frac{1}{\sqrt{2}}e^{-\widetilde{C}}\left[\sqrt{\frac{t}{\pi}}e^{-\frac{B}{t}}-\sqrt{B}\,\mathrm{erfc}\left(\sqrt{\frac{B}{t}}\right)\right.$$

$$\left.-\frac{\widetilde{D}}{3}\left(\sqrt{\frac{t}{\pi}}(t-2B)e^{-\frac{B}{t}}+2B^{3/2}\,\mathrm{erfc}\left(\sqrt{\frac{B}{t}}\right)\right)\right]+o\left(t^{5/2}e^{-\frac{B}{t}}\right)$$

where erfc is the complementary error function, equal to the cumulative of the standard normal distribution up to $\sqrt{2}$ factors:

$$\mathrm{erfc}(x)=\frac{2}{\sqrt{\pi}}\int_x^{+\infty}\mathrm{d}y\,e^{-y^2}=2\mathcal{N}\left(-\sqrt{2}x\right).$$

The asymptotic expansion of this function at $+\infty$

$$\mathrm{erfc}(x)=\frac{e^{-x^2}}{x\sqrt{\pi}}\left(1-\frac{1}{2x^2}+\frac{3}{4x^4}+o\left(\frac{1}{x^4}\right)\right)$$

with $x=\sqrt{\frac{B}{t}}$ gives the asymptotic expansion of the time value:

$$\frac{1}{2}\int_0^T \mathrm{d}t\,\mathbb{E}\Big[\sigma_F^2(F_t,V_t)\delta(F_t-K)\Big]$$
$$=\frac{T^{3/2}}{2\sqrt{2\pi}}e^{-\frac{B}{T}-\widetilde{C}-\ln(B)-\widetilde{D}T-\frac{3}{2B}T+o(T)}. \quad (20)$$

3.4 Implied Volatility

The final step consists in computing the same expansion for the Black–Scholes model, which is simpler as there is no stochastic volatility to be integrated. The metric is given by the inverse of the variance:

$$g_{FF}=\frac{1}{\sigma^2F^2}.$$

The Christoffel symbol is therefore

$$\Gamma^F_{FF} = -\frac{1}{F}.$$

The $\mathbb{R}$-connection components are computed using (4) and (5):

$$A_F = \frac{1}{2F}$$
$$Q = \frac{\sigma^2}{8}.$$

The geodesic distance is

$$d(F_0, K) = \left| \int_{F_0}^{K} \frac{\mathrm{d}F}{\sigma F} \right| = \frac{1}{\sigma} \left| \ln \frac{K}{F_0} \right|.$$

The Van Vleck–Morette determinant is simply

$$\Delta(F_0, K) = 1.$$

The parallel transport is

$$\mathcal{P}(F_0, K) = e^{-\int_{F_0}^{K} \mathrm{d}F \, A_F} = \sqrt{\frac{F}{K}}.$$

Putting all these elements together, the heat kernel expansion of $p(K, t)$ is according to (8)

$$p(K, t) = \frac{1}{\sigma K \sqrt{2\pi t}} \sqrt{\frac{F}{K}} e^{-\frac{\ln^2 \frac{K}{F}}{2\sigma^2 t}} \left(1 - \frac{\sigma^2}{8} t + o(t) \right).$$

Multiplying by the local variance, we get

$$\mathbb{E}\Big[\sigma_F^2(K)\delta(F - K)\Big] = \sigma^2 K^2 p(K, t) = \frac{1}{\sqrt{2\pi t}} e^{-\frac{B_{\mathrm{BS}}}{t} - \widetilde{C}_{\mathrm{BS}} - \widetilde{D}_{\mathrm{BS}} t + o(t)} \tag{21}$$

with

$$B_{\mathrm{BS}} = \frac{1}{2\sigma^2} \ln^2 \frac{K}{F_0}$$
$$\widetilde{C}_{\mathrm{BS}} = -\ln(\sigma) - \frac{1}{2} \ln(K F_0)$$
$$\widetilde{D}_{\mathrm{BS}} = \frac{\sigma^2}{8}.$$

(In fact formula (21) is exact: there is no $o(t)$ correction and it can be integrated exactly to get the Black–Scholes formula.)

Writing Eq. (20) for both the stochastic volatility model and the Black-Scholes model, the implied volatility is such that both quantities are equal:

$$\frac{B}{T} + \widetilde{C} + \ln(B) + \widetilde{D}T + \frac{3}{2B}T = \frac{B_{\mathrm{BS}}}{T} + \widetilde{C}_{\mathrm{BS}} + \ln(B_{\mathrm{BS}}) + \widetilde{D}_{\mathrm{BS}}T + \frac{3}{2B_{\mathrm{BS}}}T + o(T). \tag{22}$$

Expanding the implied volatility σ as a Taylor expansion

$$\sigma(K,T) = \sigma_0(K) + \sigma_1(K)T + \sigma_2(K)T^2 + o(T^2)$$

and plugging this into Eq. (22) on the Black–Scholes side, we get

$$\begin{aligned} &\frac{1}{2\sigma_0^2 T}\ln^2\left(\frac{K}{F_0}\right)\left(1 - 2\frac{\sigma_1}{\sigma_0}T - 2\frac{\sigma_2}{\sigma_0}T^2 + 3\left(\frac{\sigma_1}{\sigma_0}\right)^2 T^2\right) - \ln(\sigma_0) - \frac{\sigma_1}{\sigma_0}T \\ &-\frac{1}{2}\ln(KF_0) + \ln\left(\frac{1}{2\sigma_0^2}\ln^2\frac{K}{F_0}\right) - 2\frac{\sigma_1}{\sigma_0}T + \frac{\sigma_0^2}{8}T + \frac{3\sigma_0^2}{\ln^2\frac{K}{F_0}}T \\ &= \frac{B}{T} + \widetilde{C} + \ln(B) + \widetilde{D}\,T + \frac{3}{2B}T + o(T). \end{aligned}$$

Coefficients must be equal at each order in T, which gives our final expansion of the implied volatility.

Power -1 gives the order 0 implied volatility

$$\sigma_0 = \frac{\left|\ln\frac{K}{F_0}\right|}{\sqrt{2B}} = \frac{\left|\ln\frac{K}{F_0}\right|}{d(F_0, V_0; K, V_{\min})} \tag{23}$$

which was already obtained in [2, 8].

The first order correction is extracted from the constant term:

$$\frac{\sigma_1}{\sigma_0} = -\frac{\widetilde{C} + \ln\left(\sigma_0\sqrt{KF_0}\right)}{2B}. \tag{24}$$

Finally the $O(T)$ term gives the second order correction:

$$\frac{\sigma_2}{\sigma_0} = \frac{3}{2}\left(\frac{\sigma_1}{\sigma_0}\right)^2 - \frac{1}{2B}\left(\widetilde{D} + 3\frac{\sigma_1}{\sigma_0} - \frac{\sigma_0^2}{8}\right). \tag{25}$$

This gives our final result as the implied volatility expansion

$$\sigma = \sigma_0\left(1 + \frac{\sigma_1}{\sigma_0}T + \frac{\sigma_2}{\sigma_0}T^2 + o(T^2)\right). \tag{26}$$

We stress that this result is exact in strike: for a given strike, we have computed exactly the three first coefficients of the Taylor expansion. Moreover, contrary to other expansions, the order 1 expansion is extracted from the order 0 expansion of the probability. This technique allows us to extract a second order term for the implied volatility from the order 1 term in the probability expansion. This method can be used to compute the Taylor expansion of implied volatility up to any order, although the computation becomes more complicated and involves integrals of increasing dimension: the a_k coefficient of the heat kernel expansion involves $k+1$ integrals.

3.5 At the Money

The computation we have performed makes the implicit hypothesis that we are not exactly at the money: $K \neq F_0$. Otherwise, the dominant term in the exponential would vanish and we could not use the asymptotic expansion of the erfc function at infinity. Precisely at the money, we should use instead a Taylor expansion in 0. As the implied volatility surface is smooth, we just take the limit of formulas (23), (24) and (25) at $K \to F_0$. If we perform instead the Taylor expansion of the erfc function at 0, we find only the two first orders

$$\sigma_0(F_0) = \frac{e^{-\widetilde{C}(F_0)}}{F_0}$$
$$\frac{\sigma_1}{\sigma_0}(F_0) = \frac{1}{3}\left(\frac{\sigma_0^2}{8} - \widetilde{D}(F_0)\right).$$

Careful Taylor expansions of all quantities at the money can be used to check that this is indeed the limit of Eqs. (23) and (24). Moreover, it can be seen that the existence of these limit are conditions for formulas (24) and (25) to be convergent, as B goes to 0 at the money (at order 2 in the geodesic distance, which means that the numerators must in fact vanish at order 2).

3.6 CEV Volatility

Instead of Black volatility, the asymptotic expansion can be computed for other local volatility models. Without stochastic volatility, the SABR model reduces to the CEV model. The local volatility part of the model is thus taken into account exactly without introducing approximation besides the stochastic corrections. In view of our application to the SABR model, we will compute here a CEV implied volatility. There are closed formulas for this model, involving Bessel functions. This implied volatility can therefore be used in the CEV pricing formula in order to get the price of the option.

For a CEV model with parameter β_0 and volatility factor σ, such that

$$\mathrm{d}F = \sigma F^{\beta_0} \mathrm{d}W,$$

the function B, $\widetilde{C}$ and $\widetilde{D}$ are

$$\begin{aligned}
B_0 &= \frac{1}{2\sigma^2} \ln^2(q_0) \\
\widetilde{C}_0 &= -\ln(\sigma) - \frac{1}{2}\beta_0 \ln(K F_0) \\
\widetilde{D}_0 &= \frac{\beta_0(2-\beta_0)\sigma^2}{8K^{1-\beta_0} F_0^{1-\beta_0}}
\end{aligned}$$

with

$$q_0 = \begin{cases} \dfrac{K^{1-\beta_0} - F_0^{1-\beta_0}}{1-\beta_0} & \beta_0 < 1 \\ \ln\left(\dfrac{K}{F_0}\right) & \beta_0 = 1. \end{cases}$$

Formulas (23), (24) and (25) are modified as follows.

$$\sigma_0 = \frac{|q_0|}{\sqrt{2B}} = \frac{|q_0|}{d(F_0, V_0; K, V_{\min})} \tag{27}$$

$$\frac{\sigma_1}{\sigma_0} = -\frac{\widetilde{C} + \ln(\sigma_0) + \frac{1}{2}\beta_0 \ln(K F_0)}{2B}. \tag{28}$$

$$\frac{\sigma_2}{\sigma_0} = \frac{3}{2}\left(\frac{\sigma_1}{\sigma_0}\right)^2 - \frac{1}{2B}\left(\widetilde{D} + 3\frac{\sigma_1}{\sigma_0} - \frac{\beta_0(2-\beta_0)\sigma_0^2}{8K^{1-\beta_0} F_0^{1-\beta_0}}\right). \tag{29}$$

This gives the CEV implied volatility expansion

$$\sigma = \sigma_0 \left(1 + \frac{\sigma_1}{\sigma_0} T + \frac{\sigma_2}{\sigma_0} T^2 + o(T^2)\right). \tag{30}$$

The Black implied volatility formulas correspond to the special case $\beta_0 = 1$. The Bachelier (i.e. normal) implied volatility would correspond to $\beta_0 = 0$.

3.7 Generalization

This technique can be generalized easily to other parameterizations of the options prices. Consider a model with local volatility or stochastic volatility, for which there are closed form formulas for European option prices. It can be used as a proxy in the following way.

- Denoting by z_i the parameters of the model, compute $B_*(z_i)$, $\widetilde{C}_*(z_i)$ and $\widetilde{D}_*(z_i)$, the quantities B, $\widetilde{C}$ and $\widetilde{D}$ of the asymptotic expansion (20) for this model at a given strike.
- Find parameters $z_i^{(0)}$ such that $B_*(z_i^{(0)}) = B$ (there can be several solutions).
- Choose a one-dimensional subset of the parameters $z_i = z_i(\lambda)$ which allows a wide range of option prices at the given strike and such that $z_i(0) = z_i^{(0)}$.
- Compute derivatives of $B_*(z_i)$, $\widetilde{C}_*(z_i)$ and $\widetilde{D}_*(z_i)$ with respect to λ at $\lambda = 0$. We use the notation $B_* = B_*(z_i(0))$, $B'_* = \partial_\lambda B_*(z_i(\lambda)) \mid_{\lambda=0} \cdots$
- Write a Taylor expansion $\lambda(T) = \lambda_1 T + \lambda_2 T^2 + o(T^2)$ and write the equality of the asymptotic expansion (20) for the model and the proxy model:

$$\frac{B}{T} + \widetilde{C} + \ln(B) + \widetilde{D}T + \frac{3}{2B}T = \frac{B_* + \lambda_1 T B'_* + \lambda_2 T^2 B'_* + \frac{1}{2}\lambda_1^2 T^2 B''_*}{T}$$
$$+ \widetilde{C}_* + \lambda_1 T \widetilde{C}'_* + \ln(B_* + \lambda_1 T B'_*) + \widetilde{D}_* + \frac{3}{2B_*}T + o(T).$$

- This gives the Taylor expansion of λ:

$$\lambda_1 = \frac{\widetilde{C} - \widetilde{C}_* + \ln(B) - \ln(B_*)}{B'_*}$$
$$\lambda_2 = \frac{\widetilde{D} - \widetilde{D}_* - \lambda_1 \frac{B'_*}{B_*} - \lambda_1 \widetilde{C}'_* - \frac{1}{2}\lambda_1^2 B''_*}{B'_*}.$$

- Plug parameters $z_i(\lambda_1 T + \lambda_2 T^2)$ into the closed form option price of the proxy model to get an approximate price of the option in the real model.

The closer the models are, the better the approximation is. It is clear that if the proxy model is the real model itself, there are no corrections at all. This procedure consists in approximating only the differences between models at a given strike and not the option price itself. In the basic case of Sect. 3.4 where the proxy model is the Black-Scholes model, the approximation leverages on the fact that the volatility surface is more regular than the option price.

4 SABR Model

4.1 Model

The SABR Model [6] is a stochastic volatility model where the volatility is a local volatility function multiplied by a lognormal stochastic volatility:

$$\begin{aligned} dF &= VC(F)dW_1 \\ dV &= \nu V dW_2 \end{aligned}$$

with $\langle dW_1 dW_2 \rangle = \rho dt$. The initial value for V is the parameter[5] α:

$$\alpha = V(0).$$

$C(F)$ is a local volatility function, which is generally

$$C(F) = F^\beta$$

β is a number between 0 and 1 which controls the local skew. 0 corresponds to a normal process and 1 to a lognormal process. The implied volatility at time 0 and at the money is the local volatility αF_0^β.

Depending on the parameters, the origin $F = 0$ could be reached with finite probability in finite time. For example this happens for the CEV process (i.e. even without stochastic volatility) for $\beta \leq \frac{1}{2}$. If F models a positive variable, a boundary condition must be imposed. The asymptotic expansion does not distinguish between different boundary conditions, as the computation is local around the geodesic path. It is valid as long as this geodesic does not reach the boundary. However the maturity validity range may be reduced for low strikes, when the probability of reflection or absorbtion at the origin modifies the probability distribution at the strike considered in a significant way.

In the following sections, we compute the asymptotic expansion for the SABR model. This short maturity expansion is valid when both $\alpha^2 T$ and $\nu^2 T$ are small enough in front of 1. If ones uses CEV implied volatility instead of lognormal implied volatility, the expansion is in $\nu^2 T$ only. Numerical experiments indicates that the approximation remains very good for $\nu^2 T < 1$.

[5]We use the standard notation of α for the initial value of the volatility variable in the SABR model instead of V_0 as in the previous section.

4.2 Order 0: Metric

In order to compute the order 0 implied volatility, the only geometric object involved is the metric. According to the dictionary of Sect. 2.2, its inverse is the covariance matrix

$$\left(g^{ij}\right) = \begin{pmatrix} V^2 C(F)^2 & \rho\nu V^2 C(F) \\ \rho\nu V^2 C(F) & \nu^2 V^2 \end{pmatrix}.$$

This matrix is first simplified by changing the variable F to

$$q = \int_{F_0}^{F} \frac{\mathrm{d}F}{C(F)} \tag{31}$$

which for $C(F) = F^{\beta}$ reads for $\beta \neq 1$

$$q = \frac{F^{1-\beta} - F_0^{1-\beta}}{1-\beta}$$

and for $\beta = 1$

$$q = \ln\left(\frac{F}{F_0}\right).$$

In addition, we rescale the time such that ν disappears of the equations while keeping the same solution of the equations (the variances which are the physical quantities are not changed):

$$\begin{aligned} t &\longrightarrow \nu^2 t \\ \alpha &\longrightarrow \frac{\alpha}{\nu} \\ \nu &\longrightarrow 1 \end{aligned}$$

At the end of the computation, the inverse transformation must be applied to the implied volatility:

$$\sigma\nu \longleftarrow \sigma$$

The matrix in the set of variables (q, V) after this rescaling is

$$\left(g^{ij}\right) = V^2 \begin{pmatrix} 1 & \rho \\ \rho & 1 \end{pmatrix}.$$

This is diagonalized by going from variables (q, V) to (x, y) with

$$x = \frac{q - \rho V}{\sqrt{1 - \rho^2}}$$
$$y = V.$$

The covariance matrix becomes

$$\left(g^{ij}\right) = y^2 \begin{pmatrix} 1 & 0 \\ 0 & 1 \end{pmatrix}$$

and its inverse is the metric

$$\left(g_{ij}\right) = \frac{1}{y^2} \begin{pmatrix} 1 & 0 \\ 0 & 1 \end{pmatrix}$$

which corresponds to the infinitesimal distance

$$\mathrm{d}s^2 = \frac{\mathrm{d}x^2 + \mathrm{d}y^2}{y^2}.$$

This geometry corresponds to the hyperbolic plane, in the Poincaré half-plane representation $(y > 0)$ [7, 8]. Geodesics are vertical lines and semi-circles orthogonal to the $y = 0$ axis. The geodesic distance between two points (x_1, y_1) and (x_2, y_2) can be computed:

$$d(x_1, y_1; x_2, y_2) = \cosh^{-1}\left(1 + \frac{(x_2 - x_1)^2 + (y_2 - y_1)^2}{2y_1 y_2}\right).$$

In the (q, V) variables, going from $q = 0$, $V = \alpha$ to q, V the geodesic distance is

$$d(0, \alpha; q, V) = \cosh^{-1}\left(1 + \frac{q^2 + (V - \alpha)^2 - 2\rho q (V - \alpha)}{2(1 - \rho^2)\alpha V}\right).$$

For a given strike, i.e. a given q, it is minimized by the volatility

$$V_{\min} = \sqrt{\alpha^2 + 2\rho\alpha q + q^2}$$

and the minimal distance is

$$d(0, \alpha; q) = \cosh^{-1}\left(\frac{V_{\min} - \rho q - \rho^2 \alpha}{(1 - \rho^2)\alpha}\right) = \left|\ln\left(\frac{V_{\min} + \rho\alpha + q}{(1 + \rho)\alpha}\right)\right|.$$

Equation (23) gives the order 0 implied volatility

$$\sigma_0 = \frac{\ln\left(\frac{K}{F_0}\right)}{\ln\left(\frac{V_{\min} + \rho\alpha + q}{(1+\rho)\alpha}\right)}.$$

(We have dropped the absolute values as the numerator and the denominator have the same sign.)

Plugging the expression for $V_{\min}$ and going back to the original time, with ν factors, the order 0 implied volatility for the SABR model is

$$\sigma_0 = \frac{\nu \ln\left(\frac{K}{F_0}\right)}{\ln\left(\frac{\sqrt{\alpha^2 + 2\rho\alpha\nu q + \nu^2 q^2} + \rho\alpha + q\nu}{(1+\rho)\alpha}\right)} \tag{32}$$

with

$$q = \begin{cases} \frac{K^{1-\beta} - F_0^{1-\beta}}{1-\beta} & \beta < 1 \\ \ln\left(\frac{K}{F_0}\right) & \beta = 1. \end{cases}$$

At the money, the limit of this expression is simply

$$\sigma_0(F_0) = \alpha F_0^{\beta-1}$$

which is the local volatility.

4.3 Order 1: Connection

To compute the order 1 correction, we need the scalar factor in the time value expansion, given by $\widetilde{C}$ in Eq. (15), with C given in Eq. (12).

For the hyperbolic plane, the Van Vleck–Morette determinant can be computed as a function of the geodesic distance:

$$\Delta = \frac{d}{\sinh(d)}.$$

We need also

$$\sigma_F(K, V) = VC(K) = VK^{\beta}$$

$$g(K,V) = \frac{1}{V^4 C(K)^2(1-\rho^2)} = \frac{1}{V^4 K^{2\beta}(1-\rho^2)}$$

Using that $d' = 0$ for the minimum distance at $V = V_{\min}$, we compute

$$B'' = dd'' = \frac{d}{\alpha V_{\min}(1-\rho^2)\sinh(d)}$$

Terms simplify against each other to give at $V = V_{\min}$

$$\widetilde{C} = -\ln\left(\sqrt{\alpha V_{\min}}K^{\beta}\right) + \mathcal{M} \tag{33}$$

$\mathcal{M}$ is the integral of the connection 1-form on the geodesic. According to formula (4), the connection is given by

$$A = \frac{1}{2(1-\rho^2)}\,\mathrm{d}\ln(C(F)) - \frac{\rho C'(F)}{2(1-\rho^2)}\,\mathrm{d}V. \tag{34}$$

In fact, A can be rewritten in (x, y) variables as

$$A = \frac{C'(F)}{2\sqrt{1-\rho^2}}\,\mathrm{d}x.$$

It must be integrated from

$$\begin{pmatrix} x_1 \\ y_1 \end{pmatrix} = \begin{pmatrix} \frac{-\rho\alpha}{\sqrt{1-\rho^2}} \\ \alpha \end{pmatrix}$$

to

$$\begin{pmatrix} x_2 \\ y_2 \end{pmatrix} = \begin{pmatrix} \frac{q-\rho V}{\sqrt{1-\rho^2}} \\ V \end{pmatrix}.$$

If $\beta = 1$, the 1-form

$$A = \frac{1}{2\sqrt{1-\rho^2}}\,\mathrm{d}x$$

is exact and can be integrated directly:

$$\mathcal{M} = \frac{q-\rho V+\rho\alpha}{2\left(1-\rho^2\right)} = \frac{1}{2}\ln\left(\frac{K}{F_0}\right) + \frac{\rho}{2\left(1-\rho^2\right)}\left(\rho\ln\left(\frac{K}{F_0}\right) - V + \alpha\right).$$

We consider now the general case where $\beta < 1$. If $x_1 = x_2$, the geodesic is a vertical line. As A is along $\mathrm{d}x$, its integral is zero:

$$\mathcal{M} = 0.$$

In other cases, the first part of Eq. (34) is an exact form and can be integrated directly:

$$\int_{(F_0,\alpha)}^{(K,V)} \frac{1}{2(1-\rho^2)}\,\mathrm{d}\ln(C(F)) = \frac{1}{2(1-\rho^2)}\ln\left(\frac{C(K)}{C(F)}\right) = \frac{\beta}{2(1-\rho^2)}\ln\left(\frac{K}{F}\right). \tag{35}$$

The second part must be integrated on the geodesic path. The geodesic is a semi-circle with origin $(X, 0)$, radius R and going through (x_1, x_2) and (y_1, y_2). The origin is therefore

$$X = \frac{x_2^2 - x_1^2 + y_2^2 - y_1^2}{2(x_2 - x_1)} \tag{36}$$

and the radius

$$R = \sqrt{y_1^2 + (x_1 - X)^2}. \tag{37}$$

We parameterize the geodesic by $t = \tan(\theta/2)$ where θ is the angle on the circle:

$$x = X + R\,\frac{1-t^2}{1+t^2}$$
$$y = R\,\frac{2t}{1+t^2}.$$

In this parametrization, the geodesic distance is given by

$$\mathrm{d}s = \mathrm{d}\ln(t)$$

and we can compute

$$-\int_{\mathcal{C}} \frac{\rho\beta F^{\beta-1}}{2(1-\rho^2)}\mathrm{d}V = -\frac{\rho^2\beta}{2(1-\rho^2)}\ln\left(\frac{K}{F}\right) - \frac{\rho\beta}{(1-\beta)\sqrt{1-\rho^2}}\left[G(t_2) - G(t_1)\right] \tag{38}$$

with

$$G(t) = \tan^{-1}(t)$$
$$+\begin{cases} -\dfrac{a+bX}{\sqrt{(a+bX)^2-(1-\beta)^2R^2}}\tan^{-1}\left(\dfrac{cR+t(a+b(X-R))}{\sqrt{(a+bX)^2-(1-\beta)^2R^2}}\right) & (a+bX)^2 > (1-\beta)^2R^2 \\ \dfrac{a+bX}{cR+t(a+b(X-R))} & (a+bX)^2 = (1-\beta)^2R^2 \\ \dfrac{a+bX}{\sqrt{(1-\beta)^2R^2-(a+bX)^2}}\widetilde{\tanh^{-1}}\left(\dfrac{cR+t(a+b(X-R))}{\sqrt{(1-\beta)^2R^2-(a+bX)^2}}\right) & (a+bX)^2 < (1-\beta)^2R^2 \end{cases} \tag{39}$$

$$a = F_0^{1-\beta}$$
$$b = (1-\beta)\sqrt{1-\rho^2}$$
$$c = (1-\beta)\rho$$

$$t_i = \sqrt{\frac{R - x_i + X}{R + x_i - X}} \tag{40}$$

and

$$\widetilde{\tanh^{-1}}(z) = \frac{1}{2}\ln\left|\frac{1+z}{1-z}\right|,$$

which coincides with the inverse function of tanh on $]-1;1[$. Summing Eqs. (35) and (38), the integral of the connection is finally

$$\mathcal{M} = \frac{\beta}{2}\ln\left(\frac{K}{F}\right) - \frac{\rho\beta}{(1-\beta)\sqrt{1-\rho^2}}\left[G(t_2) - G(t_1)\right]. \tag{41}$$

Replacing $\mathcal{M}$ in Eq. (33) we get

$$\widetilde{C} = -\frac{1}{2}\ln\left(\alpha F_0^{\beta} V_{\min} K^{\beta}\right) + \begin{cases} -\dfrac{\rho\beta}{(1-\beta)\sqrt{1-\rho^2}}\left[G(t_2) - G(t_1)\right] & \beta < 1 \\ \dfrac{\rho}{2\left(1-\rho^2\right)}\left(\rho\ln\left(\dfrac{K}{F_0}\right) - V_{\min} + \alpha\right) & \beta = 1 \end{cases} \tag{42}$$

Restoring the factor ν, the order 1 correction is given by Eq. (24):

$$\frac{\sigma_1}{\sigma_0} = -\nu^2 \frac{\widetilde{C} + \ln\left(\frac{\sigma_0}{\nu}\sqrt{K F_0}\right)}{\ln^2\left(\dfrac{\sqrt{\alpha^2 + 2\rho\alpha\nu q + \nu^2 q^2} + \rho\alpha + q\nu}{(1+\rho)\alpha}\right)} \tag{43}$$

σ_0 is given by Eq. (32), $\widetilde{C}$ by Eq. (42) (with α divided by ν, also inside $V_{\min}$), where X, R, t_1 and t_2 are given in Eqs. (36), (37) and (40) from

$$\begin{pmatrix} x_1 \\ y_1 \end{pmatrix} = \begin{pmatrix} \dfrac{-\rho\alpha}{\nu\sqrt{1-\rho^2}} \\ \dfrac{\alpha}{\nu} \end{pmatrix} \text{ and } \begin{pmatrix} x_2 \\ y_2 \end{pmatrix} = \begin{pmatrix} \dfrac{\nu q - \rho\sqrt{\alpha^2 + 2\rho\alpha\nu q + \nu^2 q^2}}{\nu\sqrt{1-\rho^2}} \\ \dfrac{\sqrt{\alpha^2 + 2\rho\alpha\nu q + \nu^2 q^2}}{\nu} \end{pmatrix}$$

and $G(t)$ is defined in formula (39).

Using this expression, the first order implied volatility is

$$\sigma = \sigma_0 \left(1 + \frac{\sigma_1}{\sigma_0} t + o(t)\right)$$

which is valid for all positive strikes. Exactly at the money, the formula we give must be replaced by its limit, which can be computed by a Taylor expansion or numerically. At the money and only at the money it appears to be equal to the original HKLW formula:

$$\frac{\sigma_1}{\sigma_0}(F_0) = \frac{1}{24}\alpha^2(1-\beta)^2 F_0^{2(\beta-1)} + \frac{1}{4}\rho\alpha\nu\beta F_0^{\beta-1} + \frac{1}{12}\nu^2 - \frac{1}{8}\rho^2\nu^2. \tag{44}$$

This is not surprising as their expansion is in fact an expansion in both maturity and moneyness (eventually of order 0 in moneyness).

4.4 Order 2

To compute the second order correction to implied volatility, we need to compute $\widetilde{D}$ as defined in Eq. (16), with $D = -a_1$ defined in Eq. (9).

We have to compute a_1 as defined in Eq. (9). Most of the integration can be done analytically. We have first the integral of Q along the geodesic:

$$a_1^{(Q)} = -\frac{1}{d} \int_C Q \mathrm{d}s.$$

According to Eq. (5), Q is

$$Q = \frac{\beta}{4}\left(1 - \beta + \frac{\beta}{2(1-\rho^2)}\right) \frac{V^2}{F^{2(1-\beta)}}.$$

Using the values defined in the previous section for X, R, t_1, t_2, a, b and c, its integral along the geodesic is

$$a_1^{(Q)} = \frac{\beta}{2}\left(1 - \beta + \frac{\beta}{2(1-\rho^2)}\right) \frac{R^2}{(1-\beta)^2 R^2 - (a+bX)^2} \frac{H(t_2) - H(t_1)}{\ln(t_2) - \ln(t_1)} \tag{45}$$

with

$$H(t) = \frac{a + b(R + X) + cRt}{(a + bX)(1 + t^2) + bR(1 - t^2) + 2cRt}$$
$$+ \begin{cases} \dfrac{cR}{\sqrt{(a+bX)^2 - (1-\beta)^2 R^2}} \tan^{-1}\left(\dfrac{cR + t(a + b(X - R))}{\sqrt{(a+bX)^2 - (1-\beta)^2 R^2}} \right) & (a+bX)^2 > (1-\beta)^2 R^2 \\ -\dfrac{cR}{\sqrt{(1-\beta)^2 R^2 - (a+bX)^2}} \widetilde{\tanh^{-1}}\left(\dfrac{cR + t(a + b(X - R))}{\sqrt{(1-\beta)^2 R^2 - (a+bX)^2}} \right) & (a+bX)^2 < (1-\beta)^2 R^2 . \end{cases}$$

Note that in the denominator, the quantity $\ln(t_2) - \ln(t_1)$ is up to a sign the geodesic distance d. If $\beta = 1$, $a_1^{(Q)}$ reduces to

$$a_1^{(Q)} = -\frac{R \mid x_2 - x_1 \mid}{8(1 - \rho^2)d}. \tag{46}$$

The Laplacian on the hyperbolic plane is in (x, y) coordinates

$$D^i D_i = y^2(\partial_x^2 + \partial_y^2).$$

As the Van Vleck–Morette determinant $\Delta = \frac{s}{\sinh(s)}$ depends only on the geodesic distance s, its derivative on the orthogonal coordinate vanishes: $\partial_\perp \Delta = 0$. On the other hand, by definition the parallel transport on the geodesic curve has no covariant derivative along the curve: $\nabla_s \mathcal{P} = 0$. As a consequence, there is no crossed term and both terms decouple: we have

$$\mathcal{P}^{-1}\Delta^{-1/2}\nabla^i\nabla_i(\mathcal{P}\Delta^{1/2}) = \mathcal{P}^{-1}\nabla^i\nabla_i\mathcal{P} + \Delta^{-1/2}D^i D_i\Delta^{1/2}$$

(the $\mathbb{R}$ charge is carried only by $\mathcal{P}$).

The metric part can be integrated analytically [1, 7]:

$$a_1^{(R)} = -\frac{1}{8}\left[1 + \frac{1}{d}\left(\frac{\cosh(d)}{\sinh(d)} - \frac{1}{d}\right)\right]. \tag{47}$$

The last part to integrate is the $\mathbb{R}$ connection term $\mathcal{P}^{-1}\nabla^i\nabla_i\mathcal{P}$. As the action of gauge transformations on the heat kernel expansion is fully carried by the parallel transport term $\mathcal{P}$, a_1 can only depend on gauge-invariant quantities constructed from $F = \mathrm{d}A$. We split therefore $A = A^{(0)} + A^{(1)}$ into a pure gauge part $A^{(0)}$ and $A^{(1)}$ such that $F = \mathrm{d}A^{(1)}$:

$$A^{(0)} = \frac{1}{2}\mathrm{d}\ln(C(F))$$
$$A^{(1)} = \frac{\rho^2}{2(1-\rho^2)}\mathrm{d}\ln(C(F)) - \frac{\rho C'(F)}{2(1-\rho^2)}\mathrm{d}V.$$

Forgetting $A^{(0)}$ which is pure gauge (it can be checked by hand that $A^{(0)}$ will not contribute), we denote by $\mathcal{P}^{(1)}$ the $A^{(1)}$ part of the parallel transport:

$$\mathcal{P}^{(1)} = e^{-\mathcal{M}^{(1)}} = e^{-\int_C A^{(1)}}$$

with

$$\mathcal{M}^{(1)} = \begin{cases} -\dfrac{\rho\beta}{(1-\beta)\sqrt{1-\rho^2}} \left[G(t_2) - G(t_1)\right] & \beta < 1 \\ \dfrac{\rho}{2\left(1-\rho^2\right)} \left(\rho \ln\left(\dfrac{K}{F_0}\right) - V + \alpha\right) & \beta = 1. \end{cases} \tag{48}$$

Using $\mathcal{M}^{(1)}$ and $A^{(1)}$ which in (x, y) variables is

$$A^{(1)} = \frac{\rho C'(F)}{2}\left(\frac{\rho}{\sqrt{1-\rho^2}}\,\mathrm{d}x - \mathrm{d}y\right),$$

we can rewrite

$$\frac{1}{2}\mathcal{P}^{-1}\nabla^i\nabla_i\mathcal{P} = \frac{1}{2}y^2\Big[-\left(\partial_x^2 + \partial_y^2\right)\mathcal{M}^{(1)} + \left(\partial_x\mathcal{M}^{(1)} - A_x^{(1)}\right)^2 + \left(\partial_y\mathcal{M}^{(1)} - A_y^{(1)}\right)^2 + \partial_x A_x^{(1)} + \partial_y A_y^{(1)}\Big]$$

where the specific form of $A^{(1)}$ can be used to simplify terms

$$\partial_x A_x^{(1)} + \partial_y A_y^{(1)} = 0.$$

Computing $\mathcal{M}^{(1)}$ and $A^{(1)}$ analytically, we compute numerically the x and y derivatives and integrate numerically along the geodesic curve to get

$$a_1^{(A)} = \frac{1}{2d}\int_C \mathrm{d}s\, y^2\Big[-\left(\partial_x^2 + \partial_y^2\right)\mathcal{M}^{(1)} + \left(\partial_x\mathcal{M}^{(1)} - A_x^{(1)}\right)^2 + \left(\partial_y\mathcal{M}^{(1)} - A_y^{(1)}\right)^2\Big]. \tag{49}$$

This integral can be computed by a numerical quadrature with few points. For $\beta = 1$ the connection has in fact no curvature and therefore $a_1^{(A)} = 0$.

We compute also the following quantities (at $V = V_{\min}$):

$$\begin{aligned}
d'' &= \frac{1}{\alpha(1-\rho^2)V_{\min}\sinh(d)} \\
B'' &= dd'' \\
\frac{B^{(3)}}{B''} &= -\frac{3}{V_{\min}} \\
\frac{B^{(4)}}{B''} &= \frac{12}{V_{\min}^2} - 3d''\left(\frac{\cosh(d)}{\sinh(d)} - \frac{1}{d}\right).
\end{aligned} \tag{50}$$

We need finally $\widetilde{C}'$ and $\widetilde{C}''$. Using our decomposition $\mathcal{M} = \frac{\beta}{2}\ln\frac{K}{F} + \mathcal{M}^{(1)}$, we can write from formulas (12) and (15)

$$\widetilde{C} = -\frac{\beta}{2}\ln(KF) - \frac{1}{2}\ln\left(\frac{d}{\sinh(d)}\right) + \frac{1}{2}\ln\left(B''\right) + \ln\left(1-\rho^2\right) + \mathcal{M}_1.$$

Its first and second derivatives in V at $V = V_{\min}$ are

$$\begin{aligned}
\widetilde{C}' &= \frac{1}{2}\frac{B^{(3)}}{B''} + \mathcal{M}^{(1)\prime} \\
\widetilde{C}'' &= \frac{1}{2}d''\left(\frac{\cosh(d)}{\sinh(d)} - \frac{1}{d}\right) + \frac{1}{2}\frac{B^{(4)}}{B''} - \frac{1}{2}\left(\frac{B^{(3)}}{B''}\right)^2 + \mathcal{M}^{(1)\prime\prime}.
\end{aligned}$$

We choose to differentiate $\mathcal{M}^{(1)}$ numerically, by finite difference, as the analytical expression is very long and hard to simplify.

Simplifying $\frac{\widetilde{C}''}{2B''}$ against $-a_1^{(R)}$ we get finally

$$\widetilde{D} = -a_1^{(Q)} - a_1^{(A)} + \frac{1}{8} + \frac{1}{2B''}\left[\mathcal{M}^{(1)\prime\prime} - \mathcal{M}^{(1)\prime 2} + \frac{3}{V_{\min}}\mathcal{M}^{(1)\prime} - \frac{3}{4V_{\min}^2}\right]$$

with $a_1^{(Q)}$ given in Eq. (45), $a_1^{(A)}$ in (49), B'' in (50) and $\mathcal{M}^{(1)}$ in (48) with $G(t)$ defined in Eq. (39). For $\beta = 1$, this expression can be simplified using Eq. (46) for $a_1^{(Q)}$ and

$$\begin{aligned}
a_1^{(A)} &= 0 \\
\mathcal{M}^{(1)\prime} &= -\frac{\rho}{2(1-\rho^2)} \\
\mathcal{M}^{(1)\prime\prime} &= 0.
\end{aligned}$$

We have finally obtained the second order corrective term

$$\frac{\sigma_2}{\sigma_0} = \frac{3}{2}\left(\frac{\sigma_1}{\sigma_0}\right)^2 - \frac{1}{2B}\left(\widetilde{D} + 3\frac{\sigma_1}{\sigma_0} - \frac{\sigma_0^2}{8}\right)$$

such that we get the quadratic approximation to implied volatility

$$\sigma = \sigma_0\left(1 + \frac{\sigma_1}{\sigma_0}T + \frac{\sigma_2}{\sigma_0}T^2 + o(T^2)\right).$$

The computation has been done in redefined variables such that $\nu = 1$. To restore the ν factors, α must be replaced by $\frac{\alpha}{\nu}$, T by $\nu^2 T$ and the final implied volatility must be multiplied by ν.

At the money, the formula for $\frac{\sigma_2}{\sigma_0}$ looks divergent but its limit is well defined. We compute this limit numerically, although it could be done analytically.

4.5 CEV Volatility

The results of Sect. 3.6 can be used to invert the SABR volatility into a CEV fractional volatility. Using formulas of Sect. 3.6 the implied CEV volatility is computed and used in the closed-form option prices of the CEV model.

This appears to be useful at low strikes for $\beta < 1/2$ or with small volatility of volatility: only the corrections to the CEV model which come from the stochastic volatility are approximated, not the local volatility part. For example, at the money the first order coefficient of the Black volatility which is given by Eq. (44) becomes for the CEV volatility

$$\frac{\sigma_1}{\sigma_0}(F_0) = \frac{1}{4}\rho\alpha\nu\beta F_0^{\beta-1} + \frac{1}{12}\nu^2 - \frac{1}{8}\rho^2\nu^2.$$

The corrective term in $\alpha^2 T$ has disappeared.

4.6 Numerical Results

We present in Figs. 1 and 2 the implied volatility given by our expansion and compare it to the implied volatility computed by a two-dimensional finite difference method scheme. We also show for comparison the implied volatility given by the original formula of [6]. In this example, parameters are $F_0 = 4$, $\alpha = 30\,\%$, $\beta = 0.7$, $\nu = 40\,\%$, $\rho = -0.5$. The FDM scheme is a second order Yanenko scheme [14] with exponential fitting. We use 400 points in strike, 200 points in volatility and 30 time steps.

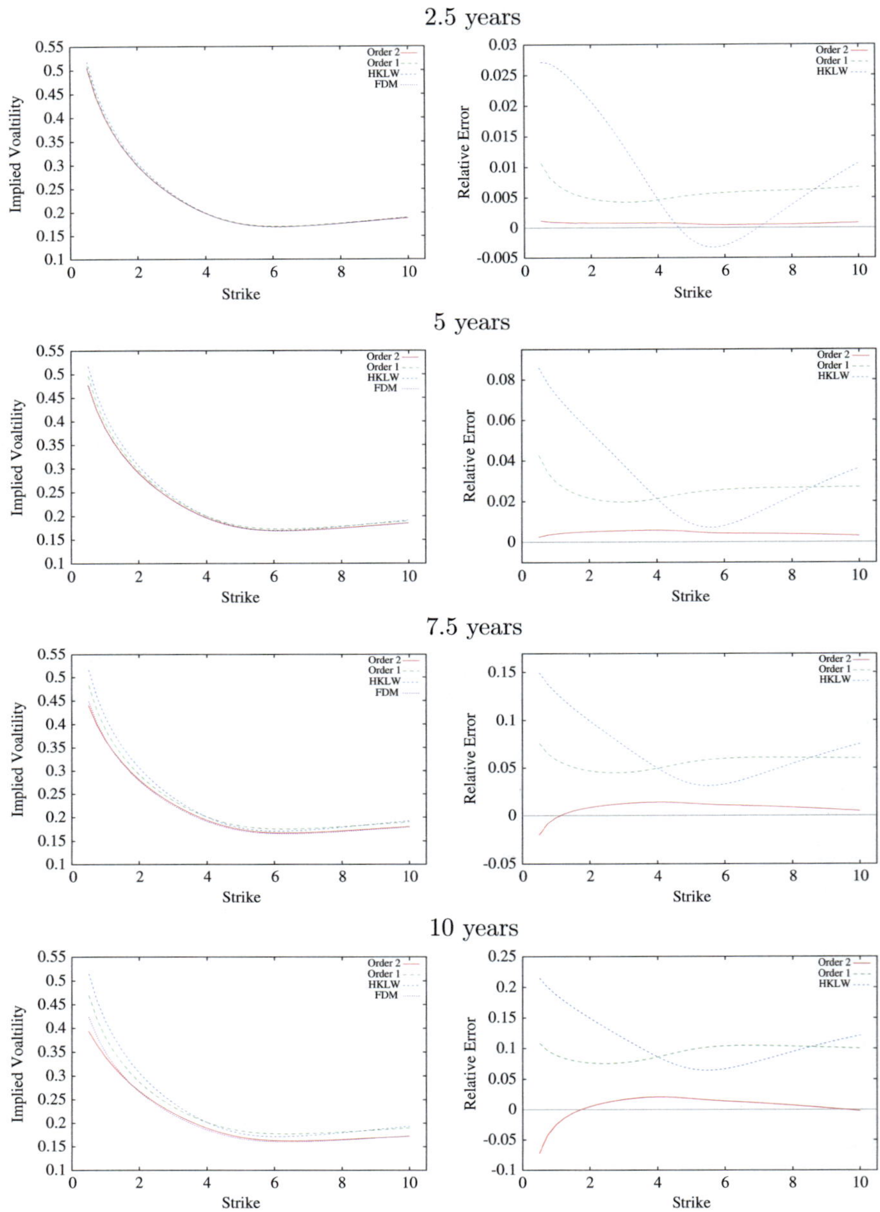

Fig. 1 Implied volatility and relative error for the SABR model with parameters $F_0 = 4$, $\alpha = 30\,\%$, $\beta = 0.7$, $\nu = 40\,\%$, $\rho = -0.5$ and maturities 2.5 yr, 5 yr, 7.5 yr and 10 yr. On the *left*, implied volatilities are plotted for our first order and second order expansions, the original formula of [6] and are compared to the result of a FDM solution. On the *right* are the relative errors with respect to this reference solution

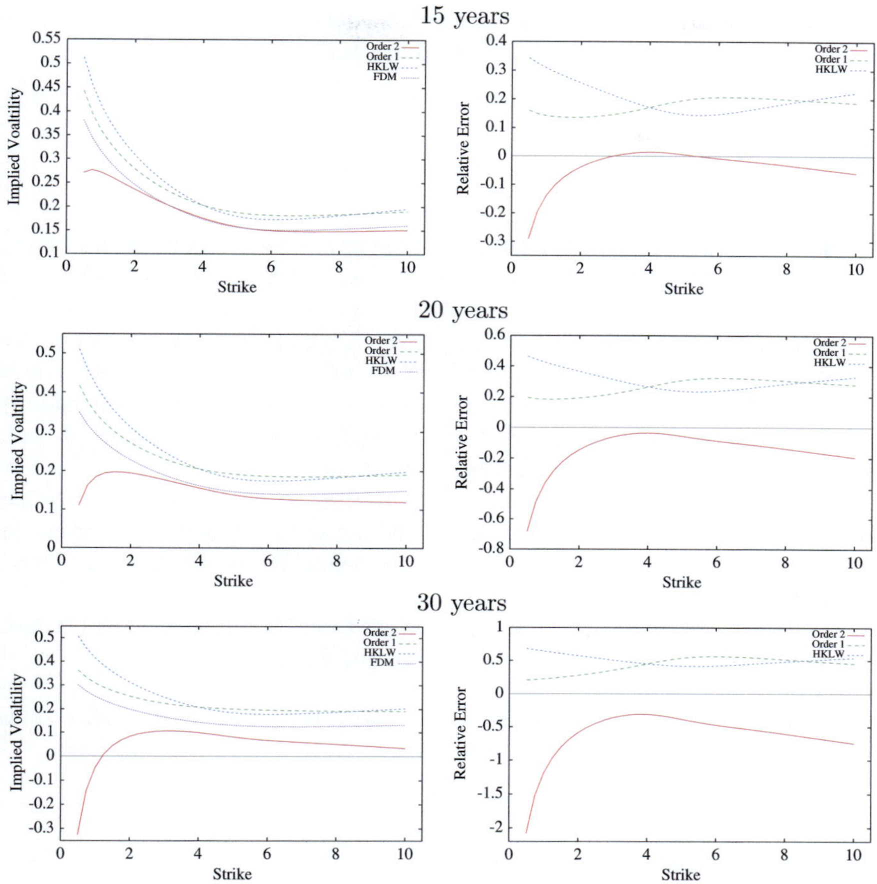

Fig. 2 Implied volatility and relative error for the SABR model with parameters $F_0 = 4, \alpha = 30\,\%$, $\beta = 0.7, \nu = 40\,\%, \rho = -0.5$ and maturities 10 yr, 20 yr and 30 yr. On the *left*, implied volatilities are plotted for our first order and second order expansions, the original formula of [6] and are compared to the result of a FDM solution. On the *right* are the relative errors with respect to this reference solution

At very short maturities, all expansions are acceptable as the expansion is dominated by the order 0 term. At first order our expansion is equal to the HKLW at the money but is more regular in strikes and is better in the wings as our computation does not involve any approximation in the moneyness. Our second order expansion is one order of magnitude more precise.[6] When maturity grows, first order expansions lose precision but the second order remain relatively good up to 10 years, where $\nu^2 T = 1.6$. At higher maturities, the second order expansion explodes quadratically and finally gives even negative volatilities at very long maturity and low strikes. At

[6]In fact at very short maturities, the FDM scheme we use is less precise and less stable than this second order expansion, especially in the wings where the probability density is very small.

long maturities, a FDM or an other numerical method must be used, unless a valid long maturity expansion could be computed more efficiently.

5 Conclusion

We have presented a general method to compute a Taylor expansion in maturity of implied volatility for stochastic volatility models. We give exact formulas for the first and second order corrections. As an application, we have computed this expansion for the SABR model and compared it to the implied volatility given by a numerical scheme and to the original HKLW formula. It appears that it gives more precise results than the usual formula and extends the domain where a short maturity expansion can be used. Outside this range of validity, other methods must be used: numerical schemes or possibly other approximations.

If a closer model with closed formulas than Black-Scholes exists, we provide a method to use this model as a proxy to extend the domain of validity of the expansion. It would be interesting to see the results of this method for the SABR model with a stochastic volatility model as a proxy.

Obtaining exact option prices at all maturities would be a non-perturbative computation, which is a longstanding issue in theoretical physics.

Acknowledgments We thank Erwan Curien for his interest in this work, Martial Millet, Xavier Lacroze and Wafaa Bennehhou for many discussions.

Appendix: Mean reversion

At long maturity, the SABR model is not realistic as the volatility process is a geometric Brownian motion. In particular the variance of the volatility increases linearly in time. A direct extension would be to add mean reversion to the volatility process, either on the volatility or on the variance. The asymptotic expansion can still be computed at order 2. However its domain of validity is usually reduced: in addition to other conditions, the maturity must be small compared to the mean reversion characteristic time.

We impose mean reversion on the volatility process as

$$\mathrm{d}V = \nu V \mathrm{d}W_2 + \kappa(\overline{V} - V)\mathrm{d}t$$

The metric is not modified as it describes the diffusion part. Expressions for A and Q are modified as follows:

$$A = A_{\mathrm{SABR}} - \frac{\rho\kappa(V - \overline{V})}{\nu V^2(1-\rho^2)F^\beta}\mathrm{d}F + \frac{\kappa(V - \overline{V})}{\nu^2 V^2(1-\rho^2)}\mathrm{d}V$$

$$Q = Q_{\mathrm{SABR}} + \frac{1}{2}\frac{\kappa^2 (V - \overline{V})^2}{\nu^2 V^2 (1-\rho^2)} + \frac{1}{2}\kappa - \kappa\frac{\overline{V}}{V} - \frac{1}{2}\frac{\rho\beta\kappa(V - \overline{V})}{\nu(1-\rho^2)F^{1-\beta}}.$$

At first order, the integral of A along the geodesic is needed. Using the same notations as in Sect. 4.3 where variables have been rescaled such that $\nu = 1$, it gets an additional term

$$\mathcal{M} = \mathcal{M}_{\mathrm{SABR}} + \frac{\rho\kappa}{\sqrt{1-\rho^2}}\left[2\left(\tan^{-1}(t_2) - \tan^{-1}(t_1)\right) - \frac{\overline{V}}{R}\ln\left(\frac{t_2}{t_1}\right)\right]$$
$$+ \kappa\left[\ln\left(\frac{V}{\alpha}\right) + \frac{\overline{V}}{V} - \frac{\overline{V}}{\alpha}\right].$$

The second order correction involves a one-dimensional integral which can be computed numerically by a quadrature with a few points, although it may be possible to get an analytical expression.

References

1. Avramidi, I.G.: Analytical and geometric methods for heat kernel applications in finance. http://infohost.nmt.edu/~iavramid/notes/hkt/hktslides7b.pdf (2007)
2. Berestycki, H., Busca, J., Florent, I.: Asymptotics and calibration of local volatility models. Quant. Financ. **2**(1), 61–69 (2002)
3. Berestycki, H., Busca, J., Florent, I.: Computing the implied volatility in stochastic volatility models. Commun. Pure Appl. Math. **57**(10), 1352–1373 (2004)
4. Cox, J.C.: Notes on option pricing I: constant elasticity of diffusions. Working paper (1975)
5. DeWitt, B.S.: Dynamical Theory of Groups and Fields. Gordon and Breach, New York (1965)
6. Hagan, P.S., Kumar, D., Lesniewski, A.S., Woodward, D.E.: Managing smile risk. Wilmott Mag. **1**(8), 84–108 (2002)
7. Hagan, P.S., Lesniewski, A.S., Woodward, D.E.: Probability distribution in the SABR model of stochastic volatility. In: Friz, P., Gatheral, J., Gulisashvili, A., Jacquier, A., Teichmann, J. (eds.) Large Deviations and Asymptotic Methods in Finance. Springer Proceedings in Mathematics and Statistics, vol. 110 (2015)
8. Henry-Labordère, P.: A general asymptotic implied volatility for stochastic volatility models. http://arxiv.org/abs/cond-mat/0504317 (2005)
9. Henry-Labordère, P.: Analysis, Geometry, and Modeling in Finance: Advanced Methods in Option Pricing. Chapman & Hall/CRC, Boca Raton (2008)
10. Heston, S.L.: A closed-form solution for options with stochastic volatility with applications to bond and currency options. Rev. Financ. Stud. **6**(2), 327–343 (1993)
11. Varadhan, S.: Diffusion processes in a small time interval. Commun. Pure Appl. Math. **20**(4), 659–685 (1967)
12. Varadhan, S.: On the behavior of the fundamental solution of the heat equation with variable coefficients. Commun. Pure Appl. Math. **20**(2), 431–455 (1967)
13. Vassilevich, D.V.: Heat kernel expansion: user's manual. Physics report 388, 279–360. http://arxiv.org/abs/hep-th/0306138 (2003)
14. Yanenko, N.N.: The Method of Fractional Steps. Springer, New York (1971)

Unifying the BGM and SABR Models: A Short Ride in Hyperbolic Geometry

Pierre Henry-Labordère

Abstract In this paper, using a geometric method introduced in (Henry-Labordère Large Deviations and Asymptotic Methods in Finance (2015) [12]) and initiated by (Avellaneda et al. Risk Mag. (2002) [4]), we derive an asymptotic swaption implied volatility at the first-order for a general stochastic volatility Libor Market Model. This formula is useful to quickly calibrate a model to a full swaption matrix. We apply this formula to a specific model where the forward rates are assumed to follow a multi-dimensional CEV process correlated to a SABR process. For a caplet, this model degenerates to the classical SABR model and our asymptotic swaption implied volatility reduces naturally to the Hagan-al formula (Hagan et al. Willmott Mag. 88–108 (2002) [11]). The geometry underlying this model is the hyperbolic manifold $\mathbb{H}^{n+1}$ with n the number of Libor forward rates.

Keywords Heat kernel expansion · Hyperbolic geometry · Asymptotic smile formula · Stochastic Libor market model

1 Introduction

The BGM model [6, 14] has recently been the focus of much attention as it gives a theoretical justification for pricing caps-floors using the classical Black-Scholes formula. The basic (physical) random variables are given by the Libor forward rates which are assumed to follow a correlated log-normal process. As the forward swap rate model implied by the BGM model is quite complicated (the swap forward rate is not log-normally distributed), the calibration to a swaption matrix is difficult. An asymptotic swaption implied volatility (at the zero-order in the swaption maturity) was initially derived by Rebonato [17], Hull and White [9] for the BGM model.

Despite its great success, the BGM model presents the same drawbacks as the classical Black-Scholes theory: as the forward rates follow a correlated log-normal

P. Henry-Labordère (✉)
Société Générale, Global Markets Quantitative Research, Paris, France
e-mail: pierre.henry-labordere@sgcib.com

P.K. Friz et al. (eds.), *Large Deviations and Asymptotic Methods in Finance*, Springer Proceedings in Mathematics & Statistics 110,
DOI 10.1007/978-3-319-11605-1_3

process, the model is not able to calibrate the full swaption matrix in/out-the money (in particular the caplets smile) and give a good dynamics to the Libor rates. The incorporation of a swaption smile can be obtained by introducing more elaborated models which should be flexible enough to calibrate caplets and a grid of swaption volatilities (not necessary at the money) across all swaption expiries and underlying swap maturities. One property that these models must still share is their ability to quickly calibrate the swaption matrix without using complicated numerical routines such as Monte-Carlo simulation which are usually noisy and time-consuming. In this context, Andersen-Andreasen introduced the CEV Libor Market Model (LMM) [1] which assumes that each Libor forward rate follows a CEV process, and showed how to obtain an asymptotic swaption smile. Their method is still based on the Rebonato "freezing" argument which consists in assuming that the ratio of a forward Libor rate over the swap rate and the derivative of the swap rate according to a forward Libor rate are almost constant (and therefore equal to their values at the spot).

Recently, for this specific model, Kawai found a more accurate asymptotic formula using the Wiener chaos expansion [15]. Although giving more flexibility than the BGM model, the CEV LMM model is still not able to calibrate the swaption matrix for in/out strikes and in this context, we are naturally led to use stochastic volatility LMM. The literature on this subject is not particularly large. Andersen-al introduced a LMM where the Libors follow a multi-dimensional correlated CEV process coupled (but uncorrelated) to a Heston model [2, 3] and recently Piterbarg modifies this model by allowing the model parameters to be time-dependent [16]. Using an averaging principle, which consists in replacing the time-dependent parameters by some effective constant parameters, Piterbarg derives an asymptotic volatility. Note that as these models are uncorrelated to the stochastic volatility, the swaption fair value is simply given by the fair price in the case of a local volatility model conditional to the stochastic volatility process as explained by the Hull and White decomposition [10]. An asymptotic expression can then be generated by approximating the moments of the volatility process [2].

For pricing exotic options (such as Bermudan swaptions for example), it is simpler or more natural to model directly the forward swap rate with a stochastic volatility process. For example, the SABR model [11] was introduced to fulfill this goal. An asymptotic swaption smile formula (at the first-order) was derived for this specific model and helps to calibrate quickly the model to liquid market data. In this context, it is natural to try to reconcile/unify both benchmark models, the BGM and SABR models. We therefore introduce a LMM where the forward rates follow a multi-dimensional CEV process (with one beta for each Libor forward rate) correlated to a SABR model. As it is the case for the SABR model, we impose that the Libors are correlated to an unique volatility and it is therefore not possible to follow the Andersen-al [3] method (i.e. the Hull and White decomposition) to derive an asymptotic swaption smile.

In this paper, we pursue our previous work on the application of the heat kernel expansion on a Riemannian manifold endowed with an Abelian connection [12] to derive an asymptotic smile formula for a swaption. The plan of this paper is as follows: in the first part, we will recall some definitions and present a list of recent

Libor Market Models. In the second part, we apply the heat kernel expansion to derive an asymptotic swaption smile formula at the first-order valid for any LMM. In the third part, we present our stochastic LMM and apply this general formula. We will prove that the geometry underlying this model is the hyperbolic manifold $\mathbb{H}^{n+1}$ with n the number of forward rates. Furthermore, we show that the "freezing" argument is no longer valid when we try to price a swaption in/out the money.

2 Libor Market Models

We denote by $F_k(t) \equiv F(t, T_{k-1}, T_k)$ with the forward rate resetting at T_{k-1} with $\tau_k = T_{k-1} - T_k$ the tenor. As the product of the bond $P(t, T_k)$ with the forward rates $F_k(t)$ is a difference of two bonds with maturity T_{k-1} and T_k, $\frac{1}{\tau_k}(P(t, T_{k-1}) - P(t, T_k))$, and therefore a traded asset, F_k is a (local) martingale under $\mathbb{Q}^k$, the (forward) measure associated with the numéraire $P(t, T_k)$. Therefore, we assume the following driftless dynamics

$$\begin{aligned} dF_k(t) &= \sigma_k(t)\Phi_k(a, F_k)dW_k \ , \ \forall t \leq T_{k-1} \ , \ k = 1, \ldots, n \\ dW_k dW_l &= \rho_{kl}(t)dt \end{aligned}$$

with the initial conditions $a(t = 0) = \alpha$ and $F_k(t = 0) = F_k^0$. Throughout this paper, W denotes a Brownian under the forward measure.

In order to achieve some flexibility, we assume that the (normal) local volatility $\Phi_k(a, F_k)$ depends on a hidden Markov process a (to be specified later) representing a stochastic volatility. We therefore assume that all the forward rates are coupled with the same stochastic volatility a. Table 1 presents a list of the different functional forms for Φ_k used in the literature. The BGM, (limited) CEV and shifted log-normal models correspond to local volatility models ($a = 1$) and the others to stochastic volatility models with a unique stochastic volatility a driven by a Heston process.

Table 1 Examples of stochastic (or local) volatility Libor models

Libor market model	SDE
BGM [6]	$dF_k = \sigma_k(t)F_k dW_k$
CEV [1]	$dF_k = \sigma_k(t)F_k^{\beta} dW_k$
Limited CEV	$dF_k = \sigma_k(t)F_k \min(F_k^{\beta-1}, \epsilon^{\beta-1})dW_k$ with ϵ a small positive number
Shifted log-normal	$dX_k = \sigma_k(t)X_k dW_k$ with $F_k = X_k + \alpha_k$
FL-SV [2]	$dF_k = \sigma_k(t)(\beta_k F_k + (1-\beta_k)F_k^0)\sqrt{v}dW_k$
	$dv = \lambda(v - \bar{\lambda})dt + \nu\sqrt{v}dZ; dW_k dZ = 0$
FL-TSS [16]	$dF_k = \sigma_k(t)(\beta_k(t)F_k + (1-\beta_k(t))F_k^0)\sqrt{v}dW_k$
	$dv = \lambda(v - \bar{\lambda})dt + \nu\sqrt{v}dZ; dW_k dZ = 0$

Note that the stochastic differential equation for the Libors rate F_k has been written in the forward measure $\mathbb{Q}^k$ and the stochastic equation for a remains the same in the forward or forward swap rate measures as a is assumed to be uncorrelated with the Libor rates. This will not be the case in our LMM.

3 Asymptotic Swaption Smile

We note $s_{\alpha\beta}$ the forward swap rate starting at T_α and expiring at T_β. The forward swap rate satisfies the following driftless dynamics in the forward-swap measure $\mathbb{Q}^{\alpha\beta}$ (associated to the numéraire $C_{\alpha\beta}(t) = \sum_{i=\alpha+1}^{\beta} \tau_i P(t, T_i)$)

$$ds_{\alpha\beta} = \sum_{k=\alpha+1}^{\beta} \frac{\partial s_{\alpha\beta}}{\partial F_k} \sigma_k(t)\Phi_k(a, F_k) dZ_k$$

Throughout this paper, W denotes a Brownian under the forward measure.

The local volatility associated to the forward swap rate ($ds_{\alpha\beta} = \sigma_{loc}^{\alpha\beta}(t, s_{\alpha\beta})dB$) is then by definition

$$\begin{aligned}(\sigma_{loc}^{\alpha\beta})^2(t, K) &\equiv \mathbb{E}^{\alpha\beta}\left[\sum_{i,j=\alpha+1}^{\beta} \rho_{ij}(t)\sigma_i(t)\sigma_j(t)\Phi_i(a, F_i)\Phi_j(a, F_j)\frac{\partial s_{\alpha\beta}}{\partial F_i}\frac{\partial s_{\alpha\beta}}{\partial F_j}|s_{\alpha\beta} = K\right] \\ &= \sum_{i,j=\alpha+1}^{\beta} \rho_{ij}(t)\sigma_i(t)\sigma_j(t)\frac{\int_{\mathbb{B}} \Phi_i(a, F_i)\Phi_j(a, F_j)\frac{\partial s_{\alpha\beta}}{\partial F_i}\frac{\partial s_{\alpha\beta}}{\partial F_j} p(da \prod_i dF_i)}{\int_{\mathbb{B}} p(da \prod_i dF_i)}\end{aligned} \tag{3.1}$$

with the submanifold $\mathbb{B} = \{(\{F_i\}_i, a)|s_{\alpha\beta} = K\}$ and $p \equiv p(F_i, a, t|F_i^0, \alpha)$ the conditional probability satisfying the (backward) Kolmogorov equation associated to the SDE for the Libors and the volatility a in the forward swap measure $\mathbb{Q}^{\alpha\beta}$. An asymptotic expression in the short time limit for the local volatility $\sigma_{\alpha\beta}(s, t)$ can be found in two steps: First by finding an asymptotic expansion for the conditional probability p (in $\mathbb{Q}^{\alpha\beta}$) and then by doing the integration over $\mathbb{B}$. The first step is achieved via the heat kernel expansion technique summarized in Appendix A. and the second step using the Laplace saddle-point method explained briefly in Appendix B.

3.1 Saddle-Point

As the conditional probability at the zero-order is proportional to $e^{-\frac{d(x,x^0)^2}{4t}}$ (see Appendix A) with $d(x, x^0)$ the geodesic distance between the points $x = (\{F_i\}_{i=1,\ldots,n}, a) \in \mathbb{R}^{n+1}$ and $x^0 = (\{F_i^0\}_{i=1,\ldots,n}, \alpha)$, the saddle-point corresponds to the point x on the submanifold $s_{\alpha\beta} = K$ which minimizes the geodesic distance $d(x, x^0)$ [4, 5]

$$(\{F_i^*\}, a^*) \equiv \text{argmin}_{\{F_i\},a|s_{\alpha\beta}=K}[d(x, x^0)^2] \tag{3.2}$$

Introducing a Lagrange multiplier λ, this is equivalent to

$$(\{F_i^*\}, a^*) \equiv \text{argmin}_{\{F_i\},a,\lambda}[d(x, x^0)^2 + \lambda(s_{\alpha\beta}(F) - K)] \tag{3.3}$$

3.2 Asymptotic Local Volatility

Plugging our asymptotic expression for the conditional probability (5.2) into (3.1) and doing the integration over $\mathbb{B}$ using the Laplace method (see Appendix B), we finally obtain the local volatility at the first-order

$$(\sigma_{\text{loc}}^{\alpha\beta})^2(t, K) = \sum_{i,j=\alpha+1}^{\beta} \rho_{ij}(t)\sigma_i(t)\sigma_j(t) f_{ij}(F^*, a^*) \Bigg\{$$
$$1 + 2t \sum_{\mu,\nu=1}^{n+1} A^{\mu\nu} \left(\frac{\partial_\mu f_{ij}(F^*, a^*)}{f_{ij}(F^*, a^*)} \left(2\frac{\partial_\nu \psi(F^*, a^*)}{\psi(F^*, a^*)} - \sum_{\gamma,\delta=1}^{n+1} A^{\gamma\delta} \partial_{\nu\gamma\delta} d^2 \right) + \frac{\partial_{\mu\nu} f_{ij}(F^*, a^*)}{f_{ij}(F^*, a^*)} \right) \Bigg\}$$

with $f_{ij}(F, a) = \Phi_i(a, F_i)\Phi_j(a, F_j)\frac{\partial s_{\alpha\beta}}{\partial F_i}\frac{\partial s_{\alpha\beta}}{\partial F_j}$, $\psi(F, a) = \sqrt{g\Delta\mathcal{P}}$ and $A^{\mu\nu} = [\partial_{\mu\nu} d^2]^{-1}$. $g, \mathcal{P}$ and Δ are defined in Appendix A and computed explicitly in Sect. 4. Note that as opposed to other asymptotic methods presented in the literature, this formula is exact at $t \to 0$. A similar zero-order formula (independent of the time t for $\sigma_i(t), \rho_{ij}(t)$ constant) was derived for a general multi-dimensional local volatility model by [4]. Moreover, in the expansion, we assumed that the time t is small but we have made no assumption that F_k is close to the spot Libor or that the volatility of volatility is small.

3.3 Asymptotic Smile

The asymptotic smile can be derived in two steps from the asymptotic local volatility: first, we have ($s_0 \equiv s_{\alpha\beta}(t=0)$)

$$ds_{\alpha\beta} = \frac{\sigma_{\mathrm{loc}}^{\alpha\beta}(t, s_{\alpha\beta})}{\sigma_{\mathrm{loc}}^{\alpha\beta}(t, s_0)} \sigma_{\mathrm{loc}}^{\alpha\beta}(t, s_0) dB_t$$

and doing a change of local time $t' = \int_0^t \sigma_{\mathrm{loc}}^{\alpha\beta}(u, s_0)^2 du$, we now obtain the associated local volatility model for the swap rate

$$ds_{\alpha\beta} = \bar{\sigma}_{\mathrm{loc}}^{\alpha\beta}(t, s_{\alpha\beta}) dB'_t$$

with $\bar{\sigma}_{\mathrm{loc}}^{\alpha\beta}(t,s) = \frac{\sigma_{\mathrm{loc}}^{\alpha\beta}(t,s)}{\sigma_{\mathrm{loc}}^{\alpha\beta}(t,s_0)}$. Secondly, we know that there is a one-to-one correspondence between this local volatility and the smile [12] given at the first-order by

$$\sigma_{\mathrm{BS}}^{\alpha\beta}(K, T_\alpha) = \sqrt{\frac{\int_0^{T_\alpha} (\sigma_{\mathrm{loc}}^{\alpha\beta})^2(u, s_0) du}{T_\alpha}} \sigma_{\mathrm{BS}}^{\alpha\beta}(K)_0$$
$$\times \left(1 + \frac{1}{2} \int_0^{T_\alpha} (\sigma_{\mathrm{loc}}^{\alpha\beta})^2(u, s_0) du \, \sigma_{\mathrm{BS}}^{\alpha\beta}(K)_1\right) \quad (3.4)$$

$$\sigma_{\mathrm{BS}}^{\alpha\beta}(K)_0 = \frac{\ln(\frac{K}{s_0})}{\int_{s_0}^K \frac{dx}{C(x)}}$$

$$\sigma_{\mathrm{BS}}^{\alpha\beta}(K)_1 = -\frac{1}{\left(\int_{s_0}^K \frac{dx}{C(x)}\right)^2} \ln\left(\frac{(\sigma_{\mathrm{BS}}^{\alpha\beta}(K)_0)^2 K s_0}{C(K)}\right) + \frac{1}{(\sigma_{\mathrm{loc}}^{\alpha\beta})^2(0, s_0)} \frac{\partial_t \sigma_{\mathrm{loc}}^{\alpha\beta}(f_{\mathrm{av}}, 0)}{C(f_{\mathrm{av}})}$$

with $C(f) \equiv \frac{\sigma_{\mathrm{loc}}^{\alpha\beta}(0,K)}{\sigma_{\mathrm{loc}}^{\alpha\beta}(0,s_0)}$, $f_{\mathrm{av}} \equiv \frac{s_0+K}{2}$.

4 SABR-LMM Model

We have seen that the asymptotic local and implied volatilities can be computed if we know the geodesic distance and a parametrization of geodesic curves on $\mathcal{M}^{n+1}$. This is the case for the hyperbolic space $\mathbb{H}^n$ for all n. This manifold has a lot of important properties. In the first part, we present our BGM-LMM-SABR model and show that the underlying geometry is $\mathbb{H}^{n+1}$ (with n the number of forward Libor rates).

Using this connection, we will find an asymptotic local volatility and an asymptotic swaption implied volatility.

4.1 Dynamics

We introduce the SABR-LMM model, given by the following SDE under the spot Libor measure $\mathbb{Q}$ (associated to the numéraire $B_m(t) = \prod_{j=1}^{\beta(t)-1}(1+\tau_j F_j(T_{j-1}))$ $P(t, T_{\beta(t)-1})$ where $\beta(t) = m$ if $T_{m-2} < t < T_{m-1}$)

$$\begin{aligned} dF_k &= a^2 B_k(t, F)dt + \sigma_k(t) a C_k(F_k) dZ_k \\ da &= \nu a dZ_{n+1}, \ dZ_i dZ_j = \rho_{ij}(t)dt \ \ i, j = 1, \dots, n+1 \end{aligned}$$

with

$$\begin{aligned} C_k(F_k) &= \phi_k F_k^{\beta_k} \\ B_k(t, F) &= \sum_{j=\beta(t)}^{k} \frac{\tau_j \rho_{jk} \sigma_k(t) \sigma_j(t) C_k(F_k) C_j(F_j)}{(1+\tau_j F_j)} \end{aligned}$$

Z is a correlated Brownian motion under the measure $\mathbb{Q}$. Here we take the local volatility term $\Phi_k(a, F_k)$ of type $a\phi_k F_k^{\beta_k}$ with $\phi_k \in \mathbb{R}$.

The functions $C_k(F_k)$ have been scaled by ϕ_k and therefore we can impose that $\sigma_k(0) = 0$. The stochastic equation for a was written in the spot Libor measure in order to get a SDE independent of a specific underlying swap $s_{\alpha\beta}$ or a forward bond. Under the forward swap measure $\mathbb{Q}^{\alpha\beta}$, we have

$$\begin{aligned} dF_k &= a^2 b_k(t, F)dt + \sigma_k(t) a C_k(F_k) dZ_k \\ da &= -\nu^2 a^2 b_a(t, F)dt + \nu a dZ_{n+1}, \ dZ_i dZ_j = \rho_{ij}(t)dt \ \ i, j = 1, \dots, n+1 \end{aligned}$$

with

$$\begin{aligned} b_k(t, F) &= \sum_{j=\alpha+1}^{\beta} (2 1_{j \le k} - 1) \tau_j \frac{P(t, T_j)}{C_{\alpha\beta}(t)} \sum_{i=\min(k+1, j+1)}^{\max(k, j)} \frac{\tau_i \rho_{ki} \sigma_i(t) \sigma_k(t) C_i(F_i) C_k(F_k)}{(1+\tau_i F_i)} \\ C_{\alpha\beta}(t) &= \sum_{i=\alpha+1}^{\beta} \tau_i P(t, T_i) \\ b_a(F, t) &= \sum_{i=\alpha+1}^{\beta} \tau_i \omega_i(t) \sum_{k=\beta(t)}^{i} \frac{\tau_k C_k(F_k) \rho_{ka} \sigma_k(t)}{1+\tau_k F_k(t)} \end{aligned}$$

and with $\omega_i(t) = \frac{\prod_{k=\beta(t)}^{i} \frac{1}{(1+\tau_k F_k)}}{\sum_{j=\alpha+1}^{\beta} \prod_{k=\beta(t)}^{j} \frac{1}{(1+\tau_k F_k)}}$. Here Z is a correlated Brownian motion under the measure $\mathbb{Q}^{\alpha\beta}$. Note that the forward-rate dynamics under the forward measure $\mathbb{Q}^k$ is much simpler and given by the following stochastic differential equations (SDE)

$$dF_k(t) = \sigma_k(t) a C_k(F_k) dW_k \ , \ dW_k dW_p = \rho_{kp}(t) dt$$

As it is the case for the BGM model, we can use a piecewise parametric form or a functional form for the serial volatilities $\sigma_i(t)$ and the correlation $\rho_{ij}(t)$ (here full rank) as

$$\sigma_i(t) = N_i[(a(T_{i-1} - t) + d)e^{-b(T_{i-1}-t)} + c] \ \forall t \leq T_{i-1}$$
$$\rho_{ij}(t = 0) = \rho_L + (1 - \rho_L)e^{-(\delta_A - \delta_B \ \min[T_{i-1}, T_{j-1}])|T_{i-1} - T_{j-1}|}$$

The constants N_i are fixed such as $\sigma_i(0) = 1$. The model depends on $9 + 3n$ parameters (see Table 2) which are calibrated on the swaption matrix. In the next subsection, we derive the metric, the geodesic distance and the Abelian connection underlying this model.

4.2 Hyperbolic Geometry

By definition (see Eq. (5.1a)), the infinitesimal distance (at $t = 0$) between the point x_α and $x_\alpha + dx_\alpha$ (5.1a) is given by ($\rho^{ij} \equiv [\rho^{-1}]_{ij}$, $(i, j) = (1, \ldots, n)$, $\rho^{ia} \equiv [\rho^{-1}]_{ia}$ and $\rho^{aa} \equiv [\rho^{-1}]_{aa}$ are the components of the inverse of the correlation matrix ρ)

$$ds^2 \equiv \sum_{\alpha,\beta=1}^{n+1} g_{\alpha\beta} dx_\alpha dx_\beta$$
$$= \frac{2}{\nu^2 a^2} \left(\sum_{i,j=1}^{n} \rho^{ij} \frac{\nu dF_i}{C_i(F_i)} \frac{\nu dF_j}{C_j(F_j)} + 2 \sum_{i=1}^{n} \rho^{ia} \frac{\nu dF_i}{C_i(F_i)} da + \rho^{aa} da^2 \right)$$

Table 2 SABR-LMM: $9 + 3n$ parameters

BGM parameters	$a, b, c, d, \phi_i, \rho_L, \delta_A, \delta_B$
CEV parameters	$\beta_i, \ i = 1, \ldots, n$
SABR parameters	$\alpha, \nu, \rho_{ia} \ i = 1, \ldots, n$

After some algebraic manipulations, we show that in the new coordinates $[x_k]_{k=1\dots n+1}$ ($\hat{L}$ is the Cholesky decomposition of the (reduced) correlation matrix: $[\rho]_{i,j=1\dots n} = [\hat{L}\hat{L}^\dagger]_{i,j=1\dots n}$)

$$x_k = \sum_{i=1}^{n} \nu \hat{L}^{ki} \int_{F_i^0}^{F_i} \frac{dF_i'}{C_i(F_i')} + \sum_{i=1}^{n} \rho^{ia} \hat{L}_{ik} a \ , \ k = 1, \dots, n$$

$$x_{n+1} = (\rho^{aa} - \sum_{i,j}^{n} \rho^{ia} \rho^{ja} \rho_{ij})^{\frac{1}{2}} a$$

the metric becomes

$$ds^2 = \frac{2(\rho^{aa} - \sum_{i,j}^{n} \rho^{ia} \rho^{ja} \rho_{ij})}{\nu^2} \frac{\sum_{i=1}^{n} dx_i^2 + dx_{n+1}^2}{x_{n+1}^2}$$

Written in the coordinates $[x_i]$, the metric is therefore the standard hyperbolic metric on $\mathbb{H}^{n+1}$ modulo a constant factor $\mu = \frac{2(\rho^{aa} - \sum_{i,j,k=1}^{n} \rho^{ia} \rho^{ja} \rho_{ij})}{\nu^2}$. This factor can be eliminated by doing a change of time $t' = \mu^{-1} t$. In order to compute our saddle-point (3.3), we need the geodesic distance which is given in [18].

Proposition 4.2.1 *The geodesic distance $d(x, x')$ on $\mathbb{H}^{n+1}$ is given by*

$$d(x, x^0) = \cosh^{-1} \left[1 + \frac{\sum_{i=1}^{n+1} (x_i - x_i^0)^2}{2 x_{n+1} x_{n+1}^0} \right] \tag{4.3}$$

Using the geodesic distance on $\mathbb{H}^{n+1}$ between the points $x = (\{F\}_k, a)$ and the initial point $x^0 = (\{F^0\}_k, \alpha)$ ($q_i \equiv \int_{F_i^0}^{F_i} \frac{dF_i'}{C_i(F_i')}$) given by

$$d(F, a | F^0, \alpha) = \cosh^{-1} \left[1 + \frac{\nu^2 \sum_{i,j=1}^{n} \rho^{ij} q_i q_j + 2\nu(a - \alpha) \sum_{j=1}^{n} \rho^{ja} q_j + (a - \alpha)^2 \rho^{aa}}{2(\rho^{aa} - \sum_{i,j=1}^{n} \rho^{ia} \rho^{ja} \rho_{ij}) a \alpha} \right]$$

In this model, the non-linear equation (3.3), satisfied by the saddle-point $a^*(s)$, $q_i^*(s)$ which implicitly depends on the swaption strike s, read:

$$a^*(s)^2 \rho^{aa} = \alpha^2 \rho^{aa} - 2\nu\alpha \sum_{i=1}^{n} \rho^{ia} q_i^* + \nu^2 \sum_{i,j=1}^{n} \rho^{ij} q_i^* q_j^* \tag{4.4}$$

$$\frac{(\rho^{ia} \frac{(a^*(s) - \alpha)}{\nu} + \sum_{j=1}^{n} \rho^{ij} q_j^*) d(q^*, a^*)}{a^*(s)(\cosh(d(a^*, \{q_i^*\}))^2 - 1)^{\frac{1}{2}}} = \frac{\lambda}{\alpha} \frac{\partial s_{\alpha\beta}}{\partial q_i} |_{(a,q)=(a^*,q^*)} \tag{4.5}$$

with

$$q_i^* = \phi_i^{-1} \int_{F_i^0}^{F_i^*} x^{-\beta_i} dx$$

The saddle-point is determined by solving these non-linear equations (4.4) and (4.5) and an approximation (which could be used as a guess solution in a numerical optimization routine) is found by linearizing these equations around the spot Libor rates (i.e. $q_i = 0$)

$$\lambda^*(s) = \frac{(s - s_0)}{\sum_{p,q=1}^n \omega_p \omega_q \tilde{\rho}_{pq}} \tag{4.6}$$

$$\frac{F_i^*}{F_i^0} = 1 + \frac{\sum_{j=1}^n \tilde{\rho}_{ij} \omega_j (s - s_0)}{\sum_{p,q=1}^n \omega_p \omega_q \tilde{\rho}_{pq}} + O((s - s_0)^2) \tag{4.7}$$

with $\omega_i \equiv \frac{\partial s_{\alpha\beta}}{\partial q_i}(q_i = 0)$ and $\tilde{\rho}^{ij} = \rho^{ij} - \frac{\rho^{ia}\rho^{ja}}{\rho^{aa}}$. Note that when the strike is close to at-the-money, the saddle-points are close to the spot Libors and $a^* = \alpha$. Moreover, by using the explicit expression for the hyperbolic distance, the Van-Vleck-Morette determinant is

$$\Delta(F, a|F^0, \alpha) = \frac{d(F, a|F^0, \alpha)}{\sqrt{\cosh^2(d(F, a|F^0, \alpha)) - 1}}$$

4.3 Connection

The Abelian connection is given by (5.1b)

$$\mathcal{A}_i = \frac{1}{C_i(F_i)} \left(\sum_{j=1}^n \rho^{ij} \left(\frac{b_j(t, F)}{C_j(F_j)} - \frac{\partial_j C_j(F_j)}{2} \right) - \nu \rho^{ia} b_a(F, t) \right)$$

$$\mathcal{A}_a = \frac{1}{\nu} \left(\sum_{j=1}^n \rho^{aj} \left(\frac{b_j(t, F)}{C_j(F_j)} - \frac{\partial_j C_j(F_j)}{2} \right) - \nu \rho^{aa} b_a(F, t) \right)$$

where we have used that

$$\sqrt{g} = \frac{2^{\frac{n+1}{2}} \det[\rho]^{-\frac{1}{2}}}{\nu a^{n+1} \prod_{i=1}^n C_i(F_i)}$$

Finally, the Abelian 1-form connection, $\mathcal{A} = \sum_{i=1}^{n} \mathcal{A}_i dF_i + \mathcal{A}_a da$, is

$$\mathcal{A} = \frac{1}{\nu} \sum_{j=1}^{n} \left(\frac{b_j(t,F)}{C_j(F_j)} - \frac{\partial_j C_j(F_j)}{2} \right) \left(\nu \sum_{i=1}^{n} \rho^{ij} dq_i + \rho^{aj} da \right)$$
$$- b_a(t,F) \left(\nu \sum_{i=1}^{n} \rho^{ia} dq_i + \rho^{aa} da \right)$$

In order to compute the log of the parallel gauge transport $\ln(\mathcal{P})(a, q|\alpha) = \int_{\mathcal{C}} \mathcal{A}$, we need to know a parametrization of the geodesic curve on $\mathbb{H}^{n+1}$. However, we can directly find $\ln(\mathcal{P})(a, q|\alpha)$ if we approximate the drifts $b_k(t, F)$ by their values at the Libor spots (and $t = 0$). A similar approximation was done in the Hagan-al formula [11] as was shown in [12]. Modulo this approximation,

$$\ln(\mathcal{P})(a, q|\alpha) \sim \frac{1}{\nu} \sum_{j=1}^{n} \left[\left(\frac{b_j(0, F^0)}{C_j(F_j^0)} - \frac{\partial_j C_j(F_j^0)}{2} \right) \left(\nu \sum_{i=1}^{n} \rho^{ij} q_i + \rho^{aj} (a - \alpha) \right) \right]$$
$$- \; b_a(0, F^0) \left(\nu \sum_{i=1}^{n} \rho^{ia} q_i + \rho^{aa} (a - \alpha) \right)$$

4.4 Asymptotic Smile—Summary

The asymptotic local volatility is given by (3.4)

$$(\sigma_{\text{loc}}^{\alpha\beta})^2(t,s) = \sum_{i,j=\alpha+1}^{\beta} \rho_{ij} \sigma_i(t) \sigma_j(t) f_{ij}(a, F)$$
$$\times \left(1 + 2t' \sum_{\mu,\nu=\alpha+1}^{\beta} A^{\mu\nu} \left\{ \frac{\partial_{\mu\nu} f_{ij}(F*, a^*)}{f_{ij}(F*, a^*)} + 2 \frac{\partial_\mu f_{ij}(F*, a^*)}{f_{ij}(F*, a^*)} \frac{\partial_\nu \psi(F*, a^*)}{\psi(F*, a^*)} \right. \right.$$
$$\left. \left. - \sum_{\gamma,\delta=1}^{n+1} A^{\gamma\delta} \frac{\partial_\mu f_{ij}(F*, a^*)}{f_{ij}(F*, a^*)} \partial_{\nu\gamma\delta} d^2(F*, a^*) \right\} \right)$$

with $(a^* \equiv a^*(s), F^* \equiv \{F_i^*(s)\}_i)$ the saddle-point satisfying the Eqs. (4.4) and (4.5) approximated by (4.6) and (4.7) and

$$
\begin{aligned}
f_{ij}(a,F) &= a^2 C_i(F_i)C_j(F_j)\frac{\partial s_{\alpha\beta}}{\partial F_i}\frac{\partial s_{\alpha\beta}}{\partial F_j}\ , \ \psi(a,F) = \sqrt{g\Delta\mathcal{P}}\ , \ A^{\alpha\beta} = [\partial_{\alpha\beta}d^2]^{-1} \\
d(F,a|F^0,\alpha) &= \cosh^{-1}\left[1 + \frac{\nu^2\sum_{i,j=1}^n \rho^{ij}q_iq_j + 2\nu(a-\alpha)\sum_{j=1}^n \rho^{ja}q_j + (a-\alpha)^2}{2(\rho^{aa} - \sum_{i,j=1}^n \rho^{ia}\rho^{ja}\rho_{ij})a\alpha}\right] \\
\ln(\mathcal{P})(a,q|\alpha) &\sim \frac{1}{\nu}\sum_{j=1}^n\left[\left(\frac{b_j(0,F^0)}{C_j(F_j^0)} - \frac{\partial_j C_j(F_j^0)}{2}\right)\left(\nu\sum_{i=1}^n \rho^{ij}q_i + \rho^{aj}(a-\alpha)\right)\right] \\
&- b_a(0,F^0)\left(\nu\sum_{i=1}^n \rho^{ia}q_i + \rho^{aa}(a-\alpha)\right) \\
\Delta(F,a|F^0,\alpha) &= \frac{d(F,a|F^0,\alpha)}{\sqrt{\cosh^2(d(F,a|F^0,\alpha)) - 1}} \\
\sqrt{g} &= \frac{2^{\frac{n+1}{2}}\det[\rho]^{-\frac{1}{2}}}{\nu a^{1+n}\prod_{i=1}^n C_i(F_i)}
\end{aligned}
$$

Note that this expression is *exact* when t goes to zero. The smile at the first-order is then obtained by plugging the above expression into (3.4) with $t' = \frac{\nu^2}{2(\rho^{aa} - \sum_{i,j,k=1}^n \rho^{ia}\rho^{ja}\rho_{ij})}t$.

Remark 4.4.1 (Libor CEV model) Note that our model reduces for ν goes to zero (and $\alpha \equiv 1$) to the Andersen-Andreasen CEV Libor model (with different CEV parameters for each Libors) and the above expressions degenerates into

$$
\begin{aligned}
f_{ij}(F) &= C_i(F_i)C_j(F_j)\frac{\partial s_{\alpha\beta}}{\partial F_i}\frac{\partial s_{\alpha\beta}}{\partial F_j} \\
d(F) &= \sqrt{2\sum_{i,j=1}^n \rho^{ij}q_iq_j} \\
\ln(\mathcal{P})(q) &= \sum_{j=1}^n\left(\frac{b_j(0,F^0)}{C_j(F_j^0)} - \frac{\partial_j C_j(F_j^0)}{2}\right)\sum_{i=1}^n \rho^{ij}q_i \\
\Delta(F,F^0) &= 1 \\
\sqrt{g} &= \frac{2^{\frac{n}{2}}\det[\rho]^{-\frac{1}{2}}}{\prod_{i=1}^n C_i(F_i)}
\end{aligned}
$$

with the saddle-points (4.4) and (4.5) satisfying the non-linear equations (modulo the constraint $s_{\alpha\beta} = s$)

$$\rho^{ij} q_j^* = \lambda \frac{\partial s_{\alpha\beta}}{\partial q_i}\bigg|_{q^*} .$$

4.5 Comments and Numerical Tests

It is interesting to note that for $n = 1$, i.e. for a caplet, the caplet asymptotic smile reduces to the classical SABR formula by construction. Moreover, the asymptotic local volatility is given at the zero-order by

$$(\sigma_{\text{loc}}^{\alpha\beta})^2(s,t) = \sum_{i,j=1}^{n} \rho_{ij}(t)\sigma_i(t)\sigma_j(t)a^{*2}(F^*)C_i(F_i^*)C_i(F_i^*)\frac{\partial s_{\alpha\beta}}{\partial F_i}(F^*)\frac{\partial s_{\alpha\beta}}{\partial F_j}(F^*)$$

with F^* depending implicitly on s via (4.4) and (4.5). At this stage, it is useful to recall how a similar asymptotic local volatility is derived using the "freezing" argument. The forward swap rate satisfies the following SDE in the forward swap numéraire $\mathbb{Q}^{\alpha\beta}$

$$ds_{\alpha\beta} = \sum_{k=1}^{n} \frac{\partial s_{\alpha\beta}}{\partial F_k}\sigma_k(t)aC_k(F_k)dZ_k \tag{4.8}$$

The "freezing" argument consists in assuming that the terms $\frac{\partial s_{\alpha\beta}}{\partial F_k}$ and $\frac{C(s)}{C(F_i)}$ are almost constant. Therefore, the SDE (4.8) can be approximated by

$$ds_{\alpha\beta} = \sum_{k=1}^{n} \frac{\partial s_{\alpha\beta}}{\partial F_k}(F^0)\sigma_k(t)a\frac{C_k(F_k^0)}{C_k(s^0)}C_k(s)dZ_k$$

and the local volatility is

$$(\sigma_{\text{loc}}^{\alpha\beta})^2(s,t) = \sum_{i,j=1}^{n} \rho_{ij}(t)\sigma_i(t)\sigma_j(t)a^{*2}(s)\frac{C_i(F_i^0)}{C_i(s^0)}\frac{C_j(F_j^0)}{C_i(s^0)}\frac{\partial s_{\alpha\beta}}{\partial F_i}$$
$$\times (F^0)\frac{\partial s_{\alpha\beta}}{\partial F_j}(F^0)C_i(s)C_j(s)$$

Table 3 Scenario A: Libor volatility $\lambda_i(t) = 5\,\%$

Swaption	Strike	MC (%)	F1	F2
5×15	4 % (ITM)	22.42	22.41 % (−1)	22.61 % (19)
	6 % (ATM)	20.33	20.41 % (8)	20.46 % (13)
	8 % (OTM)	18.92	18.93 % (1)	19.01 % (10)
10×10	4 % (ITM)	22.41	22.51 % (11)	22.67 % (26)
	6 % (ATM)	20.38	20.41 % (3)	20.50 % (12)
	8 % (OTM)	18.93	18.93 % (−1)	19.05 % (12)

Libor $L_i(0) = 6\,\%$. $\beta = 0.5$

Table 4 Scenario B: Libor volatility $dL_i = 0.25(0.17 + 0.002(T_{i-1} - t))L_i^{\beta} dW$

Swaption	Strike	MC (%)	F1	F2
5×15	5.08 % (ITM)	18.12	18.20 % (8)	18.17 % (5)
	7.26 % (ATM)	16.51	16.61 % (10)	16.63 % (12)
	9.44 % (OTM)	15.38	15.38 % (0)	15.56 % (18)
10×10	5.55 % (ITM)	17.80	17.81 % (1)	17.89 % (9)
	7.93 % (ATM)	16.26	16.33 % (7)	16.38 % (11)
	10.31 % (OTM)	15.17	15.19 % (2)	15.32 % (15)

Libor $L_i(0) = log(a + bi)$, $L_0(0) = 5\,\%$, $L_{19}(0) = 9\,\%$. $\beta = 0.5$

We can reproduce this formula for the swaption smile at-the-money[1] as the saddle-point Libor rates coincides with the spot rates. This is not the case for in/out-the-money swaption. Therefore our expression (exact at the zero-order) shows that the freezing argument is no longer correct when we try to fit a swaption implied smile in/out-the-money. In the following, we have tested our asymptotic swaption formula at the zero-order with the same beta $\beta_k = \beta$ and $\nu = 0$ (Formula F1) against the Andersen-Andreasen asymptotic formula (Formula F2) [1] in the case $\nu = 0$. The accuracy of these approximations are examined using Monte-Carlo (MC) prices as a benchmark. Following [15], we consider five scenarii (see Tables 3, 4, 5, 6 and 7). In the following tables, the implied volatility is reported and the numbers in brackets are the errors (in basis points i.e. true volatility times 10^4) corresponding to the implied volatility computed using the F1 or F2 formula minus the MC implied volatility. An $x \times y$ swaption has an option maturity of x years, a swap length of y years and a tenor of one year. We set a time-step for Monte-Carlo $\delta = 0.125$ and 2^{16} paths.[2] Our formula F1 is more accurate than F2.

[1] An at-the-money swaption (ATM) has a strike K equal to the spot rate $s_{\alpha\beta}(0)$ and an out-of-the money (OTM) (resp. in-the-money (ITM)) swaption has $K < s_{\alpha\beta}(0)$ (resp. $K > s_{\alpha\beta}(0)$).

[2] We have used a predictor-corrector scheme with a Brownian bridge.

Table 5 Scenario C: Libor volatility $dL_i = 0.25(0.17 - 0.002(T_{i-1} - t))L_i^{\beta} dW$

Swaption	Strike	MC (%)	F1	F2
5 × 15	5.08 % (ITM)	14.89	14.97 % (8)	15.08 % (19)
	7.26 % (ATM)	13.73	13.79 % (4)	13.81 % (8)
	9.44 % (OTM)	12.92	12.91 % (−1)	12.92 % (0)
10 × 10	5.55 % (ITM)	14.52	14.53 % (1)	14.64 % (12)
	7.93 % (ATM)	13.33	13.38 % (5)	13.40 % (7)
	10.31 % (OTM)	12.51	12.51 % (0)	12.54% (3)

Libor $L_i(0) = log(a + bi)$. $L_0(0) = 5\,\%$, $L_{19}(0) = 9\,\%$. $\beta = 0.5$

Table 6 Scenario D: $dL_i = 0.05L_i^{\beta}(\frac{b_{i1}(t)}{\sqrt{b_{i1}(t)^2+b_{i2}(t)^2}} dW_1 + \frac{b_{i2}(t)}{\sqrt{b_{i1}(t)^2+b_{i2}(t)^2}} dW_2)$. $b_{i1}(t) = \rho e^{-k_1(T_{i-1}-t)} + \theta e^{-k_2(T_{i-1}-t)}$, $b_{i2}(t) = \sqrt{1-\rho^2} e^{-k_1(T_{i-1}-t)}$

Swaption	Strike	MC (%)	F1	F2
5 × 15	5.08 % (ITM)	19.19	19.33 % (14)	19.38 % (19)
	7.26 % (ATM)	17.59	17.72 % (13)	17.75 % (16)
	9.44 % (OTM)	16.46	16.49 % (3)	16.61 % (15)
10 × 10	5.55 % (ITM)	18.92	18.94 % (2)	19.06 % (14)
	7.93 % (ATM)	17.31	17.39 % (8)	17.45 % (14)
	10.31 % (OTM)	16.18	16.21 % (3)	16.32 % (14)

$\rho = 0.99$, $\theta = -0.99$, $k_1 = k_2 = 0.54$. Libor $L_i(0) = log(a + bi)$. $L_0(0) = 5\,\%$, $L_{19}(0) = 9\,\%$. $\beta = 0.5$

Table 7 Scenario E: Scenario D with $\beta = 0.3$

Swaption	Strike	MC (%)	F1	F2
5 × 15	5.08 % (ITM)	33.09	33.65 % (56)	34.24 % (115)
	7.26 % (ATM)	29.47	29.96 % (49)	30.23 % (76)
	9.44 % (OTM)	26.92	27.14 % (22)	27.49 % (57)
10 × 10	5.55 % (ITM)	31.75	32.41 % (66)	33.33 % (158)
	7.93 % (ITM)	28.47	28.88 % (41)	29.37 % (91)
	10.31 % (OTM)	26.01	26.18 % (17)	26.68 % (67)

5 Conclusion

In this short note, we have introduced a LMM model coupled to a SABR stochastic volatility process. By using the heat kernel expansion technique in the short time limit, we have obtained an asymptotic swaption implied volatility at the first-order, compatible with the Hagan-al classical formula for caplets. Moreover, we have seen that this exact expression (when the expiry is very short) is incompatible with the analogous expression obtained using the freezing argument.

Appendix A: Heat Kernel Expansion

An short-time expansion of the conditional probability for a multi-dimensional Itô diffusion process can be achieved using the heat kernel expansion. In that purpose, the Kolmogorov equation is rewritten as the heat kernel equation on a (n)-dimensional Riemannian manifold $\mathcal{M}^n$ endowed with an Abelian connection as explained in [12, 13]. Let's assume that our multi-dimensional stochastic equations (in $\mathbb{Q}^{\alpha\beta}$) are written as

$$dx_\mu = b_\mu(t,x)dt + \sigma_\mu(t,x)dW_\mu$$

with $dW_\mu dW_\nu = \rho_{\mu\nu}(t)dt$. Then, the metric $g_{\mu\nu}$ depends only on the diffusion terms σ_μ and the connection $\mathcal{A}_\mu$ on the drift terms b_μ as well

$$g_{\mu\nu}(t,x) = 2\frac{\rho^{\mu\nu}(t)}{\sigma_\mu(t,x)\sigma_\nu(t,x)},\ \mu,\nu = 1,\ldots,n\ ,\ \rho^{\mu\nu} \equiv [\rho^{-1}]_{\mu\nu} \tag{5.1a}$$

$$\mathcal{A}_\alpha(t,x) = \sum_{\mu=1}^{n} g_{\alpha\mu}\frac{1}{2}\left(b_\mu(t,x) - \sum_{\nu=1}^{n+1} g^{-\frac{1}{2}}\partial_\nu\left(g^{1/2}g^{\mu\nu}(t,x)\right)\right),\ \alpha = 1,\ldots,n \tag{5.1b}$$

with $g(t,x) \equiv \det[g_{\mu\nu}(t,x)]$. In terms of these functions, the asymptotic solution to the Kolmogorov equation in the short-time limit is given by

$$p(x,t|x^0) = \frac{\sqrt{g(x)}}{(4\pi t)^{\frac{n}{2}}}\sqrt{\Delta(x,x^0)}\mathcal{P}(x,x^0)e^{-\frac{\sigma(x,x^0)}{2t}}\left(1+\sum_{n=1}^{\infty}a_n(x,x^0)t^n\right), t\to 0 \tag{5.2}$$

- Here, $\sigma(x,x^0)$ is the Synge function defined as

$$\sigma(x,x^0) = \frac{d^2(x,x^0)}{2}$$

  The distance $d(x,x^0)$ is defined as the minimizer of

$$d(x,x^0)^2 = \min_C \int_0^T g_{\mu\nu}(t=0,x)\frac{dx_\mu(t)}{dt}\frac{dx_\nu(t)}{dt}dt$$

  and t parameterizes the curve $\mathcal{C}(x,x^0)$ joining $x(t=0) \equiv x^0$ and $x(T) \equiv x$.
- $\Delta(x,x^0)$ is the so-called Van Vleck-Morette determinant

$$\Delta(x,x^0) = g(0,x)^{-\frac{1}{2}}\det\left(-\frac{\partial^2\sigma(x,x^0)}{\partial x\partial x^0}\right)g(0,x^0)^{-\frac{1}{2}} \tag{5.3}$$

with $g(x) = \det[g_{\mu\nu}(0, x)]$

- $\mathcal{P}(x, x^0)$ is the parallel transport of the Abelian connection along the geodesic $\mathcal{C}(x, x^0)$ from the point x to x^0.

$$\mathcal{P}(x, x^0) = e^{-\int_{\mathcal{C}(x^0,x)} \mathcal{A}_\mu(t=0,x)dx^\mu} \tag{5.4}$$

- The $a_i(x, x^0)$ are smooth functions on $\mathcal{M}^n$ and depend on geometric invariants such as the scalar curvature R. More details can be found in [12].

Appendix B: Saddle-Point Method

The integration over $\mathbb{B}$ is obtained by using a saddle-point method which consists in approximating at the first order the integral $\int f(x)e^{\epsilon\phi(x)}dx$ in the limit ϵ large by [8]

$$\int f(x)e^{\epsilon\phi(x)}dx \sim_{\epsilon>>1} f(x^*)e^{\epsilon\phi(x^*)}\Bigg(1 + \frac{1}{\epsilon}\Bigg(-\frac{\partial_{\alpha\beta} f}{2f}A_{\alpha\beta} + \left(\frac{\partial_\alpha f}{2f}\partial_{\beta\gamma\delta}\phi + \frac{1}{8}\partial_{\alpha\beta\gamma\delta}\phi\right)A_{\alpha\beta}A_{\gamma\delta} - \frac{\partial_{\alpha\beta\gamma}\phi\partial_{\delta\mu\nu}\phi}{72}A_{\alpha\beta}A_{\gamma\delta}A_{\mu\nu}\Bigg)\Bigg)$$

with $A^{\alpha\beta} = [\partial_{\alpha\beta}\phi]^{-1}$, $dx \equiv \prod_{i=1}^n dx_i$ and x^* the saddle-point (which minimizes $\phi(x)$). Here we have used Einstein summation convention. This expression can be obtained by developing $\phi(x)$ and $f(x)$ in series around x^*. The quadratic part in $\phi(x)$ leads to a Gaussian integration over x which can be performed.

References

1. Andersen, L., Andreasen, J.: Volatility skews and extensions of the Libor market model. Appl. Math. Financ. **7**(1), 1–32 (2000)
2. Andersen, L., Andreasen, J.: Volatile volatilities. Risk **15**(12), 163–168 (2002)
3. Andersen, L., Brotherton-Ratcliffe, R.: Extended Libor market models with stochastic volatility. J. Comput. Financ. **9**(1), 1–40 (2005)
4. Avellaneda, M., Boyer-Olson, D., Busca, J., Fritz, P.: Reconstructing the smile. Risk Mag. (2002)
5. Berestycki, H., Busca, J., Florent, I.: Asymptotics and calibration of local volatility models. Quant. Financ. **2**, 31–44 (1998)
6. Brace, A., Gatarek, D., Musiela, M.: The market model of interest rate dynamics. Math. Financ. **7**, 127–154 (1996)
7. Brigo, D., Mercurio, F.: Interest Rate Models, Theory and Practice. Springer Finance. Springer, Berlin (2015)
8. Erdély, A.: Asymptotic Expansions. Dover, New York (1956)

9. Hull, J., White, A.: Forward rate volatilities swap rate volatilities, and the implementation of the Libor market model. J. Fixed Income **3**, 46–62 (2000)
10. Hull, J., White, A.: The pricing of options on assets with stochastic volatilities. J. Financ. **42**, 281–300 (1987)
11. Hagan, P., Kumar, D., Lesniewski, A., Woodward, D.: Managing smile risk. Willmott Mag. 84–108 (2002)
12. Henry-Labordère, P.: A General Asymptotic Implied Volatility for Stochastic Volatility Models. http://arxiv.org/abs/condmat/0504317
13. Henry-Labordère, P.: Solvable local and stochastic volatility models: supersymmetric methods in option pricing. Quant. Financ. **7**(5), 525–535 (2007)
14. Jamshidian, F.: Libor and swap market models and measures. Financ. Stoch. **1**(4), 293–330 (1997)
15. Kawai, A.: A new approximate swaption formula in the Libor model market: an asymptotic approach. Appl. Math. Financ. **10**(1), 49–74 (2003)
16. Piterbarg, V.: A stochastic volatility forward Libor model with a term structure of volatility smiles, SSRN
17. Rebonato, A.: On the pricing implications of the joint lognormal assumption for the swaption and cap markets. J. Comput. Financ. **3**(3), 5–26 (1999)
18. Terras, A.: Harmonic Analysis on Symmetric Spaces and Applications, vol. I, II. Springer, New York (1985, 1988)

Second Order Expansion for Implied Volatility in Two Factor Local Stochastic Volatility Models and Applications to the Dynamic λ-Sabr Model

Gérard Ben Arous and Peter Laurence

Abstract Using an expansion of the transition density function of a two dimensional time inhomogeneous diffusion, we obtain the first and second order terms in the short time asymptotics of the local volatility function in a family of time inhomogeneous local-stochastic volatility models. With the local volatility function at our disposal, we show how recent results (Gatheral et al., Math. Financ. 22:591–620, 2012, [28]) for one dimensional diffusions can be applied to also determine expansions for call prices as well as for the implied volatility. The results are worked out in detail in the case of the dynamic Sabr model, thus generalizing earlier work by Hagan et al. (Wilmott Mag. 84–108, 2003, [31]), Hagan and Lesniewski (Springer Proceedings in Mathematics and Statistics, vol. 110, 2015, [32]) and by Henry-Labordère (Springer Proceedings in Mathematics and Statistics, vol. 110, 2015, Geometry, and Modeling in Finance. Chapman & Hall/CRC Financial Mathematics Series, 2008, [39, 40]).

Keywords Implied volatility · Local volatility · Asymptotic expansion · Heat kernels

1 Introduction

Stochastic volatility models offer a widely accepted approach to incorporating into the modeling of option markets a flexibility that accounts for the implied volatility smile or skew. From a historical perspective he first models to be introduced into the

---

Part of this work carried out while the author was a Visiting Scholar at the Courant Institute.

G. Ben Arous (✉) · P. Laurence
Courant Institute of Mathematical Sciences, New York University, 251 Mercer Street, New York, NY 10012, USA
e-mail: benarous@courant.nyu.edu

P. Laurence
Dipartimento di Matematica, Università di Roma 1 Piazzale Aldo Moro, 2, 00185 Rome, Italy

P.K. Friz et al. (eds.), *Large Deviations and Asymptotic Methods in Finance*, Springer Proceedings in Mathematics & Statistics 110,
DOI 10.1007/978-3-319-11605-1_4

literature were the Hull and White model [37], the Stein and Stein model [50] and the Heston model [33]. In three of these models the underlying asset and its volatility are driven by Brownian motions that may or may not be instantaneously correlated. The correlation coefficient is taken to be a constant ρ. Then Bates introduced the first of a series of models incorporating jumps [7]. These were followed by work by Andersen and Andreasen [2].

In recent times there has been an explosion of models using the method of stochastic time changes to produce ever more versatile models [15, 16]. However purely diffusive models have retained of their popularity. A case in point is the introduction into the literature of the dynamic SABR (stochastic alpha-beta-rho model), by Hagan, Lesniewski and Woodward [32]

$$\begin{aligned} dF_t &= \gamma(t) F_t^{\beta} y_t dW_{1t} \\ dy_t &= \nu(t) y_t dW_{2t} \\ < dW_{1t}, dW_{2t} > &= \rho(t) dt \\ F_0 &= \bar{F}_0; \, y_0 = \alpha \end{aligned}$$

A possible shortcoming of the Sabr was the lack of mean reversion in the stochastic volatility. To address this a generalization was proposes by Henry-Labordère, who in [39] introduced the λ Sabr model. In this new model the second equation is complemented by a mean-reverting term

$$dy_t = \kappa(\theta - y_t)dt + y_t dW_{2t}$$

The incorporation of a mean-reverting volatility was well received in the market place, since most practitioners think this feature is inherent in the volatility's evolution.

In this paper we will consider a model which includes both the dynamic Sabr model and the λ-Sabr models, in that we allow the volatility to be *mean reverting* and in addition for all parameters to depend explicitly on time. We explain how the approach via Laplace asymptotics needs to be adjusted in order to allow for this time dependence. The family of models considered also encompasses the so-called local volatility Heston models.

Two major tools have been developed to evaluate European option prices in stochastic volatility models in closed form. The first is the so-called mixing formula, due to Hull and White [37] and Renault and Touzi [49]. This method is particularly powerful when applied to uncorrelated stochastic volatility models, since in this case, the price of the European option is fully determined, provided the law of the time integral of the instantaneous variance is known. In the language of stochastic time changes, we need to determine the law of the time change. The second tool, and the one closest to the approach taken in this paper, was introduced into mathematical finance by Kunitomo and Takahashi [38]. It is the method of asymptotic expansions. Hagan and Woodward [30] applied this method to find asymptotic expansions for the implied volatility of European options in a local volatility setting. Hagan, Kumar,

Lesniewski and Woodward [32] used asymptotics methods to obtain approximations for the implied volatility in the two factor Sabr models. In a Courant Institute lecture André Lesniewski [43] introduced the *geometric approach* to asymptotics, by relating the underlying geometry of diffusion associated to the SABR model in the case $\beta = 0$ to the *Poincaré plane*, a model of hyperbolic space, and outlined an approach to the asymptotics in stochastic volatility models via a WKB expansion. This approach was further developed in the important unpublished working paper [31] by Hagan, Lesniewski and Woodward. These authors used changes of variables to reduce the Sabr model with $\beta \neq 0$ to a perturbed form of the Sabr model with $\beta = 0$ and then used the Hausdorff-Baker Campbell formula to find approximate solutions for the fundamental solution of the perturbed problem. In [39] Henry-Labordère's made contributions of both a theoretical and a practical nature. On the theoretical side he showed how an approach to small time asymptotics of fundamental solutions can be derived via the heat kernel expansion. This expansion was introduced by Minakshisundaram and Pleijel in [44], building on earlier work by Hadamard [29]. Using the heat kernel methods a fundamental solution for linear parabolic differential operators with variable coefficients can be obtained even in cases where the operator contains first order (i.e. drift) and zero order terms. On the practical side, as mentioned earlier, Henry-Labordère introduced the λ-Sabr model, and showed how the heat kernel method yields asymptotic formulas for the fundamental solution and for the implied volatility and local volatility in this model. Also, in recent work, Forde and Jacquier [24, 25], have begun the rigorous investigation of implied volatility in both the short and long time limit, in the Heston model. In particular they obtained the first closed form near the money expression for the implied volatility in the Heston setting. This work was further extended in joint work with Mijatović and with Lee [26, 27]. A different, i.e. fast mean reverting, regime was investigated in [23]. Also, Takahashi and collaborators have recently applied their Malliavin and Wiener-Itô iterated integral based approach to the λ-Sabr model [51], and have demonstrated its effectiveness and versatility. Benhamou and Croissant [8] examine the concept of local time in the Sabr model and show that this valuation works using a Black-Scholes like formula in which complex quantities appear. Bourgade and Croissant [13], following an approach by Molchanov [47], in a working paper, applied the latter to a generalized p-homogeneous version of the Sabr model [13].

In a recent paper Gatheral, Hsu, Laurence, Ouyang and Wang [28] have reconsidered the implied volatility expansions in local volatility models and developed highly accurate expansions up to the second order for the implied volatility. They have also pointed out the need to consider separate "regimes" in developing such expansions. A "near the money" small time limit regime was considered in an influential paper by Megvedev and Scaillet [45] based on the nonlinear PDE approach of Berestycki et al. [11], extended to a class of jump diffusions. Gatheral et al. [28] considered two complimentary regimes, an "away from the money regime" and an at the money regime. In the former, the limit is taken as the time to maturity τ goes to zero, for *fixed* value of the spot (or of the forward) and for fixed strike. This regime is associated to exponential decay of the difference between call prices and their intrinsic value as $\tau \to 0$. On the practical side, the main contribution of [28] was extending earlier asymptotic results for local volatility models to second order, so that they become capable to furnish highly accurate expansions both for time homogeneous and *time inhomogeneous* models. In addition, unlike earlier treatments (with the exception of Forde and Jacquier's treatment of the Heston model) the expansions in [28] were put on a rigorous mathematical basis. In particular it is shown that the terms in the expansion of the implied volatility can actually be interpreted as *derivatives* with respect to time to maturity, and for this reason, are in a certain sense optimal.

In Sect. 4.2 we combine the results in [28] with the Gyöngy projection technique and with the heat kernel method for time inhomogeneous diffusions to develop asymptotic expansions that are highly accurate and we show, using in part the results in [28] that these results *extend to stochastic volatility models with time dependent parameters*. That is to say, we describe in detail the *first and second term* in the expansion of the implied volatility in the limit of short times to expiration in two factor local-stochastic volatility models. The form of the resulting expansion is:

- **Second order expansion of local volatility** $\sigma_L(f_t, a_t, f, T)$:

$$\begin{aligned}\sigma_L(f_t, a_t, t, f, T) &= \sigma_L^{(0)}(f_t, a_t, t, f) + \sigma_L^{(0)}(f_t, a_t, t, f)\tau \\ &\quad + \sigma_L^{(1)}(f_t, a_t, t, f)\tau^2 + o(\tau^2) \quad \tau \to 0 \end{aligned} \tag{1.1}$$

  where we have set $\tau = T - t$ and where, in the time inhomogeneous case, we note that the coefficients $\sigma_L^{(i)}$, $i = 0, 1, 2$, will depend explicitly on the spot time t, as well.
- **Order $\tau^{5/2}$ expansion of the call prices**

$$\begin{aligned}&\left[C(s,t,K,T) - (s-K)^+\right] e^{d(K,s,t)^2/2\tau} \\ &= C^{(1)}(s,K,t)\tau^{3/2} + C^{(2)}(s,K,t)\tau^{5/2} + o(\tau^{5/2}).\end{aligned} \tag{1.2}$$

- **Second order expansion of implied volatility** $\sigma_{\mathrm{BS}}(f_t, a_t, f, T)$;

$$\begin{aligned}&\sigma_{\mathrm{BS}}((f_t, a_t, t, f, T) \\ &= \sigma_{\mathrm{BS}}^{(0)}((f_t, a_t, t, f) + \sigma_{\mathrm{BS}}^{(0)}((f_t, a_t, t, f)\tau + \sigma_{\mathrm{BS}}^{(1)}((f_t, a_t, t, f)\tau^2 + O(\tau^2)\end{aligned} \tag{1.3}$$

The key to the proof will be determining the coefficients in the first expansion (for local volatility). With the local volatility up to second order in hand, we can apply the asymptotic expansion obtained for time inhomogeneous local volatility models in [28] to derive the implied volatility. One difference between this paper and earlier ones is that we do not make simplifications for the purpose of making formulas shorter or simpler. We take the point of view that the derivation of the full length formulas should be made clear and these should be presented at a first stage. At a *second* stage, one can explore ways to simplify or shorten the formulas. Since the formulas are all in closed form, up to quadrature, this requires at most approximating a one dimensional integral by Simpson's rule and thus, even if the formulas are long their calculation is instantaneous. An illustration is given for the λ-Sabr model. We devote a section to the asymptotics obtained in [28] for the one factor (local volatility) models, since these can be readily applied in the present setting once the local volatility function has been determined.

As this work was nearing completion we learned of nice independent work by Louis Paulot [48] who also derives a second order asymptotic expansion for the implied volatility in two factor stochastic volatility models in the *time homogeneous case*. Paulot does not consider the time inhomogeneous case considered in this paper. Paulot takes a somewhat different (and very reasonable) approach to ours, in which he bypasses the computation of the local volatility function. For the local variance function (square of the local volatility), which he does not try to determine explicitly, his calculations stop at the first order (4.23) and do not seek to determine the next order (4.24). The determination of the *local* volatility in stochastic volatility models is of independent interest and is provided by our method. Once the local volatility has been determined the determination of the *call prices* is a straightforward "plug and play", using the Proposition (5.2). Alternatively using the local volatility one can use Dupire's forward equation to obtain the call prices for all strikes and maturities. The determination of the implied volatility is more involved, since the relationship between local volatility and implied volatility derived in [28] and recalled in (5.3) is correspondingly more complicated.

Another important issue concerns rigor. Two questions are often asked concerning asymptotic expansions used in mathematical finance and also in other areas of the applied sciences. The first is, are the results rigorous? The second, related question concerns the role of the boundary conditions and the influence of the boundary conditions *on* the asymptotics. In the present paper, in the style of classical asymptotic analysis, our main concern is not on full rigor, but rather to provide details of full expansions. As opposed to previous works we provide full detail of intermediate steps.

However, in order to point out the rigorous underpinnings of this work, the influence of degenerating coefficients, and of boundary conditions, we devote a section to the discussion of how (certain aspects) of how our results here can be made rigorous, based on work of Azencott and co-workers [6], the relevance of whose work seems to have not been noticed in mathematical finance heretofore. In particular, we point out, that, the Sabr and λ Sabr model are associated to an *incomplete* Riemannian manifold, when $\beta < 1$.

2 The Family of Models Considered and Application of the Heat Kernel Expansion

2.1 An Efficient Approach

In this section, we introduce the family of models to which our results will apply. It is understood that we are working under a risk neutral forward measure and are assuming zero dividends for the asset process

$$df_t = C(f_t, t)a_t dW_{1,t} \,, \tag{2.1}$$
$$da_t = \alpha(a_t, t)dt + v(a_t, t)dW_{2,t} \,, \tag{2.2}$$
$$< dW_1, dW_2 > = \rho(t)dt \tag{2.3}$$

with initial conditions at time zero given by (s_0, a_0). Here W_1 and W_2 are Brownian motions with deterministic, possibly time dependent correlation $\rho(t)$.

The associated Kolmogorov backward and forward equations for this family of models are given by:

$$p_t + \frac{1}{2}C^2 a^2 p_{ff} + \rho C a v p_{fa} + \frac{1}{2}v^2 p_{aa} + \alpha p_a = 0 \tag{2.4}$$
$$p_t - \frac{1}{2}(C^2 a^2 p)_{ff} - (\rho C a v p)_{fa} - \frac{1}{2}(v^2 p)_{aa} + (\alpha p)_a = 0 \tag{2.5}$$

The matrix **a** with entries $a_{ij}, i, j = 1, 2$

$$\mathbf{a} = \begin{pmatrix} C^2 a^2 & \rho C a v \\ \rho C a v & v^2 a^2 \end{pmatrix} \tag{2.6}$$

is the *diffusion matrix.*

The price, at time t of a European call option $C(f_t, a_t; K, T)$ struck at K, under the forward measure with maturity T is given by the expectation

$$C(f_t, a_t; K, T) = \mathbb{E}_{(f_t, a_t)}[(f_T - K)^+],$$

which can be expressed via the transition density $p(f_t, a_t, t; f_T, a, T)$ in the form

$$C(f_t, a_t, t; K, T) = \iint (f - K)^+ p(f_t, a_t, t; f, a, T) df da.$$

Here and throughout this paper we will assume the expectations are taken with respect to the a (given) risk neutral measure.

Heat Kernel: Time homogeneous case

In the time homogeneous case, we may without loss of generality assume the initial time is 0. It is well-studied in differential geometry and stochastic analysis, under certain technical conditions (see the discussion in Sect. 3), the transition density has in the time homogeneous case the following expansion as $T \to 0^+$

$$p(f_0, a_0; f, a, T) = \sqrt{g(f, a)}U(f_0, a_0; f, a, T)\frac{e^{-\frac{d^2(f_0, a_0; f, a)}{2T}}}{2\pi T} \tag{2.7}$$

where, letting $x_0 = (f_0, a_0)$ and $x = (f, a)$

- g is the volume form associated with the Riemannian metric determined by the inverse of the diffusion matrix (2.6). The inner product, denoted $< ., . >$, of two vectors A and B is given by

$$< A, B >= g_{ij}a_i b_j$$

  where the Einstein summation is used to sum over repeated indices.
- $d(x_0, x)$ is the geodesic (Riemannian) distance between x_0 and x, in the above mentioned metric g_{ij} corresponding to the inverse of the diffusion matrix.
- U is the series expansion:

$$U(x_0; x, T) = \sum_{k=0}^{n} u_k(x_0; x)T^k + O(T^{n+1}).$$

  The u_k's are called the heat kernel coefficients. In particular, $u_0(x_0; x) = \sqrt{\Delta(x_0, x)}e^{\int_\gamma \langle V, \dot{\gamma}\rangle}$, where Δ and V are defined below.
- Δ is the Van Vleck-DeWitt determinant:

$$\Delta(x_0, x) = \frac{1}{\sqrt{g(x_0)g(x)}} \det\left(-\frac{1}{2}\frac{\partial^2 d^2}{\partial x_0 \partial x}\right),$$

- $\mathcal{P} = e^{\int_\gamma \langle V, \dot{\gamma}\rangle}$ is the exponential of the work done by the vector field $\mathcal{A}$ along the geodesic γ, with $V = V^i \partial_i$ and

$$V^i = b^i - \frac{1}{\sqrt{g}}\frac{\partial}{\partial x^j}\left[\sqrt{g}g^{ij}\right]. \tag{2.8}$$

  Here the bracket "$< \cdot, \cdot >$" with vector entries V and W is defined via the metric g_{ij} as $\sum g_{ij} v_i w_j$.
- By adding and subtracting first order terms (2.4) can then be re-expressed in the form

$$u_t + \frac{1}{2}\Delta_B u + V \cdot \nabla u = 0, \tag{2.9}$$

  where Δ_B is the second order Laplace Beltrami operator $\frac{1}{\sqrt{g}}\frac{\partial}{\partial x_i}(\sqrt{g}g^{ij}\frac{\partial}{\partial x_j})$.

Note that the zeroth order heat kernel coefficient is known in closed form provided we have available in closed form both the distance function and the geodesics. The higher order on diagonal heat kernel coefficients $u_i(x, x)$ have been calculated up to order 4 in very general settings. On the other hand the efficient calculation of the *off diagonal* heat kernel coefficients U_i, for $i \geq 1$, is still an active field of research. We refer to Hsu [34] for an in-depth introduction to heat kernel expansion in stochastic analysis perspective.

Given the heat kernel expansion in (2.7) for the transition density p, the call price C as $T \to 0^+$ has the expansion

$$C(s_0, a_0; K, T) \sim \sum_{k=0}^{n} \frac{1}{2\pi T} \iint_{\{s \geq K\}} e^{-\frac{d^2(x_0,x)}{2T}} G_k(s, a) T^k ds da.$$

where for notational simplicity we have denoted by G_k the expression

$$G_k(s, a) = (s - K) u_k(s, a) \sqrt{g(s, a)}. \tag{2.10}$$

Heat kernel expansion: Time inhomogeneous case

For models of the form (2.1)–(2.3) with time dependent coefficients, the heat kernel is modified as follows: again, let $x_0 = (f_0, a_0)$ and $x = (f, a)$ and let $\tau = T - t$

$$p(x_0, t; x, T) = \frac{e^{-\frac{d^2(x_0;x,t)}{2\tau}}}{2\pi\tau} u(x_0, t; x, T) \sqrt{g(x, T)} \tag{2.11}$$

where now the distance function d, the Van Vleck-De Witt determinant Δ and the contravariant drift (2.8), depend *explicitly* on time because of the explicit dependence of the metric and the drift on time. The series expansion H in this case reads

$$u(x_0, t; x, T) = \sum_{k=0}^{n} u_k(x_0, x, t) \tau^k + O(\tau^{n+1}).$$

Moreover, the explicit form of the heat kernel coefficients is given by the following formula

$$u_0(x_0, x, t) = \sqrt{\Delta(x_0, x)} e^{\int_\gamma \langle V, \dot{\gamma} \rangle} \exp\left(- \int_0^{d(f_0, a_0, f, a, t)} \frac{\partial d(\tilde{f}(\rho), \tilde{a}(\rho), f, a)}{\partial t} d\rho \right)$$

and recursively

$$u_k(x_0, x, t) = \frac{u_0(x_0, x, t)}{d^k(x_0, x, t)} \int_0^{d(f_0, a_0, f, a, t)} \frac{\rho^{k-1}}{U_0} \left(L U_{k-1} + \frac{\partial U_{k-1}}{\partial t} \right) d\rho, \tag{2.12}$$

where $\tilde{s}(0) = f, \tilde{a}(0) = a$. Since the definition of u_i always involves u_0, it is convenient to factor the latter out and to define

$$\hat{u_i} = \frac{u_i}{u_0} \tag{2.13}$$

Comparison with Henry-Labordère approach

In [39] Henry-Labordère takes a different, in the time-homogeneous case equivalent to ours, approach to implementing the heat kernel expansion which involves a complex line bundle. Thus Eq. (2.9) is recast in the form

$$u_t + g^{-1/2}(\partial_i + \mathcal{A}_i)g^{1/2}g^{ij} + (\partial_j + \mathcal{A}_j)u + Qu = 0, \tag{2.14}$$

where

$$\mathcal{A}_i = g_{ij}\mathcal{A}^j$$

where $\mathcal{A}^j$ is given in (2.8), and where

$$Q = g^{ij}(\mathcal{A}_i\mathcal{A}_j - b_j\mathcal{A}_i - \partial_j\mathcal{A}_i)$$

While the form (2.14) is equivalent to our form (2.9) it makes it necessary to introduce the potential term Q. The motivations for introducing such a formulation are undoubtedly related to the arduous task encountered in quantum field theory of determining the higher order heat kernel coefficients. This covariant approach, introduced among others by Avramidi in [4] and discussed in his lectures on mathematical finance [5], introduces heavy machinery, complex line bundles gauge groups etc., which for instance in the case of time homogeneous parabolic operators leads to the determination of heat kernel coefficients u_n via the solution of the transport equations:

$$(1 + \frac{1}{n}\sigma_i)u_n = \mathcal{P}^{-1}\Delta^{-1/2}(D + Q)\Delta^{1/2}\mathcal{P}u_{n-1}$$

where

$$\sigma_i = \nabla_i\sigma \quad \nabla_i = \partial_i + \mathcal{A}_i$$
$$\sigma^i = g^{ij}\sigma_i$$

This covariant approach undoubtedly has computational and conceptual advantages when calculating the *higher order* dimensional heat kernel coefficients. However the method explained in the following subsection, due to Yoshida, is we think, much simpler and more effective in linking in an intuitive way the coefficients of the original parabolic operator to the u_i. Moreover it is extremely easy in Yoshida's approach to incorporate the influence of time inhomogeneity on the heat kernel coefficients.

3 The Starting Point for Rigorous Justification of Heat Kernel Method: Varadhan's Lemma

This section is meant to give the reader an intuitive grasp of some of the key concepts. We do not aspire to completeness but try to indicate original sources where various issues are dealt with in depth. This section can be skipped by those interested only in the practical results.

Varadhan's lemma, in the form he first derived it (see [52, 53]), applies to the case where the underlying operator is uniformly parabolic, with sufficient regularity on the coefficients. In its original form as obtained in [52], it relates to the unique fundamental solution in all of $\mathbb{R}^n$ associated to a diffusion

$$p_t - Lp = 0$$

and finds the small time behavior of such a diffusion to be

$$\lim_{t \to 0} -2t \log p_t(x, y) = d^2(x, y) \tag{3.1}$$

where d is the Riemannian distance introduced in Sect. 2, associated to the principle part of the diffusion operator. Varadhan proved (see Theorem 4.1.3) holds uniformly for x and y in compact sets, for which $d(x, y)$ remains bounded. When expressed in the above form, one has in mind that the domain considered does not have a boundary. Varadhan's lemma can be viewed as a weak form of the zero-th order heat kernel expansion described in Sect. 2. In fact, a general principle is that in many cases, once it is known that Varadhan's lemma holds, the full heat kernel expansion (suitably modified by a Levy parametrix, see Sect. 3.4 in [28] readily follows.

On a domain with a boundary, as is the case encountered in this paper, where the domain is $\mathbb{R}^2_+$, or the upper half plane[1] the situation is more delicate. It is natural that this be the case, since to begin with, one on a domain with boundary, one needs to deal with the possibility that the diffusion can reach the boundary. In cases where the diffusion *can* reach the boundary, in order to uniquely define the diffusion thereafter we can impose an absorbing boundary condition. This amounts to considering the transition density for the diffusion that is killed the first time it reaches the boundary of the domain.

$$\begin{aligned} &p_D(x, y, t) \\ &= \text{Prob. of reaching } y \text{ at time } t \\ &\textit{and} \text{ not reaching the boundary before time } t\text{, starting at } x \end{aligned}$$

[1] Depending on whether forward or log of forward is taken as state variable.

This leads to a general construction of the so-called *minimal heat kernel* which applies to non-compact Riemannian manifolds like $\mathbb{R}^2_+$. In such a construction, which is discussed in detail in Hsu [34], the manifold is exhausted by a sequence D_n of nested compact domains for which the Dirichlet heat kernel p_n (diffusion killed on exiting D_n), is constructed and the minimal heat kernel corresponds to the pointwise limit of these as $n \to \infty$. It is intuitively clear that at points on the boundary of the manifold where the diffusion can reach the boundary, this construction leads to a fundamental solution satisfies a Dirichlet condition in the backward variables.

As was shown by Azencott [6], a simple sufficient condition for Varadhan's lemma to hold that *does* extend to the case where the coefficients degenerate at the boundary, is the case of *complete* Riemannian manifolds. To put this geometric concept into perspective, consider the case where the underlying manifold consists of the first quadrant, corresponding to non-negative forward price and volatility. In this case, the manifold together with the natural Riemannian metric associated to the inverse of the diffusion metric, is complete, if the boundary (and the other points "at infinity", if the domain is unbounded) are at an *infinite* distance from any point in the interior of the manifold. It can be shown that a Riemannian manifold is complete if it is metrically complete, i.e., all Cauchy sequences converge to a point in M (and not on the boundary of M).

The above, with a modern proof appears as Theorem 5.2.1 in Hsu's book [34].

Theorem 3.1 *Let M be a complete Riemannian manifold and $p_M(t, x, y)$ the minimal heat kernel on M. Then, uniformly on compact subsets of $M \times M$, we have (3.1).*

When the underlying Riemannian manifold fails to be complete, Hsu [36] refined the work by Azencott by showing that a sufficient condition under which *Varadhan's lemma is still guaranteed to hold* is

$$d(x, y) \leq d(x, \infty) + d(y, \infty), \tag{3.2}$$

which in our setting amounts to requiring that the distance of x to y does not exceed the sum of the distance of x to the boundary and the distance of y to the boundary.

Below we illustrate this with a couple of examples:

Consider the well known distance function associated to the family of λ-Sabr models (7.8) in the (f, a) plane.

- When $\beta = 1$, the boundary $a = 0$ is at an infinite distance from any point in the interior (f, a). This is because the quantity q involved in the definition of the distance equals

$$|\lim_{\epsilon \to 0} \int_{\epsilon}^{f} \frac{1}{u} du| = \infty$$

Similarly it is easily seen that the $f = 0$ axis is at an infinite distance from any point in the interior. Thus in the case $\beta = 0$ the Riemannian manifold is complete and Varadhan's lemma holds for all points in the manifold.

- When $\beta < 1$, since $\frac{1}{f^\beta}$ is integrable at zero, the points on the $f = 0$ axis lie at a finite distance and those on the $a = 0$ axis at an infinite distance. Thus here we need to apply the sufficient condition (3.2) to guarantee the applicability of Varadhan's lemma.
- For the family of SV models

$$\begin{aligned} df_t &= a_t dW_{1t} + \mu_f dt \\ da_t &= \gamma a_t^q dW_{2t} + \mu_a dt \\ < dW_{1t}, dW_{2t} > &= \rho dt \end{aligned}$$

where ρ and γ are constants. Then
- The (Gaussian) curvature of the Riemannian metric naturally associated to the problem is *independent* of the correlation and of the drift and is given by.
- The curvature is equal to

$$(q-2)y^{2(q-1)} \tag{3.3}$$

Also, it is easily seen that when $q < 1$ the $a = 0$ axis lies at a finite distance from interior points (it suffices to note that vertical lines are geodesics) and also in this case the curvature of the metric blows up at $a = 0$.

A final note concerns the relationship between the remarks above and the interesting recent work by Doust [19] who in determining call prices in the Sabr model for $0 < \beta < 1$, takes into account the probability that the forward hits zero. Doust correctly points out, just as Lewis had done, in the case of the CEV model (see [42] p. 305), that the call price needs to be adjusted to allow for this possibility. There is non conflict between this result and the modified version's of Varadhan's lemma. These simply give sufficient conditions for the distance of points x, y from the boundary (here "infinity"), so that the effect of paths of the forward which reach the boundary is *exponentially negligible* in the small time limit. As Doust points out and as his numerics shows, for longer times these need to be taken into account.

We also would like to remark that the when the local volatility component $C(f, t)$ of the local-stochastic volatility models is such that $C(0, t) = 0$ (for instance $C(f, t) = \gamma(t) f^\beta$ with $0, \beta < 1$, the series constructed by means of the heat kernel *vanishes on the boundary* in the backward variables. Remarkably, this comes "for free" (see (8.1) in Sect. 8) and was not part of the heat kernel's explicit construction.

4 The Projection Method via Gyöngy Projection

A driftless diffusion (say in the forward measure)

$$df_t = \sigma_L(f_t, t)dW_t,$$

gives rise to the so-called *local volatility* models, made famous by the work of Bruno Dupire. Given a two factor local-stochastic volatility model, of the form

$$\begin{aligned} df_t &= b(t)f_t dt + C(f_t, t)a_t dW_{1,t} \\ da_t &= \mu(a, t)dt + v(a_t, t)dW_{2,t}, \\ f_0 &= f, \quad a_0 = a \\ &< dW_{1,t}, dW_{2,t} >= \rho(t)dt \end{aligned} \tag{4.1}$$

the Gyöngy projection technique yields a one factor model

$$df_t = b(t)f(t)dt + \sigma_L(f_t, t)dW_t \tag{4.2}$$

with effective local volatility $\sigma_L(f_t)$ and drift $\bar{b}$ defined by

$$\sigma_L^2(f, T) = \sigma^2(F, T)E\left[a_T^2 \mid f_T = F, f_0 = f, a_0 = a\right] \tag{4.3}$$

Note that the effective drift of f_t hasn't changed when the original drift is of the special form $b(t)f(t)$. In fact, applying Gyöngy's result in our case, the effective drift is given by

$$\bar{b}(F, T) = E\left[b(t)f_t \mid f_T = F, f_0 = f, a_0 = a\right] \tag{4.4}$$

and the latter clearly equals $b(T)F$.

The local volatility model (4.2) has the same marginals with respect to the f variable as the original two factor model (4.1). Independently of Gyöngy, this projection (sometimes also called[2] *marginalization*) technique was discovered by Bruno Dupire who was the first to apply it in mathematical finance in [21].

As pointed out and exploited in Hagan Kumar Lesniewski and Woodward, in Hagan Lesniewski and Woodward [32] and in Henry-Labordère [39], formula (4.4) can be expressed using the joint probability density $p(f, a, t, F, A, T)$ as

$$\sigma_L^2(F, T) = \frac{C^2(F, T)\int_{\mathbb{R}_+} A^2 p(f, a, F, A, T)dA}{\int_{\mathbb{R}_+} p(f, a, F, A, T)dA} \tag{4.5}$$

---

[2]This expression was coined by Marco Avellaneda, who in 1999, without prior knowledge of Gyöngy or Dupire's work, independently discovered the technique.

Therefore plugging in the expansion (2.11) into (4.5) we obtain

$$\sigma_L^2(F,T) = \frac{C^2(F,T)\int_{\mathbb{R}_+}\sum_{i=0}^{n} A^2 e^{-\frac{d^2}{2\tau}}\alpha_i(A,T)\tau^i dA}{\int_{\mathbb{R}_+}\sum_{i=0}^{n}\alpha_i(A,T)(\tau)^i dA}, \tag{4.6}$$

where

$$\alpha_0 =: \sqrt{g}(F,A,T)u_0 \tag{4.7}$$
$$=: \hat{C} \tag{4.8}$$
$$\alpha_i = \sqrt{g}(F,A,T)u_0\hat{u}_i(f,a,F,A,t), \quad i \geq 1 \tag{4.9}$$

and where we have canceled the common factor $\frac{1}{(2\pi(\tau))^{\frac{n}{2}}}$. Letting $\epsilon = \tau$ and $\phi = \frac{d^2}{2}$ the above may be expressed in the form

$$\sigma_L^2(F,T \mid f,a) = \frac{C^2(F,T)\sum_{i=0}^{n}\int_{\mathbb{R}_+} A^2\alpha_i e^{\frac{\phi}{\epsilon}} dA}{\sum_{i=0}^{n}\int_{\mathbb{R}_+}\alpha_i e^{\frac{\phi}{\epsilon}} dA} \tag{4.10}$$

Now we apply the Laplace expansion to obtain an expansion of the local volatility. It is convenient to consider separately the time homogeneous and the more involved time inhomogeneous cases:

4.1 Laplace Expansion: Time-Homogeneous Case

Proposition 4.1 *Suppose $\phi(Y)$ has a unique minimum in $(0,+\infty)$, at the point A_0. The asymptotic expansion of*

$$\int_0^{+\infty} f(A)e^{\frac{-\phi(A)}{\epsilon}} dA, \tag{4.11}$$

as $\epsilon \to 0$ is given by:

$$= \sqrt{\frac{2\pi\epsilon}{\phi''(A_0)}} e^{-\frac{\phi(A_0)}{\epsilon}}$$
$$\times \left\{ f^{(0)}(A_0) + f^{(1)}(A_0)\epsilon + f^{(2)}(A_0)\epsilon^2 \right\}$$

where

$$f^{(0)}(A_0) = f(A_0)$$

$$f^{(1)}(A_0) = \frac{f''(A_0)}{2\phi''(A_0)} - \frac{\phi^{(4)}(A_0)f(A_0)}{8(\phi''(a_0))^2} - \frac{f'(A_0)\phi^{(3)}(A_0)}{2(\phi''(A_0))^2} + \frac{5(\phi^{(3)}(A_0))^2 f(A_0)}{24(\phi''(A_0))^3} \tag{4.12}$$

$$f^{(2)}(A_0) = \tag{4.13}$$

$$\frac{1}{1152\,(\phi'')^6}\Big(-480\,(\phi''(A_0))^3\, f^{(3)}(A_0)\phi^{(3)}(A_0) + 840\,(\phi''(A_0))^2\, f''(A_0)\phi^{(3)}(A_0)^2 \tag{4.14}$$

$$- 840\phi''(A_0)f'(A_0)\phi^{(3)}(A_0)^3 + 385 f(A_0)\phi^{(3)}(A_0)^4 + 144\,(\phi''(A_0))^4\, f^{(4)}(A_0)$$
$$- 360\,(\phi''(A_0))^3\, f''(A_0)\phi^{(4)}(A_0)\,(\phi''(A_0))^2\, f'(A_0)\phi^{(3)}(A_0)\phi^{(4)}(A_0)$$
$$- 630 f(A_0)\phi''\phi^{(3)}(A_0)^2\phi^{(4)}(A_0) + 105 f(A_0)\,(\phi''(A_0))^2\,\phi^{(4)}(A_0)^2$$
$$- 144\,(\phi''(A_0))^3\, f'(A_0)\phi^{(5)}(A_0) + 168 f(A_0)\,(\phi''(A_0))^2\,\phi^{(3)}(A_0)\phi^{(5)}(A_0)$$
$$- 24 f(A_0)\,(\phi''(A_0))^3\,\phi^{(6)}(A_0)\Big) \tag{4.15}$$

where A_0 is the point that minimizes ϕ.[3]

In using the Laplace expansion in conjunction with the heat kernel form (4.16), we must deal with a more general (than (4.16) expression of the form

$$\int_0^{+\infty} \Big(f_0(A) + \epsilon f_1(A) + \epsilon^2 f_2(A)\Big)\, e^{-\frac{\phi(A)}{\epsilon}}\, dA \tag{4.16}$$

Let us introduce the notation $f_i^{(j)}$, $i, j = 1, 2$ to indicate the jth term in the Laplace expansion above applied to the function f_i. We apply the proposition to numerator and denominator of the expression defining the local volatility. This yields an expression that has the following form

$$\frac{f_0^{(0)} + \epsilon f_0^{(1)} + \epsilon^2 f_0^{(2)} + (f_1^{(0)} + \epsilon f_1^{(1)} + \epsilon^2 f_1^{(2)})\epsilon + f_2^{(0)}\epsilon^2 + o(\epsilon^2)}{\zeta_0^{(0)} + \epsilon\zeta_0^{(1)} + \epsilon^2\zeta_0^{(2)} + (\zeta_1^{(0)} + \epsilon\zeta_1^{(1)} + \epsilon^2\zeta_1^{(2)})\epsilon^2 + \zeta_2^{(0)}\epsilon^2 + o(\epsilon^2)}$$
$$= \frac{f_0^{(0)} + \epsilon(f_0^{(1)} + f_1^{(0)}) + \epsilon^2(f_0^{(2)} + f_1^{(1)}) + o(\epsilon^2)}{\zeta_0^{(0)} + \epsilon(\zeta_0^{(1)} + \zeta_1^{(0)}) + \epsilon^2(\zeta_0^{(2)} + \zeta_1^{(1)}) + o(\epsilon^2)},$$

This ratio can be expanded in powers of ϵ, and yields the following asymptotic expansion which can be applied to obtain the effective local volatility valid in a wide

[3]The expansion up to the first order (4.12) appears in several sources including in textbook form, as in Bender and Orszag [9], p. 273. On the other hand we have not been able to locate a source for the second order expansion given here as (4.15).

range of models. is:

$$\frac{f_0^{(0)}}{\zeta_0^{(0)}}$$
$$+\frac{1}{(\zeta_0^{(0)})^2}\left(-f_0^{(0)}(\zeta_0^{(1)}+\zeta_1^{(0)})+\zeta_0^{(0)}(f_0^{(1)}+f_1^{(0)})\right)\epsilon$$
$$\frac{1}{(\zeta_0^{(0)})^3}\left(f_0^{(0)}(\zeta_0^{(1)}+\zeta_1^{(0)})^2-f_0^{(0)}(f_0^{(2)}+f_1^{(1)}+f_2^{(0)})\right.$$
$$\left.+\zeta_0^{(0)}(-(\zeta_0^{(1)}+\zeta_1^{(0)})(f_0^{(1)}+f_1^{(0)})+f_0^{(0)}(f_0^{(2)}+f_1^{(1)}+f_2^{(0)})\right)\epsilon^2+o(\epsilon^2),\quad \epsilon\to 0$$

Let us introduce the following notation in relation to the family of stochastic volatility models in (4.1), in conjunction with the notation introduced in our discussion of the heat kernel expansion, following Yoshida. Recall from (4.8) the definition of $\hat{C}$

$$\hat{C}(a)=\sqrt{\det g(a,f)}\sqrt{\Delta(f_0,a_0,f,a)}\mathcal{P}(f_0,a_0,f,a)$$

where on the left hand side we suppress the dependence of $\hat{C}$ on variables other than a, since these are not relevant in the Laplace asymptotics.

Proposition 4.2 *Assume the distance function in family of local-stochastic volatility models has a unique minimum, as a function of the final value of the volatility a, at* $c=a_{min}$. *Let* $\phi=\frac{d^2}{2}$. *Then the effective local volatility in the family of local stochastic volatility models (4.1), is given, up to the second order, by*

$$\sigma_L^2(t,T,a_0,f_0,f)=:V_L(f,t)=V_L^{(0)}(f,t)+V_L^{(1)}(f,t)\tau+V_L^{(2)}(f,t)\tau^2, \tag{4.17}$$

and correspondingly, by taking square roots and expanding, the local volatility is given by[4]

$$\sigma_L^\epsilon=\sigma_L^{(0)}+\sigma_L^{(1)}\epsilon+\sigma_L^{(2)}\epsilon^2+\cdots\quad \textit{where } \epsilon=\tau \tag{4.18}$$

where

$$\sigma_L^{(0)}=\sqrt{V_L^{(0)}}\qquad (=C(f)c) \tag{4.19}$$

$$\sigma_L^{(1)}=\frac{V_L^{(1)}}{2\sigma_L^{(0)}} \tag{4.20}$$

[4]In (4.17) we once again suppressed on the right hand side the dependence of variables other than t, T and f. In the next section, when we combine the asymptotics for local volatility with the above asymptotics for implied volatility, *given* local volatility, it will be *important* that the local volatility depends on the initial time t and on the final time T in the particular way indicated.

$$\sigma_L^{(2)} = \frac{-(V_L^{(1)})^2 + 4V_L^{(0)}V_L^{(2)}}{8(V_L^{(0)})^{3/2}} = \frac{-4(\sigma_L^{(1)})^2(\sigma_L^{(0)})^2 + 4(\sigma_L^{(0)})^2 V_L^{(2)}}{8(\sigma_L^{(0)})^3} \tag{4.21}$$

where

$$V_L^{(0)}(f,t) = C^2(f)c^2 \tag{4.22}$$

$$V_L^{(1)}(f,t) = \frac{C^2(f)\left(\left(\hat{C}(c) + 2c\hat{C}'(c)\right)\phi''(c) - c\hat{C}(c)\phi^{(3)}(c)\right)}{\hat{C}(c)\phi''(c)^2} \tag{4.23}$$

$$V_L^{(2)}(f,t) = \frac{V_L^{(num,2)}(f,t)}{V_L^{(den,2)}(f,t)} \tag{4.24}$$

with

$$V_L^{(den,2)}(f,t) = \frac{1}{24\hat{C}(c)^2\,(\phi''(c))^6\,\phi''(c)^5}$$

and

$$\begin{aligned}
&V_L^{(num,2)}(f,t)\\
&= C(f)^2\Big(12c\,(\phi''(c))^2\left(-2\hat{C}'(c) + c\hat{C}(c)u_1(c)\phi''(c)\right)\Big(\phi''(c)\Big(\hat{C}''(c)\\
&\quad + 2\hat{C}(c)u_1(c)\phi''(c)\Big) - \hat{C}'(c)\phi^{(3)}(c)\Big) + \hat{C}(c)\phi''(c)\Big(2\phi''(c)^6\Big(12c\phi''(c)^2\hat{C}^{(3)}(c)\\
&\quad + 35c\hat{C}'(c)\phi^{(3)}(c)^2 + 6\hat{C}''(c)\phi''(c)\left(3\phi''(c) - 5c\phi^{(3)}(c)\right)\\
&\quad - 15\hat{C}'(c)\phi''(c)\left(2\phi^{(3)}(c) + c\phi^{(4)}(c)\right)\Big) + (\phi'')^6\Big(-12\hat{C}''(c)\phi''(c)\\
&\left(\phi''(c) + c^2u_1(c)\phi''(c)^2 - c\phi^{(3)}(c)\right) + \hat{C}(c)u_1(c)\phi''(c)\\
&\Big(-24\phi''(c)^2 - 24c^2u_1(c)\phi''(c)^3 + 5c^2\phi^{(3)}(c)^2 - 3c\phi''(c)\Big(-8\phi^{(3)}(c)\\
&\quad + c\phi^{(4)}(c)\Big)\Big) + 2\hat{C}'(c)\Big(-11c\phi^{(3)}(c)^2 + 6cu_1(c)\phi''(c)^2\left(4\phi''(c) + c\phi^{(3)}(c)\right)\\
&\quad + 3\phi''(c)\left(2\phi^{(3)}(c) + c\phi^{(4)}(c)\right)\Big)\Big)\Big) + \hat{C}(c)^2\,(u_1(c)\\
&(\phi'')^6\,\phi''(c)^2\Big(24\phi''(c)^2 - 5c^2\phi^{(3)}(c)^2 3c\phi''(c)\Big(-8\phi^{(3)}(c)\\
&\quad + c\phi^{(4)}(c)\Big)\Big) + (\phi'')^6\Big(48cu_1'(c)\phi''(c)^4 + \left(\phi''(c) - c\phi^{(3)}(c)\right)
\end{aligned}$$

$$\left(-5\phi^{(3)}(c)^2+3\phi''(c)\phi^{(4)}(c)\right)\Big)+\phi''(c)^6\left(-35c\phi^{(3)}(c)^3+35\phi''(c)\phi^{(3)}(c)\right.$$
$$\left.\left(\phi^{(3)}(c)+c\phi^{(4)}(c)\right)-3\phi''(c)^2\left(5\phi^{(4)}(c)+2c\phi^{(5)}(c)\right)\Big)\Big)\right)$$

4.2 Laplace Expansion: Time-Inhomogeneous Case

Since the heat kernel expansion for the probability transition density involves the backward (in financial terms, spot) time in all places *except* for the factor $\sqrt{g}(F, A, T)$ in Eq. 4.9, the Laplace expansion technique is essentially unchanged. As was done in [28], in Sect. 2, we need to develop $\sqrt{g}(F, A, T)$ in a power series expansion around in time around $T = t$. Ie.

$$\begin{aligned}
\sqrt{g}(F, A, T) &= \sqrt{g}(F, A, t) + \frac{\partial\sqrt{g(F, A, T)}}{\partial T}|_{T=t}(\tau) \\
&+ \frac{1}{2}\frac{\partial\sqrt{g}(F, A, T)}{\partial T}|_{T=t}(\tau)^2 + o(\tau^2) \\
&=: d_0(A, t) + d_1(A, t)\epsilon + d_2(A, t)\epsilon^2 + o(\epsilon^2), \quad \epsilon = \tau
\end{aligned} \tag{4.25}$$

The presence of powers of ϵ in the expansion above implies that letting

$$\beta_i = \frac{\alpha_i}{\sqrt{g}(F, A, T)} =: \sqrt{\Delta}\mathcal{P}u_i,$$

where α_i was defined in (4.9), the expansion in powers of ϵ will now follow from the expansion in powers of ϵ of

$$\begin{aligned}
&(\beta_i^{(0)} + \beta_i^{(1)}\epsilon + \beta_i^{(2)}\epsilon^2 + o(\epsilon^2))(c_0 + c_1\epsilon + c_2\epsilon^2 + o(\epsilon^2)) \\
&= \beta_i^{(0)}c_0 + (\beta_i^{(1)}c_0 + \beta_i^{(0)}c_1)\epsilon + (\beta_i^{(2)}c_0 + c_2\beta_i^{(0)} + \beta_i^{(1)}c_1) + o(\epsilon^2) \\
&=: \underbrace{\overset{(0)}{\underset{i}{\gamma}}}_{\alpha_i^{(0)}} + \overset{(1)}{\underset{i}{\gamma}}\epsilon + \overset{(2)}{\underset{i}{\gamma}}\epsilon^2 + o(\epsilon^2)
\end{aligned}$$

I.e. it suffices to replace in the Laplace expansion of numerator and denominator of the expansion of the local volatility function, α_i by γ_i throughout. Notice that the zero-th order $\gamma_i^{(0)}$ coincides with $\alpha_i^{(0)}$ in the time homogeneous case. Since only α_0 enters into the definition of $\sigma_L^{(1)}$, this means that the form of the latter is basically unchanged.

The new form of the $V_L^{(2)}$ is given in Appendix.

5 Coupling with the Local Volatility and Call Price Expansion

5.1 The Key Quantities in One-Dimensional Case

In [28] we obtained an optimal result for the asymptotics of the implied volatility in a local volatility model of the form[5]

$$df_t = \sigma_L(f_t, t)dW_t \tag{5.1}$$

In order to formulate our asymptotic result in the stochastic volatility setting, we will need to combine those results with the results in the previous section. We begin by recalling some of the required auxiliary quantities derived in [28].

- One dimensional (signed) distance function

$$d_1(f, K, t) = \int_K^f \frac{1}{\sigma_L(u, t)} du, \quad t \in [0, T]$$

- One dimensional heat kernel coefficients $u_{L,0}$ and $u_{1,0}$, given by

$$u_0^{(1\text{d})}(f, K, t) = \sqrt{\frac{\sigma_L(f, t)}{\sigma_L(K, t)}} \exp\left[-\int_K^s \frac{(d_1)_t(K, \eta, t)}{\sigma_L(\eta, t)} d\eta\right]. \tag{5.2}$$

and

$$u_1^{(1\text{d})} = \frac{u_0^{(1\text{d})}(f, K, t)}{d_1(K, f, t)} \int_K^f \left[\frac{\sigma_L^2}{2}\left(H^2 + H_f\right) + bH + c + \int_K^\eta H_t(\zeta, K, t)d\zeta\right] \times \frac{d\eta}{\sigma_L(\eta, t)} \tag{5.3}$$

where

$$H(f, K, t) = \frac{\partial}{\partial f}[\ln u_0(f, K, t)] = \frac{(\sigma_L)_f(f, t)}{2\sigma_L(f, t)} - \frac{(d_1)_t(K, f, t)}{\sigma_L(f, t)}.$$

Expansion for call prices

As noted by Henry-Labordère, following on the work by Dupire and Derman and Kani, it is possible to use a formula, which actually goes back even further, i.e., to the work of Carr and Jarrow [14] for the call prices $C(s, K, t, T)$ which reads

[5] Reference [28] explains how to adjust the results to allow for a non zero but constant interest (or other constant yield) rate.

$$C(s,K,t,T)=(s-K)^+ + \frac{1}{2}\int_t^T \sigma_L(K,u)^2 p(s,t,K,u)du.$$

$$C(s,K,t,T)-(s-K)^+$$
$$\sim \frac{1}{2\sqrt{2\pi}}\sum_{i=0}^{k}\left[\int_t^T \sigma_L(K,u)e^{-d(K,s,t)^2/2(u-t)}(u-t)^{i-\frac{1}{2}}du\right]u_i(s,K,t).$$

Letting

$$U_i(\omega,\tau)=\int_0^\tau u^{i-\frac{1}{2}}e^{-d^2/2u}du. \tag{5.4}$$

the expansion may, in the time inhomogeneous case, be expressed in the compact form

Proposition 5.1 *The expansion of the call prices in a driftless local volatility model, is given by:*

$$C(s,K,t,T)-(s-K)^+ \tag{5.5}$$
$$\sim\frac{1}{2\sqrt{2\pi}}\sum_{i=0}^{k}\left[U_i(d,\tau)+(\sigma_L)_t(K,t)U_{i+1}(d,\tau)+\frac{1}{2}(\sigma_L)_{tt}(K,t)U_{i+2}\right]u_i(s,K,t).$$

Using that

$$U_0(\omega,\tau)\sim 2\left[\frac{\tau^{3/2}}{d^2}-\frac{3\tau^{5/2}}{d^4}\right]e^{-\omega^2/2\tau},$$
$$U_1(\omega,\tau)\sim\left[\frac{2\tau^{5/2}}{d^2}\right]e^{-d^2/2\tau}$$

an alternate form for the call price expansion up to order $\tau^{5/2}$ is

Proposition 5.2 *The expansion of the call prices in a driftless local volatility model, is given, in the small time limit $\tau\to 0$, up to the order $\tau^{5/2}$ by*

$$C(s,K,t,T)-(s-K)^+$$

$$=\frac{1}{\sqrt{2\pi}}e^{-d^2/2\tau}\left[\left(\frac{1}{d^2}\sigma_L(K,t)u_0(s,K,t)\right)\tau^{3/2}\right.$$
$$\left.+\left(-\frac{3}{d^4}\sigma_L(K,t)u_0(s,K,t)+(\sigma_L)_t\frac{1}{d^2}u_0(s,K,t)+\frac{1}{d^2}\sigma_L(K,t)u_1(s,K,t)\right)\tau^{5/2}\right]$$

Implied volatility

For the implied volatility the following expansion was obtained in the same paper:

Proposition 5.3 *The implied volatility σ_{BS} admits the following asymptotic expansion, away from the money:*

$$\sigma_{\text{BS}} \sim \sigma_{\text{BS},0} + \sigma_{\text{BS},1}\tau + \sigma_{\text{BS},2}\tau^2 + o(\tau^2), \quad T \to t, \quad \textit{for} \quad f \neq K$$

where

$$\sigma_{\text{BS},0} = \frac{\xi}{d_1(K, f, t)} = \frac{\ln\left(\frac{f}{K}\right)}{\int_K^f \frac{d\eta}{\sigma_L(\eta,t)}}. \tag{5.6}$$

$$\sigma_{\text{BS},1} = \frac{\sigma_{\text{BS},0}^3 \ln\left[\frac{u_0^{(1d)}(f,K,t)\sigma_L(K,t)\xi^2}{\sqrt{fK}d^2\sigma_{\text{BS},0}^3}\right]}{\xi^2} = \frac{\xi \ln\left[\frac{\sigma_L(K,t)u_0^{(1d)}(f,K,t)d(K,f,t)}{\xi\sqrt{fK}}\right]}{d_1(K,f,t)^3}. \tag{5.7}$$

$$\begin{aligned}\sigma_{\text{BS},2} &= -\frac{3\sigma_{\text{BS},1}\sigma_{\text{BS},0}^2}{\xi^2} + \frac{3\sigma_{\text{BS},1}^2}{2\sigma_{\text{BS},0}} + \frac{\sigma_{\text{BS},0}^3}{\xi^2} \\ &\left[\frac{3\sigma_{\text{BS},0}^2}{\xi^2} + \frac{\sigma_{\text{BS},0}^2}{8} + \frac{(\sigma_L)_t(K,t)}{\sigma_L(K,t)} - \frac{3}{d^2(K,f,t)} + \frac{u_1^{(1d)}(f,K,t)}{u_0^{(1d)}(f,K,t)}\right] \\ &= -\frac{3\sigma_{\text{BS},1}}{d^2} + \frac{3\sigma_{\text{BS},1}^2}{2\sigma_{\text{BS},0}} + \frac{\xi^3}{8d^5} + \frac{\xi}{d^3}\left[\frac{(\sigma_L)_t(K,t)}{\sigma_L(K,t)} + \frac{u_1^{(1d)}(f,K,t)}{u_0^{(1d)}(f,K,t)}\right].\end{aligned} \tag{5.8}$$

5.2 *Plugging the Local Volatility into the Call Price and into the Implied volatility Expansion*

Given the expansion (8.4) for the local volatility in powers of $\epsilon = \tau$, we can use the 1D expansion above to determine the contributions of the higher order terms to the implied volatility. Note that the above expansion of the Black-Scholes implied volatility in powers of τ has an underlying functional dependence on the underlying local volatility, $\sigma_L \sim \sigma_L^{(0)} + \epsilon\sigma_L^{(1)} + \epsilon^2\sigma_L^{(2)}\epsilon^2$, obtained in Sect. 3. In order to emphasize this dependence below we write $\sigma_{BS,i}[\sigma_L]$. In order to combine the local volatility asymptotics (4.2) with the implied volatility asymptotics, we must plug the former into the latter and again expand in powers of $\epsilon = \tau$. Now an additional subtlety arises and taking it into account properly is *crucial* to the correct derivation of all subsequent formulas. The subtlety is that the local volatility expansion obtained in the previous section introduces an additional dependence on the backward variable f_0 which (alongside a_0) serves as an initial condition when using the call decomposition formula (4.10). Thus f enters in as a *parameter* and the "active" variable in $\sigma_L(f_0, a_0, f, t)$ is f. Every time a differentiation or integration with respect to a spatial argument needs to be carried out, we *need to use f as the active variable.* This may seem a little surprising because in deriving the 2-D heat kernel we are freezing the *forward variable f* (or K), whereas now we are freezing the backward variable f_0. This state of affairs is easily clarified when we remember that underlying Dupire's derivation of the local volatility function is a differentiation with respect to the forward variable. So, in a sense we are mixing a backward and a forward representation. Although this may seem odd, it does carry has some distinct advantages especially in the time-inhomogeneous case, since all quantities of interest there are evaluated at the spot time t, which is mostly frozen throughout.

As an example, letting $d_1^\epsilon(f, f_0, t)$ denote the $1-D$ distance function introduced in the beginning of the preceding section, *corresponding* to the expansion for the local volatility denote σ_L^ϵ obtained in (8.4)

$$d^\epsilon(f_0, f, t, T) = \int_f^{f_0} \frac{1}{\sigma_L^\epsilon(f_0, a_0, f, t, T)} du$$

So, we obtain using that by definition $\frac{\partial}{\partial T}|_{T=t} = \frac{\partial}{\partial \epsilon}$

$$= \frac{d}{d\epsilon} d(f, f_0, t, \epsilon)|_{\epsilon=0} = -\int_f^{f_0} \frac{\sigma_L^{(1)}(f_0, a_0, f, t)}{(\sigma_L^{(0)}(f_0, a_0, f, t))^2} du \tag{5.9}$$

Therefore the expression appearing in u_0 can be written

$$e^{\int_f^{f_0} \frac{1}{\sigma^{(0)}(f_0, a_0, \eta, t)} \int_K^\eta \frac{\sigma_L^{(1)}(f_0, a_0, u, t)}{(\sigma_L^{(0)}(f_0, a_0, u, t))^2} du d\eta}$$

Applying this procedure throughout, we obtain the following results:

Proposition 5.4 *The small time to maturity expansion of the call prices in the family (2.1)–(2.3) is given by*

$$C(s, K, t, T) - (s-K)^+$$

$$= \frac{1}{\sqrt{2\pi}} e^{-d^2/2\tau} \left[\left(\frac{1}{d^2} \sigma_L(K, t) u_{0,1}(s, K, t) \right) \tau^{3/2} + \left(-\frac{3}{d^4} \sigma_L^{(0)}(K, t) u_0(s, K, t) \right.\right.$$
$$\left.\left. + \sigma_L^{(1)} \frac{1}{d^2} u_{0,1}(s, K, t) + \frac{1}{d^2} \sigma_L^{((0)}(K, t) u_{1,1}(s, K, t) \right) \tau^{5/2} \right]$$

where $\sigma_L^{(0)}$ *is given by (4.19),* $\sigma_L^{(1)}$ *is given by (4.20) and where* $u_{0,1}$ *and* $u_{1,1}$ *are given by (5.14) and* $u_{1,1}$ *is given by (5.16).*

Proposition 5.5 *Given the family of local-stochastic volatility models (2.1)–(2.3), and assuming the coefficients do not depend on time, we have the following asymptotic expansion for* $\epsilon \to 0$

$$\sigma_{BS}^{SV}[\sigma_L](f_0, a_0, K, t) \equiv \sigma_{BS,0}^{SV}[\sigma_L]$$
$$+\sigma_{\mathrm{BS},1}^{SV}[\sigma_L](f_0, a_0, K, t)\tau + \sigma_{\mathrm{BS},2}^{SV}[\sigma_L](f_0, K, T)\tau^2, +o(\tau^2) \quad \tau = T - t \to 0$$

where

$$\sigma_{\mathrm{BS},0}^{SV}[\sigma_L](f_0, a_0, K, t) = \frac{\xi}{\int_K^{f_0} \frac{1}{\sigma_L^{(0)}(f_0, a_0, u, t)} du} \tag{5.10}$$

$$\sigma_{\mathrm{BS},1}^{SV}[\sigma_L](f_0, a_0, K, t) \tag{5.11}$$
$$= \frac{\xi \ln \left[\frac{\sigma_L^{(0)}(f_0, a_0, K, t) u_{0,1}(f_0, K, t) d^{(0)}(f_0, a_0, K, t)}{\xi \sqrt{f_0 K}} \right]}{d^{(0)}(f_0, K, t)^3}. \tag{5.12}$$

In the equation above

$$d^{(0)}(f_0, K, t) = \frac{1}{\int_K^{f} \frac{1}{\sigma_L^{(0)}(f_0, a_0, u, t)} du} \tag{5.13}$$

$$u_{0,1}(f, K, t)$$
$$= \sqrt{\frac{\sigma_L^{(0)}(f_0, a_0, f_0, t)}{\sigma_L^{(0)}(f_0, a_0, K, t)}} \exp \left[\int_K^f \frac{1}{\sigma_L^{(0)}(f_0, \eta, t)} \int_K^\eta \frac{\sigma_L^{(1)}(f_0, a_0, u, t)}{(\sigma_L^{(0)}(f_0, a_0, u, t)))^2} du d\eta \right]. \tag{5.14}$$

Also

$$\sigma_{\mathrm{BS},2}^{SV}(f_0, a_0, K, t)$$
$$= -\frac{3\sigma_{\mathrm{BS},1}^{SV}}{(d^{(0)})^2} + \frac{3(\sigma_{\mathrm{BS},1}^{SV})^2}{2\sigma_{\mathrm{BS},0}^{SV}} + \frac{\xi^3}{8(d^{(0)})^5} + \frac{\xi}{(d^{(0)})^3} \left[\frac{(\sigma_L^{(1)}}{\sigma_L^{(0)}} + \frac{u_{1,1}}{u_{0,1}} \right] \tag{5.15}$$

where, recalling (5.3), and, for brevity *suppressing the dependence* on the initial variables (f_0, a_0) we have

$$u_{1,1} = \frac{u_{0,1}}{d^{(0)}} \int_K^{f_0} \frac{\sigma_L^{(0)}(\eta, t)}{d^{(0)}}$$
$$\times \left[\left(\mathcal{H}_{(2)}^2 + (\mathcal{H}_f)_{(2)} \right) + \int_K^{\eta} (\mathcal{H}_t)_{(2)}(\zeta, K, t) d\zeta \right] \frac{d\eta}{\sigma_L^{(0)}(\eta, t)}, \tag{5.16}$$

where

$$\mathcal{H}_{(2)} = \frac{\sigma_L^{(1)}}{2(\sigma_L^{(0)})^2} - \frac{\int_K^{f_0} \frac{\sigma_L^{(1)}(u,t)}{(\sigma_L^{(0)}(u,t))^2} du}{\sigma_L^{(0)}},$$

$$(\mathcal{H}_f)_{(2)}$$
$$= \frac{\sigma_L^{(1)}}{(\sigma_L^{(0)})^3} - \frac{\left(\int_K^{f_0} \frac{\sigma_L^{(1)}}{(\sigma_L^{(0)})^2} du \right) \sigma_L^{(1)}}{(\sigma_L^{(0)})^2} - \frac{(\sigma_L^{(0)})_K^2}{2(\sigma_L)^{(0)})^2} + \frac{(\sigma_L^{(0)})_{KK}}{2\sigma_L^{(0)}},$$

$$(\mathcal{H}_t)_{(2)}$$
$$= \frac{1}{2(\sigma_L^{(0)})^2} \left(2 \left(\int_K^{f} \frac{-2(\sigma_L^{(1)})^2 + \sigma_L^{(0)} \sigma_L^{(2)}}{(\sigma_L^{(0)})^3} du \right) \sigma_L^{(0)} \right.$$
$$\left. + 2 \left(\int_K^{f} \frac{\sigma_L^{(0,1)}}{(\sigma_L^{(0)})^2} du \right) \sigma_L^{(1)} - \sigma_L^{(1)} (\sigma_L^{(0)})_K + \sigma_L^{(0)} (\sigma_L^{(1)})_K \right)$$

Remark 5.6 A simplified formula at the second order

A remarkable aspect of the above expression for $\sigma_{\mathrm{BS},2}^{SV}$ is that in almost the *entire expression* only $\sigma_L^{(0)}$ and $\sigma_L^{(1)}$ and their derivatives are involved. Recall that the determination of the first of these required only the heat kernel coefficient u_0 in the two heat kernel expansion, while the second of these required only the heat kernel coefficient u_1. The lengthy expression for $\sigma_L^{(2)}$, i.e. the coefficient of τ^2 in (8.4), expressed in terms of the lengthy expression for V_2, is required above, only for the first term under the integral of $(\mathcal{H}_t)_{(2)}$. Thus, we may propose as a very reasonable approximation for the implied volatility expansion at the second order to use (5.15), with all terms the same but using instead the modified $\mathcal{H}_2^m$ defined by

$$\frac{1}{2(\sigma_L^{(0)})^2} \left(2 \left(\int_K^{f} \frac{-2(\sigma_L^{(1)})^2 +}{(\sigma_L^{(0)})^3} du \right) \sigma_L^{(0)} \right.$$
$$\left. + 2 \left(\int_K^{f} \frac{\sigma_L^{(0,1)}}{(\sigma_L^{(0)})^2} du \right) \sigma_L^{(1)} - \sigma_L^{(1)} (\sigma_L^{(0)})_K + \sigma_L^{(0)} (\sigma_L^{(1)})_K \right)$$

6 Example: Dynamic λ-SABR Model

6.1 Step 1: Reduction to Laplace Beltrami + Drift Form

In this section we apply our expansions to the dynamic λ Sabr model

$$\begin{aligned} df_t &= C(f,t)a_t dW_{1,t} \\ da_t &= \kappa(t)(\bar{\lambda} - a_t) + a_t \nu(t) dW_{2,t} \end{aligned}$$

Our diffusion matrix is

$$\left(g^{ij}\right) = \begin{pmatrix} C^2a^2 & \gamma\nu\rho Ca^2 \\ C\nu\rho a^2 & \nu^2 a^2 \end{pmatrix}$$

with inverse

$$\left(g_{ij}\right) = \begin{pmatrix} \frac{1}{C^2a^2(1-\rho^2)} & -\frac{\rho}{a^2\nu(1-\rho^2)C} \\ -\frac{\rho}{a^2\nu C(1-\rho^2)} & \frac{1}{\nu^2(1-\rho^2)a^2} \end{pmatrix}$$

Recall that the Laplace Beltrami Δ_B operator is given by

$$\Delta_B = g^{-1/2}\partial_i\left(g^{1/2}g^{ij}\partial_j\right) = g^{ij}\partial_i\partial_j + g^{-1/2}\partial_i(g^{1/2}g^{ij})\partial_j$$

In the case of our metric we have

$$g^{-1} = -a^4C^2\nu^2(-1+\rho^2); \tag{6.1}$$

We now use the definition of the Laplace-Beltrami operator to rewrite the original PDE in the form

$$u_t + \frac{1}{2}\Delta_B u + V_1 u_f + V_2 u_a = 0$$

where a straightforward calculation using the explicit form of the metric coefficients yields that the "metric" part of the drift is given by

$$V_m^1 = \frac{1}{2}\left(a^2\nu\gamma f^\beta\sqrt{1-\rho^2}\left(\partial_f(\frac{1}{a^2\nu\gamma f^\beta\sqrt{1-\rho^2}}g^{11}\right) + \partial_a\left(\frac{1}{a^2\nu\gamma f^\beta\sqrt{1-\rho^2}}g^{12}\right)\right)$$
$$= \frac{1}{2}\left(a^2\nu\gamma f^\beta\sqrt{1-\rho^2}\left(\frac{\beta\gamma f^{\beta-1}}{\nu\sqrt{1-\rho^2}}\right)\right) = \frac{1}{2}\gamma^2\beta a^2 f^{2\beta-1}$$

$$V^2 = \frac{1}{2}\left(a^2\nu\gamma f^\beta\sqrt{1-\rho^2}\,\partial_f\left(\frac{1}{a^2\nu\gamma f^\beta\sqrt{1-\rho^2}}g^{12}\right) + \partial_a\left(\frac{1}{a^2\nu\gamma f^\beta\sqrt{1-\rho^2}}g^{22}\right)\right)$$

$$= \frac{1}{2}\left(a^2\nu\gamma f^\beta\sqrt{1-\rho^2}\partial_f\left(\frac{1}{a^2\nu\gamma f^\beta\sqrt{1-\rho^2}}\gamma\nu\rho f^\beta a^2\right)\right.$$

$$\left. + \partial_a\left(\frac{1}{a^2\nu\gamma f^\beta\sqrt{1-\rho^2}}\nu^2 a^2\right)\right)$$

$$= 0$$

Thus the metric part of the contravariant drift can be written

$$V_m = \frac{1}{2}\overset{2}{\gamma}\beta a^2 f^{2\beta-1}\partial_f$$

The corresponding covector field, is obtained by lowering the indices

$$V_1^m = g_{11}V_m^1 \quad V_2^m = g_{12}V_m^1$$

Hence

$$V_1^m = \frac{1}{2}\overset{2}{\gamma}\beta a^2 f^{2\beta-1}\frac{1}{\gamma^2 f^{2\beta}\left(1-\rho^2\right)a^2}$$

$$= \frac{1}{2}\frac{\beta}{f(1-\rho^2)}$$

$$V_2^m = -\frac{\rho}{\nu\gamma\left(1-\rho^2\right)f^\beta a^2}\frac{1}{2}\overset{2}{\gamma}\beta a^2 f^{2\beta-1}$$

$$= -\frac{\beta\gamma\rho}{2\nu(1-\rho^2)}f^{\beta-1}$$

Thus the covector form of the metric part of the drift is

$$\frac{\beta}{2f(1-\rho^2)}df - \frac{\beta\gamma\rho}{2(1-\rho^2)}f^{\beta-1}\frac{da}{\nu} \tag{6.2}$$

Remark: Allowing general $C(f,t)$.

Via the same procedure, when instead of $\gamma(t)f^\beta$ we consider a general form $C(f,t)$ in (2.1), we obtain, for the metric part of the drift, when C does not depend explicitly on time:

$$\frac{1}{2(1-\rho^2)}\frac{\frac{\partial C}{\partial f}}{C}df - \frac{\rho\frac{\partial C}{\partial f}}{2\nu(1-\rho^2)}da, \tag{6.3}$$

and, in the new variables defined by (6.6) and (6.7) below, this term may be expressed in the form

$$\frac{\frac{\partial C}{\partial f}}{2\sqrt{1-\rho^2}}dx,$$

a fact already pointed out by (Henry-Labordère [39] and Paulot in [48]).

Note also that the first part of (6.3) is a perfect differential $\frac{1}{2(1-\rho^2)}d\log C(f)$ and can thus be integrated directly along the original geodesic (from final point (K, a) to initial point (f_0, a_0) to get

$$\frac{1}{2(1-\rho^2)}\log\left(\frac{C(f_0,t)}{C(K,t)}\right)$$

The work done by this term is thus immediately evaluated to be

$$\left(\frac{C(f_0,t))}{C(K,t)}\right)^{\frac{1}{2(1-\rho^2)}}$$

In particular in the λ-Sabr model, we get

$$\left(\frac{F_0}{K}\right)^{\frac{\beta}{2(1-\rho^2)}} \tag{6.4}$$

To this metric part, we need to add the drift coming from the mean reverting volatility. This vector has only the ∂_a component

$$\begin{aligned}
V^{\text{meanrev}} &= \kappa(\bar{\lambda} - a)\partial_a \\
V_{\text{meanrev}} &= g_{12}\kappa(\bar{\lambda} - a)df + g_{22}\kappa(\bar{\lambda} - a)da \\
&= \frac{\rho\kappa[t](-a + \lambda[t])}{a^2\nu\left(-1+\rho^2\right)C[f,t]}df - \frac{\kappa(-a+\lambda[t])}{a^2\nu^2\left(-1+\rho^2\right)}da
\end{aligned} \tag{6.5}$$

Define

$$q = \frac{f^{1-\beta}}{\gamma(t)(1-\beta)}$$

We make the change of variables

$$x = \frac{\nu q - \rho a}{\nu\sqrt{1-\rho^2}} \tag{6.6}$$

$$y = \frac{a}{\nu} \tag{6.7}$$

$$df = \gamma\sqrt{-\rho^2}Cdx + \gamma\rho Cdy \tag{6.8}$$

This change of variables is defined in such a way that the principal part of the partial differential equation, expressed in the new coordinates, is the standard Sabr model corresponding to the (rescaled) hyperbolic plane, i.e., of the form

$$\frac{1}{2}\nu^2 y^2 (u_{xx} + u_{yy})$$

After writing the backward partial differential equation in the form $\frac{1}{2}\Delta_B + V$, each of the terms is covariant. This means that in order to determine the contravariant drift in the new coordinates, it suffices to transform the contravariant drift in the old coordinates by the recipe for the change of a contravariant vector under changes of the independent variables. However, since the changes of variables we are making (6.6) and (6.7) depend *explicitly on time* it is clear by the expression of the new dependent variable in terms of the old, i.e. $u(f, a, t) = v(x(f, a, t), y(f, a, t), t)$ that the in the drift term for the new PDE there is an extra term coming from the time derivatives of the independent variables. A simple calculation shows that the metric part of the drift (6.2) can be expressed in the compact form

$$\frac{\beta f^{\beta-1}}{2\sqrt{1-\rho^2}}dx \tag{6.9}$$

Recalling the transformation (6.6) and (6.7), the time derivatives are

$$\begin{aligned} x_t &= (\frac{1}{\gamma})' \frac{f^{1-\beta}}{(1-\beta)\sqrt{1-\rho^2}} - \frac{\rho'}{(1-\rho^2)^{3/2}} y - (\frac{1}{\nu})' \frac{\rho}{\sqrt{1-\rho^2}} \nu y \\ &= -\frac{\gamma'}{\gamma}(x + \rho \frac{y}{\sqrt{1-\rho^2}}) - \frac{\rho'}{(1-\rho^2)^{3/2}} y + (\frac{\nu'}{\nu}) \frac{\rho}{\sqrt{1-\rho^2}} y \end{aligned} \tag{6.10}$$

$$y_t = (\frac{1}{\nu})' \nu y \tag{6.11}$$

Note that from the transformations (6.6) and (6.7) we obtain the relation

$$dx = \frac{df}{\sqrt{1-\rho^2}\,\gamma\, f^\beta} - \rho \frac{dy}{\sqrt{1-\rho^2}} \tag{6.12}$$

For the non-metric part of the drift arising from the mean reversion we obtain from (6.12) and from (6.5):

$$-\frac{\rho\kappa(\bar{\lambda} - \nu y)}{\sqrt{1-\rho^2}\nu^3 y^2} dx + \left(\frac{-\rho^2 \kappa(\bar{\lambda} - \nu y)}{(1-\rho^2)\nu^3 y^2} + \frac{\kappa(\bar{\lambda} - \nu y)}{(1-\rho^2)\nu^3 y^2} \right) dy \tag{6.13}$$

All in all the PDE satisfied by v is therefore

$$v_t + \frac{1}{2}\nu^2 y^2 (v_{xx} + v_{yy}) + \tilde{V}^x v_x + \tilde{V}^y v_y = 0, \tag{6.14}$$

where

$$\tilde{V}_x dx = \left(\frac{\beta \gamma f^{\beta-1}}{2\sqrt{1-\rho^2}} - \frac{\rho\kappa(\bar{\lambda} - \nu y)}{\sqrt{1-\rho^2}\nu^3 y^2} + \frac{1}{y^2} x_t \right) dx$$

Since the first part contains an exact part, it is better to express the first expression in the equation above in the form

$$\frac{1}{2\sqrt{1-\rho^2}} d \log f^\beta - \frac{\beta\gamma\rho}{2(1-\rho^2)} f^{\beta-1} dy$$

As mentioned earlier, the first part is trivially integrated (see (6.4) and the second part of $\tilde{V}_y dy$ so we have the full drift written as

$$\tilde{\tilde{V}}_x dx + \tilde{\tilde{V}}_y dy + \frac{1}{2\sqrt{1-\rho^2}} d \log f^\beta, \tag{6.15}$$

where

$$\tilde{\tilde{V}}_x dx = \left(-\frac{\rho\kappa(\bar{\lambda} - \nu y)}{\sqrt{1-\rho^2}\nu^3 y^2} + \frac{1}{y^2} x_t \right) dx$$

$$\tilde{\tilde{V}}_y dy = \left(-\frac{\beta\gamma\rho}{2(1-\rho^2)} f^{\beta-1} dy + \frac{1}{y^2} y_t + \frac{\kappa(\bar{\lambda} - \nu y)}{\nu^3 y^2} \right) dy, \tag{6.16}$$

$$= \left(-\frac{\beta\gamma\rho}{2(1-\rho^2)} \frac{1}{(1-\beta)\gamma\sqrt{1-\rho^2}x + \rho(1-\beta)\gamma y} + \frac{1}{y^2} y_t + \frac{\kappa(\bar{\lambda} - \nu y)}{\nu^3 y^2} \right) dy \tag{6.17}$$

where, in the next to last step, we used the relation

$$f^{\beta-1} = \frac{1}{(1-\beta)\gamma\sqrt{1-\rho^2}x + \rho(1-\beta)\gamma y}$$

Recall that the difference between $\tilde{\tilde{V}}$ and $\tilde{V}$ is that in $\tilde{\tilde{V}}$ we removed the *exact* part of the drift vector. For completeness we note that, alternatively, including the exact part explicitly, we have the relations:

$$\tilde{V}_x = \left(\frac{\beta \gamma f^{\beta-1}}{2\sqrt{1-\rho^2}} - \frac{\rho\kappa(\bar{\lambda} - \nu y)}{\sqrt{1-\rho^2}\nu^3 y^2} + \frac{1}{y^2} x_t \right) dx$$
$$= \left(\frac{\beta}{2\sqrt{1-\rho^2}(1-\beta)(\sqrt{1-\rho^2}x + \rho y)} - \frac{\rho\kappa(\bar{\lambda} - \nu y)}{\sqrt{1-\rho^2}\nu^3 y^2} + \frac{1}{y^2} x_t \right) dx \tag{6.18}$$

$$\tilde{V}_y = \left(\frac{1}{y^2} y_t + \frac{\kappa(\bar{\lambda} - \nu y)}{\nu^3 y^2} \right) dy, \tag{6.19}$$

and the associated contravariant components:

$$\tilde{V}^x = \frac{\beta \nu^2 y^2}{2\sqrt{1-\rho^2}(1-\beta)(\sqrt{1-\rho^2}x + \rho y)} - \frac{\nu^2 \rho\kappa(\bar{\lambda} - \nu y)}{\sqrt{1-\rho^2}\nu^3} + \nu^2 x_t$$
$$\tilde{V}^y = \nu^2 y_t + \frac{\kappa(\bar{\lambda} - \nu y)}{\nu} \tag{6.20}$$

6.2 The Geodesics in the Poincaré Plane

The changes of variables we have made in the last section, transform the original operator into the operator

$$\partial_t + \frac{1}{2}\nu^2 y^2 (\partial_x^2 + \partial_y^2) + \tilde{V}^x \partial_x + \tilde{V}_y \partial_y$$

The factor ν^2 in front of the second order part means we are not yet in the standard Poincaré plane where $\nu = 1$. To adjust for this we further make the change of time

$$t = \frac{t'}{\nu^2} \tag{6.21}$$

so, in the t' variable the operator is now in standard form

$$\partial_{t'} + \frac{1}{2} y^2 (\partial_x^2 + \partial_y^2) + \frac{\tilde{V}^x}{\nu^2} \partial_x + \frac{\tilde{V}_y}{\nu^2} \partial_y$$

As is easily seen the covariant form of the drift remains the same (due to offsetting effects), but the contravariant form changes by a factor of ν^2.

The geodesic passing through the points (x_1, y_1) and (x_2, y_2) is known to be the semi-circle with origin at $(x_0, 0)$ where

$$x_0(x_1, y_1, x_2, y_2) = \frac{x_2^2 - x_1^2 + y_2^2 - y_1^2}{2(x_2 - x_1)}$$

and where the radius R is given by:

$$\begin{aligned} R &= \sqrt{y_1^2 + (x_1 - x_0)^2} \\ &= \sqrt{y_2^2 + (x_2 - x_0(x_1, y_1, x_2, y_2))^2} \end{aligned} \tag{6.22}$$

It will be convenient to use standard polar coordinates

$$\begin{aligned} x &= x_0 + R\cos\theta, \quad 0 \le \theta \le \pi \\ y &= R\sin\theta, \quad 0 \le \theta \le \pi \end{aligned} \tag{6.23}$$

Notice that if we denote by (x, y) the running point on the geodesic, then the polar angle corresponding to this point, issuing from a fixed point (x_2, y_2) is given by

$$\theta(x,y) = \begin{cases} \arctan\left(\frac{y}{x - \frac{x_2^2 - x^2 + y^2 - y^2}{2(x_2 - x)}}\right) & \arctan\left(\frac{y}{x - \frac{x_2^2 - x^2 + y^2 - y^2}{2(x_2 - x)}}\right) > 0 \\ \arctan\left(\frac{y}{x - \frac{x_2^2 - x^2 + y^2 - y^2}{2(x_2 - x)}}\right) + \pi, & \arctan\left(\frac{y}{x - \frac{x_2^2 - x^2 + y^2 - y^2}{2(x_2 - x)}}\right) < 0 \end{cases} \tag{6.24}$$

6.3 Calculation of the Work Done by the Drift

The integral we need to calculate is

$$\int_{\gamma((x_1,y_1),(x_2,y_2))} \tilde{V}_x dx + \tilde{V}_y dy$$

where $\gamma((x_1, y_1), (x_2, y_2)$ is the geodesic joining the points (x_1, y_1) and (x_2, y_2) in the standard hyperbolic plane, which is the image of the original f, a planed under the coordinate transformation (6.6) and (6.7), call it Λ.

Let us first carry this out for the so-called "metric part" of the drift, i.e. the part corresponding to $\kappa = 0$ and no time dependence in the coefficients.

Then, using (6.18), for the metric part of the drift we have that

$$\begin{aligned} &\int_{\gamma(x,y)} \tilde{V}^m dy \\ &= -\frac{\beta\gamma\rho}{2(1-\rho^2)} \int_{\theta_1}^{\theta_2} \frac{\cos\theta}{\sqrt{1-\rho^2}(x_0 + R\cos\theta) + R\rho\sin\theta} d\theta \end{aligned}$$

Therefore we must calculate the integral

$$\int \frac{1}{\sqrt{1-\rho^2}(x_0 + R\cos\theta) + \rho R \sin\theta} \cos\theta d\theta$$

To evaluate this integral, note that we may write the identity

$$\begin{aligned}
&\frac{\cos\theta}{\sqrt{1-\rho^2}(x_0 + R\cos\theta) + \rho R \sin\theta} \\
&= \frac{\sqrt{1-\rho^2}}{R} + \frac{\rho}{R} d\log\left(\sqrt{1-\rho^2}(x_0 + R\cos\theta) + \rho R \sin\theta\right) \\
&- \frac{x_0\left(1-\rho^2\right)}{R}\left(\frac{1}{\sqrt{1-\rho^2}(x_0 + R\cos\theta) + \rho R \sin\theta}\right)
\end{aligned}$$

and therefore the integral can be written

$$\begin{aligned}
&\frac{\sqrt{1-\rho^2}}{R}(\theta_2 - \theta_1) + \frac{\rho}{R}\log\left(\sqrt{1-\rho^2}(x_0 + R\cos\theta) + \rho R \sin\theta\right)\Big|_{\theta_1}^{\theta_2} \\
&- \frac{x_0\left(1-\rho^2\right)}{R}\int_{\theta_1}^{\theta_2} \frac{1}{\sqrt{1-\rho^2}(x_0 + R\cos\theta) + \rho R \sin\theta} d\theta
\end{aligned} \tag{6.25}$$

and it remains only to evaluate this last integral.

This may be expressed as

$$\int_{\theta_1}^{\theta_2} \frac{1}{\sqrt{1-\rho^2}x_0 + R\cos\left(\theta - \arctan(\frac{\rho}{\sqrt{1-\rho^2}})\right)} d\theta$$

Letting

$$\begin{aligned}
a_1 &= \sqrt{1-\rho^2}x_0 \\
\hat{\gamma} &= \arctan(\frac{\rho}{\sqrt{1-\rho^2}}),
\end{aligned}$$

the last integral can be expressed as:

$$\underbrace{\int_{\theta_1}^{\theta_2} \frac{1}{a_1 + R\cos(\theta - \hat{\gamma})} d(\theta - \hat{\gamma}) =}_{I_1} \tag{6.26}$$

$$= \int_{\theta_1-\hat{\gamma}}^{\theta_2-\hat{\gamma}} \frac{1}{a_1 + R\cos\theta} d\theta \tag{6.27}$$

$$\begin{cases} \left\{ \frac{2}{\sqrt{a_1^2-R^2}} \arctan\left(\sqrt{\frac{a_1-R}{a_1+R}} \tan\frac{\theta}{2}\right) \right\} |_{\theta=\theta_1-\hat{\gamma}}^{\theta=\theta_2-\hat{\gamma}} & \text{if } a_1^2 > R^2 \\ \left\{ \frac{2}{\sqrt{R^2-a_1^2}} \operatorname{arctanh}\left(\sqrt{\frac{R-a_1}{R+a_1}} \tan\frac{\theta}{2}\right) \right\} |_{\theta=\theta_1-\hat{\gamma}}^{\theta=\theta_2-\hat{\gamma}} & \text{if } a_1^2 < R^2 \\ \left\{ \frac{1}{R} \tan(\frac{\theta}{2}) \right\} |_{\theta=\theta_1-\hat{\gamma}}^{\theta=\theta_2-\hat{\gamma}} & \text{if } a_1 = R \\ \left\{ \frac{1}{R} \cot(\frac{\theta}{2}) \right\} |_{\theta=\theta_1-\hat{\gamma}}^{\theta=\theta_2-\hat{\gamma}} & \text{if } a_1 = -R \end{cases} \tag{6.28}$$

So that all told, taking into account (6.4) we obtain the following expression for the line integral corresponding to the metric part:

$$\log(\mathcal{P})_{ME}$$

$$= -\frac{\beta\gamma\rho}{2(1-\rho^2)}\left(\frac{\sqrt{1-\rho^2}}{R}(\theta_2-\theta_1) + \frac{\rho}{R}\log\left(\sqrt{1-\rho^2}(x_0 + R\cos\theta) + \rho R\sin\theta\right)|_{\theta_1}^{\theta_2} \right. \tag{6.29}$$

$$\left. - \frac{x_0(1-\rho^2)}{R} I_1 \right) + \frac{\beta}{2\sqrt{1-\rho^2}} \log(\frac{f_0}{K}) \tag{6.30}$$

λSabr part

Next we deal with the specific to λ-Sabr part, which have found above to be

$$-\frac{\rho\kappa(\bar{\lambda} - \nu y)}{\sqrt{1-\rho^2}\nu^3 y^2} dx + \frac{\kappa(\bar{\lambda} - \nu y)}{\nu^3 y^2} dy \tag{6.31}$$

Inserting this into the line integral that must be evaluated, we find

$$= \frac{\rho\kappa\bar{\lambda}}{R\sqrt{1-\rho^2}\nu^3} \int_{\theta_1}^{\theta_2} \csc\theta d\theta$$

$$- \frac{\rho\kappa}{\sqrt{1-\rho^2}\nu^2} \int_{\theta_1}^{\theta_2} 1 d\theta$$

$$+\frac{\kappa\bar{\lambda}}{\nu^3 R}\int_{\theta_1}^{\theta_2}\frac{\cos\theta}{\sin^2\theta}d\theta$$

$$-\frac{\kappa}{\nu^2}\int_{\theta_1}^{\theta_2}\frac{\cos\theta}{\sin\theta}d\theta$$

Now

$$\int \csc\theta d\theta = \log(\tan(\theta/2)) + C$$
$$\int \frac{\cos\theta}{\sin^2\theta} = -\frac{1}{\sin\theta} + C$$
$$\int \cot\theta d\theta = \log(|\sin\theta|) + C$$

Therefore, all told we obtain for the line integral corresponding to the mean reverting part:

$$\frac{\rho\kappa\bar{\lambda}}{R\sqrt{1-\rho^2}\nu^3}\log|\frac{\tan(\theta_2/2)}{\tan(\theta_1/2)}| - \frac{\rho\kappa}{\sqrt{1-\rho^2}\nu^2}(\theta_2-\theta_1) \quad (6.32)$$
$$+\frac{\kappa\lambda}{\nu^3 R}(\frac{1}{\sin\theta_1}-\frac{1}{\sin\theta_2}) - \frac{\kappa}{\nu^2}\log(\frac{|\sin\theta_2|}{|\sin\theta_1|} \quad (6.33)$$

Since θ varies between 0 and π and since in this range both $\sin\theta$ and $\tan\theta/2$ are non-negative, we can remove the absolute value signs above and arrive at

$$\log(\mathcal{P})_{MR}$$
$$= \frac{\rho\kappa\bar{\lambda}}{R\sqrt{1-\rho^2}\nu^3}\log\left(\frac{\tan(\theta_2/2)}{\tan(\theta_1/2)}\right) - \frac{\rho\kappa}{\sqrt{1-\rho^2}\nu^2}(\theta_2-\theta_1)$$
$$+\frac{\kappa\lambda}{\nu^3 R}(\frac{1}{\sin\theta_1}-\frac{1}{\sin\theta_2}) - \frac{\kappa}{\nu^2}\log(\frac{\sin\theta_2}{\sin\theta_1}) \quad (6.34)$$

Time dependent component of drift

The last integral we need to evaluate arises from the term $x_t v_x + y_t v_y$ which, as we have seen, gives rise, after lowering the indices in the hyperbolic metric to:

$$\int_{\gamma[x,y]}\frac{x_t}{y^2}dx + \int_{\gamma[x,y]}\frac{y_t}{y^2}dy$$

Using the explicit expressions (6.10) and (6.11) we find that

$$\frac{x_t}{y^2} = -\frac{\gamma'}{\gamma}(\frac{x}{y^2} + \rho\frac{1}{\sqrt{1-\rho^2}y}) - \frac{\rho'}{(1-\rho^2)^{3/2}}\frac{1}{y} + (\frac{\nu'}{\nu})\frac{\rho}{\sqrt{1-\rho^2}}\frac{1}{y}$$

$$\frac{y_t}{y^2} = -\frac{\nu'}{\nu}\frac{1}{y}$$

So the integrals to be calculated are:

$$\log(\mathcal{P}_t) = \frac{1}{\sqrt{1-\rho^2}}(-\frac{\gamma'\rho}{\gamma} - \frac{\rho'}{\sqrt{1-\rho^2}} + \frac{\nu'\rho}{\nu})\int_{\theta_1}^{\theta_2}\frac{dx}{y} - \frac{\nu'}{\nu}\int_{\theta_1}^{\theta_2}\frac{dy}{y} - \frac{\gamma'}{\gamma}\int_{\theta_1}^{\theta_2}\frac{x}{y^2}dx \tag{6.35}$$

which gives

$$\log(\mathcal{P}_t) = \frac{1}{\sqrt{1-\rho^2}}(-\frac{\gamma'\rho}{\gamma} - \frac{\rho'}{\sqrt{1-\rho^2}} + \frac{\nu'\rho}{\nu})(\theta_1 - \theta_2) - \frac{\nu'}{\nu}\log|\frac{\sin\theta_2}{\sin\theta_1}| - \frac{\gamma'}{\gamma}\left(\frac{1}{R}\log\frac{\tan\theta_2}{\tan\theta_1} - \log|\frac{\sin\theta_2}{\sin\theta_1}|\right) \tag{6.36}$$

7 Calculation of u_1 Heat Kernel Coefficient

The calculations of the heat kernel coefficient u_1 (and all other heat kernel coefficients) can be carried out either in the original variables (f, a) or in the (rescaled) standard hyperbolic place with metric $ds^2 = \frac{1}{y^2}(dx^2 + dy^2)$,[6] provided we use the transformed drift vector (6.20).

Recall, from formula (2.12) with $i = 1$ that

$$u_1 = u_0\frac{1}{d}\int_0^d u_0^{-1}(Lu_0 + u_\tau)\, ds \tag{7.1}$$

[6] Recall that a time change (6.21) gets rid of the extra factor ν^2.

where

$$u_0 = \sqrt{\Delta}\mathcal{P} \underbrace{\exp\left(- \int_0^{d(f_0,a_0,f,a,t)} \frac{\partial d(\tilde{s}(\rho), \tilde{a}(\rho), s, a)}{\partial t} \, d\rho \right)}_{u0t} \tag{7.2}$$

where we recall that using (6.4), (6.34) and (6.36), $\mathcal{P}$ is known in closed form and where in the a formula above the distance function is given by (7.8). Also recall that Δ is the Van-Vleck De Witt determinant, and we know that in the hyperbolic space of curvature -1

$$\Delta = \frac{d}{\sinh d}, \tag{7.3}$$

As noted by Willmore [54] p. 208, the Van-Vleck De Witt determinant (which Willmore calls the "discriminant function") is *independent of the coordinate system chosen*. This means that to express the VVDW determinant in the original coordinate system, it is sufficient to use the same expression (7.3) in conjunction with the distance function in *that* coordinate system, i.e. (7.8). Since all ingredients in u_0 are *explicit* we can now simply insert this expression into (7.1) and use a symbolic calculation engine like Mathematica or Maple to do the calculations.

Alternatively, we can do the calculations using the formula for u_1 in the (x, y) plane. This has the advantage that many of the terms arising during the calculation can be computed, once again, in closed form. We discuss this further in the remainder of this section.

The fact that Δ depends exclusively on the distance simplifies some of the expressions involved in u_1. In polar coordinates based at the point y, we have that the Laplace Beltrami operator can be expressed in the form

$$\frac{\partial^2}{\partial r^2} + \left(\frac{\partial}{\partial r} \log \left(\frac{r^{n-1}}{\Delta} \right) \right) \frac{\partial}{\partial r} + a(r, \theta, t) \frac{\partial^2}{\partial \theta^2} \tag{7.4}$$

$$= \frac{\partial^2}{\partial r^2} + \left(- \frac{\frac{\partial \Delta}{\partial r}}{\Delta} + \frac{n-1}{r} \right) \frac{\partial}{\partial r} + a(r, \theta, t) \frac{\partial^2}{\partial \theta^2}, \tag{7.5}$$

see Proposition G.V.3, p. 134 of Berger-Gauduchon-Mazet [12] and use the fact that $(\Delta)^{-1} = \theta(ru)$ as defined in C.III.3 in [12]. For brevity's sake, let

$$c(r, t)) = \left(- \frac{\frac{\partial \Delta}{\partial r}}{\Delta} + \frac{n-1}{r} \right)$$

Using the explicit form of Δ on calculates that, in the case of standard, hyperbolic plane,

$$c(r,t) = \text{Cotanh}(d)$$
$$a(r,t) = \frac{1}{\sinh^2(d)}$$

so that the above operator can be expressed as

$$\frac{\partial^2}{\partial r^2} + c(r,t)\frac{\partial}{\partial r} + a(r,\theta,t)\frac{\partial^2}{\partial \theta^2}$$

Letting Δ_B act on u_0, there are some simplifications due to the fact that Δ (Van Vleck De Witt determinant), depends only on r

$$\begin{aligned}
&\Delta_B u_0 = \Delta_B(\sqrt{\Delta(r)}\mathcal{P}(r,\theta)) \\
&= \left(\frac{\partial^2}{\partial r^2} + c(r,t)\frac{\partial}{\partial r} + a(r,\theta,t)\frac{\partial^2}{\partial \theta^2}\right)\Delta_B(\mathcal{P}(r,\theta)) \\
&= (\mathcal{P}(r,\theta))\left(\frac{\partial^2}{\partial r^2} + c(r,t)\frac{\partial}{\partial r}\right)(\sqrt{\Delta(r)}) + \left(a(r,\theta,t)\frac{\partial^2}{\partial \theta^2}(\mathcal{P}(r,\theta))\right)(\sqrt{\Delta(r)})
\end{aligned}$$

Therefore

$$\begin{aligned}
&Lu_0 = \\
&= \mathcal{P}L\sqrt{\Delta} + \sqrt{\Delta}L\mathcal{P} \\
&= \mathcal{P}L\sqrt{\Delta} + \sqrt{\Delta}\mathcal{P}L\mathcal{A} + \frac{1}{2}g^{ij}\mathcal{A}_i\mathcal{A}_j\mathcal{P}\sqrt{\Delta}
\end{aligned}$$

And so

$$\begin{aligned}
&u_0^{-1}Lu_0 \\
&= \frac{L\sqrt{\Delta}}{\sqrt{\Delta}} + L\mathcal{A} + \frac{1}{2}g^{ij}\mathcal{A}_i\mathcal{A}_j \\
&= \frac{L\sqrt{\Delta}}{\sqrt{\Delta}} + L\mathcal{A} + \frac{1}{2}y^2(\mathcal{A}_x^2 + \mathcal{A}_y^2)
\end{aligned}$$

So, that putting it all together and inserting into (7.1) we obtain an expression of the form

$$\sqrt{\Delta}\mathcal{P}\frac{1}{d}\int_{\gamma[x_2,y_2,x_1,y_1]}\left\{\frac{L\sqrt{\Delta}}{\sqrt{\Delta}} + L\mathcal{A} + \frac{1}{2}y^2(\mathcal{A}_x^2 + \mathcal{A}_y^2)\right\}ds$$

Since the first part $\frac{L\sqrt{\Delta}}{\sqrt{\Delta}}$ depends only of the distance, it can be integrated in closed form, and we obtain, using Mathematica, and the expression (7.7) below, that it equals

$$-\mathcal{P}\frac{(-1+d^2+d\coth d)\sqrt{d\operatorname{csch} d}}{4d^2}$$

Now

$$L\mathcal{A} = \frac{y^2}{2}(\mathcal{A}_{xx}+\mathcal{A}_{yy}) + \tilde{V}^x\mathcal{A}_x + \tilde{V}^y\mathcal{A}_y \tag{7.6}$$

$$\frac{L\sqrt{\Delta}}{\sqrt{\Delta}} = \frac{\left(-1-3d^2+(1+d^2)\cosh 2d\right)\operatorname{csch}^4 d\sqrt{\sinh d}}{8d^2(\operatorname{csch} d)^{3/2}} \tag{7.7}$$

So a key quantity that we must calculate is the action of the differential operator L on the exponent $\mathcal{A}$ in $\mathcal{P}$.

Expressing this exponent using the *standard* polar coordinates (6.23).[7] We have, with fixed point (x_2, y_2), letting for brevity $\theta_1 = \theta(x_1, y_1)$, $\theta_2 = \theta(x_2, y_2)$ with θ given by (6.24)

$$\mathcal{A}(x, y, x_2, y_2) = R(x, y, x_2, y_2)\int_{\theta_2}^{\theta_1(x,y,x_2,y_2)}(-\tilde{V}_x\sin\theta + \tilde{V}_y\cos\theta)d\theta$$

Therefore

$$\frac{\partial\mathcal{A}}{\partial x}(x_1, y_1) = \left\{R\left(-\tilde{V}_x\sin\theta_1 + \tilde{V}_y(x_1, y_1)\cos\theta_1\right)(\theta_1)_x\right\}|_{x=x_1,y=y_1}$$
$$+\left\{(R)_x\int_{\theta_2}^{\theta_1(x_1,y_2,x_2,y_2)}(-\tilde{V}_x\sin\theta + \tilde{V}_y\cos\theta)d\theta\right\}_{x=x_1,y=y_1}$$

$$\frac{\partial^2\mathcal{A}}{\partial x^2}(x_1, y_1) = \left\{(R)_{xx}\int_{\theta_2}^{\theta_1(x_1,y_2,x_2,y_2)}(-\tilde{V}_x\sin\theta + \tilde{V}_y\cos\theta)d\theta\right\}_{x=x_1,y=y_1}$$
$$+2\left\{(R)_x\left(-\tilde{V}_x\sin\theta_1 + \tilde{V}_y(x_1, y_1)\cos\theta_1\right)(\theta_1)_x\right\}|_{x=x_1,y=y_1}$$
$$+\left\{R\left(-\tilde{V}_x\sin\theta_1 + \tilde{V}_y(x_1, y_1)\cos\theta_1\right)(\theta_1)_{xx}\right\}|_{x=x_1,y=y_1}$$

and similarly for other partial order derivatives $\frac{\partial\mathcal{A}}{\partial y}$. Note that the evaluation of the terms above requires knowledge of the value of the integral which was computed in the previous section.

[7] Note this angle is *not* the same angle as that used above in geodesic polar coordinates.

Also we may calculate the second order partial derivatives $\mathcal{A}_{xx}$, $\mathcal{A}_{yy}$, $\mathcal{A}_{xy}$ and *express these in terms of* $\tilde{V}$ and derivatives of the polar angle θ_1, given as in (6.23), (6.24) and of the radius, given by (6.22).

7.1 Distance Function and Derivatives in Time Dependent Case

The distance function in the original coordinates is given by

$$d((f_0, a_0), (f, a_1) = \frac{1}{\nu}\cosh^{-1}\left(1 + \frac{q^2 - 2\rho(t)(y_1 - y_0)q + (y_1 - y_0)^2}{2(1-\rho^2)y_0 y_1}\right), \quad a = \nu(t)y \tag{7.8}$$

$$= \frac{1}{\nu}\cosh^{-1}\left(1 + \frac{\nu^2(t)q^2 - 2\rho(t)\nu(t)(a_1 - a_0)q + \nu^2(t)(a_2 - a_1)^2}{2(1-\rho^2)a_1 a_2}\right)$$

where

$$q(f, f_1, t) = \int_{f_1}^{f_0} \frac{1}{\gamma(t) f^\beta} df = \frac{1}{\gamma(1-\beta)}(f^{1-\beta} - f_1^{1-\beta})$$

Next we check the value that minimizes the geodesic distance from the hyperplane $f = f_1$:

This value is given by

$$c(t) := (a_1)_{min}(t) = a_0\sqrt{a_0^2 - 2\nu(t)q(t)a_0\rho(t) + \nu^2(t)q^2(t)} \tag{7.9}$$

After some simplification the value of the minimum distance at the minimum point this may be written in one of the two forms, one corresponding to a distance function and the second to a *signed* distance function (of the point (a_0, f_0) to the line $f = K$)

$$d_{min}^{us} = \frac{1}{\nu}\frac{q\nu\rho - a_0\rho^2 - \sqrt{a_0^2 + q^2\nu^2 + 2a_0 q\nu\rho}}{a_0\left(1-\rho^2\right)}, \quad \text{unsigned} \tag{7.10}$$

and

$$d_{min}^{s} = \frac{1}{\nu}\log\left(\frac{1}{a_0(1-\rho)}\left[\nu q - a_0\rho + \sqrt{a_0^2 - 2a_0\nu\rho q + \nu^2 q^2}\right]\right), \quad \text{signed} \tag{7.11}$$

As mentioned above, this second version of the distance function can become negative, but it's absolute value coincides with the expression above. This distinction seems to have not been emphasized in previous treatments of the subject. In calculating the local volatility expansion, especially the term V_2, we require derivatives up to the sixth order of the $\phi = \frac{d^2}{2}$. These are listed below:

Let

$$\phi^{(2)} \tag{7.12}$$

$$= \frac{\mathrm{Csch}(\nu d) d}{\nu \sqrt{a_0^2 + q^2 - 2q\nu\rho}\,(a_0 - a_0\rho^2)} \tag{7.13}$$

$$\phi^{(3)}$$

$$= \frac{3\mathrm{Csch}(\nu\, d) d}{a_0 \nu\,(a_0^2 + q^2 - 2q\nu\rho)\,(-1+\rho^2)} \tag{7.14}$$

$$\phi^{(4)}$$

$$= \frac{\left(3\mathrm{Csch}(d\nu)\left(-4da_0\nu\,(-1+\rho^2) + \sqrt{a_0^2 + q^2\nu^2 - 2qa_0\nu\rho}(1 - d\nu\mathrm{Coth}(d\nu))\mathrm{Csch}(d\nu)\right)\right)}{\left(a_0^2\nu^2\,(a_0^2 + q^2\nu^2 - 2qa_0\nu\rho)^{3/2}\,(-1+\rho^2)^2\right)} \tag{7.15}$$

$$\phi^{(5)}$$

$$= \frac{30\mathrm{Csch}(d\nu)\left(\frac{2d\nu}{(a_0^2+q^2\nu^2-2qa_0\nu\rho)^2} + \frac{(-1+d\nu\mathrm{Coth}(d\nu))\mathrm{Csch}(d\nu)}{a_0\,(a_0^2+q^2\nu^2-2qa_0\nu\rho)^{3/2}(-1+\rho^2)}\right)}{a_0\nu^2\,(-1+\rho^2)} \tag{7.16}$$

and

$$\phi^{(6)}$$

$$= -\frac{1}{\left(a^3\nu^2\,(a^2 + q^2\nu^2 - 2qa\nu\rho)^{5/2}\,(-1+\rho^2)^3\right)}$$

$$\times \left(15\mathrm{Csch}[d\nu]\left(24da^2\nu\left(-1+\rho^2\right)^2 - 3\left(a^2+q^2\nu^2-2qa\nu\rho\right)\mathrm{Coth}[d\nu]\mathrm{Csch}[d\nu]^2\right.\right.$$

$$+ d\nu\left(a^2+q^2\nu^2-2qa\nu\rho\right)(2+\mathrm{Cosh}[2d\nu])\mathrm{Csch}[d\nu]^4$$

$$\left.\left.+ 18a\sqrt{a^2+q^2\nu^2-2qa\nu\rho}\left(-1+\rho^2\right)\mathrm{Csch}[d\nu]^2(d\nu\mathrm{Cosh}[d\nu] - \mathrm{Sinh}[d\nu])\right)\right) \tag{7.17}$$

8 Probability Distribution, Local and Implied Volatility in Dynamic λ-Sabr Model

Collecting the results in this paper, we see that the probability density function in the λ-Sabr model is given, at the zero-th order, by

$$
\begin{aligned}
& p(f, a, t, K, A, T) \\
&= \frac{1}{2\pi(T-t)} \frac{1}{\sqrt{1-\rho(T)^2 C(K,T)\nu(T)A^2}} \left(\frac{\sqrt{C(f,t}}{\sqrt{C(K,t)}}\right)^{\frac{1}{1-\rho^2}} \sqrt{\frac{d}{\sinh d}} \\
&\times e^{-\frac{d^2}{2(T-t)} - \left(-\int_0^{d(f_0,a_0,f,a,t)} \frac{\partial d(\tilde{f}(\rho),\tilde{a}(\rho),f,a)}{\partial t} d\rho\right) + \log(\mathcal{P})_{MR} + \log(\mathcal{P}_t)}
\end{aligned}
$$

where the distance d in the above formula is given by (7.8). In the case of the λ-Sabr model, the results in this paper allow us to sharpen the above results and obtain the expansion for the heat kernel up to the *zero-th order* in the form

$$
\begin{aligned}
& p^{\mathrm{CEV}}(f, a, t, K, A, T) \\
&= \frac{1}{2\pi(T-t)} \frac{1}{\sqrt{1-\rho(T)^2 \nu(T)\gamma(T) K^{\beta} A^2}} \left(\frac{f^{\frac{\beta}{2}}}{K^{\frac{\beta}{2}}}\right)^{\frac{1}{1-\rho^2}} \sqrt{\frac{d}{\sinh d}} \\
&\times e^{-\frac{d^2}{2(T-t)} + \left(-\int_0^{d(f_0,a_0,f,a,t)} \frac{\partial d(\tilde{f}(\rho),\tilde{a}(\rho),f,a)}{\partial t} d\rho\right) + \log(\mathcal{P})_{MR} + \log(\mathcal{P}_t)}
\end{aligned} \tag{8.1}
$$

where $\log(\mathcal{P})_{MR}$ is explicit and defined in (6.34) and also the fully explicit $\log(\mathcal{P}_t)$ is given by (6.36). Also the derivative of the distance function appearing in the exponent of the exponential can easily calculated for any specific functional form of the time dependent parameters $\gamma(t)$, $\nu(t)$, $\rho(t)$ by differentiating (7.8) with respect to t.

Using (8.4) we see that the zero-th order local volatility σ_L, for the family of stochastic volatility models (2.1), (2.2) (for any local volatility $C(f,t)$), can be expressed, in the form

$$
\begin{aligned}
& \sigma_L^{(0)}(f_0, a_0, K, \tau) \\
&= C(K,t)(a_1)_{min} = C(K,t)\sqrt{a_0^2 - 2qa_1\nu(t)\rho(t)q + q^2\nu^2(t)}, \qquad \left(q(f_0,K) = \int_K^{f_0} \frac{1}{C(u,t)}\right)
\end{aligned}
$$

To determine the *implied* volatility we need to calculate the integral

$$
\int_K^{f_0} \frac{1}{C(u,t)\sqrt{a_1^2 + 2q(f_0,u)a_1\nu\rho q(f_0,u) + q^2(f_0,u)\nu^2}} du
$$

Note that this integration is easy to carry out because $-\frac{du}{C(u,t)} = dq$. In particular, in the λ-Sabr model case, with $C(f,t) = \gamma(t) f^\beta$ we get:

$$\sigma_{\rm BS}^{(0)} = \frac{\xi}{\nu(t)} \log\left[\frac{1}{a_0(1+\rho(t))}(\nu q + a_0\rho(t) + \sqrt{a_0^2 - 2qa_0\nu(t)\rho(t) + q^2\nu^2(t)})\right], \quad (8.2)$$

$$= \frac{\xi}{d^{(0)}}|_{amin} = \frac{\xi}{d^s_{min}} \quad (8.3)$$

where ξ was defined via (5.6), d^s_{min} is the signed distance to line $f = K$ given in (7.11) and q was defined by (7.9), and we recall that a_0 is the initial value of a_t. This zero-th order result, in the time homogeneous case, where the coefficients ν, ρ are constants and where $\gamma = 1$ agrees with that given in Berestycki et al., Henry-Labordère in his Encyclopedia article [35] but does not agree with the formula in Hagan et al. [31], a discrepancy already pointed out by Obloj in [46]

$$\sigma_L^\epsilon = \sigma_L^{(0)} + \sigma_L^{(1)}\epsilon + \sigma_L^{(2)}\epsilon^2 + \cdots \quad \text{where } \epsilon = \tau \quad (8.4)$$

$$= \sqrt{V_L^{(0)}} + \frac{V_L^{(1)}}{2\sqrt{V_L^{(0)}}}\tau + \frac{-(V_L^{(1)})^2 + 4V_L^{(0)}V_L^{(2)}}{8(V_L^{(0)})^{3/2}}\tau^2 + \cdots \quad (8.5)$$

where, recalling that $c = (a_1)_{min}$, given by (7.9), we have for a g

$$\sigma_L^{(1)}(f,t) = \frac{\gamma(t) f^\beta \left(\left(\hat{C}(c) + 2c\hat{C}'(c)\right)\phi''(c) - c\hat{C}(c)\phi^{(3)}(c)\right)}{2c\hat{C}(c)\phi''(c)^2}, \quad (8.6)$$

where recalling the definition of $\hat{C}$ from (4.8) we have

$$\hat{C}(c) = \frac{1}{a^2\nu(t)\gamma(t) f^\beta\sqrt{(1-\rho^2(t))}}\sqrt{\frac{d_{min}}{\sinh(d_{min})}} \times \left(\frac{f^{\frac{\beta}{2}}}{K^{\frac{\beta}{2}}}\right)^{\frac{1}{1-\rho^2}} e^{\log(\mathcal{P}_{MR}) + \log(\mathcal{P}_t)}$$

where the quantities in the exponent of the exponential were defined respectively in, (6.34), and (6.36), and where the derivatives of the functions ϕ were supplied in (7.13) and (7.14).

The Black-Scholes implied volatility is now given by (5.12) whose expression we recall here

$$\sigma_{\rm BS}^{(1)} = \frac{\xi \ln\left[\frac{\sigma_L^{(0)}(f_0,a_0,K,t)u_{0,1}(f_0,K,t)d_{min}(f_0,a_0,K,t)}{\xi\sqrt{f_0K}}\right]}{d_{min}^3}.$$

and where we recall that the definition of $u_{0,1}$, given by (5.14) (where we need to plug in the quantity, and *only* that term, involves $\sigma_L^{(1)}$ and hence involves the non-metric of the zero-th order heat kernel. Also, recall that d_{min} was given above (8.3).

Moreover for the second order expansion of local volatility and implied volatility respectively we have

$$\sigma_L^{(2)} = \frac{-4(\sigma_L^{(1)})^2(\sigma_L^{(0)})^2 + 4(\sigma_L^{(0)})^2 V_L^{(2)}}{8(\sigma_L^{(0)})^3}$$

where $V_L^{(2)}$, given by (4.24) in the time-homogeneous case and by (A.1) in the time *inhomogeneous case*. The ingredients going into the definition of $V_L^{(2)}$, include the derivatives up to order three of $\hat{C}$ at the minimum point, and derivatives of ϕ up to order six, which were given in (7.13)–(7.17). Lastly we obtain $\sigma_{\rm BS}^{(2)}$ by using expression (5.15).

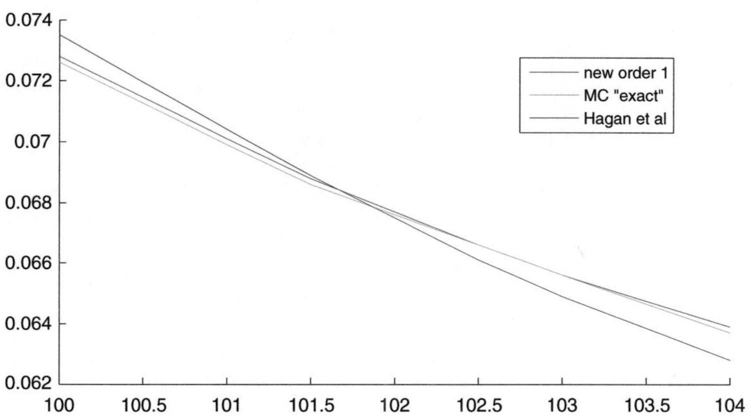

Strike	T	MC Option	Implied Vol MC	σ_0	Order 1 Hag	Order 1 new
100	1	4.004	0.0726	0.0701	0.0735	0.0728
101	1	3.356	0.0699	0.0671	0.704	0.0701
101.5	1	3.039	0.0686	0.0657	0.0689	0.0688
102.5		2.459	0.0666	0.0596	0.0661	0.0666
103	1	2.212	0.0656	0.0619	0.0649	0.0656
104	1	1.737	0.0637	0.0599	0.0628	0.0639

Fig. 1 1 Year; model parameters $S_0 = 102$, $a_0 = 0.65$, $\nu = 1$, $T = 1$, $\beta = 0.5$, $\rho = -0.5$. The third column contains the option price calculated using Monte Carlo and the fourth column contains the (benchmark) numerically inverted implied volatility

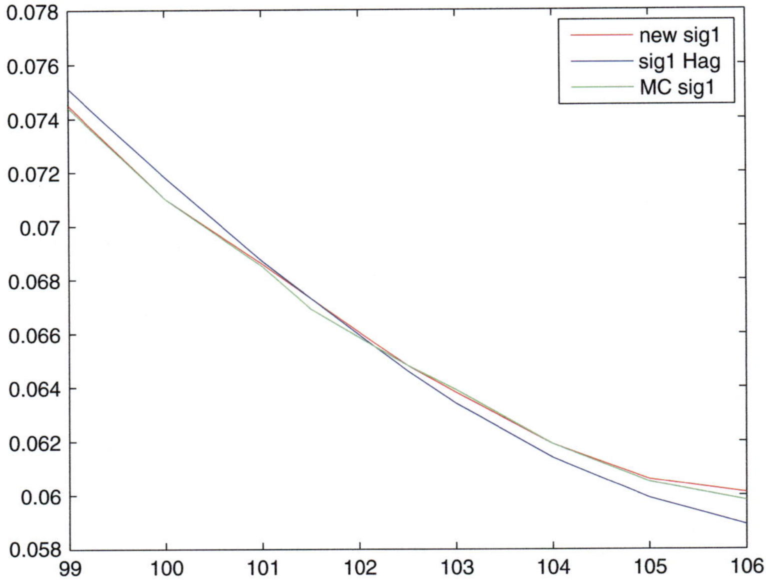

Strike	T	MC Option	Implied Vol MC	σ_0	Order 1 Hag	Order 1 new
99	0.5	3.939	0.0744	0.0678	0.0751	0.0745
100	0.5	5.340	0.071	0.07013	0.0718	0.071
101	0.5	2.499	0.0685	0.0671	0.687	0.0686
101.5	0.5	2.181	0.0669	0.0657	0.0673	0.0673
102.5	0.5	1.625	0.0648	0.0596	0.0646	0.0648
103	0.5	1.39	0.0639	0.059	0.0634	0.0638
104	0.5	0.973	0.0619	0.0580	0.0614	0.0619
105	0.5	0.657	0.0605	0.0577	0.0599	0.0606
106	0.5	0.427	0.0598	0.579	0.0589	0.0601

Fig. 2 Half a year ($T = 0.5$); model parameters $S_0 = 102$, $a_0 = 0.65$, $\nu = 1$, $T = 1$, $\beta = 0.5$, $\rho = -0.5$

9 Numerics

In this section we compare the accuracy of our first order expansion with that of Hagan-Kumar-Lesniewski-Woodward [31], who do not provide second order approximations, on three time horizons. In all of the numerics below, the benchmark prices used were obtained from a Monte Carlo simulation with 40 time steps and 300,000 paths per time step. The standard error associated, calculated by means of dividing the empirical standard deviation by the square root of the number of paths, was 0.0032. The dimensionless parameters involved in the empirical work were large, since we took vol of vol ν to be equal to 1.

What we have found is that our first order approximation is more accurate on time horizons of 0.5 and 1 year, both in and out of the money. On a 2 year time

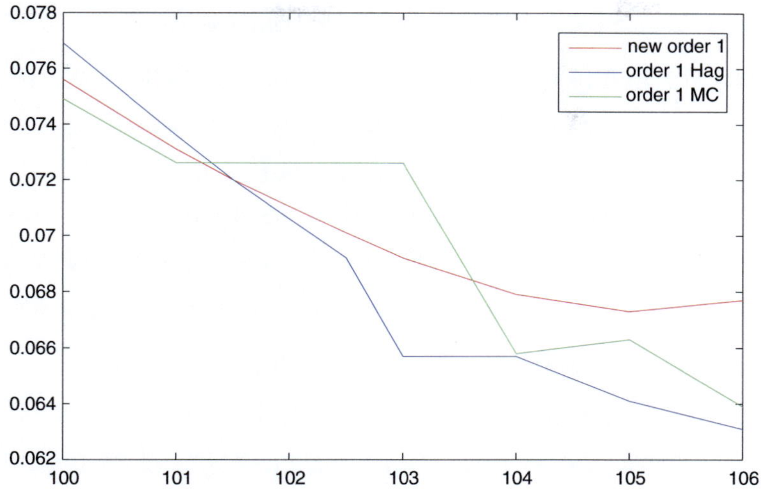

Strike	T	MC Option	Implied Vol MC	σ_0	Order 1 Hag	Order 1 new
100	2	5.3406	0.0749	0.7013	0.0769	0.0756
101	2	4.962	0.0726	0.067122	0.736	0.0731
101.5	2	4.3558	0.0726	0.0657	0.072	0.072
102.5	2	3.735	0.0726	0.0596	0.0692	0.701
103	2	2.94	0.0726	0.059	0.0657	0.0692
104	2	2.906	0.0066	0.0580	0.0657	0.0679
105	2	2.552	0.0663	0.0577	0.0641	0.0673
106	2	2.092	0.0639	0.579	0.0631	0.0677

Fig. 3 2 year; model parameters $S_0 = 102$, $a_0 = 0.65$, $\nu = 1$, $T = 1$, $\beta = 0.5$, $\rho = -0.5$

horizon, Hagan et al.'s first order approximation is sometimes better than ours out of the money and tends to be worse in the money. These results are illustrated in the figures (Figs. 1, 2 and 3).

Acknowledgments We would like to thank Tai-Ho Wang of Baruch College for his assistance in calculating the higher derivatives of ϕ in Sect. 7, and for having pointed out to us the validity of the curvature formula (3.3) in the case $q = 1$. We would also like to thank Elton Hsu for interesting discussions.

Appendix: Form of $V_L^{(2)}$ in the Time-Inhomogeneous Case

$$V_L^{(2)} =$$

$$-\frac{1}{24\hat{C}t(c)^2 d_0(c)^2 \phi''(c)^5} C(f,t)^2 \left(12c\phi''(c)^2 \left(-2d_0(c)\widetilde{Ct}'(c) + c\widetilde{Ct}(u)(d_1(u) + u_1(u))\phi''(c)\right)\right.$$

$$\left(\phi''(c)\left(-2\widetilde{Ct}'(c)d_0'(c) - d_0(c)\widetilde{Ct}''(c) - 2\widetilde{Ct}(u)(d_1(u) + u_1(u))\phi''(c)\right) + d_0(c)\widetilde{Ct}'(c)\phi^{(3)}(c)\right)$$

$$\widetilde{Ct}(c)\phi''(c)\Big(12c\phi''(c)^2\Big(2\widetilde{Ct}'(c)d_0'(c)\left(2d_0'(c)+c(d_1(c)+u_1(c))\phi''(c)\right)+$$
$$+\widetilde{Ct}(u)(d_1(u)+u_1(u))\Big(\phi''(c)\left(4d_0'(c)-cd_0''(c)+2c(d_1(c)+u_1(c))\phi''(c)\right)+cd_0'(c)\phi^{(3)}(c)\Big)\Big)$$
$$-\ d_0(c)\phi''(c)\Big(48cd_0'(c)\widetilde{Ct}''(c)\phi''(c)-24\widetilde{Ct}(u)d_1(u)\phi''(c)^2-24\widetilde{Ct}(u)u_1(u)\phi''(c)^2$$
$$-\ 12c^2d_1(c)\widetilde{Ct}''(c)\phi''(c)^2-12c^2u_1(c)\widetilde{Ct}''(c)\phi''(c)^2+24c\widetilde{Ct}(u)d_1(u)\phi''(c)\phi^{(3)}(c)$$
$$+\ 24c\widetilde{Ct}(u)u_1(u)\phi''(c)\phi^{(3)}(c)+5c^2\widetilde{Ct}(u)d_1(u)\phi^{(3)}(c)^2+5c^2\widetilde{Ct}(u)u_1(u)\phi^{(3)}(c)^2+12\widetilde{Ct}'(c)$$
$$\times\Big(4d_0'(c)\Big(\phi''(c)-c\phi^{(3)}(c)\Big)+c\phi''(c)\Big(4d_0''(c)+(d_1(c)+u_1(c))\Big(4\phi''(c)+c\phi^{(3)}(c)\Big)\Big)\Big)$$
$$-3c^2\widetilde{Ct}(u)(d_1(u)+u_1(u))\phi''(c)\phi^{(4)}(c)\Big)-24d_0(c)^2\Big(c\phi''(c)^2\widetilde{Ct}^{(3)}(c)+2c\widetilde{Ct}'(c)\phi^{(3)}(c)^2$$
$$+\widetilde{Ct}''(c)\phi''(c)\Big(\phi''(c)-2c\phi^{(3)}(c)\Big)-\widetilde{Ct}'(c)\phi''(c)\Big(2\phi^{(3)}(c)+c\phi^{(4)}(c)\Big)\Big)\Big)$$
$$+\ \widetilde{Ct}(c)^2\Big(12c\phi''(c)^2\left(2d_0'(c)+c(d_1(c)+u_1(c))\phi''(c)\right)\Big(d_0''(c)\phi''(c)-d_0'(c)\phi^{(3)}(c)\Big)$$
$$+\ d_0(c)\phi''(c)\Big(-24d_1(c)\phi''(c)^3-24u_1(c)\phi''(c)^3-48cd_1'(c)\phi''(c)^3-48cu_1'(c)\phi''(c)^3$$
$$-\ 24c\phi''(c)^2d_0^{(3)}(c)+\ 48d_0'(c)\phi''(c)\phi^{(3)}(c)+24cd_1(c)\phi''(c)^2\phi^{(3)}(c)$$
$$+\ 24cu_1(c)\phi''(c)^2\phi^{(3)}(c)-48cd_0'(c)\phi^{(3)}(c)^2+\ 5c^2d_1(c)\phi''(c)\phi^{(3)}(c)^2+5c^2u_1(c)\phi''(c)\phi^{(3)}(c)^2$$
$$-24d_0''(c)\phi''(c)\Big(\phi''(c)-2c\phi^{(3)}(c)\Big)-3c\phi''(c)\left(-8d_0'(c)+c(d_1(c)+u_1(c))\phi''(c)\right)\phi^{(4)}(c)\Big)$$
$$+\ 2d_0(c)^2\Big(15c\phi^{(3)}(c)^3-\phi''(c)\phi^{(3)}(c)\Big(15\phi^{(3)}(c)+16c\phi^{(4)}(c)\Big)$$
$$+3\phi''(c)^2\Big(2\phi^{(4)}(c)+c\phi^{(5)}(c)\Big)\Big)\Big)\Big) \qquad \text{(A.1)}$$

where all quantities appearing above where already defined in connection with (4.24) in the time-inhomogeneous case and where

$$\widetilde{Ct}=\sqrt{\Delta}\mathcal{P}$$

and d_1, d_2 where defined in (4.25). In the case of the λ-Sabr model all required derivatives of ϕ (up to order 6) were supplied in Sect. 7.1.

References

1. Andersen, L.B.G.: Option pricing with quadratic volatility: a revisit. Financ. Stoch. **15**(2), 191–219 (2011)
2. Andreasen, J., Andersen, L.B.G.: Jump diffusion pricing: volatility smile fitting and numerical methods for option pricing. Rev. Deriv. Res. **4**, 231–262 (2000)
3. Avellaneda, M., Boyer-Olson, D., Busca, J., Friz, P.: Reconstructing volatility. Risk Mag. **15**(10), 87–91 (2002)
4. Avramidi, I.: Heat Kernel and Quantum Gravity. Lecture Notes in Physics, New Series, Monographs, vol. 64. Springer, Berlin (2000)
5. Avramidi, I.: Lectures on Heat Kernel Applications in Finance. New Mexico Technology. http://infohost.nmt.edu/iavramid/notes/hkt/hkt.html
6. Azencott, R.: Géodésiques et diffusions en temps petit. Séminaire de Probabilités. pp. 84–85 (1991)

7. Bates, D.: Jumps and stochastic volatility: the exchange rate processes implicit in Deutschemark opions. Rev. Financ. Stud. **9**, 69–107 (1996)
8. Benhamou, E., Croissant, O.: Local time for the SABR model: connection with the 'complex' black scholes and application to CMS and spread options, SSRN preprint (2004)
9. Bender, C.M., Orszag, S.A.: Advanced Mathematical Methods for Scientists and Engineers. Springer, Berlin (1999)
10. Beresticky, H., Busca, J., Florent, I.: Asymptotics and calibration of local volatility models. Quant. Financ. **2**, 61–69 (2002)
11. Beresticky, H., Busca, J., Florent, I.: Computing the implied volatility in stochastic volatility models. Commun. Pure Appl. Math. **57**(10), 1352–1373 (2004)
12. Berger, M., Gauduchon, P., Mazet, E.: Le spectre d'une variété Riemanniene. Lecture Notes in Mathematics. Springer, Berlin (1970)
13. Bourgade, P., Croissant, O.: Heat kernel expansion for a family of stochastic volatility models: δ geometry, SSRN (2005)
14. Carr, P., Jarrow, R.: The stop-loss start-gain paradox and option valuation: a new decomposition into intrinsic and time value. Rev. Financ. Stud. **3**(3), 469–492 (1990)
15. Carr, P., Wu, L.: Stochastic skew for currency options. J. Financ. Econ. **86**, 213–247 (2007)
16. Carr, P., Geman, H., Madan, D., Yor, M.: From local volatility to local Lévy models. Quant. Financ. **4**, 581–588 (2004)
17. Chavel, I.: Eigenvalues in Riemannian Geometry. Academic Press, Waltham (1984)
18. Cox, J.: Notes on option pricing I: constant elasticity of diffusions. Unpublished draft. Palo Alto, CA: Stanford University, September 1975
19. Doust, P.: No arbitrage Sabr, Royal Bank of Scotland working paper (2010)
20. Dupire, B.: Pricing with a smile. Risk **7**, 18–20 (1994)
21. Dupire, B.: A unified theory of volatility, discussion paper paribas capital management. In: Peter, C. (ed.) Reprinted in Derivative Pricing: The Classic Collection. Risk Books (2004)
22. Derman, E., Kani, I.: Riding on a smile. Risk **7**, 32–39 (1994)
23. Feng, J., Forde, M., Fouque, J.-P.: Short-maturity asymptotics for a fast mean-reverting Heston stochastic volatility model. SIAM J. Financ. Math. **1**, 126–141 (2010)
24. Forde, M., Jacquier, A.: Small time asymptotics for implied volatility under the Heston model. Int. J. Theory Appl. Financ. **12**, 861–876 (2009)
25. Forde, M., Jacquier, A.: The large-maturity smile for the Heston model. Financ. Stoch. **15**(4), 755–780 (2011)
26. Forde, M., Jacquier, A., Mijatović, A.: Asymptotic formulae for implied volatility in the Heston model. Proc. R. Soc. A **466**(2124), 3593–3620 (2010)
27. Forde, M., Jacquier, A., Lee, R.: The small-time smile and term structure of implied volatility under the Heston model. SIAM J. Financ. Math. **3**(1), 690–708 (2012)
28. Gatheral, J., Hsu, E.P., Laurence, P., Ouyang, C., Wang, T.H.: Asymptotics of implied volatility in local volatility models. Math. Financ. **22**(4), 591–620 (2012)
29. Hadamard, J.: Lecons sur les equations différentielles. Dover Publications, Sewed (1952)
30. Hagan, P., Woodward, D.E.: Equivalent black volatilities. Appl. Math. Financ. **6**, 147–157 (1999)
31. Hagan, P., Kumar, P.S., Lesniewski, A., Woodward, D.E.: Managing smile risk. Wilmott Mag., 84–108 (2003)
32. Hagan, P., Lesniewski, A.: Probability distribution in the SABR model of stochastic volatility. In: Friz, P., Gatheral, J., Gulisashvili, A., Jacquier, A., Teichmann, J. (eds.) Large Deviations and Asymptotic Methods in Finance. Springer Proceedings in Mathematics and Statistics, vol. 110 (2015)
33. Heston, S.: A closed form solution for options with stochastoc volatility, with applications to bond and currency pricing. Rev. Financ. Stud. **6**, 327–342 (1993)
34. Hsu, E.: Stochatic analysis on manifolds, Graduate Studies in Mathematics. American Mathematical Society (2002)
35. Henry-Labordère, P.: SABR Model. Encyclopedia of Quantitative Finance (2010)
36. Hsu, E.: The heat kernel on non-complete manifolds. Indiana Math. J. **39**(2), 431 (1990)

37. Hull, J., White, A.: The pricing of options on assets with stochastic volatilities. J. Financ. **42**(2), 282–300 (1987)
38. Kunimoto, N., Takahashi, A.: The asymptotic expansion approach to the valuation of interest rate contingent claims. Math. Financ. **11**(1), 117–151 (2001)
39. Henry-Labordère, P.: A general asymptotic implied volatility for stochastic volatility models. http://arxiv.org/abs/cond-mat/0504317 (2005)
40. Henry-Labordère, P.: Analysis, Geometry, and Modeling in Finance. Chapman & Hall/CRC Financial Mathematics Series (2008)
41. Henry-Labordère, P.: Solvable local and stochastic volatility models: supersymmetric methods in option pricing. Quant. Financ. **7**(5), 525–535 (2007)
42. Lewis, A.L.: Option valuation under stochastic volatility: with mathematica code. CA: Finance Press (2000)
43. Lesniewski, A.: Swaption Smiles via the WKB Method. Seminar Mathematical Finance, Courant Institute of Mathematical Sciences (February 2002)
44. Minakshisundaram, S., Pleijel, A.: Some properties of the eigenfunctions of the Laplace- operator on Riemannian manifolds. Can. J. Math. **1**, 242–256 (1949)
45. Megvedev, A., Scaillet, O.: Approximation and calibration of short-term implied volatilities under jump-diffusion stochastic volatility. Rev. Financ. Stud. **20**, 427–459 (2007)
46. Obloj, J.: Fine-tuning your smile: correction to Hagan et al. Wilmott Mag. (2008)
47. Molchanov, S.: Diffusion processes and Riemannian geometry. Russ. Math. Surv. **30**, 11–63 (1975)
48. Paulot, L.: Asymptotic implied volatility at the second order with application to the SABR model. In: Friz, P., Gatheral, J., Gulisashvili, A., Jacquier, A., Teichmann, J. (eds.) Large Deviations and Asymptotic Methods in Finance. Springer Proceedings in Mathematics and Statistics, vol. 110 (2015)
49. Renault, E., Touzi, N.: Option hedging and implied volatilities in a stochastic volatility model. Math. Financ. **6**(3), 279–302 (1996)
50. Stein, E.M., Stein, J.C.: Stock distributions with stochastic volatility: an analytic approach. Rev. Financ. Stud. **4**(4), 727–752 (1991)
51. Takahashi, A., Takehara, K., Toda, M.: Computation in an asymptotic expansion method, SSRN preprint (2009)
52. Varadhan, S.: On the behavior of the fundamental solution of the heat equation with variable coefficients. Commun. Pure Appl. Math. **20**, 431–455 (1967)
53. Varadhan, S.: Diffusion processes in a small time interval. Commun. Pure Appl. Math. **20**, 659–685 (1967)
54. Willmore, T.J.: Riemannian Geometry. Oxford Science Publications, Oxford (2002)
55. Yoshida, K.: On the fundamental solution of the parabolic equation in a Riemannian space. Osaka Math. J. **1**(1), 1–52 (1953)

General Asymptotics of Wiener Functionals and Application to Implied Volatilities

Yasufumi Osajima

Abstract In the present paper, we give an asymptotic expansion of probability density for a component of general diffusion models. Our approach is based on infinite dimensional analysis on the Malliavin calculus and Kusuoka-Stroock's asymptotic expansion theory for general Wiener functionals (Kusuoka and Stroock, J. Funct. Anal. 99:1–74, 1991 [12]). The initial term of the expansion is given by the geodesic distance and we calculate it by solving Hamilton's equation. We apply our approach to obtain asymptotic expansion formulae for implied volatilities in general diffusion models, e.g. CEV and SABR model.

Keywords Wiener functional · Stochastic volatility · Hamilton equation · Malliavin calculus · Asymptotic approximation · SABR model

1 Introduction

There are many applications of asymptotic expansion theory to mathematical finance. The most popular is the singular perturbation approach. For example, Hagan and Woodward [6] gave an asymptotic expansion formula for implied volatilities of local volatility models and Hagan et al. [7] gave a formula for a stochastic volatility model (SABR model) based on Hagan et al. [8]. Their formula is well-known to practitioners. Berestycki-Busca-Florent [1] applied non-linear PDE analysis to this problem. Henry-Labordère [9] applied a heat kernel expansion method and gave an asymptotic expansion formula for a mean-reverting SABR model.

In this paper, we take an approach based on Malliavin calculus. The theory of asymptotic expansions of probability densities based on Malliavin calculus was originated by Bismut [2] and was developed by Watanabe [17] and Kusuoka and Stroock [11, 12]. Many applications of this theory to finance were given by Yoshida [18],

Y. Osajima (✉)
BNP Paribas, Fixed Income Research and Strategies, 10 Harewood Avenue, London NW1 6AA, UK
e-mail: yasufumi.osajima@uk.bnpparibas.com

P.K. Friz et al. (eds.), *Large Deviations and Asymptotic Methods in Finance*, Springer Proceedings in Mathematics & Statistics 110,
DOI 10.1007/978-3-319-11605-1_5

Takahashi-Kunitomo [10] and Siopacha and Teichmann [16]. In [14], we gave an asymptotic expansion for implied volatilities of SABR model with time-dependent coefficients. Deuschel et al. [3, 4] gave density expansions for multi dimensional hypoelliptic diffusions $(X^1, \ldots, X^d)$ at fixed time T and projected to their first l coordinates. They applied their results to short time and tail asymptotics of implied volatilities for some stochastic volatility models.

In this paper, we apply the methods of Kusuoka-Stroock [12] to the asymptotic expansion for implied volatilities of call options. The key theorem is given in [13] and also summarized in the Appendix. Finally we give explicit analytic formulae in general diffusion models.

Let $(\Omega, \mathscr{F}, P)$ be a probability space and let $\{W^1(t), \ldots, W^d(t);\ t \in [0, T]\}$ be a d-dimensional Brownian motion. Let $V_0, \ldots, V_d \in C_b^\infty([0, T] \times \mathbf{R}^N; \mathbf{R}^N)$. Here $C_b^\infty([0, T] \times \mathbf{R}^N; \mathbf{R}^N)$ denotes the space of $\mathbf{R}^N$-valued smooth functions defined in $[0, T] \times \mathbf{R}^N$ whose derivatives of any order are bounded.

Now let $X_\varepsilon(t), t \in [0, T], \varepsilon \in (0, 1]$, be the solution to the following stochastic differential equation:

$$dX_\varepsilon^i(t) = \sum_{k=1}^d \varepsilon V_k^i(t, X_\varepsilon(t))dW^k(t) + V_0^i(t, X_\varepsilon(t))dt, \quad 1 \le i \le N,$$
$$X_\varepsilon(0) = x_0 = (x_0^1, \ldots, x_0^N), \quad x_0 \in \mathbf{R}^N. \tag{1}$$

In view of financial applications, cf. below, we assume

$$V_0^1 \equiv 0, \tag{2}$$

and the ellipticity of $V_1, \ldots, V_d$ at x_0, i.e. there exists a constant $\delta > 0$ such that

$$\sum_{k=1}^d V_k(0, x_0) \otimes V_k(0, x_0) \ge \delta I, \tag{3}$$

where I denotes the identity matrix. Then there exists a unique solution to (1). Moreover, we assume that $X_\varepsilon(t)$ is continuous in t with probability one.

We investigate the distribution of $X_\varepsilon^1(T)$. From the ellipticity condition (3), the law of $X_\varepsilon^1(T)$, denoted by ν_ε, is absolutely continuous and has a smooth density $p_\varepsilon(y)$. Let H be the Cameron-Martin space of d-dimensional Wiener space. We consider the associated ordinary differential equation:

$$\frac{d}{dt}y^i(t; h) = \sum_{k=1}^d V_k^i(t, y(t; h))\dot{h}^k(t) + V_0^i(t, y(t; h)), \quad t \in [0, T], \quad h \in H,$$
$$y(0; h) = x_0, \quad x_0 \in \mathbf{R}^n. \tag{4}$$

We define the energy function $e : \mathbf{R} \to \mathbf{R}$ by

$$e(y) = \inf \left\{ \frac{1}{2} \sum_{i=1}^{d} \int_0^T |\dot{h}^i(s)|^2 ds;\ h \in H,\ y^1(T; h) = y \right\}. \tag{5}$$

Since $V_0^1 \equiv 0$, this energy function satisfies $e(x_0^1) = 0$. Let us define a flow ϕ: $[0, T] \times \mathbf{R}^N \to \mathbf{R}^N$ by

$$\frac{d}{dt}\phi(t, x) = V_0(t, \phi(t, x)), \quad t \in [0, T], \quad x \in \mathbf{R}^N, \tag{6}$$
$$\phi(0, x) = x.$$

Then the map $\phi(t, \cdot) : \mathbf{R}^N \to \mathbf{R}^N,\ t \in [0, T]$ is a diffeomorphism denoted by ϕ_t. Note that $\phi_t^1(x) = x^1$. We define

$$\tilde{V}_k^i(t, y) = \sum_{j=1}^{N} \frac{\partial \phi^i}{\partial x^j}(-t, \phi(t, y)) V_k^j(t, \phi(t, y)), \quad 1 \le i \le N, \quad 1 \le k \le d, \tag{7}$$

which is the push-forward of the vector field V by the map ϕ_t. Let us define $(g^{ij})_{1 \le i,j \le N} : [0, T] \times \mathbf{R}^N \to \mathbf{R}$ by

$$g^{ij}(t, x) = \sum_{k=1}^{d} \tilde{V}_k^i(t, x) \tilde{V}_k^j(t, x), \quad 1 \le i, j \le N.$$

From (3), the matrix $(g^{ij})_{1 \le i,j \le N}$ is positive definite corresponding to the Riemaniann metric on $\mathbf{R}^N$. We define the generating operator $L_t,\ t \in [0, T]$ by

$$(L_t f)(x) = \frac{1}{2} \sum_{i,j=1}^{N} g^{ij}(t, x) \frac{\partial^2 f}{\partial x^i \partial x^j}(x) + \sum_{i=1}^{N} \tilde{V}_0^i(t, x) \frac{\partial f}{\partial x^i}(x),$$
$$f \in C_b^\infty(\mathbf{R}^N), \quad x \in \mathbf{R}^N,\ t \in [0, T], \tag{8}$$

where $\tilde{V}_0^i \in C_b^\infty([0, T] \times \mathbf{R}^N; \mathbf{R}^N)$ is given by

$$\tilde{V}_0^i(t, y) = \frac{1}{2} \sum_{k,l=1}^{N} \sum_{m=1}^{d} \frac{\partial^2 \phi^i}{\partial x^k \partial x^l}(-t, \phi(t, y)) V_m^k(t, \phi(t, y)) V_m^l(t, \phi(t, y)),\ 1 \le i \le N. \tag{9}$$

Let us define linear operators $V : C_b^\infty([0, T] \times \mathbf{R}^N) \to C_b^\infty([0, T] \times \mathbf{R}^N)$ and $\Gamma : C_b^\infty([0, T] \times \mathbf{R}^N) \otimes C_b^\infty([0, T] \times \mathbf{R}^N) \to C_b^\infty(\mathbf{R}^N)$ by

$$(Vf)(t,x) \equiv \sum_{i=1}^{N} g^{1i}(t,x) \int_t^T \frac{\partial f}{\partial x^i}(s,x)ds, \tag{10}$$

$$\Gamma(f,g)(x) \equiv \sum_{i,j=1}^{N} g^{ij}(t,x)\Big(\int_t^T \frac{\partial f}{\partial x^i}(s,x)ds\Big)\Big(\int_t^T \frac{\partial g}{\partial x^j}(s,x)ds\Big)\,dt. \tag{11}$$

Our main result is:

Theorem 1 *There is a constant $r_0 > 0$ satisfying the following (1) and (2).*
(1) The energy function $e \in C^2([x_0^1 - r_0, x_0^1 + r_0])$ and there is a constant $C_0 > 0$ such that the asymptotic expansion of energy e satisfies

$$\Big|e(y) - \Big[\frac{1}{2b_1}(y - x_0^1)^2 - \frac{b_2}{3b_1^3}(y - x_0^1)^3 + \Big(-\frac{b_3}{4b_1^4} + \frac{b_2^2}{2b_1^5}\Big)(y - x_0^1)^4\Big]\Big|$$
$$\leq C_0|y - x_0^1|^5, \quad y \in [x_0^1 - r_0, x_0^1 + r_0], \tag{12}$$

where

$$b_1 = \int_0^T g^{11}(t,x_0)dt, \quad b_2 = \frac{3}{2}\int_0^T (Vg^{11})(t,x_0)dt, \tag{13}$$
$$b_3 = 2\int_0^T (V^2 g^{11})(t,x_0)dt + \frac{1}{2}\int_0^T \Gamma(g^{11},g^{11})(x_0).$$

(2) There are constants $C_1,\ C_2 > 0$ such that the probability density $p_\varepsilon(y)$ satisfies the following:

$$\Big|(2\pi\varepsilon^2)^{\frac{1}{2}} \exp\Big(\frac{e(y)}{\varepsilon^2}\Big) p_\varepsilon(y) - a_0(y) - \varepsilon^2 a_2(y)\Big| \leq \varepsilon^4 C_1, \ \ y \in [x_0^1 - r_0, x_0^1 + r_0]. \tag{14}$$

Here, a_0 and a_2 are continuous functions which satisfy

$$\Big|a_0(y) - \Big(\frac{\partial^2 e(y)}{\partial y^2}\Big)^{\frac{1}{2}} \exp\Big(\frac{L(y - x_0^1)^2}{2b_1^2}\Big)\Big| \leq C_2|y - x_0^1|^3, \ \ y \in [x_0^1 - r_0, x_0^1 + r_0], \tag{15}$$

and

$$a_2(x_0^1) = \frac{1}{\sqrt{b_1}}\Big(-\frac{L}{2b_1} - \frac{5}{6}\frac{b_2^2}{b_1^3} + \frac{3}{4}\frac{b_3}{b_1^2}\Big), \tag{16}$$

where

$$L = \int_{0<u<t<T} L_u(g^{11}(t,\cdot))(x_0)\,du\,dt. \tag{17}$$

Remark 1 We can restate our results (14) as the heat kernel expansion:

$$p_\varepsilon(y) \sim e^{-e(y)/\varepsilon^2} \frac{1}{\varepsilon(2\pi)^{1/2}}(a_0(y) + \varepsilon^2 a_2(y) + O(\varepsilon^4)).$$

Next, we apply our results to the asymptotic expansion of call option values and their implied volatilities. We regard X_ε^1 as the underlying of these options. Then the forward value of a call option of strike rate K and maturity T is given by

$$C_\varepsilon(T, K) = E[(X_\varepsilon^1(T) - K)^+], \quad \varepsilon \in (0, 1], \quad K > 0.$$

We define smooth functions $\varphi_n \in C_b^\infty([0, \infty))$, $n \geq 0$, by

$$\varphi_n(x) = \int_0^\infty z^n \exp\left(-xz - \frac{z^2}{2}\right) dz, \quad x \geq 0. \tag{18}$$

Some properties of φ_n are given in Lemma 7. Since (12), we can define the following function $q \in C^2([x_0^1 - r_0, x_0^1 + r_0]; \mathbf{R}_+)$ such that

$$e(x) = \frac{1}{2}\left(\int_{x_0^1}^x \frac{dy}{q(y)}\right)^2, \quad x \in [x_0^1 - r_0, x_0^1 + r_0]. \tag{19}$$

Then the asymptotic expansion of call option values are given by the following.

Theorem 2 *There are constants $K_0 < K_1$ and C_1 such that the value of the call option with strike rate K, maturity T satisfies*

$$\left|\sqrt{2\pi}\exp\left(\frac{e(K)}{\varepsilon^2}\right)C_\varepsilon(T, K) - \varepsilon a_0(K) q(K)^2 \varphi_1\left(\frac{\sqrt{2e(K)}}{\varepsilon}\right) R_2(\varepsilon, K)\right| \leq C_1 \varepsilon^4,$$
$$\varepsilon \in (0, 1], \; K \in [K_0, K_1],$$

where

$$\begin{aligned} R_2(\varepsilon, K) &= \varepsilon q(K)\left(\frac{a_0'(K)}{a_0(K)} + \frac{3}{2}\frac{q'(K)}{q(K)}\right)\frac{\varphi_2(\sqrt{2e(K)}/\varepsilon)}{\varphi_1(\sqrt{2e(K)}/\varepsilon)} \\ &\quad + \varepsilon^2 q(K)^2\left[\frac{1}{2}\frac{a_0''(K)}{a_0(K)} + 2\frac{a_0'(K)}{a_0(K)}\frac{q'(K)}{q(K)} + \frac{7}{6}\left(\frac{q'(K)}{q(K)}\right)^2 + \frac{2}{3}\frac{q''(K)}{q(K)}\right] \\ &\quad \times \frac{\varphi_3(\sqrt{2e(K)}/\varepsilon)}{\varphi_1(\sqrt{2e(K)}/\varepsilon)} + \varepsilon^2\frac{a_2(K)}{a_0(K)}. \end{aligned} \tag{20}$$

Next we calculate the asymptotic expansion of implied volatilities of call options. Let us define $f \in C^\infty(\mathbf{R}_+; \mathbf{R}_+)$ by

$$f(x) = \frac{1}{x}\frac{1}{\sqrt{2\pi}}\exp\left(-\frac{x^2}{2}\right)\varphi_1(x), \quad x > 0. \tag{21}$$

We can easily check that f is strictly decreasing and

$$f(0_+) = \infty, \quad f(\infty) = 0.$$

Therefore the inverse function $f^{-1}: \mathbf{R}_+ \to \mathbf{R}_+$ is well defined. When we consider the following normal model:

$$d\tilde{X}(t) = \sigma d\tilde{W}(t), \quad \tilde{X}(0) = x_0^1,$$

the value of the call option with strike rate K and maturity T is given by

$$C_N(T,K) = \frac{1}{\sqrt{2\pi\sigma^2 T}}\int_{-\infty}^{\infty}(z + x_0^1 - K)^+ \exp\left(-\frac{z^2}{2\sigma^2 T}\right)dx = (K - x_0^1)\cdot f\Big(\frac{K - x_0^1}{\sigma\sqrt{T}}\Big).$$

Therefore the implied normal volatility can be written as

$$\sigma_N^\varepsilon(T,K) = \frac{K - x_0^1}{f^{-1}(C_\varepsilon(T,K)/(K - x_0^1))\sqrt{T}}, \quad K > x_0^1.$$

The asymptotic expansion of the implied normal volatilities are given by the following.

Theorem 3 *The asymptotic expansion of implied normal volatilities are given by*

$$\left|\Big(\frac{\varepsilon|K - x_0^1|}{\sqrt{2e(K)T}}\Big)^{-1}\sigma_N^\varepsilon(T,K) - \exp(J)\right| \le C(\varepsilon + |K - x_0^1|)^3, \quad K \in [x_0^1, K_1], \tag{22}$$

where

$$\begin{aligned} J = {} & \frac{|K - x_0^1|^2}{b_1^2}\Big(\frac{L}{2} + \frac{1}{6}\frac{b_2^2}{b_1^2} - \frac{1}{4}\frac{b_3}{b_1}\Big)\varphi_1\left(\frac{\sqrt{2e(K)}}{\varepsilon}\right) + \frac{\varepsilon^2}{b_1}\Big(-\frac{L}{2} - \frac{5}{6}\frac{b_2^2}{b_1^2} + \frac{3}{4}\frac{b_3}{b_1}\Big)\varphi_1\left(\frac{\sqrt{2e(K)}}{\varepsilon}\right) \\ & + \frac{\varepsilon}{\sqrt{b_1}}\frac{|K - x_0^1|}{b_1}\Big(L + \frac{2}{3}\frac{b_2^2}{b_1^2} - \frac{3}{4}\frac{b_3}{b_1}\Big)\varphi_2\left(\frac{\sqrt{2e(K)}}{\varepsilon}\right) + \frac{\varepsilon^2}{b_1}\Big(\frac{L}{2} + \frac{b_2^2}{2b_1^2} - \frac{b_3}{2b_1}\Big)\varphi_3\left(\frac{\sqrt{2e(K)}}{\varepsilon}\right). \end{aligned}$$

Remark 2 Since we can give the same formula for put options, Theorem 3 still holds in the case $K < x_0^1$. The implied volatility for a put option of strike rate K and maturity T is the same as the implied volatility for a call option with the same strike rate and maturity due to the put-call parity. See Appendix 3 for the details.

2 Hamilton Equation and the Energy of Path

In this section, we investigate the correspondence between the Hamilton equation and the energy of path defined by (5). Without loss of generality, we can assume $T = 1$. Let H be a separable real Hilbert space defined by

$$H = \left\{ h \in C_0([0,1]; \mathbf{R}^d) : h \text{ is absolutely continuous and } \sum_{i=1}^{d} \int_0^1 |\dot{h}^i(t)|^2 dt < \infty \right\}.$$

The inner product is given by

$$(h, k)_H = \sum_{i=1}^{d} \int_0^1 \dot{h}^i(t) \dot{k}^i(t) dt.$$

This Hilbert space H is called the Cameron-Martin space.

Let $y(t; h)$, $t \in [0, 1]$, $h \in H$, be the solution to the ordinary differential equation:

$$\frac{d}{dt} y^i(t; h) = \sum_{k=1}^{d} \tilde{V}_k^i(t, y(t; h)) \dot{h}^k(t), \quad 1 \le i \le N, \quad t \in [0, 1],$$
$$y(0; h) = x_0, \quad x_0 \in \mathbf{R}^N.$$

Let $(g^{ij})_{1 \le i,j \le N} : [0, 1] \times \mathbf{R}^N \to \mathbf{R}$ be given by

$$g^{ij}(t, x) = \sum_{k=1}^{d} \tilde{V}_k^i(t, x) \tilde{V}_k^j(t, x).$$

We define Hamilitonian $\mathscr{H} : [0, 1] \times \mathbf{R}^N \times \mathbf{R}^N \to \mathbf{R}$ by

$$\mathscr{H}(t, x, p) = \frac{1}{2} \sum_{i,j=1}^{N} g^{ij}(t, x) p_i p_j. \tag{23}$$

Then the correspondence between Hamilton equation and the energy of path is given by the following.

Proposition 1 *Let* $J_j^i : [0, 1] \times H \to \mathbf{R}$ *be the solution to the following ordinary differential equation:*

$$
\begin{aligned}
&\frac{d}{dt}J^i_j(t;h) = \sum_{k=1}^{d}\sum_{r=1}^{N}\frac{\partial \tilde{V}^i_k}{\partial x^r}(t, y(t;h))J^r_j(t;h)\dot{h}^k(t),\\
&J^i_j(0;h) = \delta_{ij}, \quad 1 \le i, j \le N,
\end{aligned}
$$

where δ_{ij} is Kronecker's delta. Let $\bar{J}(t;h) = J^{-1}(t;h)$. We assume there is $h_0 \in H$ and $\lambda \in \mathbf{R}^N$ such that[1]

$$
h_0 = \sum_{k=1}^{N}\lambda_k Dy^k(1;h_0). \tag{24}
$$

We define $x, p \in C^\infty([0,1];\mathbf{R}^N)$ by

$$
\begin{aligned}
x(t) &= y(t;h_0),\\
p_i(t) &= \textstyle\sum_{j,k=1}^{N}\bar{J}^j_i(t;h_0)J^k_j(1;h_0)\lambda_k.
\end{aligned} \tag{25}
$$

Then (x,p) satisfies the Hamilton equation:

$$
\begin{aligned}
&\frac{d}{dt}x^i(t) = \frac{\partial}{\partial p_i}\mathscr{H}(t, x(t), p(t)),\\
&\frac{d}{dt}p_i(t) = -\frac{\partial}{\partial x^i}\mathscr{H}(t, x(t), p(t)), \quad 0 \le t \le 1, \quad 1 \le i \le N,\\
&x(0) = x_0, \quad x_0 \in \mathbf{R}^n.
\end{aligned} \tag{26}
$$

Furthermore, we have $\lambda = p(1)$ and

$$
\begin{aligned}
&\frac{d}{dt}h^k_0(t) = \sum_{i=1}^{N}p_i(t)\tilde{V}^i_k(t;x(t)), \quad 0 \le t \le 1, \quad 1 \le k \le d,\\
&\|h_0\|^2 = \sum_{i,j=1}^{N}\int_0^1 g^{ij}(t, x(t))p_i(t)p_j(t)dt.
\end{aligned} \tag{27}
$$

Proof We note that $\bar{J}^i_j : [0,1]\times H \to \mathbf{R}$ satisfies the following ordinary differential equation:

$$
\begin{aligned}
&\frac{d}{dt}\bar{J}^i_j(t;h) = -\sum_{k=1}^{d}\sum_{r=1}^{N}\frac{\partial}{\partial x^j}\tilde{V}^r_k(t, y(t;h))\bar{J}^i_r(t;h)\dot{h}^k(t),\\
&\bar{J}^i_j(0;h) = \delta_{ij}, \quad 1 \le i, j \le N.
\end{aligned}
$$

From Proposition 6.6 in Shigekawa [15], we have

---

[1] We define $Dy(\cdot;h)[k] = \frac{d}{d\varepsilon}y(\cdot;h+\varepsilon k)$.

$$Dy^i(1;h)[k] = \sum_{l=1}^{d}\sum_{r,j=1}^{N} J_r^i(1;h)\int_0^1 \bar{J}_j^r(t;h)\tilde{V}_l^j(t,y(t;h))\dot{k}^l(t)dt, \quad 1 \le i \le N, \tag{28}$$

From (25), it is easy to see $\lambda = p(1)$. Since $h_0 = \sum_{i=1}^N \lambda_i Dy^i(h_0)$, we see that

$$(h_0, k) = \sum_{i=1}^{N}\sum_{l=1}^{d}\int_0^1 p_i(t)\tilde{V}_l^i(t, y(t;h_0)\dot{k}^l(t)dt.$$

Therefore we have (27). We can check that $(x(t), p(t)),\ 0 \le t \le 1$, satisfies (26) as follows:

$$\frac{d}{dt}x^i(t) = \sum_{k=1}^{d}\tilde{V}_k^i(t,x(t))\dot{h}_0^k(t) = \sum_{j=1}^{N} g^{ij}(t,x(t))p_j(t),$$

$$\frac{d}{dt}p_i(t) = -\sum_{k=1}^{d}\sum_{j,r=1}^{N}\frac{\partial \tilde{V}_k^j}{\partial x^i}(t,x(t))p_j(t)\dot{h}_0^k(t) = -\sum_{j,r=1}^{N}\frac{\partial g^{jr}}{\partial x^i}(t,x(t))p_j(t)p_r(t).$$

□

Remark 3 We will give a remark on condition (24). We define an energy function $E : \mathbf{R}^N \to \mathbf{R}$ as

$$E(y) = \inf\left\{\frac{1}{2}\|h\|^2;\ h \in H,\ y(1;h) = y\right\}.$$

and let $h_0 \in H$ be the minimizer of the energy function. Then we can apply Lagrange's method and there is a $\lambda \in \mathbf{R}^N$ such that

$$h_0 = \sum_{k=1}^{N}\lambda_k Dy^k(1;h_0),$$

which is the condition (24). In particular, the condition (29) in the next proposition is corresponding to the energy function (5).

Let us define the following notations.

$$f \underset{k}{\sim} g \overset{\text{def}}{\Longleftrightarrow} \lim_{w\downarrow 0}\frac{f(w)-g(w)}{w^k} = 0, \quad k \ge 0, \quad f, g \in C([0,1]).$$

In the following case, we obtain the asymptotic solutions.

Proposition 2 *Let $x(t;w),\ p(t;w)$ be the solution to the Hamilton equation (26) with*

$$\lambda_i = \begin{cases} w \ \ (i=1), & w \in \mathbf{R} \\ 0 \ \ (2 \le i \le N), \end{cases} \tag{29}$$

under the boundary condition $x(0) = x_0,\ p(1) = \lambda$. Then the asymptotic expansion of $x^1(1;w)$ is given as follows:

$$x^1(1;w) \underset{3}{\sim} x_0 + b_1 w + b_2 w^2 + b_3 w^3, \tag{30}$$

where $b_1,\ b_2,\ b_3$ are defined by (13).

Proof The solution can be written as

$$x^i(t;w) = x_0^i + \sum_{j=1}^{N} \int_0^t g^{ij}(s, x(s;w)) p_j(s;w) ds, \tag{31}$$

$$p_i(t;w) = p_i(1;w) + \frac{1}{2} \sum_{j,r=1}^{N} \int_t^1 \frac{\partial g^{jr}}{\partial x^i}(s, x(s;w)) p_j(s;w) p_r(s;w) ds. \tag{32}$$

We calculate the asymptotic expansion inductively. Since $x(t;0) = x_0,\ p(t;0) = 0$, we have

$$x(t;w) \underset{0}{\sim} x_0,\ \ p(t;w) \underset{0}{\sim} 0. \tag{33}$$

Since the integral term in (32) is of the second order in w and from the boundary condition (29), we have the first order expansion of p:

$$p_i(t;w) \underset{1}{\sim} p_i(1;w) = \begin{cases} w \ \ (i=1) \\ 0 \ \ (2 \le i \le N). \end{cases} \tag{34}$$

We substitute (34) for (31), we have the first order expansion of x:

$$x^i(t;w) \underset{1}{\sim} x_0^i + \left(\int_0^t g^{i1}(s, x_0) ds\right) w. \tag{35}$$

Substituting (34) for (32), we have the second order expansion of p:

$$\begin{aligned} p_i(t;w) &\underset{2}{\sim} p_i(1;w) + \frac{1}{2} \sum_{j,r=1}^{N} \left(\int_t^1 \frac{\partial g^{jr}}{\partial x^i}(s, x(s;w)) ds\right) p_j(1;w) p_r(1;w) \\ &\underset{2}{\sim} p_i(1;w) + \frac{1}{2}\left(\int_t^1 \frac{\partial g^{11}}{\partial x^i}(s, x_0) ds\right) w^2. \end{aligned}$$

We substitute (35) for (31). Then we have the second order expansion of x:

$$x^i(t;w) \underset{2}{\sim} x_0^i + \sum_{j=1}^N \int_0^t g^{ij}(s, x(s;w))\Big\{p_j(1;w) + \frac{1}{2}\Big(\int_s^1 \frac{\partial g^{11}}{\partial x^j}(r, x_0)dr\Big)w^2\Big\}ds$$

$$\underset{2}{\sim} x_0^i + \Big(\int_0^t g^{i1}(s, x_0)ds\Big)w + \sum_{j=1}^N \Big(\int_0^t \int_0^s \frac{\partial g^{i1}}{\partial x^j}(s, x_0) g^{j1}(u, x_0)duds$$

$$+ \frac{1}{2}\int_0^t \int_s^1 g^{ij}(s, x_0)\frac{\partial g^{11}}{\partial x^j}(u, x_0)\, du\, ds\Big)w^2.$$

From the second order expansion of p and the first order expansion of x, we have third order expansion of p:

$$p_i(t;w) \underset{3}{\sim} p_i(1;w) + \frac{1}{2}\Big(\int_t^1 \frac{\partial g^{11}}{\partial x^i}(s, x_0)ds\Big)w^2$$

$$+ \frac{1}{2}\sum_{j=1}^N \Big\{\int_t^1 \frac{\partial g^{j1}}{\partial x^i}(s, x_0)\Big(\int_s^1 \frac{\partial g^{11}}{\partial x^j}(u, x_0)du\Big)ds$$

$$+ \int_t^1 \frac{\partial^2 g^{11}}{\partial x^i \partial x^j}(s, x_0)\Big(\int_0^s g^{j1}(u, x_0)du\Big)ds\Big\}w^3.$$

Finally we have the following third order expansion of x:

$$x^i(t;w) \underset{3}{\sim} x_0^i + \Big(\int_0^t g^{i1}(s, x_0)ds\Big)w$$

$$+ \sum_{j=1}^N \Big(\int_0^t \int_0^s \frac{\partial g^{i1}}{\partial x^j}(s, x_0) g^{j1}(u, x_0)duds$$

$$+ \frac{1}{2}\int_0^t \int_s^1 g^{ij}(s, x_0)\frac{\partial g^{11}}{\partial x^j}(u, x_0)duds\Big)w^2$$

$$+ \sum_{j,k=1}^N \Big[\frac{1}{2}\int_0^t g^{ij}(s, x_0)\Big(\int_s^1 \frac{\partial g^{k1}}{\partial x^j}(u, x_0)\Big(\int_u^1 \frac{\partial g^{11}}{\partial x^k}(r, x_0)dr\Big)du\Big)ds$$

$$+ \frac{1}{2}\int_0^t g^{ij}(s, x_0)\Big(\int_s^1 \frac{\partial^2 g^{11}}{\partial x^j \partial x^k}(u, x_0)\Big(\int_0^u g^{k1}(r, x_0)dr\Big)du\Big)ds$$

$$+ \frac{1}{2}\int_0^t \frac{\partial g^{ij}}{\partial x^k}(s, x_0)\Big(\int_s^1 \frac{\partial g^{11}}{\partial x^j}(u, x_0)du\Big)\Big(\int_0^s g^{k1}(r, x_0)dr\Big)ds$$

$$+ \frac{1}{2}\int_0^t \frac{\partial g^{i1}}{\partial x^j}(s, x_0)\Big(\int_0^s g^{jk}(u, x_0)\Big(\int_u^1 \frac{\partial g^{11}}{\partial x^k}(r, x_0)dr\Big)du\Big)ds$$

$$+ \int_0^t \frac{\partial g^{i1}}{\partial x^j}(s, x_0)\Big(\int_0^s \frac{\partial g^{j1}}{\partial x^k}(u, x_0)\Big(\int_0^u g^{k1}(r, x_0)dr\Big)du\Big)ds$$

$$+\frac{1}{2}\int_0^t \frac{\partial^2 g^{i1}}{\partial x^j \partial x^k}(s, x_0)\Big(\int_0^s g^{j1}(u, x_0)du\Big)\Big(\int_0^s g^{k1}(r, x_0)dr\Big)ds\Big]w^3.$$

From the definition of the linear operator V given in (10), we have

$$x^1(1; w) \underset{3}{\sim} x_0^1 + b_1 w + b_2 w^2 + b_3 w^3.$$

□

3 Proof of Theorem 1

3.1 Proof of Theorem 1 (1)

Let $\tilde{X}_\varepsilon$ be defined by $\tilde{X}_\varepsilon(t) = \phi(-t, X_\varepsilon(t))$. Then $\tilde{X}$ satisfies the following stochastic differential equation:

$$d\tilde{X}_\varepsilon^i(t) = \varepsilon \sum_{k=1}^d \tilde{V}_k^i(t, \tilde{X}_\varepsilon(t))dW^k(t) + \varepsilon^2 \tilde{V}_0^i(t, \tilde{X}_\varepsilon(t))dt, \quad 1 \le i \le N, \quad t \in [0, 1],$$
$$\tilde{X}_\varepsilon(0) = x_0, \tag{36}$$

where $\tilde{V}$ is defined as (7) and (9). The solution to the associated ordinary differential equation $\tilde{y}$ satisfies (37) in the next lemma.

Lemma 1 *Let $y(t; h) : [0, 1] \times H \to \mathbf{R}$, be the solution defined by (4). Let us define*

$$\tilde{y}(t; h) = \phi(-t, y(t; h)), \quad 1 \le i \le N, \quad t \in [0, 1],$$

then $\tilde{y}$ satisfies the ordinary differential equation:

$$\frac{d}{dt}\tilde{y}^i(t; h) = \sum_{k=1}^d \tilde{V}_k^i(t, \tilde{y}(t; h))\dot{h}^k(t), \quad 1 \le i \le N, \quad t \in [0, 1]. \tag{37}$$

Proof From the definition of ϕ given by (6), we have

$$-V_0^i(t, \phi(-t, \phi(t, y))) + \sum_{j=1}^d \nabla_j \phi^i(-t, \phi(t, y))V_0^j(t, \phi(t, y)) = 0.$$

Therefore we have our lemma. □

Proof (Theorem 1(1)) Since $V_0^1 \equiv 0$, we have $\tilde{y}^1(t; h) = y^1(t; h)$, and the energy function can be defined as follows.

$$e(x) = \frac{1}{2} \inf\Big\{\sum_{k=1}^{d} \int_0^1 |\dot{h}_0^k(t)|^2 dt : \tilde{y}^1(1;h) = x\Big\}.$$

Therefore it is enough to prove the theorem for the driftless case, i.e. $V_0 \equiv 0$.

Let h_0 be defined by

$$h_0(x) \equiv argmin\{e(h); h \in H, y^1(1;h) = x\}. \tag{38}$$

We denote $h_0(x)(t) \equiv h_0(t,x)$. Then from non-degeneracy condition, there is an $r > 0$ such that $h_0(x)$ is unique in $x \in (x_0 - r, x_0 + r)$. Using Lagrange multiplier theorem, we have

$$h_0(x) = \lambda(x) DF^1(0, h_0(x)), \tag{39}$$

where $\lambda : (x_0 - r, x_0 + r) \to \mathbf{R}$ is a smooth function. Applying Proposition 2, we have

$$\Big|x^1(1;\lambda(x)) - \Big(x_0^1 + b_1\lambda(x) + b_2\lambda(x)^2 + b_3\lambda(x)^3\Big)\Big| = O(|x - x_0|^4).$$

Therefore we have the following asymptotic expansion of λ in x:

$$\lambda(x) \underset{3}{\sim} c_1(x - x_0^1) + c_2(x - x_0^1)^2 + c_3(x - x_0^1)^3, \tag{40}$$

where

$$c_1 = \frac{1}{b_1},\ c_2 = -\frac{b_2}{b_1^3},\ c_3 = -\frac{b_3}{b_1^4} + 2\frac{b_2^2}{b_1^5}. \tag{41}$$

From [13] we have

$$\lambda(x) = \frac{\partial e(x)}{\partial x}. \tag{42}$$

Since $e(x_0^1) = 0$, we can calculate the path of energy by

$$e(x) = \int_{x_0^1}^{x} \lambda(y) dy \underset{4}{\sim} \frac{c_1}{2}(x - x_0^1)^2 + \frac{c_2}{3}(x - x_0^1)^3 + \frac{c_3}{4}(x - x_0^1)^4.$$

Therefore we have Theorem 1 (1). □

Let us define $\alpha : [0,1] \to \mathbf{R}$ by

$$\alpha(t) = c_1\Big(\int_0^t \tilde{V}_k^1(u; x_0) du\Big). \tag{43}$$

Then we have the following.

Corollary 1 *Let $h_0 \in H$ be the element defined in (38), then we have*

$$\|h_0^k(x) - \alpha(\cdot)(x - x_0^1)\|_H = O(|x - x_0^1|^2).$$

Proof From (27) and the proof of Theorem 1 (1), we have

$$\begin{aligned} h_0^k(t,x) &= \sum_{i=1}^N \int_0^t p_i(u;w)\tilde{V}_k^i(u, x(u;w))dt \\ &\underset{1}{\sim} \Big(\int_0^t \tilde{V}_k^1(u;x_0)du\Big)w \underset{1}{\sim} \Big(\int_0^t \tilde{V}_k^1(u;x_0)du\Big)c_1(x - x_0^1). \qquad \square \end{aligned}$$

3.2 Proof of Theorem 1 (2)

In this section, we will use the same notations as in [12, 13]. Let $(\Theta, \|\cdot\|_\Theta)$ be a separable Banach space and $(H, \|\cdot\|_H)$ be a separable Hilbert space such that H is a dense subspace of Θ and the inclusion map is continuous. Let μ_s, $s \in [0,\infty)$, be the (necessarily unique) probability measure on $(\Theta, \mathscr{B}_\Theta)$ with the property that

$$\int_\Theta \exp[\sqrt{-1}\langle u, \theta\rangle]\mu_s(d\theta) = \exp(-\frac{s}{2}\|u\|_H^2), \ u \in \Theta^*.$$

We can rewrite (36) replacing ε^2 by s :

$$\begin{aligned} dX_s^i(t,\theta) &= \sum_{k=1}^d V_k^i(t, X_s(t,\theta))d\theta^k(t) + sV_0^i(t, X_s(t,\theta))dt, \quad 1 \le i \le N, \quad t \in [0,1], \\ X_s(0) &= x_0. \end{aligned} \tag{44}$$

Here we replaced $\tilde{X}$ and $\tilde{V}$ in (36) by X and V respectively for simplicity.

Let us define Wiener functionals $F^i : (0,1) \times \Theta \times [x_0^1 - r_0, x_0^1 + r_0] \to \mathbb{R}$, $1 \le i \le N$, by

$$F^i(s,\theta,y) = X_s^i(1,\theta) - y. \tag{45}$$

The main theorem in [13] is summarized in Appendix 2. To apply Theorem 7, it is necessary to check the assumptions (A-1), ..., (A-5) in Appendix 2. Since $f \equiv 0$, we can check (A-1). Since $h(0) = 0$, we can check (A-2), (A-3) and (A-4) in the neighborhood of origin. Since the ellipticity condition at origin, we can check (A-5), using the same discussion given in Appendix B in [14]. Then we have the following.

For each $(s, y) \in (0, 1] \times [-r_0, r_0]$, the density function $p_s(y)$ satisfies

$$\left|(2\pi s)^{1/2} \exp(\frac{e(y)}{s}) p_s(y) - a_0(y)\right| \leq K_0 s^{1/2}, \quad (s, y) \in (0, 1] \times [-r_0, r_0].$$

The function $a_0 \in C([-r_0, r_0])$ is given by

$$a_0(y) = \Big(\frac{\partial^2 e(y)}{\partial y^2}\Big)^{\frac{1}{2}} \det{}_2(I_H - B(y))^{-\frac{1}{2}} \exp\Big(\frac{\partial e(y)}{\partial y} \mathscr{A} F^1(0, h_0(y), y)\Big). \quad (46)$$

Here $\mathscr{A}$ is called the heat operator defined by

$$\mathscr{A} f(s, \theta) = [\frac{\partial f}{\partial s} + \frac{1}{2} trace_H D^2 f](s, \theta),$$

and

$$B(y) \equiv \frac{\partial e(y)}{\partial y} D^2 F^1(0, h_0(y), y). \quad (47)$$

In this section, we calculate each terms in right hand side of (46) explicitly. First we calculate the heat operator.

Lemma 2 *There are constants $C > 0$ and $r > 0$ such that*

$$\Big|\mathscr{A} F^1(0, h_0(y), y) - \frac{(y - x_0^1)}{2b_1} \Big\{\sum_{i=1}^{N} \int_0^1 \int_0^t V_0^i(u, x_0) \nabla_i g^{11}(t, x_0) du dt$$
$$+ \sum_{i,j=1}^{N} \sum_{k=1}^{d} \int_0^1 V_k^1(t, x_0) \nabla_{i,j}^2 V_k^1(t, x_0) \Big(\int_0^t g^{ij}(u, x_0) du\Big) dt\Big\}\Big|$$
$$= O(|y - x_0^1|^2), \quad y > x_0^1.$$

Proof Since the adaptivity of X, we have

$$\mathscr{A} F^i(s, \theta, y) = \sum_{k=1}^{d} \int_0^1 \mathscr{A}[V_k^i(u, X_s(u, \theta))] d\theta_s^k(u) + \int_0^1 V_0^i(u, X_s(u, \theta)) du$$
$$+ s \int_0^1 \mathscr{A}[V_0^i(u, X_s(u, \theta))] du, \quad 1 \leq i \leq N.$$

Therefore we have

$$\begin{aligned}
&\mathscr{A}F^1(0,h_0(y),y)\\
&\quad=\sum_{j=1}^{N}\sum_{k=1}^{d}\int_0^1\nabla_j V_k^1(u,X_0(u,h_0(u;y)))\mathscr{A}X_0^j(u,h_0(u;y))\dot{h}_0^k(u;y)du\\
&\quad+\frac{1}{2}\sum_{i,j=1}^{N}\sum_{k=1}^{d}\int_0^1\nabla_{i,j}^2 V_k^1(u,X_0(u,h_0(u;y)))\langle DX_0^i(u),DX_0^j(u)\rangle\dot{h}_0^k(u;y)du.
\end{aligned}$$

Then using Corollary 1, we have the following.

$$\begin{aligned}
&\Big|\mathscr{A}F^1(0,h_0(y),y)-(y-x_0^1)\Big(\sum_{j=1}^{N}\sum_{k=1}^{d}\int_0^1\nabla_j V_k^1(u,X_0(u;0))\mathscr{A}X_0^j(u;0)\dot{\alpha}^k(u)du\\
&\quad+\frac{1}{2}\sum_{i,j=1}^{N}\sum_{k=1}^{d}\int_0^1\nabla_{i,j}^2 V_k^1(u,X_0(u;0))\langle DX_0^i(u;0),DX_0^j(u;0)\rangle\dot{\alpha}^k(u)du\Big)\Big|\\
&=O(|y-x_0^1|^2),
\end{aligned}$$

where $\mathscr{A}X_0^j(t;0)=\int_0^t V_0^j(u,X_0(u;0))du$. □

Lemma 3 *The Hilbert-Schmidt norm of D^2F^1 is given by*

$$\|D^2F^1(0,0,x_0)\|_{HS}^2=2\sum_{l_1,l_2=1}^{N}\sum_{m=1}^{d}\int_0^1\int_0^t g^{l_1l_2}(u,x_0)\nabla_{l_1}V_m^1(t,x_0)\nabla_{l_2}V_m^1(t,x_0)\,du\,dt.$$

Proof The Malliavin derivatives of X_0^i, $1\le i\le N$, to the direction $k\in H$ is given by

$$\begin{aligned}
DX_0^i(t;h)[k]&=\sum_{l=1}^{N}\sum_{m=1}^{d}\int_0^t\nabla_l V_m^i(u,X_0(u;h))DX_0^l(u;h)[k]\dot{h}^m(u)du\\
&\quad+\sum_{m=1}^{d}\int_0^t V_m^i(u,X_0(u;h))\dot{k}^m(u)du.
\end{aligned}$$

The second Malliavin derivative of F^1 to the direction $k_1, k_2 \in H$ is given by

$$\begin{aligned}
&D^2F^1(0,0,x_0)[k_1][k_2] \\
&= \sum_{l=1}^{N}\sum_{m=1}^{d}\int_0^1 \nabla_l V_m^1(u,x_0) DX_0^l(u;0)[k_1]\dot{k}_2^m(u)du \\
&+ \int_0^1 \nabla_l V_m^1(u,x_0) DX_0^l(u;0)[k_2]\dot{k}_1^m(u)du \\
&= \sum_{l=1}^{N}\sum_{m_1,m_2=1}^{d}\int_0^1 \nabla_l V_{m_1}^1(t,x_0)\Big(\int_0^t V_{m_2}^l(u,x_0)\dot{k}_2^{m_2}(u)du\Big)\dot{k}_1^{m_1}(t)dt \\
&= \sum_{l=1}^{N}\sum_{m_1,m_2=1}^{d}\int_0^1\int_0^1 (\nabla_l V_{m_1}^1(t,x_0)V_{m_2}^l(u,x_0)1_{t>u} \\
&+ \nabla_l V_{m_2}^1(u,x_0)V_{m_1}^l(t,x_0)1_{t<u})\dot{k}_1^{m_1}(t)\dot{k}_2^{m_2}(u)\,du\,dt.
\end{aligned}$$

Therefore we can calculate the Hilbert-Schmidt norm of D^2F^1 as follows:

$$\begin{aligned}
&\|D^2F^1(0,0,x_0)\|_{HS}^2 \\
&= \sum_{l=1}^{N}\sum_{m_1,m_2=1}^{d}\int_0^1\int_0^1 ((\nabla_l V_{m_1}^1(t,x_0)V_{m_2}^l(u,x_0)1_{t>u} \\
&+ \nabla_l V_{m_2}^1(u,x_0)V_{m_1}^l(t,x_0)1_{t<u})^2\,du\,dt \\
&= 2\sum_{l=1}^{N}\sum_{m_1,m_2=1}^{d}\int_0^1\int_0^t (\sum_{l=1}^{N}\nabla_l V_{m_1}^1(t,x_0)V_{m_2}^l(u,x_0))^2\,du\,dt \\
&= 2\sum_{l_1,l_2=1}^{N}\sum_{m_1,m_2=1}^{d}\int_0^1\int_0^t \nabla_{l_1}V_{m_1}^1(t,x_0)V_{m_2}^{l_1}(u,x_0)\nabla_{l_2}V_{m_1}^1(t,x_0)V_{m_2}^{l_2}(u,x_0)\,du\,dt \\
&= 2\sum_{l_1,l_2=1}^{N}\sum_{m=1}^{d}\int_0^1\int_0^t g^{l_1l_2}(u,x_0)\nabla_{l_1}V_m^1(t,x_0)\nabla_{l_2}V_m^1(t,x_0)\,du\,dt.
\end{aligned}$$

□

Finally we will complete the proof of Theorem 1.

Proof (Theorem 1 (2)) Using (46), we have

$$\log a_0(y) = -\frac{1}{2}\log\big(\det{}_2(I_H - B(y))\big) + \frac{\partial e(y)}{\partial y}\mathscr{A}F^1(0,h(y),y) + \frac{1}{2}\log\Big(\frac{\partial^2 e(y)}{\partial y^2}\Big).$$

In the right hand side, the asymptotic expansion of second term is given by Lemma 2, so we will give the asymptotic expansion of the first term.

Since B is defined by (47) and $\frac{\partial e(y)}{\partial y} \underset{1}{\sim} c_1(y - x_0^1)$, we have

$$\left|B(y) - c_1 D^2 F^1(0, 0, x_0)(y - x_0^1)\right| = O(|y - x_0^1|^2).$$

Since $B(x_0^1) = 0$, if $|y - x_0^1|$ is sufficiently small, we have

$$\det{}_2(I - B(y)) = \exp(-\sum_{n=2}^{\infty} \frac{1}{n} trace_H(B(y)^n)).$$

Therefore we have

$$\left|\log\bigl(\det{}_2(I_H - B(y))\bigr) + \frac{c_1^2(y - x_0^1)^2}{2} \|D^2 F(0, 0, x_0)\|_{HS}^2\right| = O(|y - x_0|^3). \tag{48}$$

The Hilbert-Schmidt norm of $D^2 F$ is given by Lemma 3. Therefore we have

$$\begin{aligned}
\log a_0(y) \underset{2}{\sim} &\frac{(y - x_0^1)^2}{4b_1^2} \|D^2 F^1(0, 0, x_0)\|_{HS}^2 \\
&+ \frac{(y - x_0^1)}{b_1} \mathscr{A} F^1(0, h_0(y), y) + \frac{1}{2}\log\Bigl(\frac{\partial^2 e(y)}{\partial y^2}\Bigr) \\
= &\frac{1}{2}\frac{(y - x_0^1)^2}{b_1^2} \sum_{l_1,l_2=1}^{N} \sum_{m=1}^{d} \int_0^1 \int_0^t g^{l_1 l_2}(u, x_0) \nabla_{l_1} V_m^1(t, x_0) \nabla_{l_2} V_m^1(t, x_0) du dt \\
&+ \frac{(y - x_0^1)^2}{2b_1^2} \sum_{j=1}^{N} \int_0^1 \int_0^t V_0^j(u, x_0) \nabla_j g^{11}(t, x_0) du dt \\
&+ \frac{1}{2}\frac{(y - x_0^1)^2}{b_1^2} \sum_{l_1,l_2=1}^{N} \sum_{m=1}^{d} \int_0^1 \int_0^t V_m^1(t, 0) g^{l_1 l_2}(u, x_0) \nabla_{l_1,l_2} V_m^1(t, x_0) du dt \\
&+ \frac{1}{2}\log\Bigl(\frac{\partial^2 e(y)}{\partial y^2}\Bigr).
\end{aligned}$$

From the definition of (8), we have

$$\log a_0(y) \underset{2}{\sim} \frac{(y - x_0^1)^2}{2b_1^2} \int_{0<u<t<1} L_u(g^{11}(t, \cdot)) du dt + \frac{1}{2}\log\Bigl(\frac{\partial^2 e(y)}{\partial y^2}\Bigr).$$

Then we have (15).

Finally we calculate $a_2(x_0^1)$. First we give an asymptotic expansion of the density using Hermite polynomials. Let $y = x_0^1 + \varepsilon\frac{z}{\sqrt{c_1}}$. Then the asymptotic expansion in ε up to the second order is given as follows:

$$\begin{aligned}
p_\varepsilon(y)dy &= p_\varepsilon(x_0^1 + \frac{\varepsilon z}{\sqrt{c_1}})\frac{\varepsilon dz}{\sqrt{c_1}} \\
&\underset{2}{\sim} (a_0(x_0^1 + \frac{\varepsilon z}{\sqrt{c_1}}) + \varepsilon^2 a_2(x_0^1))\frac{1}{\sqrt{2\pi}}\exp\Big[-(\frac{c_1}{2}(\frac{z}{\sqrt{c_1}})^2 + \frac{\varepsilon c_2}{3}(\frac{z}{\sqrt{c_1}})^3 \\
&\qquad + \frac{\varepsilon^2 c_3}{4}(\frac{z}{\sqrt{c_1}})^4)\Big]dz \\
&= \Big[1 - \frac{c_2}{3c_1^{3/2}}\varepsilon(z^3 - 3z) + \varepsilon^2\Big\{\frac{c_2^2}{18c_1^3}z^6 - (\frac{c_2^2}{3c_1^3} + \frac{c_3}{4c_1^2})z^4 \\
&\quad + (-\frac{c_2^2}{2c_1^3} + \frac{3c_3}{2c_1^2} + \frac{Lc_1}{2})z^2 + \varepsilon^2 a_2(x_0^1)\Big\}\Big]\phi(z)dz \\
&= \Big[1 - \varepsilon\frac{c_2}{3c_1^{3/2}}H_3(z) + \varepsilon^2\frac{c_2^2}{18c_1^3}H_6(z) + \varepsilon^2(\frac{c_2^2}{2c_1^3} - \frac{c_3}{4c_1^2})H_4(z) \\
&\quad + \varepsilon^2(\frac{c_1 L}{2})H_2(z) + \varepsilon^2(a_2(x_0^1) - \frac{c_2^2}{18c_1^3}H_6(0) - (\frac{c_2^2}{2c_1^3} - \frac{c_3}{4c_1^2})H_4(0) \\
&\quad - (\frac{c_1 L}{2})H_2(0))\Big]\cdot\sqrt{\frac{c_1}{2\pi\varepsilon^2}}\exp(-\frac{z^2}{2})dz,
\end{aligned}$$

where H_n, $n \in \mathbf{N}$, are Hermite polynomials e.g.

$$\begin{aligned}
H_2(x) &= x^2 - 1, \\
H_3(x) &= x^3 - 3x, \\
H_4(x) &= x^4 - 6x^2 + 3, \\
H_6(x) &= x^6 - 15y^4 + 45y^2 - 15.
\end{aligned}$$

Since p_ε is probability density, we have

$$1 = \int_{-\infty}^{\infty} p_\varepsilon(y)dy = \int_{-\infty}^{\infty} p_\varepsilon(\varepsilon z)\varepsilon dz.$$

The orthogonality of Hermite polynomials implies

$$\int_{-\infty}^{\infty} H_n(z)\frac{1}{\sqrt{2\pi}}\exp(-\frac{z^2}{2})dz = 0, \quad n \geq 1,$$

then we have

$$a_2(x_0^1) = \frac{c_2^2}{18c_1^3}H_6(0) - (\frac{c_2^2}{2c_1^3} - \frac{c_3}{4c_1^2})H_4(0) - (\frac{c_1 L}{2})H_2(0).$$

This completes the proof of Theorem 1 (2). □

The asymptotic expansion of the probability density in ε using Hermite polynomials is given as follows.

Corollary 2 *For each $z \in \mathbf{R}$, let $y = x_0^1 + \varepsilon\frac{z}{\sqrt{c_1}}$, $\varepsilon \in (0, 1]$. For any $r \geq 0$, there is a constant $C > 0$ such that*

$$\Big|\sqrt{\frac{2\pi\varepsilon^2}{c_1}}\exp(\frac{z^2}{2})p_\varepsilon(y) - \Big[1 - \varepsilon\frac{c_2}{3c_1^{3/2}}H_3(z) + \varepsilon^2\frac{c_2^2}{18c_1^3}H_6(z) + \varepsilon^2(\frac{c_2^2}{2c_1^3} - \frac{c_3}{4c_1^2})H_4(z) + \varepsilon^2(\frac{c_1 L}{2})H_2(z)\Big]\Big| \leq \varepsilon^3 C, \quad \varepsilon \in (0, 1], \quad z \in [-r, r].$$

4 Proof of Theorem 2

First we prove the following theorem.

Theorem 4 *We assume $X_\varepsilon^1(T)$ has a density $p_\varepsilon(y)$, $y \in \mathbf{R}$ and let*

$$a_\varepsilon(y) = (2\pi\varepsilon^2)^{1/2}\exp(\frac{e(y)}{\varepsilon^2})p_\varepsilon(y), \quad y \in \mathbf{R}.$$

We assume that there are constants $N \in \mathbf{N}$, $C_0 > 0$ and $K_0 > 0$ such that

$$\Big|a_\varepsilon(y) - \sum_{k=0}^{N} a_{2k}(y)\varepsilon^{2k}\Big| \leq C_0\varepsilon^{2N+2}, \quad y \in [x_0^1, K_0],$$

and assume that the energy function e satisfies $e'(x) > 0$, $x \in (x_0^1, K_0]$. We define $g : \mathbf{R} \to \mathbf{R}$ by

$$e(g(x)) = \frac{x^2}{2}.$$

Since e is strictly increasing, g is well defined. Then there are constants $K_1 < K_0$ and C_1, such that the value of the call option satisfies following:

$$\Big|\sqrt{2\pi}\exp(\frac{e(K)}{\varepsilon^2})C_\varepsilon(T, K) - \varepsilon\varphi_1(\frac{g^{-1}(K)}{\varepsilon})a_0(K)q(K)^2 R_N(\varepsilon, K)\Big| \leq C_1\varepsilon^{N+1}, \quad \varepsilon \in (0, 1], \; K \in [x_0^1, K_1].$$

where

$$R_N(\varepsilon, K) = \sum_{\substack{n,m\geq 0, n+m\geq 1\\ 2n+m+1\leq N}} \frac{c_{n,m}(g^{-1}(K))}{c_{0,0}(g^{-1}(K))} \frac{\varphi_{m+1}(g^{-1}(K)/\varepsilon)}{\varphi_1(g^{-1}(K)/\varepsilon)} \varepsilon^{2n+m}. \tag{49}$$

Here $c_{n,m} \in C(\mathbf{R})$ *is given by*

$$c_{n,m}(x) = \sum_{k=0}^{m} \frac{1}{(k+1)!(m-k)!} \Big(\frac{d}{dx}\Big)^{k+1} g(x) \cdot \Big(\frac{d}{dx}\Big)^{m-k} A_n(x), \tag{50}$$

where

$$A_k(x) = a_{2k}(g(x))g'(x), \quad n \in \mathbf{N}, \quad x \in [x_0^1, K_1]. \tag{51}$$

We prepare the following lemma for the proof of Theorem 4.

Lemma 4

$$A_0(x_0^1) = 1.$$

Proof Since

$$1 = \int_{-\infty}^{\infty} p_\varepsilon(y)dy = \frac{1}{(2\pi\varepsilon^2)^{1/2}} \int_{-\infty}^{\infty} a_\varepsilon(y) \exp(-\frac{e(y)}{\varepsilon^2})dy,$$

we have

$$1 = \frac{1}{(2\pi)^{1/2}} \int_{-\infty}^{\infty} a_\varepsilon(g(\varepsilon y)) \exp(-\frac{y^2}{2}) g'(\varepsilon y)dy.$$

Since the right hand side is bounded, taking the limit of $\varepsilon \downarrow 0$, we have $a_0(g(0))g'(0) = 1$. □

Proof (Proof of Theorem 4) We can divide the value of a call option into two parts:

$$C_\varepsilon(T, K) = \tilde{C}_\varepsilon(T, K) + R_\varepsilon(K_0),$$

where

$$\tilde{C}_\varepsilon(T, K) = \int_K^{K_0} (y - K)p_\varepsilon(y)dy = \int_K^{K_0} (y - K)\Big(\frac{1}{2\pi\varepsilon^2}\Big)^{\frac{1}{2}} \exp(-\frac{e(y)}{\varepsilon^2})a_\varepsilon(y)dy,$$

and

$$R_\varepsilon(K_0) = E[X_\varepsilon^1(T) - K : X_\varepsilon^1(T) > K_0].$$

Since $e(g(x)) = \frac{x^2}{2}$, we have

$$\tilde{C}_\varepsilon(T,K) = \int_{g^{-1}(K)}^{g^{-1}(K_0)} (g(x)-K)\Big(\frac{1}{2\pi\varepsilon^2}\Big)^{\frac{1}{2}} \exp\Big(-\frac{x^2}{2\varepsilon^2}\Big) a_\varepsilon(g(x)) g'(x)dx.$$

Let $A_\varepsilon(x) = a_\varepsilon(g(x))g'(x)$ and $\tilde{K}_\varepsilon = \frac{1}{\varepsilon}(g^{-1}(K_0) - g^{-1}(K))$. Putting $x = \varepsilon z + g^{-1}(K)$, we have

$$\begin{aligned}
&\exp\Big(\frac{g^{-1}(K)^2}{2\varepsilon^2}\Big)\tilde{C}_\varepsilon(T,K)\\
&= \int_0^{\tilde{K}_\varepsilon} \big(g(\varepsilon z + g^{-1}(K)) - K\big)\frac{1}{\sqrt{2\pi}}\exp\Big(-\frac{z^2}{2} - \frac{zg^{-1}(K)}{\varepsilon}\Big)A_\varepsilon(\varepsilon z + g^{-1}(K))dz.
\end{aligned}$$

We define

$$\tilde{A}_{\varepsilon,n}(x) = \bar{a}_{\varepsilon,n}(g(x))g'(x) = \sum_{k=0}^{n} A_k(x)\varepsilon^{2k}.$$

We also define

$$\begin{aligned}
&\tilde{C}_{\varepsilon,n}(T,K)\\
&\quad = \exp\Big(-\frac{g^{-1}(K)^2}{2\varepsilon^2}\Big)\int_0^{\tilde{K}_\varepsilon} \big(g(\varepsilon z + g^{-1}(K)) - K\big)\frac{1}{\sqrt{2\pi}}\exp\Big(-\frac{z^2}{2} - \frac{zg^{-1}(K)}{\varepsilon}\Big)\\
&\qquad \times \tilde{A}_{\varepsilon,n}(\varepsilon z + g^{-1}(K))dz.
\end{aligned}$$

Then there exist constants $C_1,\ C_2 > 0$ such that

$$\exp\Big(\frac{g^{-1}(K)^2}{2\varepsilon^2}\Big)\Big|\tilde{C}_\varepsilon(T,K) - \tilde{C}_{\varepsilon,n}(T,K)\Big| \le C_1\varepsilon^{2n+2}.$$

Since

$$\begin{aligned}
&\Big|(g(\varepsilon z + g^{-1}(K)) - K)\tilde{A}_{\varepsilon,n}(\varepsilon z + g^{-1}(K))\\
&\quad - \sum_{\substack{n,m\ge 0\\ 2n+m+1\le N}} c_{n,m}(g^{-1}(K))\varepsilon^{2n+m+1}z^{m+1}\Big| \le C_2\varepsilon^{N+1}, \quad K \in [x_0^1, K_1],
\end{aligned}$$

we have

$$\begin{aligned}
&\Big|\exp\Big(\frac{e(K)}{\varepsilon^2}\Big)\tilde{C}_{\varepsilon,n}(T,K) - \sum_{\substack{n,m\ge 0\\ 2n+m+1\le N}} c_{n,m}(g^{-1}(K))\varepsilon^{2n+m+1}\frac{1}{\sqrt{2\pi}}\varphi_{m+1}\Big(\frac{g^{-1}(K)}{\varepsilon}\Big)\Big|\\
&\quad \le R\varepsilon^{N+1}, \quad K \in [x_0^1, K_1].
\end{aligned}$$

For any $\delta > 0$, we have

$$R_\varepsilon(K_0) \le E[X_\varepsilon^1(T); X_\varepsilon^1(T) > K_0] \\ \le E[X_\varepsilon^1(T)^{1/\delta}]^\delta P(X_\varepsilon^1(T) > K_0)^{1-\delta}.$$

Therefore we have

$$\lim_{\varepsilon\downarrow 0} \varepsilon^2 \log R_\varepsilon(K_0) \le \lim_{\varepsilon\downarrow 0} \varepsilon^2(1-\delta)\log P(X_\varepsilon^1(T) > K_0) = -(1-\delta)e(K_0).$$

Note that $e(K_0) > e(K_1)$, we have

$$\limsup_{\varepsilon\downarrow 0} \varepsilon^2 \log R_\varepsilon(K_0) < -e(K_1).$$

The function q defined by (19) can be written as

$$q(K) = g'(g^{-1}(K)) = \big(\frac{d}{dK} g^{-1}(K)\big)^{-1}.$$

Then we have our assertion. □

Finally we prove Theorem 2.

Proof (Proof of Theorem 2) From the definition of $R_2(\varepsilon, K)$ given in (49), we have

$$R_2(\varepsilon, K) = \varepsilon \frac{c_{0,1}(g^{-1}(K))}{c_{0,0}(g^{-1}(K))} \frac{\varphi_2(g^{-1}(K)/\varepsilon)}{\varphi_1(g^{-1}(K)/\varepsilon)} + \varepsilon^2 \frac{c_{0,2}(g^{-1}(K))}{c_{0,0}(g^{-1}(K))} \frac{\varphi_3(g^{-1}(K)/\varepsilon)}{\varphi_1(g^{-1}(K)/\varepsilon)} \\ + \varepsilon^2 \frac{c_{1,0}(g^{-1}(K)/\varepsilon)}{c_{0,0}(g^{-1}(K)/\varepsilon)}.$$

The second and third derivatives of g at $g^{-1}(K)$ are given as follows:

$$\frac{d^2}{dK^2} g(g^{-1}(K)) = q(K)q'(K), \\ \frac{d^3}{dK^3} g(g^{-1}(K)) = q(K)q'(K)^2 + q(K)^2 q''(K).$$

Using the definition of $c_{n,m}$ given in (50), we can calculate $c_{0,0}$, $c_{0,1}$, $c_{1,0}$, $c_{0,2}$ explicitly as follows:

$$\begin{aligned}
c_{0,0}(g^{-1}(K)) &= a_0(K)q(K)^2,\\
c_{1,0}(g^{-1}(K)) &= a_2(K)q(K)^2,\\
c_{0,1}(g^{-1}(K)) &= a_0'(K)q(K)^3 + \frac{3}{2}a_0(K)q(K)^2q'(K),\\
c_{0,2}(g^{-1}(K)) &= \frac{1}{2}a_0''(K)q(K)^4 + 2a_0'(K)q(K)^3q'(K) + \frac{7}{6}a_0(K)q(K)^2q'(K)^2\\
&\quad + \frac{2}{3}a_0(K)q(K)^3q''(K).
\end{aligned}$$

Then we have our theorem. □

5 Proof of Theorem 3

First, we define smooth functions θ_n, $n \in \mathbf{N}$, inductively by

$$\begin{aligned}
\theta_1(x) &= \varphi_1(x),\\
\theta_{n+1}(x) &= -n\theta_n(x) + \theta_n'(x)\theta_1(x)x.
\end{aligned} \tag{52}$$

We define the function $h : [0, 1] \times \mathbf{R} \to \mathbf{R}$ by

$$h(t, y) \equiv f^{-1}(tf(y)), \tag{53}$$

where f is defined by (21). The properties of h are given in Lemma 9. Then we have the following.

Proposition 3 *The implied normal volatilities of call options are given as follows.*

$$\sigma_N^{\varepsilon}(T, K) = \frac{\varepsilon(K - x_0^1)}{g^{-1}(K)\sqrt{T}} \exp\Big(-\int_1^{1+l(\varepsilon,K)} \frac{1}{t}\varphi_1(h(t, \frac{g^{-1}(K)}{\varepsilon}))dt\Big), \quad K > x_0^1.$$

Here

$$l(\varepsilon, K) = (1 + R(\varepsilon, K)(1 + r(K)) - 1,$$

where

$$R(\varepsilon, K) = \frac{\sqrt{2\pi}\exp(\frac{e(K)}{\varepsilon^2})C_\varepsilon(T, K)}{\varepsilon c_{0,0}(g^{-1}(K))\varphi_1(g^{-1}(K)/\varepsilon)} - 1,$$

and

$$r(K) = \frac{g^{-1}(K)c_{0,0}(g^{-1}(K))}{(K - x_0^1)} - 1.$$

R and r satisfies the following respectively:

$$\left|R(\varepsilon, K) - R_N(\varepsilon, K)\right| \le C\varepsilon^N, \tag{54}$$

and

$$\lim_{K \downarrow x_0^1} r(K) = 0.$$

Proof From Theorem 4 and Lemma 7, we have (54). Using l'Hospital's rule, we have

$$\lim_{K \downarrow 0} \frac{g^{-1}(K)c_{0,0}(g^{-1}(K))}{K - x_0^1} = g'(x_0^1)a_0(x_0^1) = 1.$$

By definition of R, we can rewrite the value of call option as

$$C_\varepsilon(T, K) = f(g^{-1}(K)/\varepsilon)g^{-1}(K)c_{0,0}(g^{-1}(K))(1 + R(\varepsilon, K)).$$

On the other hand, the value of call option under the normal model is given by

$$V = (K - x_0^1)f\left(\frac{K - x_0^1}{\sigma\sqrt{T}}\right).$$

Therefore we have

$$f\left(\frac{K - x_0^1}{\sigma\sqrt{T}}\right) = (1 + r(K))(1 + R(\varepsilon, K))f\left(\frac{g^{-1}(K)}{\varepsilon}\right).$$

Using the definition of h given by (53) and Lemma 9, we have our assertion. □

Next we will give the asymptotic expansion of implied volatilities.

Theorem 5 *For any $N \in \mathbf{N}$, there is a constant $C > 0$ such that the asymptotic expansion of implied volatilities satisfies the following:*

$$\left|\left(\frac{\varepsilon(K - x_0^1)}{g^{-1}(K)\sqrt{T}}\right)^{-1}\sigma_N(T, K) - \exp\left(\sum_{n=0}^{N} \frac{l_N(\varepsilon, K)^{n+1}}{(n+1)!}\theta_{n+1}\left(\frac{g^{-1}(K)}{\varepsilon}\right)\right)\right|$$
$$< C(\varepsilon + |K - x_0^1|)^{N+1}, \quad K \in [x_0^1, K_1].$$

Here

$$l_N(\varepsilon, K) = (1 + R_N(\varepsilon, K))(1 + r(K)) - 1, \tag{55}$$

where

$$r(K) = \frac{g^{-1}(K)c_{0,0}(g^{-1}(K))}{K - x_0^1} - 1. \tag{56}$$

Proof Using Lemma 9, we have

$$\left(\frac{\partial}{\partial t}\right)^n \frac{1}{t}\varphi_1(h(t,y))\Big|_{t=1} = \theta_n(y),\ n \geq 1.$$

Therefore

$$\begin{aligned}
&\Big|\int_1^{1+l(\varepsilon,K)} \frac{1}{t}\theta_1(h(t,\frac{g^{-1}(K)}{\varepsilon}))dt - \sum_{n=0}^N \int_1^{1+l_N(\varepsilon,K)} \frac{\theta_n(y)}{n!}(t-1)^n dt\Big| \\
&\leq \Big|\int_1^{1+l(\varepsilon,K)} \frac{1}{t}\theta_1(h(t,\frac{g^{-1}(K)}{\varepsilon}))dt - \int_1^{1+l_N(\varepsilon,K)} \frac{1}{t}\theta_1(h(t,\frac{g^{-1}(K)}{\varepsilon}))dt\Big| \\
&+ \Big|\int_1^{1+l_N(\varepsilon,K)} \frac{1}{t}\theta_1(h(t,\frac{g^{-1}(K)}{\varepsilon}))dt - \sum_{n=0}^N \int_1^{1+l_N(\varepsilon,K)} \frac{\theta_n(y)}{n!}(t-1)^n dt\Big| \\
&\leq C_1|l(\varepsilon,K) - l_N(\varepsilon,K)| + C_2|l_N(\varepsilon,K)|^N \leq C(\varepsilon + |K - x_0^1|)^N. \qquad \square
\end{aligned}$$

Finally we prove Theorem 3.

Lemma 5 *The derivatives of q, a_0, a_2 at x_0 are given as follows:*

$$\begin{aligned}
&q(x_0^1) = \frac{1}{\sqrt{c_1}},\quad \frac{q'(x_0^1)}{q(x_0^1)} = -\frac{2}{3}\frac{c_2}{c_1},\quad \frac{q''(x_0^1)}{q(x_0^1)} = \frac{11}{9}\left(\frac{c_2}{c_1}\right)^2 - \frac{3}{2}\frac{c_3}{c_1},\\
&\frac{a_0'(x_0^1)}{a_0(x_0^1)} = \frac{c_2}{c_1},\quad \frac{a_0''(x_0^1)}{a_0(x_0^1)} = c_1^2 L - \left(\frac{c_2}{c_1}\right)^2 + \frac{3c_3}{c_1},\\
&\frac{a_2(x_0^1)}{a_0(x_0^1)} = \frac{1}{c_1}\left(-\frac{c_1^2 L}{2} + \frac{2}{3}\left(\frac{c_2}{c_1}\right)^2 - \frac{3}{4}\frac{c_3}{c_1}\right),
\end{aligned}$$

where c_i $(i = 1, 2, 3)$ are given by (41).

Proof Since

$$e(g(x)) = \frac{1}{2}x^2,$$

and $g'(x) > 0$, the derivatives are given by

$$\begin{aligned}
x &= e'(g(x))g'(x),\\
1 &= e''(g(x))g'(x)^2 + e'(g(x))g''(x),\\
0 &= e'''(g(x))g'(x)^3 + 3e''(g(x))g'(x)g''(x) + e'(g(x))g'''(x),\\
0 &= e^{(4)}(g(x))g'(x)^4 + 6e'''(g(x))g'(x)^2 g''(x) + 3e''(g(x))g''(x)^2\\
&\quad + 4e''(g(x))g'(x)g'''(x) + e'(g(x))g^{(4)}(x).
\end{aligned}$$

Furthermore, since

$$e'(x_0^1) = 0,\ e''(x_0^1) = \frac{1}{b_1},\ e'''(x_0^1) = -\frac{2b_2}{b_1^3},\ e^{(4)}(x_0^1) = -\frac{6b_3}{b_1^4} + \frac{12b_2^2}{b_1^5},$$

we have

$$g'(0) = \sqrt{b_1},\ g''(0) = \frac{2}{3}\frac{b_2}{b_1},\ g'''(0) = \frac{\sqrt{b_1}}{6}(9b_1b_3 - 8b_2^2). \qquad \square$$

Lemma 6

$$\left|R_2(\varepsilon, K) - R_2^0(\varepsilon, K)\right| \le C(\varepsilon + |K - x_0^1|)^3,$$

$$\left|r(K) - r^0(K)\right| \le C|K - x_0^1|^3,$$

where

$$\begin{aligned} R_2^0(\varepsilon, K) = &\frac{\varepsilon(K - x_0^1)}{\sqrt{c_1}}\Big[c_1^2 L - \frac{5}{6}\Big(\frac{c_2}{c_1}\Big)^2 + \frac{3}{4}\frac{c_3}{c_1}\Big]\frac{\varphi_2(g^{-1}(K)/\varepsilon)}{\varphi_1(g^{-1}(K)/\varepsilon)} \\ &+ \frac{\varepsilon^2}{c_1}\Big[\frac{c_1^2 L}{2} - \frac{1}{2}\Big(\frac{c_2}{c_1}\Big)^2 + \frac{1}{2}\frac{c_3}{c_1}\Big]\frac{\varphi_3(g^{-1}(K)/\varepsilon)}{\varphi_1(g^{-1}(K)/\varepsilon)} \\ &+ \frac{\varepsilon^2}{c_1}\Big[-\frac{c_1^2 L}{2} + \frac{2}{3}\Big(\frac{c_2}{c_1}\Big)^2 - \frac{3}{4}\frac{c_3}{c_1}\Big], \end{aligned}$$

and

$$r^0(K) = \Big[-\frac{1}{3}\Big(\frac{c_2}{c_1}\Big)^2 + \frac{1}{4}\frac{c_3}{c_1} + \frac{c_1^2 L}{2}\Big](K - x_0^1)^2.$$

Proof We will calculate each terms of R_2 given by (20). From Lemma 7, the functions φ_2/φ_1 and φ_3/φ_1 are bounded above. Since the first term is $O(\varepsilon)$ and other terms are $O(\varepsilon^2)$, it is enough to calculate the first order of K in the first term and 0th order in the other terms. Using Lemma 5, we have

$$\frac{c_{0,1}(x_0^1)}{c_{0,0}(x_0^1)} = 0,$$

and the first derivative is given by

$$\frac{d}{dK}\frac{c_{0,1}(g^{-1}(K))}{c_{0,0}(g^{-1}(K))} = q(K)\Big[\frac{a_0''(K)}{a_0(K)} + \Big(\frac{a_0'(K)}{a_0(K)}\Big)^2 + \frac{3}{2}\frac{q''(K)}{q(K)} + \frac{a_0'(K)}{a_0(K)}\frac{q'(K)}{q(K)}\Big].$$

Using Lemma 5 again, we have

$$\frac{c_{0,1}(g^{-1}(K))}{c_{0,0}(g^{-1}(K))} \underset{1}{\sim} \frac{(K - x_0^1)}{\sqrt{c_1}}\Big[c_1^2 L - \frac{5}{6}\Big(\frac{c_2}{c_1}\Big)^2 + \frac{3}{4}\frac{c_3}{c_1}\Big],$$

$$\frac{c_{0,2}(g^{-1}(K))}{c_{0,0}(g^{-1}(K))} \underset{0}{\sim} \frac{1}{c_1}\Big[\frac{c_1^2 L}{2} - \frac{1}{2}\Big(\frac{c_2}{c_1}\Big)^2 + \frac{1}{2}\frac{c_3}{c_1}\Big],$$

$$\frac{c_{1,0}(g^{-1}(K))}{c_{0,0}(g^{-1}(K))} \underset{0}{\sim} \frac{1}{c_1}\Big[-\frac{c_1^2 L}{2} + \frac{2}{3}\Big(\frac{c_2}{c_1}\Big)^2 - \frac{3}{4}\frac{c_3}{c_1}\Big].$$

We can calculate $r(K)$ in the same way and we have our results. □

Proof (Proof of Theorem 1.3) Using (55) we have

$$l_2(\varepsilon, K) \underset{2}{\sim} R_2^0(\varepsilon, K) + r^0(K).$$

Since R_2^0 and r^0 are of the second order in ε, K, we have

$$\sum_{n=0}^{2} \frac{l_2(\varepsilon, K)^{n+1}}{(n+1)!}\theta_{n+1}\Big(\frac{g^{-1}(K)}{\varepsilon}\Big) \underset{2}{\sim} (R_2^0(\varepsilon, K) + r^0(K))\varphi_1\Big(\frac{g^{-1}(K)}{\varepsilon}\Big).$$

Hence we have our result. □

6 Examples

In this section, we apply our results to some known models.

6.1 Local Volatility Models

We assume the following model. Let $\sigma : \mathbf{R} \to \mathbf{R}_+$ be a smooth function whose derivatives of any order are bounded. Let λ be continuous $\mathbf{R}_+$-valued functions defined on $[0, T]$.

$$dX^\varepsilon(t) = \varepsilon\lambda(t)\sigma(X^\varepsilon(t))dW_t,$$
$$X^\varepsilon(0) = x_0.$$

In this case we can solve the energy as follows:

$$e(y) = \frac{1}{2\Lambda}\Big(\int_{x_0}^{y} \frac{dx}{\sigma(x)}\Big)^2,$$

where

$$\Lambda = \int_0^T \lambda^2(t)dt.$$

The minimum energy path h is given by

$$h(t) = \frac{1}{\Lambda}\Big(\int_{x_0}^{y} \frac{dx}{\sigma(x)}\Big)\int_0^t \lambda(s)ds.$$

We can easily calculate the coefficients.

$$\begin{aligned}
b_1 &= \sigma(x_0)^2\Lambda, \quad b_2 = \frac{3}{2}\sigma(x_0)^3\sigma'(x_0)\Lambda^2,\\
b_3 &= \Big(\frac{8}{3}\sigma(x_0)^4\sigma'(x_0)^2 + \frac{2}{3}\sigma(x_0)^5\sigma''(x_0)\Big)\Lambda^3,\\
L &= \Big(\frac{1}{2}\sigma(x_0)^2\sigma'(x_0)^2 + \frac{1}{2}\sigma(x_0)^3\sigma''(x_0)\Big)\Lambda^2, \quad g^{-1}(y) = \frac{1}{\sqrt{\Lambda}}\Big(\int_{x_0}^{y} \frac{dx}{\sigma(x)}\Big).
\end{aligned}$$

Then using Theorems 1 and 3 we can calculate the density function and implied normal volatilities. We illustrate some cases.

Example 1 (CEV model) This is the case $\lambda(t) \equiv \alpha$ and

$$\sigma(x) = x^{\beta}.$$

Each terms are given by

$$\begin{aligned}
&\Lambda = \alpha^2 T, \quad b_1 = x_0^{2\beta}\Lambda, \quad b_2 = \frac{3}{2}\beta x_0^{4\beta-1}\Lambda^2, \quad b_3 = \frac{2}{3}(\beta^2 - \beta + 4)x_0^{6\beta-2}\Lambda^3,\\
&L = (\beta^2 - \frac{\beta}{2})x_0^{4\beta-2}\Lambda^2, \quad e''(y) = \frac{\beta(1+\beta)}{2\alpha^2 T y^{\beta+2}},
\end{aligned}$$

$$g^{-1}(y) = \begin{cases} \frac{1}{\sqrt{\Lambda}}\Big(\frac{y^{1-\beta} - x_0^{1-\beta}}{1-\beta}\Big) & (\beta \neq 1)\\ \frac{1}{\sqrt{\Lambda}}\log(\frac{y}{x_0}) & (\beta = 1). \end{cases}$$

Example 2 (Displaced diffusion) This is the case $\lambda(t) \equiv \sigma$ and

$$\sigma(x) = qx + (1-q)x_0.$$

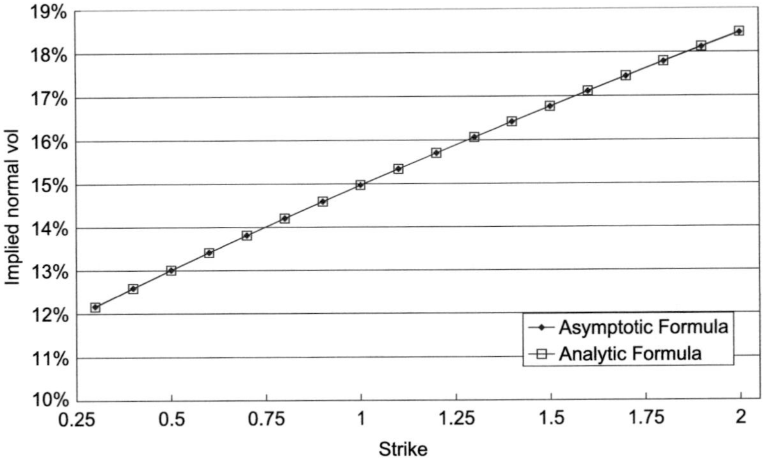

Fig. 1 Implied volatility smile of displaced diffusion, asymptotic expansion versus analytic solution with $x_0 = 1.0,\ q = 0.5,\ \sigma = 0.15,\ T = 10$

Each terms are given by (Fig. 1)

$$\Lambda = \sigma^2 T, \quad b_1 = x_0^2 \Lambda, \quad b_2 = \frac{3}{2} x_0^3 q \Lambda^2, \quad b_3 = \frac{8}{3} x_0^4 q^2 \Lambda^3, \quad L = \frac{1}{2} x_0^2 q^2 \Lambda^2,$$

$$g^{-1}(y) = \frac{1}{\sqrt{\Lambda}} \int_{x_0}^{y} \frac{dx}{qx + (1-q)x_0} = \frac{1}{q\sqrt{\Lambda}} \log\left(\frac{qy + (1-q)x_0}{x_0}\right),$$

$$e''(y) = \frac{1 + g^{-1}(y) q \sqrt{\Lambda}}{\Lambda (qy + (1-q)x_0)^2}.$$

Black-Scholes model is the case $q = 1$. We present a numerical results of the asymptotic expansion formula, comparing with analytical solution.

6.2 SABR Model

We investigate the following model which is called SABR model.

$$\begin{aligned} dX^{\varepsilon}(t) &= \varepsilon \alpha^{\varepsilon}(t) \sigma(X^{\varepsilon}(t))(\rho dW(t) + \sqrt{1-\rho^2} dZ(t)), \\ d\alpha^{\varepsilon}(t) &= \varepsilon \nu \alpha^{\varepsilon}(t) dW(t), \\ X^{\varepsilon}(0) &= x_0, \ \alpha^{\varepsilon}(0) = \alpha. \end{aligned}$$

This model was investigated in Hagan and Woodward [6, 14]. The energy function was given in Hagan et al. [8] as follows.

$$e(y) = \frac{1}{2\nu^2 T} \log\left(\frac{\sqrt{1 - 2\rho\zeta + \zeta^2} - \rho + \zeta}{1 - \rho}\right)^2 = \frac{\hat{x}(\zeta(y))^2}{2\nu^2 T},$$

where

$$\zeta(y) = -\frac{\nu}{\alpha} \int_{x_0}^{y} \frac{dz}{\sigma(z)}.$$

In Theorem 3.1 [14], we also gave the energy function by solving Hamilton equations. Then the parameters are given by (Fig. 2)

$$\begin{aligned}
b_1 &= \alpha^2 \sigma(x_0)^2 T, \quad b_2 = \frac{3}{2}\sigma(x_0)^3 \alpha^3 (\alpha\sigma'(x_0) + \nu\rho) T^2, \\
b_3 &= \Big(\frac{8}{3}\alpha^6 \sigma(x_0)^4 \sigma'(x_0)^2 + \frac{2}{3}\alpha^6 \sigma(X_0)^5 \sigma''(x_0) + 6\nu\rho\sigma(x_0)^4 \sigma'(x_0)\alpha^5 \\
&\quad + 2\nu^2\rho^2\sigma(x_0)^4\alpha^4 + \frac{2}{3}\alpha^4 \sigma(x_0)^4 \nu^2\Big) T^3, \\
L &= \frac{\alpha^2 \sigma(x_0)^2 T^2}{2}\Big(\alpha^2(\sigma'(x_0)^2 + \sigma(x_0)\sigma''(x_0)) + 4\nu\rho\alpha\sigma'(x_0) + \nu^2\Big), \\
g^{-1}(y) &= \frac{1}{\nu\sqrt{T}} \log\left(\frac{\sqrt{1 - 2\rho\zeta(y) + \zeta(y)^2} - \rho + \zeta(y)}{1 - \rho}\right).
\end{aligned}$$

We present a numerical results of the asymptotic expansion formula comparing with Monte Carlo simulation. Here we assume $\sigma(x) = x^{\beta}$.

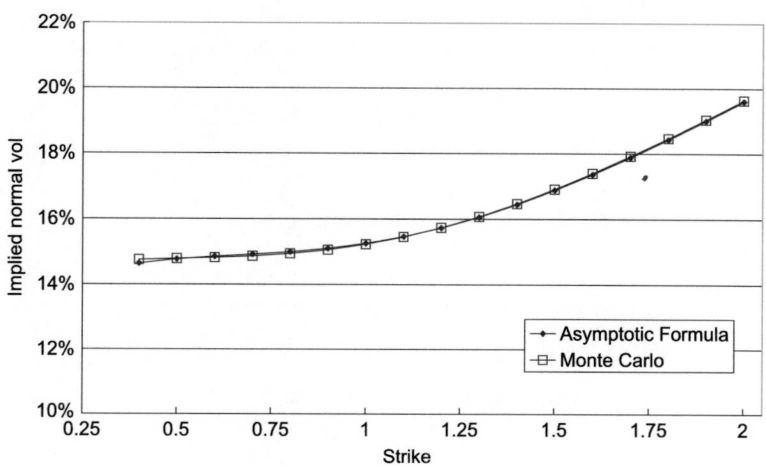

Fig. 2 Implied volatility smile of SABR model, asymptotic expansion versus Monte Carlo simulation with $x_0 = 1$, $\alpha = 0.15$, $\beta = 0.5$, $\nu = 0.2$, $\rho = -0.2$, $T = 10$.

Acknowledgments The author would like to thank Professor Shigeo Kusuoka for useful discussions.

Appendix 1

In this section, we investigate some properties of functions defined in Sect. 1. First we consider $\varphi_n,\ n \geq 0$ defined by (18).

Lemma 7 *The functions φ_n have the following properties.*
(1) $\varphi_n(x) > 0,\ x \geq 0.$
(2) $\lim_{x\to\infty} x^{n+1}\varphi_n(x) = n!.$
(3) $\sup_x \dfrac{\varphi_n(x)}{\varphi_1(x)} < \infty, \qquad n \geq 1.$

Proof (1) is easy to check. We prove (2). Putting $y = xz$

$$\varphi_n(x) = \int_0^\infty \exp(-\frac{y^2}{2x^2} - y)(\frac{y}{x})^n \frac{dy}{x} = \frac{1}{x^{n+1}} \int_0^\infty y^n \exp(-y - \frac{y^2}{2x^2})dy$$

Then we have

$$\lim_{x\to\infty} x^{n+1}\varphi_n(x) = \int_0^\infty y^n e^{-y} dy = n!.$$

(3) is an easy consequence of (1) and (2). □

The following is easy to check.

Lemma 8 *The functions $\{\varphi_n\}$ satisfy the following recurrence relations.*

$$\varphi_{n+1}(x) = -x\varphi_n(x) + n\varphi_{n-1}(x),$$
$$\varphi_n'(x) = -\varphi_{n+1}(x).$$

Example 3 φ_i $(0 \leq i \leq 3)$ are given as follows:

$$\varphi_0(x) = \exp(\frac{x^2}{2}) \int_x^\infty \exp(-\frac{z^2}{2})dz,$$
$$\varphi_1(x) = -x\varphi_0(x) + 1,$$
$$\varphi_2(x) = (x^2 + 1)\varphi_0(x) - x,$$
$$\varphi_3(x) = -(x^3 + 3x)\varphi_0(x) + x^2 + 2.$$

Next we consider the function $h \in C^\infty([0, 1] \times \mathbf{R}_+)$ defined by (53).

Lemma 9 *The n-times differentiation of* $\log h(t, y)$ *with respect to t is given as follows. We define θ in (52).*

$$\left(\frac{\partial}{\partial t}\right)^n \log h(t,y) = \frac{1}{t^n}\theta_n(h(t,y)), \quad t \in [0,1],\ y > 0,$$

where $\theta_n \in C_b[0,\infty],\ n \geq 1$ *are given inductively as follows:*

$$\theta_1(x) = \varphi_1(x),$$
$$\theta_{n+1}(x) = n\theta_n(x) + \theta_n'(x)\theta_1(x)x.$$

Proof In the case $n = 1$, since $f(h(t,y)) = tf(y)$, we have

$$\frac{\partial h}{\partial t}(t,y) = \frac{f(h(t,y))}{tf'(h(t,y))}.$$

Since

$$f'(x) = -\left(\frac{1}{x} + x + \frac{\varphi_2(x)}{\varphi_1(x)}\right)f(x) < 0,\ x > 0,$$

we have

$$\theta_1(x) = \frac{f(x)}{xf'(x)} = \left(1 + x^2 + x\frac{\varphi_2(x)}{\varphi_1(x)}\right)^{-1} = \varphi_1(x).$$

It is easy to check that $\theta_1 \in C_b([0,\infty])$ and $x\theta_1(x) \in C_b([0,\infty])$. We have

$$\frac{\partial}{\partial t}\log h(t,y) = \frac{1}{t}\theta_1(h(t,y)).$$

Since

$$\frac{\partial}{\partial t}\left(\frac{1}{t^n}\theta_n(h(t,y))\right) = \frac{1}{t^{n+1}}\left(-n\theta_n(h(t,y)) + \theta_n'(h(t,y))\theta_1(h(t,y))h(t,y)\right),$$

it is easy to prove our lemma. □

Appendix 2

In this section, we summarize the main theorem in Kusuoka and Osajima [13]. See [13] for the definitions.

Let $f,\ g \in \mathcal{G}^\infty(\mathcal{A};\mathbf{R})$ and $F \in \mathcal{G}^\infty(\mathcal{A};\mathbf{R}^N)$ be completely P-regular functions and Y be a compact subset in $\mathbf{R}^N$. We assume the following.

(A1) There is an $\alpha > 0$ such that

$$\sup_{s\in(0,1]} s\log\left(\int_\Theta \exp\left(\frac{(1+\alpha)f(s,\theta)}{s}\right)\mu_s(d\theta)\right) < \infty.$$

We define $e : \mathbf{R}^N \to [-\infty, \infty]$ by

$$e(x) \equiv \inf\{\frac{\|h\|^2}{2} - f(0,h) : F(0,h) = x\}, \qquad x \in \mathbf{R}^N.$$

We also assume the following.
(A2) For each $y \in Y$,

$$M(y) \equiv \{h \in H;\, F(0,h) = y\} \neq \emptyset$$

and that

$$e(y) = \frac{\|h(y)\|^2}{2} - f(0, h(y))$$

for precisely one $h(y) \in M(y)$.

We assume moreover the following.
(A3) $T(y) \equiv DF(0, h(y))$ has rank N for every $y \in Y$.

Let $\pi(y) = T(y)^*(T(y)T(y)^*)^{-1}T(y),\ y \in Y$. $\pi(y)$ is an orthogonal projection in H. Let $\pi(y)^\perp = I_H - \pi(y)$. Then $\pi(y)^\perp$ is also an orthogonal projection in H onto $ker\ T(y)$. Let $V(y) : H \times H \to \mathbf{R}$ be a bilinear form given by

$$\begin{aligned} &V(y)(h,h') \\ &= D^2 f(0,h(y))(\pi(y)^\perp h, \pi(y)^\perp h') \\ &+ (h(y) - Df(0,h(y)), T(y)^*(T(y)T(y)^*)^{-1} D^2 F(0,h(y))(\pi(y)^\perp h, \pi(y)^\perp h'))_H. \end{aligned}$$

We assume the following furthermore.
(A4) For all $y \in Y$ and $h \in H \setminus \{0\}$

$$V(y)(h,h) < \|h\|^2.$$

Finally we define

$$\begin{aligned} A(s,\theta) &= DF(s,\theta)DF(s,\theta)^* \\ &= ((DF_i(s,\theta), DF_j(s,\theta))_H)_{1 \leqq i,j \leqq N} \end{aligned}$$

and assume the following.
(A5) For any $p \in [1, \infty)$

$$\overline{\lim_{s \downarrow 0}}\, s \log(\int_\Theta |\det A(s,\theta)|^{-p} \mu_s(d\theta)) \leqq 0.$$

Then Kusuoka-Stroock [12] proved the following.

Theorem 6 *For each $s \in (0,1]$, a signed measure $P_s(\cdot)$ on $\mathbf{R}^N$ given by*

$$P_s(\Gamma) = \int_{F(s,\theta)\in\Gamma} g(s,\theta)\exp\left(\frac{f(s,\theta)}{s}\right)\mu_s(d\theta),\ \Gamma \in \mathscr{B}(\mathbf{R}^N),$$

admits a smooth density $p_s(\cdot)$ with respect to Lebesgue's measure. Moreover, there exist sequence $\{a_n\}_{n=0}^{\infty} \subseteq C(Y;\mathbf{R})$ and $\{K_n\}_{n=0}^{\infty} \subseteq (0,\infty)$ with the property that, for every $n \in \mathbf{N}$,

$$\left|(2\pi s)^{N/2} e^{e(y)/s} p_s(y;0) - \sum_{m=0}^{n} s^{m/2} a_m(y)\right| \leqq K_n s^{(n+1)/2},\ (s,y) \in (0,1]\times Y.$$

The main theorem in Kusuoka-Osajima [13] is the following.

Theorem 7 *e is smooth in the neighborhood of Y and*

$$a_0(y) = (\det \nabla^2 e(y))^{1/2} \det{}_2(I_H - B(y))^{-1/2} \exp\Big(\sum_{i=1}^{N} \frac{\partial e}{\partial y_i}(y)\mathscr{A}F^i(0,h(y)) + \mathscr{A}f(0,h(y))\Big)$$

for $y \in Y$, where

$$B(y) \equiv \sum_{i=1}^{N} \frac{\partial e}{\partial y_i}(y) D^2 F^i(0,h(y)) + D^2 f(0,h(y)), \qquad y \in Y.$$

Here we identify a continuous symmetric bilinear form $B : H \times H \to \mathbf{R}$ with a bounded symmetric linear operator $\tilde{B} : H \to H$ given by

$$(\tilde{B}h,k)_H = B(h,k), \qquad h,k \in H,$$

and $\det_2$ is a Carleman-Fredholm determinant (c.f. Dunford and Schwartz [5] pp.1106).

Appendix 3

In this section, we discuss about the implied volatilities for the case $K < x_0^1$. We define the forward value of a put option of strike rate K and maturity T by

$$P_\varepsilon(T,K) = E[(K - X_\varepsilon^1(T))^+]$$

Since we have put-call parity, the implied volatility of the put option is the same as the implied volatility of a call option with strike rate K and maturity T. Since

$$P_\varepsilon(T,K) = E[(-X^1_\varepsilon(T) - (-K))_+] = E[(-(X^1_\varepsilon(T) - x^1_0) - (-(K - x^1_0)))_+]$$

It is enough to discuss in the case $x^1_0 = 0$.

Let $x = (x^1, \ldots, x^n) \in \mathbf{R}^n$. We denote $\bar{x} = (-x^1, x^2, \ldots, x^n)$. We define $\bar{X}_\varepsilon(t) = \bar{X}_\varepsilon(t)$. Then we have

$$d\bar{X}^i_\varepsilon(t) = \sum_{k=1}^{d} \varepsilon \bar{V}^i_k(t, \tilde{X}_\varepsilon(t)) dW^k(t) + \bar{V}^i_0(t, \tilde{X}_\varepsilon(t))dt, \quad 1 \le i \le N,$$

where

$$\bar{V}^j_k(t,x) = \begin{cases} -V^1_k(t,\bar{x}) & (1 \le k \le d) \\ V^j_k(t,\bar{x}) & (1 \le k \le d,\ j \ne 1). \end{cases}$$

Since the associated Riemaniann metric $\bar{g}^{ij}(t,x) = \sum_{k=1}^d \bar{V}^i_k(t,x)\bar{V}^j_k(t,x)$ is given by

$$\bar{g}^{11}(t,x) = g^{11}(t,x), \quad \bar{g}^{1i}(t,x) = -g^{1i}(t,x)\ (i \ne 1), \quad \bar{g}^{ij}(t,x) = g^{ij}(t,x)\ (i,j \ne 1),$$

we have

$$\bar{b}_1 = b_1, \quad \bar{b}_2 = -b_2, \quad \bar{b}_3 = b_3, \quad \bar{L} = L.$$

Therefore Theorems 1 and 3 still hold for $K < x^1_0$.

References

1. Berestycki, H., Busca, J., Florent, I.: Computing the implied volatility in stochastic volatility models. Commun. Pure Appl. Math. **57**(10), 1352–1373 (2004)
2. Bismut, J.M.: Large Deviations and the Malliavin Calculus. Birkhauser, Boston (1984)
3. Deuschel, J.D., Friz, P.K., Jacquier, A., Violante, S.: Marginal density expansions for diffusions and stochastic volatility, part I: theoretical foundations. commun. Pure Appl. Math. **67**(1) (2014)
4. Deuschel, J.D., Friz, P.K., Jacquier, A., Violante, S.: Marginal density expansions for diffusions and stochastic volatility, part II: applications. Commun. Pure Appl. Math. **67**(2), 321–350 (2014)
5. Dunford, N., Schwartz, J.T.: Linear Operators, Part II. Wiley, New York (1988)
6. Hagan, P.S., Woodward, D.E.: Equivalent black volatilities. Appl. Math. Financ. **6**, 147–157 (1999)
7. Hagan, P.S., Kumar, D., Lesniewski, S., Woodward, D.E.: Managing smile risk. Wilmott Mag. **18**(11), 84–108 (2002)
8. Hagan, P.S., Lesniewski, S., Woodward, D.E.: Probability distribution in the SABR model of stochastic volatility. In: Friz, P., Gatheral, J., Gulisashvili, A., Jacquier, A., Teichmann, J. (eds.) Large Deviations and Asymptotic Methods in Finance. Springer Proceedings in Mathematics and Statistics, vol. 110 (2015)

9. Henry-Labordère, P.: A General Asymptotic Implied Volatility for Stochastic Volatility Models, preprint, http://arxiv.org/abs/cond-mat/0504317 (2005)
10. Kunitomo, N., Takahashi, A.: The asymptotic expansion approach to the valuation of interest rate contingent claims. Math. Financ. **11**, 117–151 (2001)
11. Kusuoka, S., Stroock, D.W.: Applications of Malliavin Calculus, Part I. In: Ito, K. (ed.) Proceedings of the Taniguchi International Symposium on Stochastic Analysis, Kyoto and Katata, 1982, pp. 271–360. Kinokuniya, Tokyo (1984)
12. Kusuoka, S., Stroock, D.W.: Precise asymptotics of certain Wiener functionals. J. Funct. Anal. **99**, 1–74 (1991)
13. Kusuoka, S., Osajima, Y.: A remark on the asymptotic expansion of density function of Wiener functionals. J. Funct. Anal. **255**, 2545–2562 (2007)
14. Osajima, Y.: The Asymptotic Expansion Formula of Implied Volatility for Dynamic SABR model and FX hybrid model, BNP Paribas, Date posted: 26 Feb 2007 SSRN working paper series
15. Shigekawa, I.: Stochastic analysis. Am. Math. Soc. (2004)
16. Siopacha, M., Teichmann, J.: Weak and strong Taylor methods for numerical solutions of stochastic differential equations. Quant. Financ. **11**(4), 517–528 (2011)
17. Watanabe, S.: Analysis of wiener functionals (Malliavin calculus) and its application to heat kernels. Ann. Probab. **15**, 1–39 (1987)
18. Yoshida, N.: Asymptotic expansions of maximum likelihood estimators for small diffusions via the theory of Malliavin-Watanabe. Probab. Theory Relat. Fields **92**, 275–311 (1992)

Implied Volatility of Basket Options at Extreme Strikes

Archil Gulisashvili and Peter Tankov

Abstract In the paper, we characterize the asymptotic behavior of the implied volatility of a basket call option at large and small strikes in a variety of settings with increasing generality. First, we obtain an asymptotic formula with an error bound for the left wing of the implied volatility, under the assumption that the dynamics of asset prices are described by the multidimensional Black-Scholes model. Next, we find the leading term of asymptotics of the implied volatility in the case where the asset prices follow the multidimensional Black-Scholes model with time change by an independent increasing stochastic process. Finally, we deal with a general situation in which the dependence between the assets is described by a given copula function. In this setting, we obtain a model-free tail-wing formula that links the implied volatility to a special characteristic of the copula called *the weak lower tail dependence function*.

Keywords Implied volatility asymptotics · Basket options · Index options · Large/ small strikes · Time change · Copula

---

We thank the anonymous reviewer for the careful reading of our manuscript and many constructive comments.

A. Gulisashvili
Department of Mathematics, Ohio University, Athens, OH, USA
e-mail: gulisash@ohio.edu

P. Tankov (✉)
Laboratoire de Probabilités et Modèles Aléatoires, Université Paris Diderot, Paris, France
e-mail: tankov@math.univ-paris-diderot.fr

P. Tankov
International Laboratory of Quantitative Finance,
National Research University "Higher School of Economics", Moscow, Russia

P.K. Friz et al. (eds.), *Large Deviations and Asymptotic Methods in Finance*, Springer Proceedings in Mathematics & Statistics 110,
DOI 10.1007/978-3-319-11605-1_6

1 Introduction

In option markets, prices of vanilla call and put options are commonly quoted in terms of their *implied volatility* $I(T, K)$, defined as the value of the volatility parameter which must be substituted into the Black-Scholes option pricing formula to obtain the quoted option price. Similarly, given a risk-neutral model, one can define the function $(T, K) \mapsto I(T, K)$ from the prices of vanilla options computed for that model. However, since in most stochastic asset price models the implied volatility function is not known explicitly, it becomes important to obtain efficient and accurate asymptotic approximations for it. Such approximations are useful for at least two reasons. First, they may shed light on the qualitative behavior of the implied volatility in the asset price model, and also on the effect of different model parameters on the shape of the model-generated implied volatility surface. Second, they allow to perform an approximate calibration of the model by comparing the market implied volatility with the asymptotic approximation. Such preliminary estimates can be used as intelligent guesses in the construction of a numerical calibration algorithm to accelerate its convergence.

Approximations to the implied volatility have been studied by many authors in a variety of asymptotic regimes, both in specific models and in model-independent settings. One of the early references on the subject is the book by Lewis [31] dealing with stochastic volatility models. Various model-free formulas describing the wing behavior of the implied volatility were obtained in the last decade. To our knowledge, celebrated Lee's moment formulas were the first model-independent asymptotic formulas for the implied volatility at extreme strikes (see [30]). Lee's results were later refined by Benaim and Friz [8, 9] and Gulisashvili [22–24]. In Gao and Lee [19], higher order asymptotic formulas for the implied volatility at extreme strikes were found, and in Tehranchi [41], uniform estimates for the implied volatility are obtained. Small-time behavior of implied volatility is analyzed, among other papers, in [11] (in local volatility models), [17] (for the Heston stochastic volatility model), [33] (for jump-diffusions), and in [2, 16, 34, 38] (for exponential Lévy models). Formulae for the implied volatility far from maturity are given in [18] (for the Heston model) and [40] (model-independent). Finally, sharp price and implied volatility approximations for various models have been obtained as "expansions around the Black-Scholes model" in [10, 21].

Implied volatility is also quoted in the market for options on a basket of stocks. Note that the Black-Scholes formula can be applied to price a vanilla option by considering the entire basket (index) as a log-normal random variable. In particular, options on stock indices or major exchange traded funds are often liquid and quoted in terms of their implied volatility. Several studies [5, 13, 29] explore the relationship between the implied volatilities of index options and those of the constituents, with the aim of designing dispersion trading strategies. Another example is provided by swaptions, which are also quite liquid, often quoted in terms of their implied volatility, and can be interpreted as basket options on the underlying Libor rates [4, 37]. A tractable relationship between swaption and caplet implied volatilities could be used to design a calibration procedure for the correlation structure of the Libor rates.

In the above cases, finding reliable asymptotic approximations to the implied volatility can be even more important, since calculating the exact value numerically can be computationally very expensive due to the large dimension of the basket. Approximations based on the *small-noise* asymptotics in multidimensional local volatility models have been developed in [5] and more recently refined in [7], but in other asymptotic regimes, much less is known about multi-asset options, than in the single-asset case.

Our main goal in the present paper is to characterize the asymptotic behavior of the implied volatility of a call option on a basket of stocks (with positive weights) for large and small strikes. Three different classes of multidimensional risk-neutral models with increasing generality are considered in the paper. In Sect. 3, we discuss the case of correlated log-normal assets, in other words, the assets which follow the multidimensional Black-Scholes model. Using a recent characterization of the tail behavior of sums of correlated log-normal random variables [27], we obtain a sharp asymptotic formula with error estimates for the implied volatility at small strikes. On the other hand, the asymptotics of the implied volatility at large strikes can be easily characterized using the results obtained in [3]. It turns out that for very large strikes, the implied volatility of a basket call option converges to the highest volatility among the stocks in the basket.

Section 4 deals with the case where the assets follow the multidimensional Black-Scholes model time-changed by an independent increasing stochastic process. It is assumed in this section that the marginal density of the time-change process decays at infinity like the function $s \mapsto s^{\alpha} e^{-\theta s}$ with $\alpha \in \mathbb{R}$ and $\theta > 0$. The class of such models, includes standard multidimensional extensions of various exponential Lévy models, for instance, of the variance gamma model, the normal inverse Gaussian model, or the generalized hyperbolic model. These extensions were previously discussed in, e.g., [15, 32, 36]. To our knowledge, for such a class of multidimensional models, the tail behavior of marginal distributions has not been studied before. In Sect. 4, we provide two-sided estimates for the distribution function of the asset price in the time-changed multidimensional Black-Scholes model, and use these estimates to find the leading term in the asymptotic expansion of the implied volatility.

Finally, in Sect. 5, we deal with the case where the assets in the basket are correlated, and the dependence structure is described by a given copula function (we refer the reader to the book [12] for details on this modeling approach). Here we obtain an asymptotic formula that can be considered as a generalization to the multidimensional setting of one of the tail-wing formulae established in [9]. The new tail-wing formula uses a special characteristic of the copula called *weak lower tail dependence function*. This notion was recently introduced in [39].

Remarks on the notation used in the paper

- Let f and g be functions defined on $\mathbb{R}$, and let $a \in [-\infty, \infty]$. Throughout the present paper, we write "$f \sim g$ as $x \to a$" provided that

$$\lim_{x \to a} \frac{f(x)}{g(x)} = 1.$$

We also use the notation "$f \lesssim g$ as $x \to a$" if

$$\limsup_{x \to a} \frac{f(x)}{g(x)} \leq 1,$$

and write "$f(x) \approx g(x)$ as $x \to a$" if there exist $c_1 > 0$ and $c_2 > 0$ such that

$$c_1 g(x) \leq f(x) \leq c_2 g(x)$$

for all x in some neighborhood of a.
- A positive function f defined in $[a, \infty)$ for some $a > 0$ is called regularly varying at infinity with index $\alpha \in \mathbb{R}$ if for any $\lambda > 0$,

$$\lim_{x \to 0} \frac{f(\lambda x)}{f(x)} = \lambda^{\alpha}.$$

for all $\alpha > 0$. The class of all regularly varying functions with index α is denoted by R_α. The elements of the class R_0 are called slowly varying functions. Regularly varying functions at zero can be defined similarly.
- The following set will be used in the paper:

$$\Delta_d := \{w \in \mathbb{R}^d : w_i \geq 0, i = 1, \ldots, d, \text{ and } \sum_{i=1}^{d} w_i = 1\}.$$

- Let $w \in \Delta_d$. We set

$$\mathcal{E}(w) := -\sum_{i=1}^{d} w_i \log w_i, \tag{1}$$

with the convention $x \log x = 0$ for $x = 0$.

2 Model-Free Formulae for the Implied Volatility

Let X_t be a non-negative martingale on a filtered probability space $(\Omega, \mathcal{F}, \{\mathcal{F}_t\}_{t\geq 0}, \mathbb{P})$. Consider a stochastic model where the process X models the price dynamics of an asset. Define the call and put pricing functions in the price model described above by

$$C(T, K) = \mathbb{E}[(X_T - K)^+] \quad \text{and} \quad P(T, K) = \mathbb{E}[(K - X_T)^+], \tag{2}$$

respectively. Here $T > 0$ is the maturity, while $K > 0$ is the strike price.

The implied volatility $(T, K) \mapsto I(T, K)$ is determined from the following equality:

$$C(K, T) = C_{BS}(T, K, \sigma = I(T, K)),$$

where the symbol C_{BS} stands for the Black-Scholes call pricing function. In the sequel, the maturity T will be fixed, and the implied volatility will be considered as a function of only the strike price.

We will next formulate two model-free asymptotic formulas, characterizing the left-wing behavior of the implied volatility in terms of the put pricing function. These formulas will be needed below. Suppose the initial condition for the price process is $X_0 = 1$. Suppose also that the asset price model does not have atoms at zero. The previous assumption means that $\mathbb{P}(X_T = 0) = 0$. Then the following asymptotic formula (a zero order formula for the implied volatility) holds:

$$I(K) = \frac{\sqrt{2}}{\sqrt{T}} \sqrt{\log \frac{1}{\widetilde{P}(K)} - \frac{1}{2} \log\log \frac{K}{\widetilde{P}(K)}} - \frac{\sqrt{2}}{\sqrt{T}} \sqrt{\log \frac{K}{\widetilde{P}(K)} - \frac{1}{2} \log\log \frac{K}{\widetilde{P}(K)}} + O\left(\left(\log \frac{K}{\widetilde{P}(K)}\right)^{-\frac{1}{2}}\right) \tag{3}$$

as $K \to 0$. Here $\widetilde{P}$ is a positive function satisfying the condition $P(K) \approx \widetilde{P}(K)$ as $K \to 0$. Formula (3) was established in [22] (see also Theorem 9.29 in [24]). The fact that the absence of atoms is a necessary condition for the validity of formula (3) was noticed in [14] (see also [25]).

The next asymptotic formula (a first-order formula for the implied volatility) can be easily deduced from the results formulated in [24, Sects. 9.6 and 9.9]:

$$I(K) = \frac{\sqrt{2}}{\sqrt{T}}\sqrt{\log\frac{1}{P(K)} - \frac{1}{2}\log\log\frac{K}{P(K)} + \log B(K)}$$
$$-\frac{\sqrt{2}}{\sqrt{T}}\sqrt{\log\frac{K}{P(K)} - \frac{1}{2}\log\log\frac{K}{P(K)} + \log B(K)}$$
$$+ O\left(\log\log\frac{K}{P(K)}\left(\log\frac{K}{P(K)}\right)^{-\frac{3}{2}}\right) \tag{4}$$

as $K \to 0$, where

$$B(K) = \frac{\sqrt{\log\frac{1}{P(K)}} - \sqrt{\log\frac{K}{P(K)}}}{2\sqrt{\pi}\sqrt{\log\frac{1}{P(K)}}}. \tag{5}$$

Formula (4) takes into account the results obtained in [19]. It provides more terms in the asymptotic expansion of the implied volatility at small strikes than formula (3) with $\widetilde{P} = P$. More information on model free formulas for the implied volatility can be found in [24].

3 Basket Options in Multidimensional Black-Scholes Model

Our goal in the present section is to characterize the asymptotic behavior of the implied volatility at small strikes in the case of a basket option of European style in the n-dimensional driftless Black-Scholes model. We assume that the interest rate is equal to zero. Let $S^1, \ldots, S^n$ be a basket of assets such that

$$\log \widetilde{S}_t = \log \widetilde{S}_0 - \frac{\mathrm{diag}(\mathfrak{B})t}{2} + \mathfrak{B}^{\frac{1}{2}} W_t,$$

where $\widetilde{S}_t = (S_t^1, \ldots, S_t^n)$, $\widetilde{S}_0 = (S_0^1, \ldots, S_0^n)$, W is an n-dimensional standard Brownian motion, $\mathfrak{B}$ is the covariance matrix, and $\mathrm{diag}(\mathfrak{B})$ stands for the main diagonal of $\mathfrak{B}$. We denote by $(\lambda_1, \ldots, \lambda_n) \in \Delta_n$ the weight vector associated with the assets in the basket.

Consider the price process of the following form:

$$S_t = \sum_{i=1}^{n} \lambda_i S_t^i, \quad t \geq 0. \tag{6}$$

The initial condition for the process S is given by $S_0 = \sum_{i=1}^n \lambda_i S_0^i$, and we will assume in the sequel that $S_0^i = 1$ for all $1 \le i \le n$. The previous condition implies that $S_0 = 1$. Therefore,

$$S_t = \sum_{i=1}^n \exp\{Y_t^i\}, \tag{7}$$

where

$$Y_t^i = \log \lambda_i - \frac{b_{ii}t}{2} + \sum_{j=1}^n \beta_{ij} W_t^j, \quad 1 \le i \le n. \tag{8}$$

In (8), the symbols β_{ij} stand for the elements of the matrix $\mathfrak{B}^{\frac{1}{2}}$. We also set

$$\mu_{i,t} = \log \lambda_i - \frac{b_{ii}t}{2}, \quad 1 \le i \le n. \tag{9}$$

It is clear that the following equality holds: $\exp\{Y_t^i\} = \lambda_i S_t^i$, $t > 0$, $1 \le i \le n$.

3.1 Asymptotics of Put Pricing Functions in Multidimensional Black-Scholes Model

The distribution density of the random variable S_T will be denoted by p_T. An asymptotic formula for p_T was recently established in [27]. Let us briefly recall the notation used in that paper. Let $\bar{w} \in \Delta_n$ be the unique vector such that

$$\bar{w}^{\perp} \mathfrak{B} \bar{w} = \min_{w \in \Delta_n} w^{\perp} \mathfrak{B} w. \tag{10}$$

The existence and uniqueness of $\bar{w}$ follows from the non-degeneracy of the matrix $\mathfrak{B}$. We let

$$\bar{n} := \mathrm{Card}\,\{i = 1, \dots, n : \bar{w}_i \neq 0\}, \quad \bar{I} := \{i = 1, \dots, n : \bar{w}_i \neq 0\} := \{\bar{k}(1), \dots, \bar{k}(\bar{n})\},$$

$\bar{\mu} \in \mathbb{R}^{\bar{n}}$ with $\bar{\mu}_i = \mu_{\bar{k}(i)}$, and $\bar{\mathfrak{B}} \in M_{\bar{n}}(\mathbb{R})$ with $\bar{\mathfrak{B}}_{ij} = \mathfrak{B}_{\bar{k}(i),\bar{k}(j)}$. The inverse matrix of $\bar{\mathfrak{B}}$ is denoted by $\bar{\mathfrak{B}}^{-1}$, and the elements and the row sums of $\bar{\mathfrak{B}}$ are denoted by $\bar{a}_{ij}$ and $\bar{A}_k := \sum_{j=1}^{\bar{n}} \bar{a}_{kj}$, respectively. Since the variables $Y_1, \dots, Y_n$ in (7) are exchangeable, we can assume with no loss of generality that for the covariance matrix $\mathfrak{B}$, $\bar{I} = \{1, \dots, \bar{n}\}$ with $\bar{n} \le n$. By the strict convexity of the objective function, the minimizer of $\min_{w \in \Delta_{\bar{n}}} w^{\perp} \bar{\mathfrak{B}} w$ coincides with the first $\bar{n}$ components of $\bar{w}$ and therefore belongs to the interior of the set $\mathbb{R}_+^{\bar{n}}$. The minimizer over $\Delta_{\bar{n}}$ then coincides with the minimizer over the set $\{w \in \mathbb{R}^{\bar{n}} : \sum_{i=1}^{\bar{n}} w_i = 1\}$, which means that

$$(\bar{w}_i)_{i=1,\dots,\bar{n}} = \frac{\bar{\mathfrak{B}}^{-1}\mathbf{1}}{\mathbf{1}^\perp \bar{\mathfrak{B}}^{-1}\mathbf{1}},$$

or, equivalently,

$$\bar{w}_k = \frac{\bar{A}_k}{\sum_{i=1}^{\bar{n}} \bar{A}_i}, \quad k = 1, \dots, \bar{n}. \tag{11}$$

Since $\sum_{i=1}^{\bar{n}} \bar{A}_i > 0$ (the matrix $\bar{\mathfrak{B}}^{-1}$ is positive definite), this implies that $\bar{A}_k > 0$ for $k = 1, \dots, \bar{n}$.

We will next formulate a condition under which the asymptotic formula for the density p_T holds.

Assumption ($\mathcal{A}$) For every $i \in \{1, \dots, n\} \setminus \bar{I}$, $(e^i - \bar{w})^\perp \mathfrak{B} \bar{w} \neq 0$, where $e^i \in \mathbb{R}^n$ satisfies $e^i_j = 1$ if $i = j$ and $e^i_j = 0$ otherwise.

Assumption ($\mathcal{A}$) is a natural nondegeneracy condition for our problem. The following straightforward equality gives a relation between the optimization problem in (10) and a similar problem without the normalization constraint:

$$\inf_{w \in \Delta_n, r \geq 0} \frac{r^2}{2} w^\perp \mathfrak{B} w - r = \inf_{v \in \mathbb{R}^n : v_i \geq 0, i=1,\dots,n} \frac{1}{2} v^\perp \mathfrak{B} v - \mathbf{1}^\perp v. \tag{12}$$

A minimizer $\bar{v}$ of the right-hand side can therefore be constructed from the minimizer $\bar{w}$ of (10) as follows:

$$\bar{v} = \frac{\bar{w}}{\bar{w}^\perp \mathfrak{B} \bar{w}}.$$

Now, introducing the vector $\lambda \in \mathbb{R}^d$ of Lagrange multipliers for the positivity constraints on the right-hand side of (12), we get the Lagrangian $\frac{1}{2} v^\perp \mathfrak{B} v - \mathbf{1}^\perp v - \lambda^\perp v$. At the extremum therefore, $\mathfrak{B}\bar{v} = \mathbf{1} + \lambda$, or in other words,

$$\frac{\mathfrak{B}\bar{w}}{\bar{w}^\perp \mathfrak{B} w} = \mathbf{1} + \lambda.$$

Therefore, Assumption ($\mathcal{A}$) simply states that for the constraints, which are saturated, the Lagrange multipliers are not equal to zero (since the constraints are inequalities, this is equivalent to the strict positivity for the multipliers). This is generally true, except when the solution of the unconstrained problem belongs to the boundary of the domain defined by the constraints. Assumption $\mathcal{A}$ is not restrictive and is satisfied in most applications. Note that if the row sums of the covariance matrix $\mathfrak{B}$ satisfy $A_i > 0$, $1 \leq i \leq n$, then Assumption $\mathcal{A}$ holds.

It was established in [27] that under Assumption ($\mathcal{A}$), the following asymptotic formula is valid for the density p_T of the price S_T of the basket:

$$p_T(x) = C_T \left(\log \frac{1}{x}\right)^{\frac{1-\bar{n}}{2}} x^{-1+\frac{1}{T}\sum_{k=1}^{\bar{n}} \bar{A}_k \left(\log \frac{\bar{A}_1+\cdots+\bar{A}_{\bar{n}}}{\bar{A}_k} + \bar{\mu}_{k,T}\right)}$$

$$\exp\left\{-\frac{1}{2T}(\bar{A}_1 + \cdots + \bar{A}_{\bar{n}}) \log^2 \frac{1}{x}\right\}\left(1 + O\left(\left(\log \frac{1}{x}\right)^{-1}\right)\right), \quad (13)$$

as $x \to 0$, where the constant C is given by

$$C_T = \frac{1}{\sqrt{2\pi T}\sqrt{|\mathfrak{B}|}} \frac{\sqrt{\bar{A}_1 + \cdots + \bar{A}_{\bar{n}}}}{\sqrt{\bar{A}_1 \cdots \bar{A}_{\bar{n}}}}$$

$$\exp\left\{-\frac{1}{2T} \sum_{i,j=1}^{\bar{n}} \bar{a}_{ij} \left(\log \frac{\bar{A}_1 + \cdots + \bar{A}_{\bar{n}}}{\bar{A}_i} + \bar{\mu}_{i,T}\right)\left(\log \frac{\bar{A}_1 + \cdots + \bar{A}_{\bar{n}}}{\bar{A}_j} + \bar{\mu}_{j,T}\right)\right\}. \quad (14)$$

Using formula (13), we can characterize the asymptotic behavior of the put pricing function P at small strikes. This can be done as follows. Consider the fractional integral of order two defined by

$$F_2 M(\sigma) = \int_\sigma^\infty (\tau - \sigma) M(\tau) d\tau, \quad (15)$$

where M is a positive function on $(0, \infty)$. Since

$$P(K) = \int_0^K (K - x) p_T(x) dx,$$

it is not hard to see that

$$P(K) = S^{-1} F_2 M(S), \quad \text{where} \quad S = K^{-1} \quad \text{and} \quad M(y) = y^{-3} p_T\left(y^{-1}\right). \quad (16)$$

Using (13), we get

$$M(y) = M_1(y)\left(1 + O\left((\log y)^{-1}\right)\right) \quad (17)$$

as $y \to \infty$, where

$$M_1(y) = C_T\,(\log y)^{\frac{1-\bar{n}}{2}}\; y^{-2-T^{-1}\sum_{k=1}^{\bar{n}} \bar{A}_k\left(\log\frac{\bar{A}_1+\cdots+\bar{A}_n}{\bar{A}_k}+\mu_{k,T}\right)}$$
$$\exp\left\{-\frac{1}{2T}(\bar{A}_1+\cdots+\bar{A}_{\bar{n}})\log^2 y\right\}, \quad y > y_0. \tag{18}$$

In (18), the constant C_T is given by (14). Set

$$M_2(y) = (\log y)^{-1}\, M_1(y). \tag{19}$$

It follows from (17) that there exist $c > 0$ and $y_1 > 0$ such that

$$|M(y) - M_1(y)| \le c M_2(y), \quad y > y_1. \tag{20}$$

In [26], a general asymptotic formula was obtained for fractional integrals (see also Theorem 5.3 in [24]). We will next formulate this general result. Suppose

$$\widetilde{M}(y) = a(y)e^{-b(y)} \quad \text{for all} \quad y \ge c$$

where $c > 0$ is some number. Suppose also that the following conditions hold:

1. $y|a'(y)| \le \gamma a(y)$ for some $\gamma > 0$ and all $y > c$.
2. $b(y) = B(\log y)$, where B is a positive increasing function on (c, ∞) such that $B''(y) \approx 1$ as $y \to \infty$.

Then as $\sigma \to \infty$,

$$F_2\widetilde{M}(\sigma) = \frac{\widetilde{M}(\sigma)}{b'(\sigma)^2}(1 + O((\log\sigma)^{-1})). \tag{21}$$

The functions $\widetilde{M} = M_1$ and $\widetilde{M} = M_2$ defined in (18) and (19) satisfy the conditions in the theorem formulated above. Applying this theorem, we obtain

$$F_2 M_i(\sigma) = \frac{M_i(\sigma)}{b'(\sigma)^2}(1 + O((\log\sigma)^{-1})) \tag{22}$$

as $\sigma \to \infty$, where $i = 1, 2$ and

$$b(u) = \frac{1}{2T}(\bar{A}_1 + \cdots + \bar{A}_{\bar{n}})\log^2 u, \tag{23}$$

It follows from (19), (20), and (22) that

$$F_2 M(\sigma) = F_2 M_1(\sigma) + O(F_2 M_2(\sigma)) = \frac{M_1(\sigma)}{b'(\sigma)^2}(1 + O((\log\sigma)^{-1})) \tag{24}$$

as $\sigma \to \infty$. Now, using (16), (23), and (24), we establish the following assertion.

Theorem 1 *Let P be the price of the put option defined in (2), and suppose Assumption $(\mathcal{A})$ holds for the covariance matrix $\mathfrak{B}$ (see [27]). Then, as $K \to 0$,*

$$P(K) = \delta_0 \left[\log\frac{1}{K}\right]^{\delta_1} \left(\frac{1}{K}\right)^{\delta_2} \exp\left\{-\delta_3 \log^2\frac{1}{K}\right\} \left(1 + O\left(\left(\log\frac{1}{K}\right)^{-1}\right)\right), \tag{25}$$

where

$$\delta_0 = \frac{C_T T^2}{(\bar{A}_1 + \cdots + \bar{A}_{\bar{n}})^2}, \quad \delta_1 = -\frac{3+\bar{n}}{2},$$

$$\delta_2 = -1 - \frac{1}{T}\sum_{k=1}^{\bar{n}} \bar{A}_k \left(\log\frac{\bar{A}_1 + \cdots + \bar{A}_{\bar{n}}}{\bar{A}_k} + \mu_{k,T}\right), \quad \delta_3 = \frac{1}{2T}(\bar{A}_1 + \cdots + \bar{A}_{\bar{n}}),$$

and C_T is given by (14).

Formula (25) will be used in the next subsection to characterize the left-wing behavior of the implied volatility associated with a basket option in the multidimensional Black-Scholes model.

3.2 Left-Wing Asymptotic Behavior of the Implied Volatility Associated with Basket Options

The next statement characterizes the asymptotic behavior of the implied volatility for small strikes.

Theorem 2 *Suppose Assumption $(\mathcal{A})$ holds for the covariance matrix $\mathfrak{B}$. Then, as $K \to 0$,*

$$I(K) = \frac{1}{\sqrt{\bar{A}_1 + \cdots + \bar{A}_{\bar{n}}}} - \frac{2\sum_{k=1}^{\bar{n}} \bar{A}_k \left(\log\frac{\bar{A}_1+\cdots+\bar{A}_{\bar{n}}}{\bar{A}_k} + \mu_{k,T}\right) + T}{2(\bar{A}_1 + \cdots + \bar{A}_{\bar{n}})^{\frac{3}{2}}} \left(\log\frac{1}{K}\right)^{-1}$$
$$- \frac{T(\bar{n}-1)}{2(\bar{A}_1 + \cdots + \bar{A}_{\bar{n}})^{\frac{3}{2}}} \log\log\frac{1}{K}\left(\log\frac{1}{K}\right)^{-2} + O\left(\left(\log\frac{1}{K}\right)^{-2}\right). \tag{26}$$

Remark 1 The leading term in the implied volatility expression above can also be written as

$$\lim_{K\downarrow 0} I(K) = \frac{1}{\sqrt{\bar{A}_1 + \cdots + \bar{A}_{\bar{n}}}} = \sqrt{\min_{w\in\Delta_n} w^{\perp}\mathfrak{B}w}. \tag{27}$$

Formula (27) for the leading term of the implied volatility holds even if assumption $(\mathcal{A})$ is not satisfied—in this case, this formula can be obtained as a corollary of Theorem 9 of this paper.

Proof It follows from (25) that as $K \to 0$,

$$\log\frac{1}{P(K)} = \log\frac{1}{\delta_0} - \delta_1\log\log\frac{1}{K} - \delta_2\log\frac{1}{K} + \delta_3\log^2\frac{1}{K} + O\left(\left(\log\frac{1}{K}\right)^{-1}\right) \tag{28}$$

and

$$\log\frac{K}{P(K)} = \log\frac{1}{\delta_0} - \delta_1\log\log\frac{1}{K} - (\delta_2+1)\log\frac{1}{K} + \delta_3\log^2\frac{1}{K} + O\left(\left(\log\frac{1}{K}\right)^{-1}\right) \tag{29}$$

where δ_0, δ_1, δ_2, and δ_3 are such as in Theorem 1. Moreover, the error term in (4) can be represented as follows:

$$O\left(\log\log\frac{1}{K}\left(\log\frac{1}{K}\right)^{-3}\right). \tag{30}$$

We will next characterize the asymptotic behavior of $\log B(K)$ as $K \to 0$. Denote the functions on the right-hand side of (28) and (29) by $V_1(K)$ and $V_2(K)$, respectively. Then, using (5), (28), and (29), we obtain

$$\log B(K) = \log\frac{1}{2\sqrt{\pi}} + \log\left[1 - \sqrt{1 - \frac{V_1(K) - V_2(K)}{V_1(K)}}\right].$$

It is easy to see that $\log(1-\sqrt{1-h}) = \log\frac{h}{2} + O(h)$ as $h \to 0$. Put $h = \frac{V_1(K)-V_2(K)}{V_1(K)}$. Then we have

$$\log B(K) = \log\frac{1}{2\sqrt{\pi}} + \log\frac{V_1(K) - V_2(K)}{2V_1(K)} + O\left(\left(\log\frac{1}{K}\right)^{-1}\right),$$

and hence

$$\log B(K) = \log \frac{1}{4\sqrt{\pi}\delta_3} - \log\log\frac{1}{K} + O\left(\left(\log\frac{1}{K}\right)^{-1}\right) \tag{31}$$

as $K \to 0$.

Our next goal is to simplify formula (4) by taking into account (28), (29), and (31), and replacing the error term by the expression in (30). We can drop the terms $O\left(\left(\log\frac{1}{K}\right)^{-1}\right)$ in (28), (29), and (31), using the mean value theorem. This will introduce an error term $O\left(\left(\log\frac{1}{K}\right)^{-2}\right)$ in the formula that follows from formula (4). Thus

$$\begin{aligned} I(K) &= \frac{\sqrt{2}}{\sqrt{T}}\sqrt{\widetilde{V}_1(K) - \frac{1}{2}\log\widetilde{V}_2(K) + \log\frac{1}{4\sqrt{\pi}\delta_3} - \log\log\frac{1}{K}} \\ &\quad - \frac{\sqrt{2}}{\sqrt{T}}\sqrt{\widetilde{V}_2(K) - \frac{1}{2}\log\widetilde{V}_2(K) + \log\frac{1}{4\sqrt{\pi}\delta_3} - \log\log\frac{1}{K}} \\ &\quad + O\left(\left(\log\frac{1}{K}\right)^{-2}\right) \end{aligned} \tag{32}$$

as $K \to 0$, where $\widetilde{V}_1(K)$ and $\widetilde{V}_2(K)$ denote the functions on the right-hand side of (28) and (29), respectively, without the terms $O\left(\left(\log\frac{1}{K}\right)^{-1}\right)$. Next, using the mean value theorem, we see that it is possible to replace $\widetilde{V}_2(K)$ in the expression $\log\widetilde{V}_2(K)$ in formula (32) by $\delta_3\log^2 K$. Now, taking into account the definitions of $\widetilde{V}_1(K)$ and $\widetilde{V}_2(K)$, we obtain

$$\begin{aligned} I(K) &= \frac{\sqrt{2}}{\sqrt{T}}\sqrt{-\log\left[4\sqrt{\pi}\delta_0\delta_3^{\frac{3}{2}}\right] - (\delta_1+2)\log\log\frac{1}{K} - \delta_2\log\frac{1}{K} + \delta_3\log^2\frac{1}{K}} \\ &\quad - \frac{\sqrt{2}}{\sqrt{T}}\sqrt{-\log\left[4\sqrt{\pi}\delta_0\delta_3^{\frac{3}{2}}\right] - (\delta_1+2)\log\log\frac{1}{K} - (\delta_2+1)\log\frac{1}{K} + \delta_3\log^2\frac{1}{K}} \\ &\quad + O\left(\left(\log\frac{1}{K}\right)^{-2}\right) \end{aligned} \tag{33}$$

as $K \to 0$. Put

$$h_1(K) = \frac{-\log\left[4\sqrt{\pi}\delta_0\delta_3^{\frac{3}{2}}\right] - (\delta_1+2)\log\log\frac{1}{K} - \delta_2\log\frac{1}{K}}{\delta_3\log^2\frac{1}{K}}$$

and

$$h_2(K) = \frac{-\log\left[4\sqrt{\pi}\delta_0\delta_3^{\frac{3}{2}}\right] - (\delta_1+2)\log\log\frac{1}{K} - (\delta_2+1)\log\frac{1}{K}}{\delta_3\log^2\frac{1}{K}}.$$

It follows from (33) that

$$I(K) = \frac{\sqrt{2}\sqrt{\delta_3}}{\sqrt{T}}\log\frac{1}{K}\left[\sqrt{1+h_1(K)} - \sqrt{1+h_2(K)}\right] + O\left(\left(\log\frac{1}{K}\right)^{-2}\right) \tag{34}$$

as $K \to 0$. Next, using the formula $\sqrt{1+h} = 1 + \frac{1}{2}h - \frac{1}{8}h^2 + O(h^3)$ as $h \to 0$ in (34), we get

$$I(K) = \frac{1}{\sqrt{2T\delta_3}} + \frac{1+2\delta_2}{4\delta_3\sqrt{2T\delta_3}}\left(\log\frac{1}{K}\right)^{-1} + \frac{\delta_1+2}{2\delta_3\sqrt{2T\delta_3}}\log\log\frac{1}{K}\left(\log\frac{1}{K}\right)^{-2} + O\left(\left(\log\frac{1}{K}\right)^{-2}\right) \tag{35}$$

as $K \to 0$. Finally, plugging the values of δ_1, δ_2, and δ_3 given in Theorem 1 into formula (35), we obtain formula (26).

This completes the proof of Theorem 2. □

Remark 2 (Implied volatility in the multidimensional Black-Scholes model for large strikes.) From Theorem 1 in [3], it follows that

$$\mathbb{P}[S_t \geq K] \sim \frac{m_n\sigma\sqrt{t}}{\sqrt{2\pi}\log K}\exp\left\{-\frac{(\log K-\mu)^2}{2\sigma^2 t}\right\}, \quad K \to \infty,$$

where $\sigma^2 = \max_{k=1,\dots,n}\mathfrak{B}_{kk}$, $\mu = \max_{\mu_{k,t}:\mathfrak{B}_{kk}=\sigma^2}$ and $m_n = \#\{k : \mathfrak{B}_{kk} = \sigma^2, \mu_{k,t} = \mu\}$. From this result, we easily deduce that

$$\mathbb{E}[(S_t-K)^+] \approx \frac{K}{\log^2 K}\exp\left\{-\frac{(\log K-\mu)^2}{2\sigma^2 t}\right\}, \quad K \to \infty.$$

Applying Corollary 2.4 in [22] (which is nothing but the right-tail version of formula (3)), we conclude that

$$I(K) = \sigma + O\left(\frac{\psi(K)}{\log K}\right)$$

as $K \to +\infty$, where ψ is any function satisfying $\psi(K) \to +\infty$ as $K \to +\infty$.

The function ψ can be removed from the error estimate in the previous formula, using Lemma 3.1, part 1, in [24]. The resulting formula is as follows:

$$I(K) = \sigma + O\left(\frac{1}{\log K}\right)$$

as $K \to \infty$.

Numerical illustration In this part of the paper we compare the theoretical left-tail limit of the implied volatility given by Formula (27) with the numerical values computed by Monte Carlo in the multidimensional Black-Scholes model. Figure 1 plots the implied volatility of two basket call options as function of the strike price with 2-standard deviation confidence intervals (for 5 million paths), as well as the horizontal line corresponding to the theoretical limit.

In the left graph, the basket contains two independent identical assets following the Black-Scholes model with volatility $\sigma = 0.3$. In the right graph, the basket contains ten identical assets following the multidimensional Black-Scholes model, where the volatility of every component is $\sigma = 0.3$ and the correlation between the log-prices of different components is $\rho = 0.5$. The maturity of the options is $T = 0.2$ years in both graphs.

We observe that in both cases the volatility is almost constant as a function of strike (note the scale on the vertical axis), and for all strikes it is very close to the theoretical limit of Formula (27). We only show the zero-order term of the expansion in Theorem 2 because the higher-order terms do not lead to an improvement of the approximation for the strikes shown in the graph. Indeed, the higher-order terms in this expansion have a singularity at $K = 1$ and have a "reasonable" value only when $\log \frac{1}{K}$ is very small.

For comparison, we also plot the implied volatility in the right wing in Fig. 2. According to Remark 2, in the right wing, the implied volatility must converge to $\sigma = 0.3$. However, from the graph in Fig. 2 we see that this convergence is very slow: for all strike values for which option prices may be computed without sophisticated

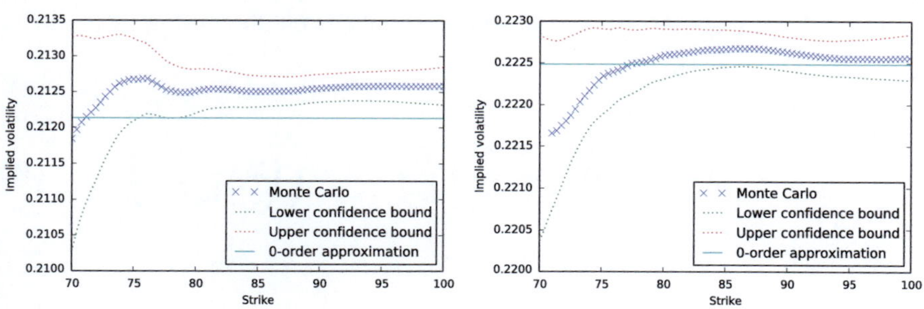

Fig. 1 Implied volatility of a basket call option in the multidimensional Black-Scholes model together with the theoretical 0-order approximation for the left wing. *Left* option on a basket of 2 identical assets. *Right* option on a basket of 10 identical assets

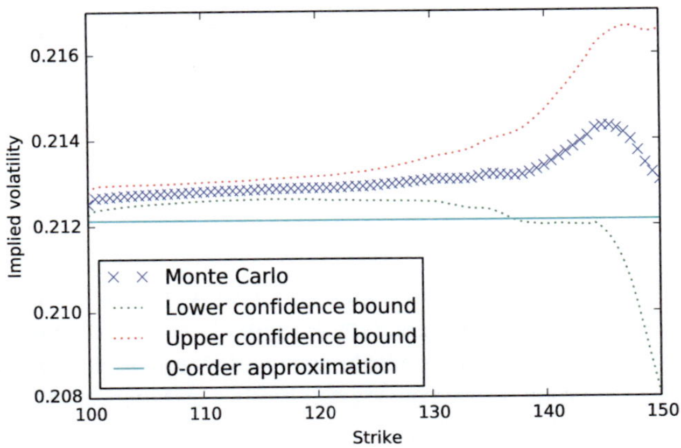

Fig. 2 Right wing of the implied volatility of a basket call option in the two-dimensional Black-Scholes model together with the theoretical 0-order left-wing approximation

variance reduction, the implied volatility, although it increases slightly with strike, remains close to its left-wing limit.

3.3 The Case Where $n = 2$

The detailed discussion of the behavior of the distribution of the sum of two log-normal variables can be found in [20, 27]. The covariance matrix in this case is as follows: $\mathfrak{B} = [b_{ij}]$, where $b_{11} = \sigma_1^2$, $b_{12} = b_{21} = \rho\sigma_1\sigma_2$, $b_{22} = \sigma_2^2$ with $\sigma_1 > 0$, $\sigma_2 > 0$, and the correlation coefficient satisfies $-1 < \rho < 1$. We will also assume $\sigma_1 \geq \sigma_2$. Note that the case where $\rho < \frac{\sigma_2}{\sigma_1}$ is a regular case, and Assumption $(\mathcal{A})$ holds. In the case where $\rho > \frac{\sigma_2}{\sigma_1}$, we have to rearrange the rows and the columns of $\mathfrak{B}$ (see the example in Sect. 2.1 of [27]). Then $\bar{\mathfrak{B}} = (\sigma_2^2)$, and Assumption $(\mathcal{A})$ holds. The case where $\rho = \frac{\sigma_2}{\sigma_1}$ is exceptional. Here Assumption $(\mathcal{A})$ does not hold.

The following asymptotic formulas for the implied volatility follow from (26):

- Suppose $\rho > \frac{\sigma_2}{\sigma_1}$. Then

$$I(K) = \sigma_2 - \sigma_2 \log \lambda_2 \left(\log \frac{1}{K} \right)^{-1} + O\left(\left(\log \frac{1}{K} \right)^{-2} \right) \tag{36}$$

as $K \to 0$.

- Suppose $\rho < \frac{\sigma_2}{\sigma_1}$. Then

$$\begin{aligned} I(K) =& \sigma_\infty - \sigma_\infty \left(\frac{T}{2}\sigma_\infty^2 + \left[\log \lambda_1 - \frac{\sigma_1^2 T}{2} - \log \bar{v} \right] \bar{v} \right. \\ & \left. + \left[\log \lambda_2 - \frac{\sigma_2^2 T}{2} - \log(1-\bar{v}) \right] (1-\bar{v}) \right) \left(\log \frac{1}{K} \right)^{-1} \\ & - \frac{T}{2}\sigma_\infty^3 \frac{\log\log \frac{1}{K}}{\log^2 \frac{1}{K}} + O\left(\left(\log \frac{1}{K} \right)^{-2} \right) \end{aligned} \tag{37}$$

as $K \to 0$, where

$$\sigma_\infty = \frac{\sigma_1 \sigma_2 \sqrt{1-\rho^2}}{\sqrt{\sigma_1^2 + \sigma_2^2 - 2\rho\sigma_1\sigma_2}} \quad \text{and} \quad \bar{v} = \frac{\sigma_2(\sigma_2 - \rho\sigma_1)}{\sigma_1^2 + \sigma_2^2 - 2\rho\sigma_1\sigma_2}.$$

Therefore, the behavior of the implied volatility experiences a qualitative change (phase transition) at $\rho^* = \frac{\sigma_2}{\sigma_1}$. Indeed, for $\rho < \rho^*$, the expression in formula (37), approximating the left wing of the implied volatility, depends on the correlation coefficient, while for $\rho > \rho^*$ the left wing is approximated by a correlation-independent expression (see (36)).

We will next discuss the asymptotic behavior of the implied volatility in the exceptional case where $n = 2$ and $\rho = \rho^*$. The following formula holds for the distribution density p_T in the exceptional case (see [20]):

$$\begin{aligned} p_T(x) \approx & x^{\frac{\mu_{2,T}}{T\sigma_2^2} - 1} \left(\log \frac{1}{x} \right)^{-\frac{1}{T(\sigma_1^2 - \sigma_2^2)}} \left(\log\log \frac{1}{x} \right)^{-\frac{1}{2}} \\ & \exp\left\{ -\frac{1}{2T\left(\sigma_1^2 - \sigma_2^2\right)} \left[\log\left(\frac{1}{\rho^2} - 1 \right) + \log\log \frac{1}{x} - \log\left(\log\left(\frac{1}{\rho^2} - 1 \right) \right. \right. \right. \\ & \left. \left. \left. + \log\log \frac{1}{x} \right) + \mu_{1,T} - \mu_{2,T} \right]^2 \right\} \\ & \exp\left\{ -\frac{\log^2 \frac{1}{x}}{2T\sigma_2^2} \right\} \end{aligned} \tag{38}$$

as $x \to 0$. Recall that we assume that $\mu = 0$. Recall also that $\mu_{1,T}$ and $\mu_{2,T}$ are defined in (9).

Remark 3 Formula (38) can be derived from formula (B20) established at the end of the proof of part (ii) of Theorem 2.3 in [20]. Note that in the present paper we assume $\sigma_1 \geq \sigma_2$, while in [20], $\sigma_1 \leq \sigma_2$.

Set

$$V_{1,T} = \log\left(\frac{1}{\rho^2} - 1\right) + \mu_{1,T} - \mu_{2,T} \quad \text{and} \quad V_2 = \log\left(\frac{1}{\rho^2} - 1\right). \tag{39}$$

It is not hard to see using the mean value theorem that

$$\log^2\left(V_2 + \log\log\frac{1}{x}\right) - \left(\log\log\log\frac{1}{x}\right)^2 = o(1)$$

as $x \to 0$. Hence

$$\exp\left\{-\frac{1}{2T(\sigma_1^2 - \sigma_2^2)}\log^2\left(V_2 + \log\log\frac{1}{x}\right)\right\} \sim \exp\left\{-\frac{1}{2T(\sigma_1^2 - \sigma_2^2)}\left(\log\log\log\frac{1}{x}\right)^2\right\}$$

as $x \to 0$. In addition,

$$\begin{aligned}
&\exp\left\{\frac{1}{T(\sigma_1^2 - \sigma_2^2)}\left(\log\log\frac{1}{x}\right)\left(\log\left(V_2 + \log\log\frac{1}{x}\right)\right)\right\} \\
&\approx \left(\log\frac{1}{x}\right)^{\frac{1}{T(\sigma_1^2-\sigma_2^2)}} \exp\left\{\frac{1}{T(\sigma_1^2 - \sigma_2^2)}\left(\log\log\frac{1}{x}\right)\left(\log\log\log\frac{1}{x}\right)\right\}
\end{aligned}$$

as $x \to 0$. Therefore, (38) implies the following estimate for the density p_T:

$$\begin{aligned}
p_T(x) \approx &\left(\frac{1}{x}\right)^{1-\frac{\mu_{2,T}}{T\sigma_2^2}} \left(\log\frac{1}{x}\right)^{-\frac{V_{1,T}}{T(\sigma_1^2-\sigma_2^2)}} \left(\log\log\frac{1}{x}\right)^{\frac{V_{1,T}}{T(\sigma_1^2-\sigma_2^2)}-\frac{1}{2}} \\
&\exp\left\{-\frac{\log^2\frac{1}{x}}{2T\sigma_2^2}\right\} \exp\left\{-\frac{1}{2T(\sigma_1^2 - \sigma_2^2)}\left(\log\log\frac{1}{x}\right)^2\right\} \\
&\exp\left\{-\frac{1}{2T(\sigma_1^2 - \sigma_2^2)}\left(\log\log\log\frac{1}{x}\right)^2\right\} \\
&\exp\left\{\frac{1}{T(\sigma_1^2 - \sigma_2^2)}\left(\log\log\frac{1}{x}\right)\left(\log\log\log\frac{1}{x}\right)\right\}
\end{aligned} \tag{40}$$

as $x \to 0$.

Our next goal is to obtain a two-sided estimate for the put pricing function P, by taking into account formula (40). We will use the ideas employed in the proof of Theorem 1. Let us set

$$B(u) = \frac{u^2}{2T\sigma_2^2} + \frac{\log^2 u}{2T(\sigma_1^2 - \sigma_2^2)} + \frac{(\log\log u)^2}{2T(\sigma_1^2 - \sigma_2^2)} - \frac{1}{T(\sigma_1^2 - \sigma_2^2)}(\log u)(\log\log u)$$

and

$$a(y) = y^{-2-\frac{\mu_{2,T}}{T\sigma_2^2}} (\log y)^{-\frac{V_{1,T}}{T(\sigma_1^2-\sigma_2^2)}} (\log\log y)^{\frac{V_{1,T}}{T(\sigma_1^2-\sigma_2^2)}-\frac{1}{2}}.$$

It is not hard to see that the restrictions, under which formula (21) is valid, are satisfied. In addition, for the function $b(x) = B(\log x)$, we have $b'(x) \approx \frac{\log x}{x}$ as $x \to \infty$. Now, reasoning as in the proof of Theorem 1, we obtain the following formula: $P(K) \approx \widetilde{P}(K)$ as $K \to 0$, where

$$\begin{aligned}\widetilde{P}(K) = &\left(\frac{1}{K}\right)^{-1-\frac{\mu_{2,T}}{T\sigma_2^2}} \left(\log\frac{1}{K}\right)^{-\frac{V_{1,T}}{T(\sigma_1^2-\sigma_2^2)}-2} \left(\log\log\frac{1}{K}\right)^{\frac{V_{1,T}}{T(\sigma_1^2-\sigma_2^2)}-\frac{1}{2}} \\ &\exp\left\{-\frac{\log^2\frac{1}{K}}{2T\sigma_2^2}\right\} \exp\left\{-\frac{1}{2T(\sigma_1^2-\sigma_2^2)}\left(\log\log\frac{1}{K}\right)^2\right\} \\ &\exp\left\{-\frac{1}{2T(\sigma_1^2-\sigma_2^2)}\left(\log\log\log\frac{1}{K}\right)^2\right\} \\ &\exp\left\{\frac{1}{T(\sigma_1^2-\sigma_2^2)}\left(\log\log\frac{1}{K}\right)\left(\log\log\log\frac{1}{K}\right)\right\}\end{aligned} \tag{41}$$

as $K \to 0$. Next, using (3) with $\widetilde{P}$ given by (41), and making numerous simplifications, we obtain the following asymptotic formula for the implied volatility in the exceptional case:

$$I(K) = \sigma_2 + O\left(\left(\log\frac{1}{K}\right)^{-1}\right) \tag{42}$$

as $K \to 0$. Comparing formula (42) with formulas (36) and (37), we see that the behavior of the implied volatility at the critical point $\rho = \frac{\sigma_2}{\sigma_1}$, where the qualitative change happens, is similar to that in the case where $\rho > \frac{\sigma_2}{\sigma_1}$.

4 Time-Changed Multidimensional Black-Scholes Model

Recall that in Sect. 3, we introduced the price process S for a basket of assets (see formula (6)). The present section deals with time changes in such processes. Suppose $\mathcal{T}_t$, $t \geq 0$, is a non-negative non-decreasing stochastic process on $(\Omega, \mathcal{F}, \{\mathcal{F}\}_{t\geq 0}, \mathbb{P})$ (a time change). Then, the time-changed process S has the following form: $t \mapsto S_{\mathcal{T}_t}$. We only consider time changes which are independent of the price process S. In the next subsections, two-sided estimates for marginal distribution functions of time-changed price processes such as above will be established. Moreover, the leading term in the asymptotic expansion of the implied volatility associated with a time-changed price process $t \mapsto S_{\mathcal{T}_t}$ in the n-dimensional Black-Scholes model will be found.

4.1 Bounds on Distribution Functions of Sums of Log-Normal Mixtures

The next assertion provides an upper bound for the distribution function of a random variable imitating the random variable $S_{\mathcal{T}_t}$ for fixed $t > 0$. The additional drift vector $\tilde{\mu}$ will be needed later to ensure the martingale property.

Theorem 3 (Upper bound) *Let Y be a centered Gaussian vector with covariance matrix $\mathfrak{B} = [b_{ij}]_{1\leq i,j\leq n}$, and let $\mu \in \mathbb{R}^n$ and $\tilde{\mu} \in \mathbb{R}^n$. Suppose Z is a random variable with values in $(0, \infty)$, which has a density $\rho(x)$ satisfying $\rho(s) \leq cs^{\alpha}e^{-\theta s}$ for $s \geq 1$, where $\theta > 0$, $c > 0$ and $\alpha \in \mathbb{R}$ are constants. Then, there exists $C > 0$ such that as $k \to +\infty$,*

$$\mathbb{P}[\sum_{i=1}^{n} e^{Y_i\sqrt{Z}+\mu_i Z+\tilde{\mu}_i} \leq e^{-k}] \lesssim Ck^{\alpha}e^{-c^*k},$$

where

$$c^* = \min_{t\geq 0} \max_{w\in\Delta_n} \left\{\theta t + \frac{(1+t\mu^{\perp}w)^2}{2w^{\perp}\mathfrak{B}wt}\right\}. \tag{43}$$

Proof In this proof, C denotes a constant which may change from line to line. For $k > 0$, set

$$F_t(k) = \mathbb{P}\left[\sum_{i=1}^{n} e^{Y_i\sqrt{kt}+\mu_i kt+\tilde{\mu}_i} \leq e^{-k}\right].$$

Fix $w \in \Delta_n$, and let t be such that $1 + t\mu^{\perp}w > 0$. Then, by Jensen's inequality,

$$
\begin{aligned}
&\mathbb{P}\left[\sum_{i=1}^{n} e^{Y_i\sqrt{kt}+\mu_i kt+\tilde{\mu}_i} \le e^{-k}\right] \\
&\le \mathbb{P}\left[\sqrt{kt}\sum_{i=1}^{n} w_i Y_i + kt\mu^{\perp}w + \tilde{\mu}^{\perp}w + \mathcal{E}(w) \le -k\right] \\
&= N\left(-\frac{k + tk\mu^{\perp}w + \tilde{\mu}^{\perp}w + \mathcal{E}(w)}{\sqrt{w^{\perp}\mathfrak{B}wkt}}\right) \\
&\le \frac{C\sqrt{t}}{(1+t\mu^{\perp}w)\sqrt{k}} \exp\left\{-\frac{(k + tk\mu^{\perp}w + \tilde{\mu}^{\perp}w + \mathcal{E}(w))^2}{2w^{\perp}\mathfrak{B}wkt}\right\} \\
&= \frac{C\sqrt{t}}{(1+t\mu^{\perp}w)\sqrt{k}} \exp\left\{-k\frac{(1+t\mu^{\perp}w)^2}{2w^{\perp}\mathfrak{B}wt}\right\} \exp\left\{-\frac{(\tilde{\mu}^{\perp}w + \mathcal{E}(w))^2}{2w^{\perp}\mathfrak{B}wkt}\right\} \\
&\qquad \times \exp\left\{-\frac{\mathcal{E}(w) + \tilde{\mu}^{\perp}w}{w^{\perp}\mathfrak{B}wt}\right\} \exp\left\{-\frac{\mu^{\perp}w(\mathcal{E}(w) + \tilde{\mu}^{\perp}w)}{w^{\perp}\mathfrak{B}w}\right\} \\
&\le \frac{C\sqrt{t}}{(1+t\mu^{\perp}w)\sqrt{k}} \exp\left\{-k\frac{(1+t\mu^{\perp}w)^2}{2w^{\perp}\mathfrak{B}wt}\right\} \exp\left\{-\frac{\tilde{\mu}^{\perp}w}{w^{\perp}\mathfrak{B}wt}\right\},
\end{aligned}
$$

where $\mathcal{E}(w)$ is defined by (1).

Consider the following function:

$$F(t, w) = \theta t + \frac{(1+t\mu^{\perp}w)^2}{2w^{\perp}\mathfrak{B}wt}.$$

The following lemma establishes some properties of this function. The proof is given in the appendix.

Lemma 1 *There exists a unique couple $(\bar{t}, \bar{w})$, with $\bar{t} \in (0, \infty)$ and $\bar{w} \in \Delta_n$ such that*

$$F(\bar{t}, \bar{w}) = \min_{t>0} \max_{w\in\Delta_n} F(t, w).$$

In addition, the function

$$f(t) = F(t, \bar{w})$$

has a unique minimum at the point $\bar{t}$.

We clearly have $1 + \bar{t}\mu^{\perp}\bar{w} > 0$. Indeed, if $1 + \bar{t}\mu^{\perp}\bar{w} < 0$ then $f(-\frac{1}{\mu^{\perp}\bar{w}}) < f(\bar{t})$ which contradicts the fact that $\bar{t}$ is the minimizer. If $1 + \bar{t}\mu^{\perp}\bar{w} = 0$ then $f'(\bar{t}) = \theta$ which also leads to a contradiction. Let

$$T' = \begin{cases} -\dfrac{1}{\mu^{\perp}\bar{w}}, & \mu^{\perp}\bar{w} < 0 \\ +\infty & \text{otherwise,} \end{cases}$$

Remark that if $T' < \infty$, then $f(T') = \theta T' > f(\bar{t})$. Let us also choose T small enough so that

$$1 - |\mu^{\perp}\bar{w}|T \geq \frac{1}{2} \quad \text{and} \quad \frac{1}{8\bar{w}^{\perp}\mathfrak{B}\bar{w}T} > f(\bar{t}).$$

and assume that k is large enough so that $k + 8\tilde{\mu}\bar{w} > 0$. We bound the distribution function of the Gaussian mixture from above as follows:

$$\mathbb{P}[\sum_{i=1}^{n} e^{Y_i\sqrt{Z}+\mu_i Z+\tilde{\mu}_i} \leq e^{-k}] = \mathbb{E}[F_{Z/k}(k)]$$

$$= \int_0^{\infty} \rho(s) F_{s/k}(k)ds = k\int_0^{\infty} \rho(tk)F_t(k)dt \tag{44}$$

$$\leq k \max_{0\leq t\leq T} F_t(k) + k\int_T^{T'} \frac{C(tk)^{\alpha}\sqrt{t}}{\sqrt{k}(\mathbf{1}+\mu t)^{\perp}w} e^{-kf(t)}dt + ck\int_{T'}^{\infty} e^{-tk\theta}(tk)^{\alpha}dt. \tag{45}$$

Now, by the choice of T, the first term on the right-hand side of the last inequality in (45) satisfies

$$k \max_{0\leq t\leq T} F_t(k) \leq C\sqrt{k}e^{-\beta k}$$

with $\beta > f(t^*)$. The second term is computed using Laplace's method. As $k \to +\infty$, up to a constant,

$$k\int_T^{T'} \frac{C(tk)^{\alpha}\sqrt{t}}{\sqrt{k}(\mathbf{1}+\mu t)^{\perp}w} e^{-kf(t)}dt \sim Ck^{\alpha}e^{-kf(t^*)}.$$

Finally, the last term is negligible by the choice of T'.

The proof of Theorem 3 is thus completed. □

Our next goal is to establish a lower estimate complementing the estimate in Theorem 3. Note that the estimates in Theorems 3 and 4 are off by the factor k^{-n}.

Theorem 4 (Lower bound) *Let Y be a centered Gaussian vector with covariance matrix $\mathfrak{B}$ and let $\mu \in \mathbb{R}^n$ and $\tilde{\mu} \in \mathbb{R}^n$. Let Z be a random variable with values in $(0, \infty)$, which has a density $\rho(x)$ satisfying $\rho(s) \geq cs^{\alpha}e^{-\theta s}$ for $s \geq 1$, where $\theta > 0$, $c > 0$ and $\alpha \in \mathbb{R}$ are constants. Then, there exists $C > 0$ such that as $k \to +\infty$,*

$$\mathbb{P}[\sum_{i=1}^{n} e^{Y_i\sqrt{Z}+\mu_i Z+\tilde{\mu}_i} \leq e^{-k}] \gtrsim Ck^{\alpha-n}e^{-c^*k},$$

where c^* is given by (43).

Proof It is clear that

$$\mathbb{P}\left[\sum_{i=1}^{n} e^{Y_i\sqrt{kt}+\mu_i kt+\tilde{\mu}_i} \le e^{-k}\right] \ge \mathbb{P}[Y_i\sqrt{kt} + \mu_i kt + \tilde{\mu}_i \le -k - \log n, i = 1, \ldots, n].$$

By Proposition 3.2 in [28], the above probability can be bounded from below (very roughly) as follows:

$$\mathbb{P}[Y_i\sqrt{kt} + \mu_i kt + \tilde{\mu}_i \le -k - \log n, i=1, \ldots, n] \ge \frac{C}{(1+k(1+t))^n} \exp\{-\alpha_t/2\},$$

where

$$\begin{aligned}
\alpha_t &= \min_{x \ge \frac{1}{\sqrt{kt}}((k+\log n)\mathbf{1}+kt\mu+\tilde{\mu})} x^{\perp}\mathfrak{B}^{-1}x \\
&= \max_{u\in\mathbb{R}^n_+} \left\{-\frac{1}{2}u^{\perp}\mathfrak{B}u + u^{\perp}\frac{1}{\sqrt{kt}}((k+\log n)\mathbf{1} + kt\mu + \tilde{\mu})\right\} \\
&= \max_{w\in\Delta_n} \frac{(k+\log n + kt\mu^{\perp}w + \tilde{\mu}^{\perp}w)^2}{2w^{\perp}\mathfrak{B}wkt} \\
&\le \max_{w\in\Delta_n} k\frac{(1+t\mu^{\perp}w)^2}{2w^{\perp}\mathfrak{B}wt} + \max_{w\in\Delta_n} \frac{(1+t\mu^{\perp}w)(\log n + \tilde{\mu}^{\perp}w)}{w^{\perp}\mathfrak{B}wt} + \max_{w\in\Delta_n} \frac{(\log n + \tilde{\mu}^{\perp}w)^2}{2w^{\perp}\mathfrak{B}wkt}.
\end{aligned}$$

Finally, we bound the distribution function of the Gaussian mixture from below as follows:

$$\mathbb{P}[\sum_{i=1}^{n} e^{Y_i\sqrt{Z}+\mu_i Z+\tilde{\mu}_i} \le e^{-k}] = k\int_0^\infty \rho(tk)F_t(k)dt \ge ck\int_{\bar{t}-1/k}^{\bar{t}+1/k} (tk)^\alpha e^{-\theta tk}F_t(k)dt$$

$$\ge \frac{Ck(\bar{t}k)^\alpha}{(1+k(1+t))^n}\int_{\bar{t}-1/k}^{\bar{t}+1/k} \exp\left\{-\theta\bar{t}k - k\max_{w\in\Delta_n}\frac{(1+t\mu^{\perp}w)^2}{2w^{\perp}\mathfrak{B}wt}\right\}dt$$

$$\ge \frac{C(\bar{t}k)^\alpha}{(1+k(1+\bar{t}))^n}\exp\left\{-\theta\bar{t}k - k\max_{w\in\Delta_n}\frac{(1+\bar{t}\mu^{\perp}w)^2}{2w^{\perp}\mathfrak{B}w\bar{t}}\right\} = \frac{Ck^\alpha e^{-kf(\bar{t})}}{(1+k(1+\bar{t}))^n}.$$

□

Remark 4 Theorems 3 and 4 show that under their assumptions, the dominating factor describing the decay of the left tail of the price of a portfolio of assets is exponential with the decay rate equal to the constant c^*. For example, for $n = 1$, we have

$$c^* = \min_{t\ge 0}\{\theta t + \frac{(1+\mu t)^2}{2\sigma^2}\} = \frac{\sqrt{2\theta\sigma^2+\mu^2}+\mu}{\sigma^2}.$$

In symmetric models with $\mu = 0$, the formula for c^* simplifies to

$$c^* = \sqrt{\frac{2\theta}{\min_{w\in\Delta_n} w^{\perp}\mathfrak{B}w}}.$$

4.2 Implied Volatility Asymptotics

Let $S^1, \ldots, S^n$ be assets such that

$$\log \widetilde{S}_t = \log \widetilde{S}_0 + \tilde{\mu}t + \mu\tau_t + \mathfrak{B}^{\frac{1}{2}} W_{\tau_t}, \tag{46}$$

where we use the same notation as in the beginning of Sect. 3. Let S denote the price process of the basket. Fix a maturity $T > 0$, and suppose the random variable τ_T has a density ρ_T. Suppose also that there exist $c_1 > 0$, $c_2 > 0$, $\theta > 0$ and $\alpha \in \mathbb{R}$ such that

$$c_1 s^{\alpha} e^{-\theta s} \leq \rho_T(s) \leq c_2 s^{\alpha} e^{-\theta s}, \quad s \geq 1. \tag{47}$$

We assume that for every $i = 1, \ldots, n$,

$$\theta > \mu_i + \frac{\mathfrak{B}_{ii}}{2}. \tag{48}$$

This assumption implies that there exists $\varepsilon > 0$ such that

$$\mathbb{E}[(S_T^i)^{1+\varepsilon}] < \infty$$

We then assume further that $\tilde{\mu}_i$ is chosen in such way that

$$\mathbb{E}[S_T^i] = S_0^i. \tag{49}$$

It follows from Theorems 3 and 4 that there exist $C_1 > 0$, $C_2 > 0$, and $y_0 > 0$ such that

$$C_1 y^{c^*}\left[\log\frac{1}{y}\right]^{\alpha-n} \leq \mathbb{P}[S_{\tau_T} \leq y] \leq C_2 y^{c^*}\left[\log\frac{1}{y}\right]^{\alpha}, \quad y < y_0. \tag{50}$$

Since we have

$$P(K) = \mathbb{E}\left[\left(K - S_{\tau_T}\right)^+\right] = \int_0^K \mathbb{P}[S_{\tau_T} \leq y]dy,$$

the estimates in (50) imply that there exist $C_3 > 0$, $C_4 > 0$, and $K_0 > 0$ such that

$$C_3\, K^{c^*+1} \left[\log \frac{1}{K}\right]^{\alpha - n} \le P(K) \le C_4\, K^{c^*+1} \left[\log \frac{1}{K}\right]^{\alpha}, \quad K < K_0. \tag{51}$$

Note that the put pricing pricing in (51) is squeezed between two regularly varying functions with the same index of regular variation at zero. Such estimates allow one to find the leading term in the asymptotic expansion of the implied volatility near zero.

Theorem 5 *Suppose condition (47) holds for the time-change process τ and that the assumptions* (48) *and* (49) *are satisfied. Then the following asymptotic formula holds for the implied volatility in time-changed n-dimensional Black-Scholes model:*

$$I(K) \sim \left(\frac{\psi(c^*)}{T}\right)^{\frac{1}{2}} \sqrt{\log \frac{1}{K}}$$

as $K \to 0$, where the function ψ is defined by

$$\psi(u) = 2 - 4(\sqrt{u^2 + u} - u), \quad u > 0 \tag{52}$$

and the constant c^* is given by Formula (43).

Proof Theorem 5 follows from (51) and Theorem 10.28 in [24]. □

Remark 5 Condition (47) holds for many processes commonly used as stochastic time changes, e.g., for the gamma process, the inverse Gaussian process, or the generalized inverse Gaussian process. The latter process is used as time change in the generalized hyperbolic Lévy model. Recall that the density of the gamma process is given by

$$\rho_t(s) = \frac{\lambda^{ct}}{\Gamma(ct)} s^{ct-1} e^{-\lambda s}, \tag{53}$$

while the density of the inverse Gaussian process is as follows:

$$\rho_t(s) = \frac{ct}{s^{3/2}} e^{2ct\sqrt{\pi\lambda} - \lambda s - \pi c^2 t^2 / s}.$$

In the previous formulas, the symbols λ and c stand for the parameters of the distributions.

We close this section with a counterpart of Theorem 5 for the right tail, which can be deduced from Theorem 10 proved in the next section.

Theorem 6 *Suppose condition (47) holds for the time-change process τ and that the assumptions* (48) *and* (49) *are satisfied. Then the following asymptotic formula holds for the implied volatility in a time-changed n-dimensional Black-Scholes model:*

$$I(K) \sim \left(\frac{\psi(c^{min}-1)}{T}\right)^{\frac{1}{2}} \sqrt{\log K}$$

as $K \to +\infty$, where

$$c^{min} = \min_{i=1,\dots,n} \frac{\sqrt{2\theta\mathfrak{B}_{ii} + \mu_i^2} - \mu_i}{\mathfrak{B}_{ii}}.$$

Proof Let $\overline{G}_i(x) = \mathbb{P}[\log S_T^i \geq x]$. By Theorems 3 and 4, there exist constants C_1 and C_2 such that

$$C_1 x^{\alpha} e^{-c_i x} \gtrsim \overline{G}_i(x) \gtrsim C_2 x^{\alpha-n} e^{-c_i x}$$

as $x \to +\infty$, where

$$c_i = \frac{\sqrt{2\theta\mathfrak{B}_{ii} + \mu_i^2} - \mu_i}{\mathfrak{B}_{ii}}.$$

Note that in the single-asset case Theorems 3 and 4 can also be applied to the right tail, by symmetry. It follows that

$$\overline{G}_i(x) \sim -c_i x$$

as $x \to +\infty$, and the proof may be completed by applying Theorem 10. □

Numerical illustration In this part of the paper we illustrate the asymptotic result of Theorem 5 with a numerical example. Figure 3 plots the squared implied volatility of

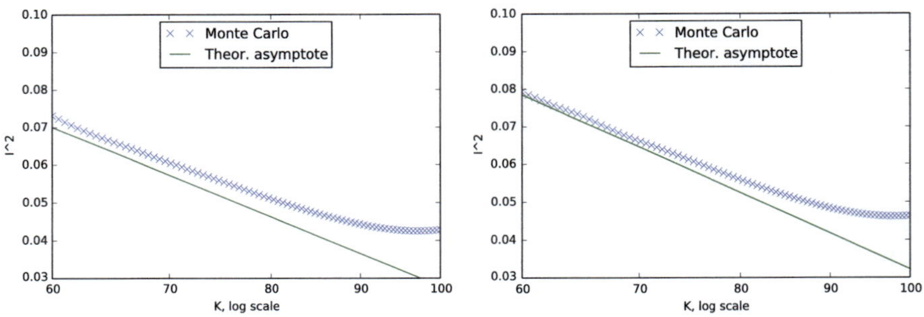

Fig. 3 Implied volatility of a basket call option together with the theoretical asymptote. *Left* option on a basket of 2 identical assets. *Right* option on a basket of 10 identical assets. The logarithms of asset prices follow the multidimensional variance gamma model

two basket call options computed by Monte Carlo as function of the strike price on logarithmic scale, as well as the theoretical asymptote with slope given by Theorem 5. Note that Theorem 5 only provides the value of the limiting slope of the squared implied volatility. Therefore, the performance of the asymptotic results should be evaluated by comparing the slope of the wing of the smile with the slope of the straight line. The intercept of the straight line has been chosen to keep the straight line relatively close to the curve, solely for the purpose of visualisation. The confidence intervals for 5 million simulated paths are very tight and not shown on the graphs.

We focus on the left wing of the smile since in the left wing the slope of the smile is correlation-dependent, and therefore can in principle be used to calibrate the correlation structure. Also, numerical experiments for the right wing (not presented in the paper) show that one needs to go much further into the tail to observe the asymptotic behavior predicted by Theorem 6.

In this numerical illustration, the time change follows the variance gamma law (53) with $\lambda = 10$ and $c = 10$. In the left graph, the basket contains two identical assets with price processes given by (46), where we take $\mu \equiv 0$, $S_0^i = 50$ for $i = 1, 2$ and the covariance matrix which satisfies $\mathfrak{B}_{ii} = \sigma^2$ with $\sigma = 0.3$ for $i = 1, 2$ and $\mathfrak{B}_{ij} = 0$ for $i \neq j$. In the right graph, the basket contains ten identical assets with price processes given by (46), where we take $\mu \equiv 0$, $S_0^i = 10$ for $i = 1, \ldots, 10$ and the covariance matrix which satisfies $\mathfrak{B}_{ii} = \sigma^2$ for $i = 1, \ldots, 10$ and $\mathfrak{B}_{ij} = \rho\sigma^2$ with $\rho = 0.5$ for $i \neq j$. The maturity of the options is $T = 0.2$ years in both graphs.

5 Assets with Dependence Structure Defined by a Copula

A popular approach to pricing European style multi-asset options is to calibrate full-fledged models for marginal distributions of asset prices, and then use a copula function from a simple parametric family to model the dependence structure. This is because information about the marginal distributions can be extracted from the prices of single asset options, which are liquidly traded, but the market quotes offer very little information about the dependence.

5.1 A Very Brief Primer on Copulas

Recall that the copula of a random vector $(X_1, \ldots, X_n)$ is a function $C : [0, 1]^n \mapsto [0, 1]$, satisfying the following conditions:

- dC is a positive measure in the sense of Lebesgue-Stieltjes integration.
- $C(u_1, \ldots, u_n) = 0$ when $u_k = 0$ for at least one k.
- $C(u_1, \ldots, u_n) = u_k$ when $u_i = 1$ for all $i \neq k$.

In addition, it is supposed that

$$\mathbb{P}[X_1 \le x_1, \dots, X_n \le x_n] = C(\mathbb{P}[X_1 \le x_1], \dots, \mathbb{P}[X_n \le x_n]), \quad (x_1, \dots, x_n) \in \mathbb{R}^n.$$

A copula exists by Sklar's theorem and is uniquely defined in the case where the marginal distributions of $X_1, \dots, X_n$ are continuous. We refer to [35] for more details on copulas.

The Gaussian copula with correlation matrix R is the unique copula of any Gaussian vector with correlation matrix R and nonconstant components (it does not depend on the mean vector and on the variances of the components).

Given a function $\phi : [0, 1] \to [0, \infty]$ which is continuous, strictly decreasing and such that its inverse ϕ^{-1} is completely monotonic, the Archimedean copula with generator ϕ is defined by

$$C(u_1, \dots, u_n) = \phi^{-1}(\phi(u_1) + \dots + \phi(u_n)).$$

Definition 1 The *weak lower tail dependence function* $\chi(\alpha_1, \dots, \alpha_n)$ of a copula C is defined by

$$\chi(\alpha_1, \dots, \alpha_n) = \lim_{u \to 0} \frac{\min_i \log u^{\alpha_i}}{\log C(u^{\alpha_1}, \dots, u^{\alpha_n})},$$

provided that the limit exists and is finite for all $\alpha_1, \dots, \alpha_n \ge 0$ such that $\alpha_k > 0$ for at least one k.

We will next formulate several known assertions (see [39]).

Theorem 7 *Let $X_1, \dots, X_n$ be random variables with state space $(0, \infty)$, marginal distribution functions $F_1, \dots, F_n$, and a copula C. Suppose that for every $k = 1, \dots, n$, the function F_k is slowly varying at zero, and there exist constants η_k, $1 \le k \le n$, and a function F such that*

$$\log F_k(x) \sim \eta_k \log F(x), \quad 1 \le k \le n.$$

Suppose also that the copula C admits a weak lower tail dependence function χ. Then,

$$\lim_{x \downarrow 0} \frac{\log \mathbb{P}[X_1 + \dots + X_n \le x]}{\min_i \log \mathbb{P}[X_i \le x]} = \frac{1}{\chi(\eta_1, \dots, \eta_n)}.$$

Theorem 8

- *Assume that a copula function C has strong tail dependence in the left tail, meaning that the limit*

$$\lambda_L = \lim_{u \downarrow 0} \frac{C(u, \dots, u)}{u},$$

  exists and satisfies $\lambda_L > 0$. Then, the weak lower tail dependence function of C satisfies $\chi(\alpha_1, \dots, \alpha_n) = 1$.

- *Let C be a Gaussian copula with correlation matrix R such that* $\det R \neq 0$. *Then,*

$$\chi(\alpha_1, \dots, \alpha_n) = \max_i \alpha_i \min_{w \in \Delta_n} w^T \Sigma w, \quad \textit{for all } \alpha_1, \dots, \alpha_n > 0,$$

where the matrix Σ has entries $\Sigma_{ij} = \frac{R_{ij}}{\sqrt{\alpha_i \alpha_j}}$, $1 \leq i, j \leq n$.

- *Let C be an Archimedean copula with a generator function ϕ such that* $\log \phi^{-1}$ *is regularly varying at ∞ with index $\lambda > 0$. Then,*

$$\chi(\alpha_1, \dots, \alpha_n) = \frac{\max(\alpha_1, \dots, \alpha_n)}{(\alpha_1^{1/\lambda} + \dots + \alpha_n^{1/\lambda})^\lambda}.$$

5.2 Copulas and the Implied Volatility Asymptotics

In this subsection, we study the left-wing behavior of the implied volatility associated with a basket call option. Recall that we denoted by $(Y_1, \dots, Y_n)$ the vector of logarithmic returns of the risky assets, and by $(\lambda_1, \dots, \lambda_n)$ the corresponding vector of weights. Let C be the copula of the vector $(Y_1, \dots, Y_n)$, and G_i be the distribution function of Y_i for $i = 1, \dots, n$. The implied volatility is considered in this section as a function $k \mapsto I(-k)$ of the variable $-k$, where k is the log-strike defined by $k = \log K$. The tail-wing formulas due to Benaim and Friz (see [9]) play an important role in the sequel.

Theorem 9 *Let $\alpha > 0$, and assume that the following are true:*

- *There exists $\varepsilon > 0$ such that $\mathbb{E}[e^{-\varepsilon Y_i}] < \infty$, $i = 1, \dots, n$.*
- *For every $1 \leq i \leq n$, the function $k \mapsto -\log G_i(-k)$, $k > k_0$, belongs to the class R_α of regularly varying functions, and there exist positive constants $\eta_1, \dots, \eta_n$ and a function G such that*

$$\log G_i(-k) \sim \eta_i \log G(-k) \quad \textit{as } k \to \infty. \tag{54}$$

- *The copula C admits a weak lower tail dependence function χ.*

Then,

$$\frac{I(-k)^2 T}{k} \sim \psi\left[-\frac{\log G(-k)}{k} \frac{\max_i \eta_i}{\chi(\eta_1, \dots, \eta_n)}\right] \tag{55}$$

as $k \to \infty$, where the function ψ is defined in (52).

Proof The distribution function F_i of the random variable $\lambda_i S_i$ is given by

$$F_i(x) = G_i(\log x - \log \lambda_i).$$

Since the function $\log G_i$ is regularly varying at $-\infty$, it is clear that $\log F_i$ is slowly varying at zero and

$$\log F_i(x) \sim \log G_i(\log x) \sim \eta_i \log G(\log x)$$

as $x \to 0$. It follows from Theorem 7 that

$$\log F(x) \sim \frac{\max_i \eta_i}{\chi(\eta_1, \ldots, \eta_n)} \log G(\log x) \quad \text{as}\, x \to 0,$$

where F is the distribution function of $\sum_{i=1}^n \lambda_i S_i$. Equivalently

$$\log F(e^{-k}) \sim \frac{\max_i \eta_i}{\chi(\eta_1, \ldots, \eta_n)} \log G(-k) \quad \text{as}\, k \to \infty,$$

and hence

$$-\frac{\log F(e^{-k})}{k} \sim -\frac{\log G(-k)}{k} \frac{\max_i \eta_i}{\chi(\eta_1, \ldots, \eta_n)} \quad \text{as}\, k \to \infty. \tag{56}$$

It follows from the assumptions in Theorem 9 that $\log G(-k) \in R_\alpha$ as $k \to \infty$. Therefore $\log F(e^{-k}) \in R_\alpha$ as well. Next, using the tail-wing formula of Benaim and Friz (see Theorem 2 in [9]), we obtain

$$\frac{I(-k)^2 T}{k} \sim \psi\left[-\frac{\log F(e^{-k})}{k}\right] \quad \text{as}\, k \to \infty. \tag{57}$$

We will need the following lemma.

Lemma 2 *Let ψ be the function defined by (52), and suppose ρ_1 and ρ_2 are positive functions on $(0, \infty)$ such that*

$$\frac{\rho_1(x)}{\rho_2(x)} \to 1 \quad as \quad x \to \infty. \tag{58}$$

Then

$$\frac{\psi(\rho_1(x))}{\psi(\rho_2(x))} \to 1 \quad as \quad x \to \infty. \tag{59}$$

Proof It is not hard to see that for all $u \geq 0$,

$$\psi(u) = \frac{2}{(\sqrt{u+1} + \sqrt{u})^2}. \tag{60}$$

The equality in (60) describes the structure of the function ψ better than the original definition.

Fix $\varepsilon > 0$. Then, using (60) and the inequality $1 < \frac{1}{1-\varepsilon}$, we get

$$\psi((1-\varepsilon)u) \le \frac{2}{(1-\varepsilon)\left(\sqrt{u+\frac{1}{1-\varepsilon}}+\sqrt{u}\right)^2} \le \frac{2}{(1-\varepsilon)(\sqrt{u+1}+\sqrt{u})^2}$$
$$= \frac{2}{(1-\varepsilon)(\sqrt{u+1}+\sqrt{u})^2} = \frac{1}{1-\varepsilon}\psi(u).$$

Similarly

$$\psi((1+\varepsilon)u) \ge \frac{1}{1+\varepsilon}\psi(u).$$

Therefore,

$$\frac{1}{1+\varepsilon}\psi(u) \le \psi((1+\varepsilon)u) \le \psi((1-\varepsilon)u) \le \frac{1}{1-\varepsilon}\psi(u). \tag{61}$$

It follows from (58) that for every $\varepsilon > 0$ there exists $x_\varepsilon > 0$ such that

$$(1-\varepsilon)\rho_2(x) \le \rho_1(x) \le (1+\varepsilon)\rho_2(x)$$

for all $x > x_\varepsilon$. Since the function ψ decreases on $(0, \infty)$, we have

$$\psi((1+\varepsilon)\rho_2(x)) \le \psi(\rho_1(x)) \le \psi((1-\varepsilon)\rho_2(x))$$

for all $x > x_\varepsilon$. Now, using (61), we obtain

$$\frac{1}{1+\varepsilon}\psi(\rho_2(x)) \le \psi(\rho_1(x)) \le \frac{1}{1-\varepsilon}\psi(\rho_2(x))$$

for all $x > x_\varepsilon$, and (59) follows. □

Finally, it is not hard to see that (56), (57), and Lemma 2 imply (55).

This completes the proof of Theorem 9. □

The next example shows that condition (54) does not prevent one from choosing different marginal laws for different components of the process $(Y_1, \dots, Y_n)$ as long as these laws have a similar tail behavior.

Example 1 Let us consider the following multidimensional extension of the example given in Sect. 5.2 of [9]. We assume that for $i = 1, \dots, n$, the distribution of the random variable Y_i is normal inverse Gaussian, more precisely, $\mathrm{NIG}(\alpha_i, \beta_i, \mu_i, \delta_i)$. It is also supposed that the parameters satisfy $\alpha_i > |\beta_i| > 0$ and $\delta_i > 0$. This means that the moment generating function of Y_i is given by

$$M_i(z) = \exp\left(\delta_i\left\{\sqrt{\alpha_i^2-\beta_i^2}-\sqrt{\alpha_i^2-(\beta_i+z)^2}\right\}+\mu_i z\right).$$

We refer the reader to [6] for more details on the normal inverse Gaussian distribution. In particular, it follows that Y_i has a density g_i which satisfies the following condition:

$$g_i(k) \sim C_i |k|^{-\frac{3}{2}} e^{-\alpha_i |k| + \beta_i k}, \quad k \to \pm\infty,$$

where C_i is a constant. Using Theorem 2 in [9], we see that $-\log G_i(-k) \in R_\alpha$ as $k \to +\infty$, and also

$$-\log G_i(-k) \sim -\log g_i(-k) \sim (\beta_i - \alpha_i)k, \quad k \to +\infty.$$

Therefore, the condition in (54) holds with $\lambda_i = \alpha_i - \beta_i$ and $G(k) = e^k$.

Assuming that the dependence structure of $(Y_1, \ldots, Y_n)$ is described by the Gaussian copula with correlation matrix R, we see that

$$\frac{I(-k)^2 T}{k} \sim \psi \left[\frac{1}{\inf_{w \in \Delta_d} w^{\perp} \Sigma w} \right], \quad k \to +\infty, \tag{62}$$

where the matrix $\Sigma = [\Sigma_{ij}]$ is such that

$$\Sigma_{ij} = \frac{R_{ij}}{\sqrt{(\alpha_i - \beta_i)(\alpha_j - \beta_j)}}.$$

In other words, the implied variance is asymptotically linear, with a correlation-dependent limiting slope, which is given by the right-hand side of (62).

Remark 6 In this remark, we compare the asymptotic formulas for the implied volatility obtained in Sects. 3 and 5 (see Theorems 2 and 9). The latter theorem is more general than the former one. It provides the leading term in the asymptotic expansion of the implied volatility under certain restrictions on the marginal distributions of log-returns and the corresponding copula, and applies to many special models. In the case of a Gaussian copula and log-normal marginal distributions, all the conditions in Theorem 9 are satisfied, and the leading term is equal to the constant $\left(\bar{A}_1 + \cdots + \bar{A}_{\bar{n}}\right)^{-\frac{1}{2}}$. This follows from Theorem 9, the second equality in formula (27), and the second statement in Theorem 8. The advantage of Theorem 9 is its generality, while the disadvantage is that the asymptotic formula for the implied volatility contains only the leading term and no error estimate. On the other hand, Theorem 2 applies only to the case of Gaussian copula and lognormal margins under a not very restrictive condition $(\mathcal{A})$, but provides a sharp asymptotic formula for the implied volatility with several terms and an error estimate.

For the sake of completeness, we include a proposition that is a counterpart of Theorem 9 in the case of the right tail. This proposition turns out to be somewhat trivial: the leading order of the implied volatility is determined by a single component with the fattest tail, and it does not depend on the copula. Let us denote by $\overline{G}_i$ the survival function of Y_i, i.e., the function $\overline{G}_i(x) = \mathbb{P}[Y_i \geq x]$.

Theorem 10 *Let $\alpha > 0$, and suppose that the following assumptions hold:*

- *There exists $\varepsilon > 0$ such that $\mathbb{E}[e^{(1+\varepsilon)Y_i}] < \infty$ for $i = 1, \ldots, n$.*
- *For each $i = 1, \ldots, n$, the function $k \mapsto -\log \overline{G}_i(k)$ belongs to the class R_α at infinity.*

Then,

$$\frac{I(k)^2 T}{k} \sim \psi\left[-1 - \frac{1}{k}\max_i \log \overline{G}_i(k)\right] \quad \text{as } k \to +\infty. \tag{63}$$

Proof Set $X_i = v_i e^{Y_i}$. Then we get

$$\mathbb{P}[X_1 + \cdots + X_n \geq x] \geq \max_i \mathbb{P}[X_i \geq x],$$

$$\mathbb{P}[X_1 + \cdots + X_n \geq x] \leq \mathbb{P}[\exists i : X_i \geq \frac{x}{n}] \leq \sum_{i=1}^n \mathbb{P}[X_i \geq \frac{x}{n}] \leq n \max_i \mathbb{P}[X_i \geq \frac{x}{n}].$$

Since for each i, the function $\log \overline{G}_i$ is regularly varying at infinity, it follows that the function $x \mapsto \log \mathbb{P}[X_i \geq x]$ is slowly varying, and therefore, for x sufficiently large and any $\varepsilon > 0$,

$$\max_i \log \mathbb{P}[X_i \geq x/n] \leq (1+\varepsilon) \max_i \log \mathbb{P}[X_i \geq x].$$

Finally,

$$\lim_{x \to +\infty} \frac{\log \mathbb{P}[X_1 + \cdots + X_n \geq x]}{\max_i \log \mathbb{P}[X_i \geq x]} = 1,$$

and formula (63) follows from Theorem 1 in [9] with a similar proof to that of Theorem 9. □

Numerical illustration In this part of the paper we illustrate the asymptotic result of Theorem 9 with a numerical experiment. Figure 4 plots the squared implied volatility of two basket call options computed by Monte Carlo as function of the strike price on logarithmic scale, as well as the theoretical asymptote with slope given by Theorem 9. Once again, we focus on the left wing of the smile since the slope of the left wing depends on the correlation of the Gaussian copula while the slope of the right wing does not.

In both graphs, the basket contains assets with price processes

$$S_T^i = S_0^i e^{\tilde{\mu}_i T + X_T^i},$$

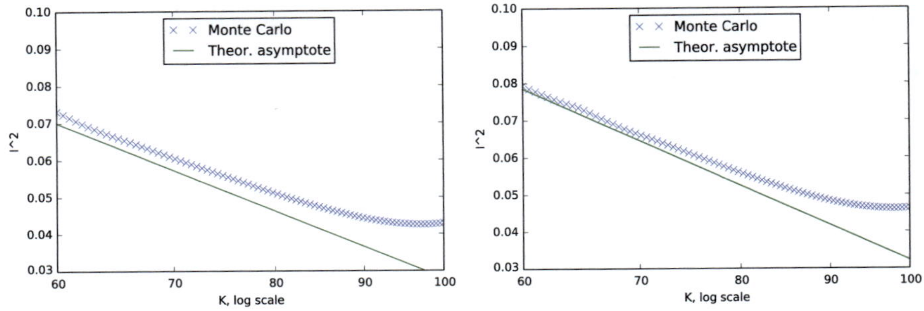

Fig. 4 Implied volatility of a basket call option together with the theoretical asymptote. *Left* option on a basket of 2 identical assets. *Right* option on a basket of 10 identical assets. The logarithms of asset prices follow the variance gamma model with dependence given by a Gaussian copula

where $\tilde{\mu}_i$ is chosen so that $\mathbb{E}[S^i_T] = S^i_0$ and X^i is the variance gamma process with characteristic function

$$E[e^{iuX^i_T}] = \left(1 - iu\kappa_i^{-1}\mu_i + \frac{\sigma_i^2 u^2 \kappa_i^{-1}}{2}\right)^{-\kappa_i T}.$$

For the numerical illustration we take $\mu_i = 0$, $\sigma_i = 0.3$ and $\kappa_i = 10$ for all i.

In the left graph, the basket contains two assets; we take $S^i_0 = 50$ for $i = 1, 2$ and assume that the assets are independent. In the right graph, the basket contains ten assets; we take $S^i_0 = 10$ for $i = 1, \dots, 10$ and assume that the dependence structure of the assets is given by the Gaussian copula with correlation matrix R with elements $R_{ij} = \rho + (1-\rho)\mathbf{1}_{i=j}$, where we took $\rho = 0.5$. The maturity of the options is $T = 0.2$ years in both graphs.

In the variance gamma model, similarly to Example 1, it can be shown (see e.g., [1]) that

$$\log G_i(k) \sim (\alpha_i + \beta_i)k, \quad k \to -\infty$$

with coefficients α_i and β_i given by

$$\alpha_i = \sqrt{\frac{\mu_i^2}{\sigma_i^4} + \frac{2\kappa_i}{\sigma_i^2}}, \quad \text{and} \quad \beta_i = \frac{\mu_i}{\sigma_i^2}.$$

Therefore, the limiting behavior of the implied volatility in this model is also given by (62). We see that the numerical illustration agrees well with the theoretical prediction. Compared to the numerical example of Sect. 4, we see that the slope of the implied volatility is steeper in the multidimensional VG model than in the copula model. This happens because the multidimensional VG model introduces additional positive dependence between the assets through the common time change process.

Proof of Lemma 1

The function F satisfies

$$F(t, w) = \max_{\lambda>0}\{\theta t + \lambda w^{\perp}(\mathbf{1} + \mu t) - \frac{\lambda^2 w^{\perp}\mathfrak{B} w t}{2}\},$$

where $\mathbf{1}$ stands for the n-dimensional vector with all elements equal to 1. Therefore,

$$\max_{w\in\Delta_n} F(t, w) = \max_{u\in\mathbb{R}^n_+} \widetilde{F}(t, u),$$

with

$$\widetilde{F}(t, u) = \{\theta t + u^{\perp}(\mathbf{1} + \mu t) - \frac{u^{\perp}\mathfrak{B} u t}{2}\}.$$

Since for every $t > 0$, $\widetilde{F}(t, u)$ is strictly concave in u, there exists a unique $\bar{u}(t) \in \mathbb{R}^n_+$ with $\bar{u}(t) \neq 0$ such that $\widetilde{F}(t, \bar{u}) = \max_{u\in\mathbb{R}^n_+} \widetilde{F}(t, u)$. This in turn implies that there exists a unique $\bar{w}(t)$ such that $F(t, \bar{w}) = \max_{w\in\Delta_n} F(t, w)$. It is also easy to see that $\bar{u}(t)$ depends continuously on t.

Let $\bar{f}(t) = \widetilde{F}(t, \bar{u}(t))$. We would like to show that $\bar{f}$ is differentiable in t and compute its derivative. $\bar{u}(t)$ may be characterized as follows: for $i = 1, \dots, n$

$$[\mathbf{1} + \mu t - t\mathfrak{B}\bar{u}(t)]_i = 0 \quad \text{if} \quad \bar{u}(t)_i > 0 \tag{64}$$

$$[\mathbf{1} + \mu t - t\mathfrak{B}\bar{u}(t)]_i \leq 0 \quad \text{if} \quad \bar{u}(t)_i = 0. \tag{65}$$

Let $I(t)$ denote the set of indices $i \in \{1, \dots, n\}$ such that $\bar{u}(t)_i > 0$, and, for a vector $x \in \mathbb{R}^n$, let $x_{I(t)}$ denote the subset of components of x with indices in $I(t)$: $x_{I(t)} = \{x_i : i \in I(t)\}$. Furthermore, let $\mathfrak{B}_{I(t),I(t)}$ denote the submatrix of the covariance matrix, containing the elements b_{ij} with $i \in I(t)$ and $j \in I(t)$. Then, the vector $\bar{u}(t)$ satisfies

$$\bar{u}(t)_{I(t)} = \frac{1}{t}\mathfrak{B}^{-1}_{I(t),I(t)}(\mathbf{1} + \mu t)_{I(t)}, \quad \bar{u}(t)_{\tilde{I}(t)} = 0,$$

where the set $\tilde{I}(t)$ contains the indices $i \in \{1, \dots, n\}$ which are not in $I(t)$.

Now, fix $t \in (0, \infty)$ and for $t' \in (0, \infty)$, define

$$v(t')_{I(t)} = \frac{1}{t'}\mathfrak{B}^{-1}_{I(t),I(t)}(\mathbf{1} + \mu t')_{I(t)}, \quad v(t)_{\tilde{I}(t)} = 0$$

First, assume that for all i such that $\bar{u}(t)_i = 0$, either $[\mathbf{1} + \mu t - t\mathfrak{B}\bar{u}(t)]_i < 0$ (with strict inequality) or

$$[\mathbf{1} + \mu t' - t'\mathfrak{B}v(t')]_i = 0$$

for all $t' \in (0, \infty)$. We shall call this Assumption 1. Then we can find $\delta > 0$, such that for every $t' \in (0, \infty)$ with $|t' - t| < \delta$, $v(t')$ satisfies the characterization (64) and (65). Therefore, $v(t') = \bar{u}(t')$. This means that

$$\bar{f}(t') = \theta t' + \frac{1}{2t'}(\mathbf{1} + \mu t')^{\perp}_{I(t)} \mathfrak{B}^{-1}_{I(t),I(t)} (\mathbf{1} + \mu t')_{I(t)}.$$

Therefore, $\bar{f}$ is differentiable at t with first derivative given by

$$\bar{f}'(t) = \theta - \frac{1}{2t^2} \mathbf{1}^{\perp}_{I(t)} \mathfrak{B}^{-1}_{I(t),I(t)} \mathbf{1}_{I(t)} + \frac{1}{2} \mu^{\perp}_{I(t)} \mathfrak{B}^{-1}_{I(t),I(t)} \mu_{I(t)} = \theta - \frac{1}{2t} \bar{u}(t)^{\perp} (\mathbf{1} - \mu t) \tag{66}$$

and second derivative

$$\bar{f}''(t) = \frac{1}{t^3} \mathbf{1}^{\perp}_{I(t)} \mathfrak{B}^{-1}_{I(t),I(t)} \mathbf{1}_{I(t)}.$$

Now assume that there exists at least one i such that $\bar{u}(t)_i = 0$ and $[\mathbf{1} + \mu t - t\mathfrak{B}\bar{u}(t)]_i = 0$, or, equivalently,

$$[\mathbf{1} + \mu t' - t'\mathfrak{B} v(t')]_i = 0$$

with $t' = t$. The case when the above equality holds for all t' is covered by Assumption 1. Since the left-hand side is linear in t', this means that for a given index set $I(t)$ and for a given i, there exists only one $t' \in (0, \infty)$ which satisfies the above equality. Since the number of possible index sets is finite, we conclude that there is at most a finite number of elements $t \in (0, \infty)$ which do not satisfy Assumption 1. But then, we can conclude by continuity that $\bar{f}$ is strictly convex (which entails uniqueness of $\bar{t}$) and differentiable for all $t \in (0, \infty)$, with the derivative given by (66) or alternatively by

$$\bar{f}'(t) = \theta - \frac{1}{2t^2 \bar{w}(t)^{\perp} \mathfrak{B} \bar{w}(t)} + \frac{(\bar{w}(t)^{\perp} \mu)^2}{2\bar{w}(t)^{\perp} \mathfrak{B} \bar{w}(t)}.$$

Comparing this with the derivative of f, which is easily computed, we see that at the point $\bar{t}$, these derivatives coincide. Since this point is characterized by the first order condition $\bar{f}'(\bar{t}) = 0$, and the function f is strictly convex, f also attains its unique minumum at $\bar{t}$.

References

1. Albin, J.M.P., Sundén, M.: On the asymptotic behaviour of Lévy processes, part I: subexponential and exponential processes. Stoch. Process. Appl. **119**, 281–304 (2009)
2. Andersen, L., Lipton, A.: Asymptotics for exponential Lévy processes and their volatility smile: survey and new results. Int. J. Theor. Appl. Financ. **16**, 1350001-1–1350001-98 (2013)
3. Asmussen, S., Rojas-Nandayapa, L.: Asymptotics of sums of lognormal random variables with Gaussian copula. Stat. Probab. Lett. **78**, 2709–2714 (2008)
4. d'Aspremont, A.: Interest rate model calibration using semidefinite programming. Appl. Math. Financ. **10**, 183–213 (2003)
5. Avellaneda, M., Boyer-Olson, D., Busca, J., Friz, P.: Reconstruction of volatility: pricing index options using the steepest-descent approximation. Risk Mag. **15**, 87–91 (2002)
6. Barndorff-Nielsen, O.: Processes of normal inverse Gaussian type. Financ. Stoch. **2**, 41–68 (1998)
7. Bayer, C., Laurence, P.: Asymptotics beats Monte Carlo: the case of correlated local vol baskets. Commun. Pure Appl. Math. **67**, 1618–1657 (2014)
8. Benaim, S., Friz, P.: Smile asymptotics II: models with known MGF. J. Appl. Probab. **45**, 16–32 (2008)
9. Benaim, S., Friz, P.: Regular variation and smile asymptotics. Math. Financ. **19**, 1–12 (2009)
10. Benhamou, E., Gobet, E., Miri, M.: Smart expansion and fast calibration for jump diffusions. Financ. Stoch. **13**, 563–589 (2009)
11. Berestycki, H., Busca, J., Florent, I.: Asymptotics and calibration of local volatility models. Quant. Financ. **2**, 61–69 (2002)
12. Cherubini, U., Luciano, E., Vecchiato, W.: Copula Methods in Finance. Wiley, Chichester (2004)
13. Cont, R., Deguest, R.: Equity correlations implied by index options: estimation and model uncertainty analysis. Math. Financ. **23**, 496–530 (2013)
14. De Marco, S., Hillairet, C., Jacquier, A.: Shapes of implied volatility with positive mass at zero (2013). arXiv:1310.1020
15. Eberlein, E., Madan, D.B.: On correlating Lévy processes. J. Risk **13**, 3–16 (2010)
16. Figueroa-López, J., Forde, M.: The small-maturity smile for exponential Lévy models. SIAM J. Financ. Math. **3**, 33–65 (2012)
17. Forde, M., Jacquier, A.: Small-time asymptotics for implied volatility under the Heston model. Int. J. Theor. Appl. Financ. **12**, 861–876 (2009)
18. Forde, M., Jacquier, A.: The large-maturity smile for the Heston model. Financ. Stoch. **15**, 755–780 (2011)
19. Gao, K., Lee, R.: Asymptotics of implied volatility to arbitrary order. Financ. Stoch. **18**, 349–392 (2014)
20. Gao, X., Xu, H., Ye, D.: Asymptotic behavior of tail density for sum of correlated lognormal variables. Int. J. Math. Math. Sci. **2009**, p. 28 (2009)
21. Gobet, E., Miri, M.: Time dependent Heston model. SIAM J. Financ. Math. **1**, 289 (2010)
22. Gulisashvili, A.: Asymptotic formulas with error estimates for call pricing functions and the implied volatility at extreme strikes. SIAM J. Financ. Math. **1**, 609–641 (2010)
23. Gulisashvili, A.: Asymptotic equivalence in Lee's moment formulas for the implied volatility, asset price models without moment explosions, and Piterbarg's conjecture. Int. J. Theor. Appl. Financ. **15**, 1250020 (2012)
24. Gulisashvili, A.: Analytically Tractable Stochastic Stock Price Models. Springer, Berlin (2012)
25. Gulisashvili, A.: Left-wing asymptotics of the implied volatility in the presence of atoms. Int. J. Theor. Appl. Finan. **18**(2) (2015)
26. Gulisashvili, A., Stein, E.M.: Asymptotic behavior of the stock price distribution density and implied volatility in stochastic volatility models. Appl. Math. Optim. **61**, 287–315 (2010)
27. Gulisashvili, A., Tankov, P.: Tail behavior of sums and differences of log-normal random variables. Bernoulli (to appear)

28. Hashorva, E., Hüsler, J.: On multivariate Gaussian tails. Ann. Inst. Stat. Math. **55**, 507–522 (2003)
29. Jourdain, B., Sbai, M.: Coupling index and stocks. Quant. Financ. **12**, 805–818 (2012)
30. Lee, R.: The moment formula for implied volatility at extreme strikes. Math. Financ. **14**, 469–480 (2004)
31. Lewis, A.: Option Valuation Under Stochastic Volatility. Finance Press, Newport Beach (2000)
32. Luciano, E., Schoutens, W.: A multivariate jump-driven financial asset model. Quant. Financ. **6**, 385–402 (2006)
33. Medvedev, A., Scaillet, O.: Approximation and calibration of short-term implied volatilities under jump-diffusion stochastic volatility. Rev. Financ. Stud. **20**, 427–459 (2007)
34. Mijatović, A., Tankov, P.: A new look at short-term implied volatility in asset price models with jumps. Math. Financ., to appear
35. Nelsen, R.: An Introduction to Copulas. Springer, New York (1999)
36. Prause, K.: The generalized hyperbolic model: estimation, financial derivatives, and risk measures, Ph.D. thesis, University of Freiburg (1999)
37. Schoenmakers, J.: Robust Libor Modelling and Pricing of Derivative Products. CRC Press, Boca Raton (2005)
38. Tankov, P.: Pricing and Hedging in Exponential Lévy Models: Review of Recent Results. Paris-Princeton Lectures on Mathematical Finance. Springer, Berlin (2010)
39. Tankov, P.: Large deviation asymptotics for the left tail of the sum of dependent positive random variables (2014). arXiv:1402.4683
40. Tehranchi, M.R.: Asymptotics of implied volatility far from maturity. J. Appl. Probab. **46**, 629–650 (2009)
41. Tehranchi, M.R.: Uniform bounds for Black-Scholes implied volatility, Pre-print (2014)

Small-Time Asymptotics for the At-the-Money Implied Volatility in a Multi-dimensional Local Volatility Model

Christian Bayer and Peter Laurence

Abstract We consider a basket or spread option based on a multi-dimensional local volatility model. Bayer and Laurence (Commun. Pure. Appl. Math., 67(10), 2014, [5]) derived highly accurate analytic formulas for prices and implied volatilities of such options when the options are not at the money. We now extend these results to the ATM case. Moreover, we also derive similar formulas for the local volatility of the basket.

Keywords Basket options · Spread options · Implied volatility · Asymptotic formulas · Heat kernel expansion

2010 Mathematics Subject Classification Primary 91G60 · Secondary 91G20, 58J90

1 Introduction

For a local volatility type model for a basket of stocks, whose forward prices are given by

---

To the memory of Peter Laurence, who passed away unexpectedly during the final stage of the preparation of this manuscript.

C. Bayer (✉)
Weierstrass Institute, Mohrenstrasse 39, 10117 Berlin, Germany
e-mail: christian.bayer@wias-berlin.de

P. Laurence
Courant Institute of Mathematical Sciences, New York University, 251 Mercer Street, New York, NY 10012, USA

P. Laurence
Dipartimento di Matematica, Università di Roma 1 Piazzale Aldo Moro, 2, 00185 Rome, Italy

P.K. Friz et al. (eds.), *Large Deviations and Asymptotic Methods in Finance*, Springer Proceedings in Mathematics & Statistics 110,
DOI 10.1007/978-3-319-11605-1_7

$$dF_i(t) = \sigma_i(F_i(t))dW_i(t), \quad i = 1, \dots, n, \tag{1.1}$$

$$d\langle W_i, W_j\rangle(t) = \rho_{ij}dt, \quad i, j = 1, \dots, n, \tag{1.2}$$

with a given correlation matrix ρ, we consider basket options with a payoff

$$P(\mathbf{F}) = \left(\sum_{i=1}^{n} w_i F_i - K\right)^{+},$$

where we generally denote in bold face a vector of the corresponding italic components, as in $\mathbf{F} = (F_1, \dots, F_n)$. Since we only assume that at least one of the weights $w_1, \dots, w_n$ is positive, we will refer to options of that type as *generalized spread options*.

The purpose of this paper is to provide an explicit first order accurate short time expansion of the price $C_{\mathcal{B}}(\mathbf{F}_0, K, T)$ of the above option using the heat kernel expansion technique (see, for instance, [12, 13, 20]) when the option is at the money. Moreover, from the asymptotic formula for the option price we also obtain an asymptotic formula for the implied and for the local volatility.[1] Thereby we complement the results obtained in [5], where a first order accurate asymptotic formula was given when the option is *not* at the money. (The zero order accurate formula is well-known, see, for instance [1]. When the option is not at the money, alternative first order accurate results can be found in [12].) Such asymptotic formulas are highly relevant, in particular when the dimension of the model is high (say $n > 3$), since then traditional (simulation or PDE) techniques to compute $C_{\mathcal{B}}$ fail or are at least very time consuming. In fact, for a wide range of different parameters, [5] show numerically that their asymptotic formula is remarkably close to the true price as given by the model, even for not so small maturities T (like 5 or even 10 years), for dimensions of up to $n = 100$ (or even more). The same holds true when the option is at the money, see Sect. 6.

We now sketch the procedure for deriving the asymptotic formulas, highlighting the differences to the non-ATM case.

- In the first step, we derive a Carr-Jarrow formula for the basket option price, separating the price into the intrinsic value of the option (which vanishes in the ATM case) and an integral over the arrival manifold $\{\sum_i w_i F_i = K\}$ with respect to the transition density $p(\mathbf{F}_0, \mathbf{F}, T)$. This is done in Sect. 2.
- The first terms in the heat kernel expansion of $p(\mathbf{F}_0, \mathbf{F}, T)$ are computed. In the non-ATM case, a zero-order heat kernel expansion was sufficient to get first order accurate formulas for the implied volatilities. At the money, we actually need to add one additional term in the heat kernel expansion. The heat kernel coefficients are computed in Lemma 3.6.
- The aforementioned integral on the arrival manifold is essentially an integral with respect to the rapidly decaying kernel $\exp\left(-d(\mathbf{F}_0, \mathbf{F})^2/(2T)\right)$, where d denotes

[1] Since we consider spread options here (for which $\sum_i w_i F_{0,i}$ may be negative), we derive implied volatilities both in the Black-Scholes and in the Bachelier sense.

the Riemannian (geodesic) distance induced by the stock price process. Hence, the integral can be approximated using Laplace's expansion for $T \to 0$, which involves the minimizer $\mathbf{F}^*$ of $\mathbf{F} \mapsto d(\mathbf{F}_0, \mathbf{F})^2$ subject to $\sum_i w_i F_i = K$. In the general case, this minimizer has to be computed numerically, while it is obviously given by $\mathbf{F}^* = \mathbf{F}_0$ when the option is at the money. On the other hand, the formulas are much longer and more complex due to the higher order heat kernel expansion used, see Proposition 3.4 together with Lemmas 3.3 and 3.7.

- In Sect. 4, we use the same Laplace's expansion technique to derive the local volatility of the basket, see Proposition 4.1.
- Finally, in Sect. 5, an asymptotic expansion for the implied volatilities is computed by a comparison of coefficients between the asymptotic expansion of the basket price derived in Proposition 5.1 and asymptotic expansions of the Black-Scholes and Bachelier formulas, respectively, see Eqs. (5.2)–(5.4).

An alternative way to derive the asymptotic expansion for at-the-money options would be to start from the non-at-the-money formulas and pass to the limit. This would involve un-determined terms "$\frac{0}{0}$", which would need to be resolved by the l'Hopital rule. In particular, we would have to compute limits of derivatives of the optimal configuration, which are not known in closed form when the option is not at-the-money. Still, one could follow that approach using similar techniques as in [4], but the derivation would hardly be any simpler than directly starting from scratch again (the course of action chosen in this article).

In Sect. 6 we present numerical examples for one particular choice of a local volatility model, namely the CEV model, corresponding to $\sigma_i(F_i) = \xi_i F_i^{\beta_i}$, $0 \le \beta_i \le 1$, $1 \le i \le n$. The numerical observations supports the claimed accuracy of the asymptotic price formulas. In fact, comparisons with highly accurate reference solutions show that the asymptotic formulas indeed have the suggested rates of convergence as $T \to 0$. Even more, they indicate that the formulas, in particular the first order formula, are highly accurate even for large maturities such as $T = 10$ years, thereby confirming the observations in [5].

Remark 1.1 In the same spirit as [5], the aim of this paper is to give *informal* derivations of fast and accurate asymptotic formulas. Indeed, there are several steps, in which our derivations are not fully rigorous. In particular, most local volatility models (like the CEV model) exhibit singular behaviour at the boundary of the domain $\mathbb{R}_+^n$ which can inhibit the validity of the heat kernel expansion, and, a fortiori, also the Laplace expansion applied later. It is clearly possible to rigorously justify both expansions under appropriate (uniform) ellipticity assumptions (see, for instance, [20] for the validity of the heat kernel expansion and [6] for a rigorous version of Laplace's expansion). An extension to general or some specific local volatility models, however, seems to be a difficult task, see also the comments in [5, Sect. 4], and, in particular, the results of [2]. Thus, we believe that a more "hands-on" approach can be justified in this particular case. For related problems see also [3, 8, 9].

2 Basket Carr-Jarrow Formula

Consider a basket $\mathcal{B} = \sum w_i F_i$ with weights $w_i \in \mathbb{R}$. Following [5, 7], we are now going to derive a Carr-Jarrow formula for the price of a generalized spread option on the basket, i.e., a decomposition of the price of the option into the intrinsic value and an integral over the arrival manifold $\{\mathcal{B} = K\}$. Take the Itô derivative of the basket's price:

$$d\sum_{i=1}^{n} w_i F_i(t) = \sum_{i=1}^{n} w_i \sigma_i(F_i(t)) dW_i(t)$$
$$= \sqrt{\underbrace{\sum_{i,j=1}^{n} w_i w_j \sigma_i(F_i(t)\sigma_j(F_j(t))\rho_{ij}}_{\sigma^2_{N,\mathcal{B}}}}\, d\bar{W}(t),$$

for a new Brownian motion $\bar{W}$. Here we have used the notation $\sigma_{N,\mathcal{B}}$ to indicate the "normal volatility" of the basket which must not be confused with the lognormal (Black) volatility $\sigma_{\mathcal{B}} = \frac{\sigma_{N,\mathcal{B}}}{\sum_{i=1}^{n} w_i F_i}$ used in reference [1]. Therefore, by the Itô-Tanaka formula we have

$$d\left(\sum_{i=1}^{n} w_i F_i(t) - K\right)^{+} = \sum_{i=1}^{n} w_i \mathbf{1}_{\sum w_i F_i(t) > K}\, dF_i(t)$$
$$+ \frac{1}{2}\delta_{\{\mathbf{F}:\sum w_i F_i(t) = K\}}\, \sigma^2_{N,\mathcal{B}}(\mathbf{F}(t)) dt.$$

Integrating we obtain

$$\left(\sum_{i=1}^{n} w_i F_i(T) - K\right)^{+} = \left(\sum_{i=1}^{n} w_i F_i(0) - K\right)^{+} + \sum_{i=1}^{n} w_i \int_0^T \mathbf{1}_{\sum w_i F_i(u) > K}\, dF_i(u) +$$
$$+ \frac{1}{2}\int_0^T \delta_{\{\mathbf{F}(u):\sum w_i F_i(u) = K\}}\, \sigma^2_{N,\mathcal{B}}(\mathbf{F}(u)) du.$$

Letting $\mathcal{E}_K = \{\mathbf{F} \in \mathbb{R}^n_+ : \sum w_i F_i = K\}$ and taking conditional expectations with respect to the filtration $\mathcal{F}_0$ at time 0, we obtain, assuming $F_i(t)$ is a martingale for each i[2]:

---

[2] In many cases of interest, $F_i(t)$ is only a local martingale and not a martingale. But the discrepancy is not "felt" for short times, since the set of paths that can reach the boundary have small probability, in this limit. This is known as the principle of "not feeling the boundary" for small times and is born out by our numerical results. More surprisingly the boundary is not felt, even for quite large times.

$$C_{\mathcal{B}}(\mathbf{F}_0, K, T) = \left(\sum_{i=1}^{n} w_i F_i(0) - K\right)^+ + \frac{1}{2}\int_0^T E\left[\sigma_{N,\mathcal{B}}^2 \, \delta_{\mathcal{E}_K}(\mathcal{B}_t)\right] dt.$$

Letting $u(\mathbf{F}) := \sum_{i=1}^n w_i F_i$ and recalling $|\nabla u| = |\mathbf{w}|$ (where $|\cdot|$ denotes the Euclidean norm) we can re-express this as

$$C_{\mathcal{B}}(\mathbf{F}_0, K, T) = \left(\sum_{i=1}^{n} w_i F_i(0) - K\right)^+ + \\ + \frac{1}{2}\frac{1}{|\mathbf{w}|}\int_0^T \int_{\mathbb{R}^n} |\nabla u(\mathbf{F})| \sigma_{N,\mathcal{B}}^2(\mathbf{F}) \delta_0\left(u(\mathbf{F}) - K\right) p(\mathbf{F}_0, \mathbf{F}, t) d\mathbf{F} dt.$$

By the *co-area formula* (see [10])

$$\int_\Omega |\nabla u(x)| g(x) dx = \int_{-\infty}^{\infty} \int_{u^{-1}(\{s\})} g(x) H_{n-1}(dx) ds$$

(where H_{n-1} denotes the Hausdorff measure on $u^{-1}(\{s\})$), we arrive at

$$\begin{aligned} C_{\mathcal{B}}(\mathbf{F}_0, K, T) &= \left(\sum_{i=1}^{n} w_i F_i(0) - K\right)^+ + \\ &\quad + \frac{1}{2}\frac{1}{|\mathbf{w}|}\int_0^T \int_{-\infty}^{\infty} \delta_0(s-K) \int_{\mathcal{E}_s} \sigma_{N,\mathcal{B}}^2(\mathbf{F}) p(\mathbf{F}_0, \mathbf{F}, t) H_{n-1}(d\mathbf{F}) ds dt \\ &= \left(\sum_{i=1}^{n} w_i F_i(0) - K\right)^+ + \\ &\quad + \frac{1}{2}\int_0^T \frac{1}{|\mathbf{w}|}\int_{\mathcal{E}_K} \sigma_{N,\mathcal{B}}^2(\mathbf{F}) p(\mathbf{F}_0, \mathbf{F}, t) H_{n-1}(d\mathbf{F}) dt. \end{aligned}$$

Note that H_{n-1} coincides with the $(n-1)$-dimensional Lebesgue measure on $\mathcal{E}_K$.

Proposition 2.1 *The value of a call option on a basket $\mathcal{B}$ is given by*

$$C_{\mathcal{B}}(\mathbf{F}_0, K, T) = \left(\sum_{i=1}^{n} w_i F_i(0) - K\right)^+ + \\ + \frac{1}{2}\int_0^T \frac{1}{|\mathbf{w}|}\int_{\mathcal{E}_K} \sum_{i,j=1}^{n} w_i w_j \sigma_i(F_i) \sigma_j(F_j) \rho_{ij} p(\mathbf{F}_0, \mathbf{F}, u) H_{n-1}(d\mathbf{F}) du. \quad (2.1)$$

Using the formula for the basket's local volatility, [1, 12], expressed in the notation introduced above, after canceling common factors we also have the

Proposition 2.2 *The local volatility of the basket option is given by:*

$$\sigma_{loc}^2(K,T)K^2 = \frac{\int\limits_{\mathcal{E}(K)} \sum\limits_{i,j=1}^{n} w_i w_j \sigma_i(F_i)\sigma_j(F_j)\rho_{ij} p(\mathbf{F}_0, \mathbf{F}, T) H_{n-1}(d\mathbf{F})}{\int\limits_{\mathcal{E}(K)} p(\mathbf{F}_0, \mathbf{F}, T) H_{n-1}(d\mathbf{F})}.$$

3 A General Asymptotic Expansion Procedure

The starting point is the basket Carr-Jarrow formula derived above for the calculation of the option prices as in Propositions 2.1 and 2.2 for the calculation of the local volatilities. The next step is to approximate the transition density there using the heat kernel. For reasons that will become clear in the course of the asymptotics, it will be necessary to use the so-called geometric expansion

$$p(\mathbf{F}_0, \mathbf{F}, t) = \frac{1}{(2\pi t)^{\frac{n}{2}}} \sqrt{\det g(\mathbf{F})} e^{-\frac{d^2(\mathbf{F}_0, \mathbf{F})}{2t}} (u_0(\mathbf{F}_0, \mathbf{F}) + t u_1(\mathbf{F}_0, \mathbf{F}) + o(t)). \quad (3.1)$$

For a detailed exposition of the geometrical underpinning of (3.1) we refer to [5, 12, 13, 17, 20]. Here, we just give a very quick reminder.

Remark 3.1 We shall assume that the process $\mathbf{F}_t$ is *locally elliptic* in the sense that ρ is invertible and $\sigma_i(F_i) > 0$ for any $F_i > 0$ and any i, i.e., for $\mathbf{F}$ in the interior of the domain of the process $\mathbf{F}_t$. A rigorous heat kernel expansion for locally elliptic diffusions is given in [2], with the restriction that the expansion is only valid for compact subsets of $\mathbb{R}_+^n$.

The state space $\mathbb{R}_+^n$ is equipped with a Riemannian metric by defining the inverse g^{-1} of the metric tensor by

$$g^{ij}(\mathbf{F}) = \sigma_i(F_i)\rho_{ij}\sigma_j(F_j), \quad 1 \le i, j \le n.$$

Hence, the metric tensor itself is given by

$$g_{ij}(\mathbf{F}) = \sigma_i(F_i)^{-1}\rho^{ij}\sigma_j(F_j)^{-1}, \quad 1 \le i, j \le n,$$

with determinant

$$\det g(\mathbf{F}) = \det\left(\rho^{-1}\right)\prod_{k=1}^{n}\sigma_k(F_k)^{-2}$$

(where ρ^{ij} denotes the (i,j)-component of the inverse matrix ρ^{-1} of the correlation matrix ρ). The (geodesic) distance between two points $\mathbf{F}_0$ and $\mathbf{F}$ is denoted by $d(\mathbf{F}_0, \mathbf{F})$.

The *specific form* of these quantities in the setting of local volatility models has no relevance in our initial asymptotic derivations, which can be obtained for generic versions of these. So, to lighten the notation and streamline the presentation, we first derive the asymptotic expansions without any specific reference to these and then plug in the specific form only at the end of the process in order to produce the required concrete asymptotic expansions.

Plugging the heat kernel expansion (3.1) into the expressions in Propositions 2.1 and 2.2, respectively, we see that we have to compute expressions of the form

$$\frac{1}{(2\pi t)^{n/2}} \int_{\mathcal{E}_K} \Psi(\mathbf{F}) \exp\left(-\frac{d(\mathbf{F}_0, \mathbf{F})^2}{2t}\right) H_{n-1}(d\mathbf{F}), \tag{3.2}$$

where

$$\Psi(\mathbf{F}) = \bar{u}_i(\mathbf{F}_0, \mathbf{F}) := \sqrt{\det g(\mathbf{F})}\sigma^2_{\mathcal{N},\mathcal{B}}(\mathbf{F}) u_i(\mathbf{F}_0, \mathbf{F}), \quad i = 0, 1, \tag{3.3}$$

for the option price and for the numerator in Proposition 2.2 and

$$\Psi(\mathbf{F}) = \hat{u}_i(\mathbf{F}_0, \mathbf{F}) := \sqrt{\det g(\mathbf{F})} u_i(\mathbf{F}_0, \mathbf{F}), \quad i = 0, 1 \tag{3.4}$$

for the denominator in Proposition 2.2.

The integral on the $n-1$ dimensional subspace $\mathcal{E}_K$ of $\mathbb{R}^n$ can be transformed into an integral over $\mathbb{R}^{n-1}$, by eliminating one of the variables. We choose to eliminate the nth one, using the payoff

$$F_n(F_1, \ldots, F_{n-1}, K) = \frac{1}{w_n}\left(K - \sum_{i=1}^{n-1} w_i F_i\right), \tag{3.5}$$

Denoting

$$\mathbf{G} = (F_1, \ldots, F_{n-1}) \in \mathbb{R}^{n-1}_+,$$

$$\mathbf{G}_K = \left\{\mathbf{G} \in \mathbb{R}^{n-1} \; : \; \sum_{i=1}^{n-1} w_i F_i < K\right\},$$

so that for our hyperplane's intersection

$$\mathcal{E}_K \cap \mathbb{R}^n_+ = \left\{\mathbf{F} \in \mathbb{R}^n_+ \; : \; \mathbf{F} = \left(\mathbf{G}, \frac{1}{w_n}\left(K - \sum_{i=1}^{n-1} w_i F_i\right)\right), \mathbf{G} \in \mathbf{G}_K\right\}.$$

Note that the set $\mathbf{G}_K$ is introduced in order to ensure that F_n in (3.5) is non-negative, as it needs to be. The set $\mathcal{E}_K$ is an $n-1$ dimensional hyperplane in $\mathbb{R}^n_+$.

Note that, when we parametrize the hyperplane $\mathcal{E}_k$ using $(F_1, \ldots, F_{n-1})$, as in (3.5)

$$F_K(F_1, \ldots, F_{n-1}) = (F_1, \ldots, F_{n-1}, F_n(F_1, \ldots, F_{n-1}, K)),$$

we will always assume that the weight multiplying F_n is *positive*. This can always be achieved by choosing as the nth asset one of the assets with a positive weight. Then for the surface measure, we have

$$dH_{n-1} = \sqrt{1 + |\nabla F_n|^2} dF_1 \ldots dF_{n-1} = \frac{|\mathbf{w}|}{|w_n|} dF_1 \ldots dF_n.$$

In this notation, with $\Lambda = \frac{d^2}{2}$, the integral (3.2) reads

$$\frac{1}{(2\pi t)^{n/2}} \frac{|\mathbf{w}|}{|w_n|} \int\limits_{\mathbf{G}_K} e^{-\frac{\Lambda(\mathbf{F}_0, \mathbf{F}_K(\mathbf{G}))}{t}} \Psi(\mathbf{F}_K(\mathbf{G})) dF_1 \ldots dF_{n-1} = \\ \frac{1}{(2\pi t)^{n/2}} \frac{|\mathbf{w}|}{|w_n|} \int\limits_{\mathbf{G}_K} e^{-\frac{\Phi(\mathbf{G})}{t}} \Psi(\mathbf{G}) d\mathbf{G}, \tag{3.6}$$

using the notation $\Phi(\mathbf{G}) := \Lambda(\mathbf{F}_0, \mathbf{F}_K(\mathbf{G}))$ and (by abuse of notation) $\Psi(\mathbf{G}) := \Psi(\mathbf{F}_K(\mathbf{G}))$. We now use *Laplace asymptotics* for multiple integrals. The main contribution comes from a neighborhood of the minimum point.

$$\begin{aligned} \mathbf{G}^* &= \arg\min_{\mathbf{G} \in \mathbf{G}_K} d^2(\mathbf{F}_0, (\mathbf{G}, F_n(\mathbf{G}, K)), \\ &= d^2(\mathbf{F}_0, \mathcal{E}_K). \end{aligned} \tag{3.7}$$

Set $\mathbf{F}_K^* = (\mathbf{G}^*, F_n(\mathbf{G}^*, K))$. (Of course, when the option is at the money, we have $\mathbf{G}^* = (F_{0,1}, \ldots, F_{0,n-1})$.)

Order Zero. The zero-th order term in the Laplace expansion of

$$\int\limits_{\mathbf{G}_K} e^{-\frac{\Phi(\mathbf{G})}{t}} \Psi(\mathbf{G}) d\mathbf{G}$$

is identical to the one in [5] except that in the present setting we have $d(\mathbf{F}_0, \mathbf{F}_K^*) = 0$. We get, as in [5]

$$t^{\frac{n-1}{2}} \Psi(\mathbf{G}^*) \times \int_{\mathbb{R}^{n-1}} e^{-\frac{z^T Q z}{2}} dz_2 \ldots dz_n = t^{\frac{n-1}{2}} \Psi(\mathbf{G}^*) \frac{(2\pi)^{\frac{n-1}{2}}}{(\det Q)^{\frac{1}{2}}},$$

where $Q = D^2\Phi(\mathbf{G}^*)$ is the Hessian of Φ at the minimum point. Thus, bringing back the missing factor and taking into account that $\mathbf{F}_K^* = \mathbf{F}_0$ in the current (ATM) setting, we see that the lowest order term in the Laplace expansion of (3.2) is

$$h_0^{\Psi} := \frac{|\mathbf{w}|}{|w_n|} \frac{1}{\sqrt{2\pi t \det Q}} \Psi(\mathbf{F}_0). \tag{3.8}$$

Order One. For obtaining first order implied or local volatility terms in the ATM regime, we need to push the Laplace expansion one step further, i.e., we need one additional term for

$$\int_{\mathbf{G}_K} e^{-\frac{\Phi(\mathbf{G})}{t}} \Psi(\mathbf{G}) d\mathbf{G}$$

Hence, we apply the (multi-variate) Taylor expansion for $\Phi(\mathbf{G}) := \Lambda(\mathbf{F}_0, \mathbf{F}_K(\mathbf{G}))$ up to order 4 around the maximizer $\mathbf{G}^*$, which can be expressed in tensor notation as

$$\Phi(\mathbf{G}) = \Phi(\mathbf{G}^*) + \underbrace{D\Phi(\mathbf{G}^*)}_{=0} \left(\mathbf{G} - \mathbf{G}^*\right) + \frac{1}{2} D^2\Phi(\mathbf{G}^*) \left(\mathbf{G} - \mathbf{G}^*\right)^{\otimes 2} + \\ + \frac{1}{6} D^3\Phi(\mathbf{G}^*) \left(\mathbf{G} - \mathbf{G}^*\right)^{\otimes 3} + \frac{1}{24} D^4\Phi(\mathbf{G}^*) \left(\mathbf{G} - \mathbf{G}^*\right)^{\otimes 4} + \cdots,$$

with

$$D^k\Phi(\mathbf{x})\mathbf{y}^{\otimes k} := \sum_{i_1,\ldots,i_k} \frac{\partial^k}{\partial x_{i_1} \cdots \partial x_{i_k}} \Phi(\mathbf{x}) y_{i_1} \cdots y_{i_k}$$

(This notation makes sense as any multi-linear map on a vector space—such as $D^k\Phi(\mathbf{x})$—corresponds to a linear map—here also denoted by $D^k\Phi(\mathbf{x})$—on the tensor product space). Of course, we are aware that when the option is at the money, the optimal configuration is the same as the initial configuration $\mathbf{F}_0$. Nonetheless, we think that using a different symbol for the optimal configuration at this stage leads to a clearer exposition of the underlying ideas. Likewise, we apply Taylor expansion up to second order for the map $\Psi(\mathbf{G})$ around $\mathbf{G}^*$,

$$\Psi(\mathbf{G}) = \Psi(\mathbf{G}^*) + D\Psi(\mathbf{G}^*) \left(\mathbf{G} - \mathbf{G}^*\right) + \frac{1}{2} D^2\Psi(\mathbf{G}^*) \left(\mathbf{G} - \mathbf{G}^*\right)^{\otimes 2} + \cdots.$$

In the end, we are interested in small-time asymptotics, so we change variables

$$\mathbf{z} := \frac{1}{\sqrt{t}} \left(\mathbf{G} - \mathbf{G}^*\right),$$

so that we can express the above Taylor expansions as expansions in t,

$$\frac{1}{t}\Phi(\mathbf{G}) = \frac{1}{t}\Phi(\mathbf{G}^*) + \frac{1}{2}D^2\Phi(\mathbf{G}^*)\mathbf{z}^{\otimes 2} + \frac{1}{6}D^3\Phi(\mathbf{G}^*)\mathbf{z}^{\otimes 3}\sqrt{t} + \frac{1}{24}D^4\Phi(\mathbf{G}^*)\mathbf{z}^{\otimes 4}t + o(t),$$

and

$$\Psi(\mathbf{G}) = \Psi(\mathbf{G}^*) + D\Psi(\mathbf{G}^*)\mathbf{z}\sqrt{t} + \frac{1}{2}D^2\Psi(\mathbf{G}^*)\mathbf{z}^{\otimes 2}t + o(t).$$

Using the above Taylor expansions, the change of variables, and

$$e^{a\sqrt{t}+bt} = 1 + a\sqrt{t} + \left(\frac{a^2}{2} + b\right)t + o(t),$$

we obtain

$$\int_{\mathbf{G}_K} e^{-\frac{\Lambda(\mathbf{F}_0,\mathbf{F}_K(\mathbf{G}))}{t}}\Psi(\mathbf{G})d\mathbf{G} = t^{(n-1)/2}e^{-\Phi(\mathbf{G}^*)/t}\int_{(\mathbf{G}_K-\mathbf{G}^*)/\sqrt{t}} e^{-\frac{1}{2}D^2\Phi(\mathbf{G}^*)\mathbf{z}^{\otimes 2}} \times\left[1 - \frac{1}{6}D^3\Phi(\mathbf{G}^*)\mathbf{z}^{\otimes 3}\sqrt{t} + \left(\frac{1}{2}\left\{-\frac{1}{6}D^3\Phi(\mathbf{G}^*)\mathbf{z}^{\otimes 3}\right\}^2 - \frac{1}{24}D^4\Phi(\mathbf{G}^*)\mathbf{z}^{\otimes 4}\right)t + o(t)\right] \times \left[\Psi(\mathbf{G}^*) + D\Psi(\mathbf{G}^*)\mathbf{z}\sqrt{t} + \frac{1}{2}D^2\Psi(\mathbf{G}^*)\mathbf{z}^{\otimes 2}t + o(t)\right]d\mathbf{z}. \tag{3.9}$$

In the next step, we approximate the integral by replacing the domain of integration $(\mathbf{G}_K - \mathbf{G}^*)/\sqrt{t}$ by $\mathbb{R}^{n-1}$. Then we can see that the integration kernel in (3.9) is Gaussian with vanishing mean, so that the integral of any odd monomial with respect to the kernel vanishes. Thus, we obtain the expansion

$$\frac{|\mathbf{w}|}{|w_n|}\frac{1}{(2\pi t)^{n/2}}\int_{\mathbf{G}_K} e^{-\frac{\Lambda(\mathbf{F}_0,\mathbf{F}_K(\mathbf{G}))}{t}}\Psi(\mathbf{G})d\mathbf{G} = \left[h_0^\Psi + h_1^\Psi t + o(t)\right], \tag{3.10}$$

with h_0^Ψ defined in (3.8) and

$$h_1^\Psi := \frac{|\mathbf{w}|}{|w_n|}\frac{1}{(2\pi t)^{n/2}}\int_{\mathbb{R}^{n-1}} e^{-\frac{1}{2}\mathbf{z}^T Q\mathbf{z}}\left[\frac{1}{2}D^2\Psi(\mathbf{G}^*)\mathbf{z}^{\otimes 2} - \frac{1}{6}D^3\Phi(\mathbf{G}^*)\mathbf{z}^{\otimes 3} \times D\Psi(\mathbf{G}^*)\mathbf{z} + \frac{1}{2}\left(\frac{1}{6}D^3\Phi(\mathbf{G}^*)\mathbf{z}^{\otimes 3}\right)^2\Psi(\mathbf{G}^*) - \frac{1}{24}D^4\Phi(\mathbf{G}^*)\mathbf{z}^{\otimes 4}\Psi(\mathbf{G}^*)\right]d\mathbf{z}. \tag{3.11}$$

Here we assume that Ψ is polynomially bounded and $\mathbf{F}_0 > 0$ (i.e., all components of $\mathbf{F}_0$ are strictly positive. Indeed, under these assumptions we observe that the error in the approximation (3.10) decays, in fact, like $e^{-1/t}$ by properties of the normal distribution.

Using Isserlis' Theorem (see [14]), the Eq. (3.11) for h_1^Ψ can be computed explicitly.

Lemma 3.2 (Isserlis' theorem for fourth and sixth moments) *For a covariance matrix* $\Sigma \in \mathbb{R}^{d \times d}$ *let* $T^2(\Sigma) \in (\mathbb{R}^d)^{\otimes 4}$ *and* $T^3(\Sigma) \in (\mathbb{R}^d)^{\otimes 6}$ *be the tensors defined by*

$$T^2(\Sigma)_{i_1,\ldots,i_4} = \Sigma_{i_1 i_2}\Sigma_{i_3 i_4} + \Sigma_{i_1 i_3}\Sigma_{i_2 i_4} + \Sigma_{i_1 i_4}\Sigma_{i_2 i_3}$$

and

$$\begin{aligned} T^3(\Sigma)_{i_1,\ldots,i_6} = {} & \Sigma_{i_1 i_2}\Sigma_{i_3 i_4}\Sigma_{i_5 i_6} + \Sigma_{i_1 i_2}\Sigma_{i_3 i_5}\Sigma_{i_4 i_6} + \Sigma_{i_1 i_2}\Sigma_{i_3 i_6}\Sigma_{i_4 i_5} \\ & + \Sigma_{i_1 i_3}\Sigma_{i_2 i_4}\Sigma_{i_5 i_6} + \Sigma_{i_1 i_3}\Sigma_{i_2 i_5}\Sigma_{i_4 i_6} + \Sigma_{i_1 i_3}\Sigma_{i_2 i_6}\Sigma_{i_4 i_5} + \Sigma_{i_1 i_4}\Sigma_{i_2 i_3}\Sigma_{i_5 i_6} \\ & + \Sigma_{i_1 i_4}\Sigma_{i_2 i_5}\Sigma_{i_3 i_6} + \Sigma_{i_1 i_4}\Sigma_{i_2 i_6}\Sigma_{i_3 i_5} + \Sigma_{i_1 i_5}\Sigma_{i_2 i_3}\Sigma_{i_4 i_6} + \Sigma_{i_1 i_5}\Sigma_{i_2 i_4}\Sigma_{i_3 i_6} \\ & + \Sigma_{i_1 i_5}\Sigma_{i_2 i_6}\Sigma_{i_3 i_5} + \Sigma_{i_1 i_6}\Sigma_{i_2 i_3}\Sigma_{i_4 i_5} + \Sigma_{i_1 i_6}\Sigma_{i_2 i_4}\Sigma_{i_3 i_5} + \Sigma_{i_1 i_6}\Sigma_{i_2 i_5}\Sigma_{i_3 i_4}, \end{aligned}$$

$1 \le i_1, \ldots, i_6 \le d$. *For* $Z \sim \mathcal{N}(0, \Sigma)$ *we have*

$$E\left[Z^{\otimes 4}\right] = T^2(\Sigma), \quad E\left[Z^{\otimes 6}\right] = T^3(\Sigma).$$

Hence, we can get an explicit formula also for h_1^Ψ in terms of derivatives of Ψ and Φ—which are easy to compute, but lead to quite long formulas that depend on the individual choice of the local volatility model. These formulas are not included here, as they essentially boil down to exercises in the product rule.

Lemma 3.3 *With the short-hand notation* $\partial_{i_1,\ldots,i_k} := \frac{\partial^k}{\partial F_{i_1} \cdots \partial F_{i_k}}$, *we have*

$$\begin{aligned} h_1^\Psi = \frac{|\mathbf{w}|}{|w_n|}\frac{1}{\sqrt{2\pi t \det Q}}\Bigg[& \frac{1}{2} D^2\Psi(\mathbf{G}^*) Q^{-1} \\ & - \frac{1}{6}\sum_{i_1,\ldots,i_4} (\partial_{i_1,i_2,i_3}\Phi)(\mathbf{G}^*)(\partial_{i_4}\Psi)(\mathbf{G}^*) T^2(Q^{-1})_{i_1,\ldots,i_4} \\ & + \frac{1}{72}\Psi(\mathbf{G}^*)\sum_{i_1,\ldots,i_6}\left(\partial_{i_1,i_2,i_3}\Phi\right)(\mathbf{G}^*)\left(\partial_{i_4,i_5,i_6}\Phi\right)(\mathbf{G}^*) T^3(Q^{-1})_{i_1,\ldots,i_6} \\ & - \frac{1}{24}\Psi(\mathbf{G}^*)\sum_{i_1,\ldots,i_4}\left(\partial_{i_1,\ldots,i_4}\Phi\right)(\mathbf{G}^*) T^2(Q^{-1})_{i_1,\ldots,i_4}\Bigg]. \end{aligned}$$

These results are summarized in

Proposition 3.4 *Assume that we have a locally elliptic local volatility model such that the heat kernel expansion* (3.1) *holds, initial stock prices* $\mathbf{F}_0$ *are strictly positive, and* Ψ *is polynomially bounded. Moreover, we assume that the minimization problem* (3.7) *has a unique solution. Then we have the Laplace expansion*

$$\frac{1}{(2\pi t)^{n/2}}\int_{\mathcal{E}_K}\Psi(\mathbf{F})\exp\left(-\frac{d(\mathbf{F}_0,\mathbf{F})^2}{2t}\right)H_{n-1}(d\mathbf{F}) = h_0^\Psi + t h_1^\Psi + o(t)$$

with h_0^Ψ given in (3.8) *and h_1^Ψ given in Lemma 3.3.*

Remark 3.5 We note that the key assumptions of Proposition 3.4 are not easy to verify in general. We refer to [5] for elements of a proof of the heat kernel expansion and to [3] for further discussion.

The last ingredient needed for the asymptotic expansions of both implied and local volatilities are the heat kernel coefficients u_0 and u_1. As we are assuming the options to be ATM, we only need the heat kernel coefficients on the diagonal.

Lemma 3.6 *For a local volatility model, we have the following formulas for the heat kernel coefficients on the diagonal:*

$$u_0(\mathbf{F}, \mathbf{F}) = 1,$$

$$u_1(\mathbf{F}, \mathbf{F}) = \frac{1}{4} \sum_{i=1}^{n} \sigma_i(F_i)\sigma_i''(F_i) - \frac{1}{8} \sum_{i,j=1}^{n} \sigma_i'(F_i)\rho^{ij}\sigma_j'(F_j),$$

where, as usual, ρ^{ij} denotes the (i, j)-component of ρ^{-1}.

Proof Note that the infinitesimal generator A of the process $\mathbf{F}(t)$ can be expressed (using the summation convention) as

$$A = \frac{1}{2}\Delta - \frac{1}{2} f_i(\mathbf{F}) \frac{\partial}{\partial F_i},$$

where

$$\Delta = \frac{1}{\sqrt{\det g}} \frac{\partial}{\partial F_i} g^{ij} \sqrt{\det g} \frac{\partial}{\partial F_j}$$

denotes the Laplace-Beltrami operator associated to g and the vector field f is given by

$$f_i(\mathbf{F}) = \sigma_i(\mathbf{F})\sigma_i'(F_i), \quad i = 1, \ldots, n.$$

As indicated in (3.1), the transition density of the process $\mathbf{F}(t)$ satisfies (under certain assumptions, see Proposition 3.4 and Remark 3.5)

$$p(\mathbf{F}_0, \mathbf{F}, T) = \frac{1}{(2\pi T)^{n/2}} \sqrt{\det g(\mathbf{F})} e^{-\frac{d(\mathbf{F}_0, \mathbf{F})^2}{2T}} \left(u_0(\mathbf{F}_0, \mathbf{F}) + T u_1(\mathbf{F}_0, \mathbf{F}) + o(T)\right),$$

where $d(\mathbf{F}_0, \mathbf{F})$ is the geodesic distance between $\mathbf{F}_0$ and $\mathbf{F}$ and u_0 and u_1 are the heat kernel coefficients.

The order zero heat kernel coefficient is given by $u_0(\mathbf{F}_0, \mathbf{F}) = \sqrt{\Delta(\mathbf{F}_0, \mathbf{F})}$ $e^{-\frac{1}{2}\int_\gamma \langle f, \dot{\gamma}_g \rangle}$, where $\int_\gamma \langle f, \dot{\gamma}_g \rangle$ is understood as integral along the geodesic γ joining $\mathbf{F}_0$ and $\mathbf{F}$ and $\Delta(\mathbf{F}_0, \mathbf{F})$ is the Van Vleck–De Witt determinant,

$$\Delta(\mathbf{F}_0, \mathbf{F}) = \frac{1}{\sqrt{\det g(\mathbf{F}_0) \det g(\mathbf{F})}} \det \left(-\frac{1}{2} \frac{\partial^2 d^2(\mathbf{F}_0, \mathbf{F})}{\partial \mathbf{F}_0 \partial \mathbf{F}} \right).$$

On the diagonal, we clearly have $\int_\gamma \langle f , \dot{\gamma}_g \rangle = 0$ and for any local volatility model we have $\Delta(\mathbf{F}_0, \mathbf{F}) \equiv 1$, as the geometry is isomorphic to the Euclidean geometry by the coordinate transformation $\mathbf{F} \mapsto L\mathbf{y}$, where $L\rho L^T = \mathrm{Id}$ and $y_i := \int_0^{F_i} \sigma_i(u)^{-1} du$. Hence, $u(\mathbf{F}, \mathbf{F}) = 1$.

For the first order heat kernel coefficient, we refer to [16, Eq. (4.1)], where it is shown that

$$u_1(\mathbf{F}, \mathbf{F}) = \frac{1}{6}\kappa + \frac{1}{4} \operatorname{div}_g f(\mathbf{F}) - \frac{1}{8} |f(\mathbf{F})|_g^2 .$$

Here, κ denotes the scalar curvature, which vanishes for local volatility model due to the isomorphism with the Euclidean geometry already used above. (Note that [16] consider the heat kernel corresponding to $\Delta + f$, whereas we consider the operator $\frac{1}{2}\Delta + \frac{1}{2} f$. Hence, we evaluate the formula obtained in [16, Eq. (4.1)] at $t/2$ instead of t.) For the remaining terms we have

$$\operatorname{div}_g f(\mathbf{F}) = \frac{1}{\sqrt{\det g(\mathbf{F})}} \frac{\partial}{\partial F_i} \left[f_i(\mathbf{F} \sqrt{\det g(\mathbf{F})} \right] = \sigma_i(F_i) \sigma_i''(F_i),$$

$$|f(\mathbf{F})|_g^2 = g_{ij}(\mathbf{F}) f_i(\mathbf{F}) f_j(\mathbf{F}) = \sigma_i'(F_i) \rho^{ij} \sigma_j'(F_j). \qquad \square$$

Finally, we can explicitly compute the determinant of the Hessian Q of Φ at $\mathbf{G}^* = (F_{0,1}, \ldots, F_{0,n-1})$ in the ATM regime.

Lemma 3.7 *The Hessian Q of Φ satisfies*

$$\det Q = \frac{\sum_{i,j=1}^n w_i \sigma_i(F_{0,i}) \rho_{ij} w_j \sigma_j(F_{0,j})}{w_n^2 \det \rho \prod_{i=1}^n \sigma_i(F_{0,i})^2} = \sigma_{N,\mathcal{B}}^2(\mathbf{F}_0) \det g(\mathbf{F}_0)/w_n^2.$$

The proof of Lemma 3.7 is deferred to the Appendix.

4 Basket Local Volatility

The numerator in the right hand side of the formula in Proposition 2.2 is given by

$$\frac{1}{(2\pi t)^{n/2}} \int_{\mathcal{E}_K} (\bar{u}_0(\mathbf{F}_0, \mathbf{F}) + t\bar{u}_1(\mathbf{F}_0, \mathbf{F})) \exp\left(-\frac{d(\mathbf{F}_0, \mathbf{F})^2}{2t} \right) H_{n-1}(d\mathbf{F}) =$$
$$h_0^{\bar{u}_0} + t \left(h_1^{\bar{u}_0} + h_0^{\bar{u}_0} \right) + o(t),$$

where, by abuse of notation, we denote the function $\mathbf{F} \mapsto \bar{u}_i$ by $\bar{u}_i$ again, $i = 0, 1$. For the denominator, we get

$$\frac{1}{(2\pi t)^{n/2}} \int_{\mathcal{E}_K} \left(\hat{u}_0(\mathbf{F}_0, \mathbf{F}) + t\hat{u}_1(\mathbf{F}_0, \mathbf{F})\right) \exp\left(-\frac{d(\mathbf{F}_0, \mathbf{F})^2}{2t}\right) H_{n-1}(d\mathbf{F}) =$$
$$h_0^{\hat{u}_0} + t\left(h_1^{\hat{u}_0} + h_0^{\hat{u}_1}\right) + o(t).$$

As

$$\frac{a_1 + b_1 t + o(t)}{a_2 + b_2 t + o(t)} = \frac{a_1}{a_2} + \frac{a_2 b_1 - a_1 b_2}{a_2^2} t + o(t),$$

we arrive at

$$\sigma_{\text{loc}}(K, T)^2 K^2 = \frac{h_0^{\bar{u}_0}}{h_0^{\hat{u}_0}} + \frac{h_0^{\hat{u}_0}(h_1^{\bar{u}_0} + h_0^{\bar{u}_1}) - h_0^{\bar{u}_0}(h_1^{\hat{u}_0} + h_0^{\hat{u}_1})}{\left(h_0^{\hat{u}_0}\right)^2} T + o(T).$$

As $\bar{u}_0 = \sigma_{N,\mathcal{B}}^2 \hat{u}_0$, we can easily simplify

$$\frac{h_0^{\bar{u}_0}}{h_0^{\hat{u}_0}} = \sigma_{N,\mathcal{B}}^2(\mathbf{F}_0).$$

For the first order term, we note that all the terms $h_i^{\bar{u}_j}$ and $h_i^{\hat{u}_j}$ have the common factor $\frac{|w|}{|w_n|} \frac{1}{\sqrt{2\pi T \det Q}}$, which, hence, cancels out in the first order term—in particular, implying that the "first order term" is really first order in T. Thus, we get

Proposition 4.1 *For $K = \overline{F}_0 = \sum_{i=1}^n w_i F_{0,i}$, the basket local volatility has the asymptotic expansion $\sigma_{loc}^2(T, K) = \sigma_{loc,0}^2(K) + \sigma_{loc,1}^2(K)T + o(T)$, with*

$$\sigma_{loc,0}^2(K) = \frac{\sigma_{N,\mathcal{B}}^2(\mathbf{F}_0)}{K^2},$$
$$\sigma_{loc,1}^2(K) = \frac{h_0^{\hat{u}_0}(h_1^{\bar{u}_0} + h_0^{\bar{u}_1}) - h_0^{\bar{u}_0}(h_1^{\hat{u}_0} + h_0^{\hat{u}_1})}{\left(h_0^{\hat{u}_0}\right)^2 K^2}.$$

We recall the definition

$$\sigma_{N,\mathcal{B}}(\mathbf{F})^2 = \sum_{i,j=1}^n w_i \sigma_i(F_i) \rho_{ij} w_j \sigma_j(F_j).$$

5 Implied Volatility

The strategy for obtaining an asymptotic expansion for the implied volatility is as follows: we first compute an asymptotic expansion of the basket option price in our local volatility model, then we compare coefficients with the short time expansion of the corresponding call option price in the Black-Scholes or Bachelier model, respectively. Hence, we first apply our general asymptotic expansion obtained in Proposition 3.4 to the Carr-Jarrow formula from Proposition 2.1, getting (for $K = \overline{F}_0$)

Now we can insert these results back into Proposition 2.1, and we obtain

$$\begin{aligned} C_{\mathcal{B}}(\mathbf{F}_0, K, T) &= \frac{1}{2|w|}\int_0^T \left(h_0^{\bar{u}_0} + t\left(h_1^{\bar{u}_0} + h_0^{\bar{u}_1}\right) + o(\sqrt{t})\right) dt \\ &= \frac{1}{2}\int_0^T \left(\frac{g_0^{\bar{u}_0}}{\sqrt{t}} + \sqrt{t}\left(g_1^{\bar{u}_0} + g_0^{\bar{u}_1}\right) + o(\sqrt{t})\right) dt \\ &= g_0^{\bar{u}_0}\sqrt{T} + \frac{1}{3}\left(g_1^{\bar{u}_0} + g_0^{\bar{u}_1}\right) T^{3/2} + o\left(T^{3/2}\right), \end{aligned}$$

where

$$g_i^{\bar{u}_j} := \frac{\sqrt{t}}{|w|} h_i^{\bar{u}_j}, \quad i, j = 0, 1 \tag{5.1}$$

is independent of t. Finally, using (3.8) together with (3.3), and Lemma 3.7, we get

$$g_0^{\bar{u}_0} = \frac{\sigma_{N,\mathcal{B}}^2(\mathbf{F}_0)\sqrt{\det g(\mathbf{F}_0)}}{|w_n|\sqrt{2\pi}\det Q} = \frac{\sigma_{N,\mathcal{B}}(\mathbf{F}_0)}{\sqrt{2\pi}}.$$

Proposition 5.1 *The expansion of the call prices (at-the-money) in drift-less local volatility models is asymptotically equivalent, to first order, to*

$$C_{\mathcal{B}}(\mathbf{F}_0, K, T) = \frac{\sigma_{N,\mathcal{B}}(\mathbf{F}_0)}{\sqrt{2\pi}} + \frac{1}{3}\left(g_1^{\bar{u}_0} + g_0^{\bar{u}_1}\right) T^{3/2} + o(T^{3/2})$$

as $T \to 0$.

In the final step, we compute an expansion of the implied volatility with respect to either Black-Scholes or Bachelier model. Let us consider the prices of call options with stock price $\overline{F}_0 = \sum_{i=1}^n w_i F_{0,i} = K$ in the Black-Scholes and Bachelier models, assuming that the respective volatilities are of the form $\sigma_{BS} = \sigma_{BS,0} + T\sigma_{BS,1}$ and $\sigma_{Bach} = \sigma_{Bach,0} + T\sigma_{Bach,1}$. We obtain the well known formulas

$$\begin{aligned} C_{BS}(\overline{F}_0, K, T) =& C_{BS}(K, K, T) = \\ & \frac{K}{\sqrt{2\pi}}\sigma_{BS,0}\sqrt{T} + \frac{K}{\sqrt{2\pi}}\left[\sigma_{BS,1} - \frac{1}{24}\sigma_{BS,0}^3\right] + o(T^{3/2}), \end{aligned}$$

$$C_{Bach}(\overline{F}_0, K, T) = C_{Bach}(K, K, T) = \frac{K}{\sqrt{2\pi}}\sigma_{Bach,0}\sqrt{T} + \frac{K}{\sqrt{2\pi}}\sigma_{Bach,1}T^{3/2} + o(T^{3/2}).$$

5.1 Zeroth Order Implied Volatility

Despite being well-known, we recall the zeroth order implied volatility coefficients and some of their properties. By comparison of coefficients, see Proposition 5.1 and the above expansions for C_{BS} and C_{Bach}, respectively, we find that

$$\sigma_{BS,0} = \sigma_{Bach,0} = \frac{1}{|w_n|K}\bar{u}_0(\mathbf{F}_0, \mathbf{F}_0)\,(\det Q)^{-\frac{1}{2}} = \frac{\sigma_{\mathcal{N},\mathcal{B}}(\mathbf{F}_0)}{\overline{F}_0}, \tag{5.2}$$

where we also used $\overline{F}_0 = K$. Note, in particular, that the basket implied volatility (5.2) can be interpreted as a weighted mean of the individual components' (ATM) implied volatilities in the sense that $(\sigma_{BS,0})^2 = \sum_{i,j=1}^{n} \rho_{ij} w_i \frac{F_{0,i}}{K}\sigma^i_{BS,0} w_j \frac{F_{0,j}}{K}\sigma^j_{BS,0}$.

Remark 5.2 The right hand side in Eq. (5.2) is nothing but the local volatility of the basket $\sum_{i=1}^{n} w_i F_i$ at $\mathbf{F}_0$ in the Black-Scholes (i.e., log-normal) sense. Hence, we have obtained that the zero order term in the small time expansion of the implied volatility of the basket is equal to its local volatility when we consider an ATM option. That result is not surprising in light of [11], where similar results were obtained (in one-dimensional models). In this sense, one could even take (5.2) as an ex-post justification of Lemma 3.7.

5.2 First Order Implied Volatility

The first order implied volatilities in the Black Scholes and the Bachelier model do not coincide any more. Indeed, we immediately have the first order correction term in the Bachelier model

$$\sigma_{Bach,1} = \frac{\sqrt{2\pi}}{3K}\left(g_1^{\bar{u}_0} + g_0^{\bar{u}_1}\right). \tag{5.3}$$

On the other hand, for the Black-Scholes model we have

$$\sigma_{BS,1} = \frac{\sqrt{2\pi}}{3K}\left(g_1^{\bar{u}_0} + g_0^{\bar{u}_1}\right) + \frac{\sigma^3_{BS,0}}{24} = \sigma_{Bach,1} + \frac{\sigma^3_{BS,0}}{24}, \tag{5.4}$$

implying that implied volatility quoted in the Black-Scholes framework is strictly larger than the implied volatility in the Bachelier framework up to first order—the prices are, of course, equal up to first order.

6 Numerical Results

6.1 The CEV Model

As in [5], we consider the CEV model for the numerical examples. The CEV model is a special case of the general local volatility model considered so far, where the local volatilities are given by

$$\sigma_i(F_i) = \xi_i F_i^{\beta_i}, \quad i = 1, \ldots, n,$$

for some parameters $\xi_i \geq 0$ and $\beta_i > 0$. In fact, the most realistic scenario here is $0 < \beta_i \leq 1$. Note that we allow $\beta_i < 1/2$, which implies degenerate densities of $\mathbf{F}_t$ at the boundary.

6.2 Implementation of the Approximate Formulas and Simulation

Implementation of the zero order terms of the implied volatilities in either Black-Scholes or Bachelier setting is, of course, easy using (5.2). On the other hand, the formulas for $\sigma_{BS,1}$ and $\sigma_{Bach,1}$ are much less straightforward to implement. While the formulas in the ATM case are fully explicit (unlike in [5]) an efficient implementation is much less trivial. The formula for h_1 in Lemma 3.3, for instance, depends on the derivatives up to order four of the squared Riemannian distance at $\mathbf{F}_0$ and on the Jacobi matrix of $\mathbf{F} \mapsto u_0(\mathbf{F}_0, \mathbf{F})$. Already the evaluation of the $(n-1) \times (n-1) \times (n-1) \times (n-1)$ tensor $D^4\Phi$ can be very time-consuming, if a naive implementation is used, which does not take into account that most derivatives actually vanish. But even when more efficient implementations are used, the sheer size of the tensor may impose limitations on the dimension of the problem. So far, we have implemented (3.11) in Mathematica using symbolic differentiation of the squared Riemannian distance and the zeroth order heat kernel coefficient u_0, which works for small dimensions, up to $n = 5$, say.

As in the paper [5], we compare the approximate prices against prices obtained from sophisticated Monte Carlo simulation. Here, the CEV-SDE is discretized using the Ninomiya-Victoir scheme [18], which is a second order weak approximation scheme based on a splitting of the generator. Strictly speaking, the CEV process violates the strong regularity assumption of that scheme, especially at the boundary of the domain, but, as often in equity modelling, we do empirically observe second order convergence for CEV-baskets, yet another beneficial effect of "not feeling the boundary". For variance reduction, we combine the discretization with the mean value Monte Carlo method, see [19]. This is a variant of the control variate technique, where a linear combination of one-dimensional geometrical Brownian motions is used as control variate. More precisely, we freeze each component but one of the basket, and replace the dynamics of the remaining basket by a corresponding Black-Scholes dynamics. In the resulting model, the true option price can be explicitly calculated. Finally, we choose a linear combination of those partially frozen model so as to minimize the variance of the Monte Carlo estimator.

The expectation of the random variable obtained by combining the Ninomiya-Victoir discretization of the CEV process and the mean value Monte Carlo method is the approximated using Sobol numbers. In some sense, this contradicts the above motivation for the variance reduction, but we do find empirically that the integration error for a Quasi Monte Carlo estimator is also reduced by the variance reduction, i.e., the variance reduction also seems to reduce the number of most relevant dimensions of the integration problem. Finally, we sacrifice some of the accuracy available by the combination of the three techniques mentioned so far by introducing a random shift of the Sobol numbers, i.e., we use the Randomized Quasi Monte Carlo technique, see L'Ecuyer [15]. In this way, we can obtain reliable computable error bounds for the integration error.

6.3 Numerical Example

We consider a three-dimensional spread option, which is determined by the following parameters:

$$\mathbf{F}_0 = \begin{pmatrix} 8 \\ 17 \\ 12 \end{pmatrix}, \quad \sigma = \begin{pmatrix} 0.4 \\ 0.8 \\ 0.7 \end{pmatrix}, \quad \beta = \begin{pmatrix} 0.7 \\ 0.5 \\ 0.3 \end{pmatrix}, \quad \mathbf{w} = \begin{pmatrix} -1 \\ 1 \\ 1 \end{pmatrix},$$

with a correlation matrix

$$\rho = \begin{pmatrix} 1 & 0.9167390 & 0.7425194 \\ 0.9167390 & 1 & 0.8099573 \\ 0.7425194 & 0.8099573 & 1 \end{pmatrix}.$$

We compute the ATM price, i.e., the option price at $K = 21$, for maturities $T \in \{0.5, 1, 2, 5, 10\}$ years, which we compare with the zeroth and first order prices in the corresponding Bachelier model. We also report $\sigma_{Bach,0} = 0.1487036$ and $\sigma_{Bach,1} = -6.72781 \times 10^{-5}$. Note that the "error bounds" reported in Tables 1 and 2 are upper estimates for the integration error (i.e., quasi Monte Carlo error) for the reference values. Hence, numbers obtained from the first order approximation formula are within the error bounds around the reference values.

In Fig. 1, we plot (linear interpolations of) the relative errors of the zeroth and first order approximate pricing formulas close to the money (as obtained in [5]) and compare them to the ATM-formulas represented by circles. We see that the accuracy is extremely good in both cases, and that our approximation formulas for ATM CEV-

Table 1 Prices

Time	Price	0th order price	1st order price	Error bound
0.5	0.88073	0.88092	0.88072	2.43e-05
1	1.24525	1.24581	1.24524	4.63e-05
2	1.76023	1.76184	1.76024	8.90e-05
5	2.77895	2.78571	2.77941	3.21e-04
10	3.91968	3.93959	3.92176	5.92e-04

Error bounds given correspond to the (quasi) Monte Carlo error in the numerical scheme. The discretization error is of higher order

Table 2 Relative errors

Time	0th order rel. error	1st order rel. error	Error bound
0.5	2.19e-04	6.85e-06	2.43e-05
1	4.49e-04	3.80e-06	4.63e-05
2	9.15e-04	9.02e-06	8.90e-05
5	2.43e-03	1.65e-04	3.21e-04
10	5.08e-03	5.33e-04	5.92e-04

Error bounds given correspond to the (quasi) Monte Carlo error in the numerical scheme. The discretization error is of higher order

basket options nicely interpolate the formulas available away from the money. Indeed, deviations from the non-ATM values only appears at very small orders of magnitude in the logarithmic scale of Fig. 1 (where the Monte Carlo error contained in the reference values probably dominates). For the sake of completeness, Fig. 2 reports the absolute errors of the respective asymptotic formulas over a wide range of strike prices, indicating that the asymptotic formulas exhibit their worst quality ATM.

Appendix A: Proof of Lemma 3.7

We present a proof of Lemma 3.7. Recall that we want to compute the determinant of the Hessian Q of the map

$$\Phi(\mathbf{G}) := \frac{1}{2} d\left(\mathbf{F}_0, (\mathbf{G}, F_N(\mathbf{G}, K))\right)^2$$

evaluated at $\mathbf{G} = \left(F_{0,1}, \ldots, F_{0,n-1}\right)$. Let $\mathfrak{S}_i(x)$ denote the anti-derivative of $1/\sigma_i$ satisfying (for simplicity) $\mathfrak{S}_i(F_{0,i}){=}0$. Now consider the change of variables $\mathbf{F} \to \mathbf{y}$ with $y_i := \mathfrak{S}_i(F_i)$, $i = 1, \ldots, n$. As verified in [5], this transformation turns the Riemannian geometry introduced above into an (almost) Euclidean geometry, with

$$d(\mathbf{F}_0, \mathbf{F})^2 = \mathbf{y}^T \rho^{-1} \mathbf{y}.$$

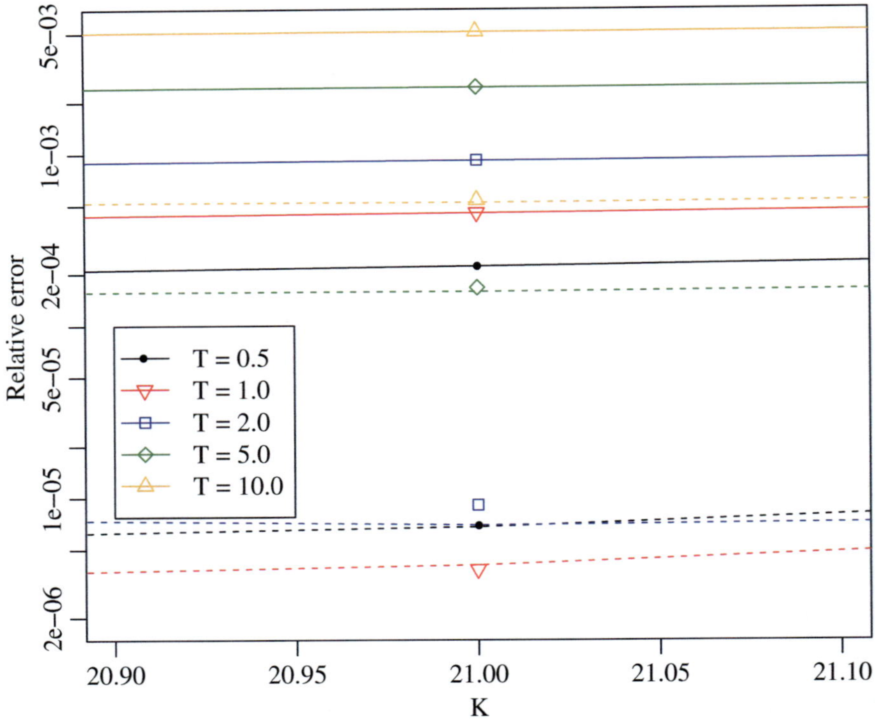

Fig. 1 Relative errors. *Solid lines* correspond to prices obtained from (non-ATM) zeroth order approximate formulas, *dashed lines* to (non-ATM) first order approximate formulas. The corresponding ATM-approximate prices are represented by *circles* and other symbols. Note that the option is ATM for $K = 21$

Of course, the constraint on $\mathbf{F}$ translates into a constraint on $\mathbf{y}$, which can be removed by eliminating one variable. Indeed, setting $\mathbf{x} := (y_1, \dots, y_{n-1})$, we get

$$y_n(\mathbf{x}) = \mathfrak{S}_n(F_n) = \mathfrak{S}_n\left(\frac{1}{w_n}\left(K - \sum_{j=1}^{n-1} w_j \mathfrak{S}_j^{-1}(y_j)\right)\right).$$

This way, we understand $\Phi(\mathbf{G})$ as a function $\varphi(\mathbf{x})$ in the new (reduced) coordinates, and obtain for the Hessian

$$H_{\mathbf{G}}\Phi(\mathbf{G}) = J(\mathbf{G})^T H_{\mathbf{x}}\varphi(\mathbf{x}) J(\mathbf{G}),$$

where $H_{\mathbf{G}}$ and $H_{\mathbf{x}}$ denote the Hessians in the $\mathbf{G}$- and $\mathbf{x}$-coordinates, respectively, and $J(\mathbf{G})$ denotes the Jacobian matrix of the change of coordinates $\mathbf{G} \to \mathbf{x}$. As $\mathfrak{S}_i' = 1/\sigma_i$, we have $J(\mathbf{G}) = \operatorname{diag}(1/\sigma_1(F_1), \dots, 1/\sigma_{n-1}(F_{n-1}))$. Regarding the matrix $H_{\mathbf{x}}\varphi$, an elementary calculation using the fact that $\mathbf{F} = \mathbf{F}_0$ corresponds to $\mathbf{y} = 0$, we obtain

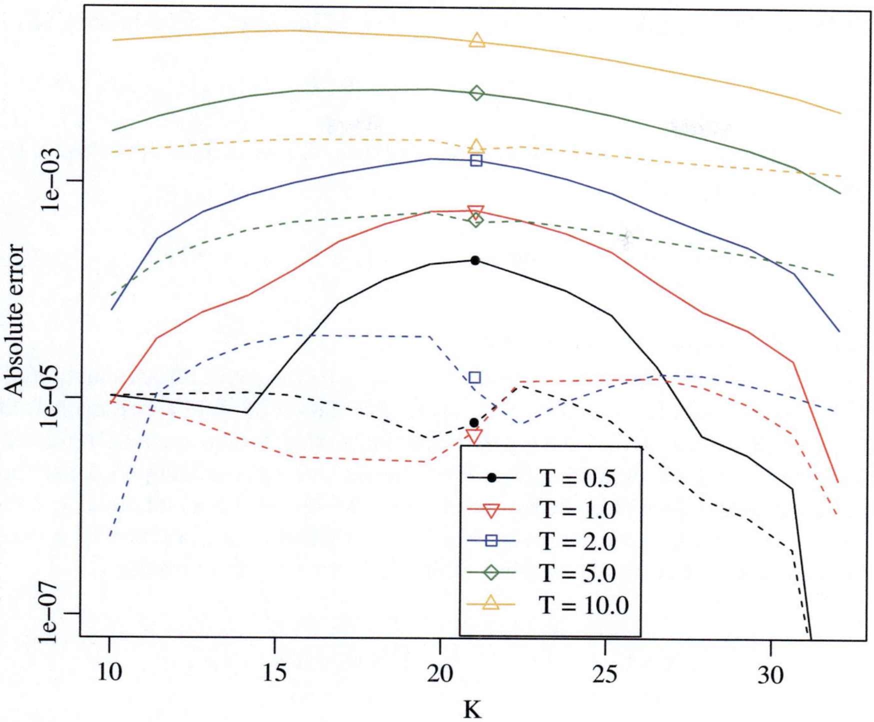

Fig. 2 Absolute errors. *Solid lines* correspond to prices obtained from (non-ATM) zeroth order approximate formulas, *dashed lines* to (non-ATM) first order approximate formulas. The corresponding ATM-approximate prices are represented by *circles* and other symbols. Note that the option is ATM for $K = 21$

$$H_{\mathbf{x}}\varphi(0) = \left(\rho^{ij} - \rho^{in}\frac{w_j\sigma_j(F_{0,j})}{w_n\sigma_n(F_{0,n})} - \rho^{jn}\frac{w_i\sigma_i(F_{0,i})}{w_n\sigma_n(F_{0,n})} + \rho^{nn}\frac{w_i\sigma_i(F_{0,i})w_j\sigma_j(F_{0,j})}{w_n^2\sigma_n(F_{0,n})^2}\right)_{i,j=1}^{n-1}.$$

From the structure of the above expression and the expression in Lemma 3.7, we see that we may assume that $w_i = 1, i = 1, \dots, n$, and $\sigma_n(F_{0,n}) = 1$. In this case, we are left to prove that the determinant of the matrix

$$A := \left(\rho^{ij} - \rho^{in}s_j - \rho^{jn}s_i + \rho^{nn}s_is_j\right)_{i,j=1}^{n-1}$$

is equal to the expression $a := \mathbf{s}^T\rho\mathbf{s}/\det\rho$, where we used the short-hand notation $s_i = \sigma_i(F_{0,i}), i = 1, \dots, n-1$, and $s_n = 1$, and $\mathbf{s} = (s_1, \dots, s_n)$.

As both $\det A$ and a are polynomials in $s_1, \dots, s_{n-1}$, we prove this equality by establishing that they have the same coefficients. Here, Cramer's rule is the essential tool:

$$B^{-1} = \frac{1}{\det B}\mathrm{Adj}(B),$$

where the *adjugate* matrix Adj B is the transpose of the matrix of co-factors, i.e.,

$$(\text{Adj } B)_{ij} = (-1)^{i+j} \det B_{\hat{j}\hat{i}},$$

with $B_{\hat{j}\hat{i}}$ being obtained from B by removing the jth row and the ith column. By symmetry, we hence have

$$\frac{\rho_{ij}}{\det \rho} = (-1)^{i+j} \det \rho^{-1}_{\hat{i}\hat{j}}, \quad \forall (i, j) \in \{1, \dots, n-1\}^2, \tag{A.1}$$

where $\rho^{-1}_{\hat{i}\hat{j}}$ is understood in the sense of $(\rho^{-1})_{\hat{i}\hat{j}}$.

Let us also establish a few notations. Let S_{n-1} be the set of all permutations of $\{1, \dots, n-1\}$ and let, similarly, $S(A; B)$ denote the set of all bijective maps from $A \subset \mathbb{N}$ to $B \subset \mathbb{N}$, with A, B having the same (finite) size. Moreover, the definition of the signature sign is extended to $S(A; B)$ in the obvious way (as being ± 1 depending on the number of inversions being even or odd). Moreover, for a monomial x in the variables $s_1, \dots, s_{n-1}$ we denote by $\pi_x p$ the coefficient of any polynomial p w.r.t. the monomial x. In order to establish Lemma 3.7, we need to prove that

$$\forall x \in \bigcup_{k=0}^{2(n-1)} \{s_1, \dots, s_{n-1}\}^k : \ \pi_x \det A = \pi_x a.$$

We distinguish different cases according to the degree.

Case 0. For $\deg x = 0$, i.e., $x = 1$, we have

$$\pi_1 \det A = \sum_{\sigma \in S_{n-1}} \text{sign}(\sigma) \prod_{i=1}^{n-1} \rho^{i\sigma(i)} = \det \rho^{-1}_{\hat{n}\hat{n}} = \text{Adj}(\rho^{-1})_{nn} = \frac{\rho_{nn}}{\det \rho} = \pi_1 a.$$

Case 1. For some fixed s_k we have

$$\begin{aligned}
\pi_{s_k} \det A = \sum_{\sigma \in S_{n-1}} \text{sign}(\sigma)(-1) \Bigg[& \rho^{\sigma^{-1}(k)n} \prod_{i \in \{1,\dots,n-1\} \setminus \{\sigma^{-1}(k)\}} \rho^{i\sigma(i)} \\
& + \rho^{\sigma(k)n} \prod_{i \in \{1,\dots,n-1\} \setminus \{k\}} \rho^{i\sigma(i)} \Bigg] \\
= -2 \sum_{\sigma \in S_{n-1}} \text{sign}(\sigma) & \rho^{\sigma(k)n} \prod_{i \in \{1,\dots,n-1\} \setminus \{k\}} \rho^{i\sigma(i)}
\end{aligned}$$

by symmetry of ρ^{-1}. There is a one-to-one correspondence between S_{n-1} and $S(\{1,\ldots,n\}\setminus\{k\};\{1,\ldots,n-1\})$ given by $\sigma\mapsto\tilde{\sigma}$ defined by

$$\tilde{\sigma}(i)=\begin{cases}\sigma(i), & i\in\{1,\ldots,n-1\}\setminus\{k\},\\ \sigma(k), & i=n.\end{cases}$$

Moreover, one can see that $\operatorname{sign}(\tilde{\sigma})=(-1)^{k+n-1}\operatorname{sign}(\sigma)$. Hence, we obtain

$$\begin{aligned}\pi_{s_k}\det A &= -2\sum_{\sigma\in S_{n-1}}\operatorname{sign}(\sigma)\rho^{n\tilde{\sigma}(n)}\prod_{i\in\{1,\ldots,n-1\}\setminus\{k\}}\rho^{i\tilde{\sigma(i)}}\\ &= 2(-1)^{k+n}\sum_{\tilde{\sigma}\in S(\{1,\ldots,n\}\setminus\{k\};\{1,\ldots,n-1\})}\operatorname{sign}(\tilde{\sigma})\rho^{n\tilde{\sigma}(n)}\prod_{i\in\{1,\ldots,n-1\}\setminus\{k\}}\rho^{i\tilde{\sigma(i)}}\\ &= 2(-1)^{k+n}\det\rho^{-1}_{\hat{k}\hat{n}}\\ &= 2\operatorname{Adj}(\rho^{-1})_{kn}=\frac{2\rho_{kn}}{\det\rho}=\pi_{s_k}a.\end{aligned}$$

Case 2. We consider $x=s_k s_l$. For simplicity, we assume $k=l$ ($k\neq l$ works analogously). We have

$$\begin{aligned}\pi_{s_k^2}\det A=\sum_{\sigma\in S_{n-1}}\operatorname{sign}(\sigma)\Bigg[&\mathbf{1}_{k=\sigma(k)}\rho^{nn}\prod_{i\in\{1,\ldots,n-1\}\setminus\{k\}}\rho^{i\sigma(i)}+\\ &+\mathbf{1}_{k\neq\sigma(k)}\rho^{\sigma(k)n}\rho^{\sigma^{-1}(k)n}\prod_{i\in\{1,\ldots,n-1\}\setminus\{k,\sigma^{-1}(k)\}}\rho^{i\sigma(i)}\Bigg].\end{aligned}$$

We construct a bijective map from S_{n-1} to $S(\{1,\ldots,n\}\setminus\{k\};\{1,\ldots,n\}\setminus\{k\})$ by mapping $\sigma\in S_{n-1}$ to $\tilde{\sigma}$ defined by

$$\tilde{\sigma}(i)=\begin{cases}\sigma(i), & i\in\{1,\ldots,n-1\}\setminus\{k\},\\ n, & i=n,\end{cases}$$

for the case $k=\sigma(k)$ and

$$\tilde{\sigma}(i)=\begin{cases}\sigma(i), & i\in\{1,\ldots,n-1\}\setminus\{k,\sigma^{-1}(k)\},\\ n, & i=\sigma^{-1}(k),\\ \sigma(k), & i=n,\end{cases}$$

else. Note that it is easy to see that $\operatorname{sign}(\sigma)=\operatorname{sign}(\tilde{\sigma})$. Hence, we have

$$\pi_{s_k^2} \det A = \sum_{\sigma \in S_{n-1}} \operatorname{sign}(\sigma) \prod_{i \in \{1,\dots,n\}\setminus\{k\}} \rho^{i\tilde{\sigma}(i)}$$
$$= \sum_{\tilde{\sigma} \in S(\{1,\dots,n\}\setminus\{k\};\{1,\dots,n\}\setminus\{k\})} \operatorname{sign}(\tilde{\sigma}) \prod_{i \in \{1,\dots,n\}\setminus\{k\}} \rho^{i\tilde{\sigma}(i)}$$
$$= \det \rho^{-1}_{\hat{k}\hat{k}} = \pi_{s_k^2} a.$$

Higher order terms. Regarding the higher order terms, we note that $\pi_x a = 0$ for any monomial of degree larger than two. Therefore, the same should be true for $\det A$, where it does not to seem to follow from an obvious argument. Note that we only need to consider polynomials where each individual variable s_k appears at most two times, as any other monomial cannot appear in $\det A$ by the definition of A and of the determinant. But any coefficient of $\det A$ with respect to such monomials can be understood as the determinant of a matrix $\widetilde{\rho^{-1}}$, which is obtained from ρ^{-1} by omitting one row and one column *and* by replacing some rows/columns by copies of other rows/columns. Of course, any such matrix $\tilde{\rho}$ has vanishing determinant, implying that $\pi_x \det A = 0$. For concreteness, we indicate this mechanism by appealing to two special cases. First, take $x = s_k^2 s_l$, $l \neq k$. Similarly to the case of $x = s_k$, one can show that

$$\pi_{s_k^2 s_l} \det A = -2 \sum_{\sigma \in S_{n-1}} \operatorname{sign}(\sigma) \Bigg[\mathbf{1}_{k=\sigma(k)} \rho^{nn} \rho^{\sigma^{-1}(l)n} \prod_{i \in \{1,\dots,n-1\}\setminus\{k,l\}} \rho^{i\sigma(i)} +$$
$$+ \mathbf{1}_{k\neq\sigma(k)} \rho^{\sigma(k)n} \rho^{\sigma^{-1}(k)n} \rho^{\sigma^{-1}(l)n} \prod_{i \in \{1,\dots,n-1\}\setminus\{k,\sigma^{-1}(k)\sigma^{-1}(l)\}} \rho^{i\sigma(i)} \Bigg],$$

which is (the multiple of) the determinant of $\widetilde{\rho^{-1}}$, which is obtained from $\rho^{-1}_{\hat{k}\hat{k}}$ by replacing the lth row by the last row. As the last row appears twice in $\widetilde{\rho^{-1}}$, the determinant, and hence $\pi_{s_k^2 s_l} \det A$, vanishes.

The mechanism is even more transparent for the most extreme monomial $x = s_1^2 \cdots s_{n-1}^2$. In this case,

$$\pi_{s_1^2 \cdots s_{n-1}^2} \det A = \sum_{\sigma \in S_{n-1}} \operatorname{sign}(\sigma) (\rho^{nn})^{n-1} = 0,$$

as the determinant of the $(n-1) \times (n-1)$ matrix with all entries being equal to ρ^{nn}.

References

1. Avellaneda, M., Boyer-Olson, D., Busca, J., Friz, P.: Application of large deviation methods to the pricing of index options in finance. C. R. Math. Acad. Sci. Paris **336**(3), 263–266 (2003)
2. Azencott, R.: Densité des diffusions en temps petit: développements asymptotiques. I. Seminar on Probability, XVIII. Lecture Notes in Mathematics, vol. 1059, pp. 402–498. Springer, Berlin (1984)
3. Bayer, C., Friz, P., Laurence, P.: On the Probability Density Function of Baskets. Springer Proceedings in Mathematics & Statistics (2014)
4. Bayer, C., Laurence, P.: Calculation of greeks for basket options. Working paper
5. Bayer, C., Laurence, P.: Asymptotics beats Monte Carlo: the case of correlated local vol baskets. Commun. Pure Appl. Math. **67**(10), 1618–1657 (2014)
6. Breitung, K., Hohenbichler, M.: Asymptotic approximations for multivariate integrals with an application to multinormal probabilities. J. Multivar. Anal. **30**, 80–97 (1989)
7. Carr, Peter P., Jarrow, Robert A.: The stop-loss start-gain paradox and option valuation: a new decomposition into intrinsic and time value. Rev. Financ. Stud. **3**(3), 469–492 (1990)
8. Deuschel, J., Friz, P., Jacquier, A., Violante, S.: Marginal density expansions for diffusions and stochastic volatility, part I: theoretical foundations. Commun. Pure Appl. Math. **67**(1), 40–82 (2013)
9. Deuschel, J., Friz, P., Jacquier, A., Violante, S.: Marginal density expansions for diffusions and stochastic volatility, part II: applications. Commun. Pure Appl. Math. **67**(2), 321–350 (2013)
10. Evans, L.C., Gariepy, R.F.: Measure Theory and Fine Properties of Functions. Studies in Advanced Mathematics. CRC Press, Boca Raton (1992)
11. Gatheral, J., Hsu, E.P., Laurence, P., Ouyang, C., Wang, T.: Asymptotics of implied volatility in local volatility models. Math. Financ. **22**(4), 591–620 (2012)
12. Henry-Labordère, P.: Analysis, Geometry, and Modeling in Finance: Advanced Methods in Option Pricing. Chapman & Hall/CRC Financial Mathematics Series. CRC Press, Boca Raton (2009)
13. Hsu, P.: Heat kernel on noncomplete manifolds. Indiana Univ. Math. J. **39**(2), 431–442 (1990)
14. Isserlis, L.: On a formula for the product-moment coefficient of any order of a normal frequency distribution in any number of variables. Biometrika **12**(1/2), 134–139 (1918)
15. L'Ecuyer, P.: Quasi-Monte Carlo methods with applications in finance. Financ. Stoch. **13**(3), 307–349 (2009)
16. McKean Jr., H.P., Singer, I.M.: Curvature and the eigenvalues of the Laplacian. J. Differ. Geom. **1**(1), 43–69 (1967)
17. Minakshisundaram, S., Pleijel, Å.: Some properties of the eigenfunctions of the Laplace-operator on Riemannian manifolds. Can. J. Math. **1**, 242–256 (1949)
18. Ninomiya, S., Victoir, N.: Weak approximation of stochastic differential equations and application to derivative pricing. Appl. Math. Financ. **15**(1–2), 107–121 (2008)
19. Pellizzari, P.: Efficient Monte Carlo pricing of European options using mean value control variates. Decis. Econ. Financ. **24**(2), 107–126 (2001)
20. Yosida, K.: On the fundamental solution of the parabolic equation in a Riemannian space. Osaka Math. J. **5**, 65–74 (1953)

A Remark on Gatheral's 'Most-Likely Path Approximation' of Implied Volatility

Martin Keller-Ressel and Josef Teichmann

Abstract We give a new proof of the representation of implied volatility as a time-average of weighted expectations of local or stochastic volatility. With this proof we clarify the question of existence of 'forward implied variance' in the original derivation of Gatheral, who introduced this representation in his book 'The Volatility Surface'.

Keywords Implied volatility · Local volatility · Most-likely path

1 Gatheral's Most-Likely Path Approximation

In his book 'The Volatility Surface—A Practitioners Guide', Jim Gatheral presents an approximation formula for the implied volatility of a European option, when the underlying stock follows a general diffusion process

$$\frac{dS_t}{S_t} = \mu(t, S_t)\, dt + \sigma(t, S_t)\, dW_t\ . \tag{1}$$

The 'most-likely path approximation' to implied Black-Scholes volatility in this model consists of two parts: The first part is the assertion that implied variance—the square of implied volatility—can be written as a time-average of weighted expectations of $\sigma^2(t, S_t)$:

M. Keller-Ressel (✉)
Fachrichtung Mathematik, Institut f. Math. Stochastik, TU Dresden, 01062 Dresden, Germany
e-mail: martin.keller-ressel@tu-dresden.de

J. Teichmann
Department of Mathematics, ETH Zürich, Rämistrasse 101, 8092 Zürich, Switzerland
e-mail: josef.teichmann@math.ethz.ch

P.K. Friz et al. (eds.), *Large Deviations and Asymptotic Methods in Finance*, Springer Proceedings in Mathematics & Statistics 110,
DOI 10.1007/978-3-319-11605-1_8

$$\sigma^2_{\text{imp}}(K,T) = \frac{1}{T}\int_0^T \mathbb{E}^{\mathbb{G}_t}\left[\sigma^2(t,S_t)\right]dt\ . \tag{2}$$

Here, the measures $\mathbb{G}_t$ are given by their Radon-Nikodym derivatives with respect to the risk-neutral measure $\mathbb{Q}$,

$$\frac{d\mathbb{G}_t}{d\mathbb{Q}} = \frac{S^2{}_t\Gamma_{\text{BS}}(S_t,\overline{\sigma}_{K,T}(t))}{\mathbb{E}\left[S^2{}_t\Gamma_{\text{BS}}(S_t,\overline{\sigma}_{K,T}(t))\right]}, \tag{3}$$

where $\overline{\sigma}_{K,T}(t)$ is a function that is yet to be specified, Γ_{BS} denotes the Black-Scholes Gamma and expectations are always taken to be under the risk-neutral pricing measure. Let us emphasize that (2) is an exact formula, and that it is the second part of the method where the approximation happens: Gatheral argues that the density (3) is concentrated (as a function of (t,S)) close to a narrow ridge connecting today's stock price S_0 to the strike price K at time T, and claims that a good approximation to (2) is to evaluate it as if the density was *entirely concentrated* on this ridge.[1] In the terminology of Gatheral this ridge is called the most-likely path and the described approximation method the most-likely path approximation. Extensions of the representation (3) have been proposed e.g. by Guyon and Henry-Labordère [2] for implied correlations.

In this note we will only be concerned with the first part of Gatheral's method, i.e. the derivation of the exact Eq. (2), and in particular the definition of the yet unknown function $\overline{\sigma}_{K,T}(t)$. Gatheral [1] defines on p. 27 first the 'Black-Scholes forward implied variance' $v_{K,T}(t)$ by

$$v_{K,T}(t) = \frac{\mathbb{E}\left[\sigma^2(t,S_t)S^2{}_t\Gamma_{\text{BS}}(S_t,\overline{\sigma}_{K,T}(t))\right]}{\mathbb{E}\left[S_t^2\Gamma_{\text{BS}}(S_t,\overline{\sigma}_{K,T}(t))\right]}, \tag{4}$$

and then, in the equation below, the quantity $\overline{\sigma}_{K,T}(t)$ by

$$\overline{\sigma}^2_{K,T}(t) = \frac{1}{T-t}\int_t^T v_{K,T}(u)du\ . \tag{5}$$

Differentiating (5) and inserting into (4) yields an ordinary differential equation for $\overline{\sigma}_{K,T}(t)$. This definition through an ODE leaves open the question whether (and under which conditions) the quantities $v_{K,T}(t)$ and $\overline{\sigma}_{K,T}(t)$ actually exist.[2] We will show that a simpler definition of $\overline{\sigma}_{K,T}(t)$ can be given, which clarifies the problem of existence, implies Eqs. (4) and (5) and finally leads to a proof of the implied volatility representation (2).

[1]See Gatheral [1, p. 29ff] for details.

[2]See also Lee [3, Sect. 2.3], who remarks that the proof in Gatheral [1] hinges upon the assumption of the existence of $v_{K,T}(t)$.

2 A New Proof of the Implied Volatility Representation

For our proof of the implied volatility representation we assume that the stock price follows an Itô-process with respect to the risk-neutral measure $\mathbb{Q}$ (with respect to which all expectations are taken) of the form

$$\frac{dS_t}{S_t} = r\,dt + \sigma_t\,dW_t\,, \tag{6}$$

such that the discounted stock price $(e^{-rt}S_t)_{0\le t\le T}$ is a square-integrable martingale. The volatility process σ is a general predictable, W-integrable process. This setup covers in particular local volatility models, where $\sigma_t = \sigma(t, S_t)$ and stochastic volatility models where $\sigma_t = \sigma(t, V_t)$ and V_t is a stochastic factor driving the volatility. We fix a terminal time T and assume that S is non-deterministic in the sense that $\mathbb{P}(S_t \neq S_T) > 0$ for all $t \in [0, T]$. Fixing also a strike price K we are ultimately interested in the implied Black-Scholes volatility $\sigma_{\text{imp}}(T, K)$ for a European option with expiry T and strike K in the above model.

2.1 A Regime-Switching Model and Implied Forward Total Variance

To start our derivation, we associate for each $u \in [0, T]$ and $\Sigma^u \geq 0$ the 'regime-switching' process S^u to S, given by

$$\begin{aligned} \frac{dS^u_t}{S^u_t} &= r\,dt + \sigma_t\,dW_t \quad t \in [0, u] \\ \frac{dS^u_t}{S^u_t} &= r\,dt + \Sigma^u\,dW_t \quad t \in [u, T]. \end{aligned} \tag{7}$$

The process S^u switches, at time $t = u$, from the dynamics (6) to Black-Scholes dynamics with constant volatility Σ^u. It should be obvious, that $S^T = S$, while S^0 is simply a Black-Scholes model with volatility Σ^0. In what follows, it will be helpful to consider the *total variance* $w_u = (T-u)(\Sigma^u)^2$ instead of Σ^u. By simple conditioning, the price of a put option on S^u with strike K and maturity T is given by

$$\begin{aligned} e^{-rT}\mathbb{E}\left[(K - S_u)_+\right] &= e^{-ru}\mathbb{E}\left[e^{-r(T-u)}\mathbb{E}\left[\,(K - S_u)_+ |\,\mathcal{F}_u\right]\right] \\ &= e^{-ru}\mathbb{E}\left[P_{\text{BS}}(u, S_u, T, K; w_u)\right], \end{aligned}$$

where $P_{\text{BS}}(u, S, T, K; w)$ is the Black-Scholes put-price parametrized by total variance, i.e.

$$P_{\text{BS}}(u, S, T, K; w) = e^{-r(T-u)} K\Phi(-d_2) - S\Phi(-d_1)$$

and

$$d_{1,2}(w) = \frac{\log\left(\frac{e^{r(T-u)}S}{K}\right)}{\sqrt{w}} \pm \frac{\sqrt{w}}{2}.$$

Definition 2.1 For $u \in [0, T)$ we define the **implied forward total variance** $\hat{w}_u = \hat{w}_u(T, K) \geq 0$ as the solution of

$$e^{-ru}\mathbb{E}\left[P_{\text{BS}}(u, S_u, T, K; \hat{w}_u)\right] = e^{-rT}\mathbb{E}\left[(K - S_T)_+\right] \tag{8}$$

i.e. $\hat{w}_u$ is the total variance $w_u = (T - u)(\Sigma^u)^2$ that has to be chosen in the regime-switching model (7) such that the resulting put-price coincides with the put-price from the original model (6).

Proposition 2.2 *There exists a unique positive deterministic function $u \mapsto \hat{w}_u$, such that the equality*

$$e^{-ru}\mathbb{E}\left[P_{BS}(u, S_u, T, K; \hat{w}_u)\right] = e^{-rT}\mathbb{E}\left[(K - S_T)_+\right] \tag{9}$$

is satisfied for all $u \in [0, T]$.

Proof For $w = 0$, the Black-Scholes price $e^{-ru}P_{\text{BS}}(u, S_u, K, T; w)$ equals $e^{-ru}(e^{-r(T-u)}K - S_u)_+$. Since $(e^{-ru}S_u)_{0\leq u\leq T}$ is a martingale, we have by Jensen's inequality that

$$e^{-ru}\mathbb{E}\left[P_{\text{BS}}(u, S_u, K, T; 0)\right] = e^{-ru}\mathbb{E}\left[(e^{-r(T-u)}K - S_u)_+\right] \leq e^{-rT}\mathbb{E}\left[(K - S_T)_+\right].$$

For $w \to \infty$ the Black-Scholes price $P_{\text{BS}}(u, S_u, K, T; w)$ approaches $e^{-r(T-u)}K$. In this case we get

$$e^{-ru}\mathbb{E}\left[P_{\text{BS}}(u, S_u, T, K; \infty)\right] = e^{-rT}K \geq e^{-rT}\mathbb{E}\left[(K - S_T)_+\right].$$

In addition $w \mapsto P_{\text{BS}}(t, S_t, T, K; w)$ is for any given S_t a continuous and strictly monotone increasing function (here we need the non-degeneracy assumption on S), hence also $w \mapsto \mathbb{E}\left[P_{\text{BS}}(t, S_t, T, K; w)\right]$ is. Therefore we conclude that (9) has a unique solution $\hat{w}_u$ for each $u \in [0, T]$. □

Remark 2.3 Notice that the previous proof holds in fact for semi-martingales S, such that $(\exp(-rt)S_t)_{0\leq t\leq T}$ is a martingale, so neither square integrability nor absence of jumps are needed. However, we do not get regularity assertions for $u \mapsto \hat{w}_u$.

2.2 Main Result

We now present our main result on the implied forward total variance $\hat{w}_u$. Here the assumption of continuous trajectories is really needed, as well as the following L^2-continuity assumption:

Assumption 2.4 We assume that σ_u is mean-square continuous, i.e. the map $[0, T] \ni u \mapsto \sigma_u^2 \in L^2(\Omega, \mathbb{Q})$ is continuous with respect to the L^2-topology.

Theorem 2.5 *Under Assumption 2.4 the mapping $u \mapsto \hat{w}_u$ is in $C^1[0, T) \cap C^0[0, T]$ and satisfies the ODE*

$$\frac{\partial \hat{w}_u}{\partial u} = -\frac{\mathbb{E}\left[\phi(d_2(\hat{w}_u))\sigma_u^2\right]}{\mathbb{E}\left[\phi(d_2(\hat{w}_u))\right]}, \qquad u \in [0, T), \tag{10}$$

with terminal condition $\lim_{u \to T} \hat{w}_T = 0$ and where ϕ denotes the standard normal density. For $u = 0$ it holds that

$$\hat{w}_0(T, K) = T\sigma_{imp}^2(T, K),$$

where $\sigma_{imp}(T, K)$ is the implied Black-Scholes volatility for time-to-maturity T and strike K in (6).

Remark 2.6 Equation (10) can be rewritten as (2). Alternatively, it can be written as

$$-\frac{\partial \hat{w}_u}{\partial u} = \mathbb{E}\left[\sigma_u^2\right] + \mathbf{Cov}\left(\frac{\phi(d_2(\hat{w}_u))}{\mathbb{E}\left[\phi(d_2(\hat{w}_u))\right]}, \sigma_u^2\right),$$

i.e., the rate of decrease in total implied variance is given by expected instantaneous stochastic volatility plus a correction term that accounts for correlation effects between σ_u and S_u in a highly non-linear way.

Proof We set

$$F(u, w) = e^{-ru}\mathbb{E}\left[P_{\mathrm{BS}}(u, S_u, T, K; w)\right].$$

Note that the derivative of P_{BS} with respect to total variance w is given by

$$\frac{\partial}{\partial w} P_{\mathrm{BS}}(u, S, T, K; w) = \frac{1}{2\sqrt{w}} S\phi(d_1),$$

which, inserting $S = S_u$, is uniformly integrable in w on each interval (ϵ, ∞), $\epsilon > 0$. Hence for $w \in (0, \infty)$,

$$\frac{\partial}{\partial w} F(u, w) = \frac{e^{-ru}}{2\sqrt{w}}\mathbb{E}\left[S_u\phi(d_1(w))\right] = \frac{e^{-rT}}{2\sqrt{w}}\mathbb{E}\left[\phi(d_2(w))\right]. \tag{11}$$

Applying Ito's formula and using the martingale property of S we obtain

$$\frac{\partial}{\partial u}F(u,w) = e^{-ru}\mathbb{E}\left[-rP_{\mathrm{BS}} + \frac{\partial}{\partial u}P_{\mathrm{BS}} + \frac{\partial}{\partial S}P_{\mathrm{BS}} r S_u + \frac{1}{2}\frac{\partial^2}{\partial S^2}P_{\mathrm{BS}} S_u^2 \sigma_u^2\right]. \quad (12)$$

Parameterized by total implied variance, the Black-Scholes put-price P_{BS} satisfies

$$-rP_{\mathrm{BS}} + \frac{\partial}{\partial u}P_{\mathrm{BS}} + rS\frac{\partial}{\partial S}P_{\mathrm{BS}} = 0\,,$$

such that (12) simplifies to

$$\frac{\partial}{\partial u}F(u,w) = e^{-ru}\frac{1}{2}\mathbb{E}\left[\frac{\partial^2}{\partial S^2}P_{\mathrm{BS}} S_u^2 \sigma_u^2\right] = \frac{1}{2}\frac{e^{-rT}K}{\sqrt{w}}\mathbb{E}\left[\phi(d_2(w))\sigma_u^2\right]. \quad (13)$$

Note that due to Assumption 2.4 both $\partial_u F(u,w)$ and $\partial_w F(u,w)$ are continuous. Furthermore, recall that $\hat{w}_u$ is given in Definition 2.1 by the implicit equation

$$F(u,\hat{w}_u) = e^{-rT}\mathbb{E}\left[(K - S_T)_+\right], \quad (14)$$

where the right hand side depends neither on u nor on $\hat{w}_u$. Let us first examine the boundary behavior of $F(u,w)$. We easily derive that

$$\begin{aligned}
\lim_{w\to 0} F(u,w) &= \mathbb{E}\left[\left(e^{-rT}K - e^{-ru}S_u\right)_+\right],\\
\lim_{w\to \infty} F(u,w) &= e^{-rT}K,\\
\lim_{u\to 0} F(u,w) &= P_{\mathrm{BS}}(0,S_0,K;w),\\
\lim_{u\to T} F(u,w) &= e^{-rT}\mathbb{E}\left[\Phi(-d_2(w))K - \Phi(-d_1(w))S_T\right].
\end{aligned}$$

By Jensen's inequality and the assumptions on the non-degeneracy of S it holds that

$$\mathbb{E}\left[\left(e^{-rT}K - e^{-ru}S_u\right)_+\right] < e^{-rT}\mathbb{E}\left[(K - S_T)_+\right] < e^{-rT}K$$

for all $u \in [0,T)$. From (11) we see that $\partial_w F(u,w) > 0$ and hence $w \mapsto F(u,w)$ is increasing for $w \in (0,\infty)$. Altogether, it follows that for each $u \in [0,T]$ a unique $\hat{w}_u$ solving (14) exists. In addition, by the implicit function theorem, $\hat{w}_u$ is in $C^1[0,T) \cap C^0[0,T]$ with derivative

$$\frac{\partial}{\partial u}\hat{w}_u = -\frac{\partial_u F(u,w)}{\partial_w F(u,w)} = -\frac{\mathbb{E}\left[\phi(d_2(w_u))\sigma_u^2\right]}{\mathbb{E}\left[\phi(d_2(w_u))\right]},$$

where we have combined (11) and (13). The initial and terminal conditions for $\hat{w}_u$ at $u = 0$ and $u = T$ can be derived from the above boundary conditions for $F(u, w)$. Indeed,

$$P_{\text{BS}}(0, S_0, K; \hat{w}_0) = C(K, T)$$

implies that $\hat{w}_0 = T\sigma_{\text{imp}}^2$, where σ_{imp} is the Black-Scholes implied volatility corresponding to the put-price $P(K, T)$. Finally

$$\mathbb{E}\left[\Phi(-d_2(w))K - \Phi(-d_1(w))S_T\right] = P(K, T) = \mathbb{E}\left[(K - S_T)_+\right]$$

implies that $w = 0$ and hence both boundary conditions for $\hat{w}_u$ follow. □

Acknowledgments MKR acknowledges funding from the Excellence Initiative of the German Research Foundation (DFG).

References

1. Gatheral, J.: The Volatility Surface. Wiley Finance (2006)
2. Guyon, J., Henry-Labordère, P.: Nonlinear Option Pricing. CRC Press, Boca Raton (2013)
3. Lee, R.: Implied volatility: statics, dynamics, and probabilistic interpretation. Recent Advances in Applied Probability. Springer, New York (2004)

Implied Volatility from Local Volatility: A Path Integral Approach

Tai-Ho Wang and Jim Gatheral

Abstract Assuming local volatility, we derive an exact Brownian bridge representation for the transition density; an exact expression for the transition density in terms of a path integral then follows. By Taylor-expanding around a certain path, we obtain a generalization of the heat kernel expansion of the density which coincides with the classical one in the time-homogeneous case, but is more accurate and natural in the time inhomogeneous case. As a further application of our path integral representation, we obtain an improved most-likely-path approximation for implied volatility in terms of local volatility.

Keywords Small time asymptotic expansion · Heat kernels expansion · Implied volatility · Local volatility model · Most likely path · Path integral

1 Introduction

Because of their consistency with the known prices of European options, and despite their unrealistic dynamical implications, local volatility models continue to be used in practice as powerful tools for risk management of equity derivatives portfolios. Under the forward measure (with no drift), local volatility models take the form

$$\frac{dS_t}{S_t} = \sigma_\ell(S_t, t)\, dB_t, \tag{1.1}$$

In memory of our long term collaborator and friend, a passionate mathematician, Peter Laurence.

T.-H. Wang (✉) · J. Gatheral
Department of Mathematics, Baruch College, CUNY 1 Bernard Baruch Way, New York, NY 10010, USA
e-mail: tai-ho.wang@baruch.cuny.edu

J. Gatheral
e-mail: jim.gatheral@baruch.cuny.edu

P.K. Friz et al. (eds.), *Large Deviations and Asymptotic Methods in Finance*, Springer Proceedings in Mathematics & Statistics 110,
DOI 10.1007/978-3-319-11605-1_9

where B_t is a Brownian motion and σ_ℓ is a local volatility function that depends only on the underlying level S and the time t.

Assume that prices of European options of all strikes K and expirations T are given or equivalently that the Black-Scholes implied volatility function $\sigma_{\rm BS}(K,T)$ is known. In that case, it is straightforward to compute the local volatility function σ_ℓ from, for example, Eq. (1.10) of Gatheral [6]:

$$\sigma_\ell{}^2(K,T) = \frac{\frac{\partial w}{\partial T}}{\left(1 - \frac{k}{2w}\frac{\partial w}{\partial k}\right)^2 - \frac{1}{4}\left(\frac{1}{4}+\frac{1}{w}\right)\left(\frac{\partial w}{\partial k}\right)^2 + \frac{1}{2}\frac{\partial^2 w}{\partial k^2}} \tag{1.2}$$

where k denotes the log-strike $k := \log K/S$ and w, the Black-Scholes implied total variance, given by $w(K,T) := \sigma_{\rm BS}^2\,(K,T)\;T$.

In practice, we observe option prices for only a finite set of strikes and expirations. Moreover (see for example Gatheral and Jacquier [7]), it is very hard if not impossible to find a functional form for implied volatility that both matches observed prices and is free from static arbitrage. One alternative approach is to assume a parameterized functional form for the local volatility function $\sigma_\ell(S,t)$ and price a finite set of European options, tuning the parameters of the function until a satisfactory fit is achieved. Such calibration of local volatility models to given option prices is in practice typically performed using numerical PDE techniques. However, numerical PDE techniques are slow and moreover are not practical in higher dimensions.

Alternatively, to achieve better understanding of the qualitative properties of local volatility models, and potentially faster calibration, both academics and practitioners have exploited asymptotic expansions of implied volatility in terms of local volatility. First, Berestycki et al. [2] solved the nonlinear PDE (1.2) for the implied total variance w in the small time to expiration limit, obtaining an exact expression for implied volatility as an integral of local volatility. Subsequently, this asymptotic approximation was extended, to first order in time to expiry $\tau = T - t$ by Henry-Labordère (see the article in this volume and also Henry-Labordère [12]), and then to second order in Gatheral et al. [9] using the heat kernel expansion. Jordan and Tier [14] apply similar methods to derive an asymptotic solution for the SABR and CEV models. In related work, the paper of Cheng et al. [5] derives an operator expansion of the density, which up to first order agrees with prior expansions obtained using the heat kernel expansion. As an earlier example of work in a similar spirit to the most-likely-path approach of our paper, Baldi and Caramellino [1] develop a small-time expansion for the hitting probability of a one-dimensional diffusion.

Our contribution in this paper is to derive an exact Brownian bridge representation for the transition density, from which an exact expression for the transition density in terms of a path integral follows. Indeed, the path integral representation of the density has often been used as a powerful tool for the derivation of improved asymptotic expansions of the transition density. For example, in the foregoing, we apply a technique from the paper of Goovaerts et al. [10]. An earlier paper by Linetsky [16] provides a more general survey of the application of path integral techniques to option pricing.

By replacing all paths that contribute to the path integral by the *most-likely-path*, the unique path that minimizes the action functional in the path integral formulation, we obtain a new approximation to the transition density which is both more accurate and natural than the classical heat kernel version. As an application, we obtain an improved most-likely-path approximation for implied volatility in terms of local volatility.

The most-likely-path (MLP) approach has been used to analyze the asymptotic behavior of implied volatility in stochastic volatility models in Gatheral [6]; this analysis is further elaborated in an article by Keller-Ressel and Teichmann [15] in these proceedings. Guyon and Henry-Labordère [11] and Reghai [17] both explore alternative definitions of the most likely path, achieving improved accuracy by considering fluctuations around the MLP. In particular, Guyon and Henry-Labordère [11] compare and contrast various approximations in a unified setting. Though the approach of Guyon and Henry-Labordère [11] differs from our path integral approach in the current paper, it is worth mentioning that their heat kernel approximation is closely related to ours. Once again however, our path integral approach leads to an unambiguously natural definition of the most-likely-path.

Our paper is organized as follows. In Sect. 2, we derive Brownian bridge and path integral representations for the transition density of one dimensional diffusions. As an application, in Sect. 3, we present a novel probabilistic derivation of the heat kernel expansion, also referred to as the WKB method in the physics literature. For time homogeneous diffusions, this new expansion recovers the conventional heat kernel expansion; however, in the time-inhomogeneous case, the two expansions differ a little. In Sect. 4, we present heuristic derivations of known small time asymptotic expansion of implied volatility to zeroth order. From the path integral perspective, these known approximations are suboptimal in the sense that they correspond to computing the optimal path of an approximate but incomplete action functional. By considering the optimal path of the exact action functional, we show how an optimal approximation may be computed. An interesting feature of the optimal approximation is that it recovers the implied volatility of the time dependent Black-Scholes model exactly, which so far, to the best of our knowledge, none of the existing small time approximations are able to achieve. Finally, in Sect. 5, we summarize and conclude.

Throughout the text, B_t denotes the standard Brownian motion defined on the filtered probability space $(\Omega, \mathcal{F}_t, \mathbb{P})$ satisfying the usual conditions. X_t denotes the Brownian motion with some drift h. $p^X(T, y|t, x)$ denotes the transition density of X from x at time t to y at time T and similarly $p^S(T, s_T|t, s_t)$ is the transition density from s_t to s_T of the process S_t. Moreover, dot will always refer to the partial derivative with respect to the time variable and prime to the space variables x or s.

2 Path Integral Representations for Transition Density

In this section, we derive path integral representations of the transition density and of the call prices under local volatility, which will in turn yield the most-likely-path approximation to implied volatility. The key ingredient in this derivation is a Brownian bridge representation for the transition density, which though straightforward, does not appear to be well-known.

We start with the case of one-dimensional Brownian motion with general but Markovian drift. We reduce the more general diffusion case which concerns us here to this one by applying the well-known Lamperti change of variable.

2.1 Brownian Bridge Representations

Two Brownian bridge representations for the transition density of Brownian motion with general but Markovian, smooth and bounded, drift are derived in Theorem 1. The first expression, (2.1), will be used in the derivation of the path integral representation for transition density in Sect. 2.2 and the second, (2.2), will be used to derive the heat kernel expansion of transition density in Sect. 3.

Theorem 1 *Let X_t be a Brownian motion with drift driven by*

$$dX_t = dB_t + h(X_t, t)dt,$$

where the drift h is assumed smooth and bounded. Let H be an antiderivative of h with respect to x, i.e., $\frac{\partial}{\partial x} H(x,t) = h(x,t)$, for all x and t. The transition density p^X of X_t has the following two equivalent Brownian bridge representations:

$$p^X(T, y|t, x) = \phi(T-t, y-x)\, \tilde{\mathbb{E}}_{x,y}\left[e^{\int_t^T h(X_s,s)dX_s - \frac{1}{2}\int_t^T h^2(X_s,s)ds}\right] \tag{2.1}$$

and

$$\begin{aligned} p^X(T, y|t, x) = \phi(T-t, y-x)\, e^{H(y,T)-H(x,t)} \times \\ \tilde{\mathbb{E}}_{x,y}\left[e^{-\frac{1}{2}\int_t^T h^2(X_s,s)+h_x(X_s,s)+2H_t(X_s,s)ds}\right], \end{aligned} \tag{2.2}$$

where ϕ is the Gaussian density $\phi(t,\xi) = \frac{1}{\sqrt{2\pi t}} e^{-\frac{\xi^2}{2t}}$. The notation $\tilde{\mathbb{E}}_{x,y}[\cdot]$ denotes the expectation under the Brownian bridge measure from x to y.

Proof Note that X_t under the original measure $\mathbb{P}$ is a Brownian motion with drift h. Define a new probability measure $\tilde{\mathbb{P}}$ through the Radon-Nikydom derivative

$$\frac{d\tilde{\mathbb{P}}}{d\mathbb{P}} = e^{-\int_t^T h(X_s,s)dB_s - \frac{1}{2}\int_t^T h^2(X_s,s)ds}.$$

By the Girsanov theorem, X_t is a Brownian motion under $\tilde{\mathbb{P}}$. Given any bounded measurable function f, we have, since $dB_t = dX_t - h(X_t, t)dt$,

$$\mathbb{E}_{t,x}[f(X_T)] = \tilde{\mathbb{E}}_{t,x}\left[f(X_T)\frac{d\mathbb{P}}{d\tilde{\mathbb{P}}}\right] = \tilde{\mathbb{E}}_{t,x}\left[f(X_T)e^{\int_t^T h(X_s,s)dX_s - \frac{1}{2}\int_t^T h^2(X_s,s)ds}\right],$$

where, for notational simplicity, $\mathbb{E}_{t,x}[\cdot]$ denotes the conditional expectation $\mathbb{E}[\cdot|X_t = x]$, and similarly for $\tilde{\mathbb{E}}_{t,x}[\cdot]$. It follows that, for any bounded measurable function f,

$$\int f(y)p^X(T, y|t, x)dy = \int f(y)\tilde{\mathbb{E}}_{x,y}\left[e^{\int_t^T h(X_s,s)dX_s - \frac{1}{2}\int_t^T h^2(X_s,s)ds}\right]$$
$$\phi(T - t, y - x)dy,$$

where $\tilde{\mathbb{E}}_{x,y}[\cdot] = \tilde{\mathbb{E}}[\cdot|X_t = x, X_T = y]$. Consequently, (2.1) follows, i.e.,

$$p^X(T, y|t, x) = \phi(T - t, y - x)\tilde{\mathbb{E}}_{x,y}\left[e^{\int_t^T h(X_s,s)dX_s - \frac{1}{2}\int_t^T h^2(X_s,s)ds}\right].$$

Furthermore, Ito's formula implies that

$$\int_t^T h(X_s, s)dX_s = H(X_T, T) - H(X_t, t) - \int_t^T \left[H_t(X_s, s) + \frac{h_x(X_s, s)}{2}\right]ds,$$

where we recall that H is an antiderivative of h with respect to x. Thus,

$$e^{\int_t^T h(X_s,s)dX_s} = e^{H(X_T,T) - H(X_t,t) - \int_t^T \left[H_t(X_s,s) + \frac{h_x(X_s,s)}{2}\right]ds}.$$

We further rewrite the transition density as

$$p^X(T, y|t, x) = \phi(T - t, y - x)e^{H(y,T) - H(x,t)}$$
$$\tilde{\mathbb{E}}_{x,y}\left[e^{-\frac{1}{2}\int_t^T h^2(X_s,s) + h_x(X_s,s) + 2H_t(X_s,s)ds}\right].$$

This completes the proof of (2.2). □

Remark 1 We remark that the conditional expectations in both (2.1) and (2.2) are under the Brownian bridge measure since X_t is a Brownian motion under $\tilde{\mathbb{P}}$. One intriguing feature of the representation (2.2) is that, if we Taylor expand the conditional expectation for small $T - t$ around the straight line connecting the initial and terminal points, we recover the heat kernel expansion in the time-homogeneous case and probably do better than the heat kernel expansion in the time-inhomogeneous case. See Sect. 3 for more detailed discussions on the heat kernel expansion.

Now for the general diffusion case, consider the process S_t driven by the stochastic differential equation (SDE)

$$dS_t = \mu(S_t, t)dt + a(S_t, t)dB_t, \quad S_0 = s_0,$$

where for simplicity, we assume the coefficients μ and a are Lipschitz and of linear growth; a is further assumed strictly away from zero. By applying the Lamperti transformation $x = \int_{s_0}^{s} \frac{d\xi}{a(\xi,t)}$, the process S_t is transformed into a Brownian motion with drift. Specifically, denote the transformation from s to x by $x = \varphi(s,t) = \int_{s_0}^{s} \frac{d\xi}{a(\xi,t)}$. Applying Ito's formula to $X_t = \varphi(S_t, t)$ yields

$$\begin{aligned}
dX_t &= d\varphi(S_t, t) \\
&= \left[\dot{\varphi}(S_t, t) + \mu(S_t, t)\varphi_s(S_t, t) + \frac{a^2(S_t, t)}{2}\varphi_{ss}(S_t, t)\right] dt + \varphi_s(S_t, t)a(S_t, t)dB_t \\
&= \left[\dot{\varphi}(S_t, t) + \frac{\mu(S_t, t)}{a(S_t, t)} - \frac{a_s(S_t, t)}{2}\right] dt + dB_t \\
&= dB_t + h(X_t, t)dt,
\end{aligned}$$

where subindices of φ and a refer to partial derivatives. The function h is defined as $h(x,t) = \dot{\varphi}(s,t) + \frac{\mu(s,t)}{a(s,t)} - \frac{a_s(s,t)}{2}$, with $s = \varphi^{-1}(x,t)$. The transition densities p^S for S_t and p^X for X_t are then related as

$$p^S(T, s_T | t, s_t) = \frac{1}{a(s_T, T)} p^X(T, x_T | t, x_t),$$

with $x_T = \varphi(s_T, T)$ and $x_t = \varphi(s_t, t)$. Thus, the transition from the Brownian bridge representation for p^X to a similar representation for p^S is straightforward by applying Theorem 1. Theorem 2 formalizes this result.

Theorem 2 *Let S_t be the diffusion process driven by the stochastic differential equation*

$$dS_t = \mu(S_t, t)dt + a(S_t, t)dB_t, \quad S_0 = s_0.$$

Denote the Lamperti transformation from s to x by $x = \varphi(s,t) = \int_{s_0}^{s} \frac{d\xi}{a(\xi,t)}$. Define the function h by $h(x,t) = \dot{\varphi}(s,t) + \frac{\mu(s,t)}{a(s,t)} - \frac{a_s(s,t)}{2}$, with $s = \varphi^{-1}(x,t)$, where subindices refer to corresponding partial derivatives. Let H be an antiderivative of h with respect to x, namely, $\frac{\partial}{\partial x} H(x,t) = h(x,t)$, for all x and t. Then the transition density p^S of S_t from (t, s_t) to (T, s_T) has the following Brownian bridge representations:

$$\begin{aligned}
p^S(T, s_T | t, s_t) = &\frac{\phi(T - t, \varphi(s_T, T) - \varphi(s_t, t))}{a(s_T, T)} \tilde{\mathbb{E}}_{\varphi(s_t,t),\varphi(s_T,T)} \\
&\times \left[e^{\int_t^T h(X_s,s)dX_s - \frac{1}{2}\int_t^T h^2(X_s,s)ds}\right]
\end{aligned} \tag{2.3}$$

and

$$p^S(T, s_T|t, s_t) = \frac{\phi(T-t, \varphi(s_T, T) - \varphi(s_t, t))}{a(s_T, T)} e^{H(\varphi(s_T,T),T)-H(\varphi(s_t,t),t)} \times$$
$$\tilde{\mathbb{E}}_{\varphi(s_t,t),\varphi(s_T,T)} \left[e^{-\frac{1}{2}\int_t^T h^2(X_s,s)+h_x(X_s,s)+2H_t(X_s,s)ds} \right], \quad (2.4)$$

where again ϕ denote the Gaussian density $\phi(t, \xi) = \frac{1}{\sqrt{2\pi t}} e^{-\frac{\xi^2}{2t}}$. As before, the notation $\tilde{\mathbb{E}}_{x,y}[\cdot]$ denotes the expectation under the Brownian bridge measure from x to y.

Note that the X_t process in both expressions (2.3) and (2.4) is a Brownian bridge from $x_T = \varphi(s_T, T)$ to $x_t = \varphi(s_t, t)$. One application of such Brownian bridge representations of transition densities is to devise more efficient simulation schemes. For example, for some given function f, we may compute numerically the expectation $\mathbb{E}_{t,s_t}[f(S_T)]$ in x-space as

$$\mathbb{E}_{t,s_t}[f(S_T)] = \mathbb{E}_{t,x_t}[f(\varphi^{-1}(X_T))] = \int f \circ \varphi^{-1}(x_T, T)\; p^X(T, x_T|t, x_t)dx_T$$
$$= \int f \circ \varphi^{-1}(x_T, T)\; \phi(T-t, x_T - x_t) e^{H(x_T,T)-H(x_t,t)}$$
$$\times\, \tilde{\mathbb{E}}_{x_t,x_T} \left[e^{-\frac{1}{2}\int_t^T h^2(X_\tau,\tau)+h_x(X_\tau,\tau)+2H_t(X_\tau,\tau)d\tau} \right] dx_T$$
$$= \mathbb{E}\left\{ f \circ \varphi^{-1}(Y, T) e^{H(Y,T)-H(x_t,t)}\, \tilde{\mathbb{E}}_{x_t,Y} \left[e^{-\frac{1}{2}\int_t^T h^2(X_\tau,\tau)+h_x(X_\tau,\tau)+2H_t(X_\tau,\tau)d\tau} \right] \right\}$$

where Y is a normal random variable with mean x_t and variance $T - t$. Therefore, if there is an efficient method to calculate or approximate the conditional expectation $\tilde{\mathbb{E}}_{x_t,Y} \left[e^{-\frac{1}{2}\int_t^T h^2(X_\tau,\tau)+h_x(X_\tau,\tau)+2H_t(X_\tau,\tau)d\tau} \right]$ in the Brownian bridge measure, the expectation $\mathbb{E}_{t,s_t}[f(S_T)]$ could potentially be computed more efficiently. Since X_t is a Brownian bridge from x_t to Y, one obvious approximation is to simply replace the integral in the exponent of $\tilde{\mathbb{E}}_{x_t,Y} \left[e^{-\frac{1}{2}\int_t^T h^2(X_\tau,\tau)+h_x(X_\tau,\tau)+2H_t(X_\tau,\tau)d\tau} \right]$ with the integrand evaluated along the straight line $x_\tau = \frac{T-\tau}{T-t} x_t + \frac{\tau-t}{T-t} Y$ for $\tau \in [t, T]$. In other words,

$$\tilde{\mathbb{E}}_{x_t,Y} \left[e^{-\frac{1}{2}\int_t^T h^2(X_t,t)+h_x(X_t,t)+2H_t(X_t,t)dt} \right]$$
$$\approx e^{-\frac{1}{2}\int_t^T h^2(x_\tau,\tau)+h_x(x_\tau,\tau)+2H_t(x_\tau,\tau)d\tau}. \quad (2.5)$$

Hence,

$$\mathbb{E}_{t,s_t}[f(S_T)] \approx e^{-\frac{1}{2}\int_t^T h^2(x_\tau,\tau)+h_x(x_\tau,\tau)+2H_t(x_\tau,\tau)d\tau}$$
$$\mathbb{E}\left[f \circ \varphi^{-1}(Y,T)\, e^{H(Y,T)-H(x_t,t)}\right],$$

after which we need only simulate the normal random variable Y.

The straight-line approximation in (2.5) seems somewhat *ad hoc*. Would we do better with another path? Why not add two extra paths to take some account of the variability of the random paths in the full integral? More generally, is there an optimal or systematic way of picking these paths? The path integral representation in Sect. 2.2 may provide a partial answer to this question. The Brownian bridge representations (2.1) and (2.3) play a key role in the derivation of this path integral representation.

2.2 The Path Integral Representation of the Density

In this section, we provide a formal derivation of the path integral representation exploiting the Brownian bridge representations of Sect. 2.1 and the Chapman-Kolmogorov equation. As in Sect. 2.1, we will use the notations $\varphi(s_t,t)$ and x_t interchangeably.

Let $\{t = t_0 < t_1 < \cdots < t_n = T\}$ be a partition of the time interval $[t,T]$ with $\Delta t_i = t_i - t_{i-1} = \frac{T}{n}$, for $i = 1,\ldots,n$. By iteratively applying the Chapman-Kolmogorov equation, the transition density $p^S(T,s_T|t,s_t)$ can be written as

$$p^S(T,s_T|t,s_t) = \int\cdots\int\prod_{i=1}^n p^S(t_i,s_i|t_{i-1},s_{i-1})\,ds_1\ldots ds_{n-1}, \tag{2.6}$$

where we set $s_0 = s_t$ and $s_n = s_T$. Recall from (2.3) that the transition density p^S of S_t from (t_{i-1},s_{i-1}) to (t_i,s_i) has the Brownian bridge representation

$$p^S(t_i,s_i|t_{i-1},s_{i-1})$$
$$= \frac{\phi(\Delta t,\varphi(s_i,t_i)-\varphi(s_{i-1},t_{i-1}))}{a(s_i,t_i)}\,\tilde{\mathbb{E}}_{\varphi(s_{i-1},t_{i-1}),\varphi(s_i,t_i)}$$
$$\left[e^{\int_{t_{i-1}}^{t_i} h(X_\tau,\tau)dX_\tau - \int_{t_{i-1}}^{t_i}\frac{1}{2}h^2(X_\tau,\tau)d\tau}\right],$$

where X_τ is a Brownian bridge from $\varphi(s_{i-1},t_{-i})$ to $\varphi(s_i,t_i)$. We next compute the limit of (2.6), as $\Delta t \to 0^+$ (or equivalently $n \to \infty$), assuming that, for $i = 1,\ldots,n$, the s_i's form a discretization of a differentiable curve s_τ, for $\tau \in [t,T]$.

We have

$$\lim_{n\to\infty}\prod_{i=1}^{n}\tilde{\mathbb{E}}_{\varphi(s_{i-1},t_{i-1}),\varphi(s_i,t_i)}\left[e^{\int_{t_{i-1}}^{t_i}h(X_\tau,\tau)dX_\tau-\frac{1}{2}\int_{t_{i-1}}^{t_i}h^2(X_\tau,\tau)d\tau}\right]$$
$$=e^{\int_t^T h(\varphi(s_\tau,\tau),\tau)\dot{x}_\tau d\tau-\frac{1}{2}\int_t^T h^2(\varphi(s_\tau,\tau),\tau)d\tau}$$

and

$$\lim_{n\to\infty}\prod_{i=1}^{n}e^{-\frac{1}{2\Delta t}[\varphi(s_i,t_i)-\varphi(s_{i-1},t_{i-1})]^2}$$
$$=\lim_{n\to\infty}e^{-\frac{1}{2\Delta t}\sum_{i=1}^{n}[\varphi(s_i,t_i)-\varphi(s_{i-1},t_{i-1})]^2}$$
$$=\lim_{n\to\infty}e^{-\frac{1}{2}\sum_{i=1}^{n}\left[\varphi'_{i-1}\frac{\Delta s_i}{\Delta t}+\dot{\varphi}_{i-1}\right]^2\Delta t+\mathcal{O}(\Delta s_i{}^2+\Delta t)^2}$$
$$=e^{-\frac{1}{2}\int_t^T[\varphi'(s_\tau,\tau)\dot{s}_\tau+\dot{\varphi}(s_\tau,\tau)]^2d\tau}.$$

Substitution into (2.6) and taking the limit $n\to\infty$ yields the following path integral representation for the transition density p^S

$$p^S(T,s_T|t,s_t)=\int_{\mathcal{C}_s}e^{-\frac{1}{2}\int_t^T[\varphi'(s_\tau,\tau)\dot{s}_\tau+\dot{\varphi}(s_\tau,\tau)-h(\varphi(s_\tau,\tau),\tau)]^2d\tau}\mathcal{D}[s],\quad(2.7)$$

where

$$\mathcal{D}[s]=\lim_{n\to\infty}\frac{1}{\sqrt{2\pi\Delta t}\,a(s_T,T)}\prod_{i=1}^{n-1}\frac{1}{\sqrt{2\pi\Delta t}}\frac{ds_i}{a(s_i,t_i)},$$

and $\mathcal{C}_s$ denotes the collection of all differentiable curves from (t,s_t) to (T,s_T). Equivalently, because $dx_i=\frac{ds_i}{a(s_i,t_i)}$, we may rewrite the path integral representation (2.7) more neatly and simply in x-space as

$$p^S(T,s_T|t,s_t)=\frac{1}{a(S_T,T)}\int_{\mathcal{C}_x}e^{-\frac{1}{2}\int_t^T[\dot{x}_\tau-h(x_\tau,\tau)]^2d\tau}\mathcal{D}[x]\quad(2.8)$$

where

$$\mathcal{D}[x]=\lim_{n\to\infty}\frac{1}{\sqrt{2\pi\Delta t}}\prod_{i=1}^{n-1}\frac{dx_i}{\sqrt{2\pi\Delta t}}$$

and $\mathcal{C}_x$ denotes the collection of all differentiable curves from (t,x_t) to (T,x_T). We shall henceforth deal mostly with the simpler expression (2.8). Heuristically, one could think of the path integral representation (2.8) of the density as an exponentially-weighted average over all possible differentiable curves connecting x_t to x_T. $\mathcal{D}[x]$ could then be regarded as the "Lebesgue" measure on the space of differentiable curves connecting x_t to x_T, though mathematically such a measure does not really exist.

Assume now that under the pricing measure (assuming zero interest rate and dividend yield), the price S_t of the underlying is driven by the SDE of local volatility type

$$dS_t = a(S_t, t)dB_t.$$

The path integral representation (2.7) of the transition density p^S in this case has the following simpler form

$$p^S(T, s_T|t, s_t) = \int_{C_s} e^{-\frac{1}{2}\int_t^T \left[\frac{\dot{s}_\tau}{a(s_\tau,\tau)} + \frac{a_S(s_\tau,\tau)}{2}\right]^2 d\tau} \mathcal{D}[s].$$

Integrating the payoff function over the transition density, the path integral representation for call price is immediate:

$$C(t, s_t, K, T) = \int_K^\infty (s_T - K) \int_{C_s} e^{-\frac{1}{2}\int_t^T \left[\frac{\dot{s}_\tau}{a(s_\tau,\tau)} + \frac{a_S(s_\tau,\tau)}{2}\right]^2 d\tau} \mathcal{D}[s] ds_T,$$

or equivalently in x-space,

$$C(t, s_t, K, T) = \int_K^\infty \frac{s_T - K}{a(s_T, T)} \int_{C_x} e^{-\frac{1}{2}\int_t^T |\dot{x}_\tau - h(x_\tau,\tau)|^2 d\tau} \mathcal{D}[x] ds_T, \quad (2.9)$$

where $h(x, t) = \dot{\varphi}(s, t) - \frac{a_S(s,t)}{2}$.

3 Probabilistic Derivation of the Heat Kernel Expansion

The heat kernel expansion is a small time asymptotic expansion of the fundamental solution of the heat equation over a Riemannian manifold. Reexpressing the transition density of a diffusion process in terms of this fundamental solution leads naturally to a small time asymptotic expansion of the transition density. This topic is well-studied in the Riemannian geometry literature, see Chavel [4] for a geometric analytical approach and Hsu [13] for a probabilistic approach. In the physics literature, the heat kernel approach to deriving small time asymptotic expansions is also known as the WKB method or the ray solution, see Jordan and Tier [14]. Deriving such expansions in one dimension is much simpler than in higher dimensions where no analogue of the Lamperti transformation exists.

Though the heat kernel expansion is very well-known, the Brownian bridge representation (2.4) of Theorem 2 leads to a novel probabilistic derivation which we will now present. To fix ideas and illustrate the methodology employed, we start with the case of Brownian motion with drift; as before, the general diffusion case follows via the Lamperti transformation. To minimize mathematical technicalities, we shall assume (at least in this section) that all functions are bounded with bounded derivatives.

3.1 Heat Kernel Expansion for Brownian Motion with Drift

Theorem 3 *Let X_t be the Brownian motion with drift h, i.e., X_t satisfies the SDE $dX_t = dB_t + h(X_t, t)dt$. Denote by H an antiderivative of h with respect to x, namely, $\frac{\partial}{\partial x}H(x,t) = h(x,t)$, for all x and t. The transition density p^X of X_t has, as $t \to T^-$, the following small time asymptotic expansion:*

$$p^X(T, y|t, x) = \phi(T-t, y-x)\, e^{H(y,T)-H(x,t)} \times \left\{1 - \frac{1}{2}\int_t^T \left[h^2(x_s^*, s) + h_x(x_s^*, s) + 2H_t(x_s^*, s)\right] ds + \mathcal{O}(T-t)^2\right\} \tag{3.1}$$

where ϕ is the Gaussian density $\phi(t,\xi) = \frac{1}{\sqrt{2\pi t}}e^{-\frac{\xi^2}{2t}}$. x_s^* denotes the straight line from (t,x) to (T,y), i.e., $x_s^* = x + \frac{s-t}{T-t}(y-x)$ for $s \in [t,T]$.

Notice that in the time-inhomogeneous case $h = h(x,t)$, the approximation (3.1) is different from the heat kernel expansion (see, for example, (3.3), (3.6), and (3.7) on page 603 of Gatheral et al. [9]) in that the approximation in (3.1) involves an integration from t to T whereas, in the classical heat kernel expansion, all quantities are evaluated at the fixed initial time t. Of course, in the time homogeneous case where the drift $h = h(x)$ has no explicit dependence on t, the expansion (3.1) coincides with the classical heat kernel expansion as formalized in the following corollary.

Corollary 1 (Heat kernel expansion for Brownian motion with drift) *For Brownian motion with time homogeneous drift $h = h(x)$, the transition density p^X of X_t from (t,x) to (T,y) has the asymptotic expansion up to first order as*

$$p(T, y|t, x) = \phi(T-t, y-x)e^{H(y)-H(x)} \left\{1 - \frac{T-t}{2(y-x)}\int_x^y \left[h^2(\xi) + h'(\xi)\right] d\xi + \mathcal{O}(T-t)^2\right\},$$

which coincides with the classical heat kernel expansion up first order (see, for instance, Gatheral et al. [9]).

Proof In this case, $H_t = 0$ because $h_t = 0$. The integral in (3.1) can be evaluated as

$$\begin{aligned}
&\int_t^T \left[h^2(x_s^*) + h'(x_s^*)\right] ds \\
&= \int_t^T \left[h^2\left(x + \frac{s-t}{T-t}(y-x)\right) + h'\left(x + \frac{s-t}{T-t}(y-x)\right)\right] ds \\
&= \frac{T-t}{y-x} \int_x^y \left[h^2(\xi) + h'(\xi)\right] d\xi,
\end{aligned}$$

where in the last equation we used the change of variable $\xi = x + \frac{s-t}{T-t}(y-x)$. □

Let Y_t denote the Brownian bridge from x at time t to y at time T and $\tilde{\mathbb{E}}_{x,y}[\cdot]$ be the expectation under the Brownian bridge measure. The proof of the asymptotic expansion (3.1) requires the following two lemmas.

Lemma 1 *For a bounded function $g = g(x,s)$, $|g| \le M$ say, we have the following estimate*

$$\tilde{\mathbb{E}}_{x,y}\left[e^{\int_t^T g(Y_s,s)ds}\right] = 1 + \int_t^T \tilde{\mathbb{E}}_{x,y}\left[g(Y_s,s)\right] ds + \mathcal{O}(T-t)^2.$$

Proof The proof is based on a clever application of the convex order for random variables first observed, to our knowledge, in the paper by Goovaerts et al. [10] (see Proposition 6.2 on p. 348). Denote by $Q_{g(Y_s,s)}(q)$ the qth quantile of the random variable $g(Y_s,s)$. Since exponential functions are convex, it follows from Proposition 6.2 of Goovaerts et al. [10] that

$$\tilde{\mathbb{E}}_{x,y}\left[e^{\int_t^T g(Y_s,s)ds}\right] \le \int_0^1 e^{\int_t^T Q_{g(Y_s,s)}(q)ds} dq.$$

We establish an upper bound for the right hand side. First we Taylor expand the integrand and rewrite the integral as

$$\int_0^1 e^{\int_t^T Q_{g(Y_s,s)}(q)ds} dq = \sum_{k=0}^{\infty} \frac{1}{k!} \int_0^1 \left(\int_t^T Q_{g(Y_s,s)}(q)ds\right)^k dq.$$

An upper bound for $\int_0^1 \left(\int_t^T Q_{g(Y_s,s)}(q)ds\right)^k dq$ is then determined as

$$\begin{aligned}
&\int_0^1 \left(\int_t^T Q_{g(Y_s,s)}(q)ds\right)^k dq \\
&\le \int_0^1 (T-t)^{k-1} \int_t^T |Q_{g(Y_s,s)}(q)|^k \, ds\, dq \quad \text{(by Hölder's inequality)}
\end{aligned}$$

$$= (T-t)^{k-1}\int_t^T \tilde{\mathbb{E}}_{x,y}|g(Y_s,s)|^k ds$$
$$\leq M^k(T-t)^k \quad (\text{since } |g| \leq M).$$

Thus,

$$\int_0^1 e^{\int_t^T Q_{g(Y_s,s)}(q)\,ds}\,dq$$
$$= 1 + \int_0^1\int_t^T Q_{g(Y_s,s)}(q)\,ds\,dq + \sum_{k=2}^{\infty}\frac{1}{k!}\int_0^1\left(\int_t^T Q_{g(Y_s,s)}(q)ds\right)^k dq$$
$$\leq 1 + \int_t^T \tilde{\mathbb{E}}_{x,y}\left[g(Y_s,s)\right]\,ds + \sum_{k=2}^{\infty}\frac{1}{k!}M^k(T-t)^k$$
$$\leq 1 + \int_t^T \tilde{\mathbb{E}}_{x,y}\left[g(Y_s,s)\right]\,ds + M^2(T-t)^2 e^{M(T-t)},$$

which completes the proof. □

Lemma 2 asserts that the time integral of the conditional expectation in Lemma 1 is approximately, up to order $(T-t)^2$, equal to the integral along a straight line connecting x at time t to y at time T.

Lemma 2 *For a bounded function $g = g(x,s)$ with bounded second partial derivative with respect to x, the following asymptotic holds.*

$$\int_t^T \tilde{\mathbb{E}}_{x,y}\left[g(Y_s,s)\right]\,ds = \int_t^T g(x_s,s)ds + \mathcal{O}(T-t)^2,$$

where x_s denotes the straight line $x_s = x + \frac{s-t}{T-t}(y-x)$ from (t,x) to (T,y).

Proof Taylor's theorem implies that

$$g(Y_s,s) = g(x_s,s) + g_x(x_s,s)(Y_s-x_s) + \frac{g_{xx}(\xi_s,s)}{2}(Y_s-x_s)^2,$$

for some ξ_s between Y_s and x_s. Since Y_s is a Brownian bridge from (t,x) to (T,y), Y_s is normally distributed: $Y_s \sim N\left(x_s, \frac{(s-t)(T-s)}{T-t}\right)$. Therefore,

$$\tilde{\mathbb{E}}_{x,y}\left[g(Y_s,s)\right] = g(x_s,s) + g_x(x_s,s)\,\tilde{\mathbb{E}}_{x,y}\left[Y_s-x_s\right] + \frac{g_{xx}(\xi_s,s)}{2}\tilde{\mathbb{E}}_{x,y}\left[(Y_s-x_s)^2\right]$$
$$= g(x_s,s) + \frac{g_{xx}(\xi_s,s)}{2}\frac{(s-t)(T-s)}{T-t}.$$

Hence, by the assumption that $|g_{xx}| \leq K$,

$$\begin{aligned}\int_t^T \tilde{\mathbb{E}}_{x,y}\left[g(Y_s,s)\right]ds &= \int_s^T g(x_s,s)ds + \int_t^T \frac{g_{xx}(\xi_s,s)}{2}\frac{(s-t)(T-s)}{T-t}ds \\ &\le \int_t^T g(x_s,s)ds + \frac{K}{2}\int_t^T \frac{(s-t)(T-s)}{T-t}ds \\ &= \int_t^T g(x_s,s)ds + \frac{K}{12}(T-t)^2. \end{aligned}$$ □

The proof of Theorem 3 is now straightforward.

Proof (Proof of Theorem 3) By combining the two asymptotics in Lemmas 1 and 2 with $g(x,s) = h^2(x,s) + h_x(x,s) + 2H_t(x,s)$, under the assumption that g is bounded with bounded second partial derivative with respect to x, we obtain

$$\begin{aligned}&\tilde{\mathbb{E}}_{x,y}\left[e^{-\frac{1}{2}\int_t^T h^2(X_s,s)+h_x(X_s,s)+2H_t(X_s,s)ds}\right] \\ &= 1 - \frac{1}{2}\int_t^T \left[h^2(x_s,s) + h_x(x_s,s) + 2H_t(x_s,s)\right]ds + \mathcal{O}(T-t)^2.\end{aligned}$$

Recall expression (2.2) for the transition density:

$$\begin{aligned}p^X(T,y|t,x) = \phi(T-t,y-x)\,e^{H(y,T)-H(x,t)} \times \\ \tilde{\mathbb{E}}_{x,y}\left[e^{-\frac{1}{2}\int_t^T h^2(X_s,s)+h_x(X_s,s)+2H_t(X_s,s)ds}\right].\end{aligned}$$

Substituting the approximation of the conditional expectation above, we obtain

$$\begin{aligned}p^X(T,y|t,x) = \phi(T-t,y-x)\,e^{H(y,T)-H(x,t)} \times \\ \left\{1 - \frac{1}{2}\int_t^T \left[h^2(x_s,s) + h_x(x_s,s) + H_t(x_s,s)\right]ds + \mathcal{O}(T-t)^2\right\}.\end{aligned}$$

□

3.2 Heat Kernel Expansion for Nondegenerate Diffusions

For general nondegenerate diffusions, consider the process S_t driven by the SDE:

$$dS_t = a(S_t,t)dB_t + \mu(S_t,t)dt. \tag{3.2}$$

Again the Lamperti transformation allows us to carry over the small time asymptotic expansion (3.1) in x-space to s-space. Specifically, recall that the Lamperti transformation $x_t = \varphi(s_t,t) = \int_{s_0}^{s_t} \frac{d\xi}{a(\xi,t)}$ transforms the SDE (3.2) into a Brownian motion with drift $dX_t = dB_t + h(X_t, t)dt$, where $h(x_t, t) = \dot{\varphi}(s_t, t) + \frac{\mu(s_t,t)}{a(s_t,t)} - \frac{a_s(s_t,t)}{2}$ and that the transition densities p^S for S_t and p^X for X_t are related by

$$p^S(T, s_T|t, s_t) = \frac{1}{a(s_T, T)} p^X(T, x_T|t, x_t),$$

with $x_T = \varphi(s_T, T)$ and $x_t = \varphi(s_t, t)$. Hence, a small time asymptotic expansion as $t \to T^-$ for p^S can be obtained by simply applying the expansion (3.1). This argument is formalized in Theorem 4.

Theorem 4 *The transition density p^S of the process S_t driven by the SDE*

$$dS_t = a(S_t, t)dB_t + \mu(S_t, t)dt$$

has the small time asymptotic expansion as $t \to T^-$

$$p^S(T, s_T|t, s_t) = \frac{\phi(T-t, \varphi(s_T, T) - \varphi(s_t, t))}{a(s_T, T)} e^{H(\varphi(s_T,T),T) - H(\varphi(s_t,t),t)} \tag{3.3}$$
$$\times \left\{1 - \frac{1}{2}\int_t^T h^2(\varphi_\tau, \tau) + h_x(\varphi_\tau, \tau) + 2H_t(\varphi_\tau, \tau)d\tau + O(T-t)^2\right\},$$

where $\varphi_\tau = \frac{T-\tau}{T-t}\varphi(s_t, t) + \frac{\tau-t}{T-t}\varphi(s_T, T)$.

We stress once again that in the time-inhomogeneous case, $a = a(s,t)$, the expansion in (3.3) is not identical to the classical heat kernel expansion as it involves an integral along the path φ_τ. On the other hand, in the time-homogeneous case $a = a(s)$, (3.3) does recover the classical heat kernel expansion. In this sense therefore, we have derived a natural generalization of the classical heat kernel expansion.

Corollary 2 (Heat kernel expansion for time-homogeneous diffusions) *The transition density p^S of the process S_t driven by the time-homogeneous SDE*

$$dS_t = a(S_t)dB_t + \mu(S_t)dt$$

has the small time asymptotic expansion as $t \to T^-$ up to first order

$$p(T, s_T|t, s_t) = \frac{\phi(T-t, \varphi(s_T) - \varphi(s_t))}{a(s_T)} e^{H\circ\varphi(s_T) - H\circ\varphi(s_t)} \tag{3.4}$$
$$\times \left\{1 - \frac{T-t}{2(\varphi(s_T) - \varphi(s_t))}\int_{s_t}^{s_T} \left[h^2(\varphi(s)) + h' \circ \varphi(s)\right]\frac{ds}{a(s)} + O(T-t)^2\right\},$$

where $\varphi(s) = \int_{s_0}^{s} \frac{d\xi}{a(\xi)}$, $h \circ \varphi(s) = \frac{\mu(s)}{a(s)} - \frac{a'(s)}{2}$, *and* H *is an antiderivative of* h. ϕ *denotes the Gaussian density* $\phi(t,\xi) = \frac{1}{\sqrt{2\pi t}} e^{-\frac{\xi^2}{2t}}$. *The small time asymptotic expansion coincides with the classical heat kernel expansion up to first order.*

Proof We verify that the expansion (3.4) is indeed the classical heat kernel expansion. The classical heat kernel expansion up to first order (see, for instance, Gatheral et al. [9]) reads in our notation

$$p(T, s_T | t, s_t) \approx \frac{\phi(T-t, \varphi(s_T) - \varphi(s_t))}{a(s_T)} u(s_t, s_T) \times \left\{ 1 + \frac{T-t}{\varphi(s_T) - \varphi(s_t)} \int_{s_t}^{s_T} \frac{\mathcal{L}u(s, s_T)}{u(s, s_T)} \frac{ds}{a(s)} \right\},$$

where $u(s, s_T) = e^{\int_s^{s_T} \frac{\mu(\eta)}{a^2(\eta)} d\eta} \sqrt{\frac{a(s)}{a(s_T)}}$ and $\mathcal{L} = \frac{a^2(s)}{2} \partial_s^2 + \mu(s)\partial_s$ is the infinitesimal generator associated with the process S_t. In this case, the asymptotic expansion (3.3) reduces to

$$\frac{\phi(T-t, \varphi(s_T) - \varphi(s_t))}{a(s_T)} e^{H\circ\varphi(s_T) - H\circ\varphi(s_t)} \times \left\{ 1 - \frac{1}{2} \int_t^T h^2(\varphi_\tau) + h'(\varphi_\tau) d\tau \right\},$$

where $\varphi_\tau = \frac{T-\tau}{T-t}\varphi(s_t) + \frac{\tau-t}{T-t}\varphi(s_T)$. Therefore, it suffices to show that

$$e^{H\circ\varphi(s_T) - H\circ\varphi(s_t)} = u(s_t, s_T) \tag{3.5}$$

and

$$-\frac{1}{2} \int_t^T h^2(\varphi_\tau) + h'(\varphi_\tau) d\tau = \frac{T-t}{\varphi(s_T) - \varphi(s_t)} \int_{s_t}^{s_T} \frac{\mathcal{L}u(s, s_T)}{u(s, s_T)} \frac{ds}{a(s)}. \tag{3.6}$$

For (3.5), since $h \circ \varphi(s) = \frac{\mu(s)}{a(s)} - \frac{a'(s)}{2}$, $\varphi'(s) = \frac{1}{a(s)}$, and H is an antiderivative of h, we have

$$\begin{aligned} H \circ \varphi(s_T) - H \circ \varphi(s_t) &= \int_{\varphi(s_t)}^{\varphi(s_T)} h(\xi) d\xi = \int_{s_t}^{s_T} h \circ \varphi(s) d\varphi(s) \\ &= \int_{s_t}^{s_T} \left[\frac{\mu(s)}{a(s)} - \frac{a'(s)}{2} \right] \frac{ds}{a(s)} = \int_{s_t}^{s_T} \frac{\mu(s)}{a^2(s)} ds - \frac{1}{2} \log \left[\frac{a(s_T)}{a(s_t)} \right]. \end{aligned}$$

Therefore,

$$e^{H\circ\varphi(s_T) - H\circ\varphi(s_t)} = e^{\int_{s_t}^{s_T} \frac{\mu(s)}{a^2(s)} ds} \sqrt{\frac{a(s_t)}{a(s_T)}} = u(s_t, s_T).$$

As for (3.6), since $\varphi_\tau = \frac{T-\tau}{T-t}\varphi(s_t) + \frac{\tau-t}{T-t}\varphi(s_T)$, we have

$$
\begin{aligned}
&\int_t^T h^2(\varphi_\tau) + h'(\varphi_\tau)d\tau \\
&= \int_t^T h^2\left(\frac{T-\tau}{T-t}\varphi(s_t) + \frac{\tau-t}{T-t}\varphi(s_T)\right) + h'\left(\frac{T-\tau}{T-t}\varphi(s_t) + \frac{\tau-t}{T-t}\varphi(s_T)\right) d\tau \\
&= \frac{T-t}{\varphi(s_T) - \varphi(s_t)} \int_{s_t}^{s_T} h^2(\varphi(s)) + h'(\varphi(s))d\varphi(s).
\end{aligned}
$$

Note that $d\varphi(s) = \frac{ds}{a(s)}$ and

$$
h' \circ \varphi(s) = \frac{1}{\varphi'(s)}\frac{d}{ds}[h \circ \varphi(s)] = a(s) \times \frac{d}{ds}\left[\frac{\mu(s)}{a(s)} - \frac{a'(s)}{2}\right],
$$

consequently,

$$
\int_{s_t}^{s_T} \left[h^2(\varphi(s)) + h'(\varphi(s))\right] d\varphi(s) = \int_{s_t}^{s_T} \left[\left(\frac{\mu}{a} - \frac{a'}{2}\right)^2 + a\left(\frac{\mu}{a} - \frac{a'}{2}\right)'\right] \frac{ds}{a(s)},
$$

where we suppressed the dependence on s for notational simplicity. On the other hand, for the right hand side of (3.6), by straightforward calculation we have

$$
\begin{aligned}
\mathcal{L}u(s, s_T) &= \frac{a^2(s)}{2}\partial_s^2 u(s, s_T) + \mu(s)\partial_s u(s, s_T) \\
&= \left[-\frac{\mu^2}{2a^2} - \frac{(a')^2}{8} - \frac{a}{2}\left(\frac{\mu}{a} - \frac{a'}{2}\right)' + \frac{a'\mu}{2a}\right] u(s, s_T) \\
&= -\frac{1}{2}\left[\left(\frac{\mu}{a} - \frac{a'}{2}\right)^2 + a\left(\frac{\mu}{a} - \frac{a'}{2}\right)'\right] u(s, s_T).
\end{aligned}
$$

It follows that

$$
\begin{aligned}
&\int_{s_t}^{s_T} \frac{\mathcal{L}u(s, s_T)}{u(s, s_T)} \frac{ds}{a(s)} = -\frac{1}{2}\int_{s_t}^{s_T} \left[\left(\frac{\mu}{a} - \frac{a'}{2}\right)^2 + a\left(\frac{\mu}{a} - \frac{a'}{2}\right)'\right] \frac{ds}{a(s)} \\
&= -\frac{1}{2}\int_{s_t}^{s_T} \left[h^2(\varphi(s)) + h'(\varphi(s))\right] d\varphi(s),
\end{aligned}
$$

which completes the proof of (3.6). □

4 Implied Volatility Approximation

The implied volatility $\sigma_{\text{BS}} = \sigma_{\text{BS}}(K, T)$ is defined implicitly by solving the nonlinear equation

$$C(s, t, K, T) = C_{\text{BS}}(s, t, K, T, \sigma_{\text{BS}}(K, T)), \tag{4.1}$$

where the function C_{BS} on the right hand side is the celebrated Black-Scholes pricing formula for call options (assuming zero interest rate and dividend yield):

$$C_{\text{BS}}(s, t, K, T, \sigma_{\text{BS}}) = sN(d_1) - KN(d_2)$$

with $d_1 = \frac{\log s - \log K}{\sigma_{\text{BS}}\sqrt{T-t}} + \frac{\sigma_{\text{BS}}\sqrt{T-t}}{2}$, $d_2 = d_1 - \sigma_{\text{BS}}\sqrt{T-t}$, and $N(\cdot)$ is the cumulative normal distribution function. The Black-Scholes formula is monotonic increasing in the volatility parameter σ_{BS}, and for this reason amongst others, it is often market practice to quote options in terms of Black-Scholes implied volatility. Moreover, practitioners often calibrate their option pricing models to implied volatilities rather than price quotes. In this regard, efficient and accurate approximations of implied volatility not only permit faster calibration of option pricing models but also help build intuition.

Conventionally, asymptotic expansions of implied volatility for small time to expiry (to lowest order) are generated by matching exponents in respectively, an asymptotic approximation for a far out-of-the-money (OTM) option under Black-Scholes, and an asymptotic approximation to the option price from direct integration over the (approximated) density. For such far out-of-the-money (OTM) options, as time approaches expiry, the event that the underlying will end up in-the-money at expiry is a rare event. According to the theory of large deviations, such a rare event has exponentially small probability, so the option price is of the form

$$\int_K^\infty e^{-\frac{d(x)}{T-t}} f(x)\,dx. \tag{4.2}$$

As $t \to T^-$, the main contribution to the integral comes from the minimum point of d, which in this case is the boundary point of the support of f because, in the OTM case, $d(x)$ is strictly increasing in x, and $f(x)$ has the payoff function as a factor (see (4.4)). To zeroth order, the Laplace asymptotic formula (for example, see (5.2.23) on p. 193 of Bleistein and Handelsman [3]) then reads

$$\int_K^\infty e^{-\frac{d(x)}{T-t}} f(x)dx \approx (T-t)^2 e^{-\frac{d(K)}{T-t}} \frac{f'(K)}{|d'(K)|^2} \tag{4.3}$$

as $t \to T^-$, provided $f'(K)$ and $d'(K)$ are nonzero. Thus, the small time asymptotic expansion of the implied volatility is obtained by applying the Laplace asymptotic formula (4.3) to both sides of (4.1) then matching the corresponding coefficients. As one might expect, the dominating term of such expansions is typically the zeroth order term. Our objective in this section is to demonstrate how to implement this matching procedure from the path integral perspective.

Recasting Eq. (4.1) for implied volatility using our path integral representation of the density, and using our earlier representation (2.9) of the call price, we obtain

$$
\begin{aligned}
&\int_K^\infty \frac{S_T - K}{a(S_T, T)} \int_{\mathcal{C}_x} e^{-\frac{1}{2}\int_t^T |\dot{x}_\tau - h(x_\tau,\tau)|^2 d\tau} \mathcal{D}[x]\, dS_T, \\
&= \int_K^\infty \frac{S_T - K}{\sigma_{\text{BS}} s_T} e^{-\frac{\sigma_{\text{BS}}}{2}(x_T - x_t) - \frac{\sigma_{\text{BS}}^2}{8}(T-t)} \int_{\mathcal{C}_x} e^{-\frac{1}{2}\int_t^T |\dot{x}_\tau|^2 d\tau} \mathcal{D}[x] dS_T \\
&= \int_K^\infty \frac{S_T - K}{\sqrt{2\pi(T-t)}\sigma_{\text{BS}} S_T} e^{-\frac{1}{2}\left(\frac{\log s_T - \log s_t}{\sigma_{\text{BS}}\sqrt{T-t}} + \frac{\sigma_{\text{BS}}\sqrt{T-t}}{2}\right)^2} dS_T.
\end{aligned}
\tag{4.4}
$$

Equation (4.4) provides an implicit expression for Black-Scholes implied volatility in terms of local volatility. In the foregoing, we first show how to recover from (4.4) the heat kernel approximations of Gatheral et al. [9] and the most-likely-path approximation of Gatheral and Wang [8]. Finally, in Sect. 4.3, we show how to improve on these approximations by adopting the path integral perspective.

4.1 Recovery of the Berestycki-Busca-Florent (BBF) Formula

To rederive the results in Berestycki et al. [2] and Gatheral et al. [9] from (4.4), we approximate both sides of (4.4) as Laplace type integrals as in (4.2). The path integral on the left hand side of (4.4) is approximated as follows:

$$
\begin{aligned}
&\int_{\mathcal{C}_x} e^{-\frac{1}{2}\int_t^T |\dot{x}_\tau - h(x_\tau,\tau)|^2 d\tau} \mathcal{D}[x] \\
&= \int_{\mathcal{C}_x} e^{-\frac{1}{2}\int_t^T |\dot{x}_\tau|^2 - 2\int_t^T h(x_\tau,\tau) dx_\tau + \int_t^T h^2(x_\tau,\tau) d\tau} \mathcal{D}[x] \\
&\approx \int_{\mathcal{C}_x} e^{-\frac{1}{2}\int_t^T |\dot{x}_\tau|^2 d\tau} \left[1 - 2\int_t^T h(x_\tau, \tau) dx_\tau + \int_t^T h^2(x_\tau, \tau) d\tau\right] \mathcal{D}[x] \\
&\approx e^{-\frac{(x_T - x_t)^2}{2(T-t)}} \left[1 + \mathcal{O}(T-t)\right],
\end{aligned}
$$

where in the last step we approximated the path integral by evaluating the integral in the exponent along a single path: the straight line connecting x_t and x_T. Recall that $x_t = \varphi(s_t, t) = \int_{s_0}^{s_t} \frac{d\xi}{a(\xi,t)}$. Substitution back into the left hand side of (4.4) gives

$$
\begin{aligned}
&\int_K^\infty \frac{S_T - K}{a(S_T, T)} \int_{\mathcal{C}_x} e^{-\frac{1}{2}\int_t^T |\dot{x}_\tau - h(x_\tau,\tau)|^2 d\tau} \mathcal{D}[x]\, dS_T \\
&\approx \int_K^\infty e^{-\frac{|\varphi(s_T,T) - \varphi(s_t,t)|^2}{2(T-t)}} \frac{S_T - K}{a(S_T, T)} \left[1 + \mathcal{O}(T-t)\right] dS_T,
\end{aligned}
$$

which is of Laplace type as in (4.2). Applying the Laplace asymptotic formula (4.3), we obtain that, up to a factor,

$$C(s,t,K,T) \approx e^{-\frac{|\varphi(K,T)-\varphi(s,t)|^2}{2(T-t)}}. \tag{4.5}$$

Likewise, the Black-Scholes price on the right hand side of (4.4) is given, up to a factor, by

$$C_{\mathrm{BS}}(s,t,K,T) \approx e^{-\frac{|\log K-\log s|^2}{2\sigma_{\mathrm{BS}}^2(T-t)}}. \tag{4.6}$$

Finally, by matching the exponents in (4.5) and (4.6), we obtain the zeroth order approximation of the implied volatility as

$$\sigma_{\mathrm{BS}} \approx \frac{\log K - \log s}{\varphi(K,T)-\varphi(s,t)}.$$

In the time homogeneous case,

$$\varphi(K)-\varphi(s) = \int_s^K \frac{d\xi}{a(\xi)}$$

and we recover the BBF formula as in Berestycki et al. [2] and Gatheral et al. [9].

4.2 Recovery of the Variational-Most-Likely-Path (vMLP) Approximation of Gatheral and Wang [8]

The path integral term in (4.4) is in x-space. Alternatively, in s-space it reads

$$\int_{\mathcal{C}_s} e^{-\frac{1}{2}\int_t^T \left[\frac{\dot{s}_T}{a(s_T,T)} + \frac{a_S(s_T,T)}{2}\right]^2 d\tau} \mathcal{D}[s],$$

where

$$\mathcal{D}[s] = \lim_{n\to\infty} \frac{1}{\sqrt{2\pi\Delta t}\,a(s_T,T)} \prod_{i=1}^{n-1} \frac{1}{\sqrt{2\pi\Delta t}} \frac{ds_i}{a(s_i,t_i)}.$$

Hence, we can rewrite the left hand side of (4.4) in s-space as

$$C(t,s_t,K,T) = \int_K^\infty (s_T-K) \int_{\mathcal{C}_s} e^{-\frac{1}{2}\int_t^T \left[\frac{\dot{s}_T}{a(s_T,T)} + \frac{a_S(s_T,T)}{2}\right]^2 d\tau} \mathcal{D}[s]\,ds_T.$$

The variational most-likely-path approximation of implied volatility developed in Gatheral and Wang [8] is obtained by dropping the second term $\frac{a_S(s_T,T)}{2}$ in the path integral and evaluating the resulting path integral along the path that minimizes the functional

$$e^{-\frac{1}{2}\int_t^T \left|\frac{\dot{s}_\tau}{a(s_\tau,\tau)}\right|^2 d\tau}.$$

In other words,

$$C(s,t,K,T) \approx \int_K^\infty (s_T - K) e^{-\frac{1}{2}\int_t^T \left|\frac{\dot{s}^*_\tau}{a(s^*_\tau,\tau)}\right|^2 d\tau} ds_T,$$

where s^*_τ is the optimal path that maximizes the action functional $\int_t^T \left|\frac{\dot{s}_\tau}{a(s_\tau,\tau)}\right|^2 d\tau$ subject to the constraints that initial and terminal points are fixed at s_t and s_T respectively. Moreover, since the resulting integral is of Laplace type, the call price is given asymptotically, up to a factor, by

$$C(s,t,K,T) \approx e^{-\frac{1}{2}\int_t^T \left|\frac{\dot{s}^*_\tau}{a(s^*_\tau,\tau)}\right|^2 d\tau},$$

where the optimal path s^*_τ has initial and terminal points s and K respectively. Finally, by matching the exponent with the Black-Scholes asymptotic as in (4.6), the zeroth order approximation of implied volatility is given by

$$\sigma_{\mathrm{BS}} \approx \frac{|\log K - \log s|}{\sqrt{T-t}} \left[\int_t^T \left|\frac{\dot{s}^*_\tau}{a(s^*_\tau,\tau)}\right|^2 d\tau\right]^{-\frac{1}{2}}$$

which recovers the variational most-likely-path approximation of the implied volatility presented in Gatheral and Wang [8].

4.3 New and Improved Most-Likely-Path (MLP) Approximation

As is obvious from our presentation, the approximations obtained in Gatheral et al. [9] and in Gatheral and Wang [8] are suboptimal from the perspective of our path integral representation (4.4) in the sense that they both drop terms. This suggests that we should define the path-integral-most-likely-path to be the path that maximizes the full action functional

$$\frac{1}{2}\int_t^T |\dot{x}_\tau - h(x_\tau,\tau)|^2 d\tau \tag{4.7}$$

or equivalently in s-space the functional

$$\frac{1}{2}\int_t^T \left[\frac{\dot{s}_\tau}{a(s_\tau, \tau)} + \frac{a_s(s_\tau, \tau)}{2}\right]^2 d\tau \tag{4.8}$$

without dropping terms. The Euler-Lagrange equation associated with the functional in (4.7) is

$$\ddot{x}_\tau = h\, h_x + h_t \tag{4.9}$$

with boundary conditions x_t and x_T at times t and T respectively. Matching exponents as before gives

$$\left|\frac{\log K - \log s}{\sigma_{\text{BS}}\sqrt{T-t}} + \frac{\sigma_{\text{BS}}\sqrt{T-t}}{2}\right|^2 = \int_t^T [\dot{x}_\tau^* - h(x_\tau^*, \tau)]^2 d\tau, \tag{4.10}$$

where x_τ^* is the optimal path which maximizes the functional (4.7) (or equivalently solves (4.9)) with initial and terminal points given by $\varphi(s,t)$ and $\varphi(K,T)$ respectively. Solving (4.10) for σ_{BS} yields our new-and-improved zeroth order approximation for implied volatility.

To illustrate the accuracy of our new approximation (4.10), consider the case of time dependent Black-Scholes, where rather pleasingly, (4.10) gives the exact solution. Note in passing that, to the best of our knowledge, none of the existing small time approximations is able to recover this very simple case.

Example 1 (Implied volatility in the time dependent Black-Scholes model) Assume the price S_t of the underlying satisfies the following under the pricing measure:

$$dS_\tau = \sigma(\tau)\, S_\tau\, dB_\tau, \quad S_t = s_t.$$

In order to apply (4.10), we proceed as follows:

(a) Transform the model into x-space.
(b) Solve the Euler-Lagrange equation (4.9) for the optimal path.
(c) Evaluate the the action functional (4.9) along the optimal path, substitute into (4.10) and solve for the implied volatility.

(a) *Transform into x -space*: In this case, $x = \varphi(s,t) = \int_{s_0}^s \frac{1}{\sigma(t)\xi} d\xi = \frac{\log s - \log s_0}{\sigma(t)}$. Dropping the explicit dependence on t for ease of notation, and applying Ito's formula to $X_t = \varphi(S_t, t)$ we obtain

$$\begin{aligned} dX_t &= \dot{\varphi}(S_t,t)dt + \varphi_s(S_t,t)dS_t + \frac{1}{2}\varphi_{ss}(S_t,t)d[S]_t \\ &= dB_t - \left(\frac{\sigma'}{\sigma}X_t + \frac{\sigma}{2}\right)dt. \end{aligned}$$

Thus $h(x,t) = -\frac{\sigma}{2} - \frac{\sigma'}{\sigma}x.$

(b) *Solve the Euler-Lagrange equation*: The associated Euler-Lagrange equation (4.9) in this case reads

$$\ddot{x} = h\,h_x + h_t = \left[\left(\frac{\sigma'}{\sigma}\right)^2 - \left(\frac{\sigma'}{\sigma}\right)'\right] x.$$

With the change of variable $x = \frac{z}{\sigma}$, the above ODE for x is transformed into the following ODE for z

$$\ddot{z} - \frac{2\sigma'}{\sigma}\dot{z} = 0 \implies \frac{d}{d\tau}\left(\frac{\dot{z}}{\sigma^2}\right) = 0.$$

With boundary conditions $z_t = \sigma_t\, x_t$ and $z_T = \sigma_T\, x_T$, the solution to the Euler-Lagrange equation is given by

$$\sigma_\tau\, x_\tau = z_\tau = \sigma_t\, x_t + \frac{\sigma_T\, x_T - \sigma_t\, x_t}{\int_t^T \sigma^2(s)ds} \int_t^\tau \sigma^2(s)ds.$$

(c) *Solve for implied volatility*: It follows that the functional (4.7) evaluated along the optimal path, taking into account that $\frac{\dot{z}}{\sigma^2} = \frac{\sigma_T x_T - \sigma_t x_t}{\int_t^T \sigma^2(s)ds}$ is a constant, is given by

$$\begin{aligned}
&\int_t^T |\dot{x}_\tau - h(x_\tau, \tau)|^2 d\tau = \int_t^T \left|\dot{x}_\tau + \frac{\sigma}{2} + \frac{\sigma'}{\sigma}x\right|^2 d\tau \\
&= \int_t^T \left|\frac{\partial_\tau(\sigma\, x)}{\sigma} + \frac{\sigma}{2}\right|^2 d\tau = \int_t^T \left|\frac{\dot{z}}{\sigma} + \frac{\sigma}{2}\right|^2 d\tau \\
&= \left(\frac{\sigma_T x_T - \sigma_t x_t}{\int_t^T \sigma^2(s)ds} + \frac{1}{2}\right)^2 \int_t^T \sigma_\tau^2 d\tau \\
&= \left(\frac{\log s_T - \log s_t}{\sqrt{\int_t^T \sigma^2(s)ds}} + \frac{1}{2}\sqrt{\int_t^T \sigma^2(s)ds}\right)^2.
\end{aligned}$$

Finally, substituting this last expression into (4.10) gives the well-known result

$$\sigma_{\mathrm{BS}}^2 = \frac{1}{T-t}\int_t^T \sigma^2(s)ds,$$

which is exact.

5 Conclusion

We have shown, up to first order in $\tau = T - t$, that the classical heat kernel expansion can be derived using a novel probabilistic approach. This new probabilistic derivation of the heat kernel expansion inspires a path integral representation of the transition density; natural definitions of the most-likely-path approximation of the transition density, the call price, and the implied volatility then follow. In the time homogeneous case, we recover well-known classical results. However, in the time inhomogeneous case, we obtain a new asymptotic expansion that generalizes the classical one. We showed how the lowest order approximation of Berestycki, Busca and Florent as well as the higher order approximations of Gatheral et al. [9] and Gatheral and Wang [8] correspond to dropping terms in our lowest order path integral representation. We further showed that by restoring the dropped terms, our new representation recovers the exact expression for Black-Scholes implied volatility in the time-dependent Black-Scholes model, which no existing asymptotic expansion technique has so far been able to achieve, to the best of our knowledge. Further applications of this promising approach to the important practical problem of accurately approximating implied volatility under local volatility is left for future research.

Acknowledgments We thank the anonymous reviewer for his helpful and constructive comments. We are also grateful for helpful discussions with the participants of the following seminars: Math Finance and PDE Seminar at Rutgers University, Probability Seminar at TU Berlin, Probability Seminar at Academia Sinica, Mathematics Colloquium at Ritsumeikan University, Mathematical Finance Seminar at Osaka University. All errors are our own responsibility.

References

1. Baldi, P., Caramellino, L.: Asymptotics of hitting probabilities for general one-dimensional diffusions. Ann. Appl. Probab. **12**, 1071–1095 (2002)
2. Berestycki, H., Busca, J., Florent, I.: Asymptotics and calibration of local volatility models. Quant. Financ. **2**, 61–69 (2002)
3. Bleistein, N., Handelsman, R.A.: Asymptotic Expansions of Integrals. Dover Publications, New York (1986)
4. Chavel, I.: Eigenvalues in Riemannian geometry. Pure and Applied Mathematics, Book 115, Academic Press (1984)
5. Cheng, W., Costanzino, N., Liechty, J., Mazzucato, A.L., Nistor, V.: Closed-form asymptotics and numerical approximations of 1D parabolic equations with applications to option pricing. SIAM J. Financ. Math. **2**(1), 901–934 (2011)
6. Gatheral, J.: The Volatility Surface: A Practitioner's Guide, Wiley Finance (2006)
7. Gatheral, J., Jacquier, A.: Arbitrage-free SVI volatility surfaces. Quant. Financ. **14**(1), 59–71 (2014)
8. Gatheral, J., Wang, T.-H.: The heat kernel most-likely-path approximation. Int. J. Theor. Appl. Financ. **15**(1), 1250001 (2012)
9. Gatheral, J., Hsu, E.P., Laurence, P., Ouyang, C., Wang, T.-H.: Asymptotics of implied volatility in local volatility models. Math. Financ. **22**(4), 591–620 (2012)
10. Goovaerts, M., De Schepper, A., Decamps, M.: Closed-form approximations for diffusion densities: a path integral approach. J. Comput. Appl. Math. **164–165**, 337–364 (2004)

11. Guyon, J., Henry-Labordère, P.: From spot volatilities to implied volatilities. Risk Mag. pp. 79–84 (2011)
12. Henry-Labordère, P.: Analysis, Geometry, and Modeling in Finance. Chapman & Hall/CRC, Financial Mathematics Series (2008)
13. Hsu, E.P.: Stochastic Analysis on Manifolds. Graduate Studies in Mathematics, American Mathematical Society (2002)
14. Jordan, R., Tier, C.: Asymptotic approximations to deterministic and stochastic volatility models. SIAM J. Financ. Math. **2**(1), 935–964 (2011)
15. Keller-Ressel, M., Teichmann, J.: A remark on Gatheral's 'most-likely path approximation' of implied volatility. In: Springer Proceedings in Mathematics & Statistics (2014)
16. Linetsky, V.: The path integral approach to financial modeling and options pricing. Comput. Econ. **11**(1–2), 129–163 (1997)
17. Reghai, A.: The hybrid most likely path. Risk Mag. 34–35 (2006)

Extrapolation Analytics for Dupire's Local Volatility

Peter Friz and Stefan Gerhold

Abstract We consider wing asymptotics of local volatility surfaces. While our recent paper in the journal Risk (De Marco et al. Risk 2:82–87, 2013, [3]) discusses our approximation formula from a practical and numerical perspective, the present paper focuses on rigorous proofs of the approximations. We apply the saddle point method (Heston model) and Hankel contour integration (variance gamma model).

Keywords Local volatility · Saddle point methods · Contour integration

1 Introduction

One of the main objectives in option pricing theory is to price exotic derivatives consistently with observed vanilla prices. According to the seminal work of Dupire [5], this can in principle be achieved, for a one-dimensional underlying, by a model with dynamics $dS_t/S_t = \sigma(S_t, t)dW_t$. As opposed to stochastic volatility models, here the volatility is a deterministic function of time and current underlying price. Any given smooth call price surface $C(K, T)$, for strikes $K > 0$ and maturities $T > 0$,

---

A preprint of this article circulated under the title "Don't stay local—extrapolation analytics for Dupire's local volatility".

P. Friz (✉)
Institut für Mathematik, Technische Universität Berlin, Berlin, Germany
e-mail: friz@math.tu-berlin.de

P. Friz
Weierstraß-Institut für Angewandte Analysis und Stochastik,
Berlin, Germany

S. Gerhold
Financial and Actuarial Mathematics, Vienna University of Technology, Wiedner Hauptstraße 8/105-1 A-1040, Vienna, Austria
e-mail: sgerhold@fam.tuwien.ac.at

P.K. Friz et al. (eds.), *Large Deviations and Asymptotic Methods in Finance*, Springer Proceedings in Mathematics & Statistics 110,
DOI 10.1007/978-3-319-11605-1_10

can be recovered by a so-called local volatility model $dS_t/S_t = \sigma_{\text{loc}}(S_t, t)dW_t$, where the volatility function is given by Dupire's formula [5]

$$\sigma^2_{\text{loc}}(K, T) = \frac{2\partial_T C}{K^2 \partial_{KK} C}. \tag{1}$$

Exotic options can then be priced by Monte Carlo simulation. Local volatility models are of considerable practical importance, and serve as building blocks for more advanced models, e.g. local-stochastic-volatility (LSV) models.

In the present paper, we consider local volatility surfaces that arise from call prices that are generated by some model for the underlying. Our aim is to turn the knowledge of that model's mgf (moment generating function; of log-spot X_T) into asymptotic results of the corresponding local volatility surface. In [3], we described two applications of such approximations. One is to the design of local volatility parametrizations, whose asymptotic behavior may be matched to our results. Another application concerns model risk. Consider pricing under an "advanced" model (affine stochastic volatility, Lévy, etc.; anything with known mgf) versus a local volatility model. The relative differences between the prices has been named "toxicity index" in [13]. Roughly speaking, it measures the distance of the trade from vanilla options. The most consistent way to calculate this index is to use the local volatility model generated by the "advanced" model, because only then all vanillas will have zero toxicity. When computing the local volatility surface, our accurate approximations can then profitably replace other numerical methods in regimes where the latter become unstable (see [3] for details).

We suppose that the underlying price process $S_t = \exp(X_t)$ is a martingale under the pricing measure $\mathbb{P}$ and write $C(K, T)$ for its call price surface. For simplicity we assume zero interest rate throughout. If C is sufficiently smooth, then the associated local volatility function is given by Dupire's formula (1). Recall the main asymptotic formula from [3]:

$$\sigma^2_{\text{loc}}(K, T) \sim \left. \frac{2\frac{\partial}{\partial T} m(s, T)}{s(s-1)} \right|_{s=\hat{s}(k,T)}, \tag{2}$$

where k denotes log-strike, and $\hat{s} = \hat{s}(k, T)$ is determined as solution of the saddle point equation

$$\frac{\partial}{\partial s} m(s, T) = k. \tag{3}$$

Here, $m(s, T) := \log M(s, T)$ is the logarithm of the moment generating function (mgf) M, which is defined by $M(s, T) := \mathbb{E} \exp(sX_T)$ and is analytic in the (maximal) strip $s_-(T) < \text{Re}(s) < s_+(T)$. The numbers s_- and s_+ are called critical exponents. In this note, we will use (2) for $K \to \infty$, but other asymptotic regimes can also be covered [3, 8]; it is thus not only a local-volatility analogue of Lee's moment formula [11], but works also for maturity (or joint) asymptotics.

As described in [3], formula (2) results from saddle point approximations of numerator and denominator of Dupire's formula, after inserting the Fourier representation of the call price:

$$\sigma_{\text{loc}}^2(K,T) = \frac{2\partial_T C}{K^2 \partial_{KK} C} = \frac{2\int_{-i\infty}^{i\infty} \frac{\partial_T m(s,T)}{s(s-1)} e^{-ks} M(s,T)ds}{\int_{-i\infty}^{i\infty} e^{-ks} M(s,T)ds}. \tag{4}$$

(The real parts of the contours are in $(1, s_+)$.)

Whereas the focus of [3] is on numerical tests and applications, the present note gives proofs for the validity of (2), in the setting of the Heston and of the variance gamma model. As regards methodology, the proof for the Heston model uses a classical saddle point approach. Its most interesting ingredient, similarly to [7], is the use of ODE comparison results to furnish the necessary tail estimates, without taking recourse to the explicit form of the Heston mgf. The analysis is thus well suited to extension towards other affine stochastic volatility models. For the variance gamma model, the saddle point method is not appropriate. We apply another classical contour integration approach, based on Hankel contours, which seems to be new in mathematical finance.

2 The Heston Model

Even though practitioners seem to prefer local-stochastic-volatility models nowadays over the classical Heston model, it might still be useful for the two applications outlined in the introduction (model risk and parametrization design; recall that the large maturity Heston smile motivates the popular SVI parametrization of *implied* volatility [9]). The dynamics of the Heston model are

$$dS_t = S_t\sqrt{Y_t}dW_t, \qquad S_0 = s_0 > 0,$$
$$dV_t = (a + bV_t)dt + c\sqrt{V_t}dZ_t, \qquad V_0 = v_0 > 0,$$

with $a \geq 0$, $b \leq 0$, $c > 0$, and $d\langle W, Z\rangle_t = \rho dt$ with $\rho \in (-1, 1)$.

Theorem 1 *In the Heston model with $\rho \leq 0$ (the relevant regime in practice, at least for equity models), the asymptotic equivalence* (2) *holds for $k \to \infty$. The explicit leading term is*

$$\sigma_{\text{loc}}^2(K,T) \sim \frac{2}{s_+(s_+ - 1)R_1/R_2} \times k, \quad k \to \infty, \tag{5}$$

where $k = \log(K/S_0)$, $s_+ \equiv s_+(T)$ *and*

$$R_1 = Tc^2 s_+(s_+ - 1)\left[c^2(2s_+ - 1) - 2\rho c(s_+\rho c + b)\right] \tag{6}$$
$$- 2(s_+\rho c + b)\left[c^2(2s_+ - 1) - 2\rho c(s_+\rho c + b)\right]$$
$$+ 4\rho c\left[c^2 s_+(s_+ - 1) - (s_+\rho c + b)^2\right],$$
$$R_2 = 2c^2 s_+(s_+ - 1)\left[c^2 s_+(s_+ - 1) - (s_+\rho c + b)^2\right]. \tag{7}$$

Proof It was shown in [3] that the right hand side of (5) asymptotically equals the right hand side of (2). It thus remains to show that (2) holds for the Heston model as $k \to \infty$.

By the exponential decay of the Heston mgf towards $\pm i\infty$, the second equality in formula (4) is correct for the Heston model. For the saddle point analysis of (4), we employ the approximate saddle point

$$\hat{s}_{\text{approx}}(k) := s_+ - \beta k^{-1/2},$$

where $\beta = \frac{\sqrt{2v_0}}{c\sqrt{\sigma}}$, σ denotes the critical slope

$$\sigma(T) = -\frac{\partial T^*}{\partial s}(s_+(T)),$$

and

$$T^*(s) = \sup\{t \geq 0 : \mathbb{E}[e^{sX_t}] < \infty\}.$$

This is the same approximate saddle point as in [7]; see there for more details on its choice, and the definition of $\sigma(T)$ and $T^*(s)$. (In [7], our $\hat{s}_{\text{approx}}$ was called simply $\hat{s}$, since the *exact* saddle point of the denominator of (4), defined in (3), did not occur.) This approximate saddle may be used for both integrals in (4). As for the denominator, this was carried out in detail in [7], where an expansion of the Heston density $\partial_{KK}C$ was determined. The analysis of the numerator in (4) is similar, except that a new tail estimate is required. But first we discuss the local expansion around the saddle point. Let us fix a number $\alpha \in (\frac{2}{3}, \frac{3}{4})$ and define $h(k) = k^{-\alpha}$. Then, in the central range $|s - \hat{s}_{\text{approx}}(k)| \leq h(k)$, we have

$$\frac{1}{s(s-1)} = \frac{1}{s_+(s_+ - 1)} + O(s_+ - s)$$
$$= \frac{1}{s_+(s_+ - 1)}\left(1 + O(k^{-1/2})\right)$$

and (cf. formula (19) in [3])

$$
\begin{aligned}
2\frac{\partial}{\partial T}m(s,T) &= \frac{2\beta^2}{\sigma(s_+ - s)^2} + O\left(\frac{1}{s_+ - s}\right) \\
&= \frac{2\beta^2}{\sigma}(\beta k^{-1/2} + O(k^{-\alpha}))^{-2} + O(k^{-1/2}) \\
&= \frac{2k}{\sigma}(1 + O(k^{1/2-\alpha})).
\end{aligned}
$$

Therefore, the local expansions of the two integrands in (4) agree, up to a factor that is given by

$$\frac{2\partial_T m(s,T)}{s(s-1)} = \frac{2k}{\sigma s_+(s_+ - 1)}(1 + O(k^{1/2-\alpha})), \tag{8}$$

where the error term holds uniformly w.r.t. the integration variable s. According to Theorem 1.2 of [7], we have

$$\frac{1}{2i\pi}\int_{\hat{s}_{\text{approx}}-ih(k)}^{\hat{s}_{\text{approx}}+ih(k)} e^{-ks}M(s,T)ds \sim A_1 e^{(1-A_3)k + A_2\sqrt{k}} k^{-3/4+a/c^2} \tag{9}$$

for certain constants A_1, $A_2 = 2\beta$, and $A_3 = s_+ + 1$. Analogously, we derive from (8) that

$$
\begin{aligned}
&\frac{1}{2i\pi}\int_{\hat{s}_{\text{approx}}-ih(k)}^{\hat{s}_{\text{approx}}+ih(k)} \frac{2\partial_T m(s,T)}{s(s-1)} e^{-ks}M(s,T)ds \\
&\qquad \sim \frac{2k}{\sigma s_+(s_+ - 1)} \times A_1 e^{(1-A_3)k + A_2\sqrt{k}} k^{-3/4+a/c^2}.
\end{aligned} \tag{10}
$$

Dividing (10) by (9) shows our claim (5), provided that the tails $|s - \hat{s}_{\text{approx}}(k)| > h(k)$ of the integrals can be discarded. For the denominator of (4), this was shown in Lemma A.3 of [7]. So we proceed with the numerator. We consider only the upper tail, as the lower one is handled by symmetry. By Lemma A.3 of [7], there is a constant $B > 0$ such that

$$\left|\int_{\hat{s}_{\text{approx}}+ih(k)}^{\hat{s}_{\text{approx}}+iB} e^{-ks}M(s,T)ds\right| \le e^{(1-A_3)k}\exp(A_2\sqrt{k} - \tfrac{1}{2}\beta^{-1}k^{3/2-2\alpha} + O(\log k)). \tag{11}$$

From formula (18) in [3] we obtain

$$\left|\frac{\partial_T m(s,T)}{s(s-1)}\right| \le const \times k$$

for all s on the contour in (11). This estimate can be absorbed into the factor $\exp(O(\log k))$ in (11), so that we conclude

$$\left| \int_{\hat{s}_{\text{approx}}+Ih(k)}^{\hat{s}_{\text{approx}}+iB} \frac{\partial_T m(s,T)}{s(s-1)} e^{-ks} M(s,T) ds \right| \\ \leq e^{(1-A_3)k} \exp(A_2\sqrt{k} - \tfrac{1}{2}\beta^{-1}k^{3/2-2\alpha} + O(\log k)). \tag{12}$$

This grows slower than the right hand side of (10) (compare the relevant factors $k^{-3/4+a/c^2}$ resp. $\exp(-\frac{1}{2}\beta^{-1}k^{3/2-2\alpha})$). As for $\text{Im}(s) > B$, it was shown in [7] (Lemma A.2) that

$$\left| \int_{\hat{s}_{\text{approx}}+iB}^{\hat{s}_{\text{approx}}+i\infty} e^{-ks} M(s,T) ds \right| = O(\exp((1-A_3)k + \beta\sqrt{k})).$$

This was deduced from the exponential decay of $M(s,T)$ for large $\text{Im}(s)$ (Lemma A.1 in [7]). The following lemma implies that the new factor $\partial_T m(s,T)/(s(s-1))$ grows only polynomially, so that the exponential decay of the integrand persists for the numerator of (4). This finishes the proof of Theorem 1. □

To state the lemma, recall that $m(s,t) = \phi(s,t) + v_0\psi(s,t)$, where ϕ and ψ satisfy the Riccati equations

$$\begin{aligned} \dot{\phi} &= a\psi, \qquad \phi(0) = 0, \\ \dot{\psi} &= \tfrac{1}{2}(s^2 - s) + \tfrac{1}{2}c^2\psi^2 + b\psi + s\rho c\psi, \qquad \psi(0) = 0. \end{aligned}$$

We have to show that $\dot{m}$ grows only polynomially as $\text{Im}(s) \to \infty$. Because of the Riccati equations, it suffices to show this for ψ. Let us write $\psi = f + ig$ and $s = \xi + iy$.

Lemma 2 *Let $T > 0$, and assume that the real part ξ of s stays bounded in some interval $1 \leq \xi \leq \xi_{\max}$. Then, there are positive constants $C_{i,T}$ ($i = 1, 2, 3$) such that for $y \geq y_0$, where y_0 depends only on $\xi_{\max}$ and the other (fixed) model parameters of the Heston model,*

$$\begin{aligned} -C_{3,T}y^2 \leq f(t) \leq -C_{1,T}y, \\ 0 \leq g(t) \leq C_{2,T}\,y. \end{aligned}$$

In fact, we can take

$$\begin{aligned} C_{1,T} &= 1/(3c), \\ C_{2,T} &= \frac{1}{2}(2\xi_{\max} - 1)\,T, \\ C_{3,T} &= T\left(1 + \frac{c^2}{2}C_{2,T}^2\right). \end{aligned}$$

Proof It follows from the proof of Lemma A.1 in [7] that (e.g. with $C_{1,T} := T\theta = \frac{1}{c}\sqrt{1/6} \le \frac{1}{3c}$)

$$f(t) \le -T\theta y = -\frac{1}{c}\sqrt{1/6}y \le -\frac{1}{3c}y =: -C_{1,T}y.$$

We next provide a similar upper estimate for g. To this end we first show that $g = g(t)$ remains ≥ 0 for all times $t > 0$. The differential equation for g,

$$\dot{g} = \frac{1}{2}(2\xi y - y) + c^2 f g - \gamma g, \qquad g(0) = 0,$$

implies the first order Euler estimate

$$\begin{aligned} g(t) &= g(0) + \left\{\frac{1}{2}(2\xi y - y) + c^2 f(0)g(0) - \gamma g(0)\right\} t + o(t) \\ &= \underbrace{\frac{1}{2}(2\xi y - y)}_{>0} t + o(t), \end{aligned}$$

and hence g is positive (even strictly so) on some interval $(0, \varepsilon_1)$. Assume this interval is maximal in the sense that $g(\varepsilon_1) = 0$ and g is (strictly) negative on some further interval $(\varepsilon_1, \varepsilon_2)$. Clearly then $\dot{g}(\varepsilon_1) \le 0$, which contradicts the information from the differential equation: indeed, using $g(\varepsilon_1) = 0$, we obtain the contradiction

$$\dot{g}(\varepsilon_1) = \underbrace{\frac{1}{2}(2\xi y - y)}_{>0}.$$

The observation that $g \geq 0$ is useful to us, since it leads, together with $f \leq -C_{1,T}y$ and $\gamma \geq 0$, to the differential inequality

$$\begin{aligned}\dot{g} &= \frac{1}{2}(2\xi y - y) + c^2 fg - \gamma g \\ &\leq \frac{1}{2}(2\xi y - y) - \left(c^2 C_{1,T} + \gamma\right) g \\ &\leq \frac{1}{2}(2\xi y - y),\end{aligned}$$

and hence to the upper estimate

$$\forall 0 \leq t \leq T : g(t) \leq \frac{1}{2}(2\xi_{\max} - 1)\, T \times y =: C_{2,T}y.$$

We can feed this upper estimate on g back in the differential equation for f to obtain a lower estimate

$$\begin{aligned}\dot{f} &= \frac{1}{2}\left(\xi^2 - y^2 - \xi\right) + \frac{c^2}{2}\left(f^2 - g^2\right) - \gamma f \\ &\geq \frac{1}{2}\left(\xi^2 - y^2 - \xi\right) + \frac{c^2}{2} f^2 - \frac{c^2}{2} C_{2,T}^2 y^2 - \gamma f \\ &= -\frac{1}{2}\left(1 + c^2 C_{2,T}^2\right) y^2 + \frac{1}{2}\left(\xi^2 - \xi\right) - \gamma f + \frac{c^2}{2} f^2 \\ &\geq -\frac{1}{2}\left(1 + c^2 C_{2,T}^2\right) y^2 + \frac{1}{2}\left(\xi^2 - \xi\right) - \gamma f \\ &\geq -\left(1 + \frac{c^2}{2} C_{2,T}^2\right) y^2 - \gamma f,\end{aligned}$$

where in the last step we assume that y is large enough so that the extra amount subtracted (at least: $\frac{1}{2}y^2$) is larger than $\frac{1}{2}\left(\xi^2 - \xi\right)$, which remains bounded. We also know that $f(t) \leq -C_{1,T}y \leq 0$ for all $0 \leq t \leq T$. It follows that $-\gamma f \geq 0$ and omission leads to our final lower bound on $\dot{f}$, namely

$$\dot{f} \geq -\left(1 + \frac{c^2}{2} C_{2,T}^2\right) y^2.$$

This entails immediately

$$f(t) \geq -T\left(1 + \frac{c^2}{2} C_{2,T}^2\right) y^2 =: -C_{3,T}y^2. \qquad \square$$

3 The Variance Gamma Model

The mgf of the variance gamma model is

$$M(s,T) = e^{Tbs}(1 - \theta\nu s - \tfrac{1}{2}\sigma^2\nu s^2)^{-T/\nu},$$

where $\sigma, \nu > 0$ and $\theta \in \mathbb{R}$. The "drift" $b = \nu^{-1}\log(1 - \theta\nu - \frac{1}{2}\sigma^2\nu)$ is chosen such that $S = e^X$ becomes a martingale (w.l.o.g., $S_0 = 1$). For fixed T with $0 < T/\nu < \frac{1}{2}$, the density $\partial_{KK}C(K,T)$ of S_T has a singularity at the origin. Indeed, it behaves as $\approx |k|^{2T/\nu-1}$, which easily follows from the integral representation of the density [1] (as always, $k = \log K$). At the money, the denominator of the Dupire formula (1) thus explodes for small T. If $T/\nu > \frac{1}{2}$, then the density is continuous. This lack of smoothness is just an additional issue on top of a common feature of jump models: The associated local volatility surface explodes as $T \to 0$, and so the local volatility SDE

$$dS/S = \sigma_{\text{loc}}(S,t)dW \tag{13}$$

does not make sense on $[0,\infty) \ni t$.

However, following [8], we can start a Monte Carlo simulation of (13) at a time $T_0 > 0$ (here, $T_0 > \nu/2$) instead of time zero. With the appropriate *stochastic* initial value, sampled from the density $\partial_{KK}C(K,T_0)$, we recover call prices from time T_0 on. (T_0 is called ε in [8].) This gives a meaning to the local volatility surface of a jump model, without appealing to the practically challenging approach of local Lévy models [2]. Our aim is not to make this fully rigorous for the variance gamma model (or other jump models), which would require to show that (13) admits a unique strong solution on $[T_0,\infty)$. Our focus, instead, is on a rigorous proof that (2) is valid in this setting. To ensure the validity of the Fourier representations of density and call price, we even assume $T/\nu > 1$ (instead of $T/\nu > \frac{1}{2}$).

Theorem 3 *In the variance gamma model, formula* (2) *holds for* $k = \log K \to \infty$. *The explicit leading term is*

$$\sigma^2_{\text{loc}}(K,T) \sim \frac{2\log(k/T)}{\nu s_+(s_+ - 1)}, \quad k \to \infty. \tag{14}$$

Note that the numerator of (14) is $\sim 2\log k$. We kept the T-dependence, because the same analysis works for fixed k and $T \to 0$, and in fact for any asymptotic regime with $k/T \to \infty$. This is a common feature of Lévy models, since the right-hand side of (2) depends on k and T only through k/T.

Proof We write the moment generating function as

$$M(s,T) = e^{bTs}\Big(\tfrac{1}{2}\sigma^2\nu(s_+ - s)(s - s_-)\Big)^{-T/\nu}$$

where the critical moments are

$$s_{\pm} = \frac{-\nu\theta \pm \sqrt{2\nu\sigma^2 + \nu^2\theta^2}}{\nu\sigma^2}.$$

We analyze the denominator of (4), i.e., the density. The arguments for the numerator are analogous (see below). The shift $k \to k + bT$ makes it clear that we may w.l.o.g. assume that $b = 0$. The main part of the saddle point equation (3) is $T/(\nu(s_+ - s)) = k$, and so

$$\hat{s} = s_+ - \frac{T}{\nu k} + O(k^{-2}).$$

The saddle point approximation of the density then is

$$\frac{1}{2i\pi}\int_{-i\infty}^{i\infty} e^{-ks} M(s,T)ds \approx \frac{\exp(m(\hat{s},t) - k\hat{s})}{\sqrt{2\pi m''(\hat{s},T)}}. \tag{15}$$

The interesting point now is that (15) is *wrong* for the variance gamma model, inasmuch as asymptotic equality does not hold. The algebraic singularity of the mgf is not pronounced enough to make the saddle point method work; see also the remark after the proof. For a correct analysis, we use an integration contour as in Fig. 1. The U-shaped notch, denoted by $\mathcal{C}(k)$, extends a bit to the right of the singularity s_+, and captures enough asymptotic information from it. By transformation into a so called Hankel path, Hankel's representation of the Gamma function can be invoked after termwise integration of a local expansion. This "Hankel contour approach" is well known in analytic combinatorics, in particular, from the so-called singularity analysis of generating functions [6].

Let us first argue that the integrals over the dashed lines in Fig. 1 can be discarded. By symmetry, it suffices to consider the upper one. The real part of s is then $\mathrm{Re}(s) = s_+ + (\log k)/k$. First suppose that s is away from the singularity, say $\mathrm{Im}(s) > 1$. The

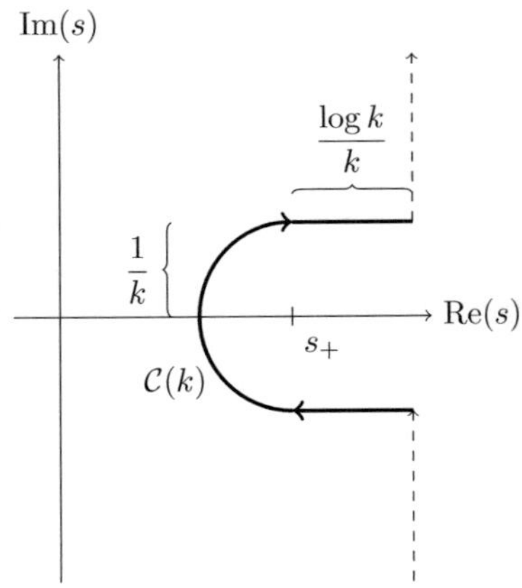

Fig. 1 The contour $\mathcal{C}(k)$, a small notch embracing the critical moment s_+

integral of $((s_+ - s)(s - s_-))^{-T/\nu}$ over this part of the contour is $O(1)$, and so we get the bound $O(e^{-k\mathrm{Re}(s)}) = O(e^{-ks_+}/k)$. Now consider s with $1/k \le \mathrm{Im}(s) < 1$. We estimate the resulting integral by the length of the contour, which is $O(1)$, times the absolute value of the integrand at the lower endpoint $s = s_+ + (\log k)/k + i/k$. The latter is easily seen to be $O(e^{-ks_+}k^{T/\nu-1}(\log k)^{-T/\nu})$.

We will now show that the integral over $\mathcal{C}(k)$ is of order $e^{-ks_+}k^{T/\nu-1}$, so that the tail estimates we have just derived are good enough. The factor $(s - s_-)$ is locally almost constant; we have, uniformly for $s \in \mathcal{C}(k)$,

$$M(s,T) \sim c_1(s_+ - s)^{-T/\nu}, \quad k \to \infty,$$

where $c_1 = c_1(T) = (\sigma^2\nu(s_+ - s_-)/2)^{-T/\nu}$. Therefore,

$$\int_{\mathcal{C}(k)} e^{-ks} M(s,T)ds \sim \int_{\mathcal{C}(k)} e^{-ks}\frac{c_1}{(s_+ - s)^{T/\nu}}ds.$$

The change of variables $s = s_+ - w/k$ transforms this into

$$\begin{aligned}\frac{e^{-ks_+}}{k}\int_{\mathcal{H}(k)} e^w c_1\left(\frac{k}{w}\right)^{T/\nu} dw &= c_1\frac{e^{-ks_+}}{k^{1-T/\nu}}\int_{\mathcal{H}(k)} e^w w^{-T/\nu}dw \\ &\sim c_1\frac{e^{-ks_+}}{k^{1-T/\nu}}\int_{\mathcal{H}(\infty)} e^w w^{-T/\nu}dw.\end{aligned}$$

The integration paths are displayed in Fig. 2. The right one, $\mathcal{H}(\infty)$, is called a Hankel contour; $\mathcal{H}(k)$ is a Hankel contour truncated at $\mathrm{Re}(s) = -\log k$. Now recall Hankel's representation for the Gamma function [12]:

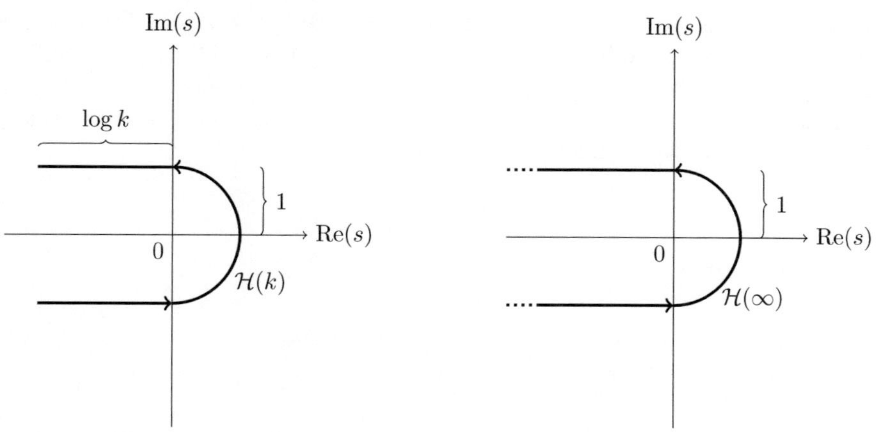

Fig. 2 The integration contours $\mathcal{H}(k)$ and $\mathcal{H}(\infty)$. The *dots* should indicate that the contour $\mathcal{H}(\infty)$ extends to $-\infty$

$$\frac{1}{2i\pi}\int_{\mathcal{H}(\infty)} e^{w} w^{-z} dw = \frac{1}{\Gamma(z)}.$$

We thus arrive at

$$\frac{1}{2i\pi}\int_{-i\infty}^{i\infty} e^{-ks} M(s,T) ds \sim \frac{c_1}{\Gamma(T/\nu)} e^{-ks_+} k^{T/\nu - 1}. \tag{16}$$

The numerator of (4) can be treated analogously, with a very similar tail estimate. The contribution of the new factor to the local expansion is

$$\begin{aligned} 2\frac{\partial_T m(s,T)}{s(s-1)} &\sim \frac{2/\nu}{s_+(s_+-1)} \log \frac{1}{s_+ - s} \\ &= \frac{2/\nu}{s_+(s_+-1)} \log \frac{k}{w} \\ &\sim \frac{2\log k}{\nu s_+(s_+-1)}, \end{aligned}$$

and so

$$2\int_{-i\infty}^{i\infty} \frac{\partial_T m(s,T)}{s(s-1)} e^{-ks} M(s,T) ds \sim \frac{2\log k}{\nu s_+(s_+-1)} \times \frac{c_1}{\Gamma(T/\nu)} e^{-ks_+} k^{T/\nu-1}. \tag{17}$$

Dividing (17) by (16) yields the desired result. □

As mentioned in the preceding proof, the saddle point formula (15) is not an asymptotic equivalence for the variance gamma model. But, as we have shown, our formula (2) is still correct. What happens is that (15), and its counterpart for the numerator of (4), are *almost* correct: They are only off by a constant factor. (This phenomenon has already been observed for similar integrals in [4].) This constant factor is the same for both integrals, and thus cancels in the quotient (4). Therefore, our asymptotic formula (2) extends well beyond models where the saddle point method is applicable. In fact, we conjecture that the formula holds whenever the mgf explodes close to the singularity s_+.

4 Other Jump Models

Without giving proofs, we briefly discuss local volatility asymptotics for two other jump models. The mgf of Kou's double exponential Lévy jump diffusion is given by

$$M(s,T) = \exp\left(T\left(bs + \frac{\sigma^2 s^2}{2} + \lambda\left(\frac{\lambda_+ p}{\lambda_+ - s} + \frac{\lambda_-(1-p)}{\lambda_- + s} - 1\right)\right)\right).$$

The critical moment is $s_+ = \lambda_+$, and the saddle point is located at

$$\hat{s} \approx s_+ - \sqrt{\frac{\lambda\lambda_+ pT}{k}}.$$

The singularity type, the same as in the Heston model, is amenable to the saddle point method. Formula (2) can thus certainly be verified, and yields

$$\sigma^2_{\text{loc}}(K,T) \sim \frac{2\sqrt{\lambda p}}{\sqrt{\lambda_+ T}(\lambda_+ - 1)} k^{1/2}, \quad k \to \infty.$$

For $T \to 0$, the blowup of local volatility is of order $T^{-1/2}$. (Just as the Hankel contour analysis in the proof of Theorem 3 can be carried out for any asymptotic regime with $k/T \to \infty$, the same is true when applying the saddle point method to the local volatility surface of a Lévy model.)

Finally, we consider the normal inverse Gaussian (NIG) model. The mgf

$$M(s,T) = \exp\left(Tbs + \delta T\left(\sqrt{\alpha^2 - \beta^2} - \sqrt{\alpha^2 - (\beta + s)^2}\right)\right)$$

has no blow-up at the critical moment

$$s_+ = \alpha - \beta,$$

but a square-root type singularity, with local expansion

$$M(s,T) \approx e^{Tbs_+ + \delta T\sqrt{\alpha^2-\beta^2}}\left(1 - \delta T\sqrt{2\alpha}\sqrt{s_+ - s}\right). \qquad (18)$$

It is still true that $\sigma^2_{\text{loc}}(K,T)$ asymptotically depends, via (4), on the local behavior of $M(s,T)$ near s_+. However, the approximation (2) hinges on the *first* term of the local expansion of $M(s,T)$. It therefore fails to capture the asymptotics of $\sigma^2_{\text{loc}}(K,T)$ here, which depend on the first *singular* term (the term $\sqrt{s_+ - s}$ in (18)). The NIG model is thus one of the few examples where (2) is wrong. (It gives the qualitatively correct result of convergence to a constant, but a wrong one.) The Hankel contour analysis in the proof of Theorem 3 can be adapted to handle this situation. The result is that local volatility tends to a constant for $k \to \infty$. This fact may be understood by comparing the NIG marginals with those of Heston's in the time $T \to \infty$ regime (this link is made precise in [10]). In particular, the result is then consistent with the Heston asymptotics (5) of local vol, given that the $O(k)$ term carries a factor $\approx 1/T$ which tends to zero as $T \to \infty$.

Acknowledgments We thank M. Drmota, J. Morgenbesser, and the referee for helpful comments, and gratefully acknowledge financial support from MATHEON (P. Friz) resp. the Austrian Science Fund (FWF) under grant P 24880-N25 (S. Gerhold).

References

1. Carr, P., Chang, E., Madan, D.: The variance gamma process and option pricing. Eur. Financ. Rev. **2**, 79–105 (1998)
2. Carr, P., Geman, H., Madan, D.P., Yor, M.: From local volatility to local Lévy models. Quant. Financ. **4**, 581–588 (2004)
3. De Marco, S., Friz, P., Gerhold, S.: Rational shapes of local volatility. Risk **2**, 82–87 (2013)
4. Drmota, M., Soria, M.: Marking in combinatorial constructions: generating functions and limiting distributions. Theor. Comput. Sci. **144**, 67–99 (1995). Special volume on mathematical analysis of algorithms
5. Dupire, B.: Pricing with a smile. Risk **7**, 18–20 (1994)
6. Flajolet, P., Odlyzko, A.: Singularity analysis of generating functions. SIAM J. Discret. Math. **3**, 216–240 (1990)
7. Friz, P., Gerhold, S., Gulisashvili, A., Sturm, S.: On refined volatility smile expansion in the Heston model. Quant. Financ. **11**, 1151–1164 (2011)
8. Friz, P.K., Gerhold, S., Yor, M.: How to make Dupire's local volatility work with jumps. Quant. Financ. **14**, 1327–1331 (2014)
9. Gatheral, J., Jacquier, A.: Convergence of Heston to SVI. Quant. Financ. **11**, 1129–1132 (2011)
10. Keller-Ressel, M.: Moment explosions and long-term behavior of affine stochastic volatility models. Math. Financ. **21**, 73–98 (2011)
11. Lee, R.W.: The moment formula for implied volatility at extreme strikes. Math. Financ. **14**, 469–480 (2004)
12. Olver, F.W.J., Lozier, D.W., Boisvert, R.F., Clark, C.W. (eds.): NIST Handbook of Mathematical Functions. U.S. Department of Commerce National Institute of Standards and Technology, Washington, DC (2010)
13. Reghai, A.: Model evolution. Presentation at the Parisian Model Validation seminar. https://sites.google.com/site/projeteuclide/les-seminaires-vmf/archives-vmf (2011)

The Gärtner-Ellis Theorem, Homogenization, and Affine Processes

Archil Gulisashvili and Josef Teichmann

Abstract We obtain a first order extension of the large deviation estimates in the Gärtner-Ellis theorem. In addition, for a given family of measures, we find a special family of functions having a similar Laplace principle expansion up to order one to that of the original family of measures. The construction of the special family of functions mentioned above is based on heat kernel expansions. Some of the ideas employed in the paper come from the theory of affine stochastic processes. For instance, we provide an explicit expansion with respect to the homogenization parameter of the rescaled cumulant generating function in the case of a generic continuous affine process. We also compute the coefficients in the homogenization expansion for the Heston model that is one of the most popular stock price models with stochastic volatility.

Keywords Affine process · Large deviation principle · Heat kernel expansion · Short time asymptotics · Laplace method · Small maturity limit in affine models

2010 Mathematics Subject Classification 60F10 · 35K08

1 Introduction

The large deviations theory has found numerous applications in mathematical finance (see, e.g., [19]). For instance, using the methods of the large deviations theory, one can estimate various important characteristics of financial models such as tails of asset price distributions, option pricing functions, and the implied volatility (see, e.g., [7–11, 13, 15] and the references therein). A popular source of information on

A. Gulisashvili (✉)
Department of Mathematics, Ohio University, Athens, OH, USA
e-mail: gulisash@ohio.edu

J. Teichmann
Department of Mathematics, ETH Zürich, Zürich, Switzerland
e-mail: jteichma@math.ethz.ch

P.K. Friz et al. (eds.), *Large Deviations and Asymptotic Methods in Finance*, Springer Proceedings in Mathematics & Statistics 110,
DOI 10.1007/978-3-319-11605-1_11

the large deviations theory is the book [4] by Dembo and Zeitouni. A useful result in the theory is the Gärtner-Ellis theorem (see [6, 12], see also [4]). This theorem allows to infer the upper and lower estimates in the large deviation principle knowing the properties of the limiting cumulant generating function.

We will next provide a brief overview of the contents of the paper. In Sect. 2, a new notion of Laplace principle equivalent expansions for families of functions and measures is introduced. This notion is motivated by the homogenization expansion of the rescaled cumulant generating function associated with an affine stochastic process X, that is, the function Λ defined by

$$\Lambda(\epsilon, u) = \epsilon \log \mathbb{E}\left[\exp\left\{-\frac{u}{\epsilon}X_\epsilon\right\}\right] = \epsilon \log \int_{\mathbb{R}} \exp\left\{-\frac{u}{\epsilon}z\right\} p_\epsilon(dz).$$

Actually, the homogenization expansion mentioned above is nothing else but the real analytic expansion of the function Λ with respect to the parameter ϵ (see Sect. 4). In Sect. 3, we gather definitions and known facts from the theory of general affine processes, while in Sect. 4, the homogenization procedure is described in all details for continuous affine processes. The main general results obtained in the paper are contained in Sect. 2 (see Theorems 2.4 and 2.7). Theorem 2.4 states that for any family of measures on the real line, satisfying the conditions in the Gärtner-Ellis theorem, and such that the homogenization expansion exists, we can find a special family of functions that is Laplace principle equivalent to the original family of measures. The structure of the function family in Theorem 2.4 resembles the first two terms in the heat kernel expansions on Riemannian manifolds (notice that we face a degenerate situation here, so we could not apply heat kernel expansion directly). Theorem 2.7 is a generalization of the Gärtner-Ellis theorem. It is shown in Theorem 2.7 that under the same conditions as in Theorem 2.4, the first order large deviation estimates are valid. Finally, in Sect. 5, we compute the coefficients in the homogenization expansion for the correlated Heston model that is one of the most popular stochastic stock price models with stochastic volatility.

2 Distributions with Equivalent Laplace Principle Expansions

Laplace's principle is an asymptotic expansion technique, which allows one to approximate integrals of the form

$$\int_a^b f(z) \exp\left\{-\frac{\phi(z)}{\epsilon}\right\} dz \tag{2.1}$$

as $\epsilon \to 0$. We will next formulate a rather general version of Laplace's principle that will be used in the sequel. Suppose the following conditions hold:

- The functions f and ϕ in (2.1) are continuous on the interval (a, b), and the integral in (2.1) converges absolutely for all $0 < \epsilon < \epsilon_0$.
- The function ϕ has a unique absolute minimum that occurs at $z = z_0$ with $a < z_0 < b$.
- The function ϕ is strictly convex in a neighborhood of z_0.
- The function ϕ is four times continuously differentiable in a neighborhood of z_0, and

$$\phi(z) = \phi(z_0) + \sum_{n=2}^{4} \frac{\partial^n \phi(z_0)}{n!}(z - z_0)^n + O\left((z - z_0)^5\right) \tag{2.2}$$

as $z \to z_0$.
- The formula in (2.2) can be differentiated. More exactly, the condition

$$\partial\phi(z) = \sum_{n=2}^{4} \frac{\partial^n \phi(z_0)}{(n-1)!}(z - z_0)^{n-1} + O\left((z - z_0)^4\right), \quad z \to z_0, \tag{2.3}$$

holds.
- The function f is twice continuously differentiable in a neighborhood of z_0, and

$$f(z) = \sum_{n=0}^{2} \frac{\partial^n f(z_0)}{n!}(z - z_0)^n + O\left((z - z_0)^3\right) \tag{2.4}$$

as $z \to z_0$.

Then, as $\epsilon \to 0$,

$$\begin{aligned}
&\int_a^b f(z) \exp\left\{-\frac{\phi(z)}{\epsilon}\right\} dz \\
&= \exp\left\{-\frac{\phi(z_0)}{\epsilon}\right\} \sqrt{\frac{2\pi\epsilon}{\partial^2\phi(z_0)}} \Bigg[f(z_0) + \epsilon \Bigg(\frac{\partial^2 f(z_0)}{2\partial^2\phi(z_0)} + \frac{5(\partial^3\phi(z_0))^2 f(z_0)}{24(\partial^2\phi(z_0))^3} \\
&- \frac{\partial^4\phi(z_0) f(z_0)}{8(\partial^2\phi(z_0))^2} - \frac{\partial^3\phi(z_0)\partial f(z_0)}{2(\partial^2\phi(z_0))^2} \Bigg) + O\left(\epsilon^2\right) \Bigg].
\end{aligned} \tag{2.5}$$

Formula (2.5) can be derived by following the proof of Theorem 8.1 in [18].

Let us next assume that weaker differentiability restrictions than those listed above are imposed on the functions f and ϕ:

- The function ϕ is twice continuously differentiable in a neighborhood of z_0, and

$$\phi(z) = \phi(z_0) + \frac{\partial^2\phi(z_0)}{2}(z - z_0)^2 + O\left((z - z_0)^3\right) \tag{2.6}$$

as $z \to z_0$.

- The formula in (2.2) can be differentiated. More exactly, the condition

$$\partial\phi(z) = \partial^2\phi(z_0)(z - z_0) + O\left((z - z_0)^2\right) \quad \text{as} \quad z \to z_0 \tag{2.7}$$

holds.
- The function f is such that

$$f(z) = f(z_0) + O(z - z_0) \quad \text{as} \quad z \to z_0. \tag{2.8}$$

Then, as $\epsilon \to 0$,

$$\int_a^b f(z) \exp\left\{-\frac{\phi(z)}{\epsilon}\right\} dz = \exp\left\{-\frac{\phi(z_0)}{\epsilon}\right\} \sqrt{\frac{2\pi\epsilon}{\partial^2\phi(z_0)}} \left[f(z_0) + O(\epsilon)\right]. \tag{2.9}$$

Remark 2.1 Using the Taylor formula, we see that (2.2), (2.3), and (2.4) hold provided that the function f is three times continuously differentiable and the function ϕ is five times continuously differentiable near z_0. Similarly, (2.6), (2.7), and (2.8) hold if f is continuously differentiable and ϕ is three times continuously differentiable near z_0.

Let $\mathbf{p} = \{p_\epsilon\}_{\epsilon>0}$ be a family of probability measures on $\mathbb{R}$. The following assumption is modeled on the behavior of the family of moment generating functions of the affine process and on the homogenization ideas (see Sect. 4 for more details):

$$\int_{\mathbb{R}} \exp\left\{-\frac{u}{\epsilon}z\right\} p_\epsilon(dz) = \exp\left(\frac{\Lambda^{(0)}(u)}{\epsilon}\right) \exp\left(\Lambda^{(1)}(u)\right)\left(1 + \epsilon\Lambda^{(2)}(u) + O(\epsilon^2)\right) \tag{2.10}$$

as $\epsilon \to 0$, where $\Lambda^{(i)}$, $0 \leq k \leq 2$, are continuous functions on the domain I. The big O estimate in (2.10) is uniform on all closed intervals contained in I.

It is not hard to see that the functions $\Lambda^{(i)}$, $0 \leq i \leq 2$, in (2.10) can be recovered from the following formulas:

$$\Lambda^{(0)}(u) = \lim_{\epsilon\to 0} \epsilon \log \int_{\mathbb{R}} \exp\left\{-\frac{u}{\epsilon}z\right\} p_\epsilon(dz), \tag{2.11}$$

$$\exp\left\{\Lambda^{(1)}(u)\right\} = \lim_{\epsilon\to 0} \exp\left\{\frac{\Lambda^{(0)}(u)}{\epsilon}\right\} \int_{\mathbb{R}} \exp\left\{-\frac{u}{\epsilon}z\right\} p_\epsilon(dz), \tag{2.12}$$

and

$$\exp\left\{\Lambda^{(1)}(u)\right\}\Lambda^{(2)}(u)$$
$$= \lim_{\epsilon\to 0}\frac{1}{\epsilon}\left[\exp\left\{\frac{\Lambda^{(0)}(u)}{\epsilon}\right\}\int_{\mathbb{R}}\exp\left\{-\frac{u}{\epsilon}z\right\}p_\epsilon(dz) - \exp\left\{\Lambda^{(1)}(u)\right\}\right]. \quad (2.13)$$

It will be assumed throughout the rest of the paper that the conditions in the Gärtner-Ellis theorem hold. More precisely, we suppose that the following are true:

- The function $\Lambda^{(0)}$ defined in (2.11) exists as an extended real number for all $u \in \mathbb{R}$. We denote by I the maximum open interval such that the number $\Lambda^{(0)}(u)$ is finite for all $u \in I$.
- The point $u = 0$ belongs to the interval I.
- The function $\Lambda^{(0)}$ is continuously differentiable on I, the derivative $\partial_u\Lambda^{(0)}$ is a strictly increasing function on I, and the range of the function $\partial_u\Lambda^{(0)}$ is $\mathbb{R}$.

The previous restrictions concern only the function $\Lambda^{(0)}$. By the Gärtner-Ellis theorem, they imply the validity of the large deviation principle for the family **p**. More information on the Gärtner-Ellis theorem can be found in [4]. The existence of the functions $\Lambda^{(1)}$ and $\Lambda^{(2)}$ (these functions are determined from (2.12) and (2.13), respectively), signals that certain refinements of large deviation results may be possible.

Remark 2.2 In the paper [16] of Jacquier and Roome, an assumption similar to that in (2.10) is imposed on the rescaled cumulant generating function (see (2.1) in [16]). Moreover, there are more similarities between the assumptions in the present section and those in Sect. 2 of [16]. Note that the main results obtained in [16] concern the asymptotic behavior of forward start options and forward smiles.

The function $\Lambda^{(0)}$ is strictly convex on I. Let us define an appropriate Legendre-Fenchel transform of $\Lambda^{(0)}$, more precisely, we put

$$\left[\Lambda^{(0)}\right]^*(z) = -\inf_{u\in I}(uz + \Lambda^{(0)}(u)), \quad z \in \mathbb{R}.$$

It is clear that there exists a unique minimizer $z \mapsto u^*(z)$ in the problem described above, satisfying the condition

$$\partial_u\Lambda^{(0)}(u^*(z)) = -z. \quad (2.14)$$

It follows that

$$\left[\Lambda^{(0)}\right]^*(z) = -zu^*(z) - \Lambda^{(0)}(u^*(z)). \quad (2.15)$$

Since $\Lambda^{(0)}(0) = 0$, we have $\left[\Lambda^{(0)}\right]^*(z) \geq 0$. It is well-known that the function $\left[\Lambda^{(0)}\right]^*$ is strictly convex on $\mathbb{R}$. The previous statements, (2.14), and (2.15) imply that $\left[\Lambda^{(0)}\right]^*(z) = 0$ if $z = -\partial_u\Lambda^{(0)}(0)$, and $\left[\Lambda^{(0)}\right]^*(z) > 0$ if $z \neq -\partial_u\Lambda^{(0)}(0)$.

Next, set

$$d(z) = \sqrt{2\left[\Lambda^{(0)}\right]^{*}(z)}. \tag{2.16}$$

It is clear that

$$\frac{d^2(z)}{2} = \left[\Lambda^{(0)}\right]^{*}(z). \tag{2.17}$$

Therefore,

$$d(z) = \sqrt{-2\left[zu^{*}(z) + \Lambda^{(0)}(u^{*}(z))\right]}. \tag{2.18}$$

By the strict convexity of the function $\Lambda^{(0)}$,

$$\inf_{z\in\mathbb{R}}\left[uz + \frac{d^2(z)}{2}\right] = -\Lambda^{(0)}(u), \quad u \in I.$$

Let $\mathbf{p}$ be a family of Borel probability measures satisfying condition (2.10). Our next goal is to find a special family of functions $\mathbf{f} = \{f_\epsilon\}_{\epsilon>0}$ on $\mathbb{R}$, for which the asymptotic behavior of rescaled moment generating functions resembles the behavior described in formula (2.10). It would be tempting to try to find an appropriate family $\mathbf{f}$ among the families of functions satisfying the following condition as $\epsilon \to 0$:

$$\int_{\mathbb{R}} \exp\left\{-\frac{u}{\epsilon}z\right\} f_\epsilon(z)dz = \exp\left\{\frac{\Lambda^{(0)}(u)}{\epsilon}\right\}\exp\left\{\Lambda^{(1)}(u)\right\} \times \left(1 + \epsilon\Lambda^{(2)}(u) + O\left(\epsilon^2\right)\right) \tag{2.19}$$

uniformly on compact subintervals of I, where the functions $\Lambda^{(k)}$, $0 \le k \le 2$, are the same as in (2.10). However, we can not always guarantee the existence of the integral on the left-hand side of formula (2.19) due to the lack of control of the tail-behavior of the function f_ϵ. The remedy here is to localize the condition in (2.19).

Definition 2.3 Let $\mathbf{p}$ be a family of Borel probability measures such that (2.10) holds. We say that a family $\mathbf{f}$ of continuous functions on $\mathbb{R}$ is Laplace principle equivalent up to order 1 to the family $\mathbf{p}$ provided that the following conditions hold:

(i) For every $n \ge 1$ there exists a proper open subinterval $J_n \subset I$ of the interval I such that as $\epsilon \to 0$,

$$\int_{-n}^{n} \exp\left\{-\frac{u}{\epsilon}z\right\} f_\epsilon(z)dz = \exp\left\{\frac{\Lambda^{(0)}(u)}{\epsilon}\right\}\exp\left\{\Lambda^{(1)}(u)\right\} \left(1 + \epsilon\Lambda^{(2)}(u) + O_{n,u}\left(\epsilon^2\right)\right)$$

for all $u \in J_n$.

(ii) The sequence of intervals $J_n, n \geq 1$, is increasing and $\bigcup_{n=1}^{\infty} J_n = I$.

The next statement explains how to construct the family $\mathbf{f}$. The ansatz, defining the structure of the function f_ϵ in formula (2.20), is based on the classical theory of heat kernel expansions.

Theorem 2.4 *Let* $\mathbf{p}$ *be a family of Borel probability measures on* $\mathbb{R}$ *satisfying* (2.10), *and suppose the conditions in the Gärtner-Ellis theorem hold. Suppose also that the function* $\Lambda^{(0)}$ *is five times continuously differentiable on* I, *the function* $\Lambda^{(1)}$ *is three times continuously differentiable on* I, *and the function* $\Lambda^{(2)}$ *is continuously differentiable on* I. *Define a family* $\mathbf{f}$ *of functions as follows:*

$$f_\epsilon(z) = \frac{1}{\sqrt{2\pi\epsilon}} \exp\left\{-\frac{d^2(z)}{2\epsilon}\right\} (C_0(z) + \epsilon C_1(z)), \quad \epsilon > 0, \tag{2.20}$$

where d *is given by* (2.18),

$$C_0(z) = \sqrt{\partial_u^2 \Lambda^{(0)}(u^*(z))} \exp\left\{\Lambda^{(1)}(u^*(z))\right\},$$

and

$$\begin{aligned} C_1(z) = C_0(z)\,\Lambda^{(2)}(u^*(z)) &- \frac{\partial^2 C_0(z)\,\partial_u^2 \Lambda^{(0)}(u^*(z))}{2} - \frac{5C_0(z)\left[\partial_u^3 \Lambda^{(0)}(u^*(z))\right]^2}{24\left[\partial_u^2 \Lambda^{(0)}(u^*(z))\right]^3} \\ &+ \frac{C_0(z)\left(3\left[\partial_u^3 \Lambda^{(0)}(u^*(z))\right]^2 - \partial_u^2 \Lambda^{(0)}(u^*(z))\,\partial_u^4 \Lambda^{(0)}(u^*(z))\right)}{8\left[\partial_u^2 \Lambda^{(0)}(u^*(z))\right]^3} \\ &+ \frac{\partial C_0(z)\,\partial_u^3 \Lambda^{(0)}(u^*(z))}{2\partial_u^2 \Lambda^{(0)}(u^*(z))}. \end{aligned}$$

Then the family $\mathbf{f}$ *is Laplace principle equivalent up to order* 1 *to the family* $\mathbf{p}$.

Proof The differentiability restrictions on the functions $\Lambda^{(i)}$, $0 \leq i \leq 2$, in the formulation of Theorem 2.4 are imposed because otherwise the functions C_0 and C_1 are not defined. Note that the function $z \mapsto u^*(z)$ is three times continuously differentiable on the real line. The previous statement easily follows from (2.14).

The proof of Theorem 2.4 is based on the following construction, which uses Laplace's principle. For every $n \geq 1$, we have

$$\begin{aligned} &\int_{-n}^{n} \exp\left\{-\frac{u}{\epsilon} z\right\} f_\epsilon(z) dz \\ &= \frac{1}{\sqrt{2\pi\epsilon}} \int_{-n}^{n} \exp\left\{-\frac{1}{\epsilon}\left(uz + \frac{d^2(z)}{2}\right)\right\} (C_0(z) + \epsilon C_1(z)) dz. \end{aligned} \tag{2.21}$$

Set

$$\phi_u(z) = uz + \frac{d^2(z)}{2}. \tag{2.22}$$

Laplace's principle will be applied to the family of integrals appearing on the right-hand side of (2.21) twice. The first time, formula (2.5) with $f = C_0$ and $\phi = \phi_u$ will be used, while for the second time, formula (2.9) will be used with $f = C_1$ and $\phi = \phi_u$.

The critical point $z^*(u)$ of the function ϕ_u given by (2.22) is the solution to the equation $\partial_z \left[\Lambda^{(0)}\right]^*(z) = u$. It is not hard to see that $z = z^*(u)$ if and only if $u = u^*(z)$. It follows from (2.14) that

$$z^*(u) = -\partial\Lambda^{(0)}(u), \quad u \in I. \tag{2.23}$$

The next formulas can be derived using (2.17), (2.15), (2.22), and (2.23). We have

$$\partial_z^2\phi_u(z^*(u)) = \frac{1}{\partial^2\Lambda^{(0)}(u)}, \tag{2.24}$$

$$\partial_z^3\phi_u(z^*(u)) = \frac{\partial^3\Lambda^{(0)}(u)}{[\partial^2\Lambda^{(0)}(u)]^3}, \tag{2.25}$$

and

$$\partial_z^4\phi_u(z^*(u)) = \frac{3[\partial^3\Lambda^{(0)}(u)]^2 - \partial^2\Lambda^{(0)}(u)\,\partial^4\Lambda^{(0)}(u)}{[\partial^2\Lambda^{(0)}(u)]^5}. \tag{2.26}$$

Let us define the intervals J_n appearing in Definition 2.3 as follows:

$$J_n = \{u \in I : z^*(u) \in (-n, n)\}, \quad n \geq 1.$$

It is not hard to see that condition (ii) in Definition 2.3 is satisfied. Next, using (2.5) and (2.21), we obtain

$$\int_{-n}^{n} \exp\left\{-\frac{u}{\epsilon}z\right\} f_\epsilon(z)dz = \exp\left\{-\frac{\phi_u(z^*(u))}{\epsilon}\right\}\sqrt{\frac{1}{\partial_z^2\phi(z^*(u))}}$$
$$\left[C_0(z^*(u)) + \epsilon\left(C_1(z^*(u)) + \frac{\partial_z^2 C_0(z^*(u))}{2\partial_z^2\phi_u(z^*(u))} + \frac{5\left(\partial_z^3\phi_u(z^*(u))\right)^2 C_0(z^*(u))}{24\left(\partial_z^2\phi_u(z^*(u))\right)^3} - \right.\right.$$
$$\left.\left. - \frac{\partial_z^4\phi_u(z^*(u))C_0(z^*(u))}{8\left(\partial_z^2\phi_u(z^*(u))\right)^2} - \frac{\partial_z^3\phi_u(z^*(u))\partial_z C_0(z^*(u))}{2\left(\partial_z^2\phi_u(z^*(u))\right)^2}\right) + O_{n,u}\left(\epsilon^2\right)\right] \tag{2.27}$$

as $\epsilon \to 0$. Note that the differentiability conditions in Theorem 2.4 allow us to use formulas (2.5) and (2.9) with the functions f and ϕ chosen above.

We will next compare the formulas in (2.10) and (2.27). Note that

$$\phi_u(z^*(u)) = u\, z^*(u) - z^*(u)\, u^*(z^*(u)) - \Lambda^{(0)}(u^*(z^*(u))) = -\Lambda^{(0)}(u).$$

This shows that if we choose the function d as in (2.16), then the first factors in formulas (2.10) and (2.27) coincide. Moreover, the functions C_0 and C_1 have to be chosen so that

$$C_0(z^*(u)) = \sqrt{\partial_u^2 \Lambda^{(0)}(u)} \exp(\Lambda^{(1)}(u)) \tag{2.28}$$

and

$$\begin{aligned} C_1(z^*(u)) &= C_0(z^*(u))\, \Lambda^{(2)}(u) - \frac{\partial_z^2 C_0(z^*(u))}{2\partial_z^2 \phi_u(z^*(u))} - \frac{5(\partial_z^3 \phi_u(z^*(u)))^2\, C_0(z^*(u))}{24(\partial_z^2 \phi_u(z^*(u)))^3} \\ &\quad + \frac{\partial_z^4 \phi_u(z^*(u))\, C_0(z^*(u))}{8(\partial_z^2 \phi_u(z^*(u)))^2} + \frac{\partial_z^3 \phi_u(z^*(u))\, \partial_z C_0(z^*(u))}{2(\partial_z^2 \phi_u(z^*(u)))^2}. \end{aligned} \tag{2.29}$$

The representations of the functions C_0 and C_1 given in Theorem 2.4 can be obtained by plugging $u = u^*(z)$ into (2.28) and (2.29), and simplifying the resulting formulas. Equalities (2.23)–(2.26) are taken into account in the simplifications.

This completes the proof of Theorem 2.4. □

Remark 2.5 We have already established that $\left[\Lambda^{(0)}\right]^*(y) \geq 0$ for all $y \in \mathbb{R}$. Since (2.14) and (2.15) hold, we have

$$\partial \left[\Lambda^{(0)}\right]^*(y) = -u^*(y)$$

for all $y \in \mathbb{R}$. Hence the infimum of the function $\left[\Lambda^{(0)}\right]^*$ on the real line is attained at the point y such that $u^*(y) = 0$. This point is given by $y = z^*(0) = \partial \Lambda^{(0)}(0)$. Moreover,

$$\inf_{y \in \mathbb{R}} \left[\Lambda^{(0)}\right]^*(y) = -\Lambda^{(0)}(0) = 0.$$

Remark 2.6 A heuristic conclusion that can be reached using Theorem 2.4 is that the family $\mathbf{f}$ is a small-time approximation to the family $\mathbf{p}$ in a certain very weak sense. Finding such approximations is an important problem. We consider our results as first modest steps in going beyond the celebrated Gärtner-Ellis theorem.

The next assertion provides a first order large deviation estimate in the Gärtner-Ellis theorem for families of measures satisfying condition (2.10). Higher order estimates can also be found, but we do not include them in the present paper. Let A

be a bounded Borel set. Denote by $\overline{A}$ the closure of the set A, and let

$$a^{+} = \sup_{z \in A}\{z\} \quad \text{and} \quad a^{-} = \inf_{z \in A}\{z\}.$$

Then we have $z^{+},\ z^{-} \in \overline{A}$.

Theorem 2.7 *Let* **p** *be a family of probability Borel measures on* $\mathbb{R}$ *such that* (2.10) *holds. Suppose also that the function* $\Lambda^{(0)}$ *is twice continuously differentiable on* I *and the conditions in the Gärtner-Ellis theorem hold* (*see the conditions listed after formula* (2.13)). *Suppose also that* $A \subset \mathbb{R}$ *is a bounded Borel set, and* $x \in A$. *Then the following are true:*

(i) If $x \geq \partial\Lambda^{(0)}(0)$, *then as* $\epsilon \to 0$,

$$p_{\epsilon}(A) \leq \exp\left\{-\frac{\left[\Lambda^{(0)}\right]^{*}(x) - u^{*}(x)\left(a^{+} - x\right)}{\epsilon}\right\}\exp\left\{\Lambda^{(1)}(u^{*}(x))\right\} \times \left(1 + \epsilon\Lambda^{(2)}(u^{*}(x)) + \mathcal{O}(\epsilon^{2})\right). \tag{2.30}$$

(ii) If $x < \partial\Lambda^{(0)}(0)$, *then as* $\epsilon \to 0$,

$$p_{\epsilon}(A) \leq \exp\left\{-\frac{\left[\Lambda^{(0)}\right]^{*}(x) - |u^{*}(x)|\left(x - a^{-}\right)}{\epsilon}\right\}\exp\left\{\Lambda^{(1)}(u^{*}(x))\right\} \times \left(1 + \epsilon\Lambda^{(2)}(u^{*}(x)) + \mathcal{O}(\epsilon^{2})\right). \tag{2.31}$$

The big O estimates in (2.30) *and* (2.31) *are uniform with respect to* $x \in A$.

Remark 2.8 The conditions $x \geq \partial\Lambda^{(0)}(0)$ and $x < \partial\Lambda^{(0)}(0)$ are equivalent to $u^{*}(x) \geq 0$ and $u^{*}(x) < 0$, respectively.

Theorem 2.9 *Let* **p** *be a family of probability Borel measures on* $\mathbb{R}$ *such that* (2.10) *holds. Suppose also that the function* $\Lambda^{(0)}$ *is twice continuously differentiable on* I *and the conditions in the Gärtner-Ellis theorem hold* (*see the conditions listed after formula* (2.13)). *Suppose also that* $A \subset \mathbb{R}$ *is a bounded open set, and* $x \in A$. *Then the following are true:*

(i) Let $x \geq \partial\Lambda^{(0)}(0)$. *Then there exists a constant* $\gamma_{A} > 0$ *depending on the set* A *such that as* $\epsilon \to 0$,

$$p_\epsilon(A) \geq \exp\left\{-\frac{\left[\Lambda^{(0)}\right]^*(x) + u^*(x)(x - a^-)}{\epsilon}\right\} \exp\left\{\Lambda^{(1)}(u^*(x))\right\}$$
$$\times \left(1 - \exp\left\{-\frac{\gamma_A}{\epsilon}\right\}\right)\left(1 + \epsilon\Lambda^{(2)}(u^*(x)) + \mathcal{O}(\epsilon^2)\right). \tag{2.32}$$

(ii) If $x < \partial\Lambda^{(0)}(0)$*, then as* $\epsilon \to 0$*,*

$$p_\epsilon(A) \geq \exp\left\{-\frac{\left[\Lambda^{(0)}\right]^*(x) + |u^*(x)|(a^+ - x)}{\epsilon}\right\} \exp\left\{\Lambda^{(1)}(u^*(x))\right\}$$
$$\times \left(1 - \exp\left\{-\frac{\gamma_A}{\epsilon}\right\}\right)\left(1 + \epsilon\Lambda^{(2)}(u^*(x)) + \mathcal{O}(\epsilon^2)\right). \tag{2.33}$$

The constant γ_A *in* (2.33) *is the same as in* (2.32)*, and the big O estimates in* (2.32) *and* (2.33) *are uniform with respect to* $x \in A$*.*

Remark 2.10 Note that performing the transformation $\limsup_{\epsilon\to 0} \epsilon \log p_\epsilon(A)$ in the upper estimates in Theorem 2.7, we obtain the upper estimate in the large deviation principle for any bounded Borel set A. This gives a little more than the upper estimate in the Gärtner-Ellis theorem. However, we should not forget that formula (2.30) was derived under a stronger restriction (2.10), than in the Gärtner-Ellis theorem.

Proof of Theorem 2.7 We borrow some ideas from the proofs of Cramer's theorem and the Gärtner-Ellis theorem given in [4]. The proofs of the upper estimates in those theorems use Chebyshev's inequality. In our case, due to a special structure of the problem, we can provide a slightly more direct proof.

Suppose the conditions in Theorem 2.7 hold, and let $u \in I$ and $\epsilon > 0$. Then we have

$$\int_A \exp\left\{-\frac{uz}{\epsilon}\right\} p_\epsilon(dz) \geq p_\epsilon(A) \inf_{z\in\overline{A}}\left[\exp\left\{-\frac{uz}{\epsilon}\right\}\right]. \tag{2.34}$$

It follows from (2.34) that for every $u \in I$ there exists $\xi(u) \in \overline{A}$ such that

$$\begin{aligned} p_\epsilon(A) &\leq \exp\left\{\frac{u\xi(u)}{\epsilon}\right\}\int_A \exp\left\{-\frac{uz}{\epsilon}\right\} p_\epsilon(dz) \\ &= \exp\left\{-\frac{\Lambda^{(0)}(u)}{\epsilon}\right\}\int_A \exp\left\{-\frac{u}{\epsilon}z\right\} p_\epsilon(dz) \\ &\quad\times \exp\left\{\frac{\Lambda^{(0)}(u) + xu + u(\xi(u) - x)}{\epsilon}\right\}. \end{aligned}$$

Indeed, we can take $\xi(u) = a^+$ if $u \geq 0$ and $\xi(u) = a^-$ if $u < 0$.

Next, by plugging $u = u^*(x)$ into the previous equalities and taking into account condition (2.10), we get

$$\begin{aligned} p_\epsilon(A) \leq &\exp\left\{-\frac{\Lambda^{(0)}(u^*(x))}{\epsilon}\right\} \int_A \exp\left\{-\frac{u^*(x)}{\epsilon} z\right\} p_\epsilon(dz) \\ &\times \exp\left\{-\frac{\left[\Lambda^{(0)}\right]^*(x) - u^*(x)(\xi(u^*(x)) - x)}{\epsilon}\right\} \\ \leq &\exp\left\{\Lambda^{(1)}(u^*(x))\right\} \exp\left\{-\frac{\left[\Lambda^{(0)}\right]^*(x) - u^*(x)(\xi(u^*(x)) - x)}{\epsilon}\right\} \\ &\times \left(1 + \epsilon\Lambda^{(2)}(u^*(x)) + \mathcal{O}(\epsilon^2)\right) \end{aligned} \tag{2.35}$$

as $\epsilon \to 0$. Now, it is not hard to see that (2.35) implies Theorem 2.7.

Proof of Theorem 2.9 The lower bounds given in Theorem 2.9 are more delicate. Here we start with the estimate

$$\int_A \exp\left\{-\frac{uz}{\epsilon}\right\} p_\epsilon(dz) \leq p_\epsilon(A) \sup_{z \in \overline{A}} \left[\exp\left\{-\frac{uz}{\epsilon}\right\}\right]$$

instead of the estimate in (2.34). This implies that

$$\begin{aligned} p_\epsilon(A) \geq &\exp\left\{\frac{u\eta(u)}{\epsilon}\right\} \int_A \exp\left\{-\frac{uz}{\epsilon}\right\} p_\epsilon(dz) \\ = &\exp\left\{-\frac{\Lambda^{(0)}(u)}{\epsilon}\right\} \int_A \exp\left\{-\frac{uz}{\epsilon}\right\} p_\epsilon(dz) \\ &\times \exp\left\{\frac{\Lambda^{(0)}(u) + xu + u(\eta(u) - x)}{\epsilon}\right\}, \end{aligned}$$

for all $u \in I$, where $\eta(u) = a^-$ if $u \geq 0$ and $\eta(u) = a^+$ if $u < 0$. Therefore

$$\begin{aligned} p_\epsilon(A) \geq &\exp\left\{-\frac{\Lambda^{(0)}(u^*(x))}{\epsilon}\right\} \int_A \exp\left\{-\frac{u^*(x)}{\epsilon} z\right\} p_\epsilon(dz) \\ &\times \exp\left\{-\frac{\left[\Lambda^{(0)}\right]^*(x) - u^*(x)(\eta(u^*(x)) - x)}{\epsilon}\right\}. \end{aligned} \tag{2.36}$$

Our next goal is to use the change of measure method. Consider a new family $\widetilde{\mathbf{p}}$ of probability measures defined by

$$\widetilde{p}_\epsilon(dz) = \frac{\exp\left\{-\frac{u^*(x)z}{\epsilon}\right\} p_\epsilon(dz)}{\int_{\mathbb{R}} \exp\left\{-\frac{u^*(x)z}{\epsilon}\right\} p_\epsilon(dz)}, \quad \epsilon > 0.$$

Note that the family $\widetilde{\mathbf{p}}$ depends on x. Then inequality (2.36) and condition (2.10) imply that

$$\begin{aligned} p_\epsilon(A) &\geq \exp\left\{-\frac{\Lambda^{(0)}(u^*(x))}{\epsilon}\right\} \int_{\mathbb{R}} \exp\left\{-\frac{u^*(x)}{\epsilon} z\right\} p_\epsilon(dz)\, \widetilde{p}_\epsilon(A) \\ &\times \exp\left\{-\frac{[\Lambda^{(0)}]^*(x) - u^*(x)(\eta(u^*(x)) - x)}{\epsilon}\right\} \\ &= \exp\left\{\Lambda^{(1)}(u^*(x))\right\} \left(1 + \epsilon\Lambda^{(2)}(u^*(x)) + \mathcal{O}(\epsilon^2)\right) \widetilde{p}_\epsilon(A) \\ &\times \exp\left\{-\frac{[\Lambda^{(0)}]^*(x) - u^*(x)(\eta(u^*(x)) - x)}{\epsilon}\right\} \end{aligned} \tag{2.37}$$

as $\epsilon \to 0$.

We will next estimate the quantity

$$\widetilde{p}_\epsilon(A) = 1 - \widetilde{p}_\epsilon(A^c) \tag{2.38}$$

from below. This will be done using the upper estimate in the Gärtner-Ellis theorem. Let us denote by $\widetilde{\Lambda}^{(0)}$ the function defined by (2.11) for the family $\widetilde{\mathbf{p}}$ instead of the family $\mathbf{p}$. Then it is not hard to see that

$$\widetilde{\Lambda}^{(0)}(v) = \Lambda^{(0)}(v + u^*(x)) - \Lambda^{(0)}(u^*(x)), \quad v \in \widetilde{I}, \tag{2.39}$$

where $\widetilde{I} = I - u^*(x)$. The function $\widetilde{\Lambda}^{(0)}$ and the interval $\widetilde{I}$ depend on x. It is clear that $0 \in \widetilde{I}$. Moreover,

$$\left[\widetilde{\Lambda}^{(0)}\right]^*(y) = -\inf_{v\in\widetilde{I}} \left\{yv + \widetilde{\Lambda}^{(0)}(v)\right\} \geq 0$$

Next, taking into account that A^c is a closed set, and using the upper large deviations estimate in the Gärtner-Ellis theorem (see Theorem 2.3.6 in [4]), we obtain

$$\limsup_{\epsilon\to 0} \left[\epsilon \log \widetilde{p}_\epsilon(A^c)\right] \leq -\inf_{y\in A^c} \left[\widetilde{\Lambda}^{(0)}\right]^*(y).$$

Set $\delta_A = \inf_{y\in A^c} \left[\widetilde{\Lambda}^{(0)}\right]^*(y)$. Using Remark 2.5 and (2.39), we see that the unique infimum of the function $\left[\widetilde{\Lambda}^{(0)}\right]^*$ on the real line is attained at the point

$$y = \partial\left[\widetilde{\Lambda}^{(0)}\right]^*(0) = \Lambda^{(0)}(u^*(x)) = x,$$

and is equal to zero. Since $x \notin A^c$, and the set A^c is closed, we have $\delta_A > 0$. Therefore, for every $\tau > 0$, there exists $\epsilon_\tau > 0$ such that

$$\widetilde{p}_\epsilon(A^c) \leq \exp\left\{\frac{-\delta_A + \tau}{\epsilon}\right\}, \quad 0 < \epsilon < \epsilon_\tau. \tag{2.40}$$

Fix any number $\tau > 0$ with $0 < \tau < \delta_A$, and set $\gamma_A = \delta_A - \tau$. Then (2.38) and (2.40) imply the following estimate:

$$\widetilde{p}_\epsilon(A) \geq 1 - \exp\left\{\frac{-\gamma_A}{\epsilon}\right\}, \quad 0 < \epsilon < \epsilon_\tau. \tag{2.41}$$

Finally, using (2.37) and (2.41), we establish estimate (2.32).

The proof of Theorem 2.7 is thus completed.

3 Affine Processes

Let D be a non-empty Borel subset of the real Euclidian space $\mathbb{R}^d$, equipped with the Borel σ-algebra $\mathcal{D}$, and assume that the affine hull of D is the full space $\mathbb{R}^d$. To D we add a point δ that serves as a 'cemetery state'. Define

$$\widehat{D} = D \cup \{\delta\}, \qquad \widehat{\mathcal{D}} = \sigma(\mathcal{D}, \{\delta\}),$$

and equip $\widehat{D}$ with the Alexandrov topology, in which any open set with a compact complement in D is declared an open neighborhood of δ.[1] Any continuous function f defined on D is extended to $\widehat{D}$ by setting $f(\delta) = 0$.

Let $(\Omega, \mathcal{F}, \mathbb{F})$ be a filtered measurable space, on which a family $(\mathbb{P}^x)_{x \in \widehat{D}}$ of probability measures is defined, and assume that $\mathcal{F}$ is $\mathbb{P}^x$-complete for all $x \in \widehat{D}$ and that the filtration $\mathbb{F}$ is right continuous. Finally, let X be a càdlàg process taking values in $\widehat{D}$, whose transition kernel

$$p_t(x, A) = \mathbb{P}^x(X_t \in A), \qquad (t \geq 0, x \in \widehat{D}, A \in \widehat{\mathcal{D}})$$

is a normal time-homogeneous Markov kernel, for which δ is absorbing. That is, $p_t(x, .)$ satisfies the following conditions:

(a) $x \mapsto p_t(x, A)$ is $\widehat{\mathcal{D}}$-measurable for each $(t, A) \in \mathbb{R}_{\geqslant 0} \times \widehat{\mathcal{D}}$.
(b) $p_0(x, \{x\}) = 1$ for all $x \in \widehat{D}$,
(c) $p_t(\delta, \{\delta\}) = 1$ for all $t \geq 0$
(d) $p_t(x, \widehat{D}) = 1$ for all $(t, x) \in \mathbb{R}_{\geqslant 0} \times \widehat{D}$, and
(e) the Chapman-Kolmogorov equation

[1]Note that the topology of $\widehat{\mathcal{D}}$ enters our assumptions in a subtle way: We require later that X is càdlàg on $\widehat{\mathcal{D}}$, which is a property for which the topology matters.

$$p_{t+s}(x, d\xi) = \int p_t(y, d\xi)\, p_s(x, dy)$$

holds for each $t, s \geq 0$ and $(x, d\xi) \in \widehat{D} \times \widehat{\mathcal{D}}$.

We equip $\mathbb{R}^d$ with the canonical inner product $\langle , \rangle$, and associate to D the set $\mathcal{U} \subseteq \mathbb{C}^d$ defined by

$$\mathcal{U} = \left\{ u \in \mathbb{C}^d : \sup_{x \in D} \operatorname{Re} \langle u, x \rangle < \infty \right\}.$$

Note that the set $\mathcal{U}$ is the set of complex vectors u such that the exponential function $x \mapsto e^{\langle u, x \rangle}$ is bounded on D. It is easy to see that $\mathcal{U}$ is a convex cone and always contains the set of purely imaginary vectors $i\mathbb{R}^d$.

Definition 3.1 (*Affine processes*) A stochastic process X is called *affine* with state space D, if the transition kernel $p_t(x, d\xi)$ of X satisfies the following conditions:

(i) It is stochastically continuous, i.e., $\lim_{s \to t} p_s(x, .) = p_t(x, .)$ weakly for all $t \geq 0, x \in D$.
(ii) The Fourier-Laplace transform of the kernel depends on the initial state in the following way: there exist functions $\Phi : \mathbb{R}_{\geqslant 0} \times \mathcal{U} \to \mathbb{C}$ and $\psi : \mathbb{R}_{\geqslant 0} \times \mathcal{U} \to \mathbb{C}^d$, such that

$$\int_D e^{\langle \xi, u \rangle} p_t(x, d\xi) = \Phi(t, u) \exp(\langle x, \psi(t, u) \rangle) \tag{3.1}$$

for all $t \in \mathbb{R}_{\geqslant 0}$, $x \in D$, and $u \in \mathcal{U}$.

Remark 3.2 Note that the previous definition does not specify $\psi(t, u)$ in a unique way. However, there is a natural unique choice for ψ that will be discussed in Proposition 3.3. Also note that as long as $\Phi(t, u)$ is non-zero, there exists $\phi(t, u)$ such that $\Phi(t, u) = e^{\phi(t,u)}$, and equality (3.1) becomes

$$\int_D e^{\langle \xi, u \rangle} p_t(x, d\xi) = \exp\left\{\phi(t, u) + \langle x, \psi(t, u) \rangle\right\}. \tag{3.2}$$

This is the essentially the definition that was used in [5]. Condition (3.2) means that the Fourier-Laplace transform of the transition function is the exponential of an *affine function* of x. This fact is usually interpreted as the reason for the name 'affine process', even though affine functions also appear in other aspects of affine processes, e.g., in the coefficients of the infinitesimal generator, or in the differentiated semi-martingale characteristics. We prefer to use equality (3.1) instead of equality (3.2), since the former equality leads to a slightly more general definition that avoids the necessity of the a-priori assumption that the left hand side of (3.1) is non-zero for all t and u.

Before we start exploring the first simple consequences of Definition 3.1, additional notation will be introduced. For any $u \in \mathcal{U}$, set $\sigma(u) := \inf\{t \geq 0 \colon \Phi(t, u) = 0\}$

and $\mathcal{Q} := \left\{(t, u) \in \mathbb{R}_{\geqslant 0} \times \mathcal{U} : t < \sigma(u)\right\}$, and let ϕ be a function on $\mathcal{Q}$ such that

$$\Phi(t, u) = e^{\phi(t,u)} \quad \text{for all} \quad (t, u) \in \mathcal{Q}.$$

The uniqueness of ϕ will be discussed below. The functions ϕ and ψ have the following properties (see [17]):

Proposition 3.3 *Let X be an affine process on D. Then*

(i) The condition $\sigma(u) > 0$ holds for any $u \in \mathcal{U}$.
(ii) The functions ϕ and ψ are uniquely defined on $\mathcal{Q}$ under the restriction that they are jointly continuous and satisfy $\phi(0, 0) = \psi(0, 0) = 0$.
(iii) The function ψ maps $\mathcal{Q}$ into $\mathcal{U}$.
(iv) The functions ϕ and ψ satisfy the semi-flow *property. For any $u \in \mathcal{U}$ and $t, s \geq 0$ with $t + s \leq \sigma(u)$, the following conditions hold:*

$$\begin{aligned} \phi(t+s, u) &= \phi(t, u) + \phi(s, \psi(t, u)), \quad & \phi(0, u) &= 0 \\ \psi(t+s, u) &= \psi(t, \psi(s, u)), & \psi(0, u) &= u \end{aligned}$$

Remark 3.4 In the sequel, the functions ϕ and ψ will always be chosen according to Proposition 3.3.

We now introduce the important notion of *regularity*.

Definition 3.5 An affine process X is called *regular* if the derivatives

$$F(u) = \left.\frac{\partial \phi(t, u)}{\partial t}\right|_{t=0+}, \qquad R(u) = \left.\frac{\partial \psi(t, u)}{\partial t}\right|_{t=0+}$$

exist for all $u \in \mathcal{U}$ and are continuous at $u = 0$.

The next statement illustrates why the regularity is a crucial property. This statement was originally established by [5] for affine processes on the state-space $\mathbb{R}^n \times \mathbb{R}^m_{\geqslant 0}$.

Proposition 3.6 *Let X be a* regular *affine process. Then there exist $\mathbb{R}^d$-vectors $b, \beta^1, \ldots, \beta^d$; $d \times d$-matrices $a, \alpha^1, \ldots, \alpha^d$; real numbers $c, \gamma^1, \ldots, \gamma^d$, and signed Borel measures $m, \mu^1, \ldots, \mu^d$ on $\mathbb{R}^d \setminus \{0\}$ such that the functions $F(u)$ and $R(u)$ can be represented as follows:*

$$F(u) = \frac{1}{2} \langle u, au \rangle + \langle b, u \rangle - c + \int_{\mathbb{R}^d \setminus \{0\}} \left(e^{\langle \xi, u \rangle} - 1 - \langle h(\xi), u \rangle\right) m(d\xi), \tag{3.3a}$$

$$R_i(u) = \frac{1}{2} \left\langle u, \alpha^i u \right\rangle + \left\langle \beta^i, u \right\rangle - \gamma^i + \int_{\mathbb{R}^d \setminus \{0\}} \left(e^{\langle \xi, u \rangle} - 1 - \langle h(\xi), u \rangle\right) \mu^i(d\xi). \tag{3.3b}$$

In the previous formulas, $h(x) = x\mathbf{1}_{\{\|x\|\leq 1\}}$ is a truncation function. In addition, for all $x \in D$, the quantities

$$A(x) = a + x_1\alpha^1 + \cdots + x_d\alpha^d, \tag{3.4a}$$

$$B(x) = b + x_1\beta^1 + \cdots + x_d\beta^d, \tag{3.4b}$$

$$C(x) = c + x_1\gamma^1 + \cdots + x_d\gamma^d, \tag{3.4c}$$

$$\nu(x, d\xi) = m(d\xi) + x_1\mu^1(d\xi) + \cdots + x_d\mu^d(d\xi) \tag{3.4d}$$

have the following properties: $A(x)$ is positive semidefinite, $C(x) \leq 0$, and

$$\int_{\mathbb{R}^d\setminus\{0\}} \left(\|\xi\|^2 \wedge 1\right)\nu(x, d\xi) < \infty.$$

Moreover, for $u \in \mathcal{U}$ and $t \in [0, \sigma(u))$, the functions ϕ and ψ satisfy the following ordinary differential equations:

$$\frac{\partial}{\partial t}\phi(t, u) = F(\psi(t, u)), \quad \phi(0, u) = 0 \tag{3.5a}$$

$$\frac{\partial}{\partial t}\psi(t, u) = R(\psi(t, u)), \quad \psi(0, u) = u. \tag{3.5b}$$

Remark 3.7 The Eq. (3.5) are called *generalized Riccati equations*, since they are classical Riccati equations when $m(d\xi) = \mu^i(d\xi) = 0$. Moreover, Eqs. (3.3) and (3.4) imply that $u \mapsto F(u) + \langle R(u), x\rangle$ is a function of Lévy-Khintchine form for each $x \in D$.

Proof See [17]. □

In general, the parameters $(a, \alpha^i, b, \beta^i, c, \gamma^i, m, \mu^i)_{i\in\{1,\ldots,d\}}$ appearing in the representations of F and R in (3.5a) and (3.5b) have to satisfy additional conditions, called the *admissibility* conditions. These conditions guarantee the existence of an affine Markov process X with state space D and with prescribed F and R. It is clear that such conditions should depend strongly on the geometry of the (boundary of the) state space D. Finding such (necessary and sufficient) conditions on the parameters for different types of state spaces has been the focus of several publications. For $D = \mathbb{R}^m_{\geqslant 0} \times \mathbb{R}^n$, the admissibility conditions were derived in [5]. For the cone of semi-definite matrices $D = S_d^+$, such conditions were found in [2], and for symmetric irreducible cones, the admissibility conditions were found in [3]. Finally, for affine diffusions ($m = \mu^i = 0$) on polyhedral cones and on quadratic state spaces, the admissibility conditions were given in [20].

Definition 3.8 We call the state space $D = \mathbb{R}^m_{\geqslant 0} \times \mathbb{R}^n$ with $m, n \geq 0$ the *canonical state space*.

Affine processes on canonical state spaces are completely characterized in [5] in terms of the admissibility conditions imposed on F and R. Affine processes on canonical state spaces have continuous trajectories (such processes are called continuous affine processes) if and only if the functions F and R satisfy the admissibility conditions and are polynomials of degree at most 2 (see Proposition 3.6).

4 Homogenization Procedure

In this section, we consider continuous, affine processes on the canonical state space $D = \mathbb{R}^m_{\geqslant 0} \times \mathbb{R}^n$. We will next introduce a natural homogenization procedure, which allows to analyze the short-time asymptotics of the law of continuous affine processes. In the case of affine processes, the homogenization leads in fact to real analytic expansions with respect to the homogenization parameter.

The following lemmas introduce the homogenization procedure.

Lemma 4.1 *Let $\psi : \mathcal{U} \times \mathbb{R}_{\geq 0} \to \mathcal{U}$ be the unique solution of the equation*

$$\frac{\partial}{\partial t}\psi(u,t) = R\big(\psi(u,t)\big), \quad \psi(u,0) = u \in \mathcal{U},$$

where $R : \mathcal{U} \to \mathbb{C}^d$ is a quadratic polynomial. Then, for every $\epsilon > 0$, the function

$$\psi^\epsilon(u,t) := \epsilon\psi\Big(\frac{u}{\epsilon}, \epsilon t\Big)$$

solves the equation

$$\frac{\partial}{\partial t}\psi^\epsilon(u,t) = R^\epsilon\big(\psi^\epsilon(u,t)\big), \quad \psi^\epsilon(u,0) = u$$

with $R^\epsilon(u) := \epsilon^2 R\big(\epsilon^{-1}u\big)$ for $u \in \mathcal{U}$.

Analogously, let $\phi : \mathcal{U} \times \mathbb{R}_{\geq 0} \to \mathbb{C}$ be the unique solution of the equation

$$\frac{\partial}{\partial t}\psi(u,t) = F\big(\psi(u,t)\big), \quad \phi(u,0) = 0.$$

Then, for every $\epsilon > 0$, the function

$$\phi^\epsilon(u,t) := \epsilon\phi\Big(\frac{u}{\epsilon}, \epsilon t\Big)$$

solves the equation

$$\frac{\partial}{\partial t}\phi^\epsilon(u,t) = F^\epsilon\big(\psi^\epsilon(u,t)\big), \quad \phi^\epsilon(u,0) = 0$$

with $F^\epsilon(u) := \epsilon^2 F\big(\epsilon^{-1}u\big)$ for $u \in \mathcal{U}$.

The proof of Lemma 4.1 is simple, and we leave it as an exercise for the reader.

Lemma 4.2 *Under the previous assumptions, the limit* $\lim_{\epsilon \to 0} \psi^\epsilon = \psi^{(0)}$ *exists uniformly on compact sets in* $\mathcal{U} \times \mathbb{R}_{\geq 0}$. *Furthermore,*

$$\psi^\epsilon(u,t) = \psi^{(0)}(u,t) + \epsilon \psi^{(1)}(u,t) + \sum_{n \geq 2} \epsilon^n \psi^{(n)}(u,t) \tag{4.1}$$

is a convergent power series expansion for small $\epsilon > 0$. *The coefficient functions in* (4.1) *satisfy certain ordinary differential equations, i.e., in particular,*

$$\frac{\partial}{\partial t} \psi^{(0)}(u,t) = R^{(0)}\big(\psi^{(0)}(u,t)\big), \quad \psi^{(0)}(u,0) = u\,,$$

and

$$\frac{\partial}{\partial t} \psi^{(1)}(u,t) = \frac{\partial}{\partial \epsilon}\bigg|_{\epsilon=0} R^\epsilon\big(\psi^{(0)}(u,t)\big)\psi^{(1)}(u,t), \quad \psi^{(1)}(u,0) = 0.$$

For $n \geq 2$, *the equations for the coefficient functions involve higher order derivatives. In complete analogy, the limit* $\lim_{\epsilon \to 0} \phi^\epsilon = \phi^{(0)}$ *exists uniformly on compact sets in* $\mathcal{U} \times \mathbb{R}_{\geq 0}$. *Furthermore*

$$\phi^\epsilon(u,t) = \phi^{(0)}(u,t) + \epsilon \phi^{(1)}(u,t) + \sum_{n \geq 2} \epsilon^u \phi^{(n)}(u,t)\,,$$

for small enough values of ϵ.

Proof Observe that $R^\epsilon = R^{(0)} + \epsilon R^{(1)} + \frac{\epsilon^2}{2} R^{(2)}$ and $F^\epsilon = F^{(0)} + \epsilon F^{(1)} + \frac{\epsilon^2}{2} F^{(2)}$. Hence, the vector fields appearing in the equation in Lemma 4.2 are polynomial in u and ϵ. Standard results on differential equations with polynomial vector fields yield the assertions in Lemma 4.2, in particular, the real analyticity of the solution with respect to ϵ. □

Let X be an affine diffusion process with the corresponding functions F and R. We can extend the solutions of the Riccati equations described above to maximal domains for $u \in \mathbb{R}^d$, i.e., consider maximal local flows on $\mathbb{R}^d$ with the vector fields F^ϵ and R^ϵ. By $\hat{\Lambda}^{(i)}$, $i \geq 0$, are denoted the functions appearing in the following power series expansion in ϵ:

$$\hat{\Lambda}^{(0)}(u) + \epsilon \hat{\Lambda}^{(1)}(u) + \ldots := \phi^\epsilon(-u,1) + \langle x, \psi^\epsilon(-u,1) \rangle\,, \tag{4.2}$$

They are the solutions of the Riccati equations appearing in the previous lemmas. Note that we suppress the dependence on the initial value x on the left-hand side of (4.2). The functions $\hat{\Lambda}^{(i)}$ exist as extended real numbers for $u \in \mathbb{R}^d$.

Remark 4.3 If the expression on right-hand side of (4.2) is finite, then the power series on the left-hand side converges absolutely for sufficiently small values of ϵ.

Remark 4.4 For continuous affine processes, the homogenization procedure leads to the following representation:

$$E\left[\exp\left\{-\langle\frac{u}{\epsilon}, X_\epsilon\rangle\right\}\right] = \int_D \exp\left\{-\langle\frac{u}{\epsilon}, z\rangle\right\} p_\epsilon(dz)$$
$$= \exp\left\{\frac{\hat{\Lambda}^{(0)}(u)}{\epsilon} + \hat{\Lambda}^{(1)}(u) + \ldots\right\}, \tag{4.3}$$

where u is such that the expressions on both sides of (4.3) are finite for small enough values of ϵ.

The representation in (4.3) valid for any continuous affine process was a motivation for us for introducing condition (2.10) used in the previous sections. However, the expansion in (4.3) is a little different from that in (2.10).

5 Example: The Heston Model

In this section, we find explicit formulas for the functions $\Lambda^{(i)}$, $0 \le i \le 2$, associated with the log-price process in the Heston model. Let us consider the following correlated Heston model:

$$dX_t = (r + kV_t)dt + \sqrt{V_t}dW_{1,t},$$
$$dV_t = (a - bV_t)dt + \sigma\sqrt{V_t}dW_{2,t}, \tag{5.1}$$

where $r, k \in \mathbb{R}$, $a, b \ge 0$, $\sigma > 0$, and $W_{1,t}$ and $W_{2,t}$ are standard Brownian motions with $d\langle W_1, W_2\rangle_t = \rho dt$. We assume that the correlation coefficient ρ satisfies the condition $-1 < \rho < 1$. In (5.1), X is the log-price process, and V is the variance process. The initial conditions for the processes X and V are denoted by x_0 and v_0, respectively. The Heston model was introduced in [14]. Note that in the present paper we consider the Heston model in which both the log-price and the variance equations contain drift terms generated by affine functions. Very often, e.g., in [7–10, 16], a special Heston model where $k = -\frac{1}{2}$ and $r = 0$ is studied. An extended Heston model, in which the defining equations contain affine drift terms, is discussed in [15].

The process X is not an affine process. It is a projection of the two-dimensional affine process (X, V) onto the first coordinate. The moment generating function of X_t is given by $M_t(u) = \mathbb{E}\left[\exp\{uX_t\}\right] = \exp\{C(u,t) + D(u,t)v_0 + ux_0\}$, where

$$C(u,t) = rut + \frac{a}{\sigma^2}\left[(b - \rho\sigma u + d(u))t - 2\log\left(\frac{1 - g(u)e^{d(u)t}}{1 - g(u)}\right)\right],$$

$$D(u,t) = \frac{b + d(u) - \rho\sigma u}{\sigma^2}\left(\frac{1 - e^{d(u)t}}{1 - g(u)e^{d(u)t}}\right),$$

$$g(u) = \frac{b - \rho\sigma u + d(u)}{b - \rho\sigma u - d(u)},$$

and

$$d(u) = \sqrt{(\rho\sigma u - b)^2 - \sigma^2(2ku + u^2)}$$

(see [1]). Here and in the sequel, the symbol $\sqrt{\cdot}$ stands for the principal square root function. We will explain below the meaning of the logarithmic function appearing in the expression for the function C (see the discussion after formula (5.7)). Note that for $u = 0$, the expressions for the functions C and D should be understood in the limiting sense. More precisely,

$$C(0,t) = \lim_{u \to 0} C(u,t) = 0 \quad \text{and} \quad D(0,t) = \lim_{u \to 0} D(u,t) = 0$$

for all $t > 0$.

It is clear that

$$\mathbb{E}\left[\exp\{-\frac{u}{t}X_t\}\right] = \exp\left\{C(-\frac{u}{t}, t) + D(-\frac{u}{t}, t)v_0 - \frac{u}{t}x_0\right\}.$$

Denote $\Lambda(u,t) = t \log \mathbb{E}\left[\exp\{-\frac{u}{t}X_t\}\right]$. Then

$$\Lambda(u,t) = tC(-\frac{u}{t}, t) + tD(-\frac{u}{t}, t)v_0 - ux_0. \tag{5.2}$$

Next, set $A(u) = b - \rho\sigma u$. It is not hard to see that

$$\begin{aligned} D(u,t) &= \frac{1}{\sigma^2}(A(u) + d(u))\frac{1 - e^{d(u)t}}{1 - \frac{A(u)+d(u)}{A(u)-d(u)}e^{d(u)t}} \\ &= \frac{1}{\sigma^2}(A(u)^2 - d(u)^2)\frac{\sinh\frac{d(u)t}{2}}{d(u)\cosh\frac{d(u)t}{2} + A(u)\sinh\frac{d(u)t}{2}}. \end{aligned}$$

Moreover,

$$\begin{aligned} C(u,t) &= rut + \frac{a}{\sigma^2}\left[(A(u) + d(u))t - 2\log\left(\frac{1 - \frac{A(u)+d(u)}{A(u)-d(u)}e^{d(u)t}}{1 - \frac{A(u)+d(u)}{A(u)-d(u)}}\right)\right] \\ &= rut + \frac{a}{\sigma^2}\left[A(u)t - 2\log\frac{d(u)\cosh\frac{d(u)t}{2} + A(u)\sinh\frac{d(u)t}{2}}{d(u)}\right]. \end{aligned}$$

Using the previous formula, we obtain

$$C\left(-\frac{u}{t},t\right)=-ru$$
$$+\frac{a}{\sigma^2}\left[bt+\rho\sigma u-2\log\frac{d(-\frac{u}{t})t\cosh\frac{d(-\frac{u}{t})t}{2}+(bt+\rho\sigma u)\sinh\frac{d(-\frac{u}{t})t}{2}}{d(-\frac{u}{t})t}\right]. \quad (5.3)$$

We also have

$$A\left(-\frac{u}{t}\right)=b+\rho\sigma\frac{u}{t},$$

$$A^2\left(-\frac{u}{t}\right)=b^2+2b\rho\sigma\frac{u}{t}+\rho^2\sigma^2\frac{u^2}{t^2},$$

$$d^2\left(-\frac{u}{t}\right)=-\frac{u^2(1-\rho^2)\sigma^2}{t^2}+\frac{2\sigma u(k\sigma+b\rho)}{t}+b^2,$$

$$\frac{1}{\sigma^2}\left(A^2\left(-\frac{u}{t}\right)-d^2\left(-\frac{u}{t}\right)\right)=\frac{u^2}{t^2}-\frac{2ku}{t},$$

and

$$D\left(-\frac{u}{t},t\right)=\left(\frac{u^2}{t^2}-\frac{2ku}{t}\right)\frac{\sinh\frac{d(-\frac{u}{t})t}{2}}{d\left(-\frac{u}{t}\right)\cosh\frac{d(-\frac{u}{t})t}{2}+A\left(-\frac{u}{t}\right)\sinh\frac{d(-\frac{u}{t})t}{2}}. \quad (5.4)$$

Let us denote by Z the set of such real numbers u that the expressions on the right-hand side of (5.3) and (5.4) are finite for all small enough values of t, and put

$$\hat{S}(u,t)=d\left(-\frac{u}{t}\right)\frac{t}{2}.$$

It is easy to see that

$$\hat{S}(u,t)=\frac{1}{2}\sqrt{-u^2(1-\rho^2)\sigma^2+2tu(k\sigma^2+b\rho\sigma)+t^2b^2}.$$

In the previous formula, t is a real number. Therefore, for every real number $u\neq 0$, $\hat{S}(u,t)$ is purely imaginary for all numbers t with $|t|$ small enough. For such u and t, $\hat{S}(u,t)=iS(u,t)$, where

$$S(u,t)=\frac{1}{2}\sqrt{u^2(1-\rho^2)\sigma^2-2tu(k\sigma^2+b\rho\sigma)-t^2b^2} \quad (5.5)$$

is a real number. It follows that

$$tC\Big(-\frac{u}{t},t\Big) = -tru + t\frac{a}{\sigma^2}\left[bt+\rho\sigma u - 2\log\frac{2S(u,t)\cos S(u,t) + (bt+\rho\sigma u)\sin S(u,t)}{2S(u,t)}\right] \tag{5.6}$$

and

$$tD\Big(-\frac{u}{t},t\Big) = \Big(u^2 - 2tku\Big)\frac{\sin S(u,t)}{2S(u,t)\cos S(u,t) + (bt+\rho\sigma u)\sin S(u,t)}. \tag{5.7}$$

Our next goal is to introduce an additional condition under which the logarithmic function appearing in formula (5.6) exists, and the expressions on the right-hand sides of (5.6) and (5.7) are finite. Recall that we have assumed that $u \neq 0$ and $|t|$ is small enough. Set

$$\widetilde{S}(u) = \lim_{t\to 0}\,[2S(u,t)\cos S(u,t) + (bt+\rho\sigma u)\sin S(u,t)]\,.$$

Then, we have

$$\lim_{t\to 0} S(u,t) = \frac{1}{2}|u|\sigma\sqrt{1-\rho^2}$$

and

$$\widetilde{S}(u) = |u|\sigma\sqrt{1-\rho^2}\cos\frac{|u|\sigma\sqrt{1-\rho^2}}{2} + u\sigma\rho\sin\frac{|u|\sigma\sqrt{1-\rho^2}}{2}.$$

Let $\rho \neq 0$, and assume that

$$-\frac{2}{\sigma\sqrt{1-\rho^2}}\arctan\frac{\sqrt{1-\rho^2}}{\rho} < u < \frac{2}{\sigma\sqrt{1-\rho^2}}\left(\pi - \arctan\frac{\sqrt{1-\rho^2}}{\rho}\right). \tag{5.8}$$

The restriction in (5.8) means that the variable u is bounded from below by the largest negative root of the function

$$\widehat{S}(u) = \sqrt{1-\rho^2}\cos\frac{u\sigma\sqrt{1-\rho^2}}{2} + \rho\sin\frac{u\sigma\sqrt{1-\rho^2}}{2},$$

and from above by the smallest positive root of the same function. Note that $\widehat{S}(0) > 0$. Therefore, we have $\widehat{S}(u) > 0$, for all u satisfying the condition in (5.8).

It is easy to see that $\widetilde{S}(u) = \sigma|u|\widehat{S}(u)$ for all $u \neq 0$, satisfying the condition in (5.8). Hence, $\widetilde{S}(u) > 0$, under the same restrictions on u. It follows from (5.7) that for all $u \neq 0$ such that (5.8) holds, the right-hand side of (5.7) is eventually finite as $t \to 0$, and moreover

$$\lim_{t\to 0} tD\Big(-\frac{u}{t},t\Big)=\frac{u\sin\frac{u\sigma\sqrt{1-\rho^2}}{2}}{\sigma\left(\sqrt{1-\rho^2}\cos\frac{u\sigma\sqrt{1-\rho^2}}{2}+\rho\sin\frac{u\sigma\sqrt{1-\rho^2}}{2}\right)}. \tag{5.9}$$

In addition, the expression under the logarithm sign in (5.6) is eventually positive, and

$$\lim_{t\to 0} tC\Big(-\frac{u}{t},t\Big)=0. \tag{5.10}$$

In the case where $\rho=0$, the condition in (5.8) becomes

$$-\frac{\pi}{\sigma}<u<\frac{\pi}{\sigma}. \tag{5.11}$$

The analysis here proceeds similarly to that in the previous case.

The next statement provides explicit expressions for the function $\Lambda^{(0)}$. This statement was obtained in [8] (see formula (2) in [8], see also [10]) in a special case where $k=-\frac{1}{2}$ and $r=0$.

Theorem 5.1 *Suppose* $\rho\neq 0$ *and condition* (5.8) *holds. Then* $u\in Z$ *and the following formula is valid:*

$$\Lambda^{(0)}(u)=\frac{v_0 u\sin\frac{u\sigma\sqrt{1-\rho^2}}{2}}{\sigma\left(\sqrt{1-\rho^2}\cos\frac{u\sigma\sqrt{1-\rho^2}}{2}+\rho\sin\frac{u\sigma\sqrt{1-\rho^2}}{2}\right)}-x_0u. \tag{5.12}$$

If $\rho=0$ *and condition* (5.11) *holds, then* $u\in Z$ *and*

$$\Lambda^{(0)}(u)=\frac{v_0u}{\sigma}\tan\frac{u\sigma}{2}-x_0u.$$

Theorem 5.1 follows from (5.2), (5.9), and (5.10).

Recall that for $x\in\mathbb{R}$, the critical point $u^*(x)$ is the solution of the equation $\partial_u\Lambda^{(0)}(u)=-x$. Put $\theta=\frac{\sigma\sqrt{1-\rho^2}}{2}$. Then, using (5.12), we obtain

$$\partial_u\Lambda^{(0)}(u)=\frac{v_0}{2\sigma}\frac{\rho[1-\cos(2\theta u)]+\sqrt{1-\rho^2}\sin(2\theta u)+\sigma(1-\rho^2)u}{(\sqrt{1-\rho^2}\cos(\theta u)+\rho\sin(\theta u))^2}-x_0.$$

In a special case where $\rho=0$, we have

$$\partial_u\Lambda^{(0)}(u)=\frac{v_0}{\sigma}\frac{\sin(2\theta u)+\sigma u}{1+\cos(2\theta u)}-x_0.$$

In the following two statements, we provide formulas for the critical point $u^*(x)$ and the second derivative of the function $\Lambda^{(0)}(u)$. These results can be used in the asymptotic formulas established in the previous sections in the case of the Heston model.

Lemma 5.2 *Suppose $\rho \neq 0$ and condition* (5.8) *holds. Then, for every $x \in \mathbb{R}$, the critical point $u^*(x)$ is the unique solution to the equation*

$$\frac{\rho[1-\cos(2\theta u)] + \sqrt{1-\rho^2}\sin(2\theta u) + \sigma(1-\rho^2)u}{(\sqrt{1-\rho^2}\cos(\theta u) + \rho\sin(\theta u))^2} = \frac{2\sigma}{v_0}(x_0 - x).$$

If $\rho = 0$ and condition (5.11) *holds, then for every $x \in \mathbb{R}$, $u^*(x)$ is the unique solution to the equation*

$$\frac{\sin(2\theta u) + \sigma u}{1 + \cos(2\theta u)} = \frac{\sigma}{v_0}(x_0 - x).$$

Lemma 5.3 *Suppose $\rho \neq 0$ and condition* (5.8) *holds. Then*

$$\partial^2 \Lambda^{(0)}(u) = \frac{v_0 S(u)}{2\sigma[\sqrt{1-\rho^2}\cos(\theta u) + \rho\sin(\theta u)]^3}$$

where

$$\begin{aligned} S(u) = {} & (2\theta + \sigma\sqrt{1-\rho^2})[\rho\sqrt{1-\rho^2}\sin(\theta u) + (1-\rho^2)\cos(\theta u)] \\ & + 2\sigma\theta(1-\rho^2)u[\sqrt{1-\rho^2}\sin(\theta u) - \rho\cos(\theta u)]. \end{aligned}$$

If $\rho = 0$ and condition (5.11) *holds, then*

$$\partial^2 \Lambda^{(0)}(u) = \frac{v_0}{2\sigma}\frac{(2\theta + \sigma)\cos(\theta u) + 2\theta\sigma u \sin(\theta u)}{\cos^3(\theta u)}.$$

Lemmas 5.2 and 5.3 are straightforward, and their proofs are omitted.
We will next compute the functions $\Lambda^{(1)}$ and $\Lambda^{(2)}$. Recall that

$$\int \exp\left(-\frac{u}{t}z\right)p_t(dz) = \exp\left(\frac{\Lambda^{(0)}(u)}{t}\right)\exp\left(\Lambda^{(1)}(u)\right)\left(1 + t\Lambda^{(2)}(u) + \dots\right).$$

Therefore,

$$\Lambda(u,t) = \Lambda^{(0)}(u) + t\Lambda^{(1)}(u) + t\log(1 + t\Lambda^{(2)}(u) + \dots).$$

By differentiating the previous formula with respect to t, we obtain

$$\Lambda^{(1)}(u) = \lim_{t\to 0} \frac{\partial \Lambda}{\partial t}(u,t)$$

and

$$\Lambda^{(2)}(u) = \frac{1}{2} \lim_{t\to 0} \frac{\partial^2 \Lambda}{\partial t^2}(u,t). \tag{5.13}$$

Let us fix $u \neq 0$ such as in Theorem 5.1. Then the function $t \mapsto S(u,t)$ defined by (5.5) is real analytic in t in a small neighborhood of $t = 0$, depending on u. Using the Taylor formula, we obtain

$$S(u,t) = c_0(u) + c_1(u)t + \frac{1}{2}c_2(t)t^2 + O\left(t^3\right) \tag{5.14}$$

as $t \to 0$, where the O-estimate depends on u, and the coefficients are given by

$$c_0(u) = \frac{|u|\sigma}{2}\sqrt{1-\rho^2}, \tag{5.15}$$

$$c_1(u) = -\frac{|u|}{u}\frac{k\sigma + b\rho}{2\sqrt{1-\rho^2}}, \tag{5.16}$$

and

$$c_2(u) = -\frac{|u|}{u^2}\frac{b^2(1-\rho^2) + (k\sigma + b\rho)^2}{2\sigma(1-\rho^2)^{\frac{3}{2}}}.$$

Our next goal is to expand the functions $t \mapsto \sin S(u,t)$ and $t \mapsto \cos S(u,t)$. Using the Taylor formula and (5.14), we get

$$\sin S(u,t) = U_0(u) + U_1(u)t + \frac{1}{2}U_2(u)t^2 + O(t^3) \tag{5.17}$$

as $t \to 0$, where

$$U_0(u) = \sin c_0(u), \tag{5.18}$$

$$U_1(u) = c_1(u)\cos c_0(u), \tag{5.19}$$

and

$$U_2(u) = c_2(u)\cos c_0(u) - c_1(u)^2 \sin c_0(u).$$

Similarly,

$$\cos S(u,t) = W_0(u) + W_1(u)t + \frac{1}{2}W_2(u)t^2 + O(t^3) \tag{5.20}$$

as $t \to 0$, where

$$W_0(u) = \cos c_0(u),$$

$$W_1(u) = -c_1(u) \sin c_0(u),$$

and

$$W_2(u) = -[c_2(u) \sin c_0(u) + c_1(u)^2 \cos c_0(u)].$$

We will next expand the functions $t \mapsto tD(-\frac{u}{t}, t)$ and $t \mapsto tC(-\frac{u}{t}, t)$. It follows from (5.14), (5.17), and (5.20) that

$$2S(u,t)\cos S(u,t) + (bt + \rho\sigma u)\sin S(u,t) = V_0(u) + V_1(u)t + \frac{1}{2}V_2(u)t^2 + O(t^3) \tag{5.21}$$

as $t \to 0$, where

$$V_0(u) = 2c_0(u)W_0(u) + \rho\sigma u U_0(u),$$

$$V_1(u) = 2c_0(u)W_1(u) + 2c_1(u)W_0(u) + bU_0(u) + \rho\sigma u U_1(u),$$

and

$$V_2(u) = 2c_0(u)W_2(u) + 4c_1(u)W_1(u) + 2c_2(u)W_0(u) + 2bU_1(u) + \rho\sigma u U_2(u).$$

It is not hard to see that

$$V_0(u) = 2c_0(u)\cos c_0(u) + \rho\sigma u \sin c_0(u), \tag{5.22}$$

$$\begin{aligned} V_1(u) = &(2 + \rho\sigma u)c_1(u)\cos c_0(u) \\ &+ (b - 2c_0(u)c_1(u))\sin c_0(u), \end{aligned} \tag{5.23}$$

and

$$\begin{aligned} V_2(u) = &[2c_2(u) + 2bc_1(u) + \rho\sigma u c_2(u) - 2c_0(u)c_1(u)^2]\cos c_0(u) \\ &- [2c_0(u)c_2(u) + 4c_1(u)^2 + \rho\sigma u c_1(u)^2]\sin c_0(u). \end{aligned}$$

Therefore,

$$tD(-\frac{u}{t}, t) = (u^2 - 2tku)\frac{U_0(u) + U_1(u)t + \frac{1}{2}U_2(u)t^2 + O(t^3)}{V_0(u) + V_1(u)t + \frac{1}{2}V_2(u)t^2 + O(t^3)} \tag{5.24}$$

as $t \to 0$ (see (5.7), (5.17), and (5.21)).

Set

$$tD(-\frac{u}{t}, t) = T_0(u) + T_1(u)t + \frac{1}{2}T_2(u)t^2 + O(t^3) \tag{5.25}$$

as $t \to 0$. Then, (5.24) and (5.25) give

$$T_0(u)V_0(u) = u^2U_0(u),$$

$$T_0(u)V_1(u) + T_1(u)V_0(u) = u^2U_1(u) - 2kuU_0(u),$$

and

$$\frac{1}{2}T_0(u)V_2(u) + T_1(u)V_1(u) + \frac{1}{2}T_2(u)V_0(u) = \frac{1}{2}u^2U_2(u) - 2kuU_1(u).$$

It follows from the previous equalities that

$$T_0(u) = \frac{u^2U_0(u)}{V_0(u)},$$

$$T_1(u) = \frac{u^2U_1(u)V_0(u) - 2kuU_0(u)V_0(u) - u^2U_0(u)V_1(u)}{V_0(u)^2},$$

and

$$T_2(u) = \frac{Q(u)}{V_0(u)^3},$$

where

$$\begin{aligned} Q(u) =& u^2U_2(u)V_0(u)^2 - 4kuU_1(u)V_0(u)^2 - u^2U_0(u)V_0(u)V_2(u) \\ &- 2u^2U_1(u)V_0(u)V_1(u) + 4kuU_0(u)V_0(u)V_1(u) + 2u^2U_0(u)V_1(u)^2. \end{aligned} \tag{5.26}$$

Therefore, the following asymptotic formula holds:

$$\begin{aligned} tD(-\frac{u}{t},t) =& \frac{u^2U_0(u)}{V_0(u)} + t\frac{u^2U_1(u)V_0(u) - 2kuU_0(u)V_0(u) - u^2U_0(u)V_1(u)}{V_0(u)^2} \\ &+ \frac{t^2}{2}\frac{Q(u)}{V_0(u)^3} + O(t^3) \end{aligned} \tag{5.27}$$

as $t \to 0$.

Now, we turn our attention to the function $t \mapsto tC(-\frac{u}{t},t)$. Using (5.6), we see that

$$tC(-\frac{u}{t},t) = -tru + t\frac{a}{\sigma^2}\left[bt + \rho\sigma u - 2\log\frac{V_0(u) + V_1(u)t + O(t^2)}{2c_0(u) + 2c_1(u)t + O\left(t^2\right)}\right]. \tag{5.28}$$

Set

$$\frac{V_0(u) + V_1(u)t + O(t^2)}{2c_0(u) + 2c_1(u)t + O\left(t^2\right)} = L_0(u) + L_1(u)t + O(t^2)$$

as $t \to 0$. It is not hard to see that

$$L_0(u) = \frac{V_0(u)}{2c_0(u)}, \tag{5.29}$$

$$L_1(u) = \frac{c_0(u)V_1(u) - c_1(u)V_0(u)}{2c_0(u)^2}. \tag{5.30}$$

We also have

$$\log[L_0(u) + L_1(u)t + O(t^2)] = \log L_0(u) + \frac{L_1(u)}{L_0(u)}t + O(t^2) \tag{5.31}$$

as $t \to 0$. It follows from (5.28)–(5.31) that

$$tC(-\frac{u}{t}, t) = \left[\frac{a\rho u}{\sigma} - ru - \frac{2a}{\sigma^2}\log\frac{V_0(u)}{2c_0(u)}\right]t \\ + \left[\frac{ab}{\sigma^2} - \frac{2a}{\sigma^2}\frac{c_0(u)V_1(u) - c_1(u)V_0(u)}{c_0(u)V_0(u)}\right]t^2 + O(t^3) \tag{5.32}$$

as $t \to 0$.

Next, we will find explicit expressions for the functions $\Lambda^{(1)}$ and $\Lambda^{(2)}$. Suppose $\rho \neq 0$, $u \neq 0$, and condition (5.8) holds. Then

$$\Lambda^{(1)}(u) = \left(\frac{a\rho}{\sigma} - r\right)u - \frac{2a}{\sigma^2}\log\frac{V_0(u)}{2c_0(u)} \\ + v_0\frac{u^2U_1(u)V_0(u) - 2kuU_0(u)V_0(u) - u^2U_0(u)V_1(u)}{V_0(u)^2}. \tag{5.33}$$

Formula (5.33) can be established, using (5.27) and (5.32).

The next statement provides an explicit expression for the function $\Lambda^{(1)}$ in terms of the Heston model parameters.

Theorem 5.4 *Suppose $\rho \neq 0$, $u \neq 0$, and condition* (5.8) *holds. Then*

$$\Lambda^{(1)}(u) = \left(\frac{a\rho}{\sigma} - r\right)u - \frac{2a}{\sigma^2}\log\frac{\sqrt{1-\rho^2}\cos\frac{u\sigma\sqrt{1-\rho^2}}{2} + \rho\sin\frac{u\sigma\sqrt{1-\rho^2}}{2}}{\sqrt{1-\rho^2}} \\ + v_0\frac{E_1(u)\cos^2\frac{u\sigma\sqrt{1-\rho^2}}{2} + E_2(u)\cos\frac{u\sigma\sqrt{1-\rho^2}}{2}\sin\frac{u\sigma\sqrt{1-\rho^2}}{2} + E_3(u)\sin^2\frac{u\sigma\sqrt{1-\rho^2}}{2}}{\sigma^2\left(\sqrt{1-\rho^2}\cos\frac{u\sigma\sqrt{1-\rho^2}}{2} + \rho\sin\frac{u\sigma\sqrt{1-\rho^2}}{2}\right)^2}, \tag{5.34}$$

where

$$E_1(u) = -\frac{u\sigma(k\sigma + b\rho)}{2},$$

$$E_2(u) = -2k\sigma\sqrt{1-\rho^2} + \frac{k\sigma + b\rho}{\sqrt{1-\rho^2}},$$

and

$$E_3(u) = -\left(2k\rho\sigma + b + u\sigma\frac{k\sigma + b\rho}{2}\right).$$

If $\rho = 0$, *then formula* (5.34) *holds for all* u *satisfying condition* (5.11).

Proof Taking into account (5.33), (5.15), (5.16), (5.18), (5.19), (5.22), (5.23),we obtain

$$\Lambda^{(1)}(u) = \left(\frac{a\rho}{\sigma} - r\right)u - \frac{2a}{\sigma^2}\log\frac{|u|\sigma\sqrt{1-\rho^2}\cos\frac{|u|\sigma\sqrt{1-\rho^2}}{2} + \rho\sigma u\sin\frac{|u|\sigma\sqrt{1-\rho^2}}{2}}{|u|\sigma\sqrt{1-\rho^2}}$$
$$+ v_0\frac{\widetilde{E}_1(u)\cos^2\frac{|u|\sigma\sqrt{1-\rho^2}}{2} + \widetilde{E}_2(u)\cos\frac{|u|\sigma\sqrt{1-\rho^2}}{2}\sin\frac{|u|\sigma\sqrt{1-\rho^2}}{2} + \widetilde{E}_3(u)\sin^2\frac{|u|\sigma\sqrt{1-\rho^2}}{2}}{\left(|u|\sigma\sqrt{1-\rho^2}\cos\frac{|u|\sigma\sqrt{1-\rho^2}}{2} + \rho\sigma u\sin\frac{|u|\sigma\sqrt{1-\rho^2}}{2}\right)^2},$$

where

$$\widetilde{E}_1(u) = -\frac{u^3}{2}\sigma(k\sigma + b\rho),$$

$$\widetilde{E}_2(u) = u\left[-\rho\sigma u|u|\frac{k\sigma + b\rho}{2\sqrt{1-\rho^2}} - 2k|u|\sigma\sqrt{1-\rho^2} + (2+\rho\sigma u)|u|\frac{k\sigma + b\rho}{2\sqrt{1-\rho^2}}\right],$$

and

$$\widetilde{E}_3(u) = -u^2\left[2k\rho\sigma + b + u\sigma\frac{k\sigma + b\rho}{2}\right].$$

□

Next, replacing $|u|$ by u in the previous formulas (it is not hard to see that this can be done) and making several cancellations, we obtain formula (5.34).

This completes the proof of Theorem 5.4.

Our final goal in the present section is to find an explicit formula for the function $\Lambda^{(2)}$ in terms of the Heston model parameters. It follows from (5.2), (5.13), (5.27), and (5.32) that

$$\Lambda^{(2)}(u) = \frac{ab}{\sigma^2} - \frac{2a}{\sigma^2}\frac{c_0(u)V_1(u) - c_1(u)V_0(u)}{c_0(u)V_0(u)} + \frac{v_0}{2}\frac{Q(u)}{V_0(u)^3}, \tag{5.35}$$

where $Q(u)$ is given by (5.26). Now, it is clear how to obtain an explicit expression for the function $\Lambda^{(2)}$, expressed in terms of the Heston model parameters. It suffices to transform the formula in (5.35), using the explicit expressions for the functions c_i, U_i, V_i with $i = 0, 1, 2$, and the function Q. Let us also note that the value of the function on the right-hand of formula (5.35) does not change if we replace $|u|$ by u. Taking into account what was said above, and making long but straightforward computations, we see that the following statement holds.

Theorem 5.5 *Suppose $\rho \neq 0$, $u \neq 0$, and condition* (5.8) *holds. Then*

$$\Lambda^{(2)}(u) = \frac{ab}{\sigma^2} - \frac{a}{\sigma^3(1-\rho^2)u}\frac{I_0(u)}{\sqrt{1-\rho^2}\cos\frac{u\sigma\sqrt{1-\rho^2}}{2} + \rho\sin\frac{u\sigma\sqrt{1-\rho^2}}{2}}$$
$$+ \frac{v_0}{2}\frac{I_1(u)}{\sigma^2\left[\sqrt{1-\rho^2}\cos\frac{u\sigma\sqrt{1-\rho^2}}{2} + \rho\sin\frac{u\sigma\sqrt{1-\rho^2}}{2}\right]^2}$$
$$+ \frac{v_0}{2}\frac{I_2(u)I_3(u)}{u\sigma^3\left[\sqrt{1-\rho^2}\cos\frac{u\sigma\sqrt{1-\rho^2}}{2} + \rho\sin\frac{u\sigma\sqrt{1-\rho^2}}{2}\right]^3}, \tag{5.36}$$

where

$$I_0(u) = \left[-u\rho\sigma\sqrt{1-\rho^2}(k\sigma + b\rho)\right]\cos\frac{u\sigma\sqrt{1-\rho^2}}{2}$$
$$+ \left[2b + 2\rho k\sigma + u\sigma(k\sigma + b\rho)(1-\rho^2)\right]\sin\frac{u\sigma\sqrt{1-\rho^2}}{2},$$

$$I_1(u) = -\frac{u\left[b^2(1-\rho^2) + (k\sigma + b\rho)^2\right]}{2(1-\rho^2)} + \frac{(k\sigma + b\rho)^2}{1-\rho^2}\sin^2\frac{u\sigma\sqrt{1-\rho^2}}{2}$$
$$+ \left[\frac{b^2(1-\rho^2) + (k\sigma + b\rho)^2}{\sigma(1-\rho^2)^{\frac{3}{2}}} + \frac{b(k\sigma + b\rho)}{\sqrt{1-\rho^2}}\right]\sin\frac{u\sigma\sqrt{1-\rho^2}}{2}\cos\frac{u\sigma\sqrt{1-\rho^2}}{2},$$

$$I_2(u) = 2\left[2k\sigma\sqrt{1-\rho^2} - \frac{(2+\rho\sigma u)(k\sigma + b\rho)}{2\sqrt{1-\rho^2}}\right]\cos\frac{u\sigma\sqrt{1-\rho^2}}{2}$$
$$+ 2\left[2k\rho\sigma + b + \frac{u\sigma(k\sigma + b\rho)}{2}\right]\sin\frac{u\sigma\sqrt{1-\rho^2}}{2},$$

and

$$I_3(u) = -u\sigma\sqrt{1-\rho^2} + b\sin^2\frac{u\sigma\sqrt{1-\rho^2}}{2} - \frac{k\sigma + b\rho}{\sqrt{1-\rho^2}}\sin\frac{u\sigma\sqrt{1-\rho^2}}{2}\cos\frac{u\sigma\sqrt{1-\rho^2}}{2}.$$

If $\rho = 0$*, then formula* (5.36) *holds for all* u *satisfying condition* (5.11).

Proof The second term on the right-hand side of (5.36) can be obtained from the corresponding term in (5.35) by taking into account (5.15), (5.16), (5.22), and (5.23). Next, using (5.26), we see that

$$\frac{Q(u)}{V_0(u)^3} = \frac{u^2\left[U_2(u)V_0(u) - U_0(u)V_2(u)\right]}{V_0(u)^2} + \frac{\left[4kuV_0(u) + 2u^2V_1(u)\right]\left[U_0(u)V_1(u) - U_1(u)V_0(u)\right]}{V_0(u)^3}. \tag{5.37}$$

□

Moreover

$$U_2(u)V_0(u) - U_0(u)V_2(u) = 2c_0(u)c_2(u) + 4c_1(u)^2\sin^2 c_0(u) - 2\left[c_2(u) + bc_1(u)\right]\sin c_0(u)\cos c_0(u),$$

$$4kuV_0(u) + 2u^2V_1(u) = 2u\left[4kc_0(u) + u(2 + \rho\sigma u)c_1(u)\right]\cos c_0(u) + 2u^2\left[2k\rho\sigma + b - 2c_0(u)c_1(u)\right]\sin c_0(u),$$

and

$$U_0(u)V_1(u) - U_1(u)V_0(u) = b\sin^2 c_0(u) - 2c_0(u)c_1(u) + 2c_1(u)\sin c_0(u)\cos c_0(u).$$

Set

$$I_1(u) = U_2(u)V_0(u) - U_0(u)V_2(u),$$

$$I_2(u) = u^{-2}\left[4kuV_0(u) + 2u^2V_1(u)\right],$$

and

$$I_3(u) = U_0(u)V_1(u) - U_1(u)V_0(u).$$

Next, taking into account (5.35), (5.37), and using the explicit expressions for the functions c_i, U_i, and V_1 with $i = 0, 1, 2$, which were found above, we obtain (5.36).

This completes the proof of Theorem 5.5.

Remark 5.6 The present remark concerns the continuity of the functions $\Lambda^{(i)}$ with $i = 0, 1, 2$ on their domain. Recall that $\Lambda^{(i)}(0) = 0$. It follows from Theorems 5.1, 5.4, and 5.5 that the functions $\Lambda^{(i)}$ are continuous on their domain with a possible exception of the point $u = 0$. However, it is not hard to see, using the explicit expressions for the functions $\Lambda^{(i)}$, provided in the theorems mentioned above, that

$$\lim_{u \to 0} \Lambda^{(i)}(u) = 0 \quad \text{for} \quad i = 0, 1, 2.$$

Acknowledgments We gratefully acknowledge the support by the Institute for Mathematical Research (FIM) and the ETH foundation. We would also like to thank Antoine Jacquier for reading the paper and making very helpful comments.

References

1. Aït-Sahalia, Y., Yu, J.: Saddlepoint approximations for continuous-time Markov processes. J. Econom. **134**, 507–551 (2006)
2. Cuchiero, C., Filipovic, D., Mayerhofer, E., Teichmann, J.: Affine processes on positive semi-definite matrices. Ann. Appl. Probab. **21**, 397–463 (2011)
3. Cuchiero, C., Keller-Ressel, M., Mayerhofer, E., Teichmann, J.: Affine processes on symmetric cones. J. Theor. Probab. doi:10.1007/s10959-014-0580-x (2014)
4. Dembo, A., Zeitouni, O.: Large Deviations Techniques and Applications. Jones and Bartlett Publishers Inc., Boston (1993)
5. Duffie, D., Filipovic, D., Schachermayer, W.: Affine processes and applications in finance. Ann. Appl. Probab. **13**, 984–1053 (2003)
6. Ellis, R.S.: Large deviations for a general class of random vectors. Ann. Probab. **12**, 1–12 (1984)
7. Forde, M., Jacquier, A.: Small-time asymptotics for implied volatility under the Heston model. IJTAF **12**, 861–876 (2009)
8. Forde, M., Jacquier, A.: The large-maturity smile for the Heston model. Financ. Stoch. **15**, 755–780 (2011)
9. Forde, M., Jacquier, A., Mijatović, A.: Asymptotic formulae for implied volatility in the Heston model. Proc. R. Soc. A **466**, 3593–3620 (2010)
10. Forde, M., Jacquier, A., Lee, R.: The small-time smile and term structure of implied volatility under the Heston model. SIAM J. Financ. Math. **3**, 690–708 (2012)
11. Forde, M., Kumar, R.: Large-time option pricing for a general stochastic volatility model with a stochastic interest rate, using the Donsker-Varadhan LDP, Preprint (2013)
12. Gärtner, J.: On large deviations from the invariant measure. Theory Probab. Appl. **22**, 24–39 (1977)
13. Gulisashvili, A., Laurence, P.: The Heston Riemannian distance function. Journal de Mathématiques Pures et Appliquées **101**, 303–329 (2014)
14. Heston, S.L.: A closed-form solution for options with stochastic volatility, with applications to bond and currency options. Rev. Financ. Stud. **6**, 327–343 (1993)
15. Jacquier, A., Mijatović, A.: Large deviations for the extended Heston model: the large-time case. Asia-Pacific Finan. Markets **21**(3), 263–280 (2014)

16. Jacquier, A., Roome, P.: Asymptotics of forward implied volatility. SIAM J. Finan. Math. **6**(1), 307–351 (2015)
17. Keller-Ressel, M., Teichmann, J., Schachermayer, W.: Regularity of affine processes on general state spaces. Electron. J. Probab. **18**, 1–17 (2013)
18. Olver, F.W.J.: Asymptotics and Special Functions. A K Peters Ltd., Wellesley (1997)
19. Pham, H.: Some applications and methods of large deviations in finance and insurance. Paris-Princeton Lectures on Mathematical Finance 2004. Lecture Notes in Mathematics, vol. 1919, pp. 191–244 (2007)
20. Spreij, P., Veerman, E.: Affine diffusions with non-canonical state space. Stoch. Anal. Appl. **30**, 605–641 (2012)

Asymptotics for d-Dimensional Lévy-Type Processes

Matthew Lorig, Stefano Pagliarani and Andrea Pascucci

Abstract We consider a general d-dimensional Lévy-type process with killing. Combining the classical Dyson series approach with a novel polynomial expansion of the generator $\mathcal{A}(t)$ of the Lévy-type process, we derive a family of asymptotic approximations for transition densities and European-style options prices. Examples of stochastic volatility models with jumps are provided in order to illustrate the numerical accuracy of our approach. The methods described in this paper extend the results from Corielli et al. (SIAM J Financ Math 1:833–867, 2010, [4]), Pagliarani and Pascucci (Int. J. Theor. Appl. Financ. 16(8):1–35, 2013, [20]) to Lorig et al. (Analytical expansions for parabolic equations, 2013, [13]) for Markov diffusions to Markov processes with jumps.

Keywords Multi-dimensional Lévy-type process with killing · Asymptotic approximation · Integro-differential equation · Levy processes · Parametrix

1 Introduction

In a multi-dimensional Markovian setting, the time evolution of a market model is usually described by the solution X of a Lévy-Itô stochastic differential equation (SDE). Such a model allows for features commonly seen in markets, such as stochastic volatility, jumps, default, co-integration and correlation. Many quantities of interest (e.g., option prices, net present values) can be expressed as expectations of

M. Lorig
Department of Applied Mathematics, University of Washington, Seattle, WA, USA
e-mail: mlorig@uw.edu

S. Pagliarani
Ecole Polytechnique, CMAP, Route de Saclay, 91128 Palaiseau Cedex, France
e-mail: pagliarani@cmap.polytechnique.fr

A. Pascucci (✉)
Dipartimento di Matematica, Università di Bologna, Bologna, Italy
e-mail: andrea.pascucci@unibo.it

P.K. Friz et al. (eds.), *Large Deviations and Asymptotic Methods in Finance*, Springer Proceedings in Mathematics & Statistics 110,
DOI 10.1007/978-3-319-11605-1_12

the form $u(t, x) := \mathbb{E}[\varphi(X_T)|X_t = x]$. Under mild conditions, the function $u(t, x)$ is the unique classical solution of a partial integro-differential equation (PIDE). Unfortunately, closed form and even semi-closed form solutions of these PIDEs are available only in rare cases. As such, it is important to develop general methods for finding analytical approximations for the solutions of these PIDEs.

Within the mathematical finance literature, a number of different approaches have been taken for finding approximate transition densities and option prices for markets described by Markov processes. Most of these techniques involve expansions that exploit a small parameter or a limiting case. For example, Benhamou et al. [1] develop analytical approximations for models with local volatility and Gaussian jumps in the small diffusion and small jump frequency/size limits (see also the recent review paper by Bompis and Gobet [2]). Deuschel et al. [5] obtain densities for diffusion processes in a small noise limit. Fouque et al. [6] find option prices for Black-Scholes-like multiscale models where volatility is driven by two factors, one running on a fast scale, one running on a slow scale. Lorig [11], Lorig and Lozano-Carbassé [12] extend these multiscale techniques to more general diffusions and to the exponential Lévy setting.

Recently, Pagliarani and Pascucci [19] introduce a method for finding asymptotic solutions of parabolic PDEs. The approach, called the *adjoint expansion method*, is extended by Lorig et al. [15], Pagliarani et al. [21] to models with jumps and it was further generalized by Lorig et al. [13] to a family of asymptotic expansions for a d-dimensional market described by an Itô SDE (i.e., a Markov market with no jumps). The method consists of expanding the pricing PDE in polynomial basis functions, which results in a nested sequence of Cauchy problems, and deriving analytical solutions for these nested Cauchy problems. In this paper, we extend the results of Lorig et al. [13], Lorig et al. [15], Pagliarani et al. [21] to the PIDEs that arise when markets are described by a d-dimensional Lévy-Itô SDE. Results presented here also simplify results from Lorig et al. [13], Lorig et al. [15], Pagliarani et al. [21].

The rest of this paper proceeds as follows. In Sect. 2 we present a general d-dimensional market model. We also describe the kinds of derivative-assets we wish to price, and we relate the price of such derivative-assets to the solution of a parabolic PIDE. In Sect. 3 we introduce the idea of polynomial expansions of the pricing PIDE and in Sect. 4, we derive a family of analytical price approximations—one for each polynomial expansion of the pricing PIDE. Lastly, in Sect. 5 we provide a numerical example, illustrating the versatility and accuracy of our methods.

2 Market Model

We take, as given, an equivalent martingale measure $\mathbb{Q}$ defined on a complete filtered probability space $(\Omega, \mathcal{F}, \{\mathcal{F}_t, t \geq 0\}, \mathbb{Q})$. All stochastic processes defined below live on this probability space and all expectations are taken with respect to $\mathbb{Q}$. The risk-neutral dynamics of our market are described by the following d-dimensional Markov Lévy-type process

$$\mathrm{d}X_t = \mu(t, X_t)\,\mathrm{d}t + \sigma(t, X_t)\,\mathrm{d}W_t + \int_{\mathbb{R}^d} z\,\mathrm{d}\widetilde{N}(t, X_{t-}, \mathrm{d}t, \mathrm{d}z).$$

Here W is a standard m-dimensional Brownian motion, and $\widetilde{N}(\cdot, \cdot, \mathrm{d}t, \mathrm{d}z)$, given by

$$\widetilde{N}(t, x, \mathrm{d}t, \mathrm{d}z) = N(t, x, \mathrm{d}t, \mathrm{d}z) - \nu(t, x, \mathrm{d}z)\mathrm{d}t, \qquad (t, x) \in \mathbb{R}^+ \times \mathbb{R}^d,$$

is a family of compensated Poisson measures on $\mathcal{B}(\mathbb{R}) \otimes \mathcal{B}(\mathbb{R}^d)$. The drift vector μ and volatility matrix σ map $\mu : \mathbb{R}_+ \times \mathbb{R}^d \to \mathbb{R}^d$ and $\sigma : \mathbb{R}_+ \times \mathbb{R}^d \to \mathbb{R}^{d\times m}$, respectively. We assume the Lévy kernel ν satisfies

$$\int_{\mathbb{R}^d} \min\left\{|z|, |z|^2\right\} \bar{\nu}(\mathrm{d}z) < \infty, \qquad \bar{\nu}(\mathrm{d}z) := \sup_{(t,x)\in\mathbb{R}_+\times\mathbb{R}^d} \nu(t, x, \mathrm{d}z), \qquad (2.1)$$

which is rather standard for Lévy-type models. The components of X could represent a number of things such as e.g., economic factors, asset prices, indices, or functions of these quantities. We assume a risk-free interest rate of the form $r(t, X_t)$ where $r : \mathbb{R}_+ \times \mathbb{R}^d \to \mathbb{R}$. We also introduce a random time ζ, which is given by

$$\zeta = \inf\left\{t \geq 0 : \int_0^t \gamma(s, X_s)\mathrm{d}s \geq \mathcal{E}\right\}, \qquad \gamma : \mathbb{R}_+ \times \mathbb{R}^d \to \mathbb{R}_+,$$

with $\mathcal{E}$ exponentially distributed and independent of X. The random time ζ could represent the default time of an asset, the arrival of an economic shock, etc.

Denote by V the no-arbitrage price of a European derivative expiring at time T with payoff

$$H(X_T)\,\mathbb{I}_{\{\zeta>T\}} + G(X_T)\,\mathbb{I}_{\{\zeta\leq T\}} = \big(H(X_T) - G(X_T)\big)\,\mathbb{I}_{\{\zeta>T\}} + G(X_T).$$

It is well known (see, for instance, Jeanblanc et al. [9]) that

$$\begin{aligned} V_t = \mathbb{E}&\left[\mathrm{e}^{-\int_t^T r(s,X_s)\mathrm{d}s} G(X_T)|X_t\right] + \\ &\mathbb{I}_{\{\zeta>t\}}\mathbb{E}\left[\mathrm{e}^{-\int_t^T (r(s,X_s)+\gamma(s,X_s))\mathrm{d}s}\Big(H(X_T) - G(X_T)\Big)|X_t\right], \quad t < T. \end{aligned} \qquad (2.2)$$

Thus, to value a European-style option, one must compute functions of the form

$$u(t, x) := \mathbb{E}\left[\mathrm{e}^{-\int_t^T \lambda(s,X_s)\mathrm{d}s}\varphi(X_T) \mid X_t = x\right]. \qquad (2.3)$$

Under mild assumptions (see, for instance, Pascucci [22]), the function u, defined by (2.3), satisfies the Kolmogorov backward equation

$$(\partial_t + \mathcal{A}(t))u = 0, \qquad u(T, x) = \varphi(x), \qquad x \in \mathbb{R}^d, \qquad (2.4)$$

where the operator $\mathcal{A}(t)$ is given explicitly by

$$\mathcal{A}(t) = \int_{\mathbb{R}^d} \nu(t,x,\mathrm{d}z)\left(\mathrm{e}^{\langle z,\nabla_x\rangle} - 1 - \langle z,\nabla_x\rangle\right) + \frac{1}{2}\sum_{i,j=1}^{d}\left(\sigma\sigma^{\mathrm{T}}\right)_{ij}(t,x)\partial_{x_i}\partial_{x_j} + \sum_{i=1}^{d}\mu_i(t,x)\partial_{x_i} - \lambda(t,x), \tag{2.5}$$

with

$$\langle z,x\rangle := \sum_{i=1}^{d} z_i x_i, \qquad \nabla_x := (\partial_{x_1},\partial_{x_2},\ldots,\partial_{x_d}), \qquad \mathrm{e}^{\langle z,\nabla_x\rangle} f(x) := f(x+z).$$

The formal representation of the *shift operator* $\mathrm{e}^{\langle z,\nabla_x\rangle}$ is motivated by the fact that its Taylor expansion applied to the function $f(x)$ gives the Taylor expansion of $f(x+z)$ about the point x. As in Øksendal and Sulem [18, Chap. 1], we regard the domain of $\mathcal{A}(t)$ to be all functions $f\colon \mathbb{R}^d \to \mathbb{R}$ such that $\mathcal{A}(t)f(x)$ exists and is finite for all $x \in \mathbb{R}^d$.

Remark 2.1 (Martingale property) Let us denote by $X^{(i)}$ the ith component of the vector X and assume that

$$\int_{|z|\geq 1} \mathrm{e}^{z_i}\,\bar{\nu}(\mathrm{d}z) < \infty,$$

for some $i \leq d$, with $\bar{\nu}$ as in (2.1). If $S_t := \mathbb{I}_{\{\zeta>t\}}\mathrm{e}^{X_t^{(i)}}$ is supposed to be a traded asset then, in order for S to be a martingale, the drift μ_i must satisfy

$$\mu_i(t,x) = \gamma(t,x) - \int_{\mathbb{R}^d} \nu(t,x,\mathrm{d}z)(\mathrm{e}^{z_i} - 1 - z_i) - \frac{1}{2}\left(\sigma\sigma^{\mathrm{T}}\right)_{ii}(t,x),$$

To see this, set $H(x) = \mathrm{e}^{x_i}$, $G(x) = 0$ and impose $V_t = S_t$ in (2.2).

3 General Expansion Basis

Let us start by rewriting the differential operator (2.5) in the more compact form

$$\mathcal{A}(t) := \int_{\mathbb{R}^d} \nu(t,x,\mathrm{d}z)\left(\mathrm{e}^{\langle z,\nabla_x\rangle} - 1 - \langle z,\nabla_x\rangle\right) + \sum_{|\alpha|\leq 2} a_\alpha(t,x)D_x^\alpha, \quad t\in\mathbb{R},\ x\in\mathbb{R}^d,$$

where by standard notations

$$\alpha = (\alpha_1, \dots, \alpha_d) \in \mathbb{N}_0^d, \qquad |\alpha| = \sum_{i=1}^d \alpha_i, \qquad D_x^{\alpha *} = \partial_{x_1}^{\alpha_1} \cdots \partial_{x_d}^{\alpha_d}.$$

In this section we introduce a family of expansion schemes for $\mathcal{A}(t)$, which we shall use to construct closed-form approximate solutions (one for each family) of (2.4).

Definition 3.1 For $|\alpha| \leq 2$ and $n \leq N \in \mathbb{N}_0$, let $a_{\alpha,n} = a_{\alpha,n}(t,x)$ and $\nu_n = \nu_n(t,x,\mathrm{d}z)$ be such that the following hold:

(i) For any $t \in [0,T]$, $a_{\alpha,n}(t,\cdot)$ are polynomial functions with $a_{\alpha,0}(t,x) \equiv a_{\alpha,0}(t)$, and for any $x \in \mathbb{R}^d$ the functions $a_{\alpha,n}(\cdot,x)$ belong to $L^\infty([0,T])$.
(ii) For any $t \in [0,T]$, $x \in \mathbb{R}^d$, we have

$$\nu_n(t,x,\mathrm{d}z) = \sum_{|\beta| \leq M_n} x^\beta \nu_{n,\beta}(t,\mathrm{d}z), \qquad M_n \in \mathbb{N}_0, \tag{3.1}$$

where each $\nu_{n,\beta}(t,\mathrm{d}z)$ satisfies condition (2.1). Moreover, $M_0 = 0$, $\nu_0 \geq 0$ and

$$\int_{|z|\geq 1} \mathrm{e}^{\lambda |z|} \nu_0(t,\mathrm{d}z) < \infty, \qquad t \in [0,T], \tag{3.2}$$

for some positive λ.

Then we say that $(\mathcal{A}_n(t))_{0 \leq n \leq N}$, defined by

$$\begin{aligned}\mathcal{A}_n(t) f(x) &= \sum_{|\beta| \leq M_n} x^\beta \int_{\mathbb{R}^d} \nu_{n,\beta}(t,\mathrm{d}z) \left(\mathrm{e}^{\langle z, \nabla_x\rangle} - 1 - \langle z, \nabla_x \rangle\right) f(x) + \sum_{|\alpha| \leq 2} a_{\alpha,n}(t,x) D_x^\alpha f(x) \\ &\equiv \int_{\mathbb{R}^d} \nu_n(t,x,\mathrm{d}z) \left(\mathrm{e}^{\langle z, \nabla_x\rangle} - 1 - \langle z, \nabla_x \rangle\right) f(x) + \sum_{|\alpha| \leq 2} a_{\alpha,n}(t,x) D_x^\alpha f(x),\end{aligned} \tag{3.3}$$

is an *Nth order polynomial expansion* of $\mathcal{A}(t)$.

Definition 3.1 allows for very general polynomial specifications. The idea is to choose an expansion $(\mathcal{A}_n(t))$ that closely approximates $\mathcal{A}(t)$. The precise sense of this approximation will depend on the application. Below, we present three polynomial expansions. The first two expansion schemes provide an accurate approximation $\mathcal{A}(t)$ in a pointwise local sense, under the assumption of smooth coefficients. The last expansion scheme approximates $\mathcal{A}(t)$ in a global sense and can be applied even in the case of discontinuous coefficients.

Example 3.2 (Taylor polynomial expansion) Assume the coefficients $a_\alpha(t,\cdot) \in C^N(\mathbb{R}^d)$ and that the compensator ν takes the form

$$\nu(t,x,\mathrm{d}z) = h(t,x,z)\bar{\nu}(\mathrm{d}z)$$

where $h(t,\cdot,z) \in C^N(\mathbb{R}^d)$ with $h \geq 0$, and $\bar{\nu}$ is a Lévy measure. Then, for any fixed $\bar{x} \in \mathbb{R}^d$ and $n \leq N$, we define ν_n and $a_{\alpha,n}$ as the nth order term of the Taylor expansions of ν and a_α respectively in the spatial variables x around the point $\bar{x}$. That is, we set

$$\nu_n(t,x,\mathrm{d}z) = \sum_{|\beta|=n} \frac{D_x^\beta h(t,\bar{x},z)}{\beta!}(x-\bar{x})^\beta \bar{\nu}(\mathrm{d}z),$$

$$a_{\alpha,n}(t,x) = \sum_{|\beta|=n} \frac{D_x^\beta a_\alpha(t,\bar{x})}{\beta!}(x-\bar{x})^\beta, \qquad |\alpha| \leq 2,$$

where as usual $\beta! = \beta_1! \cdots \beta_d!$ and $x^\beta = x_1^{\beta_1} \cdots x_d^{\beta_d}$. The expansion proposed in Lorig et al. [14, 17] is the particular case when $\nu \equiv 0$, whereas the expansion proposed in Lorig et al. [15, 16] is a particular case when $d = 1$.

Example 3.3 (Time-dependent Taylor polynomial expansion) Under the assumptions of Example 3.2, fix a trajectory $\bar{x} : \mathbb{R}_+ \to \mathbb{R}^d$. We then define $\nu_n(t,x,\mathrm{d}z)$ and $a_{\alpha,n}(t,x)$ as the nth order term of the Taylor expansions of $\nu(t,x,\mathrm{d}z)$ and $a_\alpha(t,x)$ respectively around $\bar{x}(t)$. More precisely, we set

$$\nu_n(t,x,\mathrm{d}z) = \sum_{|\beta|=n} \frac{D_x^\beta h(t,\bar{x}(t),z)}{\beta!}(x-\bar{x}(t))^\beta \bar{\nu}(\mathrm{d}z),$$

$$a_{\alpha,n}(t,x) = \sum_{|\beta|=n} \frac{D_x^\beta a_\alpha(t,\bar{x}(t))}{\beta!}(x-\bar{x}(t))^\beta, \qquad |\alpha| \leq 2.$$

This expansion for the coefficients allows the expansion point $\bar{x}$ of the Taylor series to evolve in time according to the evolution of the underlying process X_t. For instance, one could choose $\bar{x}(t) = \mathbb{E}[X_t]$. In Lorig et al. [14] this choice results in a highly accurate approximation for option prices and implied volatility in the Heston [8] model.

Example 3.4 (Hermite polynomial expansion) Hermite expansions can be useful when the diffusion coefficients are discontinuous. A remarkable example in financial mathematics is given by the Dupire's local volatility formula for models with jumps (see Friz et al. [7]). In some cases, e.g., the well-known Variance-Gamma model, the fundamental solution (i.e., the transition density of the underlying stochastic model) has singularities. In such cases, it is natural to approximate it in some L^p norm rather than in the pointwise sense. For the Hermite expansion centered at $\bar{x}$, one sets

$$\nu_n(t,x,\mathrm{d}z) = \sum_{|\beta|=n} \langle \mathbf{H}_\beta(\cdot - \bar{x}), \nu(t,\cdot,\mathrm{d}z)\rangle_\Gamma \mathbf{H}_\beta(x-\bar{x}),$$

$$a_{\alpha,n}(t,x) = \sum_{|\beta|=n} \langle \mathbf{H}_\beta(\cdot - \bar{x}), a_\alpha(t,\cdot)\rangle_\Gamma \mathbf{H}_\beta(x-\bar{x}), \qquad |\alpha| \le 2,$$

where the inner product $\langle\cdot,\cdot\rangle_\Gamma$ is an integral over $\mathbb{R}^d$ with a Gaussian weighting centered at $\bar{x}$ and the functions $\mathbf{H}_\beta(x) = H_{\beta_1}(x_1)\cdots H_{\beta_d}(x_d)$ where H_n is the nth one-dimensional Hermite polynomial (properly normalized so that $\langle \mathbf{H}_\alpha, \mathbf{H}_\beta\rangle_\Gamma = \delta_{\alpha,\beta}$ with $\delta_{\alpha,\beta}$ being the Kronecker's delta function).

4 Formal Solution Via Dyson Series

In this section we present a heuristic argument to pass from an expansion of the operator $\mathcal{A}(t)$ in (2.5) to an expansion for u, the solution of problem (2.4). The following argument is not intended to be rigorous. Rather, the computations that follow provide motivation for the price expansion given in Definition 4.1. Throughout this section, we will generally omit x-dependence, except where it is needed for clarity. To begin, we presume that the operator $\mathcal{A}(t)$ can be formally written as

$$\mathcal{A}(t) = \mathcal{A}_0(t) + \mathcal{B}(t), \qquad \mathcal{B}(t) = \sum_{n=1}^{\infty} \mathcal{A}_n(t). \tag{4.1}$$

We insert expansion (4.1) for $\mathcal{A}(t)$ into Cauchy problem (2.4) and find

$$(\partial_t + \mathcal{A}_0(t))u(t) = -\mathcal{B}(t)u(t), \qquad u(T) = \varphi.$$

Note that, by construction, $\mathcal{A}_0(t)$ is the generator of an additive process. Therefore, by Duhamel's principle, we have

$$u(t) = \mathcal{P}_0(t,T)\varphi + \int_t^T \mathrm{d}t_1\, \mathcal{P}_0(t,t_1)\mathcal{B}(t_1)u(t_1), \tag{4.2}$$

where $\mathcal{P}_0(t,T)$ is the semigroup of operators generated by $\mathcal{A}_0(t)$. Inserting expression (4.2) for u into the right-hand side of (4.2) and iterating we obtain

$$
\begin{aligned}
u(t) &= \mathcal{P}_0(t,T)\varphi + \int_t^T \mathrm{d}t_1\, \mathcal{P}_0(t,t_1)\mathcal{B}(t_1)\mathcal{P}_0(t_1,T)\varphi \\
&\quad + \int_t^T \mathrm{d}t_1 \int_{t_1}^T \mathrm{d}t_2\, \mathcal{P}_0(t,t_1)\mathcal{B}(t_1)\mathcal{P}_0(t_1,t_2)\mathcal{B}(t_2)u(t_2) \\
&= \cdots \\
&= \mathcal{P}_0(t,T)\varphi + \sum_{k=1}^{\infty} \int_t^T \mathrm{d}t_1 \int_{t_1}^T \mathrm{d}t_2 \cdots \int_{t_{k-1}}^T \mathrm{d}t_k \\
&\quad \mathcal{P}_0(t,t_1)\mathcal{B}(t_1)\mathcal{P}_0(t_1,t_2)\mathcal{B}(t_2)\cdots\mathcal{P}_0(t_{k-1},t_k)\mathcal{B}(t_k)\mathcal{P}_0(t_k,T)\varphi \qquad (4.3)\\
&= \mathcal{P}_0(t,T)\varphi + \sum_{n=1}^{\infty}\sum_{k=1}^{n} \int_t^T \mathrm{d}t_1 \int_{t_1}^T \mathrm{d}t_2 \cdots \int_{t_{k-1}}^T \mathrm{d}t_k \\
&\quad \sum_{i\in I_{n,k}} \mathcal{P}_0(t,t_1)\mathcal{A}_{i_1}(t_1)\mathcal{P}_0(t_1,t_2)\mathcal{A}_{i_2}(t_2)\cdots\mathcal{P}_0(t_{k-1},t_k)\mathcal{A}_{i_k}(t_k)\mathcal{P}_0(t_k,T)\varphi,
\end{aligned}
\tag{4.4}
$$

$$
I_{n,k} = \{i = (i_1,i_2,\ldots,i_k) \in \mathbb{N}^k \mid i_1+i_2+\cdots+i_k = n\}. \tag{4.5}
$$

The second-to-last equality (4.3) is known as the *Dyson series expansion* of u (see, for instance, Sect. 5.7 of Sakurai and Tuan [23] or Chap. IX.2.6 of Kato [10]). To obtain (4.4) from (4.3) we have used (4.1) to replace $\mathcal{B}(t)$ by the infinite sum $\sum_{n=1}^{\infty}\mathcal{A}_n(t)$, and we have partitioned on the sum of the subscripts of the $(\mathcal{A}_{i_k})$. Expansion (4.4) motivates the following definition.

Definition 4.1 For a fixed Nth order polynomial expansion $(\mathcal{A}_n(t))_{0\le n\le N}$ satisfying Definition 3.1, we define $\bar{u}_N$, the *Nth order price approximation* of u, as

$$
\bar{u}_N := \sum_{n=0}^{N} u_n, \tag{4.6}
$$

where

$$
\begin{aligned}
u_0(t) &:= \mathcal{P}_0(t,T)\varphi, \\
u_n(t) &:= \sum_{k=1}^{n} \int_t^T \mathrm{d}t_1 \int_{t_1}^T \mathrm{d}t_2 \cdots \int_{t_{k-1}}^T \mathrm{d}t_k \\
&\quad \sum_{i\in I_{n,k}} \mathcal{P}_0(t,t_1)\mathcal{A}_{i_1}(t_1)\mathcal{P}_0(t_1,t_2)\mathcal{A}_{i_2}(t_2)\cdots\mathcal{P}_0(t_{k-1},t_k)\mathcal{A}_{i_k}(t_k)\mathcal{P}_0(t_k,T)\varphi, \qquad n\ge 1.
\end{aligned}
\tag{4.7}
$$

Here, $\mathcal{P}_0(t,T)$ is the semigroup generated by $\mathcal{A}_0(t)$ and $I_{n,k}$ is as given in (4.5).

In Sects. 4.1 and 4.2 we will provide explicit expressions for u_0 and $(u_n)_{n\geq 1}$ respectively.

4.1 Expression for u_0

In what follows, it will be helpful to recall the definition of the Fourier and inverse Fourier transforms. For any function φ in the Schwartz class, we define

$$\text{Fourier transform:} \qquad \mathcal{F}[\varphi](\xi) = \hat{\varphi}(\xi) = \int_{\mathbb{R}^d} \mathrm{d}x\, \varphi(x) \mathrm{e}^{\mathrm{i}\langle \xi, x\rangle},$$

$$\text{Inverse transform:} \qquad \mathcal{F}^{-1}[\hat{\varphi}](x) = \varphi(x) = \frac{1}{(2\pi)^d} \int_{\mathbb{R}^d} \mathrm{d}\xi\, \hat{\varphi}(\xi) \mathrm{e}^{-\mathrm{i}\langle \xi, x\rangle}.$$

Recall that by construction $M_0 = 0$ (cf. Definition 3.1) and therefore the operator $\mathcal{A}_0(t)$ has time-dependent coefficients which are independent of x. Then the action of the semigroup of operators $\mathcal{P}_0(t,T)$ of $\mathcal{A}_0(t)$ is well-known:

$$u_0(t) := \mathcal{P}_0(t,T)\varphi = \frac{1}{(2\pi)^d} \int_{\mathbb{R}^d} \hat{P}_0(t,x,T,\xi) \hat{\varphi}(-\xi)\, \mathrm{d}\xi \tag{4.8}$$

where

$$\hat{P}_0(t,x,T,\xi) := \mathrm{e}^{\mathrm{i}\langle \xi, x\rangle + \Phi_0(t,T,\xi)} \tag{4.9}$$

with

$$\Phi_0(t,T,\xi) = \sum_{|\alpha|\leq 2} (\mathrm{i}\xi)^\alpha \int_t^T \mathrm{d}s\, a_{\alpha,0}(s) + \Psi_0(t,T,\xi), \tag{4.10}$$

and

$$\Psi_0(t,T,\xi) = \int_t^T \int_{\mathbb{R}^d} \left(\mathrm{e}^{\mathrm{i}\langle \xi, z\rangle} - 1 - \mathrm{i}\langle \xi, z\rangle \right) \nu_0(s, \mathrm{d}z) \mathrm{d}s.$$

Remark 4.2 We introduce $\hat{P}$ and e_ξ, the *characteristic function* and *oscillating exponential*, respectively

$$\hat{P}(t,x,T,\xi) := \mathbb{E}\left[\mathrm{e}^{\int_t^T a_{0,0}(s,X_s)\mathrm{d}s} \mathrm{e}^{\mathrm{i}\langle \xi, X_T\rangle} | X_t = x \right], \quad \mathrm{e}_\xi(x) = \mathrm{e}^{\mathrm{i}\langle \xi, x\rangle}, \tag{4.11}$$

where $a_{0,0}$ is short-hand for $a_{(0,0,\ldots,0),0}$. From (2.3) we observe that $\hat{P}(t,x,T,\xi)$ is obtained as the special case $\varphi = \mathrm{e}_\xi$. We note that $\hat{P}_0(t,x,T,\xi)$ in (4.9) represents the

0th order approximation of $\hat{P}(t, x, T, \xi)$. More generally, we denote by $\hat{P}_n(t, x, T, \xi)$ the nth order approximation of $\hat{P}(t, x, T, \xi)$, obtained by setting $\varphi = \mathrm{e}_\xi$ in (4.7).

4.2 Expression for u_n

Remarkably, as the following proposition shows, every $u_n(t)$ can be expressed as a pseudo-differential operator $\mathcal{L}_n(t, T)$ acting on $u_0(t)$.

Proposition 4.3 *Assume that φ belongs to the Schwartz class, and that Φ_0 in* (4.10) *is a smooth function of the variable ξ. Then the function u_n defined in* (4.7) *is given explicitly by*

$$u_n(t) = \mathcal{L}_n(t, T) u_0(t), \tag{4.12}$$

where u_0 is given by (4.8) *and*

$$\mathcal{L}_n(t, T) = \sum_{k=1}^{n} \int_t^T \mathrm{d}t_1 \int_{t_1}^T \mathrm{d}t_2 \cdots \int_{t_{k-1}}^T \mathrm{d}t_k \sum_{i \in I_{n,k}} \mathcal{G}_{i_1}(t, t_1) \mathcal{G}_{i_2}(t, t_2) \cdots \mathcal{G}_{i_k}(t, t_k), \tag{4.13}$$

with $I_{n,k}$ as defined in (4.5) *and*

$$\begin{aligned} \mathcal{G}_j(t, t_k) &:= \mathcal{A}_j(t_k, \mathcal{M}(t, t_k)) \\ &= \int_{\mathbb{R}^d} \nu_j(t_k, \mathcal{M}(t, t_k), \mathrm{d}z) \left(\mathrm{e}^{\langle z, \nabla_x \rangle} - 1 - \langle z, \nabla_x \rangle \right) + \sum_{|\alpha| \le 2} a_{\alpha, j}(t_k, \mathcal{M}(t, t_k)) D_x^\alpha, \end{aligned} \tag{4.14}$$

$$\mathcal{M}(t, t_k) := x + \int_{\mathbb{R}^d} \int_t^{t_k} z \left(\mathrm{e}^{\langle z, \nabla_x \rangle} - 1 \right) \nu_0(s, \mathrm{d}z) \mathrm{d}s + \int_t^{t_k} m(s) \mathrm{d}s + \int_t^{t_k} C(s) \nabla_x \mathrm{d}s, \tag{4.15}$$

$$m(s) = \left(a_{(1,0,\dots,0),0}(s) \quad a_{(0,1,\dots,0),0}(s) \quad \dots \quad a_{(0,0,\dots,1),0}(s) \right),$$

$$C(s) = \begin{pmatrix} 2a_{(2,0,\dots,0),0}(s) & a_{(1,1,\dots,0),0}(s) & \dots & a_{(0,0,\dots,1),0}(s) \\ a_{(1,1,\dots,0),0}(s) & 2a_{(0,2,\dots,0),0}(s) & \dots & a_{(0,1,\dots,1),0}(s) \\ \vdots & \vdots & \ddots & \vdots \\ a_{(1,0,\dots,1),0}(s) & a_{(0,1,\dots,1),0}(s) & \dots & 2a_{(0,0,\dots,2),0}(s) \end{pmatrix}.$$

Moreover, the components of $\mathcal{M}(t, t_k)$ commute. Therefore the operators $(\mathcal{G}_j(t, t_k))$, which are polynomials in $\mathcal{M}(t, t_k)$ by construction, are well defined.

Proof The proof consists in showing that the operator $\mathcal{G}_j(t, t_k)$ in (4.14) satisfies

$$\mathcal{P}_0(t, t_k) \mathcal{A}_j(t_k) = \mathcal{G}_j(t, t_k) \mathcal{P}_0(t, t_k). \tag{4.16}$$

Assuming (4.16) holds, we can use the fact that $\mathcal{P}_0(t_k, t_{k+1})$ is a semigroup

$$\mathcal{P}_0(t,T) = \mathcal{P}_0(t,t_1)\mathcal{P}_0(t_1,t_2)\cdots\mathcal{P}_0(t_{k-1},t_k)\mathcal{P}_0(t_k,T), \quad t \le t_1 \le \cdots \le t_k \le T,$$

and we can re-write (4.7) as

$$u_n(t) = \sum_{k=1}^{n} \int_t^T \mathrm{d}t_1 \int_{t_1}^T \mathrm{d}t_2 \cdots \int_{t_{k-1}}^T \mathrm{d}t_k \sum_{i \in I_{n,k}} \mathcal{G}_{i_1}(t,t_1)\mathcal{G}_{i_2}(t,t_2)\cdots\mathcal{G}_{i_k}(t,t_k)\mathcal{P}_0(t,T)\varphi,$$

from which (4.12) and (4.13) follows directly. Thus, we only need to show that $\mathcal{G}_j(t,t_k)$ satisfies (4.16). It is sufficient to investigate how the operator $\mathcal{P}_0(t,t_k)\mathcal{A}_j(t_k)$ acts on the oscillating exponential in (4.11). First, we note that

$$\mathcal{P}_0(t,t_k)\mathrm{e}_\xi(x) = \mathrm{e}^{\Phi_0(t,t_k,\xi)}\mathrm{e}_\xi(x), \tag{4.17}$$

where $\Phi_0(t,t_k,\xi)$, as given in (4.10), is a smooth function by condition (3.2). Next, we observe that the operator $\mathcal{M}(t,t_k)$ in (4.15) can be written

$$\mathcal{M}(t,t_k) = M(t,t_k,-\mathrm{i}\nabla_x), \quad M(t,t_k,\xi) = -\mathrm{i}\nabla_\xi\left(\Phi_0(t,t_k,\xi) + \mathrm{i}\langle\xi,x\rangle\right). \tag{4.18}$$

Denote by $\mathcal{M}_j$ and M_j the jth component of $\mathcal{M}$ and M respectively. Then, using (4.18) we have

$$\begin{aligned}(-\mathrm{i}\partial_{\xi_i})(-\mathrm{i}\partial_{\xi_j})\mathrm{e}^{\Phi_0(t,t_k,\xi)}\mathrm{e}_\xi(x) &= (-\mathrm{i}\partial_{\xi_i})M_j(t,t_k,\xi)\mathrm{e}^{\Phi_0(t,t_k,\xi)}\mathrm{e}_\xi(x)\\ &= \mathcal{M}_j(t,t_k)(-\mathrm{i}\partial_{\xi_i})\mathrm{e}^{\Phi_0(t,t_k,\xi)}\mathrm{e}_\xi(x)\\ &= \mathcal{M}_j(t,t_k)M_i(t,t_k,\xi)\mathrm{e}^{\Phi_0(t,t_k,\xi)}\mathrm{e}_\xi(x)\\ &= \mathcal{M}_j(t,t_k)\mathcal{M}_i(t,t_k)\mathrm{e}^{\Phi_0(t,t_k,\xi)}\mathrm{e}_\xi(x).\end{aligned} \tag{4.19}$$

More generally for any multi-index β we have

$$(-\mathrm{i}\nabla_\xi)^\beta\mathrm{e}^{\Phi_0(t,t_k,\xi)}\mathrm{e}_\xi(x) = (\mathcal{M}(t,t_k))^\beta\mathrm{e}^{\Phi_0(t,t_k,\xi)}\mathrm{e}_\xi(x). \tag{4.20}$$

From (4.19) we deduce that operators $\mathcal{M}_i$ and $\mathcal{M}_j$ commute when applied to $\mathrm{e}^{\Phi_0(t,t_k,\xi)}\mathrm{e}_\xi(x)$, because so do ∂_{ξ_i} and ∂_{ξ_j}. Consequently, $\mathcal{M}_i$ and $\mathcal{M}_j$ also commute when applied to $\mathrm{e}_\xi(x)$ or any function that admits a representation as a Fourier transform. To see this observe that

$$\mathcal{M}_j(t,t_k)\mathcal{M}_i(t,t_k)\mathrm{e}^{\Phi_0(t,t_k,\xi)}\mathrm{e}_\xi(x) = \mathcal{M}_i(t,t_k)\mathcal{M}_j(t,t_k)\mathrm{e}^{\Phi_0(t,t_k,\xi)}\mathrm{e}_\xi(x).$$

Therefore, since $\mathcal{M}_j(t,t_k)$ acts on x and not ξ we have

$$\mathcal{M}_j(t,t_k)\mathcal{M}_i(t,t_k)\mathrm{e}_\xi(x) = \mathcal{M}_i(t,t_k)\mathcal{M}_j(t,t_k)\mathrm{e}_\xi(x).$$

Finally, we compute

$$\begin{aligned}
\mathcal{P}_0(t,t_k)\mathcal{A}_j(t_k)\mathrm{e}_\xi(x) &= \mathcal{P}_0(t,t_k)\int_{\mathbb{R}^d}\nu_j(t_k,x,\mathrm{d}z)(\mathrm{e}^{\langle z,\nabla_x\rangle}-1-\langle z,\nabla_x\rangle)\mathrm{e}_\xi(x)\\
&\quad+\sum_{|\alpha|\le 2}\mathcal{P}_0(t,t_k)a_{\alpha,j}(t_k,x)D_x^\alpha\mathrm{e}_\xi(x) && \text{(by (3.3))}\\
&= \mathcal{P}_0(t,t_k)\int_{\mathbb{R}^d}(\mathrm{e}^{\mathrm{i}\langle z,\xi\rangle}-1-\mathrm{i}\langle z,\xi\rangle)\nu_j(t_k,x,\mathrm{d}z)\mathrm{e}_\xi(x)\\
&\quad+\sum_{|\alpha|\le 2}(\mathrm{i}\xi)^\alpha\mathcal{P}_0(t,t_k)a_{\alpha,j}(t_k,x)\mathrm{e}_\xi(x)\\
&= \int_{\mathbb{R}^d}(\mathrm{e}^{\mathrm{i}\langle z,\xi\rangle}-1-\mathrm{i}\langle z,\xi\rangle)\nu_j(t_k,-\mathrm{i}\nabla_\xi,\mathrm{d}z)\mathcal{P}_0(t,t_k)\mathrm{e}_\xi(x)\\
&\quad+\sum_{|\alpha|\le 2}(\mathrm{i}\xi)^\alpha a_{\alpha,j}(t_k,-\mathrm{i}\nabla_\xi)\mathcal{P}_0(t,t_k)\mathrm{e}_\xi(x)\\
&= \int_{\mathbb{R}^d}(\mathrm{e}^{\mathrm{i}\langle z,\xi\rangle}-1-\mathrm{i}\langle z,\xi\rangle)\nu_j(t_k,-\mathrm{i}\nabla_\xi,\mathrm{d}z)\mathrm{e}^{\Phi_0(t,t_k,\xi)}\mathrm{e}_\xi(x)\\
&\quad+\sum_{|\alpha|\le 2}(\mathrm{i}\xi)^\alpha a_{\alpha,j}(t_k,-\mathrm{i}\nabla_\xi)\mathrm{e}^{\Phi_0(t,t_k,\xi)}\mathrm{e}_\xi(x) && \text{(by (4.17))}\\
&= \int_{\mathbb{R}^d}(\mathrm{e}^{\mathrm{i}\langle z,\xi\rangle}-1-\mathrm{i}\langle z,\xi\rangle)\nu_j(t_k,\mathcal{M}(t,t_k),\mathrm{d}z)\mathrm{e}^{\Phi_0(t,t_k,\xi)}\mathrm{e}_\xi(x)\\
&\quad+\sum_{|\alpha|\le 2}(\mathrm{i}\xi)^\alpha a_{\alpha,j}(t_k,\mathcal{M}(t,t_k))\mathrm{e}^{\Phi_0(t,t_k,\xi)}\mathrm{e}_\xi(x) && \text{(by (4.20))}\\
&= \int_{\mathbb{R}^d}\nu_j(t_k,\mathcal{M}(t,t_k),\mathrm{d}z)(\mathrm{e}^{\langle z,\nabla_x\rangle}-1-\langle z,\nabla_x\rangle)\mathrm{e}^{\Phi_0(t,t_k,\xi)}\mathrm{e}_\xi(x)\\
&\quad+\sum_{|\alpha|\le 2}a_{\alpha,j}(t_k,\mathcal{M}(t,t_k))D_x^\alpha\mathrm{e}^{\Phi_0(t,t_k,\xi)}\mathrm{e}_\xi(x)\\
&= \int_{\mathbb{R}^d}\nu_j(t_k,\mathcal{M}(t,t_k),\mathrm{d}z)(\mathrm{e}^{\langle z,\nabla_x\rangle}-1-\langle z,\nabla_x\rangle)\mathcal{P}_0(t,t_k)\mathrm{e}_\xi(x)\\
&\quad+\sum_{|\alpha|\le 2}a_{\alpha,j}(t_k,\mathcal{M}(t,t_k))D_x^\alpha\mathcal{P}_0(t,t_k)\mathrm{e}_\xi(x) && \text{(by (4.17))}\\
&= \mathcal{G}_j(t,t_k)\mathcal{P}_0(t,t_k)\mathrm{e}_\xi(x), && \text{(by (4.14))}
\end{aligned}$$

which concludes the proof. □

Remark 4.4 Error bounds for the Taylor approximation $\bar{u}_N$ in the scalar case $d=1$ can be found in Lorig et al. [15, 16].

4.3 Fourier Representation for u_n

Using (4.8), (4.9) and (4.12) we have

$$u_n(t,x) = \mathcal{L}_n(t,T)u_0(t,x) = \frac{1}{(2\pi)^d}\int_{\mathbb{R}^d} \mathrm{e}^{\Phi_0(t,T,\xi)}\left(\mathcal{L}_n(t,T)\mathrm{e}^{\mathrm{i}\langle\xi,x\rangle}\right)\hat{\varphi}(-\xi)\mathrm{d}\xi.$$

The term in parenthesis $\mathcal{L}_n(t,T)\mathrm{e}^{\mathrm{i}\langle\xi,x\rangle}$ can be computed explicitly. However, $\mathcal{L}_n(t,T)$ is, in general, an *integro-differential* operator (when X is a diffusion $\mathcal{L}_n(t,T)$ is simply a differential operator). Thus, for models with jumps, computing $\mathcal{L}_n(t,T)\mathrm{e}^{\mathrm{i}\langle\xi,x\rangle}$ is a challenge. Remarkably, we will show that there exists a first order *differential* operator $\hat{\mathcal{L}}_n^{\xi}(t,T)$ such that

$$\mathcal{L}_n^x(t,T)\mathrm{e}^{\mathrm{i}\langle\xi,x\rangle} = \hat{\mathcal{L}}_n^{\xi}(t,T)\mathrm{e}^{\mathrm{i}\langle\xi,x\rangle}, \tag{4.21}$$

where, for clarity, we have explicitly indicated using superscripts that $\mathcal{L}_n^x(t,T)$ acts on x and $\hat{\mathcal{L}}_n^{\xi}(t,T)$ acts on ξ. With a slight abuse of terminology, we call $\hat{\mathcal{L}}_n^{\xi}$ the *symbol*[1] of the operator $\mathcal{L}_n^x(t,T)$ in (4.13).

Let us consider the operator $\mathcal{M}^x(t,t_k) \equiv \mathcal{M}(t,t_k)$ in (4.15) and denote by $\mathcal{M}_i^x(t,t_k)$ its ith component. The symbol $\widehat{\mathcal{M}}_i^{\xi}(t,t_k)$ of $\mathcal{M}_i^x(t,t_k)$ is defined analogously to (4.21), that is

$$\mathcal{M}_i^x(t,t_k)\mathrm{e}^{\mathrm{i}\langle\xi,x\rangle} = \widehat{\mathcal{M}}_i^{\xi}(t,t_k)\mathrm{e}^{\mathrm{i}\langle\xi,x\rangle}.$$

Explicitly, we have

$$\widehat{\mathcal{M}}_i^{\xi}(t,t_k) = F_i(\xi,t,t_k) - \mathrm{i}\partial_{\xi_i}, \qquad i = 1,\dots,d,$$

where the function F is defined as

$$F_i(\xi,t,t_k) = \int_{\mathbb{R}^d}\int_t^{t_k} z_i\left(\mathrm{e}^{\mathrm{i}\langle z,\xi\rangle} - 1\right)\nu_0(s,\mathrm{d}z)\mathrm{d}s + \int_t^{t_k} m_i(s)\mathrm{d}s + \mathrm{i}\int_t^{t_k}(C(s)\xi)_i\,\mathrm{d}s.$$

We note that, while $\mathcal{M}^x$ is a first order *integro-differential* operator, its symbol $\widehat{\mathcal{M}}^{\xi}$ is a first order *differential* operator. For this reason, it is more convenient to use the symbol $\widehat{\mathcal{M}}^{\xi}$ instead of the operator $\mathcal{M}^x$. Note also that

$$\begin{aligned}
\mathcal{M}_i^x(t,t_k)\mathcal{M}_j^x(t,t_k)\mathrm{e}^{\mathrm{i}\langle\xi,x\rangle} &= \mathcal{M}_i^x(t,t_k)\widehat{\mathcal{M}}_j^{\xi}(t,t_k)\mathrm{e}^{\mathrm{i}\langle\xi,x\rangle}\\
&= \widehat{\mathcal{M}}_j^{\xi}(t,t_k)\mathcal{M}_i^x(t,t_k)\mathrm{e}^{\mathrm{i}\langle\xi,x\rangle}\\
&= \widehat{\mathcal{M}}_j^{\xi}(t,t_k)\widehat{\mathcal{M}}_i^{\xi}(t,t_k)\mathrm{e}^{\mathrm{i}\langle\xi,x\rangle}.
\end{aligned}$$

[1]The operator $\hat{\mathcal{L}}_n^{\xi}$ is not a function as in the classical theory of pseudo-differential calculus. However $\mathrm{e}^{-\mathrm{i}\langle\xi,x\rangle}\hat{\mathcal{L}}_n^{\xi}\mathrm{e}^{\mathrm{i}\langle\xi,x\rangle}$ is the symbol of $\mathcal{L}_n^x(t,T)$.

Since $\mathcal{M}_i^x$ and $\mathcal{M}_j^x$ commute when applied to a function that admits a Fourier representation, then $\widehat{\mathcal{M}}_j^\xi$ and $\widehat{\mathcal{M}}_i^\xi$ also commute when applied to such functions. In particular, the operator $\left(\widehat{\mathcal{M}}^\xi(t, t_k)\right)^\beta$, for $\beta \in \mathbb{N}_0^d$, is well defined and we have

$$\left(\widehat{\mathcal{M}}^\xi(t, t_k)\right)^\beta \mathrm{e}^{\mathrm{i}\langle \xi, x\rangle} = (\mathcal{M}(t, t_k))^\beta \, \mathrm{e}^{\mathrm{i}\langle \xi, x\rangle}. \tag{4.22}$$

From identity (4.22) we obtain directly the expression of the symbol of $\mathcal{G}_j$ in (4.14). Indeed, recalling the expression (3.1) of ν_j we have

$$\begin{aligned}\hat{\mathcal{G}}_j^\xi(t, t_k) = \sum_{|\beta| \le M_j} \int_{\mathbb{R}^d} \left(\mathrm{e}^{\mathrm{i}\langle z, \xi\rangle} - 1 - \mathrm{i}\langle z, \xi\rangle\right) \nu_{j,\beta}(t_k, \mathrm{d}z) \left(\widehat{\mathcal{M}}^\xi(t, t_k)\right)^\beta \\ + \sum_{|\alpha| \le 2} (\mathrm{i}\xi)^\alpha \, a_{\alpha,j}\left(t_k, \widehat{\mathcal{M}}^\xi(t, t_k)\right).\end{aligned}$$

Thus we have proved the following lemma.

Lemma 4.5 *We have*

$$\hat{\mathcal{L}}_n^\xi(t, T) = \sum_{k=1}^n \int_t^T \mathrm{d}t_1 \int_{t_1}^T \mathrm{d}t_2 \cdots \int_{t_{k-1}}^T \mathrm{d}t_k \sum_{i \in I_{n,k}} \hat{\mathcal{G}}_{i_1}^\xi(t, t_1) \hat{\mathcal{G}}_{i_2}^\xi(t, t_2) \cdots \hat{\mathcal{G}}_{i_k}^\xi(t, t_k), \tag{4.23}$$

where $I_{n,k}$ as defined in (4.5).

The following theorem extends the Fourier pricing formula (4.8) to higher order approximations.

Theorem 4.6 *Under the assumptions of Proposition 4.3, for any $n \ge 1$ we have*

$$u_n(t) = \frac{1}{(2\pi)^d} \int_{\mathbb{R}^d} \hat{P}_n(t, x, T, \xi) \hat{\varphi}(-\xi) \, \mathrm{d}\xi, \tag{4.24}$$

where $\hat{P}_n(t, x, T, \xi)$ is the nth order term of the approximation of the characteristic function of X (cf. Remark 4.2). Explicitly, we have

$$\hat{P}_n(t, x, T, \xi) := \hat{P}_0(t, x, T, \xi) \left(\mathrm{e}^{-\mathrm{i}\langle \xi, x\rangle} \hat{\mathcal{L}}_n^\xi(t, T) \mathrm{e}^{\mathrm{i}\langle \xi, x\rangle}\right)$$

where $\hat{P}_0(t, x, T, \xi)$ is the 0th order approximation in (4.9) *and $\hat{\mathcal{L}}_n^\xi(t, T)$ is the differential operator defined in* (4.23).

Proof We first note that, since the approximating operator $\mathcal{L}_n^x$ acts in the x variables, then it commutes[2] with the Fourier pricing operator (4.8). Thus, by (4.12) combined with (4.8), we get

$$u_n(t) = \mathcal{L}_n^x(t,T)u_0(t) = \frac{1}{(2\pi)^d}\int_{\mathbb{R}^d} \mathcal{L}_n^x(t,T)\mathrm{e}^{\mathrm{i}\langle\xi,x\rangle+\Phi_0(t,T,\xi)}\hat{\varphi}(-\xi)\,\mathrm{d}\xi$$

$$= \frac{1}{(2\pi)^d}\int_{\mathbb{R}^d} \hat{P}_0(t,x,T,\xi)\left(\mathrm{e}^{-\mathrm{i}\langle\xi,x\rangle}\hat{\mathcal{L}}_n^{\xi}(t,T)\mathrm{e}^{\mathrm{i}\langle\xi,x\rangle}\right)\hat{\varphi}(-\xi)\,\mathrm{d}\xi,$$

and the thesis follows from (4.21). □

Remark 4.7 Computing the term in parenthesis above $\left(\mathrm{e}^{-\mathrm{i}\langle\xi,x\rangle}\hat{\mathcal{L}}_n^{\xi}(t,T)\mathrm{e}^{\mathrm{i}\langle\xi,x\rangle}\right)$ is a straightforward exercise since the symbol $\hat{\mathcal{L}}_n^{\xi}(t,T)$, given in (4.23), is a differential operator.

Remark 4.8 In case of non-integrable payoffs (e.g. Call and Put options), the Fourier representation (4.24) can be easily extended by considering the Fourier transform on the imaginary line $\xi = \xi_r + \mathrm{i}\xi_{\mathrm{i}}$. For instance, since the Call option payoff $\varphi(x) = \left(\mathrm{e}^x - \mathrm{e}^k\right)^+$ is not integrable, its Fourier transform $\hat{\varphi}(-\xi)$ must be computed in a generalized sense by fixing an imaginary component of the Fourier variable $\xi_{\mathrm{i}} < -1$.

Remark 4.9 Observe that the Nth order approximation (4.6)–(4.24) requires only a single Fourier inversion

$$\bar{u}_N(t,x) = \sum_{n=0}^{N} u_n(t,x) = \frac{1}{(2\pi)^d}\sum_{n=0}^{N}\int_{\mathbb{R}^d} \hat{P}_n(t,x,T,\xi)\hat{\varphi}(-\xi)\,\mathrm{d}\xi.$$

Moreover, when evaluating the inverse transform, the number of dimensions over which one must integrate numerically is equal to the number of components of x that appear in the option payoff φ. This is due to the fact that the Fourier transform of a constant is a Dirac delta function. In particular, let $\varphi(x) \equiv \bar{\varphi}(\bar{x})$ with $\bar{x} = (x_1,\ldots,x_{d'})$, for some $d' < d$. Then we have $\hat{\varphi}(\xi) = (2\pi)^{d-d'}\hat{\bar{\varphi}}\left(\bar{\xi}\right)\delta_0(\xi_{d'+1})\cdots\delta_0(\xi_d)$ with $\bar{\xi} = (\xi_1,\ldots,\xi_{d'})$, and thus

$$\bar{u}_N(t,x) = \frac{1}{(2\pi)^{d'}}\sum_{n=0}^{N}\int_{\mathbb{R}^{d'}} \hat{P}_n\left(t,x,T,\left(\bar{\xi},0\right)\right)\hat{\bar{\varphi}}\left(-\bar{\xi}\right)\,\mathrm{d}\bar{\xi}.$$

---

[2]This was one of the main points of the *adjoint expansion method* proposed by Pagliarani et al. [21].

5 Example: Heston Model with Stochastic Jump-Intensity

Consider the following model for an asset $S = \mathrm{e}^X$, written under the pricing measure $\mathbb{Q}$ assuming zero interest rates

$$\mathrm{d}X_t = \left(-\frac{1}{2} - \int_{\mathbb{R}} \nu(\mathrm{d}\zeta)(\mathrm{e}^{\zeta} - 1 - \zeta)\right) Z_t \mathrm{d}t + \sqrt{Z_t}\mathrm{d}W_t + \int_{\mathbb{R}} \zeta \mathrm{d}\widetilde{N}(t, Z_t, \mathrm{d}t, \mathrm{d}\zeta),$$
$$\mathrm{d}Z_t = \kappa(\theta - Z_t)\mathrm{d}t + \delta\sqrt{Z_t}\mathrm{d}B_t, \qquad \mathrm{d}\langle W, B\rangle_t = \rho \mathrm{d}t.$$

Note that, just as in the Heston model, the instantaneous volatility of X is given by $\sqrt{Z_t}$, where Z is a CIR process. Likewise, the instantaneous arrival rate of jumps of size $\mathrm{d}\zeta$ is given by $Z_t \nu(\mathrm{d}\zeta)$, where ν is a Lévy measure satisfying all of the usual integrability conditions. The generator $\mathcal{A}$ of the process (X, Z) is given by

$$\begin{aligned}\mathcal{A} = z &\left(\mu \partial_x + \frac{1}{2}\partial_x^2 + \int_{\mathbb{R}} \nu(\mathrm{d}\zeta)(\mathrm{e}^{\zeta \partial_x} - 1 - \zeta \partial_x)\right) \\ &+ \kappa(\theta - z)\partial_z + \frac{1}{2}\delta^2 z \partial_z^2 + \rho\delta z \partial_x \partial_y, \\ \mu = &-\frac{1}{2} - \int_{\mathbb{R}} \nu(\mathrm{d}\zeta)(\mathrm{e}^{\zeta} - 1 - \zeta).\end{aligned}$$

The characteristic function $\hat{P}(t, x, z, T, \xi) := \mathbb{E}[\mathrm{e}^{\mathrm{i}\xi X_T} | X_t = x, Z_t = z]$ is obtained in Carr and Wu [3] by expressing the process X as a time-changed Lévy process. One can also obtain the characteristic function by solving for the Fourier transform of the fundamental solution corresponding to the operator $(\partial_t + \mathcal{A})$. We have

$$\begin{aligned}\hat{P}(t, x, z, T, \xi) &= \mathrm{e}^{\mathrm{i}\xi x + C(T-t,\xi) + z D(T-t,\xi)}, \\ C(\tau, \xi) &= \frac{\kappa\theta}{\delta^2}\left((\kappa - \rho\delta\mathrm{i}\xi + d(\xi))\tau - 2\log\left[\frac{1 - f(\xi)\mathrm{e}^{d(\xi)\tau}}{1 - f(\xi)}\right]\right), \\ D(\tau, \xi) &= \frac{\kappa - \rho\delta\mathrm{i}\xi + d(\xi)}{\delta^2} \frac{1 - \mathrm{e}^{d(\xi)\tau}}{1 - f(\xi)\mathrm{e}^{d(\xi)\tau}}, \\ f(\xi) &= \frac{\kappa - \rho\delta\mathrm{i}\xi + d(\xi)}{\kappa - \rho\delta\mathrm{i}\xi - d(\xi)}, \\ d(\xi) &= \sqrt{-\delta^2 2\psi(\xi) + (\kappa - \rho\mathrm{i}\xi\delta)^2}, \\ \psi(\xi) &= \mathrm{i}\mu\xi - \tfrac{1}{2}\xi^2 + \int_{\mathbb{R}} \nu(\mathrm{d}\zeta)(\mathrm{e}^{\mathrm{i}\xi\zeta} - 1 - \mathrm{i}\xi\zeta).\end{aligned}$$

With an explicit expression for $\hat{P}(t, x, z, T, \xi)$ available, the price of a European call option can be computed using standard Fourier methods

$$u(t,x,z) = \frac{1}{2\pi}\int_{\mathbb{R}} \mathrm{d}\xi_r \, \hat{P}(t,x,z,T,\xi)\hat{\varphi}(-\xi), \qquad \hat{\varphi}(\xi) = \frac{-\mathrm{e}^{k-\mathrm{i}k\xi}}{\mathrm{i}\xi+\xi^2},$$
$$\xi = \xi_r + \mathrm{i}\xi_i, \quad \xi_i < -1. \tag{5.1}$$

Note that, since the call option payoff $\varphi(x) = (\mathrm{e}^x - \mathrm{e}^k)^+$ is not in $L^1(\mathbb{R})$, its Fourier transform $\hat{\varphi}(\xi)$ must be computed in a generalized sense by fixing an imaginary component of the Fourier variable $\xi_i < -1$.

Also of interest are sensitivities of option prices or *Greeks*. In particular, consider the Δ and the Γ, which are defined as

$$\Delta(t,x,z) := \partial_s u(t, x(s), z) = \mathrm{e}^{-x}\partial_x u(t,x,z), \tag{5.2}$$
$$\Gamma(t,x,z) := \partial_s^2 u(t, x(s), z) = \mathrm{e}^{-2x}(\partial_x^2 - \partial_x)u(t,x,z), \tag{5.3}$$

where we have used $x(s) = \log s$. When computing terms of the form $\partial_x^m u(t,x,z)$, observe that the differential operator ∂_x^m acts only on the characteristic function $\hat{P}$ appearing in (5.1) and not on the Fourier transform $\hat{\varphi}$ of the payoff φ. Likewise, when using Theorem 4.6 to compute $\partial_x^m \bar{u}_n(t,x,z) = \sum_{i=0}^n \partial_x^m u_i(t,x,z)$ the differential operator ∂_x^m acts only on $\hat{P}_i$ in (4.24).

Now, we specialize to the case where jumps are normally distributed

$$\nu(\mathrm{d}\zeta) = \frac{\lambda}{\sqrt{2\pi s^2}} \exp\left(\frac{-(\zeta-m)^2}{2s^2}\right).$$

In Fig. 1 we plot the implied volatility σ corresponding to the exact price u as well as the implied volatility $\bar{\sigma}_2$ corresponding to our second order approximation $\bar{u}_2$. To compute σ we first compute option prices using (5.1); we then invert the Black-Scholes equation numerically in order to obtain the implied volatility σ. To compute our second order approximation of implied volatility $\bar{\sigma}_2$ we first compute our second order approximation for prices $\bar{u}_2$ using Theorem 4.6; we then invert the Black-Scholes equation numerically in order to obtain $\bar{\sigma}_2$. Values from Fig. 1 can be found in Table 1. In Fig. 2 we plot the exact Δ as well as our second order approximation $\bar{\Delta}_2$. In Fig. 3 we plot the exact Γ as well as our second order approximation $\bar{\Gamma}_2$. Values from Figs. 2 and 3 are given in Tables 2 and 3 respectively. Exact Greeks are computed by combining (5.1)–(5.3). Approximate Greeks are computed by combining Theorem 4.6 and Eqs. (5.2) and (5.3).

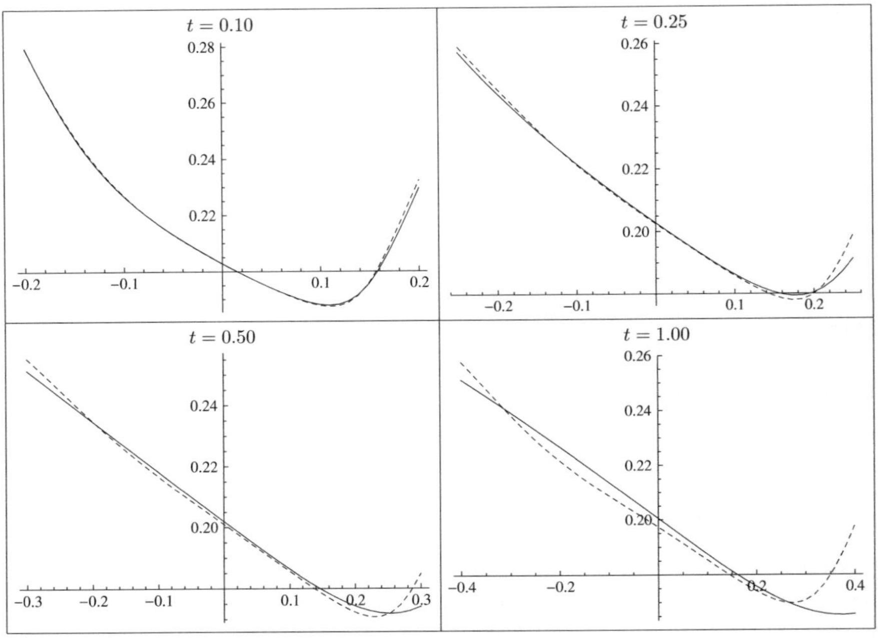

$$\nu(\mathrm{d}\zeta) = \frac{\lambda}{\sqrt{2\pi s^2}} \exp\left(\frac{-(\zeta - m)^2}{2s^2}\right).$$

Fig. 1 For the model considered in Sect. 5, we plot the implied volatility σ corresponding to the exact option price u (*solid black*) as well as the implied volatility $\bar{\sigma}_2$ corresponding to our second order option price approximation $\bar{u}_2$ (*dashed black*). The units of the horizontal axis are log strike $k := \log K$. Approximate prices are computed using the Taylor series expansion of $\mathcal{A}(t)$ as described in Example 3.2. We assume the Lévy measure ν is as parametrized above. The following parameters are used in all four plots: $\kappa = 1.15$, $\theta = 0.04$, $\delta = 0.2$, $\rho = -0.7$, $z = \theta$, $x = 0$, $m = -0.1$, $s = 0.2$, $\lambda = 2.0$

Table 1 Exact implied vols σ, second order approximation $\bar{\sigma}_2$ and relative error $|(\bar{\sigma}_2 - \sigma)/\sigma|$

	$k - x$	−0.2	−0.15	−0.1	−0.05	0.00	0.05	0.1	0.15	0.2
t = 0.10	σ	0.2797	0.2478	0.2269	0.2133	0.2028	0.1940	0.1881	0.1960	0.2296
	$\bar{\sigma}_2$	0.2795	0.2483	0.2271	0.2132	0.2028	0.1939	0.1877	0.1963	0.2324
	rel. err.	0.0006	0.0018	0.0009	0.0003	0.0002	0.0001	0.0020	0.0018	0.0120
t = 0.25	σ	0.2441	0.2323	0.2217	0.2120	0.2028	0.1941	0.1863	0.1805	0.1803
	$\bar{\sigma}_2$	0.2456	0.2328	0.2215	0.2116	0.2025	0.1939	0.1859	0.1793	0.1799
	rel. err.	0.0059	0.0018	0.0013	0.0020	0.0013	0.0009	0.0021	0.0067	0.0027
t = 0.50	σ	0.2348	0.2266	0.2183	0.2101	0.202	0.1940	0.1864	0.1796	0.1743
	$\bar{\sigma}_2$	0.2350	0.2254	0.2168	0.2088	0.201	0.1933	0.1856	0.1783	0.1723
	rel. err.	0.0005	0.0049	0.0069	0.0063	0.004	0.0037	0.0040	0.0070	0.0116
t = 1.00	σ	0.2268	0.2204	0.2138	0.2072	0.2005	0.1939	0.1875	0.1813	0.1757
	$\bar{\sigma}_2$	0.2217	0.2149	0.2089	0.2031	0.1973	0.1914	0.1854	0.1794	0.1740
	rel. err.	0.0227	0.0246	0.0230	0.0197	0.0160	0.0130	0.0111	0.0103	0.0096

Parameters are the same as those in Fig. 1

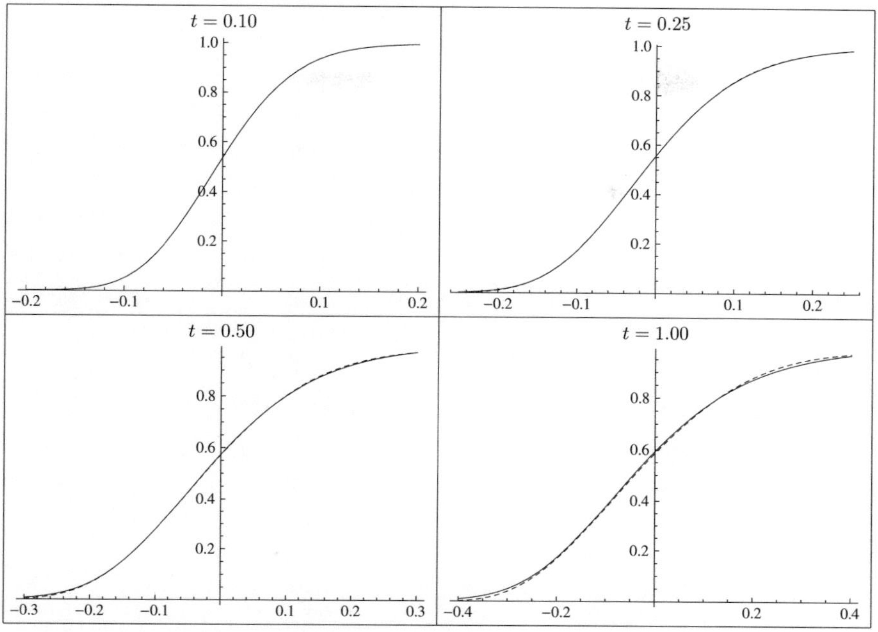

Fig. 2 For the model considered in Sect. 5, we plot the Delta Δ corresponding to the exact option price u (*solid black*) as well as the Delta $\bar{\Delta}_2$ corresponding to our second order option price approximation $\bar{u}_2$ (*dashed black*). The units of the horizontal axis are x. Approximate prices are computed using the Taylor series expansion of $\mathcal{A}(t)$ as described in Example 3.2. We assume the Lévy measure ν is as given in Fig. 1. The following parameters are used in all four plots: $\kappa = 1.15$, $\theta = 0.04$, $\delta = 0.2$, $\rho = -0.7$, $z = \theta$, $k = 0$, $m = -0.1$, $s = 0.2$, $\lambda = 2.0$

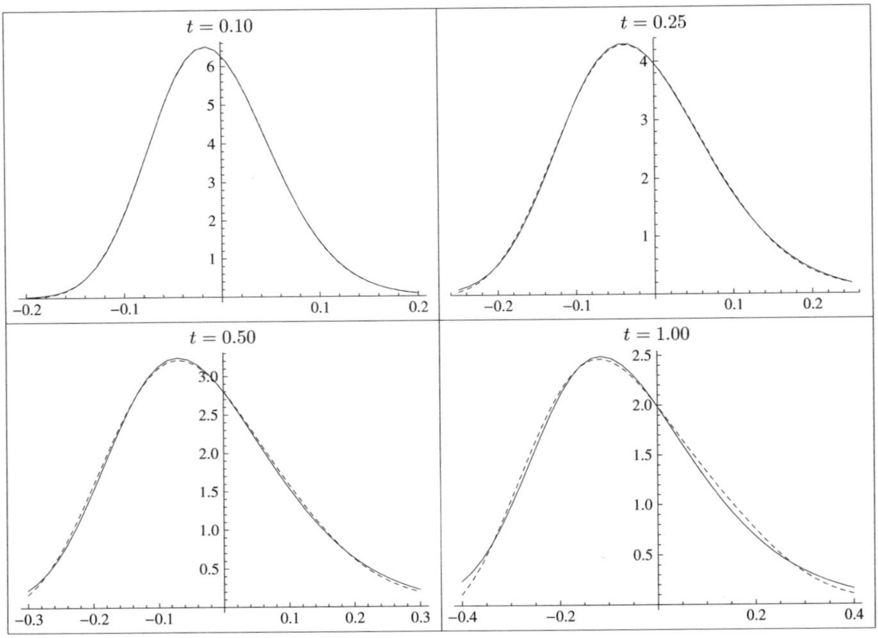

Fig. 3 For the model considered in Sect. 5, we plot the Gamma Γ corresponding to the exact option price u (*solid black*) as well as the Gamma $\bar{\Gamma}_2$ corresponding to our second order option price approximation $\bar{u}_2$ (*dashed black*). The units of the horizontal axis are x. Approximate prices are computed using the Taylor series expansion of $\mathcal{A}(t)$ as described in Example 3.2. We assume the Lévy measure ν is as given in Fig. 1. The following parameters are used in all four plots: $\kappa = 1.15$, $\theta = 0.04$, $\delta = 0.2$, $\rho = -0.7$, $z = \theta$, $k = 0$, $m = -0.1$, $s = 0.2$, $\lambda = 2.0$

Table 2 Exact Delta Δ, second order approximation $\bar{\Delta}_2$ and relative error $|(\bar{\Delta}_2 - \Delta)/\Delta|$

	x	−0.2	−0.15	−0.1	−0.05	0.00	0.05	0.1	0.15	0.2
t = 0.10	Δ	0.0008	0.00516	0.05084	0.2312	0.5370	0.8024	0.9385	0.9845	0.9959
	$\bar{\Delta}_2$	0.0009	0.00478	0.05081	0.2313	0.5368	0.8026	0.9387	0.9843	0.9958
	rel. err.	0.1309	0.07358	0.00048	0.0006	0.0003	0.0002	0.0002	0.0002	0.0000
t = 0.25	Δ	0.01311	0.05708	0.1690	0.3503	0.5559	0.7329	0.8563	0.9293	0.9672
	$\bar{\Delta}_2$	0.0114	0.05674	0.1696	0.3502	0.5552	0.7330	0.8576	0.9306	0.9673
	rel. err.	0.1305	0.00585	0.0035	0.0004	0.0012	0.0000	0.0014	0.0014	0.0000
t = 0.50	Δ	0.06608	0.1506	0.2767	0.4260	0.5739	0.7018	0.8014	0.8731	0.9215
	$\bar{\Delta}_2$	0.06425	0.1508	0.2766	0.4246	0.5719	0.7007	0.8027	0.8766	0.9256
	rel. err.	0.02773	0.0014	0.0003	0.0032	0.0034	0.0015	0.0015	0.0040	0.0044
t = 1.00	Δ	0.1708	0.2667	0.3760	0.4878	0.5927	0.6849	0.7618	0.8234	0.8713
	$\bar{\Delta}_2$	0.1662	0.2627	0.3710	0.4814	0.5857	0.6791	0.7595	0.8262	0.8789
	rel. err.	0.0268	0.01496	0.0131	0.0130	0.0117	0.0084	0.0030	0.0033	0.0088

Parameters are the same as those in Fig. 2

Table 3 Exact Gamma Γ, second order approximation $\bar{\Gamma}_2$ and relative error $|(\bar{\Gamma}_2 - \Gamma)/\Gamma|$

	x	−0.2	−0.15	−0.1	−0.05	0.00	0.05	0.1	0.15	0.2
t = 0.10	Γ	0.01828	0.2978	2.159	5.539	6.288	3.831	1.446	0.3779	0.0780
	$\bar{\Gamma}_2$	0.01197	0.2897	2.1760	5.5300	6.288	3.841	1.437	0.3748	0.0821
	rel. err.	0.3452	0.0273	0.0077	0.0015	0.0001	0.0025	0.0061	0.0082	0.0518
t = 0.25	Γ	0.5185	1.705	3.337	4.275	3.967	2.884	1.738	0.906	0.4229
	$\bar{\Gamma}_2$	0.5267	1.747	3.334	4.255	3.969	2.907	1.754	0.8925	0.4016
	rel. err.	0.0157	0.024	0.0009	0.0046	0.0003	0.0079	0.0094	0.0149	0.0503
t = 0.50	Γ	1.514	2.488	3.135	3.206	2.802	2.174	1.54	1.017	0.635
	$\bar{\Gamma}_2$	1.585	2.508	3.109	3.182	2.804	2.208	1.588	1.045	0.6244
	rel. err.	0.0468	0.0079	0.0081	0.0076	0.0007	0.015	0.0309	0.0279	0.0167
t = 1.00	Γ	2.095	2.425	2.483	2.306	1.985	1.612	1.251	0.9364	0.6814
	$\bar{\Gamma}_2$	2.134	2.418	2.452	2.280	1.988	1.656	1.331	1.028	0.7511
	rel. err.	0.0183	0.0032	0.0124	0.0110	0.0015	0.0276	0.0644	0.097	0.1023

Parameters are the same as those in Fig. 3

6 Conclusion

In this paper we derive a family of asymptotic expansions for European option prices when the underlying is modeled as a d-dimensional time inhomogeneous Lévy-type process. By combining the classical Dyson series expansion with a novel polynomial expansion of the generator, we obtain two equivalent representations for approximate option price: (i) as an integro-differential operator acting on the order zero price, and (ii) as a Fourier transform. We implement our pricing approximation on a Heston-like model which allows for both stochastic volatility and stochastic jump intensity. We find that our second order expansion provides and excellent approximation for prices (as seen through corresponding implied volatilities), as well as for the Greeks Δ and Γ.

References

1. Benhamou, E., Gobet, E., Miri, M.: Smart expansion and fast calibration for jump diffusions. Financ. Stoch. **13**(4), 563–589 (2009)
2. Bompis, R., Gobet, E.: Asymptotic and non asymptotic approximations for option valuation. Recent Developments in Computational Finance. Foundations, Algorithms and Applications, pp. 159–241. World Scientific, Hackensack (2013)
3. Carr, P., Wu, L.: Time-changed Lévy processes and option pricing. J. Financ. Econ. **71**(1), 113–141 (2004)
4. Corielli, F., Foschi, P., Pascucci, A.: Parametrix approximation of diffusion transition densities. SIAM J. Financ. Math. **1**, 833–867 (2010)
5. Deuschel, J.-D., Friz, P., Jacquier, A., Violante, S.: Marginal density expansions for diffusions and stochastic volatility, part i: theoretical foundations. Commun. Pure Appl. Math. **67**(1), 40–82 (2014)

6. Fouque, J.-P., Papanicolaou, G., Sircar, R., Solna, K.: Multiscale Stochastic Volatility for Equity, Interest Rate, and Credit Derivatives. Cambridge University Press, Cambridge (2011)
7. Friz, P. K., Gerhold, S., Yor, M.: How to make Dupire's local volatility work with jumps. Quant. Financ. **14**(8), 1327–1331 (2014)
8. Heston, S.: A closed-form solution for options with stochastic volatility with applications to bond and currency options. Rev. Financ. Stud. **6**(2), 327–343 (1993)
9. Jeanblanc, M., Yor, M., Chesney, M.: Mathematical Methods for Financial Markets. Springer, London (2009)
10. Kato, T.: Perturbation Theory for Linear Operators. Classics in Mathematics. Springer, Berlin (1995). Reprint of the 1980 edition
11. Lorig, M.: Pricing derivatives on multiscale diffusions: an eigenfunction expansion approach. Math. Finance **24**(2), 331–363 (2014)
12. Lorig, M., Lozano-Carbassé O.: Multiscale Exponential Lévy models. Quant. Finance **15**(1), 91–100 (2015)
13. Lorig, M., Pagliarani, S., Pascucci, A.: Analytical expansions for parabolic equations. SIAM J. Appl. Math. **75**(2), 468–491 (2015)
14. Lorig, M., Pagliarani, S., Pascucci, A.: Explicit implied volatilities for multifactor local-stochastic volatility models. Math. Finance (to appear) (2015). ArXiv preprint arXiv:1306.5447
15. Lorig, M., Pagliarani, S., Pascucci, A.: A family of density expansions for Lévy-type processes with default. Ann. Appl. Probab. **25**(1), 235–267 (2015)
16. Lorig, M., Pagliarani, S., Pascucci, A.: Pricing approximations and error estimates for local Lévy-type models with default. Comp. Math. App. **69**(10), 1189–1219 (2015)
17. Lorig, M., Pagliarani, S., Pascucci, A.: A Taylor series approach to pricing and implied vol for LSV models. J. Risk **17**(2), 1–17 (2014)
18. Øksendal, B., Sulem, A.: Applied Stochastic Control of Jump Diffusions. Springer, Berlin (2005)
19. Pagliarani, S., Pascucci, A.: Analytical approximation of the transition density in a local volatility model. Cent. Eur. J. Math. **10**(1), 250–270 (2012)
20. Pagliarani, S., Pascucci, A.: Local stochastic volatility with jumps: analytical approximations. Int. J. Theor. Appl. Financ. **16**(8), 1–35 (2013)
21. Pagliarani, S., Pascucci, A., Riga, C.: Adjoint expansions in local Lévy models. SIAM J. Financ. Math. **4**, 265–296 (2013)
22. Pascucci, A.: PDE and Martingale Methods in Option Pricing. Bocconi & Springer Series, vol. 2. Springer, Milan (2011)
23. Sakurai, J.J., Tuan, S.F.: Modern Quantum Mechanics, vol. 104. Addison-Wesley, Reading (Mass.) (1994)

Asymptotic Expansion Approach in Finance

Akihiko Takahashi

Abstract This paper provides a survey on an asymptotic expansion approach to valuation and hedging problems in finance. The asymptotic expansion is a widely applicable methodology for analytical approximations of expectations of certain Wiener functionals. Hence not only academic researchers but also practitioners have been applying the scheme to a variety of problems in finance such as pricing and hedging derivatives under high-dimensional stochastic environments. The present note gives an overview of the approach.

Keywords Asymptotic expansion · Derivatives · Option pricing · Hedge · Greeks · Stochastic volatility · Interest rate · Term structure model · Malliavin calculus · Watanabe theory

1 Introduction

Let $(\Omega, \mathcal{F}, \{\mathcal{F}_t\}_{t\in[0,T]}, P)$ denote a probability space with filtration, on which a r-dimensional standard Wiener process W is defined, where P is an appropriate pricing measure (a risk neutral measure) in finance, and T denotes some positive constant. Now, let $F(\omega)$ be a Wiener functional and then $\mathbf{V}$, the security or portfolio value can be expressed as $\mathbf{V} = \mathbf{E}[F(\omega)]$ under certain conditions. Evaluating this expectation is one of the main issues in finance. Moreover, if F depends on the parameter θ, computation of $\frac{\partial \mathbf{V}}{\partial \theta} = \frac{\partial}{\partial \theta}\mathbf{E}[F(\omega; \theta)]$, the sensitivity of the security value with respect to the change in this parameter (so called *Greeks*) is also an important task in practice.

I dedicate this note to the late Professor Peter Laurence and Koji Takahashi.
I am very grateful to Professor Fujii, Professor Shiraya, Professor Takehara, Dr. Toda, Dr. Tsuzuki and Professor Yamada, my coauthors in the original articles, which are main bases for this survey.

A. Takahashi (✉)
Graduate School of Economics, The University of Tokyo, 7-3-1, Hongo,
Tokyo, Bunkyo-ku 113-0033, Japan
e-mail: akihikot@e.u-tokyo.ac.jp

P.K. Friz et al. (eds.), *Large Deviations and Asymptotic Methods in Finance*, Springer Proceedings in Mathematics & Statistics 110,
DOI 10.1007/978-3-319-11605-1_13

As an example, let us consider a d-dimensional diffusion process $X^{(\epsilon)}$ which is obtained as a strong solution to the stochastic differential equation;

$$dX_t^{(\epsilon)} = V_0(X_t^{(\epsilon)}, \epsilon)dt + V(X_t^{(\epsilon)}, \epsilon)dW_t,\ t \in [0, T];\ X_0^{(\epsilon)} = x_0,$$

where $\epsilon \in [0, 1]$ is a known parameter. Here, the coefficients are assumed to satisfy some regularity conditions. In finance, many problems of pricing derivatives and evaluating the portfolios in investment theories are reduced to the problems of computing $\mathbf{E}[f(X_T^{(\epsilon)})]$, the expectation of $f(X_T^{(\epsilon)})$, that is a function of $X_T^{(\epsilon)}$.

In finance applications, it is important to deal with not only a smooth function $f(x)$ but also non-smooth one. For example, when various options are evaluated, f is expressed as $f = T \circ g$, where $T(x) = \max\{x, 0\}$ and g stands for a smooth function of $\mathbf{R}^d \mapsto \mathbf{R}$. In general, it is difficult to represent this expectation explicitly except for special cases. Hence, numerical methods such as Monte Carlo simulations or numerical solutions of partial differential equations (PDEs) are employed and various speeding up techniques are developed, since fast and precise computation is required in practice.

As a different approach, an approximation of the expectation by an asymptotic expansion of the stochastic differential equation around $\epsilon = 0$ may be considered. Furthermore, because $\frac{\partial}{\partial x_0}\mathbf{E}[f(X_T^{(\epsilon)})]$ and $\frac{\partial}{\partial \epsilon}\mathbf{E}[f(X_T^{(\epsilon)})]$, the sensitivities of the security value with respect to the changes in the initial value x_0 and in the parameter ϵ are important indicators for practical purposes, the approximations with high accuracies are so valuable. Moreover, some schemes that combine Monte Carlo simulations with asymptotic expansions with low orders are developed, since the asymptotic expansion up to the first or second order can be easily evaluated. Those schemes are able to improve the efficiencies of Monte Carlo simulations and the accuracies of approximations obtained by the asymptotic expansions.

An asymptotic expansion approach in finance has been developed for the past two decades, which is mathematically justified by Watanabe theory (Watanabe [111]) in Malliavin calculus (e.g. Malliavin [64], Chap. V-8 in Ikeda and Watanabe [39], Nualart [72]). To the best of our knowledge, the asymptotic expansion technique is firstly applied to finance for evaluation of average options that are popular derivatives in commodity markets. Kunitomo and Takahashi [48] and Takahashi [85] derive approximation formulas for average options by an asymptotic expansion method based on log-normal approximations for average prices distributions, when the underlying asset prices follow geometric Brownian motions. Yoshida [119] derives an asymptotic expansion of an average option price around a normal distribution for a general diffusion model, which is a byproduct of his result in statistics [118] based on the Watanabe theory.

Thereafter, the asymptotic expansion approach have been applied to a broad class of valuation problems in finance, which includes pricing options with stochastic volatility models, pricing options under Heath-Jarrow-Morton (HJM) models [37] or Libor market models (LMM) (Brace, Gatarek and Musiela [7], Jamshidian [43]) of interest rates, and pricing so called exotic-type options such as basket and barrier options in addition to average options.

For instance, please see Kawai [44], Kobayashi, Takahashi and Tokioka [45], Kunitomo and Takahashi [49–51], Li [59], Matsuoka, Takahashi and Uchida [66], Muroi [67], Nishiba [71], Osajima [75], Shiraya and Takahashi [78–80], Shiraya, Takahashi and Toda [81], Shiraya, Takahashi and Yamada [83], Shiraya, Takahashi and Yamazaki [82], Takahashi and Matsushima [88], Takahashi and Saito [89], Takahashi and Takehara [90–94], Takahashi, Takehara and Toda [90, 91], Takahashi and Tsuzuki [98], Takahashi and Uchida [99], Takahashi and Yamada [100–104], Takahashi and Yoshida [106, 107], Takahashi and Takehara [92, 93], Violante [110], Xu and Zheng [112, 113], and Takahashi [86, 87].

We briefly introduce some of above works in Sect. 3.6. Moreover, we remark that the asymptotic expansion approach is employed by Yamanobe [116, 117] in physics for analyses of the impulse-driven stochastic biological oscillator and global dynamics of a stochastic neuronal oscillator.

We also note that there exist many other types of the expansion/perturbation methods which have turned out to be so useful for applications in finance. For example, see Bayer and Laurence [2], Ben Arous and Laurence [3], Benaim, Friz and Lee [4], Col, Gnoatto and Grasselli [9], Davydov and Linetsky [11], Deuschel, Friz, Jacquier and Violante [12, 13], Forde and Jacquier [18], Forde, Jacquier and Lee [17], Foschi, Pagliarani, Pascucci [19], Fouque, Papanicolaou and Sircar [20, 21], Fujii [24], Fujii and Takahashi [25–27, 29], Gatheral, Hsu, Laurence, Ouyang, and Wang [30], Gnoatto and Grasselli [31], Gulisashvili [32], Hagan, Kumar, Lesniewski and Woodward [33], Henry-Labordère [38], Kato, Takahashi and Yamada [46, 47], Kusuoka and Osajima [57], Lee [58], Lipton [60], Linetsky [61], Osajima [76], Pagliarani and Pascucci [77], Siopacha and Teichmann [84], Yamamoto, Sato and Takahashi [114], Yamamoto and Takahashi [115], and references therein.

The organization of the paper is as follows. The next section describes the outline of the asymptotic expansion approach in a general diffusion setting. Then, Sect. 3 explains a computational scheme for the expansion method. Section 4 provides an extension of the general computational scheme in the previous section, and Sect. 5 briefly introduces two improvement scheme for the expansion method. Section 6 extends the approach to non-diffusion Wiener functionals by using an instantaneous forward rates model as an example. Sections 7 and 8 introduce an asymptotic expansion in jump-diffusion models and a perturbation scheme in forward backward stochastic differential equations (FBSDEs). Section 9 concludes.

2 Asymptotic Expansion in General Diffusion Setting

Following [87, 96], this section briefly describes an asymptotic expansion method in a general diffusion setting.

Let us consider a d-dimensional diffusion process $X_t^{(\epsilon)} = (X_t^{(\epsilon),1}, \ldots, X_t^{(\epsilon),d})^\top$ which is the solution to the following stochastic differential equation:

$$dX_t^{(\epsilon),j} = V_0^j(X_t^{(\epsilon)}, \epsilon)dt + \epsilon V^j(X_t^{(\epsilon)})dW_t \quad (j = 1, \ldots, d) \tag{1}$$
$$X_0^{(\epsilon)} = x_0 \in \mathbf{R}^d,$$

where $W = (W^1, \ldots, W^r)^\top$ is a r-dimensional standard Wiener process, and $\epsilon \in (0, 1]$ is a known parameter. Here, $x^\top$ denotes the transpose of x. Next, let us define $V_0 = (V_0^1, \ldots, V_0^d)^\top : \mathbf{R}^d \times (0, 1] \mapsto \mathbf{R}^d$ and $V: \mathbf{R}^d \mapsto \mathbf{R}^d \otimes \mathbf{R}^r$ whose jth row is V^j, $j = 1, \ldots, d$. Suppose also that V_0 and V satisfy some regularity conditions. (For example, V_0 and V are smooth functions with bounded derivatives of all orders.)

Next, let a function $g : \mathbf{R}^d \mapsto \mathbf{R}$ be smooth and all of its derivatives have polynomial growth. Then, a smooth Wiener functional $g(X_T^{(\epsilon)})$ has its asymptotic expansion:

$$g(X_T^{(\epsilon)}) \sim g_{0T} + \epsilon g_{1T} + \epsilon^2 g_{2T} + \cdots$$

in $\mathbf{D}^\infty$ as $\epsilon \downarrow 0$ where $g_{0T}, g_{1T}, g_{2T}, \ldots \in \mathbf{D}^\infty$. For any $k \in \mathbf{N}$, $q \in (1, \infty)$ and $s > 0$, this expansion means that

$$\frac{1}{\epsilon^k}\|g(X_T^{(\epsilon)}) - (g_{0T} + \epsilon g_{1T} + \cdots + \epsilon^{k-1} g_{k-1,T})\|_{q,s} = O(1) \ (as\ \epsilon \downarrow 0),$$

where $\|G\|_{q,s}$ represents the sum of L^q-norms of Malliavin derivatives of a Wiener functional G up to the sth order. Further, a Banach space $\mathbf{D}_{q,s} = \mathbf{D}_{q,s}(\mathbf{R})$ can be regarded as the totality of random variables bounded with respect to (q, s)-norm $\|\cdot\|_{q,s}$, and $\mathbf{D}^\infty = \cap_{s>0}\cap_{1<q<\infty}\mathbf{D}_{q,s}$. The coefficients $g_{nT} \in \mathbf{D}^\infty (n = 0, 1, \ldots)$ in the expansion can be obtained by Taylor's formula and represented based on multiple Wiener-Itô integrals. For the details of definitions and proofs above, please consult Watanabe [111], Chap. V of Ikeda and Watanabe [39], Malliavin [64], or Chap. 7 of Malliavin and Thalmaier [65].

Remark 1 As an example of applications in finance, $X^{(\epsilon)}$ consists of n stocks, $X^{(\epsilon)} = (S_1^{(\epsilon)}, \ldots, S_n^{(\epsilon)})^\top$ and $g(\cdot)$ is those weighted sum $g(x) = w_1 x_1 + \cdots + w_n x_n$ for $x = (x_1, \ldots, x_n)^\top$ with constant weights $w_i (i = 1, \ldots, n)$. Then, $g(x)$ would represent the spread, the average or the basket price of the stock prices.

As another example, we can set $X^{(\epsilon)}$ is a vector of N forward Libor rates, $X^{(\epsilon)} = (L_1^{(\epsilon)}, \ldots, L_N^{(\epsilon)})^\top$, and

$$g(X_T^{(\epsilon)}) = SR_T^{(\epsilon)} = \frac{1 - \prod_{j=0}^{N-1} \frac{1}{1+\tau L_{jT}^{(\epsilon)}}}{\tau \sum_{i=0}^{N-1} \prod_{j=0}^{i} \frac{1}{1+\tau L_{jT}^{(\epsilon)}}},$$

that is a swap rate with inception date T and maturity date $T_N = T + N\tau$. Here, L_{jT} stands for the forward Libor rate at T fixing at $T + j\tau$ with tenor τ.

Let $A_{kt} = \frac{1}{k!}\frac{\partial^k X_t^{(\epsilon)}}{\partial \epsilon^k}|_{\epsilon=0}$ and A_{kt}^j, $j = 1, \ldots, d$ denote the jth elements of A_{kt}. In particular, A_{1t} is represented by

$$A_{1t} = \int_0^t Y_t Y_u^{-1} \left(\partial_\epsilon V_0(X_u^{(0)}, 0)du + V(X_u^{(0)})dW_u \right), \tag{2}$$

where Y denotes the solution to the ordinary differential equation:

$$dY_t = \partial V_0(X_t^{(0)}, 0)Y_t dt; \; Y_0 = I_d.$$

Here, ∂V_0 denotes the $d \times d$ matrix whose (j,k)-element is $\partial_k V_0^j = \frac{\partial V_0^j(x,\epsilon)}{\partial x_k}$, V_0^j is the jth element of V_0, and I_d denotes the $d \times d$ identity matrix.

For $k \geq 2$, A_{kt}^j, $j = 1, \ldots, d$ is recursively determined by the following equation:

$$\begin{aligned} A_{kt}^j &= \frac{1}{k!}\int_0^t \partial_\epsilon^k V_0^j(X^{(0)}, 0)du \\ &+ \sum_{l=1}^{k} \sum_{\vec{l}_\beta, \vec{d}_\beta}^{(l)} \frac{1}{(k-l)!}\frac{1}{\beta!}\int_0^t \left(\prod_{j=1}^{\beta} A_{l_j u}^{d_j} \right) \partial_{\vec{d}_\beta}^\beta \partial_\epsilon^{k-l} V_0^j(X_u^{(0)}, 0)du \\ &+ \sum_{\vec{l}_\beta, \vec{d}_\beta}^{(k-1)} \frac{1}{\beta!}\int_0^t \left(\prod_{j=1}^{\beta} A_{l_j u}^{d_j} \right) \partial_{\vec{d}_\beta}^\beta V^j(X_u^{(0)})dW_u, \end{aligned} \tag{3}$$

where $\partial_\epsilon^l = \frac{\partial^l}{\partial \epsilon^l}$, $\partial_{\vec{d}_\beta}^\beta = \frac{\partial^\beta}{\partial x_{d_1} \cdots \partial x_{d_\beta}}$,

$$\sum_{\vec{l}_\beta, \vec{d}_\beta}^{(l)} := \sum_{\beta=1}^{l} \sum_{\vec{l}_\beta \in L_{l,\beta}} \sum_{\vec{d}_\beta \in \{1,\ldots,d\}^\beta} \tag{4}$$

for $l \geq 1$,

$$L_{l,\beta} := \left\{ \vec{l}_\beta = (l_1, \ldots, l_\beta); \sum_{j=1}^{\beta} l_j = l; \; (l, l_j, \beta \in \mathbf{N}) \right\}, \tag{5}$$

and for $l = 0$,

$$\sum_{\vec{l}_\beta, \vec{d}_\beta}^{(0)} = \sum_{\beta=0} \sum_{\vec{l}_0=(\emptyset)} \sum_{\vec{d}_0=(\emptyset)} .$$

Then, g_{0T} and g_{1T} can be written as

$$g_{0T} = g(X_T^{(0)}),$$
$$g_{1T} = \sum_{j=1}^{d} \partial_j g(X_T^{(0)}) A_{1T}^j,$$

where $\partial_j g(x) = \frac{\partial}{\partial x_j} g(x)$, $j = 1, \ldots, d$.

For $n \geq 2$, g_{nT} is expressed as follows:

$$g_{nT} = \sum_{\vec{l}_\beta, \vec{d}_\beta}^{(n)} \frac{1}{\beta!} \partial^\beta_{\vec{d}_\beta} g(X_T^{(0)}) A_{l_1 T}^{d_1} \cdots A_{l_\beta T}^{d_\beta}. \tag{6}$$

Here, we note that each $A_{lt}^i (i = 1, \ldots, d, l = 1, 2, \ldots, k, 0 \leq t \leq T)$ has all finite moments due to a *grading* structure. We describe the definition of the grading structure by following pp. 45–47 in Bichteler, Gravereaux and Jacod [5]: Consider the stochastic differential equation of the form:

$$dS_t = \mu(S_t, t)dt + \sigma(S_t, t)dW_t; \; S_0 = s_0 \in \mathbf{R}^d, \tag{7}$$

where $\mu : \mathbf{R}^d \times \mathbf{R}^+ \to \mathbf{R}^d$ and $\sigma : \mathbf{R}^d \times \mathbf{R}^+ \to \mathbf{R}^d \otimes \mathbf{R}^r$.

Definition 1 A grading of $\mathbf{R}^d$ is a decomposition $\mathbf{R}^d = \mathbf{R}^{d_1} \times \cdots \times \mathbf{R}^{d_q}$ with $d = d_1 + \cdots + d_q$. The coordinates of a point in $\mathbf{R}^d$ are always arranged in an increasing order along the subspace $\mathbf{R}^{d_i}$, and we set $M_0 = 0$ and $M_l = d_1 + \cdots + d_l$ for $1 \leq l \leq q$. We say that the coefficients μ and σ are graded according to the grading $\mathbf{R}^d = \mathbf{R}^{d_1} \times \cdots \times \mathbf{R}^{d_q}$ if $\mu^i(x, t)$ and $\sigma^i_j(x, t)$, $j = 1, \ldots, r$ depend upon only through the coordinates $(x^k)_{1 \leq k \leq M_p}$ when $M_{p-1} \leq i \leq M_p$.

Theorem 1 *We assume that the coefficients μ and σ in (7) are graded according to $\mathbf{R}^d = \mathbf{R}^{d_1} \times \cdots \times \mathbf{R}^{d_q}$. Moreover for $F(x, t) = \mu(x, t)$ or $\sigma_j(x, t)$, $j = 1, \ldots, r$, we assume that F is differentiable in x on $\mathbf{R}^d$ and*

1. $|F^i(0, t)| \leq Z_t$ *for* $i = 1, \ldots, d$
2. $|\frac{\partial}{\partial x^j} F^i(x, t)| \leq \hat{Z}_t(1 + |x|^\theta)$ *for all* i, j
3. $|\frac{\partial}{\partial x^j} F^i(x, t)| \leq \zeta$ *if* $M_{p-1} \leq i, j \leq M_p$ *for some* $p \leq q$,

where $\zeta, \theta \geq 0$ are constants, and Z, $\hat{Z}$ are predictable processes such that $\|Z\|_p$ and $\|\hat{Z}\|_p$ are finite for all $p \geq 1$ where $\|Z\|_p = \left\{\int_0^T E[|Z_t|^p]dt\right\}^{1/p}$. Then (7) have a unique solution S, and for every $p \geq 1$ there are constants c_p and γ_p depending only upon $(\zeta, \theta, \{||\hat{Z}||_{p'}\}_{p' \geq 1})$, such that

$$|| \sup_{0 \leq t \leq T} S_t ||_{L^p} \leq c_p(s_0 + ||Z||_{\gamma_p}).$$

For the detail of the definition and theorem above, see pp. 45–47 in Bichteler, Gravereaux and Jacod [5].

Applying Theorem 1 to the system of stochastic differential equations consisting of $A^i_{lt}(i = 1, \ldots, d, l = 1, \ldots, k, 0 \leq t \leq T)$ as well as any products of them, we obtain the following lemma.

Lemma 1 *Each coefficient in the expansion,* $A^i_{lt}(i = 1, \ldots, d, l = 1, \ldots, k, 0 \leq t \leq T)$ *has all finite moments.*

(Proof) We consider the system of stochastic differential equations (SDEs) for $A^1_1, \ldots, A^d_1, A^1_1 A^1_1, \ldots, A^d_1 A^d_1, A^1_2, \ldots, A^d_2, \ldots$. Then, the coefficients of the SDEs are represented by the derivatives at $\epsilon = 0$ of $\tilde{V}_0(X^{(\epsilon)}_u, \epsilon)$ and $\tilde{V}(X^{(\epsilon)}_u)$, which are bounded in $[0, T]$. Moreover, it is easily shown that the coefficients of the equation are graded and satisfy the conditions in Theorem 1. Hence each coefficient in the expansion, A^i_{kt} has all finite moments. □

Next, let normalize $g(X^{(\epsilon)}_T)$ to

$$G^{(\epsilon)} = \frac{g(X^{(\epsilon)}_T) - g_{0T}}{\epsilon}$$

for $\epsilon \in (0, 1]$. Then, we have

$$G^{(\epsilon)} \sim g_{1T} + \epsilon g_{2T} + \cdots$$

in $\mathbf{D}^\infty$.

Next, for $h \in H$, where H denotes the Cameron-Martin subspace of the r-dimensional Wiener space, the H-derivative of $G^{(\epsilon)}$ is expressed as

$$\mathbf{D}_h G^{(\epsilon)} = \frac{1}{\epsilon} \sum_{i=1}^{d} \partial_i g(X^{(\epsilon)}_T) \mathbf{D}_h X^{(\epsilon),i}_T = \sum_{i=1}^{d} \partial_i g(X^{(\epsilon)}_T) \int_0^T [Y^{(\epsilon)}_T (Y^{(\epsilon)}_t)^{-1} V(X^{(\epsilon)}_t) \dot{h}_t]_i \, dt,$$

where $Y^{(\epsilon)}$ is the $\mathbf{R}^d \otimes \mathbf{R}^d$-valued stochastic process which is the solution to the stochastic differential equation:

$$dY^{(\epsilon)}_t = \partial V_0(X^{(\epsilon)}_t, \epsilon) Y^{(\epsilon)}_t dt + \epsilon \sum_{i=1}^{m} \partial V^i(X^{(\epsilon)}_t) Y^{(\epsilon)}_t dw_{it}; \; Y^{(\epsilon)}_0 = I_d,$$

In fact, $Y_t = Y^{(0)}_t$. Here, $\partial V^i (i = 0, 1, \ldots, m)$ denotes the $d \times d$ matrix whose (j, k)-element is $\partial_k V^i_j$. ($\partial_k = \frac{\partial}{\partial x_k}$.)

Moreover, with a notation $\hat{V}^{(\epsilon)}_t$ that is defined by

$$\hat{V}^{(\epsilon)}_t = \left(\partial g(X^{(\epsilon)}_T)\right)^\top \left[Y^{(\epsilon)}_T (Y^{(\epsilon)}_t)^{-1} V(X^{(\epsilon)}_t)\right],$$

where $\left(\partial g(X_T^{(\epsilon)})\right)^\top = (\partial_1 g(X_T^{(\epsilon)}), \ldots, \partial_d g(X_T^{(\epsilon)}))$, the Malliavin (co)variance of $G^{(\epsilon)}$ is given by

$$\sigma_{G^{(\epsilon)}} = \int_0^T \hat{V}_t^{(\epsilon)} (\hat{V}_t^{(\epsilon)})^\top dt. \tag{8}$$

Moreover, let

$$\hat{V}_t := \hat{V}_t^{(0)} = \left(\partial g(X_T^{(0)})\right)^\top \left[Y_T Y_t^{-1} V(X_t^{(0)})\right]$$

and make the following assumption:

$$\text{(Assumption 1)} \quad \Sigma_T = \int_0^T \hat{V}_t \hat{V}_t^\top dt > 0.$$

Note that g_{1T} follows a normal distribution with variance Σ_T, and the density function of g_{1T} denoted by $f_{g_{1T}}(x)$ is given as

$$f_{g_{1T}}(x) = \frac{1}{\sqrt{2\pi\Sigma_T}} \exp\left(-\frac{(x-C)^2}{2\Sigma_T}\right)$$

where

$$C := \left(\partial g(X_T^{(0)})\right)^\top \int_0^T Y_T Y_t^{-1} \partial_\epsilon V_0(X_t^{(0)}, 0) dt. \tag{9}$$

Since Σ_T is the variance of the random variable g_{1T}, which follows a normal distribution, (Assumption 1) means the condition that the distribution of g_{1T} does not degenerate. In application, as it is easy to check this condition in most cases, it plays an important role for practical purposes.

Next, let us briefly introduce a truncated version of the Watanabe theory [111] based on Yoshida [118, 119]. Under (Assumption 1), $\sigma_{G^{(\epsilon)}}$ is uniformly non-degenerate for $\{|\eta_c^{(\epsilon)}| \leq 1\}$; that is, it can be shown that there exists a positive real number $c_0 > 0$ such that for any $c > c_0$ and $p > 1$,

$$\sup_{\epsilon \in (0,1]} \mathbf{E}[1_{\{|\eta_c^{(\epsilon)}| \leq 1\}} (|\sigma_{G^{(\epsilon)}}|)^{-p}] < \infty, \tag{10}$$

where $\eta_c^{(\epsilon)} = c \int_0^T |\hat{V}_t^{(\epsilon)} - \hat{V}_t| dt$.

Let $\mathcal{S}$ be the real Schwartz space of rapidly decreasing C^∞-functions on $\mathbf{R}$ and $\mathcal{S}'$ be its dual space. Then, for $\Phi : \mathbf{R} \mapsto \mathbf{R}$, $\Phi \in \mathcal{S}'$, a composite function $\psi(\eta_c^{(\epsilon)})\Phi \circ G^{(\epsilon)} = \psi(\eta_c^{(\epsilon)})\Phi(G^{(\epsilon)})$ is well-defined as an element of $\tilde{\mathbf{D}}^{-\infty} = \cup_{s<0} \cap_{1<p<\infty} \mathbf{D}_{p,s}$. Here, $\psi(x)$, $x \in R$ denotes a smooth function $0 \leq \psi(x) \leq 1$, defined as $\psi(x) = 1$ for $|x| \leq 1/2$ and $\psi(x) = 0$ for $|x| \geq 1$. Here, a Banach space $\mathbf{D}_{p,s}$, $s < 0$ is the dual space of $\mathbf{D}_{q,-s}(\mathbf{R})(q = p/(p-1))$.

Moreover, the coupling with the function 1 is well-defined, which is called as generalized expectation and is written as $\mathbf{E}[\psi(\eta_c^{(\epsilon)})\Phi \circ G^{(\epsilon)}]$. Further, $\psi(\eta_c^{(\epsilon)})\Phi \circ G^{(\epsilon)}$ can be expanded in $\tilde{\mathbf{D}}^{-\infty}$.

In addition, it can be shown that $\{\eta_c^{(\epsilon)}(w); \epsilon \in (0, 1]\} \subset \mathbf{D}^{\infty}$, $\eta_c^{(\epsilon)}(w)$ is $O(1)$ in $\mathbf{D}^{\infty}$ as $\epsilon \downarrow 0$, and that for any $a_0 > 0$ there exist positive constants $a_i, i = 1, 2, 3$ such that $P(\{|\eta_c^{(\epsilon)}| > a_0\}) \leq a_1 \exp(-a_2 \epsilon^{-a_3})$. Hence, for any $k = 1, 2, \ldots,$ we have

$$\lim_{\epsilon \downarrow 0} \frac{P(|\eta_c^{(\epsilon)}| > \frac{1}{2})}{\epsilon^k} < \infty.$$

This means that the probability of the events truncated by $\psi(\eta_c^{(\epsilon)})$ is smaller than any polynomial orders of ϵ. Then, in the expansion of $\psi(\eta_c^{(\epsilon)})\Phi \circ G^{(\epsilon)}$, the coefficients expressed as generalized Wiener functionals belonging to $\tilde{\mathbf{D}}^{-\infty}$ can be written by applying Taylor's formula to $\Phi(g_{0T} + \epsilon g_{1T} + \epsilon^2 g_{2T} + \cdots)$. Therefore, the asymptotic expansion of the expectation $\mathbf{E}[\Phi(G^{(\epsilon)})]$ can be obtained relatively easily. For the details of Watanabe theory and its truncated version above, please consult Watanabe [111] and Yoshida [118, 119]. For its application to valuation problems in finance, please also see [50].

In particular, if we take the delta function at $y \in \mathbf{R}$, δ_y as Φ, that is $\Phi(x) = \delta_y(x)$, we obtain an asymptotic expansion of the density function of $G^{(\epsilon)}$. Moreover, because functions such as $\Phi(x) = \max\{x, 0\}$ that is measurable but not smooth, frequently appear in finance, the framework mentioned above is necessary for the asymptotic expansion.

For instance, when we take $\max\{x, 0\}$, $\min\{x, 0\}$ or $\delta_y(x)$ as $\Phi(x)$ for a useful application in finance, the expectation of $\Phi(G^{(\epsilon)})$ is expanded as follows: for $N = 0, 1, 2, \ldots,$

$$\begin{aligned}
\mathbf{E}[\Phi(G^{(\epsilon)})] &= \sum_{n=0}^{N} \epsilon^n \sum_{\vec{k}_m}^{(n)} \frac{1}{m!} \mathbf{E}\left[\Phi^{(m)}(g_{1T}) \left(\prod_{j=1}^{m} g_{(k_j+1)T}\right)\right] + o(\epsilon^N) \\
&= \sum_{n=0}^{N} \epsilon^n \sum_{\vec{k}_m}^{(n)} \frac{1}{m!} \mathbf{E}\left[\Phi^{(m)}(g_{1T}) X^{\vec{k}_m}\right] + o(\epsilon^N) \\
&= \sum_{n=0}^{N} \epsilon^n \sum_{\vec{k}_m}^{(n)} \frac{1}{m!} \int_{-\infty}^{\infty} \Phi^{(m)}(x) \mathbf{E}[X^{\vec{k}_m} | g_{1T} = x] f_{g_{1T}}(x) dx + o(\epsilon^N) \\
&= \sum_{n=0}^{N} \epsilon^n \sum_{\vec{k}_m}^{(n)} \frac{1}{m!} \int_{-\infty}^{\infty} \Phi(x)(-1)^m \\
&\quad \times \frac{d^m}{dx^m} \left\{\mathbf{E}[X^{\vec{k}_m} | g_{1T} = x] f_{g_{1T}}(x)\right\} dx + o(\epsilon^N)
\end{aligned} \tag{11}$$

where $\Phi^{(m)}(g_{1T}) = \left.\frac{d^m\Phi(x)}{dx^m}\right|_{x=g_{1T}}$, $\sum_{\vec{k}_m}^{(n)} = \sum_{m=1}^{n}\sum_{\vec{k}_m\in L_{n,m}}$, and

$$X^{\vec{k}_m} := \prod_{j=1}^{m} g_{(k_j+1)T}. \tag{12}$$

In order to compute the asymptotic expansion (11), we need to evaluate the conditional expectations of the form:

$$E\left[\tilde{X}^{\vec{k}_m}\middle| g_{1T} = x\right],$$

where $\tilde{X}^{\vec{k}_m}$ is represented by a product of multiple Wiener-Itô integrals.

In the preceding works on application of the asymptotic expansion, the conditional expectations in (11) were directly computed with some formulas including multi-dimensional ones given for example, in [85, 86]. Recently, while the formulas up to the third order are given in the works, [95] has developed a high-order computation scheme for the conditional expectations by using the fact that each of these $\{A^j_{k,t}\}_{j,k}$, $\{g_{nT}\}_n$ and also $\{X^{\vec{k}_m}\}_{\vec{k}_m}$ can be decomposed into a finite sum of iterated multiple Wiener-Itô integrals by applications of the Itô's formula with certain properties of iterated multiple Wiener-Itô integrals. (Please see Sect. 4 of [95] for the detail.)

On the other hand, as shown in the next section, we can develop an alternative method which does not evaluate the conditional expectations directly.

3 Computational Scheme

This section follows [96] to introduce a computational scheme for the asymptotic expansion, which is an alternative to the direct calculation method for the conditional expectations given in [95].

3.1 Preparation

To compute the conditional expectations on the right hand side of (11), we use the following lemma which can be derived from a property of Hermite polynomials and leads us to compute the unconditional expectations instead of the conditional ones.

Lemma 2 *Let (Ω, F, P) be a probability space. Suppose that $X \in L^2(\Omega, P)$ and Z is a random variable with Gaussian distribution with mean 0 and variance Σ. Then, the conditional expectation $E[X|Z = x]$ has the following expansion in $L^2(\mathbf{R}, \mu)$ where μ is the Gaussian measure on $\mathbf{R}$ with mean 0 and variance Σ:*

$$E[X|Z = x] = \sum_{n=0}^{\infty} \frac{a_n}{\Sigma^n} H_n(x; \Sigma) \tag{13}$$

where $H_n(x; \Sigma)$ is the Hermite polynomial of degree n which is defined as

$$H_n(x; \Sigma) = (-\Sigma)^n e^{x^2/2\Sigma} \frac{d^n}{dx^n} e^{-x^2/2\Sigma}$$

and the coefficients a_n are given by

$$a_n = \frac{1}{n!} \frac{1}{i^n} \frac{\partial^n}{\partial \xi^n}\bigg|_{\xi=0} \left\{ e^{\frac{\xi^2}{2}\Sigma} \mathbf{E}[e^{i\xi Z} X] \right\}, \; (i = \sqrt{-1}). \tag{14}$$

(Proof) Since the system of Hermite polynomials $\{H_n(x; \Sigma)\}$ is an orthogonal basis of $L^2(\mathbf{R}, \mu)$, and $E[X|Z = x] \in L^2(\mathbf{R}, \mu)$, we have the following unique expansion of $E[X|Z = x]$ in $L^2(\mathbf{R}, \mu)$:

$$E[X|Z = x] = \sum_{n=0}^{\infty} \frac{a_n}{\Sigma^n} H_n(x; \Sigma).$$

Since we have another Taylor expansion

$$e^{i\xi x} = e^{-\frac{\xi^2}{2}\Sigma} \sum_{n=0}^{\infty} \frac{H_n(x; \Sigma)}{n!} (i\xi)^n,$$

then,

$$\begin{aligned} e^{\frac{\xi^2}{2}\Sigma} \mathbf{E}[e^{i\xi Z} X] &= e^{\frac{\xi^2}{2}\Sigma} \int_{\mathbf{R}} e^{i\xi x} \mathbf{E}[X|Z = x] \mu(dx) \\ &= \int_{\mathbf{R}} \sum_{m=0}^{\infty} \frac{H_m(x; \Sigma)}{m!} (i\xi)^m \sum_{n=0}^{\infty} a_n \frac{H_n(x; \Sigma)\mu(dx)}{\Sigma^n} \\ &= \sum_{n=0}^{\infty} a_n (i\xi)^n. \end{aligned}$$

Comparing to the coefficients of the Taylor series of $e^{\frac{\xi^2}{2}\Sigma} \mathbf{E}[e^{i\xi Z} X]$ around 0 with respect to ξ, we see that a_n can be written as (14). □

Next, we write $\hat{V}_t = (\partial g(X_T^{(0)}))^{\top} Y_T Y_t^{-1} V(X_t^{(0)})$ as $\hat{V}(X_t^{(0)})$. Then, we define $\hat{g}_1 = \{\hat{g}_{1t}; t \in \mathbf{R}^+\}$ and $Z^{\langle \xi \rangle} = \{Z_t^{\langle \xi \rangle}; t \in \mathbf{R}^+\}$ as the stochastic processes

$$\hat{g}_{1t} = \int_0^t \hat{V}(X_u^{(0)}) dW_u$$

and

$$Z_t^{\langle \xi \rangle} = \exp\left(i\xi \hat{g}_{1t} + \frac{\xi^2}{2}\Sigma_t\right),$$

respectively, where $\Sigma_t := \int_0^t \hat{V}(X_u^{(0)})\hat{V}(X_u^{(0)})^\top du$.

Then, from Lemma 2, the conditional expectations appearing on the right hand side of the Eq. (11) is expressed as

$$\begin{aligned}\mathbf{E}[X^{\vec{k}_m}|g_{1T} = x] &= \mathbf{E}[X^{\vec{k}_m}|\hat{g}_{1T} = x - C] \\ &= \sum_{l=0}^{\infty} \frac{a_l^{\vec{k}_m}}{\Sigma_T^l} H_l(x - C; \Sigma_T)\end{aligned} \tag{15}$$

where

$$a_l^{\vec{k}_m} = \frac{1}{l!}\frac{1}{i^l}\frac{\partial^l}{\partial \xi^l}\bigg|_{\xi=0} \left\{\mathbf{E}[X^{\vec{k}_m} Z_T^{\langle \xi \rangle}]\right\}. \tag{16}$$

Here it is noted that with this expression we now need to compute unconditional expectations $\mathbf{E}\left[X^{\vec{k}_\delta} Z_T^{\langle \xi \rangle}\right]$ instead of the conditional expectations.

3.2 Asymptotic Expansion of Density Function

In this subsection, we explain a new computational method through deriving a general formula for the expansion (11) with an arbitrary specification of its order N. In particular, we show that the coefficients in the expansion are obtained through a system of ordinary differential equations that is solved easily.

First, we define $\eta_{\vec{l}_\beta}^{\vec{d}_\beta}(t; \xi)$ for $\vec{l}_\beta \in L_{n,\beta}$ and $\vec{d}_\beta \in \{1, \ldots, d\}^\beta$ $(n \geq \beta \geq 1)$ as

$$\eta_{\vec{l}_\beta}^{\vec{d}_\beta}(t; \xi) = \mathbf{E}\left[\left(\prod_{j=1}^{\beta} A_{l_j t}^{d_j}\right) Z_t^{\langle \xi \rangle}\right], \tag{17}$$

and for $n = 0$ as

$$\eta_{(\emptyset)}^{(\emptyset)}(t; \xi) = \mathbf{E}\left[Z_t^{\langle \xi \rangle}\right]. \tag{18}$$

Then, by using (6) we write the unconditional expectations $\mathbf{E}[X^{\vec{k}_m} Z_T^{\langle \xi \rangle}]$ in (16) in terms of η as follows:

$$\mathbf{E}[X^{\vec{k}_m} Z_T^{\langle\xi\rangle}] = \mathbf{E}\left[\left(\prod_{j=1}^{m} g_{(k_j+1)T}\right) Z_T^{\langle\xi\rangle}\right]$$

$$= \mathbf{E}\left[\left(\prod_{j=1}^{m}\left\{\sum_{\vec{l}^j_{\beta_j},\vec{d}^j_{\beta_j}}^{(k_j+1)} \frac{1}{\beta_j!}\partial^{\beta_j}_{\vec{d}^j_{\beta_j}} g(X_T^{(0)}) A^{d^j_1}_{l^j_1 T}\cdots A^{d^j_{\beta_j}}_{l^j_{\beta_j}T}\right\}\right) Z_T^{\langle\xi\rangle}\right]$$

$$= \sum_{\vec{l}^1_{\beta_1},\vec{d}^1_{\beta_1}}^{(k_1+1)} \cdots \sum_{\vec{l}^m_{\beta_m},\vec{d}^\delta_{\beta_m}}^{(k_m+1)} \left(\prod_{j=1}^{m} \frac{1}{\beta_j!}\partial^{\beta_j}_{\vec{d}^j_{\beta_j}} g(X_T^{(0)})\right) \eta^{\vec{d}^1_{\beta_1}\otimes\cdots\otimes\vec{d}^m_{\beta_m}}_{\vec{l}^1_{\beta_1}\otimes\cdots\otimes\vec{l}^\delta_{\beta_\delta}}(T;\xi) \tag{19}$$

where

$$\vec{d}^i_{\beta_i} \otimes \vec{d}^j_{\beta_j} := (d^i_1,\ldots,d^i_{\beta_i}, d^j_1,\ldots,d^j_{\beta_j}),$$
$$\vec{l}^i_{\beta_i} \otimes \vec{l}^j_{\beta_j} := (l^i_1,\ldots,l^i_{\beta_i}, l^j_1,\ldots,l^j_{\beta_j}).$$

So, we have to calculate $\eta^{\vec{d}_\beta}_{\vec{l}_\beta}(T;\xi)$ to evaluate the asymptotic expansion (11).

In the following, we derive a system of ODEs satisfied by these $\{\eta^{\vec{d}_\beta}_{\vec{l}_\beta}\}$. Before showing a general result, we first derive the ODEs for a few leading-low-order terms explicitly to give a better intuition of a key idea of our method. Particularly, let us consider the evaluation of $\eta^j_{(2)}(T;\xi) = E[A^j_{2T} Z_T^{\langle\xi\rangle}]$ which appears in the ϵ-order. Here, for simplicity, we assume that V_0 does not depend on ϵ, and write $V_0(x,\epsilon)$ as $V_0(x)$. In this case, we first note that the SDEs of A^j_{1t} and A^j_{2t} $(j = 1,\ldots,d)$ are given as follows:

$$dA^j_{1t} = \sum_{j'=1}^{d} A^{j'}_{1t}\partial_{j'} V^j_0(X_t^{(0)})dt + V^j(X_t^{(0)})dW_t \tag{20}$$

$$dA^j_{2t} = \left[\sum_{j'=1}^{d} A^{j'}_{2t}\partial_{j'} V^j_0(X_t^{(0)}) + \frac{1}{2}\sum_{j',k'=1}^{d} A^{j'}_{1t}A^{k'}_{1t}\partial_{j'}\partial_{k'} V^j_0(X_t^{(0)})\right]dt$$
$$+ \sum_{j'=1}^{d} A^{j'}_{1t}\partial_{j'} V^j(X_t^{(0)})dW_t. \tag{21}$$

Also, the SDE of $Z_t^{\langle\xi\rangle}$ is expressed as:

$$dZ_t^{\langle\xi\rangle} = (i\xi)\hat{V}(X^{(0)})Z_t^{\langle\xi\rangle}dW_t. \tag{22}$$

Then, applying Itô's formula to $A^j_{2t}Z^{\langle\xi\rangle}_t$, we have

$$
\begin{aligned}
d(A^j_{2t}Z^{\langle\xi\rangle}_t) &= A^j_{2t}dZ^{\langle\xi\rangle}_t + Z^{\langle\xi\rangle}_t dA^j_{2t} + d\langle A^j_2, Z^{\langle\xi\rangle}\rangle_t \\
&= \left\{(i\xi)\sum_{j'=1}^{d} A^{j'}_{1t}Z^{\langle\xi\rangle}_t \hat{V}(X^{(0)}_t)\partial_{j'}V^j(X^{(0)}_t)' + \sum_{j'=1}^{d} A^{j'}_{2t}Z^{\langle\xi\rangle}_t \partial_{j'}V^j_0(X^{(0)}_t)\right. \\
&\quad \left. +\frac{1}{2}\sum_{j',k'=1}^{d} A^{j'}_{1t}A^{k'}_{1t}Z^{\langle\xi\rangle}_t \partial_{j'}\partial_{k'}V^j_0(X^{(0)}_t)\right\}dt \\
&\quad + \left\{(i\xi)A^j_{2t}Z^{\langle\xi\rangle}_t \hat{V}(X^{(0)}_t) + \sum_{j'=1}^{d} A^{j'}_{1t}Z^{\langle\xi\rangle}_t \partial_{j'}V^j(X^{(0)}_t)\right\} dW_t.
\end{aligned}
$$

Since the last term is a martingale, taking expectation on both sides, we have the following ordinary differential equation for $\eta^j_{(2)}$:

$$
\begin{aligned}
\frac{d}{dt}\eta^j_{(2)}(t;\xi) &= (i\xi)\sum_{j'=1}^{d} \eta^{j'}_{(1)}(t;\xi)\hat{V}(X^{(0)}_t)\partial_{j'}V^j(X^{(0)}_t)^\top \\
&\quad + \sum_{j'=1}^{d} \eta^{j'}_{(2)}(t;\xi)\partial_{j'}V^j_0(X^{(0)}_t) + \frac{1}{2}\sum_{j',k'=1}^{d} \eta^{j',k'}_{(1,1)}(t;\xi)\partial_{j'}\partial_{k'}V^j_0(X^{(0)}_t).
\end{aligned}
$$

Here, $\eta^j_{(1)}(j = 1,\ldots,d)$ appearing in the right hand side of the above ODE are evaluated in the similar manner:

$$
\begin{aligned}
d(A^j_{1t}Z^{\langle\xi\rangle}_t) &= A^j_{1t}dZ^{\langle\xi\rangle}_t + Z^{\langle\xi\rangle}_t dA^j_{1t} + d\langle A^j_1, Z^{\langle\xi\rangle}\rangle_t \\
&= \left\{(i\xi)Z^{\langle\xi\rangle}_t \hat{V}(X^{(0)}_t)V^j(X^{(0)}_t)^\top + \sum_{j'=1}^{d} A^{j'}_{1t}Z^{\langle\xi\rangle}_t \partial_{j'}V^j_0(X^{(0)}_t)\right\} dt \\
&\quad + \left\{(i\xi)A^j_{1t}Z^{\langle\xi\rangle}_t \hat{V}(X^{(0)}_t) + Z^{\langle\xi\rangle}_t V^j(X^{(0)}_t)\right\} dW_t,
\end{aligned}
$$

hence, we have

$$
\frac{d}{dt}\eta^j_{(1)}(t;\xi) = (i\xi)\hat{V}(X^{(0)}_t)V^j(X^{(0)}_t)^\top + \sum_{j'=1}^{d} \eta^{j'}_{(1)}(t;\xi)\partial_{j'}V^j_0(X^{(0)}_t).
$$

$\eta^{j,k}_{(1,1)}$ and other higher-order terms can be evaluated in the same way. The key observation is that each ODE does not involve any higher-order terms, and only lower- or the same order-terms appear in the right hand side of the ODE. So, one can easily solve (analytically or numerically) the system of ODEs and evaluate the expectations.

The following proposition provides a way to calculate general $\eta_{\vec{l}_\beta}^{\vec{d}_\beta}(T;\xi)$ as a solution to the system of the ordinary differential equations:

Proposition 1 *For $\eta_{\vec{l}_\beta}^{\vec{d}_\beta}(t;\xi)$ defined in* (17), *the following system of ordinary differential equations is satisfied:*

$$\begin{aligned}
\frac{d}{dt}\left\{\eta_{\vec{l}_\beta}^{\vec{d}_\beta}(t;\xi)\right\} &= \sum_{k=1}^{\beta}\frac{1}{l_k!}\left\{\eta_{\vec{l}_{\beta/k}}^{\vec{d}_{\beta/k}}(t;\xi)\right\}\left\{\partial_\epsilon^{l_k}V_0^{d_k}(X_t^{(0)},0)\right\}\\
&+\sum_{k=1}^{\beta}\sum_{l=1}^{l_k}\sum_{\vec{m}_\gamma,\vec{\tilde{d}}_\gamma}^{(l)}\frac{1}{(l_k-l)!}\frac{1}{\gamma!}\left\{\eta_{(\vec{l}_{\beta/k})\otimes\vec{m}_\gamma}^{(\vec{d}_{\beta/k})\otimes\vec{\tilde{d}}_\gamma}(t;\xi)\right\}\\
&\times\left\{\partial_{\vec{\tilde{d}}_\gamma}^{\gamma}\,\partial_\epsilon^{l_k-l}V_0^{d_k}(X_t^{(0)},0)\right\}\\
&+\sum_{\substack{k,m=1\\k<m}}^{\beta}\sum_{\vec{m}_\gamma,\vec{\tilde{d}}_\gamma}^{(l_k-1)}\sum_{\vec{m}_\delta,\vec{\hat{d}}_\delta}^{(l_m-1)}\frac{1}{\gamma!\delta!}\left\{\eta_{(\vec{l}_{\beta/k,m})\otimes\vec{m}_\gamma\otimes\vec{m}_\delta}^{(\vec{d}_{\beta/k,m})\otimes\vec{\tilde{d}}_\gamma\otimes\vec{\hat{d}}_\delta}(t;\xi)\right\}\left\{\partial_{\vec{\tilde{d}}_\gamma}^{\gamma}\,V^{d_k}(X_t^{(0)})\right\}\\
&\times\left\{\partial_{\vec{\hat{d}}_\delta}^{\delta}\,V^{d_m}(X_t^{(0)})\right\}\\
&+(i\xi)\sum_{k=1}^{\beta}\sum_{\vec{m}_\gamma,\vec{\tilde{d}}_\gamma}^{(l_k-1)}\frac{1}{\gamma!}\left\{\eta_{(\vec{l}_{\beta/k})\otimes\vec{m}_\gamma}^{(\vec{d}_{\beta/k})\otimes\vec{\tilde{d}}_\gamma}(t;\xi)\right\}\\
&\times\left\{\partial_{\vec{\tilde{d}}_\gamma}^{\gamma}\,V^{d_k}(X_t^{(0)})\right\}\hat{V}(X_t^{(0)}) \qquad (23)
\end{aligned}$$

where $\sum_{\vec{m}_\gamma,\vec{\tilde{d}}_\gamma}^{(l)}$ is defined in (4), *and*

$$\begin{aligned}
\vec{l}_{\beta/k} &:= (l_1,\ldots,l_{k-1},l_{k+1},\ldots,l_\beta)\\
\vec{l}_{\beta/k,n} &:= (l_1,\ldots,l_{k-1},l_{k+1},\ldots,l_{n-1},l_{n+1},\ldots,l_\beta),\ 1\le k<n\le\beta\\
\vec{l}_\beta\otimes\vec{m}_\gamma &:= (l_1,\ldots,l_\beta,m_1,\ldots,m_\gamma)
\end{aligned}$$

for $\vec{l}_\beta=(l_1,\ldots,l_\beta)$ and $\vec{m}_\gamma=(m_1,\ldots,m_\gamma)$.

(Proof) We firstly apply Itô's formula to $\left(\prod_{j=1}^{\beta}A_{l_jt}^{d_j}\right)$ by using (3) to obtain the following:

$$
\begin{aligned}
d\left(\prod_{j=1}^{\beta} A_{l_j t}^{d_j}\right) &= \sum_{k=1}^{\beta}\left(\prod_{\substack{j=1\\ j\neq k}}^{\beta} A_{l_j t}^{d_j}\right) dA_{l_k t}^{d_k} + \sum_{\substack{k,m=1\\ k<m}}^{\beta}\left(\prod_{\substack{j=1\\ j\neq k,m}}^{\beta} A_{l_j t}^{d_j}\right) d\langle A_{l_k}^{d_k}, A_{l_m}^{d_m}\rangle_t \\
&= \sum_{k=1}^{\beta}\left(\prod_{\substack{j=1\\ j\neq k}}^{\beta} A_{l_j t}^{d_j}\right) \frac{1}{l_k!}\partial_{\epsilon}^{l_k} V_0^{d_k}(X_t^{(0)}, 0)dt \\
&\quad + \sum_{k=1}^{\beta}\left(\prod_{\substack{j=1\\ j\neq k}}^{\beta} A_{l_j t}^{d_j}\right) \sum_{l=1}^{l_k}\sum_{\vec{m}_\gamma, \vec{\tilde{d}}_\gamma}^{(l)} \frac{1}{(l_k-l)!}\frac{1}{\gamma!}\left(\prod_{j'=1}^{\gamma} A_{m_{j'} t}^{\tilde{d}_{j'}}\right) \\
&\quad \times \partial_{\tilde{d}_\gamma}^{\gamma}\partial_{\epsilon}^{l_k - l} V_0^{d_k}(X_t^{(0)}, 0)dt \\
&\quad + \sum_{k=1}^{\beta}\left(\prod_{\substack{j=1\\ j\neq k}}^{\beta} A_{l_j t}^{d_j}\right) \sum_{\vec{m}_\gamma, \vec{\tilde{d}}_\gamma}^{(l_k-1)} \frac{1}{\gamma!}\left(\prod_{j'=1}^{\gamma} A_{m_{j'} t}^{\tilde{d}_{j'}}\right) \partial_{\tilde{d}_\gamma}^{\gamma} V^{d_k}(X_t^{(0)})dW_t \\
&\quad + \sum_{\substack{k,m=1\\ k<m}}^{\beta}\left(\prod_{\substack{j=1\\ j\neq k,m}}^{\beta} A_{l_j t}^{d_j}\right) \sum_{\vec{m}_\gamma, \vec{\tilde{d}}_\gamma}^{(l_k-1)} \sum_{\vec{m}_\delta, \vec{\hat{d}}_\delta}^{(l_m-1)} \frac{1}{\gamma!\delta!} \\
&\quad \times \left(\prod_{j'=1}^{\gamma} A_{m_{j'} t}^{\tilde{d}_{j'}}\right) \partial_{\tilde{d}_\gamma}^{\gamma} V^{d_k}(X_t^{(0)}) \left(\prod_{j'=1}^{\delta} A_{m_{j'} t}^{\hat{d}_{j'}}\right) \partial_{\hat{d}_\delta}^{\delta} V^{d_m}(X_t^{(0)})dt.
\end{aligned}
\tag{24}
$$

Note also that $dZ_t^{\langle\xi\rangle} = (i\xi)\hat{V}(X_t^{(0)})Z_t^{\langle\xi\rangle}dW_t$. Then, applying Itô's formula again to $\left(\prod_{j=1}^{\beta} A_{l_j t}^{d_j} Z_t^{\langle\xi\rangle}\right)$ and take expectations on both sides to obtain the result. □

Remark 2 Due to $\eta_{(\emptyset)}^{(\emptyset)}(t;\xi) = \mathbf{E}[Z_t^{\langle\xi\rangle}] = 1$, and the hierarchical structure of the ODEs with respect to $n = \sum_{j=1}^{\beta} l_j$, one can easily solve these ODEs successively from lower-order terms to higher-order terms with initial conditions $\eta_{\vec{l}_\beta}^{\vec{d}_\beta}(0;\xi) = 0$ for $(\vec{l}_\beta, \vec{d}_\beta) \neq (\emptyset, \emptyset)$.

Remark 3 Further, due to the structure of the system of the differential equations, it is easily shown by induction that each $\eta_{\vec{l}_\beta}^{\vec{d}_\beta}(t;\xi)$ is expressed as a polynomial of degree $n = \sum_{j=1}^{\beta} l_j$ with respect to $(i\xi)$. Then, we can also show that $\mathbf{E}[X^{\vec{k}_m} Z_T^{\langle\xi\rangle}]$

is a polynomial of degree $(n+m)$ with respect to $(i\xi)$, and thus $a_l^{\vec{k}_m} = 0 (l > n+m)$ for $\vec{k}_m \in L_{n,m}$. This ensures a convergence of the infinite sum in (15).

Then, from Lemma 2 and (11), we have the following expression of $\mathbf{E}[\Phi(G^{(\epsilon)})]$:

$$\begin{aligned}\mathbf{E}[\Phi(G^{(\epsilon)})] &= \sum_{n=0}^{N} \epsilon^n \sum_{\vec{k}_m}^{(n)} \frac{1}{m!} \int_{\mathbf{R}} \Phi(x)(-1)^m \frac{d^m}{dx^m} \\ &\quad \times \left\{ \sum_{l=0}^{n+m} \frac{a_l^{\vec{k}_m}}{\Sigma_T^l} H_l(x-C;\Sigma_T) f_{g_{1T}}(x) \right\} dx + o(\epsilon^N) \\ &= \sum_{n=0}^{N} \epsilon^n \sum_{\vec{k}_m}^{(n)} \frac{1}{m!} \int_{\mathbf{R}} \Phi(x) \\ &\quad \times \left\{ \sum_{l=0}^{n+m} \frac{a_l^{\vec{k}_m}}{\Sigma_T^{l+m}} H_{l+m}(x-C;\Sigma_T) f_{g_{1T}}(x) \right\} dx + o(\epsilon^N)\end{aligned}$$

Here we have used the well-known property of the Hermite polynomial:

$$\frac{d^m}{dx^m} \left\{ H_l(x-C;\Sigma_T) f_{g_{1T}}(x) \right\} = \left(\frac{-1}{\Sigma_T} \right)^m H_{l+m}(x-C;\Sigma_T) f_{g_{1T}}(x).$$

In particular, let Φ be the delta function at $x \in \mathbf{R}$, δ_x, we obtain the asymptotic expansion of the density of $G^{(\epsilon)}$:

$$\begin{aligned} f_{G^{(\epsilon)}}(x) &= \mathbf{E}[\delta_x(G^{(\epsilon)})] \\ &= \sum_{n=0}^{N} \epsilon^n \sum_{\vec{k}_m}^{(n)} \frac{1}{m!} \sum_{l=0}^{n+m} \frac{a_l^{\vec{k}_m}}{\Sigma_T^{l+m}} H_{l+m}(x-C;\Sigma_T) f_{g_{1T}}(x) + o(\epsilon^N). \quad (25)\end{aligned}$$

We summarize the discussion above as the following theorem:

Theorem 2 *Let $X^{(\epsilon)}$ be the solution to the stochastic differential equation* (1). *Suppose a function $g : \mathbf{R}^d \mapsto \mathbf{R}$ is smooth and all of its derivatives have polynomial growth. Then, the asymptotic expansion of the density function of $G^{(\epsilon)} = \frac{g(X_T^{(\epsilon)}) - g(X_T^{(0)})}{\epsilon}$ up to ϵ^N-order is given by*

$$\begin{aligned} f_{G^{(\epsilon)}}(x) &= f_{g_{1T}}(x) \\ &\quad + \sum_{n=1}^{N} \epsilon^n \left(\sum_{m=0}^{3n} C_{nm} H_m(x-C;\Sigma_T) \right) f_{g_{1T}}(x) + o(\epsilon^N), \end{aligned}$$

(26)

where

$$f_{g_{1T}}(x) = \frac{1}{\sqrt{2\pi\Sigma_T}} \exp\left(-\frac{(x-C)^2}{2\Sigma_T}\right) \tag{27}$$

with

$$\begin{aligned} C &= \left(\partial g(X_T^{(0)})\right)^\top \int_0^T Y_T Y_t^{-1} \partial_\epsilon V_0(X_t^{(0)}, 0) dt, \\ \Sigma_T &= \int_0^T \hat{V}(X_t^{(0)}) \hat{V}(X_t^{(0)})^\top dt > 0, \\ \hat{V}(X_t^{(0)}) &= (\partial g(X_T^{(0)}))^\top Y_T Y_t^{-1} V(X_t^{(0)}). \end{aligned}$$

$H_n(x;\Sigma)$ *is the Hermite polynomial of degree n with parameter* Σ*, which is defined as*

$$H_n(x;\Sigma) = (-\Sigma)^n e^{x^2/2\Sigma} \frac{d^n}{dx^n} e^{-x^2/2\Sigma}, \tag{28}$$

and

$$\begin{aligned} C_{nm} = &\frac{1}{\Sigma_T^m} \sum_{\vec{k}_\delta}^{(m)} \sum_{\vec{l}^1_{\beta_1}, \vec{d}^1_{\beta_1}}^{(k_1+1)} \cdots \sum_{\vec{l}^\delta_{\beta_\delta}, \vec{d}^\delta_{\beta_\delta}}^{(k_\delta+1)} \frac{1}{\delta!(m-\delta)!} \\ &\times \left(\prod_{j=1}^{\delta} \frac{1}{\beta_j!} \partial^{\beta_j}_{\vec{d}^j_{\beta_j}} g(X_T^{(0)}) \right) \\ &\times \frac{1}{i^{m-\delta}} \frac{\partial^{m-\delta}}{\partial \xi^{m-\delta}} \bigg|_{\xi=0} \left\{ \eta^{\vec{d}^1_{\beta_1} \otimes \ldots \otimes \vec{d}^\delta_{\beta_\delta}}_{\vec{l}^1_{\beta_1} \otimes \ldots \otimes \vec{l}^\delta_{\beta_\delta}} (T;\xi) \right\}, \ \left(i = \sqrt{-1}\right). \end{aligned} \tag{29}$$

$\eta^{\vec{d}_\beta}_{\vec{l}_\beta}(T;\xi)$ *are obtained as a solution to the following system of ODEs:*

$$\begin{aligned} \frac{d}{dt}\left\{ \eta^{\vec{d}_\beta}_{\vec{l}_\beta}(t;\xi) \right\} = &\sum_{k=1}^{\beta} \frac{1}{l_k!} \eta^{\vec{d}_{\beta/k}}_{\vec{l}_{\beta/k}}(t;\xi) \partial_\epsilon^{l_k} V_0^{d_k}(X_t^{(0)}, 0) \\ &+ \sum_{k=1}^{\beta} \sum_{l=1}^{l_k} \sum_{\vec{m}_\gamma, \vec{\tilde{d}}_\gamma}^{(l)} \frac{1}{(l_k-l)!} \frac{1}{\gamma!} \eta^{(\vec{d}_{\beta/k}) \otimes \vec{\tilde{d}}_\gamma}_{(\vec{l}_{\beta/k}) \otimes \vec{m}_\gamma}(t;\xi) \partial^\gamma_{\vec{\tilde{d}}_\gamma} \partial_\epsilon^{l_k - l} V_0^{d_k}(\tilde{X}_t^{(0)}, 0) \\ &+ \sum_{\substack{k,m=1 \\ k<m}}^{\beta} \sum_{\vec{m}_\gamma, \vec{\tilde{d}}_\gamma}^{(l_k-1)} \sum_{\vec{m}_\delta, \vec{\hat{d}}_\delta}^{(l_m-1)} \frac{1}{\gamma!\delta!} \eta^{(\vec{d}_{\beta/k,m}) \otimes \vec{\tilde{d}}_\gamma \otimes \vec{\hat{d}}_\delta}_{(\vec{l}_{\beta/k,m}) \otimes \vec{m}_\gamma \otimes \vec{m}_\delta}(t;\xi) \end{aligned}$$

$$\times \partial^{\gamma}_{\vec{\tilde{d}}_\gamma} V^{d_k}(X_t^{(0)})\partial^{\delta}_{\vec{\hat{d}}_\delta} V^{d_m}(X_t^{(0)})$$

$$+ (i\xi)\sum_{k=1}^{\beta}\sum_{\vec{m}_\gamma,\vec{\tilde{d}}_\gamma}^{(l_k-1)} \frac{1}{\gamma!}\eta^{(\vec{d}_{\beta/k})\otimes\vec{\tilde{d}}_\gamma}_{(\vec{l}_{\beta/k})\otimes\vec{m}_\gamma}(t;\xi)\partial^{\gamma}_{\vec{\tilde{d}}_\gamma} V^{d_k}(X_t^{(0)})\hat{V}(X_t^{(0)},t)$$

$$\eta^{\vec{d}_\beta}_{\vec{l}_\beta}(0;\xi) = 0\, for\ (\vec{l}_\beta,\vec{d}_\beta) \neq (\emptyset,\emptyset),\ \eta^{(\emptyset)}_{(\emptyset)}(t;\xi) = 1\ for\ (\vec{l}_\beta,\vec{d}_\beta) = (\emptyset,\emptyset). \tag{30}$$

Here, we use the following notations:

$$\vec{l}_{\beta/k} := (l_1,\dots,l_{k-1},l_{k+1},\dots,l_\beta)$$
$$\vec{l}_{\beta/k,n} := (l_1,\dots,l_{k-1},l_{k+1},\dots,l_{n-1},l_{n+1},\dots,l_\beta),\ 1 \le k < n \le \beta$$
$$\vec{l}_\beta \otimes \vec{m}_\gamma := (l_1,\dots,l_\beta,m_1,\dots,m_\gamma)$$

for $\vec{l}_\beta = (l_1,\dots,l_\beta)$ *and* $\vec{m}_\gamma = (m_1,\dots,m_\gamma)$.

Remark 4 Particularly, in order to calculate the expansion above up to the ϵ^2-order, we need the Hermite polynomials $H_n(x;\Sigma)$ up to $n = 6$, which are given as follows:

$$H_0(x;\Sigma) = 1,$$
$$H_1(x;\Sigma) = x,$$
$$H_2(x;\Sigma) = x^2 - \Sigma,$$
$$H_3(x;\Sigma) = x^3 - 3\Sigma x,$$
$$H_4(x;\Sigma) = x^4 - 6\Sigma x^2 + 3\Sigma^2,$$
$$H_5(x;\Sigma) = x^5 - 10\Sigma x^3 + 15\Sigma^2 x,$$
$$H_6(x;\Sigma) = x^6 - 15\Sigma x^4 + 45\Sigma^2 x^2 - 15\Sigma^3.$$

3.3 Remarks on the Asymptotic Expansion for Multi-dimensional Density Functions

We can also apply the conditional expectation formulas for the multi-dimensional case in Lemma 1.1′ of [85] and Lemma 2.1 of [86] to derive an asymptotic expansion up to the third order of the multi-dimensional density functions. This is particularly useful for pricing exotic-type options such as barrier options with discrete monitoring (e.g. [83]), and pricing Bermudan-type or approximate American-type derivatives (e.g. Nishiba [71]).

Moreover, we obtain the following result as an extension of Lemma 2, which easily leads to an asymptotic expansion of a multi-dimensional density function in the similar manner as in the one dimensional case in Theorem 2.

Lemma 3 *Let $(\Omega, \mathcal{F}, P)$ be a probability space. Suppose that $X \in L^2(\Omega, P)$ and $\vec{Z}$ is a d-dimensional random variable with Gaussian distribution with mean $\vec{0}$ (d-dimensional zero vector) and variance-covariance matrix $\underline{\Sigma}$. Then, the conditional expectation $\mathbf{E}[X|\vec{Z}=\vec{x}]$ for $\vec{x} \in \mathbf{R}^d$ has the following expansion in $L^2(\mathbf{R}^d, \boldsymbol{\mu})$ where $\boldsymbol{\mu}$ is the Gaussian measure on $\mathbf{R}^d$ with mean $\vec{0}$ and variance-covariance matrix $\underline{\Sigma}$:*

$$\mathbf{E}[X|\vec{Z}=\vec{x}] = \sum_{|\vec{n}|=0}^{\infty} a_{\vec{n}!} H_{\vec{n}}(\vec{x}; \underline{\Sigma}), \tag{31}$$

where $\vec{n} = (n_1, n_2, \ldots, n_d)$, $|\vec{n}| = n_1 + n_2 + \cdots + n_d$, $\vec{n}! = n_1! n_2! \cdots n_d!$ and

$$a_{\vec{n}} = \frac{1}{\vec{n}} \frac{1}{i^{|\vec{n}|}} \frac{\partial^{\vec{n}}}{\partial \vec{\xi}}\bigg|_{\vec{\xi}=\vec{0}} \left\{ e^{\frac{1}{2}(\vec{\xi})^{\top} \underline{\Sigma} \vec{\xi}} \mathbf{E}\left[e^{i(\vec{\xi})^{\top} \vec{Z}} X \right] \right\}, \ \left(i = \sqrt{-1} \right). \tag{32}$$

Here, $(\vec{\xi})^{\top}$ denotes the transpose of $\vec{\xi}$. $H_{\vec{n}}(\vec{x}; \underline{\Sigma})$ stands for the d-dimensional multiple Hermite polynomial of degree $|\vec{n}|$ with $\vec{n} = (n_1, n_2, \ldots, n_d)$:

$$\begin{aligned} H_{\vec{n}}(\vec{x}; \underline{\Sigma}) &= \frac{1}{n[\vec{x} : \underline{\Sigma}]} \left(-\frac{\partial}{\partial_{x_1}} \right) \left(-\frac{\partial}{\partial_{x_2}} \right) \cdots \left(-\frac{\partial}{\partial_{x_d}} \right) n[\vec{x} : \underline{\Sigma}]; \\ \vec{x} &= (x_1, x_2, \ldots, x_d) \end{aligned} \tag{33}$$

where

$$n[\vec{x} : \underline{\Sigma}] = \frac{1}{(2\pi)^{d/2} |\underline{\Sigma}|^{1/2}} \exp\left\{ -\frac{1}{2} \vec{x}^{\top} \underline{\Sigma}^{-1} \vec{x} \right\}. \tag{34}$$

(Proof) Basically, we can make a similar discussion as in the proof of Lemma 2. Indeed we first note that the system of the following Hermite polynomials is a complete biorthogonal system in $L^2(\mathbf{R}^d, \boldsymbol{\mu})$:

$$\begin{aligned} &\{H_{\vec{n}}(\vec{x} : \underline{\Sigma}) : \vec{n} = (n_1, n_2, \ldots, n_d); n_i = 0, 1, 2, \ldots, (i = 1, 2, \ldots, d)\}, \\ &\{\tilde{H}_{\vec{n}}(\vec{x} : \underline{\Sigma}) : \vec{n} = (n_1, n_2, \ldots, n_d); n_i = 0, 1, 2, \ldots, (i = 1, 2, \ldots, d)\}, \end{aligned}$$

where $H_{\vec{n}}(\vec{x} : \underline{\Sigma})$ is given by (33) and $\tilde{H}_{\vec{n}}(\vec{x} : \underline{\Sigma})$ is defined as follows:

$$\begin{aligned} \tilde{H}_{\vec{n}}(\vec{x}; \underline{\Sigma}) &= \frac{1}{n[\vec{x} : \underline{\Sigma}]} \left(-\frac{\partial}{\partial_{y_1}} \right) \left(-\frac{\partial}{\partial_{y_2}} \right) \cdots \left(-\frac{\partial}{\partial_{y_d}} \right) n[\vec{x} : \underline{\Sigma}], \\ \vec{y} &= (y_1, y_2, \ldots, y_d)^{\top} = \underline{\Sigma}^{-1} \vec{x}. \end{aligned} \tag{35}$$

Thus, we have the following expansion of $E[X|\vec{Z}=\vec{x}]$ in $L^2(\mathbf{R}^d, \mu)$:

$$\mathbf{E}[X|\vec{Z}=\vec{x}] = \sum_{|\vec{n}|=0}^{\infty} a_{\vec{n}} H_{\vec{n}}(\vec{x}; \underline{\Sigma}).$$

On the other hand, we know the relation:

$$\sum_{|\vec{j}|=0}^{\infty} \frac{(i\vec{\xi})^{\vec{j}}}{\vec{j}!} \tilde{H}_{\vec{j}}(\vec{x}; \underline{\Sigma}) = e^{i\vec{\xi}^{\top}\vec{x}} e^{\frac{1}{2}\vec{\xi}^{\top}\underline{\Sigma}\vec{\xi}}, \tag{36}$$

where $(i\vec{\xi})^{\vec{j}} = (i\xi_1)^{j_1}(i\xi_2)^{j_2}\cdots(i\xi_d)^{j_d}$. Hence,

$$e^{i\vec{\xi}^{\top}\vec{x}} = e^{-\frac{1}{2}\vec{\xi}^{\top}\underline{\Sigma}\vec{\xi}} \sum_{|\vec{j}|=0}^{\infty} \frac{(i\vec{\xi})^{\vec{j}}}{\vec{j}!} \tilde{H}_{\vec{j}}(\vec{x} : \underline{\Sigma}).$$

It is also well known that

$$\int_{\mathbf{R}^d} H_{\vec{m}}(\vec{x} : \underline{\Sigma}) \tilde{H}_{\vec{n}}(\vec{x} : \underline{\Sigma}) n[\vec{x} : \underline{\Sigma}] d\vec{x} = \begin{cases} \vec{m}! \text{ (if } \vec{m}=\vec{n}), \\ 0 \text{ (if } \vec{m}\neq\vec{n}). \end{cases} \tag{37}$$

Therefore,

$$\begin{aligned} e^{\frac{1}{2}\vec{\xi}^{\top}\underline{\Sigma}\vec{\xi}} \mathbf{E}\left[e^{i\vec{\xi}^{\top}\vec{Z}} X\right] &= e^{\frac{1}{2}\vec{\xi}^{\top}\underline{\Sigma}\vec{\xi}} \mathbf{E}\left[e^{i\vec{\xi}^{\top}\vec{Z}} \mathbf{E}\left[X|\vec{Z}\right]\right] \\ &= \int_{\mathbf{R}^d} \left\{ \sum_{|\vec{j}|=0}^{\infty} \tilde{H}_{\vec{j}}(\vec{x} : \underline{\Sigma}) \frac{(i\vec{\xi})^{\vec{j}}}{\vec{j}!} \right\} \\ &\quad \times \left\{ \sum_{|\vec{n}|=0}^{\infty} a_{\vec{n}} H_{\vec{n}}(\vec{x} : \underline{\Sigma}) \right\} \mu(d\vec{x}) \qquad (38) \\ &= \sum_{|\vec{n}|=0}^{\infty} a_{\vec{n}} i^{|\vec{n}|} (\vec{\xi})^{\vec{n}}; \ ((\vec{\xi})^{\vec{n}} = \xi_1^{n_1}\xi_2^{n_2}\cdots\xi_d^{n_d}), \qquad (39) \end{aligned}$$

and making $\vec{n} = (n_1, \ldots, n_d)$th order differentiation of both sides in the equation above with respect to $\vec{\xi} = (\xi_1, \ldots, \xi_d)$ at $\vec{\xi} = \vec{0}$, we obtain (32) and hence the result, (31)–(34).

3.4 Expansion of Option Prices

Now, we apply the approximate density function in Theorem 2 obtained by the asymptotic expansion technique to option pricing.

In particular, we consider a plain vanilla option on the underlying asset price process $(g(X_t^{(\epsilon)}))_{t\in[0,T]}$, where $(X_t^{(\epsilon)})_{t\in[0,T]}$ is the solution to the stochastic differential equation expressed as the Eq. (1). As an example, we obtain an approximation of a call option price as follows.

Theorem 3 *An asymptotic expansion up to the $\epsilon^{(N+1)}$-order of a call option price at time* 0 *with maturity T and strike price K where $K = g(X_T^{(0)}) - \epsilon y$ for arbitrary $y \in \mathbf{R}$ is given as follows:*

$$
\begin{aligned}
C(K,T) &= \epsilon P(0,T)\left[\sqrt{\Sigma_T}\, n\left(\frac{y+C}{\sqrt{\Sigma_T}}\right) + CN\left(\frac{y+C}{\sqrt{\Sigma_T}}\right) + yN\left(\frac{y+C}{\sqrt{\Sigma_T}}\right)\right] \qquad (40)\\
&+\sum_{n=1}^{N}\epsilon^{n+1}P(0,T)C_{n0}\left[\sqrt{\Sigma_T}\, n\left(\frac{y+C}{\sqrt{\Sigma_T}}\right) + CN\left(\frac{y+C}{\sqrt{\Sigma_T}}\right)\right]\\
&+\sum_{n=1}^{N}\epsilon^{n+1}P(0,T)C_{n1}\left[\Sigma_T\, N\left(\frac{y+C}{\sqrt{\Sigma_T}}\right) - \sqrt{\Sigma_T}\, y\, n\left(\frac{y+C}{\sqrt{\Sigma_T}}\right)\right]\\
&+\sum_{n=1}^{N}\epsilon^{n+1}P(0,T)\sum_{m=2}^{3n}C_{nm}\left[-y\sqrt{\Sigma_T}\, H_{m-1}\left(-(y+C);\Sigma_T\right)\, n\left(\frac{y+C}{\sqrt{\Sigma_T}}\right)\right.\\
&+\left.\Sigma_T^{\frac{3}{2}}\, H_{m-2}\left(-(y+C);\Sigma_T\right)\, n\left(\frac{y+C}{\sqrt{\Sigma_T}}\right)\right]\\
&+y\sum_{n=1}^{N}\epsilon^{n+1}P(0,T)\, C_{n0}N\left(\frac{y+C}{\sqrt{\Sigma_T}}\right)\\
&+y\sum_{n=1}^{N}\epsilon^{n+1}P(0,T)\sum_{m=1}^{3n}C_{nm}\,\sqrt{\Sigma_T}H_{m-1}\left(-(y+C);\Sigma_T\right)\, n\left(\frac{y+C}{\sqrt{\Sigma_T}}\right) + o(\epsilon^{(N+1)}).
\end{aligned}
$$

Here, C_{nm} is given by (29), *and $H_n(x;\Sigma)$ is the Hermite polynomial of degree n with parameter Σ, which is defined as*

$$
H_n(x;\Sigma) = (-\Sigma)^n e^{x^2/2\Sigma}\frac{d^n}{dx^n}e^{-x^2/2\Sigma}.
$$

C and Σ_T are given respectively by

$$
C = \left(\partial g(X_T^{(0)})\right)^{\top}\int_0^T Y_T Y_t^{-1}\partial_\epsilon V_0(X_t^{(0)},0)dt
$$

and

$$\Sigma_T = \int_0^T \hat{V}(X_t^{(0)})\hat{V}(X_t^{(0)})^\top dt,$$

where

$$\hat{V}(X_t^{(0)}) = (\partial g(X_T^{(0)}))^\top Y_T Y_t^{-1} V(X_t^{(0)}).$$

Also, $P(0,T)$ denotes the price at time 0 of a zero coupon bond with maturity T. $N(x)$ stands for the standard normal distribution function, and its density function is given by $n(x) = \frac{1}{\sqrt{2\pi}}e^{-x^2/2}$.

(Proof) We firstly note that the call price is expanded as follows:

$$\begin{aligned}
C(K,T) &= P(0,T)\mathbf{E}[\max\{g(X_T^{(\epsilon)}) - K, 0\}] \\
&= \epsilon P(0,T)\mathbf{E}\left[\max\left\{\left(\frac{g(X_T^{(\epsilon)}) - g(X_T^{(0)})}{\epsilon}\right) + \left(\frac{g(X_T^{(0)}) - K}{\epsilon}\right), 0\right\}\right] \\
&= \epsilon P(0,T)\mathbf{E}\left[\max\left\{G^{(\epsilon)} + y, 0\right\}\right] \\
&= \epsilon P(0,T)\int_{-y}^{\infty} (x+y) f_{G^{(\epsilon)},N}(x)dx + o(\epsilon^{(N+1)}).
\end{aligned} \quad (41)$$

Here, $f_{G^{(\epsilon)},N}$ is the asymptotic expansion of the density of $G^{(\epsilon)}$ up to ϵ^N-order, which is given by the first two terms on the right hand side of (26) in Theorem 2:

$$f_{G^{(\epsilon)},N}(x) = f_{g_{1T}}(x) + \sum_{n=1}^{N} \epsilon^n \left(\sum_{m=0}^{3n} C_{nm} H_m(x - C; \Sigma_T)\right) f_{g_{1T}}(x), \quad (42)$$

where

$$f_{g_{1T}}(x) = \frac{1}{\sqrt{2\pi\Sigma_T}} \exp\left(-\frac{(x-C)^2}{2\Sigma_T}\right).$$

Next, we note the well-known properties of the Hermite polynomials:

$$\begin{aligned}
&\frac{d}{dx} H_n(x;\Sigma) = n H_{n-1}(x;\Sigma) \quad (43) \\
&\frac{d^m}{dx^m}\{H_n(x;\Sigma)n(x;\Sigma)\} = \left(\frac{-1}{\Sigma}\right)^m H_{n+m}(x;\Sigma)n(x;\Sigma) \\
&H_{n+1}(x;\Sigma) = xH_n(x;\Sigma) - \Sigma n H_{n-1}(x;\Sigma),
\end{aligned}$$

where $n(x;\Sigma) = \frac{1}{\sqrt{2\pi\Sigma}}e^{-\frac{x^2}{2\Sigma}}$.

Then, we can obtain the following expressions for the Integrals appearing on the right hand side of (41):

$$\int_{-y}^{\infty} f_{g_{1T}}(x)dx = N\left(\frac{y+C}{\sqrt{\Sigma_T}}\right), \tag{44}$$

$$\int_{-y}^{\infty} x f_{g_{1T}}(x)dx = \sqrt{\Sigma_T}\, n\left(\frac{y+C}{\sqrt{\Sigma_T}}\right) + CN\left(\frac{y+C}{\sqrt{\Sigma_T}}\right),$$

$$\int_{-y}^{\infty} H_m(x-C;\Sigma_T)\, f_{g_{1T}}(x)dx = \sqrt{\Sigma_T} H_{m-1}(-(y+C);\Sigma_T)\, n\left(\frac{y+C}{\sqrt{\Sigma_T}}\right);\ m \geq 1,$$

$$\int_{-y}^{\infty} x H_m(x-C;\Sigma_T)\, f_{g_{1T}}(x)dx = -\sqrt{\Sigma_T} y\, H_{m-1}(-(y+C);\Sigma_T)\, n\left(\frac{y+C}{\sqrt{\Sigma_T}}\right)$$

$$+ \Sigma_T^{\frac{3}{2}} H_{m-2}(-(y+C);\Sigma_T)\, n\left(\frac{y+C}{\sqrt{\Sigma_T}}\right);\ m \geq 2.$$

□

Remark 5 In practical applications, usually the underlying model is given as a non-perturbed form:

$$d\hat{X}_t^j = \hat{V}_0^j(\hat{X}_t)dt + \hat{V}^j(\hat{X}_t)dW_t \quad (j = 1,\ldots,d) \tag{45}$$
$$\hat{X}_0 = x_0 \in \mathbf{R}^d.$$

Then, in order to apply the asymptotic expansion method, we may rewrite the model for instance, as

$$dX_t^{(\epsilon),j} = V_0^j(X_t^{(\epsilon)})dt + \epsilon V^j(X_t^{(\epsilon)})dW_t \quad (j = 1,\ldots,d) \tag{46}$$
$$X_0^{(\epsilon)} = x_0 \in \mathbf{R}^d,$$

where by rescaling $\hat{V}^j(x)$ we set $V^j(x)$ so that $\hat{V}^j(x) = \epsilon V^j(x)$ for some $\epsilon \in (0,1]$. Consequently, an approximate call price under the original model (45) is obtained by (40) without $o(\epsilon^{N+1})$.

3.5 Application to Computation of Greeks

We already have a so called closed form approximate formula (40) for the option price, and hence are able to obtain approximations of its Greeks (that is, sensitivities to the changes in parameters in a model) as closed forms as well (or at least with easy numerical method such as the difference quotient method with the approximate option pricing formula).

For instance, [68] implements direct differentiations of the approximate formulas for option values under a time-homogeneous general local volatility model, and obtains closed form approximate formulas for the Deltas and Vegas. Moreover, [68] applies the similar technique to computing the Deltas and Vegas for average options with continuous monitoring, and gets their closed form approximate formulas as well. They also confirms the validity of the approximations through numerical experiments in the CEV model.

By deriving asymptotic expansions of characteristic functions of option values, [93, 94] propose a new expansion scheme for pricing options on long-term currencies under a Libor market model (LMM) and a general diffusion stochastic volatility model with jump of spot exchange rates. Furthermore, applying the approximate formulas, they provide analytical (closed form) approximations for the Deltas and Gammas of the options. Please see [93, 94] for the detail.

Alternatively, for a parameter θ, the sensitivity of a call price $C(K,T)$ with respect to the change in θ is expressed as follows:

$$\begin{aligned}
\frac{\partial}{\partial\theta}C(K,T) &= P(0,T)\mathbf{E}[\max\{g(X_T^{(\epsilon)})-K,0\}] \\
&= \frac{\partial}{\partial\theta}\left(\epsilon P(0,T)\mathbf{E}\left[\max\left\{G^{(\epsilon)}+y,0\right\}\right]\right) \\
&= \left(\frac{\partial}{\partial\theta}\{\epsilon P(0,T)\}\right)\mathbf{E}\left[\max\left\{G^{(\epsilon)}+y,0\right\}\right] \\
&\quad +\epsilon P(0,T)\left(\frac{\partial}{\partial\theta}\mathbf{E}\left[\max\left\{G^{(\epsilon)}+y,0\right\}\right]\right) \\
&= \left(\frac{\partial\{\epsilon P(0,T)\}}{\partial\theta}\right)\frac{C_{AE}(K,T)}{\epsilon P(0,T)} \\
&\quad +\epsilon P(0,T)\mathbf{E}\left[\left(\frac{\partial G^{(\epsilon)}}{\partial\theta}+\frac{\partial y}{\partial\theta}\right)1_{\{G^{(\epsilon)}>-y\}}\right], \qquad (47)
\end{aligned}$$

where $C_{AE}(K,T)$ stands for the approximate call price with strike K and maturity T, which is obtained by the asymptotic expansion.

Then, we are able to obtain an approximation of the sensitivity by a direct application of the asymptotic expansion to the above equation, particularly, the second term in the last equation. For example, under one dimensional diffusion setting, that is a general time homogeneous local volatility model, [66] successfully applies the expansion technique to computation of the Deltas and the Vegas with numerical experiments.

More generally, we note that the similar method as in option pricing in the previous subsection can be applied in Greeks, since we can take $\Phi \in \mathcal{S}'$ for $\mathbf{E}[\Phi(G^{(\epsilon)})]$ in (11) and apply the integration-by-parts method in Malliavin calculus. Recently, [103] takes this approach and derives asymptotic expansions of Greeks around the Black-Scholes model in stochastic volatility environment, and develop a unified method for precise estimates of the expansion errors. Particularly, they make use of the so called Kusuoka-Stroock functions introduced by Kusuoka [52], which is a powerful tool to clarify the order of a Wiener functional with respect to the time parameter t

in a unified manner. Then, they estimate the error bounds for the Malliavin weights of both the coefficient and the residual terms in the expansions.

3.6 Approximations of Asset Values Under Diffusion Processes

The framework of the asymptotic expansion can be applied not only to the simple cases mentioned above, but also to evaluation of much broader range of asset and security values. In particular, there are many cases where the asymptotic expansion can be applied to approximate their values when the underlying asset prices of financial securities, cash flows and interest rates are expressed as some functions of a random vector $X^{(\epsilon)}$ that follows a diffusion process. The method is almost the same as the one illustrated above and hence it is omitted. In this subsection, we only review how to represent the values of financial assets.

First, just as in the previous subsections, we consider a d-dimensional diffusion process $X^{(\epsilon)}$ defined as the strong solution to the stochastic differential equation (1). As an example, the present value V of a financial asset which generates a cash flow at the maturity date T is represented as

$$V = \mathbf{E}\left[e^{-\int_0^T R_2(X_u^{(\epsilon)})du} F(g(X_T^{(\epsilon)}))\right], \tag{48}$$

where g denotes the underlying asset price and F is the cash flow which characterizes the asset to be evaluated. Note that the underlying asset price g follows a diffusion process, whose drift term (the coefficient of the dt term) is $R_1(X_t^{(\epsilon)})g - D(X_t^{(\epsilon)})$ under an equivalent martingale measure. Moreover, R_1 at time $t \in [0, T]$ is represented as

$$R_1(X_t^{(\epsilon)}) = r(X_t^{(\epsilon)}) + \sum_{j=1}^{J_1} s_{1j}(X_t^{(\epsilon)}),$$

where r denotes the risk-free interest rate and s_{1j}, $j = 1, \ldots, J_1$ stand for various spreads (the differences from the risk-free rate) such as credit spreads and liquidity spreads. Suppose also that those are expressed as functions of the variable $X^{(\epsilon)}$. Further, $D(X_t^{(\epsilon)})$ denotes a payoff generated by the underlying asset such as a dividend or an interest rate and is also represented as a function of the variable $X^{(\epsilon)}$. Meanwhile, the discount rate at time t that is, $R_2(X_t^{(\epsilon)})$ of the target asset F to be evaluated is also expressed as

$$R_2(X_t^{(\epsilon)}) = r(X_t^{(\epsilon)}) + \sum_{j=1}^{J_2} s_{2j}(X_t^{(\epsilon)}),$$

where s_{2j}, $j = 1, \ldots, J_2$ are various spreads related to the objective asset or security. We again assume that those are expressed as some functions of the variable $X^{(\epsilon)}$.

As an example, let $F = 1$ in (48) for a zero-coupon bond with the face value 1 and the maturity date T. Also, let V_i denote the price of the zero-coupon bond with the maturity T_i. Then, V, the value of a coupon bond with the maturity T_N and coupon (and principal) payments c_i at $T_i (i = 1, \ldots, N, T_1 < \cdots < T_N)$ is represented by the equation $V = \sum_{i=1}^{N} c_i V_i$. Moreover, the present value of a call option on the coupon bond with the option maturity $T (< T_1)$ can be evaluated if we set $F(x) = (x - K)^+$ and $g(X_T^{(\epsilon)}) = \sum_{i=1}^{N} c_i g_i(X_T^{(\epsilon)})$ in the equation (48), where $g_i(X_T^{(\epsilon)}), i = 1, \ldots, N$ are given by

$$g_i(X_T^{(\epsilon)}) = \mathbf{E}\left[e^{-\int_T^{T_i} R_1(X_u^{(\epsilon)})du} | X_T^{(\epsilon)} \right].$$

Finally, we briefly review applications of the asymptotic expansion technique to numerical problems in finance, which can not be introduced in the present note due to the limitation of the space.

Takahashi and Yoshida [106] applies an asymptotic expansion to a dynamic investment problem with utility maximization for the asset at the end of the investment period, and derives an approximation formula for evaluating the optimal portfolio. Although the optimal portfolio has been numerically evaluated as a function of derivatives of the solution to some Bellman equation except for special cases, it is a hard task to implement it when the number of assets is large. Takahashi and Yoshida [106] provides its approximation based on the representation which Ocone-Karatzas [74] derives by using the so called Clark-Ocone formula. Moreover, [45] applies this method to a dynamic bond portfolio problem.

In evaluation of the expectation of a Wiener functional based on Monte Carlo simulations, [107] proposes a new estimator with a control variate which has its expectation explicitly obtained by an asymptotic expansion, and has a high correlation with the target Wiener functional. The convergence of the simulation based on this estimator becomes much faster and the approximation error with the asymptotic expansion up to a low order such as the first or second order is decreased. As for the extension of this method, please see [51, 88, 99].

For pricing American options, [89] extends a well-know decomposition formula for an American option value by Carr-Jarrow-Myneni [8], and proposes an approximation of the value by making use of the approximate density function of the underlying asset, which is obtained by the asymptotic expansion.

Moreover, because of its generality and unified nature of this approach with analytical (so called closed from) formulas, the asymptotic expansion method has been applied to broad class of valuation models which have become popular recently in practice. Especially, comparing to other numerical approximation schemes such as the Monte Carlo simulations and numerically solving methods for the partial differential equations (PDEs), it has an advantage in high dimensional problems. We list the following works as examples.

Applying the framework described above to default risk models, Muroi [67] derives asymptotic expansions for approximations of CDS (credit default swap) spreads.

Shiraya et al. [82] applies the expansion technique to obtain an approximation of swaption values under the Libor market model(LMM) of interest rates (Brace, Gatarek and Musiela [7], Jamshidian [43]) with local-stochastic volatility models.

Takahashi and Takehara [90–92] develop asymptotic expansion formulas for pricing long-term currency options with a Libor market model(LMM) of interest rates and diffusion or jump-diffusion stochastic volatility processes of spot exchange rates. Moreover, [92] presents a new characteristic-function-based Monte Carlo simulation scheme with the asymptotic expansion as a control variate.

Takahashi and Takehara [96] develops a general computation scheme for a high-order expansion method explained in this section, and applies it to the SABR model (Hagan, Kumar, Lesniewski, and Woodward [33]). They derives the expansions of the option prices up to the fifth order to show that the higher order expansion improves the approximations.

Takahashi and Takehara [108] and Takahashi et al. [109] also apply this scheme to the long-term currency options such as the 10 year maturity one under a Libor market model (LMM) of interest rates and stochastic volatility processes of spot exchange rates. Again, they confirm that the fourth or the fifth order expansion provides the better approximations than the lower order ones.

Furthermore, we are able to apply the expansion method to pricing the so called *exotic* type options. For instance, [78] derives expansions of average options with discrete monitoring under stochastic volatility models in order to obtain approximate prices of commodities average options. Moreover, they implement calibration to real futures plain-vanilla option prices of the underlying commodities, and evaluate average options based on the parameters obtained by the calibration.

Shiraya et al. [83] develops new approximation formulas for pricing single and double barrier options with discrete monitoring under stochastic volatility models. In addition, they demonstrate its validity through numerical experiments.

Shiraya et al. [81] presents a new approximation scheme for pricing continuous barrier options in stochastic volatility environment. Particularly, they make use of a static hedging scheme and the fifth order expansions of the vanilla options to obtain accurate approximate prices. Further, they derives the fifth order expansions for pricing average options with continuous monitoring under stochastic volatility models to achieve very precise approximations.

Shiraya and Takahashi [79] develops a general scheme for evaluation of the so called multi-asset cross currency options. In particular, they derive the expansions of basket option prices with 100 underlying assets (200 state variables with their stochastic volatilities), and cross currency average/basket options with discrete monitoring under stochastic volatility models to obtain accurate approximations.

Kato et al. [46, 47] develop a new expansion scheme for solutions of Cauchy-Dirichlet problems for second order parabolic partial differential equations (PDEs) and apply it to pricing down-and-out/up-and-out barrier options with continuous monitoring under stochastic volatility models.

4 Extension

This section follows [97] which presents an extension of the general computational scheme of the asymptotic expansion described in the previous section. In particular, by a change of variable technique and by various ways of setting the perturbation parameters in the expansion, we are able to provide the flexibility of setting the benchmark distribution around which the expansion is made, and an automatic way for computation up to any order in the expansion. For instance we introduce expansions, called the log-normal expansion and the CEV expansion.

4.1 Change of Variable and Perturbation

We consider a d-dimensional diffusion process $X_t = (X_t^1, \ldots, X_t^d)$ which is the solution to the following stochastic differential equation:

$$\begin{aligned} dX_t^j &= V_0^j(X_t)dt + V^j(X_t)dW_t \quad (j = 1, \ldots, d) \\ X_0 &= x_0 \in \mathbf{R}^d \end{aligned} \tag{49}$$

where $W = (W^1, \ldots, W^r)$ is an r-dimensional standard Wiener process; $V_0^j : \mathbf{R}^d \mapsto \mathbf{R}$ and $V^j : \mathbf{R}^d \mapsto \mathbf{R}^d$ are smooth functions with bounded derivatives of all orders.

Next, let $C : \mathbf{R}^d \mapsto \mathbf{R}^d$ be a $\mathbf{C}^2$-function which has the unique inverse function, C^{-1}, and define $\tilde{X}_t$ as $\tilde{X}_t = C(X_t)$. Then, the dynamics of $\tilde{X}$ is given by

$$\begin{aligned} d\tilde{X}_t^j &= \tilde{V}_0^j(\tilde{X}_t)dt + \tilde{V}^j(\tilde{X}_t)dW_t \quad (j = 1, \ldots, d), \\ \tilde{X}_0 &= \tilde{x}_0, \end{aligned} \tag{50}$$

where

$$\begin{aligned} \tilde{V}_0^j(\tilde{x}) &:= \sum_{j'=1}^{d} \partial_{j'} C^j(C^{-1}(\tilde{x})) V_0^{j'}(C^{-1}(\tilde{x})) \\ &\quad + \frac{1}{2} \sum_{j',k'=1}^{d} \partial_{j'k'} C^j(C^{-1}(\tilde{x})) V^{j'}(C^{-1}(\tilde{x})) V^{k'}(C^{-1}(\tilde{x}))^\top, \\ \tilde{V}^j(\tilde{x}) &:= \sum_{j'=1}^{d} \partial_{j'} C^j(C^{-1}(\tilde{x})) V^{j'}(C^{-1}(\tilde{x})), \end{aligned}$$

and $\tilde{x}_0 = C(x_0)$. ($(C^{-1}(\tilde{x}))^\top$ denotes the transpose of $(C^{-1}(\tilde{x}))$.)

Next, we introduce a perturbation parameter $\epsilon \in (0, 1]$ as follows:

$$
\begin{aligned}
\tilde{X}_t &\mapsto \tilde{X}_t^{(\epsilon)} \\
\tilde{V}_0^j(\tilde{x}) &\mapsto \tilde{V}_0^{(\epsilon),j}(\tilde{x}, \epsilon) \\
\tilde{V}^j(\tilde{x}) &\mapsto \epsilon \tilde{V}^j(\tilde{x}),
\end{aligned}
$$

and hence, the dynamics of $\tilde{X}^{(\epsilon)}$ is expressed as

$$
d\tilde{X}_t^{(\epsilon),j} = \tilde{V}_0^{(\epsilon),j}(\tilde{X}_t^{(\epsilon)}, \epsilon)dt + \epsilon \tilde{V}^j(\tilde{X}_t^{(\epsilon)})dW_t \quad (j = 1, \ldots, d). \tag{51}
$$

Then, we are able to apply the technique developed in the previous section to the transformed SDE (51).

4.2 *Applications to Option Pricing Under Local-Stochastic Volatility Model*

We assume that the underlying process is the unique solution to the following SDE:

$$
\begin{aligned}
dS_t &= \sigma(X_t)h(S_t)dW_t \\
dX_t^j &= V_0^j(X_t)dt + V^j(X_t)dW_t \quad (j = 2, \ldots, d) \\
S_0 &= s_0 \in \mathbf{R}, \quad X_0 = x_0 \in \mathbf{R}^{d-1},
\end{aligned} \tag{52}
$$

where $\sigma : \mathbf{R}^{d-1} \to \mathbf{R}^r$, $h : \mathbf{R} \to \mathbf{R}$, and W is a r-dimensional Brownian motion. Then, we evaluate a call option with strike K and maturity T, whose underlying price process is given by S. Under the zero discount interest rate for simplicity, the call price $Call(K, T)$ with strike price K and maturity T is obtained by

$$
Call(K, T) = \mathbf{E}[(S_T - K)_+]. \tag{53}
$$

First, for $x = (x^1, x^2, \ldots, x^d)$, let

$$
C(x) = (C_1(x^1), x^2, \ldots, x^d),
$$

where $C_1 : \mathbf{R} \to \mathbf{R}$ is an invertible $\mathbf{C}^2$-function. Then, $\tilde{S}_t = C_1(S_t)$, which $\tilde{S}$ follows a process of the solution to the following SDE:

$$
\begin{aligned}
d\tilde{S}_t = & \frac{1}{2}|\sigma(X_t)|^2 h(C_1^{-1}(\tilde{S}_t))^2 C_1^{('')}(C_1^{-1}(\tilde{S}_t))dt \\
& + \sigma(X_t)h(C_1^{-1}(\tilde{S}_t))C_1^{(')}(C_1^{-1}(\tilde{S}_t))dW_t, \quad \tilde{s}_0 = C_1(s_0),
\end{aligned} \tag{54}
$$

where $C_1^{(')}(x) := \frac{d}{dx}C_1(x)$ and $C_1^{('')}(x) := \frac{d^2}{dx^2}C_1(x)$.

Next, we introduce a perturbation parameter ϵ as follows:

$$\begin{aligned} d\tilde{S}_t^{(\epsilon)} &= \frac{\eta(\epsilon)}{2}|\sigma(X_t^{(\epsilon)})|^2 h(C_1^{-1}(\tilde{S}_t^{(\epsilon)}))^2 C_1^{('')}(C_1^{-1}(\tilde{S}_t^{(\epsilon)}))dt \\ &\quad + \epsilon\sigma(X_t^{(\epsilon)})h(C_1^{-1}(\tilde{S}_t))C_1^{(')}(C_1^{-1}(\tilde{S}_t^{(\epsilon)}))dW_t, \\ dX_t^{(\epsilon),j} &= V_0^j(X_t^{(\epsilon)},\epsilon)dt + \epsilon V^j(X_t^{(\epsilon)})dW_t \quad (j = 2, \ldots, d), \end{aligned} \tag{55}$$

where $\eta(\epsilon) = \epsilon^k$ and k is a nonnegative integer such as $k = 0, 1, 2, \ldots$. Note that

$$S_t = C_1^{-1}(\tilde{S}_t) = C_1^{-1}(\tilde{S}_t^{(1)}),$$

where $\tilde{S}_t^{(1)} = \tilde{S}_t^{(\epsilon)}|_{\epsilon=1}$.

According to Theorem 2 in the previous section, we have already an asymptotic expansion of the density function of $G^{(\epsilon)} = \frac{\tilde{S}_T^{(\epsilon)} - \tilde{S}_T^{(0)}}{\epsilon}$ up to ϵ^N -order, denoted by $f_{G^{(\epsilon)},N}(x)$.

Therefore, an approximation formula of the call price is given as follows:

$$Call(K,T) = \mathbf{E}[(S_T - K)_+] = \mathbf{E}\left[\left(C_1^{-1}\left(\tilde{S}_T^{(1)}\right) - K\right)_+\right] \tag{56}$$

$$\approx \int_y^\infty \left(C_1^{-1}(x + \tilde{S}_T^{(0)}) - K\right) f_{G^{(1)},N}(x)dx, \tag{57}$$

where $y = C_1(K) - \tilde{S}_T^{(0)}$.

A simple example is the following. Set the local volatility function to be linear:

$$\begin{aligned} dS_t &= \sigma(X_t)S_t dW_t \\ dX_t^j &= V_0^j(X_t)dt + V^j(X_t)dW_t \quad (j = 2, \ldots, d). \end{aligned} \tag{58}$$

For $x = (x^1, x^2, \ldots, x^d)$, let

$$C(x) = (\log x^1, x^2, \ldots, x^d),$$

and set $\eta(\epsilon) = \epsilon^k$ where k is 0, 1 or 2. Then, we have $\tilde{S}_t^{(\epsilon)} = \log S_t^{(\epsilon)}$, where

$$\begin{aligned} d\tilde{S}_t^{(\epsilon)} &= -\frac{\epsilon^k}{2}\sigma(X_t^{(\epsilon)})^2 dt + \epsilon\sigma(X_t^{(\epsilon)})dW_t, \\ dX_t^{(\epsilon),j} &= V_0^j(X_t^{(\epsilon)},\epsilon)dt + \epsilon V^j(X_t^{(\epsilon)})dW_t \quad (j = 2, \ldots, d). \end{aligned} \tag{59}$$

This case corresponds to some existing researches. (e.g. [91, 92, 95, 96, 100])

4.3 Examples

This subsection shows more specific examples in the local-stochastic volatility model.

4.3.1 CEV Model

The first example is on the well-known CEV (Constant Elasticity of Variance) model (Cox [10]):

$$dS_t = \sigma(S_t^{\beta} S_0^{1-\beta})dW_t, \ \sigma \text{ and } S_0 \text{ are positive constants}, \ \beta \in [0, 1], \quad (60)$$

where the term $S_0^{1-\beta}$ makes the level of σ is of the same order for different β. For $x > 0$, let us take the change of variable function to be $C(x) = \log(x/S_0)$, that is $x = C^{-1}(\tilde{x}) = S_0 \exp(\tilde{x})$. Hence, $\tilde{S}_t = \log \frac{S_t}{S_0}$ and we have

$$d\tilde{S}_t = -\frac{1}{2}\sigma^2 e^{2(\beta-1)\tilde{S}_t}dt + \sigma e^{(\beta-1)\tilde{S}_t}dW_t; \ \tilde{S}_0 = 0. \quad (61)$$

Next, we introduce a perturbation $\epsilon \in [0, 1]$, again as follows:

$$d\tilde{S}_t^{(\epsilon)} = -\frac{\eta(\epsilon)}{2}\sigma^2 e^{2(\beta-1)\tilde{S}_t^{(\epsilon)}}dt + \epsilon\sigma e^{(\beta-1)\tilde{S}_t^{(\epsilon)}}dW_t; \ \tilde{S}_0 = 0. \quad (62)$$

where $\eta(\epsilon) = \epsilon^j$ and j is a nonnegative integer.

Because

$$S_T = C^{-1}\left(\tilde{S}_T^{(1)}\right) = S_0 \exp\left(\tilde{S}_T^{(1)}\right) = S_0 \exp\left(G^{(1)} + S_T^{(0)}\right),$$

an approximation formula of the call price with strike K and maturity T is given as follows:

$$Call(K,T) = \mathbf{E}[(S_T - K)_+] = \mathbf{E}\left[\left(S_0 \exp\left(G^{(1)} + \tilde{S}_T^{(0)}\right) - K\right)_+\right]$$

$$\approx \int_y^{\infty} \left(S_0 \exp\left(x + \tilde{S}_T^{(0)}\right) - K\right) f_{G^{(1)},N}(x)dx; \quad (63)$$

$$y = C(K) - \tilde{S}_T^{(0)} = \log\frac{K}{S_0} - \tilde{S}_T^{(0)}. \quad (64)$$

Note that $f_{g_{1T}}$, the first term in the asymptotic expansion of the density $f_{G^{(\epsilon)}}$ is a normal density and hence, the underlying asset price is expanded around a log-normal distribution. Thus, we could call this case a log-normal asymptotic expansion. We also remark that the case of $\eta(\epsilon) = \epsilon^0 = 1$ is harder to be evaluated than the other cases, which is essentially due to difficulty in computation of $\tilde{S}_t^{(0)}$ for $\eta(\epsilon) = 1$.

- **On the Validity of the Asymptotic Expansion for CEV model**
  Previous works such as [85, 86, 107] have considered an asymptotic expansion of (average and vanilla) option prices based on the following type of a perturbed process: For $\beta \in [1/2, 1)$,

$$dS_t^{(\epsilon)} = \epsilon (S_t^{(\epsilon)} \vee 0)^{\beta} dW_t; \; S_0^{(\epsilon)} = s_0. \tag{65}$$

Although the coefficient function in this model is not smooth at 0, the asymptotic expansion method is still applicable. For instance, we could use a smooth modification technique (e.g. [106, 107]). That is, let us take a modified process $(\tilde{S}_t^{(\epsilon)})_{t\in[0,T]}$ of $(S_t^{(\epsilon)})_{t\in[0,T]}$ as follows:

$$d\tilde{S}_t^{(\epsilon)} = \epsilon g(\tilde{S}_t^{(\epsilon)}) dW_t. \tag{66}$$

Here, $g(x)$ is a smooth modification of $g(x) = (x \vee 0)^{\beta}$ such that $g(x) = x^{\beta}$ when $x \geq a_1$ for some small $a_1 \in (0, a)$ for $a = \frac{1}{2}s_0$ and $g(x) = 0$ when $x \leq a_2$ for some $a_2 \in (0, a_1)$. Specifically, we may set $g(x)$ as follows. For $t \in [0, T]$,

$$\begin{aligned}
g(x) &= h(x)x^{\beta} \\
h(x) &= \frac{\psi(x-a_2)}{\psi(x-a_2)+\psi(a_1-x)}, \quad 0 < a_2 < a_1 \\
\psi(x) &= e^{-1/x} \text{ for } x > 0, \; \psi(x) = 0 \text{ for } x \leq 0.
\end{aligned}$$

Suppose that for a $\mathbf{R}$-valued function f, $E\left[\left|f(S^{(\epsilon)})\right|^2\right] < \infty$ and $E\left[\left|f(\tilde{S}^{(\epsilon)})\right|^2\right] < \infty$. (e.g. we can take option payoff functions as f in our setting.) Then, we have

$$\begin{aligned}
E\left[\left|f(S^{(\epsilon)}) - f(\tilde{S}^{(\epsilon)})\right| 1_{\{S^{(\epsilon)} \neq \tilde{S}^{(\epsilon)}\}}\right] &\leq \left(E\left[|f(S^{(\epsilon)})|^2\right]^{\frac{1}{2}} + E\left[|f(\tilde{S}^{(\epsilon)})|^2\right]^{\frac{1}{2}}\right) \\
&\times P\left(\{S^{(\epsilon)} \neq \tilde{S}^{(\epsilon)}\}\right)^{\frac{1}{2}}.
\end{aligned}$$

It also holds that

$$\begin{aligned}
&P\left(\{S^{\epsilon} \neq \tilde{S}^{\epsilon}\}\right) = P\left(\{S_t^{\epsilon} \leq a_1 \text{ for some } t \in [0,T]\}\right) \\
&\leq P\left(\{\sup_{0\leq t\leq T} |S_t^{\epsilon} - S_t^0| > a\}\right) \\
&+ P\left(\{S_t^{\epsilon} \leq a_1 \text{ for some } t \in [0,T]\} \cap \{\sup_{0\leq t\leq T} |S_t^{\epsilon} - S_t^0| \leq a\}\right).
\end{aligned}$$

We can easily see that the second term after the last inequality is 0. The first term is smaller than any ϵ^n for $n = 1, 2, \ldots$ by the following lemma of a large deviation inequality:

Lemma 4 *Suppose that* $Z_t^\epsilon, t \in [0, T]$ *follows a process of the solution to the SDE:*

$$dZ_t^\epsilon = \mu(Z_t^\epsilon)dt + \epsilon\sigma(Z_t^\epsilon)dW_t.$$

where $\mu(z)$ *satisfies the Lipschitz and linear growth conditions, and* $\sigma(z)$ *satisfies the linear growth condition. We assume that the unique strong solution exists. Then, there exists positive constants* c_1 *and* c_2 *independent of* ϵ *such that*

$$P(\{\sup_{0 \le s \le T} |Z_s^\epsilon - Z_s^0| > c\}) \le c_1 \exp(-c_2\epsilon^{-2}) \tag{67}$$

for all $c > 0$.

The lemma can be proved by slight modification of the Lemma 5.3 in [119] or the Lemma 7.1 in [50]. Note also that S^ϵ and $\tilde{S}^\epsilon$ satisfy the conditions in the lemma above.

Hence,

$$E\left[\left|f(S^{(\epsilon)}) - f(\tilde{S}^{(\epsilon)})\right|\right] = o(\epsilon^n), \; n = 1, 2, \ldots. \tag{68}$$

Therefore, the difference between $f(S^{(\epsilon)})$ and $f(\tilde{S}^{(\epsilon)})$ is negligible in a *small disturbance asymptotic theory*, and hence we could apply an asymptotic expansion to $E\left[f(\tilde{S}^{(\epsilon)})\right]$ instead of $E\left[f(S^{(\epsilon)})\right]$.

In particular, [107] considered the case that $\beta = 1/2$ and $f(x) = \left(\frac{1}{T}\int_0^T x_t dt - K\right)^+$, $x = S^{(\epsilon)}, \tilde{S}^{(\epsilon)}$ (an average call option's payoff). The similar modification could be applied to the asymptotic expansions for transformed processes in this section. Please also see [88] for numerical experiments under the smooth and bounded modification of this kind for volatility functions in a HJM-type model of interest rates.

4.3.2 SABR Model

Next, let us consider a stochastic volatility model so called SABR [33] (or λ-SABR [38]) Model:

$$\begin{aligned} dS_t &= \sigma_t(S_t^\beta S_0^{1-\beta})dW_t^1; \; S_0 > 0, \\ d\sigma_t &= \lambda(\theta - \sigma_t)dt + \nu\sigma_t dW_t^2; \; \sigma_0 > 0 \end{aligned} \tag{69}$$

where $\beta \in [0, 1]$, $\lambda \ge 0$, $\theta > 0$, $\nu > 0$, and $W = (W^1, W^2)$ is a two dimensional Wiener process with correlation $\rho \in [0, 1]$.

- **Log-normal Asymptotic Expansion**
Let us take a log-normal asymptotic expansion for the underlying asset price S, that is for $x_1 > 0$, set $C(x_1, x_2) = (\log(x_1/S_0), x_2)$ and $\tilde{S}_t = \log \frac{S_t}{S_0}$:

$$d\tilde{S}_t = -\frac{1}{2}\sigma_t^2 e^{2(\beta-1)\tilde{S}_t} dt + \sigma_t e^{(\beta-1)\tilde{S}_t} dW_t^1; \ \tilde{S}_0 = 0 \tag{70}$$
$$d\sigma_t = \lambda(\theta - \sigma_t)dt + \nu\sigma_t dW_t^2; \ \sigma_0 > 0.$$

Next, we introduce a perturbation $\epsilon \in [0, 1]$, again as follows:

$$d\tilde{S}_t^{(\epsilon)} = -\frac{\eta_1(\epsilon)}{2}\sigma^2 e^{2(\beta-1)\tilde{S}_t^{(\epsilon)}} dt + \epsilon\sigma e^{(\beta-1)\tilde{S}_t^{(\epsilon)}} dW_t; \ \tilde{S}_0 = 0, \tag{71}$$
$$d\sigma_t^{(\epsilon)} = \eta_2(\epsilon)\lambda(\theta - \sigma_t^{(\epsilon)})dt + \epsilon\nu\sigma_t^{(\epsilon)} dW_t^2; \ \sigma_0^{(\epsilon)} = \sigma_0,$$

where $\eta_i(\epsilon) = \epsilon^{j_i}$, $i = 1, 2$ and j_i is a nonnegative integer.
For instance, typical cases are $\eta_2(\epsilon) = \epsilon^0 = 1$ with $\eta_2(\epsilon) = \epsilon$ (an extension of *the log-normal asymptotic expansion* in [95, 100]), or $\eta_2(\epsilon) = \epsilon^2$ (an extension of [90] to the CEV-type local volatility).
An approximation formula of the call price with strike K and maturity T is given as follows:

$$Call(K, T) = \mathbf{E}[(S_T - K)_+] = \mathbf{E}\left[\left(S_0 \exp\left(G^{(1)} + \tilde{S}_T^{(0)}\right) - K\right)_+\right]$$
$$\approx \int_y^\infty \left(S_0 \exp\left(x + \tilde{S}_T^{(0)}\right) - K\right) f_{G^{(1)},N}(x)dx; \tag{72}$$
$$y = C(K) - \tilde{S}_T^{(0)} = \log \frac{K}{S_0} - \tilde{S}_T^{(0)}. \tag{73}$$

Again, we note that the case of $\eta(\epsilon) = \epsilon^0 = 1$ is harder to be evaluated than the other cases, which results from difficulty in computation of $\tilde{S}_t^{(0)}$ for $\eta(\epsilon) = 1$.
- **CEV Asymptotic Expansion**
Let us take change of variable function C as $C(x_1, x_2) = (C_1(x_1), x_2)$ for (x_1, x_2), where for $x > 0$ and $\beta \in [0, 1)$,

$$C_1(x) = \frac{1}{1-\beta}\frac{x^{1-\beta}}{S_0^{1-\beta}} \left(= \int^x \frac{dz}{z^\beta S_0^{1-\beta}}\right). \tag{74}$$

That is,

$$C_1^{-1}(\tilde{x}) = S_0(1-\beta)^{\frac{1}{(1-\beta)}} \tilde{x}^{\frac{1}{(1-\beta)}}. \tag{75}$$

Then, as $\tilde{S}_t = C_1(S_t)$, we have

$$d\tilde{S}_t = -\frac{1}{2}\frac{\beta}{1-\beta}\sigma_t^2\frac{1}{\tilde{S}_t}dt + \sigma_t dW_t^1; \ \tilde{S}_0 = \frac{1}{1-\beta} > 0 \tag{76}$$
$$d\sigma_t = \lambda(\theta - \sigma_t)dt + \nu\sigma_t dW_t^2; \ \sigma_0 > 0.$$

Again, we set a perturbed process as follows:

$$d\tilde{S}_t^{(\epsilon)} = -\frac{\eta_1(\epsilon)}{2}\frac{\beta}{1-\beta}(\sigma_t^{(\epsilon)})^2\frac{1}{\tilde{S}_t^{(\epsilon)}}dt + \epsilon\sigma_t^{(\epsilon)}dW_t^1; \ \tilde{S}_0^{(\epsilon)} = \frac{1}{1-\beta} \tag{77}$$
$$d\sigma_t^{(\epsilon)} = \eta_2(\epsilon)\lambda(\theta - \sigma_t^{(\epsilon)})dt + \epsilon\nu\sigma_t^{(\epsilon)}dW_t^2; \ \sigma_0^{(\epsilon)} = \sigma_0,$$

where $\eta_i(\epsilon) = \epsilon^{j_i}$, $i = 1, 2$ and j_i is a nonnegative integer.
For illustrative purpose, let us set $\eta_1(\epsilon) = \eta_2(\epsilon) = \epsilon$. That is,

$$d\tilde{S}_t^{(\epsilon)} = -\frac{\epsilon}{2}\frac{\beta}{1-\beta}(\sigma_t^{(\epsilon)})^2\frac{1}{\tilde{S}_t^{(\epsilon)}}dt + \epsilon\sigma_t^{(\epsilon)}dW_t^1; \ \tilde{S}_0^{(\epsilon)} = \frac{1}{1-\beta}, \tag{78}$$
$$d\sigma_t^{(\epsilon)} = \epsilon\lambda(\theta - \sigma_t^{(\epsilon)})dt + \epsilon\nu\sigma_t^{(\epsilon)}dW_t^2; \ \sigma_0^{(\epsilon)} = \sigma_0.$$

In this case, as $\tilde{S}_t^{(0)} = \frac{1}{1-\beta}$ and $\sigma_t^{(0)} = \sigma_0$ for all $t \in [0, T]$, the first two terms in the asymptotic expansion, $\tilde{g}_{1t} = \frac{1}{1-\beta} + \frac{\partial}{\partial\epsilon}\Big|_{\epsilon=0}\tilde{S}_t^{(\epsilon)}$ follows a Gaussian process:

$$d\tilde{g}_{1t} = \frac{-\beta\sigma_0^2}{2}dt + \sigma_0 dW_t^1; \ \tilde{g}_{10} = \frac{1}{1-\beta}. \tag{79}$$

Then, by applying *Itô*'s formula to

$$\hat{g}_{1t} := C_1^{-1}(\tilde{g}_{1t}) = S_0(1-\beta)^{\frac{1}{(1-\beta)}}\tilde{g}_{1t}^{\frac{1}{(1-\beta)}}, \tag{80}$$

and using

$$\tilde{g}_{1t} = \frac{1}{1-\beta}\frac{\hat{g}_{1t}^{1-\beta}}{S_0^{1-\beta}}, \tag{81}$$

we formally obtain the SDE of $\hat{g}_{1t}$ though it is generally well-defined only for $\tilde{g}_{1t} \geq 0$:

$$d\hat{g}_{1t} = \frac{\beta\sigma_0^2 S_0^{1-\beta}}{2}\hat{g}_{1t}^{\beta}\left[-1 + S_0^{1-\beta}\hat{g}_{1t}^{\beta-1}\right]dt + \sigma_0 S_0^{1-\beta}\hat{g}_{1t}^{\beta}dW_t^1; \ \hat{g}_{10} = S_0. \tag{82}$$

Here, because the diffusion coefficient of $\hat{g}_{1t}$ is given by $\sigma_0 S_0^{1-\beta}(\hat{g}_{1t})^{\beta}$ and we may think that S is expanded around $\hat{g}_1$, we call this case a CEV asymptotic expansion (though $\hat{g}_1$ is not exactly a CEV process).
In particular, when $\beta = 1/2$,

$$d\hat{g}_{1t} = \frac{\sigma_0^2}{4}\left[-\sqrt{S_0\hat{g}_{1t}} + S_0\right]dt + \sigma_0\sqrt{S_0\hat{g}_{1t}}dW_t^1;\ \hat{g}_{10} = S_0, \tag{83}$$

and because

$$\hat{g}_{1T} = \frac{S_0}{4}\tilde{g}_{1T}^2, \tag{84}$$

$\hat{g}_{1T}/(S_0\sigma_0^2 T/4)$ follows a non-central χ^2 distribution, around which the original underlying asset price S_T is expanded.
Finally, for $\eta_i(\epsilon) = \epsilon^{j_i}$, $i = 1, 2$ and j_i is a nonnegative integer, an approximation formula of the call price with strike K and maturity T is obtained as follows:

$$\begin{aligned}
Call(K,T) &= \mathbf{E}[(S_T - K)_+] = \mathbf{E}\left[\left(C_1^{-1}(\tilde{S}_T) - K\right)_+\right]\\
&= \mathbf{E}\left[\left(\left\{S_0(1-\beta)^{\frac{1}{(1-\beta)}}(\tilde{S}_T)^{\frac{1}{(1-\beta)}}\right\} - K\right)_+\right]\\
&= \mathbf{E}\left[\left(\left\{S_0(1-\beta)^{\frac{1}{(1-\beta)}}(\tilde{S}_T^{(1)})^{\frac{1}{(1-\beta)}}\right\} - K\right)_+\right]\\
&= \mathbf{E}\left[\left(\left\{S_0(1-\beta)^{\frac{1}{(1-\beta)}}(G^{(1)} + \tilde{S}_T^{(0)})^{\frac{1}{(1-\beta)}}\right\} - K\right)_+\right]\\
&\approx \int_y^{\infty}\left(\left\{S_0(1-\beta)^{\frac{1}{(1-\beta)}}(x + \tilde{S}_T^{(0)})^{\frac{1}{(1-\beta)}}\right\} - K\right) f_{G^{(1)},N}(x)dx;
\end{aligned} \tag{85}$$

$$y = C_1(K) - \tilde{S}_T^{(0)} = \frac{1}{1-\beta}\left(\frac{K}{S_0}\right)^{1-\beta} - \tilde{S}_T^{(0)}. \tag{86}$$

As numerical examples, [97] examines *normal*, *log-normal* and *CEV expansions* up to the third order for approximations of option prices under SABR model, which implies that *CEV expansion* provides the most stable approximations. We also observe that *CEV expansion* becomes more precise with the same level of absolute errors across the whole range of β along the higher order expansions. Thus, we expect a higher order *CEV expansion* will produce the better and more stable approximation than the other expansions, though further investigation seems necessary. Please see the original paper [97] for the detail of the numerical experiment.

Remark 6 If necessary, applying a similar technique as mentioned in Sect. 4.3.1, we could use the asymptotic expansion for a model with smooth (and bounded) modification of the underlying processes. For a concrete example please see Remark 3 in [97].

5 Improvement Scheme for Asymptotic Expansion

Although the asymptotic expansion up to the fifth order is known to be sufficiently accurate for option pricing (e.g. [81, 95, 96, 108, 109]), one of the main criticisms against the method would be that the approximate density function admits negative values typically at its tails that is, some region of the deep Out-of-The-Money (OTM), which could create an arbitrage opportunity in option trading. Also, even if the domain of a true density is restricted to be positive, the domain of its approximation may include negative values unless an appropriate boundary condition is assigned. To overcome the problems, we briefly introduce two recent researches related to the present asymptotic expansion approach.

5.1 New Improvement Scheme for Approximation Methods of Probability Density Functions

Takahashi and Tsuzuki [98] develops a new scheme for improving density approximation methods, which also provides precise approximations of option values. Specifically, the scheme is inspired by the idea in the Hilbert space projection theorem, and so called "Dykstra's cyclic projections algorithm" is applied for its implementation. (Please consult Deutsch [14] for the detail of the algorithm.) We also remark that the scheme can be easily implemented in practice, where we need only market data used for usual calibration such as option prices with strikes.

Furthermore, numerical experiments for vanilla option pricing under SABR model demonstrate the validity of the scheme. In fact, in terms of approximation accuracies this scheme improves the third and fifth order asymptotic expansions preserving the required conditions such as nonnegative densities under an appropriate forward measure.

We finally remark that the scheme is general and flexible enough to include a set of conditions and information as one would like to put on an approximate density, and it can be applied to approximation methods other than the asymptotic expansion method. For example, a number of researches have been going on in order to extend SABR model with fixing the problem of the negative densities in the method of [33]. (For instance, see Doust [15].) We note that the scheme is also a candidate for handling this issue. Also, the estimate of the absorption probability based on Monte Carlo simulations as in [15] can be consistently incorporated in the scheme.

5.2 A Weak Approximation with Asymptotic Expansion and Multidimensional Malliavin Weights

Takahashi and Yamada [105] develops a new weak approximation scheme for expectations of functions of the solutions to SDEs. In particular, the scheme connects approximate operators constructed based on the asymptotic expansion. More concretely, a diffusion semigroup is defined as the expectation of an appropriate function of the solution to a certain SDE: for example, $P_t^{\epsilon} f(x) = E[f(X_t^{x,\epsilon})]$ with the solution $X_t^{x,\epsilon}$ of a SDE with perturbation parameter ϵ and a function f. Then, we approximate P_t^{ϵ} by an operator $Q_t^{\epsilon,m}$ which is constructed based on the asymptotic expansion up to a certain order m. Thus, given a partition of $[0, T]$, $\pi = \{(t_0, t_1, \ldots, t_n) : 0 = t_0 < t_1 < \cdots < t_n = T\}$, we are able to approximate $P_T^{\epsilon} f(x)$ by connecting the expansion-based approximations with the use of multi-dimensional Malliavin weights sequentially: that is, roughly speaking, with $s_k = t_k - t_{k-1}, k = 1, \ldots, n$,

$$P_T^{\epsilon} f(x) \simeq Q_{s_n}^{\epsilon,m} Q_{s_{n-1}}^{\epsilon,m} \cdots Q_{s_1}^{\epsilon,m} f(x).$$

The present research justifies this idea by applying Malliavin calculus, particularly, theories developed by Watanabe [111] and Kusuoka [52–54]. In computation, in order to evaluate the Malliavin weights, the paper makes use of conditional expectation formulas for multi-dimensional asymptotic expansions in [86].

Moreover, the paper shows through numerical examples for option pricing under local and stochastic volatility models that very few partition such as $n = 2$ is mostly enough to substantially improve the errors at deep OTMs of expansions with the first or second order ($m = 1, 2$).

6 Asymptotic Expansion in an Instantaneous Forward Rates Model

Among main stochastic models in finance, there exist models in which the stochastic processes of the underlying variables do not belong to the class of diffusion processes. This section illustrates an instantaneous forward rates model as a typical example.

6.1 Asymptotic Expansion for General Wiener Functionals

Watanabe [111] derives an asymptotic expansion for general Wiener functionals. As an example of the Watanabe's expansion, [100] shows the following result:

Theorem 4 *Let us consider a family of smooth Wiener functionals* $F^{\epsilon} = (F_1^{\epsilon}, \ldots, F_n^{\epsilon})$, $F_i^{\epsilon} \in \mathbf{D}^{\infty}$ $(i = 1, \ldots, n)$ *such that* F_i^{ϵ} *has an asymptotic expansion in* $\mathbf{D}^{\infty}$.

Moreover, F^{ϵ} satisfies the uniformly non-degenerate condition:

$$\limsup_{\epsilon\downarrow 0} \|(\det \sigma_{F^{\epsilon}})^{-1}\|_{L^p} < \infty, \text{ for all } p < \infty, \tag{87}$$

where $\sigma_{F^{\epsilon}}$ stands for the Malliavin covariance matrix of F^{ϵ}. Then, for a Schwartz distribution $T \in \mathcal{S}'(\mathbf{R}^n)$, we have an asymptotic expansion in $\mathbf{R}$:

$$\Bigg| E[T(F^{\epsilon})] - \Bigg\{ \int_{\mathbf{R}^n} T(x) p^{F^0}(x)dx + \sum_{j=1}^{N} \epsilon^j \int_{\mathbf{R}^n} T(x) E \\ \times \left[\sum_{k}^{(j)} H_{\alpha^{(k)}}\left(F^0, \prod_{l=1}^{k} F_{\alpha_l}^{0,\beta_l} \right) | F^0 = x \right] p^{F^0}(x)dx \Bigg\} \Bigg| = O(\epsilon^{N+1}), \tag{88}$$

Equivalently,

$$\Bigg| E[T(F^{\epsilon})] - \Bigg\{ \int_{\mathbf{R}^n} T(x) p^{F^0}(x)dx \\ + \sum_{j=1}^{N} \epsilon^j \sum_{k}^{(j)} (-1)^k \int_{\mathbf{R}^n} T(x) \partial^k_{\alpha^{(k)}} \\ \times \left\{ E\left[\prod_{l=1}^{k} F_{\alpha_l}^{0,\beta_l} | F^0 = x \right] p^{F^0}(x) \right\} dx \Bigg\} \Bigg| = O(\epsilon^{N+1}), \tag{89}$$

where $F_i^{0,k} := \frac{1}{k!}\frac{d^k}{d\epsilon^k} F_i^{\epsilon}|_{\epsilon=0}$, $k \in \mathbf{N}$ $(i = 1, \ldots, n)$, $\alpha^{(k)}$ denotes a multi-index, $\alpha^{(k)} = (\alpha_1, \ldots, \alpha_k)$ and

$$\sum_{k}^{(j)} \equiv \sum_{k=1}^{j} \sum_{\beta_1+\cdots+\beta_k=j, \beta_i\geq 1} \sum_{\alpha^{(k)}\in\{1,\ldots,n\}^k} \frac{1}{k!}.$$

$p^{F^0}(x)$ stands for the density function of F^0. The Malliavin weight $H_{\alpha^{(k)}}$ is recursively defined as follows:

$$H_{\alpha^{(k)}}(F, G) = H_{(\alpha_k)}(F, H_{\alpha^{(k-1)}}(F, G)), \tag{90}$$

where

$$H_{(l)}(F, G) = D^*\left(\sum_{i=1}^{n} G \gamma_{li}^F DF_i \right). \tag{91}$$

Here, $F_i \in D^\infty$, $G \in D^\infty$, $D^*\left(\sum_{i=1}^n G\gamma_{li}^F DF_i\right)$ is the divergence of $\sum_{i=1}^n G\gamma_{li}^F DF_i$, DF_i is the Malliavin derivative of F_i, and $\gamma^F = \left(\gamma_{ij}^F\right)_{1\le i,j\le n}$ denotes the inverse matrix of the Malliavin covariance matrix of F. Moreover, we use the notation $\int T(x)g(x)dx$ for $T \in \mathcal{S}'(\mathbf{R}^n)$ and $g \in \mathcal{S}(\mathbf{R}^n)$ meaning that ${}_{\mathcal{S}'}\langle T, g\rangle_{\mathcal{S}}$. (See the Sect. 2 of [100] for the details of those definitions.)

Remark 7 The asymptotic expansion formula (89) is the formula developed by Watanabe [111]. Hence, this theorem shows the expansion (88) based on *push down* (conditional expectation) of Malliavin weights (divergences) is equivalent to the Watanabe's formula.

(Proof) We use α as an abbreviation of $\alpha^{(k)}$ in the proof, and the notation $\langle\cdot,\cdot\rangle_{p^{F^0}(x)dx}$ is defined as follows:

$$\langle T, E^{F^0}[\cdot]\rangle_{p^{F^0}(x)dx} := {}_{\mathcal{S}'}\langle T, E^{F^0}[\cdot]\, p^{F^0}\rangle_{\mathcal{S}}.$$

Under the uniformly non-degenerate condition of $F^\epsilon \in \mathbf{D}^\infty(\mathbf{R}^n)$, the lifting up of $T \in \mathcal{S}'(\mathbf{R}^n)$ that is, $(E^{F^\epsilon})^*T$, has the asymptotic expansion in distributions on the Wiener space $\mathbf{D}^{-\infty}$, that is for $N \in \mathbf{N}$, there exists $s \in \mathbf{N}$ such that

$$\begin{aligned}&\left\|(E^{F^\epsilon})^*T - \left\{T\circ F^0 + \sum_{j=1}^N \epsilon^j \sum_k^{(j)} (\partial_\alpha^k T)\circ F^0 \prod_{l=1}^k F_{\alpha_l}^{0,\beta_l}\right\}\right\|_{\mathbf{D}_{q,-s}}\\ &\quad = O(\epsilon^{N+1}),\ \epsilon\in(0,1],\ q<\infty.\end{aligned} \tag{92}$$

Then, there exists an asymptotic expansion of $\langle (E^{F^\epsilon})^*T, \mathbf{1}\rangle_{\mathbf{D}^{-\infty}\times\mathbf{D}^\infty}$.

The push-down of the divergence are computed as follows:

$$\begin{aligned}\left\langle \partial_\alpha^k T(F^0), \prod_{l=1}^k F_{\alpha_l}^{0,\beta_l}\right\rangle_{\mathbf{D}^{-\infty}\times\mathbf{D}^\infty} &= \left\langle T(F^0), H_\alpha\left(F^0, \prod_{l=1}^k F_{\alpha_l}^{0,\beta_l}\right)\right\rangle_{\mathbf{D}^{-\infty}\times\mathbf{D}^\infty}\\ &= \left\langle T, E^{F^0}\left[H_\alpha\left(F^0, \prod_{l=1}^k F_{\alpha_l}^{0,\beta_l}\right)\right]\right\rangle_{p^{F^0}(x)dx}\\ &= \int_{\mathbf{R}^n} T(x)E\left[H_\alpha\left(F^0, \prod_{l=1}^k F_{\alpha_l}^{0,\beta_l}\right)|F^0 = x\right]p^{F^0}(x)dx.\end{aligned} \tag{93}$$

On the other hand,

$$\begin{aligned}\left\langle \partial_\alpha^k T(F^0), \prod_{l=1}^k F_{\alpha_l}^{0,\beta_l}\right\rangle_{\mathbf{D}^{-\infty}\times\mathbf{D}^\infty} &= \left\langle \partial_\alpha^k T, E^{F^0}\left[\prod_{l=1}^k F_{\alpha_l}^{0,\beta_l}\right]\right\rangle_{p^{F^0}(x)dx}\\ &= \left\langle T, (\partial^*)_\alpha^k E^{F^0}\left[\prod_{l=1}^k F_{\alpha_l}^{0,\beta_l}\right]\right\rangle_{p^{F^0}(x)dx}\end{aligned}$$

$$= (-1)^k \int_{\mathbf{R}^n} T(x) \partial_\alpha^k \left\{ E\left[\prod_{l=1}^{k} F_{\alpha_l}^{0,\beta_l} | F^0 = x \right] p^{F^0}(x) \right\} dx. \tag{94}$$

Here, $(\partial^*)_\alpha^k$ means $(\partial^*)_\alpha^k = \partial_\alpha^* \cdots \partial_\alpha^* (k \text{ times})$, and ∂_α^* denotes the divergence operator on the space $\left(\mathbf{R}^n, p^{F^0}(x)dx\right)$. □

Corollary 1 *The asymptotic expansion of the density function of F^ϵ, $p^{F^\epsilon}(y)$ is expressed with the push-down of the Malliavin weights as the follows:*

$$p^{F^\epsilon}(y) = p^{F^0}(y) + \sum_{j=1}^{m} \epsilon^j E\left[\sum_{k}^{(j)} H_{\alpha^{(k)}}\left(F^0, \prod_{l=1}^{k} F_{\alpha_l}^{0,\beta_l} \right) | F^0 = y \right] p^{F^0}(y) + O(\epsilon^{m+1}), \tag{95}$$

where $p^{F^0}(y)$ is the density function of F^0. An alternative expression is given as follows:

$$p^{F^\epsilon}(y) = p^{F^0}(y) + \sum_{j=1}^{m} \epsilon^j \sum_{k}^{(j)} (-1)^k \partial_{\alpha^{(k)}}^k \left\{ E\left[\prod_{l=1}^{k} F_{\alpha_l}^{0,\beta_l} | F^0 = y \right] p^{F^0}(y) \right\} + O(\epsilon^{m+1}). \tag{96}$$

(*Proof*) Take a delta function $\delta_y \in \mathcal{S}'(\mathbf{R}^n)$ in the theorem above. □

6.2 Instantaneous Forward Rates Model

As a typical stochastic model for pricing the interest rate derivatives, there exists a model developed by Heath-Jarrow-Morton [37], the so called HJM model, which is formulated based on the forward rates with infinitesimal terms of the interest rates, that is the instantaneous forward rates $\{f(s,t) : 0 \le s \le t \le T\}$. Here, s is the time when the forward rate is fixed and t denotes the inception time when the forward rate is applied.

The stochastic processes for the instantaneous forward rates are considered in the framework of the asymptotic expansion by introducing a parameter $\epsilon \in [0, 1]$. For example, let W be a m-dimensional standard Wiener process and let $f(0, t)$, $t \in [0, T]$ be a given Lipschitz continuous function of t. Then, under the equivalent martingale measure, the stochastic processes of $\{f^{(\epsilon)}(s,t) : 0 \le s \le t \le T\}$ are solutions to the following stochastic integral equations:

$$f^{(\epsilon)}(s,t) = f(0,t) + \epsilon^2 \int_0^s \sum_{i=1}^m \left[\sigma_i(f^{(\epsilon)}(v,t), v, t) \int_v^t \sigma_i(f^{(\epsilon)}(v,y), v, y)dy \right] dv$$
$$+ \epsilon \sum_{i=1}^m \int_0^s \sigma_i(f^{(\epsilon)}(v,t), v, t)dW_i(v) \; ; \epsilon \in [0,1], \tag{97}$$

where the volatility functions $\{\sigma_i(x,s,t); i = 1,\ldots,m\}$ are smooth and satisfy the regularity conditions which guarantee that the equation (97) has its unique strong solution. It is to be noted that the drift term (the coefficient of the dv term) of $f^{(\epsilon)}(s,t)$ depends on $\{f^{(\epsilon)}(v,y); 0 \leq v < s, v \leq y < t\}$. Moreover, the stochastic process of the instantaneous short-term interest rate $r^{(\epsilon)}(t)$ is determined by the relation, $r^{(\epsilon)}(t) = f^{(\epsilon)}(t,t)$.

For this model, the approximations of the values for interest rate derivatives can still be considered in a unified framework with derivation of asymptotic expansions of the instantaneous forward rates when $\epsilon \downarrow 0$ and with use of the relation between the instantaneous forward rates and a zero-coupon bond price:

$$P^{(\epsilon)}(t,T) = \exp\left\{ -\int_t^T f^{(\epsilon)}(t,u)du \right\}. \tag{98}$$

As an example, we consider pricing an option on a coupon bond (or a swaption), which is a standard interest rate derivative. The payoff at maturity of a call option is given by

$$V_c(T) = \max\left\{ \sum_{i=1}^n c_i P^{(\epsilon)}(t, T_i) - K, 0 \right\},$$

where $0 \leq T \leq T_1 < \cdots < T_n$, c_i $(i = 1,\ldots,n)$ are positive constants and $K (> 0)$ is a strike price. Then, its present value is given by

$$V_c(0) = \mathbf{E}\left[e^{-\int_0^T r_u^{(\epsilon)}du} V_c(T) \right]. \tag{99}$$

When $\epsilon \downarrow 0$, the forward rate $f^{(\epsilon)}(s,t)$ is expanded around $f(0,t)$ as

$$f^{(\epsilon)}(s,t) \sim f(0,t) + \epsilon f_1(s,t) + \epsilon^2 f_2(s,t) + \cdots \; in \; \mathbf{D}^\infty, \tag{100}$$

where the coefficients of ϵ^n, $n = 1, 2, \ldots$, that is $f_1(t,u)$, $f_2(t,u) \cdots$ are also in $\mathbf{D}^\infty$.

As a result, we obtain an expansion of the zero-coupon bond price $P^{(\epsilon)}(t,T)$ around the current forward bond price $\frac{P(0,T)}{P(0,t)}$, and an expansion of the discount factor $\exp\left\{ -\int_0^T r^{(\epsilon)}(t)dt \right\}$ around the current zero-coupon bond price $P(0,T)$ as follows:

$$P^{(\epsilon)}(t,T) \sim \frac{P(0,T)}{P(0,t)}\left[1-\epsilon\int_t^T f_1(t,u)du - \epsilon^2\int_t^T f_2(t,u)du \right.$$
$$\left. + \epsilon^2\frac{1}{2}\left\{\int_t^T f_1(t,u)du\right\}^2\right] + \cdots \; in\; \mathbf{D}^\infty, \tag{101}$$

$$e^{-\int_0^T r^{(\epsilon)}(s)ds} \sim P(0,T)\left[1-\epsilon\int_0^T f_1(t,t)dt - \epsilon^2\int_0^T f_2(t,t)dt \right.$$
$$\left. + \epsilon^2\frac{1}{2}\left\{\int_0^T f_1(t,t)dt\right\}^2\right] + \cdots \; in\; \mathbf{D}^\infty, \tag{102}$$

where $f_i(s,t), i = 1, 2$ are given by

$$f_1(s,t) = \tfrac{\partial f^{(\epsilon)}(s,t)}{\partial\epsilon}|_{\epsilon=0} = \int_0^s \sum_{i=1}^m \sigma_i^{(0)}(v,t)dW_i(v),$$
$$f_2(s,t) = \tfrac{1}{2}\tfrac{\partial^2 f^{(\epsilon)}(s,t)}{\partial^2\epsilon}|_{\epsilon=0}$$
$$= \int_0^s b^{(0)}(v,t)dv + \int_0^s \sum_{i=1}^m \partial\sigma_i^{(0)}(v,t) f_1(v,t)dW_i(v).$$

Here, $\sigma_i^{(0)}(v,t) = \sigma_i(f^{(0)}(v,t), v, t)$, and $b^{(0)}(v,t)$ and $\partial\sigma_i^{(0)}(v,t)$ are defined as

$$b^{(0)}(v,t) = \sum_{i=1}^n \sigma_i(f^{(0)}(v,t), v, t)\int_v^t \sigma_i(f^{(0)}(v,y), v, y)dy,$$
$$\partial\sigma_i^{(0)}(v,t) = \frac{\partial\sigma_i(x,v,t)}{\partial x}|_{x=f(0,t)}.$$

Therefore, in a similar way as in the framework for diffusion cases in the previous sections, we define $X_t^{(\epsilon),1}$ and $X_t^{(\epsilon),i}$ $(i = 2,\ldots,n)$ as

$$X_t^{(\epsilon),1} = \exp\left\{-\int_0^t r^{(\epsilon)}(u)du\right\} \tag{103}$$
$$X_t^{(\epsilon),i} = P^{(\epsilon)}(t,T_i) = \exp\left\{-\int_t^{T_i} f^{(\epsilon)}(t,u)du\right\},\; i = 2,\ldots,n. \tag{104}$$

Then, the payoff at maturity of the call option on a coupon bond is written as

$$V_c(T) = \max\left\{\sum_{i=2}^n c_i X_t^{(\epsilon),i} - K, 0\right\}. \tag{105}$$

Moreover, let $x = (x_1, x_2, \ldots, x_n)$ and define $g(x)$ as

$$g(x) = x_1 \left(\sum_{i=2}^{n} c_i x_i - K \right). \tag{106}$$

In this way, we are able to employ a similar technique to pricing derivatives as in the case of diffusion processes. For example, with redefinition of variables such as Σ_T, the approximation of the option price $V_c(0)$ in (99) can be obtained based on the almost same asymptotic expansion method as in the previous sections. In fact, by using the above expansions of instantaneous forward rates, zero-coupon bond prices and the discount factor, we can apply the expansion to $\mathbf{E}[\max\{g(X_T^{(\epsilon)}), 0\}]$, where $X_T^{(\epsilon)} = (X_T^{(\epsilon),1}, X_T^{(\epsilon),2} \ldots, X_T^{(\epsilon),n})$.

For the details and numerical examples, please see [49, 50, 88]. In particular, [88] implements numerical experiments under a smooth and bounded modification of two factor CEV-type volatility functions (as explained in Sect. 4.3.1), and the variance reduction technique in proposed in [107] to demonstrate the effectiveness of the method. We remark that the boundedness of the volatility functions $\{\sigma_i(x, s, t); i = 1, \ldots, m\}$ for the instantaneous forward rates $f^{(\epsilon)}(s, t)$ is one of the sufficient conditions that guarantee the existence of the unique strong solution of the stochastic integral Eq. (97).

For evaluation of other various interest rate derivatives, approximations based on the asymptotic expansion approach can be derived in the similar manner. Moreover, an example of an approximate formula for derivative prices dependent on the instantaneous forward rates in the HJM model and other variables following general diffusion processes is given by [85].

7 Asymptotic Expansion in Jump and Jump-Diffusion Models

So far, we have used stochastic models whose randomnesses are generated by only Wiener processes. However, we are also able to apply the asymptotic expansion approach to stochastic processes including jumps in their sample paths. This section provides its very brief review. For the details, please see the cited papers.

In terms of the mathematical viewpoint, Yoshida [120] presented an extension of Watanabe theory to develop a framework for providing a validity of asymptotic expansions in Wiener-Poisson spaces, which can be applied to jump-diffusion models under some regularity conditions. Hayashi [34] applied a Malliaivin calculus of jump-type to prove an asymptotic expansion theorem for functionals of a Poisson random measure, and Hayashi [35] derived the coefficients in the expansion of a call option price under a pure jump model. Moreover, Hayashi and Ishikawa [36] proved an asymptotic expansion formula for the compositions of a smooth Wiener-Poisson functional with Schwartz distributions.

In direct applications to finance problems, [51, 87] derived asymptotic expansion to approximate bond prices or/and plain-vanilla option prices under jump-diffusion with local volatility models.

Subsequently, [93, 94] found a new expansion scheme for pricing long-term European currency options under a Libor market model (LMM) and a general diffusion stochastic volatility model with jumps of spot exchange rates. Particularly, thanks to a linear structure of the underlying asset price process in their model, they separated the jump component with a known characteristic function to apply the expansion technique developed in the diffusion models. Also, [100] took a Malliavin calculus approach to derive asymptotic expansions of vanilla option prices in a jump-diffusion with stochastic volatility model.

Recently, [80] has generalized the preceding researches such as [51, 87] and [100] in the asymptotic expansion approach, and developed a new approximation formula for pricing basket options in a local-stochastic volatility model with jumps. In particular, the model admits local volatility functions and jump components in not only the underlying asset price processes, but also the volatility processes. Moreover, they implemented some numerical experiments to confirm the validity of the method. Please see the paper for the details.

As an example of asymptotic expansions of option prices under jump-diffusion models, the next subsection describes the outline of the method by using a simplified version of [80].

7.1 Pricing Basket Options Under Local Stochastic Volatility with Jumps

In the first place, we define the model of the underlying asset prices and its volatility processes, which is used for pricing the European type basket options. In particular, suppose that the filtered probability space $(\Omega, \mathcal{F}, P, \{\mathcal{F}_t\}_{t\geq 0})$ is given, where P is an equivalent martingale measure and the filtration satisfies the usual conditions. The risk-free interest rate is assumed to be a nonnegative constant r for simplicity. Then, $(S_t^i)_{t\in[0,T]}$ and $(\sigma_t^i)_{t\in[0,T]}$, $i = 1, \ldots, d$ represent the underlying asset prices and their volatilities for $t \in [0, T]$, respectively. Particularly, let us assume that S_T^i and σ_T^i are given by the solutions of the following stochastic integral equations:

$$S_T^i = s_0^i + \int_0^T \alpha^i S_{t-}^i dt + \int_0^T \phi_{S^i}\left(\sigma_{t-}^i, S_{t-}^i\right) dW_t^{S^i} + \sum_{l=1}^n \left(\sum_{j=1}^{N_{l,T}} h_{S^i,l,j} S_{\tau_{j,l}-}^i - \int_0^T \Lambda_l S_{t-}^i \mathbf{E}[h_{S^i,l,j}] dt \right), \tag{107}$$

$$\sigma_T^i = \sigma_0^i + \int_0^T \lambda^i(\theta^i - \sigma_{t-}^i) dt + \int_0^T \phi_{\sigma^i}\left(\sigma_{t-}^i\right) dW_t^{\sigma^i}$$

$$+\sum_{l=1}^{n}\left(\sum_{j=1}^{N_{l,T}} h_{\sigma^i,l,j}\sigma^i_{\tau_{j,l}-} - \int_0^T \Lambda_l \sigma^i_{t-}\mathbf{E}[h_{\sigma^i,l,j}]dt\right), \quad (108)$$

where s_0^i and σ_0^i, $i = 1, \ldots, d$ are given as some constants. The notations are defined as follows:

- α^i $(i = 1, \ldots, d)$ are constants.
  λ^i and θ^i $(i = 1, \ldots, d)$ are nonnegative constants.
- $\phi_{S^i}(x, y)$ and $\phi_{\sigma^i}(x)$ are some functions with appropriate regularity conditions.
- W^{S^i} and W^{σ^i}, $(i = 1, \ldots, d)$ are correlated Brownian motions.
- Each N_l, $(l = 1, \ldots, n)$ is a Poisson process with constant intensity Λ_l. N_l, $l = 1, \ldots, n$ are independent, and also independent of all W^{S^i} and W^{σ^i}.
- $\tau_{j,l}$ stands for the jth jump time of N_l.
- For each $l = 1, \ldots, n$ and $i = 1, \ldots, d$, both $\left(\sum_{j=1}^{N_{l,t}} h_{S^i,l,j}\right)_{t\geq 0}$ and $\left(\sum_{j=1}^{N_{l,t}} h_{\sigma^i,l,j}\right)_{t\geq 0}$ are compound Poisson processes. ($\sum_{j=1}^{N_{l,t}} \equiv 0$ when $N_{l,t} = 0$.)
- For each l and x^i, $h_{x^i,l,j}$ $(j \in \mathbf{N})$ are independent and identically distributed random variables, where x^i stands for one of S^i and σ^i $(i = 1, \ldots, d)$.
  - for the log-normal jump case, $h_{x^i,l,j} = e^{Y_{x^i,l,j}} - 1$, where $Y_{x^i,l,j}$ is a random variable which follows a normal distribution with mean $m_{x^i,l}$ and variance $\gamma^2_{x^i,l}$ that is, $N(m_{x^i,l}, \gamma^2_{x^i,l})$ for all j.
- $h_{x^i,l,j}$ and $h_{x^{i'},l',j'}$ $(l \neq l')$ are independent.
  $h_{x^i,l,j}$ and $h_{x^{i'},l',j'}$ $(j \neq j')$ are independent.
  N_l and $h_{x^i,l',j}$ are independent.
  For the same l and j, $h_{S^i,l,j}$ and $h_{\sigma^{i'},l,j}$ $(i, i' = 1, \ldots, d)$ are allowed to be dependent, that is $Y_{S^i,l,j}$ and $Y_{\sigma^{i'},l,j}$ $(i, i' = 1, \ldots, d)$ are generally correlated.

Remark 8 By specifying the functions ϕ_S and ϕ_σ, we can express various types of local-stochastic volatility models. For example, the model with $\phi_S(\sigma, S) = (aS^2 + bS + c)\sqrt{\sigma}$ and $\phi_\sigma(\sigma) = \sqrt{\sigma}$ corresponds to an extension of the Quadratic Heston model. The model with $\phi_S(\sigma, S) = S^{\beta_S}\sigma$ and $\phi_\sigma(\sigma) = \sigma$ corresponds to an extended SABR (λ-SABR) model, and the one with $\phi_S(\sigma, S) = S^{\beta_S}\sigma$ and $\phi_\sigma(\sigma) = \sigma^{\beta_\sigma}$ corresponds to a local volatility on volatility with jumps model.

Next, we introduce perturbations to the model (107) and (108). That is, for a known parameter $\epsilon \in [0, 1]$ we consider the following stochastic integral equations: for $i = 1, \ldots, d$,

$$S_T^{i,(\epsilon)} = s_0^i + \int_0^T \alpha^i S_{t-}^{i,(\epsilon)}dt + \epsilon\int_0^T \phi_{S^i}\left(\sigma_{t-}^{i,(\epsilon)}, S_{t-}^{i,(\epsilon)}\right) dW_t^{S^i}$$

$$+\sum_{l=1}^{n}\left(\sum_{j=1}^{N_{l,T}} h^{(\epsilon)}_{S^i,l,j} S^{i,(\epsilon)}_{\tau_{j,l}-} - \int_0^T \Lambda_l S^{i,(\epsilon)}_{t-}\mathbf{E}[h^{(\epsilon)}_{S^i,l,1}]dt\right), \quad (109)$$

$$\sigma_T^{i,(\epsilon)} = \sigma_0^i + \int_0^T \lambda^i(\theta^i - \sigma_{t-}^{i,(\epsilon)})dt + \epsilon \int_0^T \phi_{\sigma^i}\left(\sigma_{t-}^{i,(\epsilon)}\right) dW_t^{\sigma^i}$$
$$+ \sum_{l=1}^{n} \left(\sum_{j=1}^{N_{l,T}} h_{\sigma^i,l,j}^{(\epsilon)} \sigma_{\tau_{j,l}-}^{i,(\epsilon)} - \int_0^T \Lambda_l \sigma_{t-}^{i,(\epsilon)} \mathbf{E}[h_{\sigma^i,l,1}^{(\epsilon)}]dt \right), \tag{110}$$

where $h_{x^i,l,j}^{(\epsilon)} = e^{\epsilon Y_{x^i,l,j}} - 1$, that is, we assume that the jump size follows a log-normal distribution, $\epsilon Y_{x^i,l,j} \sim N(\epsilon m_{x^i,l}, \epsilon^2 \gamma_{x^i,l}^2)$.

We assume the asymptotic expansions of $S_T^{i,(\epsilon)}$ and $\sigma_T^{i,(\epsilon)}$ around $\epsilon = 0$ as follows:

$$S_T^{i,(\epsilon)} = S_T^{i,(0)} + \epsilon S_T^{i,(1)} + \frac{\epsilon^2}{2!} S_T^{i,(2)} + \cdots, \tag{111}$$

$$\sigma_T^{i,(\epsilon)} = \sigma_T^{i,(0)} + \epsilon \sigma_T^{i,(1)} + \frac{\epsilon^2}{2!} \sigma_T^{i,(2)} + \cdots, \tag{112}$$

$$h_{x^i,l,j}^{(\epsilon)} = h_{x^i,l,j}^{(0)} + \epsilon h_{x^i,l,j}^{(1)} + \frac{\epsilon^2}{2!} h_{x^i,l,j}^{(2)} + \cdots, \tag{113}$$

where $S_t^{i,(\iota)} := \left.\frac{\partial^\iota S_t^{i,(\epsilon)}}{\partial \epsilon^\iota}\right|_{\epsilon=0}$, $\sigma_t^{i,(\iota)} := \left.\frac{\partial^\iota \sigma_t^{i,(\epsilon)}}{\partial \epsilon^\iota}\right|_{\epsilon=0}$, $h_{x^i,l,j}^{(\iota)} := \left.\frac{\partial^\iota h_{x^i,l,j}^{(\epsilon)}}{\partial \epsilon^\iota}\right|_{\epsilon=0}$.

We also suppose that $(W^{S^1}, \ldots, W^{S^d}, W^{\sigma^1}, \ldots, W^{\sigma^d})' = \varrho \cdot Z$ where ϱ is a $2d \times 2d$ correlation matrix, and Z is a $2d$-dimensional (independent) Brownian motion.

For ease of the expressions we introduce the following notations:

- $\Phi_{S^i,j} := \phi_{S^i}(\sigma^i, S^i)(\varrho)_{i,j}$ and $\Phi_{\sigma^i,j} := \phi_{\sigma^i}(\sigma^i)(\varrho)_{d+i,j}$, where $(\varrho)_{i,j}$ denotes the (i, j)-element of ϱ.
- $\Phi_{S^i} := (\Phi_{S^i,1}, \ldots, \Phi_{S^i,2d})$ and $\Phi_{\sigma^i} := (\Phi_{\sigma^i,1}, \ldots, \Phi_{\sigma^i,2d})$ are $2d$-dimensional vectors.
- $\Phi_S := (\Phi_{S^1}, \ldots, \Phi_{S^d})'$ and $\Phi_\sigma := (\Phi_{\sigma^1}, \ldots, \Phi_{\sigma^d})'$ are $d \times 2d$ matricies.
- We define a operator "$*$" as follows: When A and B are $d \times 2d$ matrices,

$$A * B := \begin{bmatrix} (A)_{1,1}(B)_{1,1} & \cdots & (A)_{1,2d}(B)_{1,2d} \\ \vdots & \ddots & \vdots \\ (A)_{d,1}(B)_{d,1} & \cdots & (A)_{d,2d}(B)_{d,2d} \end{bmatrix}. \tag{114}$$

When A is a $d \times 2d$ matrix and B is a d-dimensional vector,

$$A * B = B * A := \begin{bmatrix} (A)_{1,1}(B)_1 & \cdots & (A)_{1,2d}(B)_1 \\ \vdots & \ddots & \vdots \\ (A)_{d,1}(B)_d & \cdots & (A)_{d,2d}(B)_d \end{bmatrix}. \tag{115}$$

When A and B are d-dimensional vectors,

$$A * B := \begin{bmatrix} (A)_1(B)_1 \\ \vdots \\ (A)_d(B)_d \end{bmatrix}. \tag{116}$$

- We also define $\partial_x \Phi_S$ ($x = S$ or σ) as

$$\partial_x \Phi_S := \begin{bmatrix} \frac{\partial}{\partial x^1}(\Phi_S)_{1,1} & \cdots & \frac{\partial}{\partial x^1}(\Phi_S)_{1,2d} \\ \vdots & \ddots & \vdots \\ \frac{\partial}{\partial x^d}(\Phi_S)_{d,1} & \cdots & \frac{\partial}{\partial x^d}(\Phi_S)_{d,2d} \end{bmatrix}, \tag{117}$$

where $(\Phi_S)_{i,j}$ denotes the (i, j)-element of the $d \times 2d$ matrix Φ_S.
- Let us introduce the following notations:
$S_t = (S_t^1, \dots, S_t^d)$, $\sigma_t = (\sigma_t^1, \dots, \sigma_t^d)$,
$h_{S,l,j}^{(i)} = (h_{S^1,l,j}^{(i)}, \dots, h_{S^d,l,j}^{(i)})$, $h_{\sigma,l,j}^{(i)} = (h_{\sigma^1,l,j}^{(i)}, \dots, h_{\sigma^d,l,j}^{(i)})$,
$e^{\alpha t} = (e^{\alpha^1 t}, \dots, e^{\alpha^d t})$ and $e^{\lambda t} = (e^{\lambda^1 t}, \dots, e^{\lambda^d t})$.

Based on these preparations, we obtain the next proposition.

Proposition 2 *The coefficients,* $S_T^{(i)}$, $\sigma_T^{(i)}$ *and* $h_{x,l,j}^{(i)}$($x =, S, \sigma$), $i = 0, 1, 2$ *in the expansions* (111), (112) *and* (113) *are given as follows:*

$$S_T^{(0)} = e^{\alpha T} * s_0, \tag{118}$$

$$\sigma_T^{(0)} = \theta + (\sigma_0 - \theta) * e^{-\lambda T}, \tag{119}$$

$$h_{x,l,j}^{(0)} = 0, \tag{120}$$

$$S_T^{(1)} = \int_0^T e^{\alpha(T-t)} * \Phi_S\left(\sigma_{t-}^{(0)}, S_{t-}^{(0)}\right) dZ_t + \sum_{l=1}^{n}\left(\sum_{j=1}^{N_{l,T}} h_{S,l,j}^{(1)} \cdot - \Lambda_l T \mathbf{E}\left[h_{S,l,1}^{(1)}\right]\right) * S_T^{(0)}, \tag{121}$$

$$\sigma_T^{(1)} = \int_0^T e^{-\lambda(T-t)} * \Phi_\sigma\left(\sigma_{t-}^{(0)}\right) dZ_t + \sum_{l=1}^{n}\left(\sum_{j=1}^{N_{l,T}} h_{\sigma,l,j}^{(1)} * e^{-\lambda(T-\tau_{j,l})} * \sigma_{\tau_{j,l}-}^{(0)} - \Lambda_l \mathbf{E}\left[h_{\sigma,l,1}^{(1)}\right] * e^{-\lambda T} * \int_0^T e^{\lambda t} * \sigma_{t-}^{(0)} dt\right), \tag{122}$$

$$h_{x,l,j}^{(1)} = Y_{x,l,j} := (Y_{x^1,l,j}, \dots, Y_{x^d,l,j}), \tag{123}$$

$$S_T^{(2)} = \int_0^T e^{\alpha(T-t)} * \partial_S \Phi_S\left(\sigma_{t-}^{(0)}, S_{t-}^{(0)}\right) * S_{t-}^{(1)} dZ_t + \int_0^T e^{\alpha(T-t)} * \partial_\sigma \Phi_S\left(\sigma_{t-}^{(0)}, S_{t-}^{(0)}\right) * \sigma_{t-}^{(1)} dZ_t$$

$$+\sum_{l=1}^{n}\left(\left\{\sum_{j=1}^{N_{l,T}} h^{(2)}_{S,l,j} - \Lambda_l T\mathbf{E}\left[h^{(2)}_{S,l,1}\right]\right\} * S^{(0)}_T\right.$$

$$+\sum_{j=1}^{N_{l,T}} h^{(1)}_{S,l,j} * e^{\alpha(T-\tau_{j,l})} * S^{(1)}_{\tau_{j,l}-}$$

$$\left.-\Lambda_l\mathbf{E}\left[h^{(1)}_{S,l,1}\right] * e^{\alpha T} * \int_0^T e^{-\alpha t} * S^{(1)}_{t-}dt\right), \tag{124}$$

$$h^{(2)}_{x,l,j} = Y_{x,l,j} * Y_{x,l,j}. \tag{125}$$

Next, let us define the payoff of a basket call option with strike price K as

$$(g(x) - K)^+ \; (:= \max\{g(x) - K, 0\}), \tag{126}$$

$$g(x) := w \cdot x = \sum_{i=1}^{d} w_i x^i,$$

where $g(x)$ represents a weighted sum of the underlying asset prices of $x_1, \ldots, x_d$ with the constant weights $w_1, \ldots, w_d$. Here, we set $x := (x^1, \ldots, x^d)$ and $w := (w_1, \ldots, w_d)$.

For an approximation of a basket option price, we firstly note that $g\left(S^{(\epsilon)}_T\right)$ is expanded around $\epsilon = 0$ as:

$$g\left(S^{(\epsilon)}_T\right) = g\left(S^{(0)}_T\right) + \epsilon g\left(S^{(1)}_T\right) + \frac{\epsilon^2}{2} g\left(S^{(2)}_T\right) + o(\epsilon^2). \tag{127}$$

Then, for a strike price $K = g(S^0_T) - \epsilon y$ for an arbitrary $y \in \mathbf{R}$, the payoff of the call option with maturity T is expanded as follows:

$$\begin{aligned}\left(g\left(S^{(\epsilon)}_T\right) - K\right)^+ &= \epsilon\left(\frac{g(S^{(\epsilon)}_T) - g(S^{(0)}_T)}{\epsilon} + y\right)^+ \\ &= \epsilon\left(g\left(S^{(1)}_T\right) + \frac{\epsilon}{2} g\left(S^{(2)}_T\right) + y + o(\epsilon)\right)^+ \\ &= \epsilon\left(g(S^{(1)}_T) + y\right)^+ + \frac{\epsilon^2}{2} 1_{\left\{g(S^{(1)}_T) > -y\right\}} g\left(S^{(2)}_T\right) + o(\epsilon^2).\end{aligned} \tag{128}$$

We next note that when the number of jumps is k_l $(l = 1, \ldots, n)$, that is on $\{N_l = k_l\} := \left\{N_{1,T} = k_1, \ldots, N_{n,T} = k_n\right\}$, $S^{(1)}_T$ in the equation (121) becomes

$$\xi_{\{k_l\}} + \hat{S}_T, \tag{129}$$

where

$$\xi_{\{k_l\}} := \sum_{l=1}^{n}(k_l - \Lambda_l T)m_{S,l} * S_T^{(0)} \tag{130}$$

and

$$\hat{S}_T := \int_0^T e^{\alpha(T-t)} * \Phi_S\left(\sigma_t^{(0)}, S_t^{(0)}\right) dZ_t + \sum_{l=1}^{n}\left(\sum_{j=1}^{k_l} \gamma_{S,l} * \zeta_{S,j,l} * S_T^{(0)}\right). \tag{131}$$

Here, we use the following notations:

- $\gamma_{S,l} = (\gamma_{S^1,l}, \ldots, \gamma_{S^d,l})$
- $\zeta_{S,j,l} = (\zeta_{S^1,j,l}, \ldots, \zeta_{S^d,j,l})$ is a vector of random variables, where $\zeta_{S^i,j,l}$ follows $N(0, 1)$, that is the standard normal distribution.

We remark that the distribution of $g(\hat{S}_T)$ is $N\left(0, \Sigma_T^{\{k_l\}}\right)$, that is the normal distribution with mean zero and variance $\Sigma_T^{\{k_l\}}$ whose density function is expressed as

$$n\left(x; 0, \Sigma_T^{\{k_l\}}\right) := \frac{1}{\sqrt{2\pi\Sigma_T^{\{k_l\}}}} \exp\left\{\frac{-x^2}{2\Sigma_T^{\{k_l\}}}\right\}. \tag{132}$$

Here, $\Sigma_T^{\{k_l\}}$ is defined as follows:

$$\begin{aligned}\Sigma_T^{\{k_l\}} &:= \int_0^T \left(w * e^{\alpha(T-t)} * \Phi_S\left(\sigma_t^{(0)}, S_t^{(0)}\right)\right)^\top \left(w * e^{\alpha(T-t)} * \Phi_S\left(\sigma_t^{(0)}, S_t^{(0)}\right)\right) dt \\ &\quad + \sum_{l=1}^{n} k_l (w * \gamma_{S,l} * S_T^{(0)})^\top \vartheta_{\zeta_{S,l}} (w * \gamma_{S,l} * S_T^{(0)}),\end{aligned} \tag{133}$$

where $\vartheta_{\zeta_{S,l}}$ stands for the correlation matrix of $\zeta_{S,j,l} = (\zeta_{S^1,j,l}, \ldots, \zeta_{S^d,j,l})$, and $x^\top$ denotes the transpose of x.

Next, we define

$$\eta_2(x, \{k_l\}) = \mathbf{E}\left[g\left(S_T^{(2)}\right) \Big| g(\hat{S}_T) = x, \{N_l = k_l\}\right]. \tag{134}$$

With those preparations, we approximate the expectation of the basket call payoff under an equivalent martingale measure in the following way:

$$\mathbf{E}\left[\left(g\left(S_T^{(\epsilon)}\right)-K\right)^+\right]$$
$$=\epsilon\mathbf{E}\left[\mathbf{E}\left[\left(g(S_T^{(1)})+y\right)^+\Big|g(\hat{S}_T)=x,\{N_l=k_l\}\right]\right]$$
$$+\frac{\epsilon^2}{2}\mathbf{E}\left[\mathbf{E}\left[1_{\left\{g(S_T^{(1)})>-y\right\}}g\left(S_T^{(2)}\right)\Big|g(\hat{S}_T)=x,\{N_l=k_l\}\right]\right]+o(\epsilon^2). \quad (135)$$

We also note that the probability of $\{N_l=k_l\}:=\left\{N_{1,T}=k_1,\ldots,N_{n,T}=k_n\right\}$ is expressed as

$$p_{\{k_l\}}:=\prod_{l=1}^{n}\frac{(\Lambda_l T)^{k_l}e^{-\Lambda_l T}}{k_l!}, \quad (136)$$

which is the product of the k_l times of the jump probabilities of $N_{l,T}$ $(l=1,\ldots,n)$, that is $\prod_{l=1}^n P(\{N_{l,T}=k_l\})$, thanks to the independence of $N_{l,T}$ $(l=1,\ldots,n)$.

Then, we calculate the coefficients of ϵ and $\frac{\epsilon^2}{2}$ on the right hand of (135) as follows: The coefficient of ϵ is given by:

$$\mathbf{E}\left[\mathbf{E}\left[\left(g\left(S_T^{(1)}\right)+y\right)^+\Big|g(\hat{S}_T)=x,\{N_l=k_l\}\right]\right]$$
$$=\sum_{k=0}^{\infty}\sum_{\sum_{l=1}^n k_l=k}p_{\{k_l\}}\int_{-(g(\xi_{\{k_l\}})+y)}^{\infty}\left(x+g(\xi_{\{k_l\}})+y\right)n(x;0,\Sigma_T^{\{k_l\}})dx, \quad (137)$$

and the coefficient of $\frac{\epsilon^2}{2}$ is given by:

$$\mathbf{E}\left[\mathbf{E}\left[1_{\left\{g(S_T^{(1)})>-y\right\}}g\left(S_T^{(2)}\right)\Big|g(\hat{S}_T)=x,\{N_l=k_l\}\right]\right]$$
$$=\sum_{k=0}^{\infty}\sum_{\sum_{l=1}^n k_l=k}p_{\{k_l\}}\int_{-(g(\xi_{\{k_l\}})+y)}^{\infty}\eta_2(x,\{k_l\})n(x;0,\Sigma_T^{\{k_l\}})dx. \quad (138)$$

Then, the initial value, $C(K,T)$ of the basket call option with maturity T and strike K is expanded around $\epsilon=0$ as follows:

$$C(K,T)=$$
$$\sum_{k=0}^{\infty}\sum_{\sum_{l=1}^n k_l=k}p_{\{k_l\}}e^{-rT}\left\{\epsilon\int_{-y_{\{k_l\}}}^{\infty}(x+y_{\{k_l\}})\,n(x;0,\Sigma_T^{\{k_l\}})dx\right.$$
$$\left.+\epsilon^2\int_{-y_{\{k_l\}}}^{\infty}\eta_2(x,\{k_l\})n(x;0,\Sigma_T^{\{k_l\}})dx\right\}+o(\epsilon^2), \quad (139)$$

where $y_{\{k_l\}} := g(\xi_{\{k_l\}}) + y$, and r is a constant risk-free rate.

In order to evaluate $\eta_2(x, \{k_l\})$, that is the conditional expectation defined in (134), we apply some formulas derived in Lemma 3.2 of [80].

Consequently, with $\epsilon = 1$ we obtain an approximate pricing formula for a basket call option, which corresponds to an asymptotic expansion of the basket option price up to the ϵ^2-order.

Theorem 5 *An approximation formula for the initial value $C(K, T)$ of an basket call option with maturity T and strike price K is given by the following equation:*

$$\sum_{k=0}^{\infty} \sum_{\sum_{l=1}^{n} k_l = k} p_{\{k_l\}} e^{-rT} \left\{ y_{k_l} N\left(\frac{y_{k_l}}{\sqrt{\Sigma_T^{\{k_l\}}}} \right) + \left(\Sigma_T^{\{k_l\}} + C_1 \frac{H_1\left(y_{k_l}; \Sigma_T^{\{k_l\}}\right)}{\Sigma_T^{\{k_l\}}} \right.\right.$$
$$\left.\left. + C_2 \frac{H_2\left(y_{k_l}; \Sigma_T^{\{k_l\}}\right)}{\left(\Sigma_T^{\{k_l\}}\right)^2} + C_3 \right) n\left(y_{k_l}; 0, \Sigma_T^{\{k_l\}}\right) \right\}, \tag{140}$$

where $p_{\{k_l\}} = \prod_{l=1}^{n} \frac{(\Lambda_l T)^{k_l} e^{-\Lambda_l T}}{k_l!}$, r is a constant risk-free rate, $y = g(S_T^{(0)}) - K$, $y_{\{k_l\}} = g(\xi_{\{k_l\}}) + y$, $N(x)$ denotes the standard normal distribution function and $n(x; 0, \Sigma) = \frac{1}{\sqrt{2\pi\Sigma}} \exp\left(\frac{-x^2}{2\Sigma}\right)$. Here, $\Sigma_T^{\{k_l\}}$ is given by (133), and $\xi_{\{k_l\}}$ is defined by (130). C_1, C_2 and C_3 are some constants, which are given with the derivations in Appendix B of [80]. Moreover, $H_k\left(x; \Sigma_T^{\{k_l\}}\right)$ denotes the kth order Hermite polynomial: particularly, $H_1\left(x; \Sigma_T^{\{k_l\}}\right) = x$ and $H_2\left(x; \Sigma_T^{\{k_l\}}\right) = x^2 - \Sigma_T^{\{k_l\}}$.

8 Perturbation Scheme in Forward Backward Stochastic Differential Equations (FBSDEs)

The FBSDEs have become quite popular in finance community since El Karoui, Peng and Quenez [16], especially after the recent financial crises and the subsequent quite volatile markets, which leads us to recognize the importance of counter party risk management, particularly the credit value adjustments (CVA).

However, an explicit solution for a FBSDE has been known only for a simple linear or quadratic example. Although several techniques have been proposed in the last decade, they seem very limited in practical applications since they rely on numerical methods for non-linear partial differential equations (PDEs) or regression based Monte Carlo simulations, which are generally very difficult to implement or quite time-consuming especially for high-dimensional and long-horizon problems.

Recently, [25] has developed a simple analytical approximation scheme for the nonlinear FBSDEs, notably for not only the so called decoupled cases but also the coupled cases. Fujii and Takahashi [25] has introduced a perturbation parameter to the generator of a backward stochastic differential equation (BSDE) to expand recursively the non-linear terms around a relevant linear FBSDE. In the computation of each order, [25] explicitly represents the backward elements as the functions of the forward components and take those expectations. Hence, except the cases that the distributions of the forward process are explicitly known, we need to apply some approximations of the distributions, and so, again, the asymptotic expansion technique for the forward stochastic differential equation (FSDE) is useful in the approximations. Section 8.1 below illustrates the scheme briefly. Fujii and Takahashi [25] also provided two numerical examples, where the second-order analytic approximations work quite well compared to numerical techniques such as the finite difference method and the regression-based Monte Carlo simulation. Please see the paper for the detail.

Moreover, their subsequent work [26] has applied this scheme to the optimal portfolio problem in an incomplete market with stochastic volatility, and demonstrated the accurate approximations even for long maturities such as 10 years, as opposed to the regression based Monte Carlo simulation which works well only up to short maturities such as one year.

We also note that the method has a great advantage of deriving explicit expressions of the optimal portfolios and hedging strategies, that is very important in practice. Furthermore, we can employ the method for the general multi-dimensional cases.

In order to achieve further reduction of computational burdens in this method, the scheme with an interacting particle method has been recently developed. Section 8.2 describes the outline. Please also see [29] as an application of the method to American option pricing.

Furthermore, [104] provides a mathematical foundation for the original scheme in the decoupled case proposed in [25]. (The justification for the coupled case seems an important and interesting research topic.) It mainly consisted of two parts. That is, for the BSDE expansion with a perturbed generator they have obtained the coefficients up to an arbitrary order as the solution to a system of the associated BSDEs with the base FSDE, and present the error estimate of the expansion. Accordingly, they showed a concrete representation for each expansion coefficient of the volatility component, that is the martingale integrand in the BSDE. For the FSDE expansion, they derived an expansion formula with its sharp error estimate for the expectation of the solution to the base FSDE in terms of a small diffusion. Then, they combine the both results, particularly applying the FSDE expansion formula to the BSDE expansion coefficients to obtain a main result, that is an asymptotic expansion of FBSDEs with a perturbed generator. In the proofs, [104] effectively applied the representation results in Ma and Zhang [63] for the BSDE expansion and the properties of the Kusuoka-Stroock functions in Kusuoka [52] for the FSDE expansion.

In a different stream, [102] has proposed a new semi closed-form approximation for the solutions of FBSDEs. In particular, applying the asymptotic expansion method in [100] and [103] to the forward SDEs with a Picard-type iteration scheme for the BSDEs, they have obtained an error estimate for the approximation. Moreover, they demonstrated the effectiveness of the method through numerical examples for pricing options with counter party risk under the local and stochastic volatility models, where the credit value adjustment (CVA) is taken into account. Roughly speaking, considering a perturbed forward SDE X^{ε}, $\varepsilon \in (0,1]$ and an associated backward SDE $(Y^{\varepsilon}, Z^{\varepsilon})$, they have the following recursive asymptotic expansion around some non-degenerate gaussian model $\bar{X}^0$. That is, for $k \geq 0$, $N \geq 1$

$$
\begin{aligned}
Y_t^{\varepsilon,t,x} \simeq u^{\varepsilon,k+1,N}(t,x) &= E[g(\bar{X}_T^{0,t,x})] \\
&+ E\left[\int_t^T f(s, \bar{X}_s^{0,t,x}, Y_s^{\varepsilon,k,N,t,x}, Z_s^{\varepsilon,k,N,t,x})ds\right] \\
&+ \sum_{i=1}^N \varepsilon^i E[g(\bar{X}_T^{0,t,x})\pi_{i,T}^{0,t}] \\
&+ \sum_{i=1}^N \varepsilon^i E\left[\int_t^T f(s, \bar{X}_s^{0,t,x}, Y_s^{\varepsilon,k,N,t,x}, Z_s^{\varepsilon,k,N,t,x})\pi_{i,s}^{0,t} ds\right],
\end{aligned}
\tag{141}
$$

$$
\begin{aligned}
Z_t^{\varepsilon,t,x} \simeq (\nabla u^{\varepsilon,k+1,N}\sigma)(t,x) &= \Bigg\{ E[g(\bar{X}_T^{0,t,x})N_{0,T}^{0,t}] \\
&+ E\left[\int_t^T f(s, \bar{X}_s^{0,t,x}, Y_s^{\varepsilon,k,N,t,x}, Z_s^{\varepsilon,k,N,t,x})N_{0,s}^{0,t} ds\right] \\
&+ \sum_{i=1}^N \varepsilon^i E[g(\bar{X}_T^{0,t,x})N_{i,T}^{0,t}] \\
&+ \sum_{i=1}^N \varepsilon^i E\left[\int_t^T f(s, \bar{X}_s^{0,t,x}, Y_s^{\varepsilon,k,N,t,x}, Z_s^{\varepsilon,k,N,t,x})N_{i,s}^{0,t} ds\right]\Bigg\} \varepsilon\sigma(t,x),
\end{aligned}
\tag{142}
$$

where $Y_s^{\varepsilon,k,N,t,x} = u^{\varepsilon,k,N}(s, \bar{X}_s^{0,t,x})$ and $Z_s^{\varepsilon,k,N,t,x} = (\nabla_x u^{\varepsilon,k,N}\sigma)(s, \bar{X}_s^{0,t,x})$. Here, the processes $\pi_{i,t}^0$ and $N_{i,t}^0$, $i = 1, \ldots, N$ are the *Malliavin weights* and in particular, $N_{0,t}^0$ corresponds to the weight appeared in a representation theorem in Ma and Zhang [63].

8.1 Expansion with Perturbed Generator in BSDE

This subsection briefly describes the perturbation method following [25]. Firstly, let us consider the following decoupled FBSDE:

$$
\begin{aligned}
dV_t &= -f(X_t, V_t, Z_t)dt + Z_t \cdot dW_t \\
V_T &= \Phi(X_T),
\end{aligned}
\tag{143}
$$

where V takes the value in $\mathbf{R}$, W is a r-dimensional Wiener process, and X_t valued in $\mathbf{R}$ is assumed to follow a diffusion process, which is the solution to the (forward)

SDE:

$$dX_t = \gamma_0(X_t)dt + \gamma(X_t) \cdot dW_t; \; X_0 = x \,. \tag{144}$$

Hereafter, we assume the appropriate regularity conditions that guarantee the mathematical validity. For example, pleases see [104] on this point.

In order to approximate the pair of (V_t, Z_t) in terms of X_t, we extract the linear term from the generator f and treat the residual non-linear term as a perturbation to the linear FBSDE. That is, let us introduce a perturbation parameter ϵ, and then write the equation as

$$\begin{aligned} dV_t^{(\epsilon)} &= c(X_t)V_t^{(\epsilon)}dt - \epsilon g(X_t, V_t^{(\epsilon)}, Z_t^{(\epsilon)})dt + Z_t^{(\epsilon)} \cdot dW_t \\ V_T^{(\epsilon)} &= \Phi(X_T). \end{aligned} \tag{145}$$

Here, the above equation with $\epsilon = 1$ corresponds to the original model:

$$f(X_t, V_t, Z_t) = -c(X_t)V_t + g(X_t, V_t, Z_t) \,. \tag{146}$$

We remark that as in the previous asymptotic expansion cases, the residual part g should be small for a precise approximation. Hence, one should choose the linear term $c(X_t)V_t^{(\epsilon)}$ in such a way that the residual non-linear term g becomes as small as possible.

Now, we are going to expand the solution of BSDE (145) with respect to ϵ. That is, suppose $V_t^{(\epsilon)}$ and $Z_t^{(\epsilon)}$ are expanded as follows:

$$V_t^{(\epsilon)} = V_t^{(0)} + \epsilon V_t^{(1)} + \epsilon^2 V_t^{(2)} + \cdots \tag{147}$$

$$Z_t^{(\epsilon)} = Z_t^{(0)} + \epsilon Z_t^{(1)} + \epsilon^2 Z_t^{(2)} + \cdots \,. \tag{148}$$

For illustrative purpose, let us show a first few steps of the expansion. For the zeroth order of ϵ, it is easily seen that $V_t^{(0)}$ is a solution to the following equation:

$$dV_t^{(0)} = c(X_t)V_t^{(0)}dt + Z_t^{(0)} \cdot dW_t \tag{149}$$

$$V_T^{(0)} = \Phi(X_T) \,. \tag{150}$$

Then, $V_t^{(0)}$ can be represented as follows:

$$V_t^{(0)} = E\left[e^{-\int_t^T c(X_s)ds} \Phi(X_T) \middle| \mathcal{F}_t \right], \tag{151}$$

which is equivalent to the value of a standard European contingent claim with the terminal payoff $\Phi(X_T)$ and the discount rate $c(X_t)$ under a suitable pricing measure. Clearly, $V_t^{(0)}$ is a function of X_t due to the Markovian nature of the model. Moreover, applying Itô's formula (or the Malliavin derivative), we are able to obtain $Z_t^{(0)}$ as a function of X_t as well.

Next, let us consider the process $V^{(\epsilon)} - V^{(0)}$:

$$
\begin{aligned}
d\big(V_t^{(\epsilon)} - V_t^{(0)}\big) &= c(X_t)\big(V_t^{(\epsilon)} - V_t^{(0)}\big)dt \\
&\quad - \epsilon g(X_t, V_t^{(\epsilon)}, Z_t^{(\epsilon)})dt + \big(Z_t^{(\epsilon)} - Z_t^{(0)}\big) \cdot dW_t \\
V_T^{(\epsilon)} - V_T^{(0)} &= 0\,.
\end{aligned} \tag{152}
$$

Now, by extracting the ϵ-first order term, we can once again recover the linear FBSDE:

$$
\begin{aligned}
dV_t^{(1)} &= c(X_t)V_t^{(1)}dt - g(X_t, V_t^{(0)}, Z_t^{(0)})dt + Z_t^{(1)} \cdot dW_t \\
V_T^{(1)} &= 0\,,
\end{aligned} \tag{153}
$$

which leads to

$$
V_t^{(1)} = E\left[\int_t^T e^{-\int_t^u c(X_s)ds} g(X_u, V_u^{(0)}, Z_u^{(0)})du \,\Big|\, \mathcal{F}_t\right]. \tag{154}
$$

Because $V_u^{(0)}$ and $Z_u^{(0)}$ are some functions of X_u, we obtain $V_t^{(1)}$ as a function of X_t, and also $Z_t^{(1)}$ through Itô's formula (or Malliavin derivative).

In exactly the same way, we are able to derive an arbitrarily higher order correction. Particularly, due to the ϵ in front of the non-linear term g, the system remains to be linear in every order of the approximation. For example, $V_t^{(2)}$ that is the ϵ^2-order's coefficient of the expansion of $V_t^{(\epsilon)}$ is the solution to the following equation:

$$
\begin{aligned}
dV_t^{(2)} &= c(X_t)V_t^{(2)}dt - \left(\frac{\partial}{\partial v} g(X_t, V_t^{(0)}, Z_t^{(0)})V_t^{(1)}\right. \\
&\quad \left. + \nabla_z g(X_t, V_t^{(0)}, Z_t^{(0)}) \cdot Z_t^{(1)}\right) dt + Z_t^{(2)} \cdot dW_t \qquad (155) \\
V_T^{(2)} &= 0.
\end{aligned}
$$

In general, suppose that we have succeeded to represent backward components (V_t, Z_t) in terms of X_t up to the $(i-1)$th order. Then, in order to proceed to a higher order approximation, we need to obtain the following form of expressions with some deterministic function $G(\cdot)$ in terms of the forward components X_t.

$$
V_t^{(i)} = E\left[\int_t^T e^{-\int_t^u c(X_s)ds} G(X_u)du \,\Big|\, \mathcal{F}_t\right]. \tag{156}
$$

Even if it seems impossible to get the exact result, we can still have an analytic approximation for $(V_t^{(i)}, Z_t^{(i)})$. through again, the asymptotic expansion method.

As an example, [26] has explicitly derived an approximation formula for the dynamic optimal portfolio in an incomplete market setting, and confirmed its accuracy comparing with the exact result by the Cole-Hopf transformation (Zariphopoulou [121]).

Finally, let us provide a brief remark on an approximation of coupled FBSDEs. Let us consider the following generic *coupled* non-linear FBSDE:

$$\begin{aligned} dV_t &= -f(t, X_t, V_t, Z_t)dt + Z_t \cdot dW_t \\ V_T &= \Phi(X_T) \\ dX_t &= \gamma_0(t, X_t, V_t, Z_t)dt + \gamma(t, X_t, V_t, Z_t) \cdot dW_t;\ X_0 = x\ . \end{aligned} \tag{157}$$

We are able to treat this case in the similar way as in the decoupled case by introducing perturbations to the forward SDE in addition to the one in BSDE:

$$\begin{aligned} dV_t^{(\epsilon)} &= c(t, X_t^{(\epsilon)})V_t^{(\epsilon)}dt - \epsilon g\left(t, X_t^{(\epsilon)}, V_t^{(\epsilon)}, Z_t^{(\epsilon)}\right)dt + Z_t^{(\epsilon)} \cdot dW_t \\ V_T^{(\epsilon)} &= \Phi\left(X_T^{(\epsilon)}\right) \\ dX_t^{(\epsilon)} &= \left(r\left(t, X_t^{(\epsilon)}\right) + \epsilon\mu\left(t, X_t^{(\epsilon)}, V_t^{(\epsilon)}, Z_t^{(\epsilon)}\right)\right)dt \\ &\quad + \left(\sigma\left(t, X_t^{(\epsilon)}\right) + \epsilon\eta\left(t, X_t^{(\epsilon)}, V_t^{(\epsilon)}, Z_t^{(\epsilon)}\right)\right) \cdot dW_t \end{aligned}$$

We also note that the similar method can be applied to the coupled case under a PDE (partial differential equation) formulation based on the so called *four step scheme* (e.g. Ma-Yong [62].) Please see [25] for the details. Developing a mathematical validity of the scheme for the coupled case will be one of the research topics in the future.

8.2 Perturbation Scheme with Interacting Particle Method

This subsection briefly introduces a new scheme proposed by Fujii and Takahashi [27]. Except the cases that we are able to obtain fully closed form expressions, the high orders' expansions of perturbed FBSDEs generally contain multi-dimensional time integrations of expectation values due to a convoluted nature of the scheme, which makes standard Monte Carlo simulations too time consuming. To avoid nested simulations, one can applies a particle representation inspired by the ideas of branching diffusion models (e.g. Fujita [23], Ikeda, Nagasawa and Watanabe [40–42], McKean [69], Nagasawa and Sirao [70]). Then, we are able to provide a straightforward simulation scheme to solve nonlinear FBSDEs at each order of the approximation based on the perturbation. In particular, comparing to the direct application of the branching diffusion method, the method is expected to be less numerically intensive, because thanks to expansions of the perturbed generator, the interested system is already decomposed into a set of linear problems. We illustrate the outline of the method by following [27].

Again, let us introduce a perturbation parameter ϵ in the generator of a BSDE as follows:

$$\begin{cases} dV_s^{(\epsilon)} = -\epsilon f(X_s, V_s^{(\epsilon)}, Z_s^{(\epsilon)})ds + Z_s^{(\epsilon)} \cdot dW_s \\ V_T^{(\epsilon)} = \Psi(X_T), \end{cases} \tag{158}$$

where $X_t \in \mathbf{R}$ is assumed to follow a generic Markovian forward SDE:

$$dX_s = \gamma_0(X_s)ds + \gamma(X_s) \cdot dW_s; \; X_t = x_t. \tag{159}$$

Next, let us fix the initial time as t. We denote the Malliavin derivative of X_u $(u \geq t)$ at time t as

$$\mathcal{D}_t X_u \in \mathbf{R}^{r \times d}. \tag{160}$$

Let us also note that in terms of the future time u, the SDE of $(Y_{t,u})^i_j$ defined by $(Y_{t,u})^i_j = \partial_{x_t^j} X_u^i$ is given in the following:

$$\begin{aligned} d(Y_{t,u})^i_j &= \partial_k \gamma_0^i(X_u)(Y_{t,u})^k_j du + \partial_k \gamma_a^i(X_u)(Y_{t,u})^k_j dW_u^a \\ (Y_{t,t})^i_j &= \delta^i_j, \end{aligned} \tag{161}$$

where ∂_k denotes the partial differentiation with respect to the kth component of X, and δ^i_j stands for the Kronecker delta. Here, i and j run through $\{1, \ldots, d\}$ and $\{1, \ldots, r\}$ for a, and we adopt the Einstein notation which assumes the summation of all the paired indexes. Then, it is well-known that

$$(\mathcal{D}_t X_u^i)_a = (Y_{t,u}\gamma(x_t))^i_a,$$

where $a \in \{1, \ldots, r\}$ is the index of r-dimensional Wiener process.

First, for the ϵ-zeroth order, it is easy to see

$$V_t^{(0)} = \mathbf{E}\Big[\Psi(X_T)\Big|\mathcal{F}_t\Big] \tag{162}$$

$$Z_t^{(0),a} = \mathbf{E}\Big[\partial_i \Psi(X_T)(Y_{tT}\gamma(X_t))^i_a\Big|\mathcal{F}_t\Big]. \tag{163}$$

Then, it is clear that they can be evaluated by standard Monte Carlo simulations. However, for their use in higher order approximations, it is crucial to obtain analytical (closed form) approximate expressions for these two quantities, for example based on the asymptotic expansion technique as before.

In the following, let us suppose that we have obtained the solutions up to a given order of the asymptotic expansion, and write each of them as a function of x_t:

$$\begin{cases} V_t^{(0)} = v^{(0)}(x_t) \\ Z_t^{(0)} = z^{(0)}(x_t). \end{cases} \tag{164}$$

Next, for the ϵ-first order's coefficient $V_t^{(1)}$, we obtain an expression as

$$\begin{aligned} V_t^{(1)} &= \int_t^T \mathbf{E}\Big[f(X_u, V_u^{(0)}, Z_u^{(0)})\Big|\mathcal{F}_t\Big]du \\ &= \int_t^T \mathbf{E}\Big[f\Big(X_u, v^{(0)}(X_u), z^{(0)}(X_u)\Big)\Big|\mathcal{F}_t\Big]du. \end{aligned} \tag{165}$$

Then, we define the new process for $(s > t)$ by introducing a deterministic positive process λ_t as follows:

$$\hat{V}_{ts}^{(1)} = e^{\int_t^s \lambda_u du} V_s^{(1)}, \tag{166}$$

Here, λ_t can be a positive constant for the simplest case. Then, for the fixed initial time t, its SDE is given by

$$d\hat{V}_{ts}^{(1)} = \lambda_s \hat{V}_{ts}^{(1)} ds - \lambda_s \hat{f}_{ts}(X_s, v^{(0)}(X_s), z^{(0)}(X_s))ds + e^{\int_t^s \lambda_u du} Z_s^{(1)} \cdot dW_s \ ,$$

where

$$\hat{f}_{ts}(x, v^{(0)}(x), z^{(0)}(x)) = \frac{1}{\lambda_s} e^{\int_t^s \lambda_u du} f(x, v^{(0)}(x), z^{(0)}(x)).$$

Since we have $\hat{V}_{tt}^{(1)} = V_t^{(1)}$, one can easily see the following relation holds:

$$V_t^{(1)} = \mathbf{E}\left[\int_t^T e^{-\int_t^u \lambda_s ds} \lambda_u \hat{f}_{tu}(X_u, v^{(0)}(X_u), z^{(0)}(X_u))du \middle| \mathcal{F}_t\right] \tag{167}$$

Similarly to the cases of the standard credit risk modeling (e.g. Bielecki-Rutkowski [6]), it is the present value of default payment where the default intensity is λ_s with the default payoff at $s(> t)$ as $\hat{f}_{ts}(X_s, v^{(0)}(X_s), z^{(0)}(X_s))$. Thus, we obtain the following proposition.

Proposition 3 *The $V_t^{(1)}$ in* (165) *can be equivalently expressed as*

$$V_t^{(1)} = 1_{\{\tau > t\}} \mathbf{E}\left[1_{\{\tau < T\}} \hat{f}_{t\tau}\Big(X_\tau, v^{(0)}(X_\tau), z^{(0)}(X_\tau)\Big) \middle| \mathcal{F}_t\right]. \tag{168}$$

Here τ is the interaction time where the interaction is drawn independently from the Poisson distribution with an arbitrary deterministic positive intensity process λ_t. $\hat{f}$ is defined as

$$\hat{f}_{ts}(x, v^{(0)}(x), z^{(0)}(x)) = \frac{1}{\lambda_s} e^{\int_t^s \lambda_u du} f(x, v^{(0)}(x), z^{(0)}(x)) \,. \tag{169}$$

Now, let us consider the ϵ-order's coefficient of $Z^{(\epsilon)}$, that is the component $Z^{(1)}$. It can be expressed as

$$Z_t^{(1)} = \int_t^T \mathbf{E}\left[\mathcal{D}_t f\left(X_u, v^{(0)}(X_u), z^{(0)}(X_u)\right) \middle| \mathcal{F}_t \right] du \tag{170}$$

Firstly, we observe that the SDE of the Malliavin derivative of $V^{(1)}$ is given as follows:

$$\begin{aligned} d(\mathcal{D}_t V_s^{(1)}) &= -(\mathcal{D}_t X_s^i) \nabla_i (x, v^{(0)}, z^{(0)}) f(x, v^{(0)}, z^{(0)}) + (\mathcal{D}_t Z_s^{(1)}) \cdot dW_s; \\ \mathcal{D}_t V_t^{(1)} &= Z_t^{(1)}, \end{aligned} \tag{171}$$

where

$$\nabla_i (x, v^{(0)}, z^{(0)}) \equiv \partial_i + \partial_i v^{(0)}(x) \partial_v + \partial_i z^{a(0)}(x) \partial_{z^a}, \tag{172}$$

$$f(x, v^{(0)}, z^{(0)}) \equiv f(x, v^{(0)}(x), z^{(0)}(x)). \tag{173}$$

Then, we define for $(s > t)$, $\widehat{\mathcal{D}_t V_s^{(1)}}$ as

$$\widehat{\mathcal{D}_t V_s^{(1)}} = e^{\int_t^s \lambda_u du} (\mathcal{D}_t V_s^{(1)}), \tag{174}$$

and its SDE can be written as

$$\begin{aligned} d(\widehat{\mathcal{D}_t V_s^{(1)}}) &= \lambda_s (\widehat{\mathcal{D}_t V_s^{(1)}}) ds - \lambda_s (\mathcal{D}_t X_s^i) \nabla_i (X_s, v^{(0)}, z^{(0)}) \hat{f}_{ts}(X_s, v^{(0)}, z^{(0)}) ds \\ &\quad + e^{\int_t^s \lambda_u du} (\mathcal{D}_t Z_s^{(0)}) \cdot dW_s. \end{aligned} \tag{175}$$

Then, we again have

$$\widehat{\mathcal{D}_t V_t^{(1)}} = Z_t^{(1)}. \tag{176}$$

Hence,

$$Z_t^{(1)} = \mathbf{E}\left[\int_t^T e^{-\int_t^u \lambda_s ds} \lambda_u (\mathcal{D}_t X_u^i) \nabla_i (X_u, v^{(0)}, z^{(0)}) \hat{f}_{tu}(X_u, v^{(0)}, z^{(0)}) du \middle| \mathcal{F}_t \right]. \tag{177}$$

Thus, following the same argument as for the previous proposition, we have the next result:

Proposition 4 $Z_t^{(1)}$ *in* (170) *is equivalently expressed as*

$$Z_t^{(1),a} = 1_{\{\tau>t\}}\mathbf{E}\left[1_{\{\tau<T\}}(Y_{t,\tau}\gamma(X_\tau))_a^i \nabla_i(X_\tau, v^{(0)}, z^{(0)})\hat{f}_{t\tau}(X_\tau, v^{(0)}, z^{(0)})\Big|\mathcal{F}_t\right], \quad (178)$$

where the definitions of random time τ *and the positive deterministic process* λ *are the same as those in the previous proposition.*

Now, we are able to obtain a new Monte Carlo scheme. That is, we have a new particle interpretation of $(V^{(1)}, Z^{(1)})$ as follows:

$$V_t^{(1)} = 1_{\{\tau>t\}}\mathbf{E}\left[1_{\{\tau<T\}}\hat{f}_{t\tau}\left(X_\tau, v^{(0)}, z^{(0)}\right)\Big|\mathcal{F}_t\right] \quad (179)$$

$$Z_t^{(1)} = 1_{\{\tau>t\}}\mathbf{E}\left[1_{\{\tau<T\}}(Y_{t,\tau}\gamma(X_\tau))^i \nabla_i(X_\tau, v^{(0)}, z^{(0)})\hat{f}_{t\tau}(X_\tau, v^{(0)}, z^{(0)})\Big|\mathcal{F}_t\right], \quad (180)$$

which allows an efficient time integration with the following Monte Carlo scheme:

- Run the diffusion processes of X and Y.
- Carry out Poisson draw with probability $\lambda_s \Delta s$ at each time s and if "one" is drawn, set that time as τ.
- Then stores the relevant quantities at τ, or in the case of $(\tau > T)$ stores 0.
- Repeat the above procedures and take their expectation.

Finally, we remark that the higher order coefficients in the expansions are evaluated in the similar way. Please see [27] for the details.

9 Conclusion

The present note has reviewed an asymptotic expansion approach in finance, particularly in terms of computational problems arising in practice of financial derivatives. in finance. However, due to the limitation of the space, we have not provided thorough explanations especially for recent progress such as improvement schemes in Sect. 5, expansion methods in jump and jump-diffusion models in Sect. 7 and perturbation schemes in forward backward stochastic differential equations (FBSDEs) in Sect. 8. Please see the cited papers for the details.

Moreover, we have not introduced an application of the method to mean-variance hedging problems in partially observable markets, which is an interesting topic as an application of stochastic filtering problems in finance. Please see [29] for the detail.

References

1. Alòs, E., Eydeland, A., Laurence, P.: A Kirk's and a Bachelier's formula for three asset spread options. Energy Risk **09**(2011), 52–57 (2011)
2. Bayer, C., Laurence, P.: Asymptotics beats Monte Carlo: the case of correlated local vol baskets. Commun. Pure Appl. Math. (2013). Published online 9 October
3. Ben Arous, G., Laurence, P.: Second order expansion for implied volatility in two factor local stochastic volatility models and applications to the dynamic λ-SABR model. In: Friz, P., Gatheral, J., Gulisashvili, A., Jacquier, A., Teichmann, J. (eds.) Large Deviations and Asymptotic Methods in Finance. Springer Proceedings in Mathematics and Statistics, vol. 110. Springer, Berlin (2009)
4. Benaim, S., Friz, P., Lee, R.: On Black-Scholes implied volatility at extreme strikes. In: Cont, R. (ed.) Frontiers in Quantitative Finance: Volatility and Credit Risk Modeling. Wiley, Hoboken (2008)
5. Bichteler, K., Gravereaux, J.-B., Jacod, J.: Malliavin Calculus for Processes with Jumps. Stochastic Monographs. Gordon and Breach Science Publishers, New York (1987)
6. Bielecki, T., Rutkowski, M.: Credit Risk: Modeling, Valuation and Hedging. Springer, Berlin (2000)
7. Brace, A., Gatarek, D., Musiela, M.: The market model of interest rate dynamics. Math. Financ. **7**, 127–155 (1997)
8. Carr, P., Jarrow, R., Myneni, R.: Alternative characterizations of American put options. Math. Financ. **2**, 87–106 (1992)
9. Col, A.D., Gnoatto, A., Grasselli, M.: Smiles all around: FX joint calibration in a multi-Heston model. J. Bank. Financ. **37**(10), 3799–3818 (2013)
10. Cox, J.: Notes on option pricing I: constant elasticity of diffusions. Unpublished draft, Stanford University (1975)
11. Davydov, D., Linetsky, V.: Pricing options on scalar diffusions: an eigenfunction expansion approach. Oper. Res. **51**, 185–209 (2003)
12. Deuschel, J.D., Friz, P.K., Jacquier, A., Violante, S.: Marginal density expansions for diffusions and stochastic volatility I: theoretical foundations. Commun. Pure Appl. Math. **67–1**, 321–350 (2014)
13. Deuschel, J.D., Friz, P.K., Jacquier, A., Violante, S.: Marginal density expansions for diffusions and stochastic volatility II: applications. Commun. Pure Appl. Math. **67–2**, 40–82 (2014)
14. Deutsch, F.: Best Approximation in Inner Product Spaces. Springer, New York (2001)
15. Doust, P.: No-arbitrage SABR. J. Comput. Financ. **15**(3), 3–31 (2012)
16. El Karoui, N., Peng, S.G., Quenez, M.C.: Backward stochastic differential equations in finance. Math. Financ. **7**, 1–71 (1997)
17. Forde, M., Jacquier, A., Lee, R.: The small-time smile and term structure of implied volatility under the Heston model. SIAM J. Financ. Math. **3**, 690–708 (2012)
18. Forde, M., Jacquier, A.: Small-time asymptotics for implied volatility under the Heston model. Int. J. Theor. Appl. Financ. **12**(6), 861–876 (2009)
19. Foschi, P.P., Pagliarani, S., Pascucci, A.: Approximations for Asian options in local volatility models. J. Comput. Appl. Math. **237**, 442–459 (2013)
20. Fouque, J.-P., Papanicolaou, G., Sircar, K.R.: Financial modeling in a fast mean-reverting stochastic volatility environment. Asia-Pac. Financ. Mark. **6**(1), 37–48 (1999)
21. Fouque, J.-P., Papanicolaou, G., Sircar, K.R.: Derivatives in Financial Markets with Stochastic Volatility. Cambridge University Press, Cambridge (2000)
22. Friz, P., Gerhold, S., Gulisashvili, A., Sturm, S.: On refined volatility smile expansion in the Heston model. Quant. Financ. **11**(8), 1151–1164 (2011)
23. Fujita, H.: On the blowing up of solutions of the Cauchy problem for $u_t = \Delta u + u^{1+\alpha}$. J. Fac. Sci. Univ. Tokyo **13**, 109–124 (1966)
24. Fujii, M.: Momentum-space approach to asymptotic expansion for stochastic filtering. Ann. Inst. Stat. Math. **66**(1) (2012)

25. Fujii, M., Takahashi, A.: Analytical approximation for non-linear FBSDEs with perturbation scheme. Int. J. Theor. Appl. Financ. **15**(5) (2012)
26. Fujii, M., Takahashi, A.: Perturbative expansion of FBSDE in an incomplete market with stochastic volatility. Q. J. Financ. **2**(3) (2012)
27. Fujii, M., Takahashi, A.: Perturbative expansion technique for non-linear FBSDEs with interacting particle method. Asia-Pacific Finan. Markets (2015)
28. Fujii, M., Sato, S., Takahashi, A.: An FBSDE approach to American option pricing with an interacting particle method. CARF-F-302 (2012)
29. Fujii, M., Takahashi, A.: Making mean-variance hedging implementable in a partially observable market. Quant. Financ. **14**(10), 1709–1724 (2014)
30. Gatheral, J., Hsu, E.P., Laurence, P., Ouyang, C., Wang, T.-H.: Asymptotics of implied volatility in local volatility models. Math. Financ. **22**(4), 591–620 (2012)
31. Gnoatto, A., Grasselli, M.: An affine multi-currency model with stochastic volatility and stochastic interest rates. SIAM J. Financ. Math. **5**(1), 493–531 (2014)
32. Gulisashvili, A.: Asymptotic formulas with error estimates for call pricing functions and the implied volatility at extreme strikes. SIAM J. Financ. Math. **1**(1), 609–641 (2011)
33. Hagan, P.S., Kumar, D., Lesniewski, A.S., Woodward, D.E.: Managing smile risk. Willmott Mag. **15**, 84–108 (2002)
34. Hayashi, M.: Asymptotic expansions for functionals of a Poisson random measure. J. Math. Kyoto Univ. **48**(1), 91–132 (2008)
35. Hayashi, M.: Coefficients of asymptotic expansions of SDE with jumps. Asia-Pac. Financ. Mark. **17**(4), 373–380 (2010)
36. Hayashi, M., Ishikawa, Y.: Composition with distributions of Wiener-Poisson variables and its asymptotic expansion. Mathematische Nachrichten **285**(5–6), 619–658 (2011)
37. Heath, D., Jarrow, R., Morton, A.: Bond pricing and the term structure of interest rates: a new methodology for contingent claims valuation. Econometrica **60**, 77–105 (1992)
38. Henry-Labordère, P.: Analysis, Geometry and Modeling in Finance: Advanced Methods in Options Pricing. Chapman and Hall, Boca Raton (2008)
39. Ikeda, N., Watanabe, S.: Stochastic Differential Equations and Diffusion Processes, 2nd edn. North-Holland/Kodansha, Tokyo (1989)
40. Ikeda, N., Nagasawa, M., Watanabe, S.: Branching Markov processes. Proc. Jpn. Acad. **41**, 816–821 (1965)
41. Ikeda, N., Nagasawa, M., Watanabe, S.: Branching Markov processes. Proc. Jpn. Acad. **42**, 252–257, 370–375, 380–384, 719–724, 1016–1021, 1022–1026 (1966)
42. Ikeda, N., Nagasawa, M., Watanabe, S.: Branching Markov processes I(II). J. Math. Kyoto Univ. **8**, 233–278, 365–410 (1968)
43. Jamshidian, F.: LIBOR and Swap market models and measures. Financ. Stoch. **1**, 293–330 (1997)
44. Kawai, A.: A new approximate Swaption formula in the LIBOR market model: an asymptotic expansion approach. Appl. Math. Financ. **10**, 49–74 (2003)
45. Kobayashi, T., Takahashi, A., Tokioka, N.: Dynamic optimality of yield curve strategies. Int. Rev. Financ. **4**, 49–78 (2003) (published in 2005)
46. Kato, T., Takahashi, A., Yamada. T.: A semi-group expansion for pricing barrier options. Int. J. Stoch. Anal. **2014**(268086) (2014)
47. Kato, T., Takahashi, A., Yamada. T.: An asymptotic expansion formula for up-and-out barrier option price under stochastic volatility model. JSIAM Lett. **5**, 17–20 (2013)
48. Kunitomo, N., Takahashi, A.: Pricing average options. Jpn. Financ. Rev. **14**, 1–20 (1992). (in Japanese)
49. Kunitomo, N., Takahashi, A.: The asymptotic expansion approach to the valuation of interest rate contingent claims. Math. Financ. **11**, 117–151 (2001)
50. Kunitomo, N., Takahashi, A.: On validity of the asymptotic expansion approach in contingent claim analysis. Ann. Appl. Probab. **13**(3), 914–952 (2003)
51. Kunitomo, N., Takahashi, A.: Applications of the asymptotic expansion approach based on Malliavin-Watanabe calculus in financial problems. Stochastic Processes and Applications to Mathematical Finance, pp. 195–232 (2004)

52. Kusuoka, S.: Malliavin calculus revisited. J. Math. Sci. Univ. Tokyo **10**, 261–277 (2003)
53. Kusuoka, K.: Approximation of expectation of diffusion process and mathematical finance. Taniguchi Conference on Mathematics, Nara, 1998. Advanced Studies in Pure Mathematics, vol. 31, pp. 147–165. Mathematical Society of Japan, Tokyo (2001)
54. Kusuoka, K.: Approximation of expectation of diffusion process based on Lie algebra and Malliavin calculus. Adv. Math. Econ. **6**, 69–83 (2004)
55. Kusuoka S., Stroock, D.: Applications of the Malliavin Calculus Part I. Stochastic Analysis (Katata/Kyoto 1982), pp. 271–306 (1984)
56. Kusuoka, S., Strook, D.: Precise asymptotics of certain Wiener functionals. J. Funct. Anal. **99**, 1–74 (1991)
57. Kusuoka, S., Osajima, Y.: A remark on the asymptotic expansion of density function of Wiener functionals. J. Funct. Anal. **255**(9), 2545–2562 (2007)
58. Lee, R.: The moment formula for implied volatility at extreme. Math. Financ. **14**(3), 469–480 (2004)
59. Li, C.: Closed-form expansion, conditional expectation, and option valuation. Math. Oper. Res. **39**(2), 487–516 (2014)
60. Lipton, A.: Mathematical Methods for Foreign Exchange: A Financial Engineer's Approach. World Scientific Publication, Singapore (2001)
61. Linetsky, V.: Spectral expansions for Asian (average price) options. Oper. Res. **52**, 856–867 (2004)
62. Ma, J., Yong, J.: Forward-Backward Stochastic Differential Equations and Their Applications. Springer, Berlin (2000)
63. Ma, J., Zhang, J.: Representation theorem of backward stochastic differential equations. Ann. Appl. Probab. **12**(4), 1390–1418 (2002)
64. Malliavin, P.: Stochastic Analysis. Springer, Berlin (1997)
65. Malliavin, P., Thalmaier, A.: Stochastic Calculus of Variations in Mathematical Finance. Springer, Berlin (2006)
66. Matsuoka, R., Takahashi, A., Uchida, Y.: A new computational scheme for computing greeks by the asymptotic expansion approach. Asia-Pac. Financ. Mark. **11**, 393–430 (2004)
67. Muroi, Y.: Pricing contingent claims with credit risk: asymptotic expansion approach. Financ. Stoch. **9**(3), 415–427 (2005)
68. Matsuoka, R., Takahashi, A.: An asymptotic expansion approach to computing Greeks. FSA Res. Rev. **2005**, 72–108 (2005)
69. McKean, H.P.: Application of Brownian motion to the equation of Kolmogorov-Petrovskii-Piskunov. Commun. Pure Appl. Math. **28**, 323–331 (1975)
70. Nagasawa, M., Sirao, T.: Probabilistic treatment of the blowing up of solutions for a nonlinear integral equation. Trans. Am. Math. Soc. **139**, 301–310 (1969)
71. Nishiba, M.: Pricing exotic options and American options: a multidimensional asymptotic expansion approach. Asia-Pac. Financ. Mark. **20**(2), 147–182 (2013)
72. Nualart, D.: The Malliavin Calculus and Related Topics. Springer, Berlin (1995)
73. Nualart, D., Üstünel, A.S., Zakai, M.: On the moments of a multiple Wiener-Itô integral and the space induced by the polynomials of the integral. Stochastics **25**, 233–340 (1988)
74. Ocone, D., Karatzas, I.: A generalized clark representation formula, with application to optimal portfolios. Stoch. Stoch. Rep. **34**, 187–220 (1991)
75. Osajima, Y.: The asymptotic expansion formula of implied volatility for dynamic SABR model and FX hybrid model. Preprint, Graduate School of Mathematical Sciences, The University of Tokyo (2006)
76. Osajima, Y.: General asymptotics of wiener functionals and application to mathematical finance. In: Friz, P., Gatheral, J., Gulisashvili, A., Jacquier, A., Teichmann, J. (eds.) Large Deviations and Asymptotic Methods in Finance Springer Proceedings in Mathematics and Statistics, vol. 110 (2015)
77. Pagliarani, S., Pascucci, A.: Local stochastic volatility with jumps. Int. J. Theor. Appl. Financ **16**(8), 1350050 (2013)

78. Shiraya, K., Takahashi, A.: Pricing average options on commodities. J. Futures Mark. **31**(5), 407–439 (2011)
79. Shiraya, K., Takahashi, A.: Pricing multi-asset cross currency options. J. Futures Mark. **34**(1), 1–19 (2014)
80. Shiraya, K., Takahashi, A.: Pricing basket options under local stochastic volatility with jumps. CARF-F-336 (2013)
81. Shiraya, K., Takahashi, A., Toda, M.: Pricing barrier and average options under stochastic volatility environment. J. Comput. Financ. **15**(2), 111–148 (2011)
82. Shiraya, K., Takahashi, A., Yamazaki, A.: Pricing swaptions under the LIBOR market model of interest rates with local-stochastic volatility models. Wilmott **2011**(54), 61–73 (2011)
83. Shiraya, K., Takahashi, A., Yamada, T.: Pricing discrete barrier options under stochastic volatility. Asia-Pac. Financ. Mark. **19**(3), 205–232 (2012)
84. Siopacha, M., Teichmann, J.: Weak and strong Taylor methods for numerical solutions of stochastic differential equations. Quant. Financ. **11**(4), 517–528 (2011)
85. Takahashi, A.: Essays on the valuation problems of contingent claims. Unpublished Ph.D. Dissertation, Haas School of Business, University of California, Berkeley (1995)
86. Takahashi, A.: An asymptotic expansion approach to pricing contingent claims. Asia-Pac. Financ. Mark. **6**, 115–151 (1999)
87. Takahashi, A.: On an asymptotic expansion approach to numerical problems in finance. Selected Papers on Probability and Statistics, pp. 199–217. American Mathematical Society (2009)
88. Takahashi, A., Matsushima, S.: Monte Carlo simulation with an asymptotic expansion in HJM framework. FSA Research Review 2004, pp. 82–103. Financial Services Agency (2004)
89. Takahashi, A., Saito, T.: An asymptotic expansion approach to pricing American options. Monet. Econ. Stud. **22**, 35–87 (2003). (in Japanese)
90. Takahashi, A., Takehara, K.: An asymptotic expansion approach to currency options with a market model of interest rates under stochastic volatility processes of spot exchange rates. Asia-Pac. Financ. Mark. **14**, 69–121 (2007)
91. Takahashi, A., Takehara, K.: Fourier transform method with an asymptotic expansion approach: an applications to currency options. Int. J. Theor. Appl. Financ. **11**(4), 381–401 (2008)
92. Takahashi, A., Takehara, K.: A hybrid asymptotic expansion scheme: an application to currency options. Working paper, CARF-F-116, The University of Tokyo, http://www.carf.e.u-tokyo.ac.jp/workingpaper/ (2008)
93. Takahashi, A., Takehara, K.: A hybrid asymptotic expansion scheme: an application to long-term currency options. Int. J. Theor. Appl. Financ. **13**(8), 1179–1221 (2010)
94. Takahashi, A., Takehara, K.: Asymptotic expansion approaches in finance: applications to currency options. Finance and Banking Developments, pp. 185–232. Nova Science Publishers, New York (2010)
95. Takahashi, A., Takehara, K., Toda, M.: Computation in an asymptotic expansion method. CARF-F-149 (2009)
96. Takahashi, A., Takehara, K., Toda, M.: A general computation scheme for a high-order asymptotic expansion method. Int. J. Theor. Appl. Financ. **15**(6) (2012)
97. Takahashi, A., Toda, M.: Note on an extension of an asymptotic expansion scheme. Int. J. Theor. Appl. Financ. **16**(5), 1350031-1–1350031-23 (2013)
98. Takahashi, A., Tsuzuki, Y.: A new improvement scheme for approximation methods of probability density functions. CARF-F-350. Forthcoming in J. Comput. Financ. (2013)
99. Takahashi, A., Uchida, Y.: New acceleration schemes with the asymptotic expansion in Monte Carlo simulation. Adv. Math. Econ. **8**, 411–431 (2006)
100. Takahashi, A., Yamada, T.: An asymptotic expansion with push-down of Malliavin weights. SIAM J. Financ. Math. **3**, 95–136 (2012)
101. Takahashi, A., Yamada, T.: A remark on approximation of the solutions to partial differential equations in finance. Recent Adv. Financ. Eng. **2011**, 133–181 (2011)

102. Takahashi, A., Yamada, T.: An asymptotic expansion for forward-backward SDEs: a Malliavin calculus approach. CARF-F-296 (2012)
103. Takahashi, A., Yamada, T.: On error estimates for asymptotic expansions with Malliavin weights—application to stochastic volatility model-. CARF-F-324. Forthcoming in Math. Oper. Res. (2013)
104. Takahashi, A., Yamada, T.: An asymptotic expansion for forward-backward SDEs with a perturbed driver. CARF-F-326 (2013)
105. Takahashi, A., Yamada, T.: A weak approximation with asymptotic expansion and multidimensional Malliavin weights. CARF-F-335. Forthcoming in Ann. Appl. Probab. (2013)
106. Takahashi, A., Yoshida, N.: An asymptotic expansion scheme for optimal investment problems. Stat. Inference Stoch. Process. **7**(2), 153–188 (2004)
107. Takahashi, A., Yoshida, N.: Monte Carlo simulation with asymptotic method. J. Jpn. Stat. Soc. **35**(2), 171–203 (2005)
108. Takehara, K., Takahashi, A., Toda, M.: New unified computation algorithm in a high-order asymptotic expansion scheme. In: Recent Advances in Financial Engineering (The Proceedings of KIER-TMU International Workshop on Financial Engineering 2009), pp. 231–251 (2010)
109. Takehara, K., Toda, M., Takahashi, A.: Application of a high-order asymptotic expansion scheme to long-term currency options. Int. J. Bus. Financ. Res. **5**(3), 87–100 (2011)
110. Violante, S.P.N.: Asymptotics of Wiener functionals and applications to mathematical finance. Ph.D. Thesis, Department of Mathematics, Imperial College London (2012)
111. Watanabe, S.: Analysis of Wiener functionals (Malliavin calculus) and its applications to heat kernels. Ann. Probab. **15**, 1–39 (1987)
112. Xu, G., Zheng, H.: Basket options valuation for a local volatility jump-diffusion model with the asymptotic expansion method. Insur. Math. Econ. **47**(3), 415–422 (2010)
113. Xu, G., Zheng, H.: Lower bound approximation to basket option values for local volatility jump-diffusion models. Int. J. Theor. Appl. Financ. **17**, 1–15 (2014)
114. Yamamoto, K., Sato, S., Takahashi, A.: Probability distribution and option pricing for drawdown in a stochastic volatility environment. Int. J. Theor. Appl. Financ. **13**(2), 335–354 (2010)
115. Yamamoto, K., Takahashi, A.: A remark on a singular perturbation method for option pricing under a stochastic volatility model. Asia-Pac. Financ. Mark. **16**(4), 333–345 (2009)
116. Yamanobe, T.: Stochastic phase transition operator. Phys. Rev. E **84**, 011924 (2011)
117. Yamanobe, T.: Global dynamics of a stochastic neuronal oscillator. Phys. Rev. E **88**, 052709 (2013)
118. Yoshida, N.: Asymptotic expansion for small diffusions via the theory of Malliavin-Watanabe. Probab. Theor. Relat. Fields **92**, 275–311 (1992)
119. Yoshida, N.: Asymptotic expansions for statistics related to small diffusions. J. Jpn. Stat. Soc. **22**, 139–159 (1992)
120. Yoshida, N.: Conditional expansions and their applications. Stoch. Process. Appl. **107**, 53–81 (2003)
121. Zariphopoulou, T.: A solution approach to valuation with unhedgeable risks. Financ. Stoch. **5**, 61–82 (2001)

On Small Time Asymptotics for Rough Differential Equations Driven by Fractional Brownian Motions

Fabrice Baudoin and Cheng Ouyang

In memory of Peter Laurence

Abstract We survey existing results concerning the study in small times of the density of the solution of a rough differential equation driven by fractional Brownian motions. We also slightly improve existing results and discuss some possible applications to mathematical finance.

Keywords Small maturity limit · Mathematical foundations in non-Markovian situations · Rough differential equations

1991 Mathematics Subject Classification 28D05 · 60D58

1 Introduction

In this paper, our main goal is to survey some existing results concerning the small-time asymptotics of the density of rough differential equations driven by fractional Brownian motions. Even though we do not claim any new results, we slightly improve

The first author of this research was supported in part by NSF Grant DMS 0907326.

F. Baudoin (✉)
Department of Mathematics, Purdue University, West Lafayette, IN 47907, USA
e-mail: fbaudoin@math.purdue.edu

C. Ouyang
Department of Mathematics, Statistics and Computer Science, University of Illinois at Chicago, Chicago, IL 60607, USA
e-mail: couyang@math.uic.edu

P.K. Friz et al. (eds.), *Large Deviations and Asymptotic Methods in Finance*, Springer Proceedings in Mathematics & Statistics 110,
DOI 10.1007/978-3-319-11605-1_14

some of the existing ones and also point out some possible connections to finance. We also hope, it will be useful for the reader to have, in one place, the most recent results concerning the small-time asymptotics questions related to rough differential equations driven by fractional Brownian motions. Our discussion will mainly be based on one hand on the papers [5–7] by the two present authors and on the other hand on the papers [27, 28] by Inahama.

Random dynamical systems are a well established modeling tool for a variety of natural phenomena ranging from physics (fundamental and phenomenological) to chemistry and more recently to biology, economy, engineering sciences and mathematical finance. In many interesting models the lack of any regularity of the external inputs of the differential equation as functions of time is a technical difficulty that hampers their mathematical analysis. The theory of rough paths has been initially developed by T. Lyons [31] in the 1990s to provide a framework to analyze a large class of driven differential equations and the precise relations between the driving signal and the output (that is the state, as function of time, of the controlled system).

Rough paths theory provides a perfect framework to study differential equations driven by Gaussian processes (see [19]). In particular, using rough paths theory, we may define solutions of stochastic differential equations driven by a fractional Brownian motion with a parameter $H > 1/4$ (see [15]). Let us then consider the equation

$$X_t^x = x + \int_0^t V_0(X_s^x)ds + \sum_{i=1}^d \int_0^t V_i(X_s^x)dB_s^i, \tag{1.1}$$

where $x \in \mathbb{R}^n$, $V_0, V_1, \ldots, V_d$ are bounded smooth vector fields and $(B_t)_{t\geq 0}$ is a d-dimensional fractional Brownian motion with Hurst parameter $H \in (\frac{1}{4}, 1)$. A first basic question is the existence of a smooth density with respect to the Lebesgue measure for the random variable $X_t^x, t > 0$. After multiple works, it is now understood that the answer to this question is essentially the same as the one for stochastic differential equations driven by Brownian motions: the random variable X_t^x admits a smooth density with respect to the Lebesgue measure if Hörmander's condition is satisfied at x. More precisely, if $I = (i_1, \ldots, i_k) \in \{0, \ldots, d\}^k$, we denote by V_I the Lie commutator defined by

$$V_I = [V_{i_1}, [V_{i_2}, \ldots, [V_{i_{k-1}}, V_{i_k}] \ldots]],$$

and

$$d(I) = k + n(I),$$

where $n(I)$ is the number of 0 in the word I. The basic and fundamental result concerning the existence of a density for stochastic differential equations driven by fractional Brownian motions is the following:

Theorem 1.1 ([4, 10, 12, 24]) *Assume $H > \frac{1}{4}$ and assume that, at some $x \in \mathbb{R}^n$, there exists N such that*

$$\mathbf{span}\{V_I(x), d(I) \le N\} = \mathbb{R}^n. \tag{1.2}$$

Then, for any $t > 0$, the law of the random variable X_t^x has a smooth density $p_t(x, y)$ with respect to the Lebesgue measure on $\mathbb{R}^n$.

Once the existence and smoothness of the density is established, it is natural to study properties of this density. In particular, we are interested here in small-time asymptotics, that is the analysis of $p_t(x, y)$ when $t \to 0$. Based on the results in the Brownian motion case [1, 2], and taking into account the scaling property of the fractional Brownian motion, the following expansion (in particular when $n = d$) is somehow expected when x, y are close enough to each other:

$$p_t(x, y) = \frac{1}{(t^H)^d} e^{-\frac{d^2(x,y)}{2t^{2H}}} \left(\sum_{i=0}^{N} c_i(x, y) t^{2iH} + r_{N+1}(t, x, y) t^{2(N+1)H} \right). \tag{1.3}$$

Our goal is to discuss here the various assumptions under which such expansion is known to be true and also discuss possible variations. The approach to study the problem is similar to the case of Brownian motion, the main difficulty to overcome is to study the Laplace method on the path space of the fractional Brownian motion (see [3] for the Brownian case).

The paper is organized as follows. In Sect. 2 we give some basic results of the theory of rough paths and of the Malliavin calculus tools that will be needed. In Sect. 3, we prove a Varadhan's type small time asymptotics for $\ln p_t(x, y)$. The discussion is mainly based on [7]. In Sect. 4, we study sufficient conditions under which the above expansion (1.3) is valid. Our discussion is based on [5, 27, 28]. Finally, in Sect. 5, we discuss some models in mathematical finance where the asymptotics of the density for rough differential equations may play an important role.

2 Preliminary Material

For some fixed $H > \frac{1}{4}$, we consider $(\Omega, \mathcal{F}, \mathbb{P})$ the canonical probability space associated with the fractional Brownian motion (in short fBm) with Hurst parameter H. That is, $\Omega = \mathcal{C}_0([0, 1])$ is the Banach space of continuous functions vanishing at zero equipped with the supremum norm, $\mathcal{F}$ is the Borel sigma-algebra and $\mathbb{P}$ is the unique probability measure on Ω such that the canonical process $B = \{B_t = (B_t^1, \ldots, B_t^d),\ t \in [0, 1]\}$ is a fractional Brownian motion with Hurst parameter H. In this context, let us recall that B is a d-dimensional centered Gaussian process, whose covariance structure is induced by

$$R(t, s) := \mathbb{E}\left[B_s^j\, B_t^j\right] = \frac{1}{2}\left(s^{2H} + t^{2H} - |t - s|^{2H}\right), \quad s, t \in [0, 1] \text{ and } j = 1, \ldots, d. \tag{2.1}$$

In particular it can be shown, by a standard application of Kolmogorov's criterion, that B admits a continuous version whose paths are γ-Hölder continuous for any $\gamma < H$.

2.1 Rough Paths Theory

In this section, we recall some basic results in rough paths theory. More details can be found in the monographs [20] and [32]. For $N \in \mathbb{N}$, recall that the truncated algebra $T^N(\mathbb{R}^d)$ is defined by

$$T^N(\mathbb{R}^d) = \bigoplus_{m=0}^{N} (\mathbb{R}^d)^{\otimes m},$$

with the convention $(\mathbb{R}^d)^{\otimes 0} = \mathbb{R}$. The set $T^N(\mathbb{R}^d)$ is equipped with a straightforward vector space structure plus an multiplication $\otimes$. Let π_m be the projection on the mth tensor level. Then $(T^N(\mathbb{R}^d), +, \otimes)$ is an associative algebra with unit element $\mathbf{1} \in (\mathbb{R}^d)^{\otimes 0}$.

For $s < t$ and $m \geq 2$, consider the simplex $\Delta^m_{st} = \{(u_1, \ldots, u_m) \in [s,t]^m;\ u_1 < \cdots < u_m\}$, while the simplices over $[0,1]$ will be denoted by Δ^m. A continuous map $\mathbf{x} : \Delta^2 \to T^N(\mathbb{R}^d)$ is called a multiplicative functional if for $s < u < t$ one has $\mathbf{x}_{s,t} = \mathbf{x}_{s,u} \otimes \mathbf{x}_{u,t}$. An important example arises from considering paths x with finite variation: for $0 < s < t$ we set

$$\mathbf{x}^m_{s,t} = \sum_{1 \leq i_1, \ldots, i_m \leq d} \left(\int_{\Delta^m_{st}} dx^{i_1} \cdots dx^{i_m} \right) e_{i_1} \otimes \cdots \otimes e_{i_m}, \tag{2.2}$$

where $\{e_1, \ldots, e_d\}$ denotes the canonical basis of $\mathbb{R}^d$, and then define the truncated *signature* of x as

$$S_N(x) : \Delta^2 \to T^N(\mathbb{R}^d), \qquad (s,t) \mapsto S_N(x)_{s,t} := 1 + \sum_{m=1}^{N} \mathbf{x}^m_{s,t}.$$

The function $S_N(x)$ for a smooth function x will be our typical example of multiplicative functional. Let us stress the fact that those elements take values in the strict subset $G^N(\mathbb{R}^d) \subset T^N(\mathbb{R}^d)$, called free nilpotent group of step N, and is equipped with the classical Carnot-Caratheodory norm which we simply denote by $|\cdot|$. For a path $\mathbf{x} \in \mathcal{C}([0,1], G^N(\mathbb{R}^d))$, the p-variation norm of $\mathbf{x}$ is defined to be

$$\|\mathbf{x}\|_{p-\mathrm{var};[0,1]} = \sup_{\Pi \subset [0,1]} \left(\sum_i |\mathbf{x}_{t_i}^{-1} \otimes \mathbf{x}_{t_{i+1}}|^p \right)^{1/p}$$

where the supremum is taken over all subdivisions Π of $[0,1]$.

With these notions in hand, let us briefly define what we mean by geometric rough path (we refer to [20, 32] for a complete overview): for $p \geq 1$, an element $x : [0,1] \to G^{\lfloor p \rfloor}(\mathbb{R}^d)$ is said to be a geometric rough path if it is the p-var limit of a sequence $S_{\lfloor p \rfloor}(x^m)$. In particular, it is an element of the space

$$\mathcal{C}^{p-\text{var};[0,1]}([0,1], G^{\lfloor p \rfloor}(\mathbb{R}^d)) = \{\mathbf{x} \in \mathcal{C}([0,1], G^{\lfloor p \rfloor}(\mathbb{R}^d)) : \|\mathbf{x}\|_{p-\text{var};[0,1]} < \infty\}.$$

Let $\mathbf{x}$ be a geometric p-rough path with its approximating sequence x^m, that is, x^m is a sequence of smooth functions such that $\mathbf{x}^m = S_{\lfloor p \rfloor}(x^m)$ converges to $\mathbf{x}$ in the p-var norm. Fix any $1 \leq q \leq p$ so that $p^{-1} + q^{-1} > 1$ and pick any $h \in \mathcal{C}^{q-\text{var}}([0,1], \mathbb{R}^d)$. One can define the translation of $\mathbf{x}$ by h, denoted by $T_h(\mathbf{x})$ by

$$T_h(\mathbf{x}) = \lim_{n \to \infty} S_{\lfloor p \rfloor}(x^m + h).$$

It can be shown that $T_h(\mathbf{x})$ is an element in $\mathcal{C}^{p-\text{var}}([0,1], G^{\lfloor p \rfloor}(\mathbb{R}^d))$. Moreover, one can show that $T_h(\mathbf{x})$ uniformly continuous in h and $\mathbf{x}$ on bounded sets.

Remark 2.1 A typical situation of the above translation of $\mathbf{x}$ by h in the present paper is when $\mathbf{x} = \mathbf{B}$, the fractional Brownian motion lifted as a rough path, and h is a Cameron-Martin element of B. In this case, we simply denote $T_h(\mathbf{B}) = B + h$.

According to the considerations above, in order to prove that a lift of a d-dimensional fBm as a geometric rough path exists it is sufficient to build enough iterated integrals of B by a limiting procedure. Towards this aim, a lot of the information concerning B is encoded in the rectangular increments of the covariance function R (defined by (2.1)), which are given by

$$R_{uv}^{st} \equiv \mathbb{E}\left[(B_t^1 - B_s^1)(B_v^1 - B_u^1)\right].$$

We then call 2-dimensional ρ-variation of R the quantity

$$V_\rho(R) \equiv \sup\left\{\left(\sum_{i,j}\left|R_{s_i s_{i+1}}^{t_j t_{j+1}}\right|^\rho\right)^{1/\rho} ; (s_i), (t_j) \in \Pi\right\},$$

where Π stands again for the set of partitions of $[0,1]$. It is know that (see, for example [20]) if a process has a covariance function with finite ρ-variation for $\rho \in [1,2)$, it admits a lift to a geometric p-rough path for all $p > 2\rho$. As a consequence, we have the following for fractional Brownian motions:

Proposition 2.2 *For a fractional Brownian motion with Hurst parameter H, we have $V_\rho(R) < \infty$ for all $\rho \geq 1/(2H)$. Consequently, for $H > 1/4$ the process B admits a lift $\mathbf{B}$ as a geometric rough path of order p for any $p > 1/H$.*

2.2 Malliavin Calculus

We introduce the basic framework of Malliavin calculus in this subsection. The reader is invited to read the corresponding chapters in [33] for further details. Let $\mathcal{E}$ be the space of $\mathbb{R}^d$-valued step functions on [0, 1], and $\mathcal{H}$ the closure of $\mathcal{E}$ for the scalar product:

$$\langle(\mathbf{1}_{[0,t_1]}, \cdots, \mathbf{1}_{[0,t_d]}), (\mathbf{1}_{[0,s_1]}, \cdots, \mathbf{1}_{[0,s_d]})\rangle_{\mathcal{H}} = \sum_{i=1}^{d} R(t_i, s_i).$$

We denote by K_H^* the isometry between $\mathcal{H}$ and $L^2([0, 1])$. When $H > \frac{1}{2}$ it can be shown that $\mathbf{L}^{1/H}([0, 1], \mathbb{R}^d) \subset \mathcal{H}$, and when $\frac{1}{4} < H < \frac{1}{2}$ one has

$$C^\gamma \subset \mathcal{H} \subset L^2([0, 1])$$

for all $\gamma > \frac{1}{2} - H$.

We remark that $\mathcal{H}$ is the reproducing kernel Hilbert space for B. Let $\mathscr{H}_H$ be the Cameron-Martin space of B, one proves that the operator $\mathcal{R} := \mathcal{R}_H : \mathcal{H} \to \mathscr{H}_H$ given by

$$\mathcal{R}\psi := \int_0^{\cdot} K_H(\cdot, s)[K_H^* \psi](s)\, ds \tag{2.3}$$

defines an isometry between $\mathcal{H}$ and $\mathscr{H}_H$. Let us now quote from [20, Chap. 15] a result relating the 2-d regularity of R and the regularity of $\mathscr{H}_H$.

Proposition 2.3 *Let B be a fBm with Hurst parameter $\frac{1}{4} < H < \frac{1}{2}$. Then one has $\mathscr{H}_H \subset C^{\rho-\text{var}}$ for $\rho > (H + 1/2)^{-1}$. Furthermore, the following quantitative bound holds:*

$$\|h\|_{\mathscr{H}_H} \geq \frac{\|h\|_{\rho-\text{var}}}{(V_\rho(R))^{1/2}}.$$

Remark 2.4 The above proposition shows that for fBm we have $\mathscr{H}_H \subset C^{\rho-\text{var}}$ for $\rho > (H + 1/2)^{-1}$. Hence an integral of the form $\int h\, dB$ can be interpreted in the Young sense by means of p-variation techniques.

Remark 2.5 Under the same conditions, the above embedding can be sharpened to $\mathscr{H}_H \subset C^{\rho-\text{var}}$ for all $\rho \geq (H + 1/2)^{-1}$. We refer interested readers to [17] for more details.

A $\mathcal{F}$-measurable real valued random variable F is then said to be cylindrical if it can be written, for a given $n \geq 1$, as

$$F = f\left(B(\phi^1), \ldots, B(\phi^n)\right) = f\Big(\int_0^1 \langle \phi_s^1, dB_s\rangle, \ldots, \int_0^1 \langle \phi_s^n, dB_s\rangle\Big),$$

where $\phi^i \in \mathcal{H}$ and $f : \mathbb{R}^n \to \mathbb{R}$ is a C^∞ bounded function with bounded derivatives. The set of cylindrical random variables is denoted $\mathcal{S}$.

The Malliavin derivative is defined as follows: for $F \in \mathcal{S}$, the derivative of F is the $\mathbb{R}^d$ valued stochastic process $(\mathbf{D}_t F)_{0 \le t \le 1}$ given by

$$\mathbf{D}_t F = \sum_{i=1}^{n} \phi^i(t) \frac{\partial f}{\partial x_i} \left(B(\phi^1), \dots, B(\phi^n) \right).$$

More generally, we can introduce iterated derivatives. If $F \in \mathcal{S}$, we set

$$\mathbf{D}^k_{t_1,\dots,t_k} F = \mathbf{D}_{t_1} \dots \mathbf{D}_{t_k} F.$$

For any $p \ge 1$, it can be checked that the operator $\mathbf{D}^k$ is closable from $\mathcal{S}$ into $\mathbf{L}^p(\Omega; \mathcal{H}^{\otimes k})$. We denote by $\mathbb{D}^{k,p}$ the closure of the class of cylindrical random variables with respect to the norm

$$\|F\|_{k,p} = \left(\mathbb{E}\left(F^p\right) + \sum_{j=1}^{k} \mathbb{E}\left(\left\| \mathbf{D}^j F \right\|^p_{\mathcal{H}^{\otimes j}} \right) \right)^{\frac{1}{p}},$$

and

$$\mathbb{D}^\infty = \bigcap_{p \ge 1} \bigcap_{k \ge 1} \mathbb{D}^{k,p}.$$

Definition 2.6 Let $F = (F^1, \dots, F^n)$ be a random vector whose components are in $\mathbb{D}^\infty$. Define the Malliavin matrix of F by

$$\gamma_F = (\langle \mathbf{D}F^i, \mathbf{D}F^j \rangle_{\mathcal{H}})_{1 \le i,j \le n}.$$

Then F is called *non-degenerate* if γ_F is invertible $a.s.$ and

$$(\det \gamma_F)^{-1} \in \cap_{p \ge 1} L^p(\Omega).$$

It is a classical result that the law of a non-degenerate random vector $F = (F^1, \dots, F^n)$ admits a smooth density with respect to the Lebesgue measure on $\mathbb{R}^n$. Furthermore, the following integration by parts formula allows to get more quantitative estimates:

Proposition 2.7 *Let $F = (F^1, \dots, F^n)$ be a non-degenerate random vector whose components are in $\mathbb{D}^\infty$, and γ_F the Malliavin matrix of F. Let $G \in \mathbb{D}^\infty$ and φ be a function in the space $C_p^\infty(\mathbb{R}^n)$. Then for any multi-index $\alpha \in \{1, 2, \dots, n\}^k$, $k \ge 1$, there exists an element $H_\alpha = H_\alpha(F, G) \in \mathbb{D}^\infty$ depending on F and G such that*

$$\mathbb{E}[\partial_\alpha \varphi(F) G] = \mathbb{E}[\varphi(F) H_\alpha].$$

Moreover, the elements H_α are recursively given by

$$H_{(i)} = \sum_{j=1}^{d} \delta\left(G(\gamma_F^{-1})^{ij}\mathbf{D}F^j\right)$$
$$H_\alpha = H_{(\alpha_k)}(H_{(\alpha_1,\ldots,\alpha_{k-1})}),$$

and for $1 \le p < q < \infty$ we have

$$\|H_\alpha\|_{L^p} \le C_{p,q}\|\gamma_F^{-1}\mathbf{D}F\|^k_{k,2^{k-1}r}\|G\|_{k,q},$$

where $\frac{1}{p} = \frac{1}{q} + \frac{1}{r}$.

Remark 2.8 By the estimates for H_α above, one can conclude that there exist constants $\beta, \gamma > 1$ and integers m, r such that

$$\|H_\alpha\|_{L^p} \le C_{p,q}\|\det \gamma_F^{-1}\|^m_{L^\beta}\|\mathbf{D}F\|^r_{k,\gamma}\|G\|_{k,q}.$$

2.3 Differential Equations Driven by Fractional Brownian Motions

Let B be a d-dimensional fractional Brownian motion with Hurst parameter $H > \frac{1}{4}$. Fix a small parameter $\varepsilon \in (0, 1]$, and consider the solution X_t^ε to the stochastic differential equation

$$X_t^\varepsilon = x + \varepsilon\sum_{i=1}^{d}\int_0^t V_i(X_s^\varepsilon)dB_s^i + \int_0^t V_0(\varepsilon, X_s^\varepsilon)ds, \tag{2.4}$$

where the vector fields $V_1, \ldots, V_d$ are C^∞-bounded vector fields on $\mathbb{R}^n$ and $V_0(\varepsilon, \cdot)$ is C^∞-bounded uniform in $\varepsilon \in [0, 1]$.

Proposition 2.2 ensures the existence of a lift of B as a geometrical rough path. The general rough paths theory (see e.g. [20, 22]) together with some integrability results (see e.g. [12, 18]) allow us to state the following proposition:

Proposition 2.9 *Consider Eq. (2.4) driven by a d-dimensional fBm B with Hurst parameter $H > \frac{1}{4}$, and assume that the vector fields V_is are C^∞-bounded. Then*
(i) *For each $\varepsilon \in (0, 1]$, Eq. (2.4) admits a unique finite p-var continuous solution X^ε in the rough paths sense, for any $p > \frac{1}{H}$.*
(ii) *There exists $\lambda > 0$ such that*

$$\mathbb{E}\left[\exp\lambda\left(\sup_{t\in[0,1],\epsilon\in(0,1]}|X_t^\varepsilon|^{(2H+1)\wedge 2}\right)\right] < \infty. \tag{2.5}$$

Once Eq. (2.4) is solved, the vector X_t^ε is a typical example of random variable which can be differentiated in the Malliavin sense. We shall express this Malliavin derivative in terms of the Jacobian $\mathbf{J}^\varepsilon$ of the equation, which is defined by the relation

$$\mathbf{J}_t^{\varepsilon,ij} = \partial_{x_j} X_t^{\varepsilon,i}.$$

Setting DV_j for the Jacobian of V_j seen as a function from $\mathbb{R}^n$ to $\mathbb{R}^n$, let us recall that $\mathbf{J}^\varepsilon$ is the unique solution to the linear equation

$$\mathbf{J}_t^\varepsilon = \mathrm{Id}_n + \varepsilon \sum_{j=1}^d \int_0^t DV_j(X_s^\varepsilon)\, \mathbf{J}_s^\varepsilon\, dB_s^j, \tag{2.6}$$

and that the following results hold true (see [10, 11, 34] for further details):

Proposition 2.10 *Let X^ε be the solution to Eq. (2.4) and suppose the V_i's are C^∞-bounded. Then for every $i = 1, \ldots, n$, $t > 0$, and $x \in \mathbb{R}^n$, we have $X_t^{\varepsilon,i} \in \mathbb{D}^\infty$ and*

$$\mathbf{D}_s^j X_t^\varepsilon = \mathbf{J}_{st}^\varepsilon V_j(X_s^\varepsilon), \quad j = 1, \ldots, d, \quad 0 \le s \le t,$$

where $\mathbf{D}_s^j X_t^{\varepsilon,i}$ is the jth component of $\mathbf{D}_s X_t^{\varepsilon,i}$, $\mathbf{J}_t^\varepsilon = \partial_x X_t^\varepsilon$ and $\mathbf{J}_{st}^\varepsilon = \mathbf{J}_t^\varepsilon (\mathbf{J}_s^\varepsilon)^{-1}$.

Let us now quote the recent result [12], which gives a useful estimate for moments of the Jacobian of rough differential equations driven by Gaussian processes.

Proposition 2.11 *Consider a fractional Brownian motion B with Hurst parameter $H > \frac{1}{4}$ and $p > \frac{1}{H}$. Then for any $\eta \ge 1$, there exists a finite constant c_η such that the Jacobian $\mathbf{J}^\varepsilon$ defined at Proposition 2.10 satisfies:*

$$\mathbb{E}\left[\sup_{\varepsilon \in [0,1]} \|\mathbf{J}^\varepsilon\|_{p-\text{var};[0,1]}^\eta \right] = c_\eta. \tag{2.7}$$

Proof The integrability of $\mathbf{J}^\varepsilon$ is only proved in [12] when $\varepsilon = 1$. On the other hand, the estimates of $\mathbf{J}$ in [12] only depends on the supremum norm of the vector fields and their derivatives. In our case, the vector fields in Eq. (2.4) are $\varepsilon V_i' s$ whose derivatives together with themselves are bounded uniform in $\varepsilon \in (0, 1)$. Hence the uniform integrability of $\mathbf{J}^\varepsilon$ (in ε) follows. □

Finally, we close the discussion of this section by the following large deviation principle that will be needed later. Let $\Phi : \mathscr{H}_H \to C([0, 1], \mathbb{R}^n)$ be given by solving the ordinary differential equation

$$\Phi_t(h) = x + \sum_{i=1}^d \int_0^t V_i(\Phi_s(h))dh_s^i + \int_0^t V_0(0, \Phi_s(h))ds. \tag{2.8}$$

Theorem 2.12 *Let Φ be given in (2.8), which is a differentiable mapping from $\mathcal{H}_H$ to $C([0,1],\mathbb{R}^n)$. Introduce the following function on $\mathbb{R}^n$*

$$I(y) = \inf_{\Phi_1(h)=y} \frac{1}{2}\|h\|^2_{\mathcal{H}_H}.$$

Recall that X_1^ε is the solution to Eq. (2.4). Then X_1^ε satisfies a large deviation principle with rate function $I(y)$.

Proof Fix any $p > \frac{1}{H}$. It is known (see [20]) that $\varepsilon\mathbf{B}$ as a $G^{\lfloor p \rfloor}(\mathbb{R}^d)$-valued rough path satisfies a large deviation principle in p-variation topology with good rate function given by

$$J(h) = \begin{cases} \frac{1}{2}\|h\|^2_{\mathcal{H}} & \text{if } h \in \mathcal{H} \\ +\infty & \text{otherwise.} \end{cases}$$

It is clear $\Phi_1(\cdot) : G^{\lfloor p \rfloor}(\mathbb{R}^d) \to \mathbb{R}^n$ is continuous. Now that $X_1^\varepsilon = \Phi_1(\varepsilon\mathbf{B})$, the claimed result follows from the contraction principle. □

3 Varadhan Asymptotics

In this section, we are interested in a family of stochastic differential equations driven by fractional Brownian motions B (with Hurst parameter $H > \frac{1}{4}$) of the following form

$$X_t^\varepsilon = x + \varepsilon \sum_{i=1}^d \int_0^t V_i(X_s^\varepsilon) dB_s^i.$$

We define a map $\Phi : \mathcal{H}_H \to C[0,1]$ by solving the ordinary differential equation

$$\Phi_t(h) = x + \sum_{i=1}^d \int_0^t V_i(\Phi_s(h)) dh_s^i.$$

Clearly, we have $X_t^\varepsilon = \Phi_t(\varepsilon\mathbf{B})$. Denote by $\gamma_{\Phi_1(h)}$ the deterministic Malliavin matrix of $\Phi_1(h)$, i.e.,

$$\gamma^{ij}_{\Phi_1(h)} = \langle \mathbf{D}\Phi_1^i(h), \mathbf{D}\Phi_1^j(h) \rangle_{\mathcal{H}}.$$

Introduce the following functions on $\mathbb{R}^n$, which depend on Φ

$$d^2(y) = I(y) = \inf_{\Phi_1(h)=y} \frac{1}{2}\|h\|^2_{\mathcal{H}_H}, \quad \text{and} \quad d_R^2(y) = \inf_{\Phi_1(h)=y, \det \gamma_{\Phi_1(h)}>0} \frac{1}{2}\|h\|^2_{\mathcal{H}_H}.$$

In the absence of the drift term ($V_0 = 0$) in our setting in this section, one can show that the above two distances coincide.

Lemma 3.1 *For every $y \in \mathbb{R}^n$, we have $d(y) = d_R(y)$.*

Proof We follow an argument of Léandre (see [30]). By using Theorem I.2 in [30] and the isometry between the Cameron-Martin space of the fractional Brownian motion and the Cameron-Martin space of the Brownian motion, we see that for every $\varepsilon > 0$, there exists $h \in \mathcal{H}$ such that $\|h\|_{\mathscr{H}} \leq \varepsilon$ and $\det \gamma_{\Phi_1(h)} > 0$. Then arguing as in the Remark after Proposition II.1 in [30], we can for every $\eta > 0$ and $y \in \mathbb{R}^n$ construct $h \in \mathcal{H}$ such that $\Phi_1(h) = y$, $\det \gamma_{\Phi_1(h)} > 0$ and

$$\frac{1}{2}\|h\|^2_{\mathscr{H}} \leq d^2(y) + \eta.$$

□

Throughout the section, we assume that the following assumption Hypothesis 3.2 is satisfied. Let us first introduce some notations. Let $\mathcal{A} = \{\emptyset\} \cup \bigcup_{k=1}^{\infty} \{1, 2, \cdots, n\}^k$ and $\mathcal{A}_1 = A \setminus \{\emptyset\}$. We say that $I \in \mathcal{A}$ is a word of length k if $I = (i_1, \cdots, i_k)$ and we write $|I| = k$. If $I = \emptyset$, then we denote $|I| = 0$. For any integer $l \geq 1$, we denote by $\mathcal{A}(l)$ the set $\{I \in \mathcal{A}; |I| \leq l\}$ and by $\mathcal{A}_1(l)$ the set $\{I \in \mathcal{A}_1; |I| \leq l\}$. We also define an operation $*$ on $\mathcal{A}$ by $I * J = (i_1, \cdots, i_k, j_1, \cdots, j_l)$ for $I = (i_1, \cdots, i_k)$ and $J = (j_1, \cdots, j_l)$ in $\mathcal{A}$. We define vector fields $V_{[I]}$ inductively by

$$V_{[j]} = V_j, \quad V_{[I*j]} = [V_{[I]}, V_j], \quad j = 1, \cdots, d$$

Hypothesis 3.2 *(Uniform hypoelliptic condition)* The vector fields $V_1, \cdots, V_d$ are in $C_b^{\infty}(\mathbb{R}^n)$ and they form a uniform hypoelliptic system in the sense that there exist an integer l and a constant $\lambda > 0$ such that

$$\sum_{I \in \mathcal{A}_1(l)} \langle V_{[I]}(x), u \rangle^2_{\mathbb{R}^n} \geq \lambda \|u\|^2 \tag{3.1}$$

holds for any $x, u \in \mathbb{R}^n$

Under this assumption the main result proved in [7] is the following Varadhan's type estimate:

Theorem 3.3 *Let us denote by $p_\varepsilon(y)$ the density of X_1^ε. Then*

$$\liminf_{\varepsilon \downarrow 0} \varepsilon^2 \log p_\varepsilon(y) = -d^2(y). \tag{3.2}$$

The two key ingredients in proving Theorem 3.3 are an estimate for the Malliavin derivative $\mathbf{D}X_1^\varepsilon$ and an estimate of the Malliavin matrix $\gamma_{X_1^\varepsilon}$ of X_1^ε. Building on previous results from [8], the following estimates were obtained in [7]:

Lemma 3.4 *Assume Hypothesis 3.2. For* $H > \frac{1}{4}$, *we have*

(1) $\sup_{\varepsilon \in (0,1]} \|X_1^\varepsilon\|_{k,r} < \infty$ *for each* $k \geq 1$ *and* $r \geq 1$.
(2) $\|\gamma_{X_1^\varepsilon}^{-1}\|_r \leq c_r \varepsilon^{-2l}$ *for any* $r \geq 1$.

Proof of Theorem 3.3 We first show that

$$\liminf_{\varepsilon \downarrow 0} \varepsilon^2 \log p_\varepsilon(y) \geq -d_R^2(y). \tag{3.3}$$

Fix $y \in \mathbb{R}^n$. We only need to show for $d_R^2(y) < \infty$, since if $d_R^2(y) = \infty$ the statement is trivial. Fix any $\eta > 0$ and let $h \in \mathscr{H}_H$ be such that $\Phi_1(h) = y$, $\det_{\gamma_\Phi}(h) > 0$, and $\|h\|_{\mathscr{H}_H}^2 \leq d_R^2(y) + \eta$. Let $f \in C_0^\infty(\mathbb{R}^n)$. By the Cameron-Martin theorem for fractional Brownian motions, we have

$$\mathbb{E} f(X_1^\varepsilon) = e^{-\frac{\|h\|^2_{\mathscr{H}_H}}{2\varepsilon^2}} \mathbb{E} f(\Phi_1(\varepsilon B + h)) e^{\frac{B(h)}{\varepsilon}}.$$

Consider then a function $\chi \in C^\infty(\mathbb{R}), 0 \leq \chi \leq 1$, such that $\chi(t) = 0$ if $t \notin [-2\eta, 2\eta]$, and $\chi(t) = 1$ if $t \in [-\eta, \eta]$. Then, if $f \geq 0$, we have

$$\mathbb{E} f(X_1^\varepsilon) \geq e^{-\frac{\|h\|_{\mathscr{H}_H} + 4\eta}{2\varepsilon^2}} \mathbb{E} \chi(\varepsilon B(h)) f(\Phi_1(\varepsilon B + h)).$$

Hence, we obtain

$$\varepsilon^2 \log p_\varepsilon(y) \geq -\left(\frac{1}{2}\|h\|^2_{\mathscr{H}_H} + 2\eta\right) + \varepsilon^2 \log \mathbb{E}\big(\chi(\varepsilon B(h))\delta_y(\Phi_1(\varepsilon B + h))\big). \tag{3.4}$$

On the other hand, we have

$$\mathbb{E}\big(\chi(\varepsilon B(h))\delta_y(\Phi_1(\varepsilon B + h))\big) = \varepsilon^{-n} \mathbb{E}\left(\chi(\varepsilon B(h))\delta_0\left(\frac{\Phi_1(\varepsilon B + h) - \Phi_1(h)}{\varepsilon}\right)\right).$$

Note that

$$Z_1(h) = \lim_{\varepsilon \downarrow 0} \frac{\Phi_1(\varepsilon B + h) - \Phi_1(h)}{\varepsilon}$$

is a n-dimensional random vector in the first Wiener chaos with variance $\gamma_{\Phi_1}(h) > 0$. Hence $Z_1(h)$ is non-degenerate and we can then prove that we obtain

$$\lim_{\varepsilon \downarrow 0} \mathbb{E}\left(\chi(\varepsilon B(h))\delta_0\left(\frac{\Phi_1(\varepsilon B + h) - \Phi_1(h)}{\varepsilon}\right)\right) = \mathbb{E}\delta_0(Z_1(h)).$$

Therefore,

$$\lim_{\varepsilon\downarrow 0}\varepsilon^2\log\mathbb{E}\big(\chi(\varepsilon B(h))\delta_y(\Phi_1(\varepsilon B+h))\big)=0.$$

Letting $\varepsilon\downarrow 0$ in (3.4) we obtain

$$\liminf_{\varepsilon\downarrow 0}\varepsilon^2\log p_\varepsilon(y)\geq-\left(\frac{1}{2}\|h\|^2_{\mathscr{H}_H}+2\eta\right)\geq-\left(d_R^2(y)+3\eta\right).$$

Since $\eta>0$ is arbitrary, this completes the proof of (3.3).

Next, we show that

$$\limsup_{\varepsilon\downarrow 0}\varepsilon^2\log p_\varepsilon(y)\leq -d^2(y). \tag{3.5}$$

Fix a point $y\in\mathbb{R}^n$ and consider a function $\chi\in C_0^\infty(\mathbb{R}^n)$, $0\leq\chi\leq 1$ such that χ is equal to one in a neighborhood of y. The density of X_1^ε at point y is given by

$$p_\varepsilon(y)=\mathbb{E}(\chi(X_1^\varepsilon)\delta_y(X_1^\varepsilon)).$$

By the integration by parts formula of Proposition 2.7, we can write

$$\begin{aligned}\mathbb{E}\chi(X_1^\varepsilon)\delta_y(X_1^\varepsilon)&=\mathbb{E}\left(\mathbf{1}_{\{X_1^\varepsilon>y\}}H_{(1,2,\dots,n)}(X_1^\varepsilon,\chi(X_1^\varepsilon))\right)\\&\leq\mathbb{E}|H_{(1,2,\dots,n)}(X_1^\varepsilon,\chi(X_1^\varepsilon))|\\&=\mathbb{E}\big(|H_{(1,2,\dots,n)}(X_1^\varepsilon,\chi(X_1^\varepsilon))|\mathbf{1}_{\{X_1^\varepsilon\in\mathrm{supp}\chi\}}\big)\\&\leq\mathbb{P}(X_1^\varepsilon\in\mathrm{supp}\chi)^{\frac{1}{q}}\|H_{(1,\dots,n)}(X_1^\varepsilon,\chi(X_1^\varepsilon))\|_p,\end{aligned}$$

where $\frac{1}{p}+\frac{1}{q}=1$. By Remark 2.8 we know that

$$\|H_{(1,\dots,n)}(X_1^\varepsilon,\chi(X_1^\varepsilon))\|_p\leq C_{p,q}\|\gamma^{-1}_{X_1^\varepsilon}\|^m_\beta\|\mathbf{D}X_1^\varepsilon\|^r_{k,\gamma}\|\chi(X_1^\varepsilon)\|_{k,q},$$

for some constants $\beta,\gamma>0$ and integers k,m,r. Thus, by Lemma 3.4 we have

$$\lim_{\varepsilon\downarrow 0}\varepsilon^2\log\|H_{(1,\dots,n)}(X_1^\varepsilon,\chi(X_1^\varepsilon))\|_p=0.$$

Finally by Theorem 2.12, a large deviation principle for X_1^ε ensures that for small ε we have

$$\mathbb{P}(X_1^\varepsilon\in\mathrm{supp}\chi)^{\frac{1}{q}}\leq e^{-\frac{1}{q\varepsilon^2}(\inf_{y\in\mathrm{supp}\chi}d^2(y)+o(1))}.$$

This gives us (3.5).

Combining Lemma 3.1, (3.3) and (3.5), the proof of Theorem 3.3 is thus completed. □

4 Small-Time Kernel Expansion

4.1 Laplace Approximation

Fix $H > \frac{1}{4}$ and consider Eq. (2.4). For the convenience of our discussion, in what follows, we write the above equation in the following form

$$X_t^\varepsilon = x + \varepsilon \int_0^t \sigma(X_s^\varepsilon) dB_s + \int_0^t b(\varepsilon, X_s^\varepsilon) ds,$$

where σ is a smooth $d \times d$ matrix and b a smooth function from $\mathbb{R}^+ \times \mathbb{R}^d$ to $\mathbb{R}^d$. We also assume that σ and b have bounded derivatives to any order.

Fix $p > \frac{1}{H}$. Let F and f be two bounded infinitely Fréchet differentiable functionals on $C^{p-\text{var};[0,1]}([0,1], \mathbb{R}^d)$ with bounded derivatives (as linear operators) to any order. We are interested in studying the asymptotic behavior of

$$J(\varepsilon) = \mathbb{E}\big[f(X^\varepsilon) \exp\{-F(X^\varepsilon)/\varepsilon^2\}\big], \qquad \text{as } \varepsilon \downarrow 0.$$

Recall for each $k \in \mathscr{H}_H$, $\Phi(k)$ is the deterministic Itô map defined in (2.8). Set

$$\Lambda(\phi) = \inf\left\{\frac{1}{2}\|k\|_{\mathscr{H}_H}, \phi = \Phi(k), k \in \mathscr{H}_H\right\}.$$

Throughout our discussion we make the following assumptions:

Assumption 4.1
- H 1: $F + \Lambda$ attains its minimum at finite number of paths $\phi_1, \phi_2, \ldots, \phi_n$ on $P(\mathbb{R}^d)$.
- H 2: For each $i \in \{1, 2, \ldots, n\}$, we have $\phi_i = \Phi(\gamma_i)$ and γ_i is a non-degenerate minimum of the functional $F \circ \Phi + 1/2\|\cdot\|^2_{\mathscr{H}_H}$, i.e.:

$$\forall k \in \mathscr{H}_H \backslash \{0\}, \quad d^2(F \circ \Phi + 1/2\|\cdot\|^2_{\mathscr{H}_H})(\gamma_i)k^2 > 0.$$

The following theorem is the main result of this section.

Theorem 4.2 *Under the assumptions H 1 and H 2 above, we have*

$$J(\varepsilon) = e^{-\frac{a}{\varepsilon^2}} e^{-\frac{c}{\varepsilon}} \left(\alpha_0 + \alpha_1 \varepsilon + \cdots + \alpha_N \varepsilon^N + O(\varepsilon^{N+1})\right).$$

Here

$$a = \inf\{F + \Lambda(\phi), \phi \in P(\mathbb{R}^d)\} = \inf\{F \circ \Phi(k) + 1/2|k|^2_{\mathscr{H}_H}, k \in \mathscr{H}_H\}$$

and

$$c = \inf\left\{dF(\phi_i)Y_i, i \in \{1, 2, \ldots, n\}\right\},$$

where Y_i is the solution of

$$dY_i(s) = \partial_x\sigma(\phi_i(s))Y_i(s)d\gamma_i(s) + \partial_\varepsilon b(0, \phi_i(s))ds + \partial_x b(0, \phi_i(s))Y_i(s)ds$$

with $Y_i(0) = 0$.

In what follows, we sketch the proof of the above Laplace approximation in the case $H > \frac{1}{2}$. Remarks on the rough case $\frac{1}{4} < H < \frac{1}{2}$ will be provided afterwards.

Without loss of generality, we may assume that $F + \Lambda$ attains its minimum at a unique path ϕ. There exists a $\gamma \in \mathscr{H}_H$ such that

$$\phi = \Phi(\gamma), \qquad \text{and } \Lambda(\phi) = \frac{1}{2}\|\gamma\|^2_{\mathscr{H}_H},$$

and

$$a \stackrel{\text{def}}{=} \inf\{F + \Lambda(\phi), \phi \in P(\mathbb{R}^d)\} = \inf\left\{F \circ \Phi(k) + \frac{1}{2}\|k\|^2_{\mathscr{H}_H}, k \in \mathscr{H}_H\right\}.$$

Moreover by assumption H 2, for all non zero $k \in \mathscr{H}_H$:

$$d^2(F \circ \Phi + \frac{1}{2}\| \ \|^2_{\mathscr{H}_H})(\gamma)k^2 > 0.$$

Consider the following stochastic differential equation

$$Z_t^\varepsilon = x + \int_0^t \sigma(Z_s^\varepsilon)(\varepsilon dB_s + d\gamma_s) + \int_0^t b(\varepsilon, Z_s^\varepsilon)ds.$$

It is clear that $Z^0 = \phi$. Denote $Z_t^{m,\varepsilon} = \partial_\varepsilon^m Z_t^\varepsilon$ and consider the Taylor expansion with respect to ε near $\varepsilon = 0$, we obtain

$$Z^\varepsilon = \phi + \sum_{j=0}^{N} \frac{g_j\varepsilon^j}{j!} + \varepsilon^{N+1}R^\varepsilon_{N+1},$$

where $g_j = Z^{j,0}$. Explicitly, we have

$$dg_1(s) = \sigma(\phi_s)dB_s + \partial_x\sigma(\phi_s)g_1(s)d\gamma_s + \partial_x b(0, \phi_s)g_1(s)ds + \partial_\varepsilon b(0, \phi_s)ds.$$

Now the proof is divided into the following steps.

Step 1: By the large deviation principle, the sample paths that contribute to the asymptotics of $J(\varepsilon)$ lie in the neighborhoods of the minimizers of $F + \Lambda$. More precisely, for $\rho > 0$, denote by $B(\phi, \rho)$ the open ball (under λ-Hölder topology for a fixed $\lambda < H$) centered at ϕ with radius ρ. There exist $d > a$ and $\varepsilon_0 > 0$ such that for all $\varepsilon \le \varepsilon_0$

$$\left|J(\varepsilon) - \mathbb{E}\left[f(X_T^\varepsilon)e^{-F(X_T^\varepsilon)/\varepsilon^2}, X^\varepsilon \in B(\phi,\rho)\right]\right| \le e^{-d/\varepsilon^2}.$$

Hence, letting

$$J_\rho(\varepsilon) = \mathbb{E}\left[f(X_T^\varepsilon)e^{-F(X_T^\varepsilon)/\varepsilon^2}, X^\varepsilon \in B(\phi,\rho)\right],$$

to study the asymptotic behavior of $J(\varepsilon)$ as $\varepsilon \downarrow 0$, it suffices to study that of $J_\rho(\varepsilon)$.
Step 2: Let $\theta(\varepsilon) = F(Z^\varepsilon)$ and write

$$\theta(\varepsilon) = \theta(0) + \varepsilon\theta'(0) + \frac{1}{2}\varepsilon^2\theta''(0) + \varepsilon^3 R(\varepsilon).$$

By the Cameron-Martin theorem for fractional Brownian motions, we have

$$\begin{aligned} &J_\rho(\varepsilon) \qquad (4.1)\\ &= \mathbb{E}\left\{ f(Z^\varepsilon)\exp\left(-\frac{F(Z^\varepsilon)}{\varepsilon^2}\right)\exp\left(-\frac{1}{\varepsilon}\int_0^T ((K_H^*)^{-1}(K_H^{-1}\dot{\gamma}))_s dB_s - \frac{\|\gamma\|^2_{\mathscr{H}_H}}{2\varepsilon^2}\right); Z^\varepsilon \in B(\phi,\rho)\right\}\\ &= \mathbb{E}\left\{ \exp\left[-\frac{1}{\varepsilon^2}\left(F(\phi) + \frac{1}{2}\|\gamma\|^2_{\mathscr{H}_H}\right)\right]\exp\left[-\frac{\theta(0)' + \int_0^T ((K_H^*)^{-1}(K_H^{-1}\dot{\gamma}))_s dB_s}{\varepsilon}\right]\right.\\ &\qquad \left.\exp\left[-\frac{1}{2}\theta''(0)\right]\cdot\left[f(Z^\varepsilon)e^{-\varepsilon R(\varepsilon)}\right]; Z^\varepsilon \in B(\phi,\rho)\right\}. \end{aligned}$$

Step 3: It is clear that to prove Theorem 4.2, it suffices to analyze the four terms in the expectation above. First of all, it is apparent that the first term (of order-2) is

$$\exp\left[-\frac{1}{\varepsilon^2}\left(F(\phi) + \frac{1}{2}\|\gamma\|^2_{\mathscr{H}_H}\right)\right] = e^{-\frac{a}{\varepsilon^2}}, \tag{4.2}$$

which gives the leading term the Varadhan asymptotics.

The second term (of order-1) is deterministic. Indeed, since γ is a critical point of $F \circ \Phi + 1/2\|\cdot\|^2_{\mathscr{H}_H}$ and note $\|k\|_{\mathscr{H}_H} = \|K_H^{-1}k\|_{\mathscr{H}}$, we have

$$dF(\phi)(d\Phi(\gamma)k) = -\int_0^T ((K_H^*)^{-1}(K_H^{-1}\dot{\gamma}))_s dk_s.$$

By the continuity of Young's integral with respect to the driving path, the above extends to

$$dF(\phi)(d\Phi(\gamma)B) = -\int_0^T ((K_H^*)^{-1}(K_H^{-1}\dot{\gamma}))_s dB_s.$$

On the other hand, note

$$\theta'(0) = dF(\phi)g_1,$$

and

$$g_1 = d\Phi(\gamma)B + Y.$$

Here Y is the solution of

$$dY_s = \partial_x \sigma(\phi_s) Y_s d\gamma_s + \partial_\varepsilon b(0, \phi_s) ds + \partial_x b(0, \phi_s) Y_s ds, \quad Y(0) = 0.$$

We obtain

$$\exp\left[-\frac{\theta(0)' + \int_0^T \left((K_H^*)^{-1}(K_H^{-1}\dot{\gamma})\right)_s dB_s}{\varepsilon} \right] = \exp\left[-\frac{dF(\phi)Y}{\varepsilon} \right]. \tag{4.3}$$

For the third term (of order 0), one can show that there exists a $\beta > 0$ such that

$$\mathbb{E} \exp\left\{ -(1+\beta)\left[\frac{1}{2}\theta''(0) \right] \right\} < \infty. \tag{4.4}$$

Let us emphasize that in order to show the above integrability of $\theta''(0)$, one needs to use assumption H2 and prove that $d^2 F \circ \Phi(\gamma)(k^1, k^2)$ is Hilbert-Schmidt. For more details, we refer the reader to [5] for the case when $H > \frac{1}{2}$, and to [27] when $\frac{1}{4} < H < \frac{1}{2}$. Moreover, one can prove the following integrability of $R(\varepsilon)$.

Lemma 4.3 *There exist $\alpha > 0$ and $\varepsilon_0 > 0$ such that*

$$\sup_{0 \le \varepsilon \le \varepsilon_0} \mathbb{E}\left(e^{(1+\alpha)|\varepsilon R(\varepsilon)|}; Z^\varepsilon \in B(\phi, \rho) \right) < \infty.$$

Lemma 4.3 and (4.4) allows us to analyze the third and forth terms and show

$$\mathbb{E}\left[f(Z^\varepsilon) e^{-\frac{1}{2}\theta''(0) - \varepsilon R(\varepsilon)}; Z^\varepsilon \in B(\phi, \rho) \right] = \sum_{m=0}^{N} \alpha_m \varepsilon^m + O(\varepsilon^{N+1}). \tag{4.5}$$

Finally, combining (4.1)–(4.3), and (4.5), the proof of Theorem 4.2 is complete. □

Remark 4.4 In application (see the next section), one may also be interested in an SDE which involves a fractional order term of ε,

$$X_t^\varepsilon = x + \varepsilon \int_0^t \sigma(X_s^\varepsilon) dB_s + \varepsilon^{\frac{1}{H}} \int_0^t b(\varepsilon, X_s^\varepsilon) ds. \tag{4.6}$$

For this purpose, let us first introduce

$$\Lambda_1 = \left\{ n_1 + \frac{n_2}{H} | n_1, n_2 = 0, 1, 2, \ldots \right\}, \tag{4.7}$$

the set of fractional orders. Let $0 = \kappa_0 < \kappa_1 < \kappa_2 < \cdots$ be all elements of Λ_1 in increasing order. When $H > \frac{1}{2}$, we have

$$(\kappa_0, \kappa_1, \kappa_2, \kappa_3, \kappa_4, \ldots) = \left(0, 1, \frac{1}{H}, 2, 1 + \frac{1}{H}, \ldots \right). \tag{4.8}$$

Set

$$\Lambda_2 = \{\kappa - 2 | \kappa \in \Lambda_1 \backslash \{0\}\},$$

and define

$$\Lambda_3 = \{a_1 + a_2 + \cdots + a_m | m \in \mathbb{N}_+ \text{ and } a_1, \ldots, a_m \in \Lambda_1\}$$

and

$$\Lambda_3' = \{a_1 + a_2 + \cdots + a_m | m \in \mathbb{N}_+ \text{ and } a_1, \ldots, a_m \in \Lambda_2\}.$$

Finally let

$$\Lambda_4 = \{a + b | a \in \Lambda_3, b \in \Lambda_3'\}$$

and denote by $\{0 = \lambda_0 < \lambda_1 < \lambda_2 < \ldots\}$ all the elements of Λ_4 in increasing order. Let us note that the set Λ_3 characterizes the powers of ε coming from the term $f(Z^\varepsilon)$ in (4.1) and Λ_3' characterizes that of $e^{-\varepsilon R(\varepsilon)}$.

Similar as before, we consider

$$Z_t^\varepsilon = x + \int_0^t \sigma(Z_s^\varepsilon)(\varepsilon d B_s + d\gamma_s) + \varepsilon^{\frac{1}{H}} \int_0^t b(\varepsilon, Z_s^\varepsilon) ds. \tag{4.9}$$

It can be proved that Z^ϵ has the following expansion in ε,

$$Z^\varepsilon = \phi + \sum_{j=0}^{N} g_{\kappa_j} \varepsilon^{\kappa_j} + \varepsilon^{\kappa_N + 1} R_{\kappa_N + 1}^\varepsilon.$$

Note that in (4.8), indices up to degree two are $(0, 1, 1/H, 2)$. There is an extra term $1/H$ compared to the case without fractional order. Hence when plugging (4.9) into Step 2 of the proof of Theorem 4.2, there is an extra (but deterministic) term

$$\exp\left\{ -\frac{dF(\Phi) g_{\kappa 2}}{\varepsilon^{2-\frac{1}{H}}} \right\},$$

where g_{κ_2} satisfies

$$dg_{\kappa_2}(s) = \partial_x\sigma(\phi_s)g_{\kappa_2}(s)d\gamma_s + b(0, \phi_s)ds, \quad g_{\kappa_2}(0) = 0.$$

It is not hard to see that the other terms up to degree two remain the same, and that although higher order terms are different they could be handled similarly as before. Hence we obtain

Theorem 4.5 *Let X^ε satisfy (4.6). We have*

$$\mathbb{E}\left[f(X^\varepsilon)e^{-F(X^\varepsilon)/\varepsilon^2}\right] = e^{-\frac{a}{\varepsilon^2}}e^{-\frac{c}{\varepsilon}}\exp\left\{-\frac{d}{\varepsilon^{2-\frac{1}{H}}}\right\}\Big(\alpha_{\lambda_0} + \alpha_{\lambda_1}\varepsilon^{\lambda_1} + \cdots + \alpha_{\lambda_N}\varepsilon^{\lambda_N} + O(\varepsilon^{\lambda_{N+1}})\Big).$$

Here

$$a = \inf\{F \circ \Phi(k) + 1/2|k|^2_{\mathscr{H}_H}, k \in \mathscr{H}_H\},$$
$$c = dF(\phi)Y, \quad \text{and} \quad d = dF(\phi)g_{\kappa_2},$$

where Y and g_{κ_2} satisfy

$$dY(s) = \partial_x\sigma(\phi_i(s))Y(s)d\gamma(s) + \partial_\varepsilon b(0, \phi(s))ds + \partial_x b(0, \phi(s))Y(s)ds, \quad Y(0) = 0,$$

and

$$dg_{\kappa_2}(s) = \partial_x\sigma(\phi_s)g_{\kappa_2}(s)d\gamma_s + b(0, \phi_s)ds, \quad g_{\kappa_2}(0) = 0.$$

Remark 4.6 Theorem 4.2 for the rough case $\frac{1}{4} < H < \frac{1}{2}$ was proved by Inahama [27]. In this case, equation is understood in the rough path sense. Thanks to Proposition 2.3, equations for g_i and R_i are understood as Young's paring.

In [27] the author also discussed RDEs with fractional orders of ε, in which the index set Λ_1 was introduced. The main idea of the proof for the rough case is the same as that outlined above. But the major difficulty is to show that $d^2F \circ \Phi(\gamma)(k^1, k^2)$ is Hilbert-Schmidt. This is easier when $H > \frac{1}{2}$, since in this case $\partial_t K(t, s)$ is integrable, and one can easily obtain a nice representation for $d^2F \circ \Phi(\gamma)(k^1, k^2)$.

4.2 Expansion of the Density Function

Consider

$$X_t = x + \sum_{i=1}^{d}\int_0^t V_i(X_s)dB_s^i + \int_0^t V_0(X_s)ds. \tag{4.10}$$

We are interested in studying the small-time asymptotic behavior of X_t. It is clear that by the self-similarity of B, this is equivalent to studying the asymptotic behavior of X_1^ε (for small ε) which satisfies

$$X_t^\varepsilon = x + \sum_{i=1}^d \varepsilon \int_0^t V_i(X_s) dB_s^i + \varepsilon^{\frac{1}{H}} \int_0^t V_0(X_s) ds.$$

In what follows, we use the Laplace approximation to obtain a short time asymptotic expansion for the density of X_1^ε in the case when $H > \frac{1}{2}$. For this purpose, we need the following assumption.

Assumption 4.7 • A 1: For every $x \in \mathbb{R}^d$, the vectors $V_1(x), \cdots, V_d(x)$ form a basis of $\mathbb{R}^d$.

• A 2: There exist smooth and bounded functions ω_{ij}^l such that:

$$[V_i, V_j] = \sum_{l=1}^d \omega_{ij}^l V_l,$$

and

$$\omega_{ij}^l = -\omega_{il}^j.$$

Assumption A1 is the standard ellipticity condition. Due to the second assumption A2, the geodesics are easily described. If $k : \mathbb{R}_{\geq 0} \to \mathbb{R}$ is a α-Hölder path with $\alpha > 1/2$ such that $k(0) = 0$, we denote by $\Phi(x, k)$ the solution of the ordinary differential equation:

$$x_t = x + \sum_{i=1}^d \int_0^t V_i(x_s) dk_s^i.$$

Whenever there is no confusion, we always suppress the starting point x and denote it simply by $\Phi(k)$ as before. Then we have (see Lemma 4.2 in [5])

Lemma 4.8 *$\Phi(x, k)$ is a geodesic if and only if $k(t) = tu$ for some $u \in \mathbb{R}^d$.*

As a consequence of the previous lemma, we then have the following key result (Proposition 4.3 in [5]):

Proposition 4.9 *Let $T > 0$. For $x, y \in \mathbb{R}^d$,*

$$\inf_{k \in \mathcal{H}_H, \Phi_T(x,k) = y} \|k\|_{\mathcal{H}_H}^2 = \frac{d^2(x, y)}{T^{2H}}.$$

Lemma 4.10 *For any $x \in \mathbb{R}^d$, there exists a neighborhood V of x and a bounded smooth function $F(x, y, z)$ on $V \times V \times \mathbb{R}^d$ such that:*

(1) For any $(x, y) \in V \times V$ the infimum

$$\inf\left\{F(x,y,z)+\frac{d(x,z)^2}{2},\, z\in M\right\}=0$$

is attained at the unique point y. Moreover, it is a non-degenerate minimum. Hence there exists a unique $k^0 \in \mathcal{H}_H$ *such that (a):* $\Phi_1(x_0,k^0) = y_0$*; (b):* $d(x_0,y_0)=\|k^0\|_{\mathcal{H}_H}$*; and (c):* k^0 *is a non-degenerate minimum of the functional:* $k \to F(\Phi_1(x_0,k)) + 1/2\|k\|^2_{\mathcal{H}_H}$ *on* $\mathcal{H}_H$.

(2) For each $(x,y) \in V \times V$*, there exists a ball centered at y with radius r independent of x, y such that* $F(x,y,\cdot)$ *is a constant outside of the ball.*

Let F be in the above lemma and $p_\varepsilon(x,y)$ the density function of X_1^ε. By the inversion of Fourier transformation we have

$$\begin{aligned}
p_\varepsilon(x,y)e^{-\frac{F(x,y,y)}{\varepsilon^2}} &= \frac{1}{(2\pi)^d}\int e^{-i\zeta\cdot y}d\zeta\int e^{i\zeta\cdot z}e^{-\frac{F(x,y,z)}{\varepsilon^2}}p_\varepsilon(x,z)dz\\
&= \frac{1}{(2\pi\varepsilon)^d}\int e^{-i\frac{\zeta\cdot y}{\varepsilon}}d\zeta\int e^{i\frac{\zeta\cdot z}{\varepsilon}}e^{-\frac{F(x,y,z)}{\varepsilon^2}}p_\varepsilon(x,z)dz\\
&= \frac{1}{(2\pi\varepsilon)^d}\int d\zeta\mathbb{E}_x\left(e^{\frac{i\zeta\cdot(X_1^\varepsilon-y)}{\varepsilon}}e^{-\frac{F(x,y,X_1^\varepsilon)}{\varepsilon^2}}\right). \qquad (4.11)
\end{aligned}$$

It is clear that by applying Laplace approximation to the expectation in the last equation above and switching the order of integration (with respect to ζ) and summation, we obtain an asymptotic expansion for the density function $p_\varepsilon(x,y)$.

Remark 4.11 One might wonder why not constructing, for each fixed x, y, a function F which minimizes (at $z=y$)

$$F(x,y,z)+\frac{D(x,z)^2}{2}$$

in Lemma 4.10, where

$$D^2(x,y)=\inf_{k\in\mathcal{H}_H,\Phi_1(x,k)=y}\|k\|^2_{\mathcal{H}_H}.$$

After all $D(x,y)$ seems the natural "distance" for the system (4.10), instead of the Riemannian distance $d(x,y)$. The problem with $D(x,y)$ is that it is not clear weather it is differentiable, while the construction of F in Lemma 4.10 needs some differentiability of $D(x,y)$. This is indeed one of the reasons why we impose the structure assumption A2 so that $D(x,y)=d(x,y)$ (content of Proposition 4.9). With this identification, we know $D(x,y)$ is smooth for all $x\neq y$.

Remark 4.12 In order to show Proposition 4.9, we used the fact that $\partial K(t,s)/\partial t$ is integrable, which is only true for the smooth case $H>\frac{1}{2}$. Hence although Inahama proved the Laplace approximation for $\frac{1}{4}<H<\frac{1}{2}$ in [27], we can not repeat the proof in this section to produce an expansion of the density function for the rough case.

Recall the definition of Λ_1 in Remark 4.4 and similarly set

$$\Lambda_2 = \{\kappa - 1 | \kappa \in \Lambda_1 \backslash \{0\}\}$$

and

$$\Lambda_2' = \{\kappa - 2 | \kappa \in \Lambda_1 \backslash \{0\}\}.$$

Next define

$$\Lambda_3 = \{a_1 + a_2 + \cdots + a_m | m \in \mathbb{N}_+ \text{ and } a_1, \ldots, a_m \in \Lambda_2\}.$$

and

$$\Lambda_3' = \{a_1 + a_2 + \cdots + a_m | m \in \mathbb{N}_+ \text{ and } a_1, \ldots, a_m \in \Lambda_2'\}.$$

Finally, set

$$\Lambda_4 = \{a + b | a \in \Lambda_3, b \in \Lambda_3'\}$$

and denote by $\{0 = \lambda_0 < \lambda_1 < \lambda_2 < \ldots\}$ all the elements of Λ_4 in increasing order. Similar as before, powers of ϵ in the index set Λ_3 comes from the term $\exp\left\{i\zeta \cdot (X_1^{\varepsilon} - y)/\varepsilon\right\}$ in (4.11) and powers in Λ_3' comes from $\exp\{-F(x, y, X_1^{\varepsilon})/\varepsilon^2\}$.

Our main result of this section is the following (by letting $\varepsilon = t^H$).

Theorem 4.13 *Fix $x \in \mathbb{R}^d$. Suppose the Assumption 4.7 is satisfied, then in a neighborhood V of x, the density function $p(t; x, y)$ of X_t in (4.10) has the following asymptotic expansion near $t = 0$*

$$p(t; x, y) = \frac{1}{(t^H)^d} e^{-\frac{d^2(x,y)}{2t^{2H}} + \frac{\beta}{t^{2H-1}}} \left(\sum_{i=0}^{N} c_i(x, y) t^{\lambda_i H} + r_{N+1}(t, x, y) t^{\lambda_{N+1} H} \right), \ y \in V.$$

Here β is some constant, $d(x, y)$ is the Riemannian distance between x and y determined by $V_1, \ldots, V_d$. Moreover, we can chose V such that $c_i(x, y)$ are C^∞ in $V \times V \subset \mathbb{R}^d \times \mathbb{R}^d$, and for all multi-indices α and β

$$\sup_{t \le t_0} \sup_{(x,y) \in V \times V} |\partial_x^\alpha \partial_y^\beta r_{N+1}(t, x, y)| < \infty$$

for some $t_0 > 0$.

Remark 4.14 Differentiability of $c_i(x, y)$, r_{N+1} in the above theorem and legitimacy of Fourier inversion in (4.11) is obtained by Malliavin calculus and some uniform estimates of the coefficients in the Laplace approximation. We refer the reader to [5] for details.

Remark 4.15 Our result assumes the ellipticity condition and a strong structure condition (Assumption 4.7). Later Inahama [28, 29] proved the kernel expansion (for $H > \frac{1}{3}$) under some mild conditions on the vector fields. He takes a different approach and uses Watanabe distribution theory. Hence he is able to work with $D(x, y)$ introduced in Remark 4.11 directly and avoids the technical assumption A2 of Assumption 4.7. On the other hand, the smoothness of coefficient and the uniform estimate for the remainder terms in the expansion are not provided in [28, 29].

5 Application to Mathematical Finance

Fractional Brownian motions have been used in financial models to introduce memory. In this section, we give two examples of such models and remark on how the methods and results in the previous sections could be applied to the study of such models.

5.1 One Dimensional Models

Memories can be introduced to stock price process directly. In particular, the so-called fractional Black and Scholes model is given by

$$S_t = S_0 \exp\left(\mu t + \sigma B_t^H - \frac{\sigma^2}{2} t^{2H}\right), \tag{5.1}$$

where B^H is a fractional Brownian motion with Hurst parameter H, μ the mean rate of return and $\sigma > 0$ the volatility. Let r be the interest rate. The price for the risk-free bond is given by e^{rt}.

More generally, one can also consider a fractional local volatility model

$$dS_t = S_t(\mu dt + \sigma(S_t) dB_t^H).$$

Here the stochastic integration with respect to B^H could be understood in the sense of rough path theory. After a simple change of variable $X_t = \log S_t$, one obtains

$$dX_t = \mu dt + \sigma(e^{X_t}) dB_t^H.$$

There has been an intensive study recently of option prices and implied volatilities for options with short maturity (e.g. [9, 16, 21]). Since the above equation is a special case of (4.10), we can use the results obtained in the previous sections to obtain short-time asymptotic behavior of such models.

A drawback of the finance models discussed above is that they lead to the existence of arbitrage opportunities. For example, let the couple $(\alpha_t, \beta_t), t \in [0, T]$ be a portfolio with α_t the amount of bonds and β_t the amount of stocks at time t. One can construct an arbitrage in the fractional Black and Scholes model by (for simplicity, we assume $\mu = r = 0$)

$$\beta_t = S_t - S_0, \quad \text{and } \alpha_t = \int_0^t \beta_t dS_t - \beta_t S_t.$$

Let V_t be the value of the portfolio at time t. It is not hard to see that this is a self-financing portfolio that satisfies $V_0 = 0$ and $V_t = (S_t - S_0)^2$ for all $t > 0$, and hence it is an arbitrage. For more discussion on arbitrage in models given by fractional Brownian motions, we refer the reader to [35].

5.2 Stochastic Volatility Models

Stochastic volatility models were introduced to capture both the volatility smile and the correct dynamics of the volatility smile (see [23] for instance). For these models, modeling the volatility process is one of the key factors. In [14], the authors proposed a long memory specification of the volatility process in order to capture the steepness of long term volatility smiles without over increasing the short run persistence.

The following stochastic volatility model based on the fractional Ornstein-Uhlenbeck process provides another way introducing long memory to the volatility process:

$$dS_t = \mu S_t dt + \sigma_t S_t dW_t,$$

where $\sigma_t = f(Y_t)$ and Y_t is a fractional Ornstein-Uhlenbeck process:

$$dY_t = \alpha(m - Y_t)dt + \beta_t dB_t^H.$$

In the above W_t is a standard Brownian motion and B_t^H an independent (of W_t) fractional Brownian motion with Hurst parameter $H > \frac{1}{2}$. Examples of functions f are $f(x) = e^x$ and $f(x) = |x|$.

Comte and Renault [13] studied this type of stochastic volatility models which introduces long memory and mean reversion in the Hull and White setting [26]. The long memory property allows this model to capture the well-documented evidence of persistence of the stochastic feature of Black and Scholes implied volatilities when time to maturity increases.

Unlike one dimensional models mentioned above, the fractional Ornstein-Uhlenbeck model is arbitrage free since the stock price process is driven by a standard Brownian motion. In [25], Hu has proved that for this model, market is incomplete and the martingale measures are not unique. If we set $\gamma_t = (r - \mu)/\sigma_t$ and

$$\frac{d\mathbb{Q}}{d\mathbb{P}} = \exp\left(\int_0^T \gamma_t dW_t - \frac{1}{2}\int_0^T |\gamma_t|^2 dt\right).$$

Then $\mathbb{Q}$ is the minimal martingale measure associated with $\mathbb{P}$. Moreover, the risk minimizing-hedging price at $t = 0$ of an European call option with payoff $(S_T - K)^+$ is given by

$$C_0 = e^{-rT}\mathbb{E}_{\mathbb{Q}}(S_T - K)^+.$$

The fractional Ornstein-Uhlenbeck model takes a generalized form of Eq. (4.10) that is studied in the previous sections. It is a system of SDEs driven by fractional Brownian motions, but with varying Hurst parameter H. We believe that the methods discussed above can be extended to study small-time asymptotics of these models.

References

1. Azencott, R.: Densité des diffusions en temps petit: développements asymptotiques. I. Seminar on probability, XVIII. Lecture Notes in Mathematics, vol. 1059, pp. 402-498. Springer, Berlin (1984)
2. Ben Arous, G.: Développement asymptotique du noyau de la chaleur hypoelliptique hors du cut-locus. Ann. Sci. École Norm. Sup. (4) **21**(3), 307–331 (1988)
3. Ben Arous, G.: Méthode de Laplace et de la phase stationnaire sur l'espace de Wiener, Stochastics **25**(3), 125–153 (1988)
4. Baudoin, F., Hairer, M.: A version of Hörmander's theorem for the fractional Brownian motion. Probab. Theory Relat. Fields **139**, 373–395 (2007)
5. Baudoin, F., Ouyang, C.: Small-time kernel expansion for solutions of stochastic differential equations driven by fractional Brownian motions. Stoch. Process. Appl. **121**(4), 759–792 (2011)
6. Baudoin, F., Ouyang, C.: Gradient bounds for solutions of stochastic differential equations driven by fractional Brownian motions. Malliavin Calculus and Stochastic Analysis: A Festschrift in Honor of David Nualart. Springer, Berlin (2012)
7. Baudoin, F., Ouyang, C., Zhang, X.: Varadhan estimates for RDEs driven by fractional Brownian motions. Stoch. Proc. Appl. **125**(2), 634–652 (2015)
8. Baudoin, F., Ouyang, C., Zhang, X.: Smoothing effect of rough differential equations driven by fractional Brownian motions. Ann. Inst. Henri Poincare Probab. Statist. (2013)
9. Berestyki, H., Busca, J., Florent, I.: Computing the implied volatility in stochastic volatility models. Commun. Pure Appl. Math., Vol. **LVII**, 1352–1373 (2004)
10. Cass, T., Friz, P.: Densities for rough differential equations under Hörmander condition. Ann. Math. **171**(3), 2115–2141 (2010)
11. Cass, T., Friz, P., Victoir, N.: Non-degeneracy of Wiener functionals arising from rough differential equations. Trans. Am. Math. Soc. **361**, 3359–3371 (2009)
12. Cass, T., Litterer, C., Lyons, T.: Integrability and tail estimates for Gaussian rough differential equations. Ann. Probab. **41**(4), 3026–3050 (2013)
13. Comte, F., Renault, E.: Long memory in continuous-time stochastic volatility models. Math. Financ. **8**, 291–323 (1998)
14. Comte, F., Coutin, L., Renault, E.: Affine fractional stochastic volatility models. Ann. Financ. **8**(2–3), 337–378 (2012)
15. Coutin, L., Qian, Z.M.: Stochastic analysis, rough path analysis and fractional Brownian motions. Probab. Theory Relat. Fields **122**(1), 108–140 (2002)
16. Feng, J., Forde, M., Fouque, J.P.: Short maturity asymptotics for a fast mean reverting Heston stochastic volatility model. SIAM J. Financ. Math. **1**, 126–141 (2010)
17. Friz, P., Gess, B., Gulisashvili, A., Riedel, S.: The Jain-Monrad criterion for rough paths and applications to random Fourier series and non-Markovian Hörmander theory. Ann. Probab. (2013)
18. Friz, P., Riedel, S.: Integrability of (non-)linear rough differential equations and integrals. Stoch. Anal. Appl. **31**(2), 336–358 (2013)
19. Friz, P., Victoir, N.: Differential equations driven by Gaussian signals. Ann. Inst. Henri Poincare Probab. Stat. **46**(2), 369–413 (2010)

20. Friz, P., Victoir, N.: Multidimensional Dimensional Processes seen as Rough Paths. Cambridge University Press, Cambridge (2010)
21. Gatheral, J., Hsu, E., Laurence, P., Ouyang, C., Wang, T.-H.: Asymptotics of implied volatility in local volatility models. Math. Financ. **22**, 591–620 (2012)
22. Gubinelli, M.: Controlling rough paths. J. Funct. Anal. **216**, 86–140 (2004)
23. Hagan, P., Kumar, D., Lesniewski, A., Woodward, D.: Managing Smile Risk. Wilmott Mag. (2003)
24. Hairer, M., Pillai, N.S.: Regularity of laws and ergodicity of hypoelliptic SDEs driven by rough paths. Ann. Inst. Henri Poincaré Probab. Stat. **47**(2), 601–628 (2011)
25. Hu, Y.: Integral transformations and anticipative calculus for fractional Brownian motions. Mem. Am. Math. Soc. **175**(825), 324 (2005)
26. Hull, J., White, A.: The pricing of options on assets with stochastic volatilities. J. Financ. **3**, 281–300 (1987)
27. Inahama, Y.: Laplace approximation for rough differential equation driven by fractional Brownian motion. Ann. Probab. **41**(1), 170–205 (2013)
28. Inahama, Y.: Short time kernel asymptotics for young SDE by means of Watanabe distribution theory. To appear in J. Math. Soc. Jpn. (2013)
29. Inahama, Y.: Short time kernel asymptotics for rough differential equation driven by fractional Brownian motion. Preprint (2014)
30. Léandre, R.: Minoration en temps petit de la densité d'une diffusion dégénérée. J. Funct. Anal. **74**, 399–414 (1987)
31. Lyons, T.: Differential equations driven by rough signals. Rev. Mat. Iberoam. **14**(2), 215–310 (1998)
32. Lyons, T., Qian, Z.: System Control and Rough Paths. Oxford University Press, Oxford (2002)
33. Nualart, D.: The Malliavin Calculus and Related Topics, 2nd edn. Probability and its Applications. Springer, Berlin (2006)
34. Nualart, D., Saussereau, B.: Malliavin calculus for stochastic differential equations driven by a fractional Brownian motion. Stoch. Process. Appl. **119**(2), 391–409 (2009)
35. Rogers, L.C.G.: Arbitrage with fractional Brownian motion. Math. Financ. **7**(1), 95–105 (1997)

On Singularities in the Heston Model

Vladimir Lucic

Abstract In this note we provide characterization of the singularities of the Heston characteristic function. In particular, we show that all the singularities are pure imaginary.

Keywords Heston · Complex singularities

1 Problem Formulation

Consider the Heston stochastic volatility model, which under risk-neutral measure and with zero drift has the following dynamics

$$\begin{aligned} dS_t &= S_t\sqrt{v_t}\,dW_t^{(1)}, \\ dv_t &= \lambda(\bar{v} - v_t)\,dt + \eta\sqrt{v_t}\,(\rho\,dW_t^{(1)} + \sqrt{1-\rho^2}\,dW_t^{(2)}), \end{aligned}$$

where the parameters λ, η, and $\bar{v}$ are nonnegative, $\rho \in [-1, 1]$, and the initial values S_0 and v_0 are positive.

The Heston characteristic function is defined as

$$\phi_H(u,\tau) = \mathbb{E}\big[e^{iu\log(S_\tau/S_0)}\big], \quad \alpha < \Im(u) < \beta.$$

Results of Heston [5] and Lewis [7] show that on the strip of convergence $\alpha < \Im(u) < \beta$ the Heston characteristic function coincides with

$$\phi(u,\tau) = e^{C(u,\tau)\bar{v}+D(u,\tau)v_0}, \quad u \in \mathbb{Z},$$

V. Lucic (✉)
Quantitative Analytics, Barclays, 5 The North Colonnade, Canary Wharf, London E14 4BB, UK
e-mail: vladimir.lucic@barclays.com

P.K. Friz et al. (eds.), *Large Deviations and Asymptotic Methods in Finance*, Springer Proceedings in Mathematics & Statistics 110,
DOI 10.1007/978-3-319-11605-1_15

where

$$D(u,\tau) = r_-\frac{1-e^{-d\tau}}{1-ge^{-d\tau}}, \quad C(u,\tau) = \lambda\left[r_-\tau - \frac{2}{\eta^2}\log\left(\frac{1-ge^{-d\tau}}{1-g}\right)\right],$$

$$r_\pm = \frac{\beta \pm d}{\eta^2}, \quad d = \sqrt{\beta^2 + 2\alpha\eta^2}, \quad g = \frac{r_-}{r_+},$$

$$\alpha = \frac{u^2}{2} - \frac{iu}{2}, \quad \beta = \lambda + \rho\eta i u.$$

With the customary abuse of terminology, we'll refer to $\phi(u,\tau)$, $u \in \mathbb{Z}$ as the Heston characteristic function.

Using a result[1] of Lukacs [8], Lewis [7] points out that $\phi(u,\tau)$ has singularities on the imaginary axis at the boundaries of the strip of convergence. Whether there are any other singularities (necessarily complex-conjugate) on that boundary could not be readily established. Furthermore, no conclusions can be made about singularities outside of the strip of convergence. The purpose of this note is to provide full characterization of the singularities of $\phi(u,\tau)$.

2 Main Result

The following theorem, although presented as an existence result, allows for construction of the singularities of $\phi(u,\tau)$ via standard numerical methods.

Theorem 2.1 *All singularities of $\phi(u,\tau)$ are pure imaginary.*

Proof Assume $\eta > 0$, as for $\eta = 0$ we have the Black-Scholes model whose characteristic function is free of singularities (see, e.g., Lewis [7]).

To simplify notation we put $is = u$ and show that the (essential) singularities of $\phi(is,\tau)$ are real. To this end, we show that the transcendental equation

$$\frac{r_+}{r_-} = e^{-d\tau}, \tag{2.1}$$

where

$$\beta = \lambda - \rho\eta s \tag{2.2a}$$

$$d = \sqrt{\beta^2 - \eta^2 s(s-1)} \tag{2.2b}$$

$$r_\pm = \frac{\beta \pm d}{\eta^2} \tag{2.2c}$$

has only real roots.

---

[1] As noted in Lukacs [8], this is a corollary of a more general result on Laplace transforms, e.g. Theorem II.5b of Widder [9].

We consider (2.1) and (2.2) as a system in d and s. Equation (2.1) can be written as

$$d = (-\lambda + \rho\eta s)\tanh(\tau d/2). \tag{2.3}$$

From (2.2a) and (2.2b) we get

$$-(1-\rho^2)\eta^2 s^2 + s(\eta^2 - 2\rho\eta\lambda) + \lambda^2 - d^2 = 0, \tag{2.4}$$

so, with $q := \sqrt{1-\rho^2}$, we can express s in terms of d: for $q \neq 0$

$$s_{1/2} = \frac{\eta - 2\rho\lambda \pm \sqrt{(\eta - 2\rho\lambda)^2 + 4q^2(\lambda^2 - d^2)}}{2q^2\eta}, \tag{2.5}$$

and for $q = 0$ and $2\rho\lambda - \eta \neq 0$

$$s = \frac{d^2 - \lambda^2}{\eta^2 - 2\lambda\rho\eta}. \tag{2.6}$$

If $q = 0$ and $2\rho\lambda - \eta = 0$ from from (2.1), (2.2), and (2.4) we obtain $d = \lambda$, $\rho = 1$, $\eta = 2\lambda$, which implies that the only singularity is

$$s = \frac{1}{1 - e^{-\lambda\tau}}.$$

If $d = 0$ we have equality in (2.3), while from (2.5) and (2.6) it follows that the roots in s are real.

For $d \neq 0$ substituting (2.5) in (2.3) yields

$$d = \left(-\lambda + \rho\frac{\eta - 2\rho\lambda \pm \sqrt{(\eta - 2\rho\lambda)^2 + 4q^2(\lambda^2 - d^2)}}{2q^2}\right)\tanh(\tau d/2),$$

while substituting (2.6) in (2.3) gives

$$d = \left(-\lambda + \rho\frac{d^2 - \lambda^2}{\eta - 2\rho\lambda}\right)\tanh(\tau d/2).$$

which imply, respectively,

$$\left(2dq^2\coth(\tau d/2) + 2\lambda - \rho\eta\right)^2 = \rho^2((\eta - 2\rho\lambda)^2 + 4q^2(\lambda^2 - d^2)), \tag{2.7}$$

and

$$d\coth(\tau d/2) + \lambda = \frac{\rho}{\eta - 2\rho\lambda}(\lambda^2 - d^2). \tag{2.8}$$

With

$$\frac{\tau d}{2} = iz, \quad a = \frac{\tau(\eta - 2\lambda\rho)}{4q}\,\mathrm{sgn}(\rho), \quad b = \frac{\tau\lambda}{2}, \quad c = \frac{|\rho|}{q}$$

Lemma 2.2 implies that the roots of (2.7) are either real or pure imaginary. For the special case (2.8), Lemma 2.3 with

$$\frac{\tau d}{2} = iz, \quad b = \frac{\tau\lambda}{2}, \quad c = \frac{2\rho}{\tau(\eta - 2\rho\lambda)}$$

implies that the corresponding roots are also either real or pure imaginary.

Therefore, it follows that for $d \neq 0$ the expression in the brackets in (2.3) is real (being ratio of either real or imaginary numbers), which in turn implies that the solutions of the transcendental equation (2.1) are real in s. □

Lemma 2.2 *For real a and real nonnegative b, c the roots of the equation*

$$(z\cot(z) + b - ac)^2 = c^2(a^2 + b^2 + z^2), \quad z \in \mathbb{Z} \tag{2.9}$$

are real or pure imaginary.

Proof For $c = 0$ the result follows from Lemma A.6. If $c > 0$ from Lemma 2.4 we have that for sufficiently large N equation (2.9) has $4N + 2$ roots inside the square with vertices $(N + 1/2)(\pm\pi, \pm i\pi)$. On the other hand, from Lemmas A.1 and A.3 it follows that there are $4N + 2$ real or pure imaginary roots inside the same square, so the result follows. □

Lemma 2.3 *For real nonnegative b and real c the roots of the equation*

$$z\cot(z) + b = c(b^2 + z^2), \quad z \in \mathbb{Z} \tag{2.10}$$

are real or pure imaginary.

Proof For $c = 0$ the result follows from Lemma A.6. Putting $a = 0$ in Lemma 2.4 we conclude that for every $c \neq 0$ and sufficiently large N equation

$$(z\cot(z) + b)^2 = c^2(b^2 + z^2)^2, \quad z \in \mathbb{Z}$$

has $4N + 4$ roots inside the square with vertices $(N + 1/2)(\pm\pi, \pm i\pi)$. On the other hand, from Lemmas A.2 and A.4 it follows that both equations

$$z\cot(z) + b = \pm c(b^2 + z^2)$$

have $2N + 2$ real or pure imaginary roots inside the same square, whence the result follows. □

In the next lemma[2] we make repeated use of the Rouché's theorem (e.g., Hille [6] [Theorem 9.2.3]).

Lemma 2.4 *Let C_N, $N \in \mathbb{N}$ denote the square in complex plane with vertices at $(N+\frac{1}{2})(\pm\pi, \pm i\pi)$. Then for real a, nonnegative b, c, and $d = 1, 2$ there exists $N_0 \in \mathbb{N}$ such that for every integer $N > N_0$ the equation*

$$(z\cot(z) + b - ac)^2 = c^2(a^2 + b^2 + z^2)^d, \quad z \in \mathbb{Z} \tag{2.11}$$

has $4N + 2d$ roots inside C_N.

Proof Consider the case $d = 1$, $c > 1$ and the case $d = 2$, $c > 0$ together. On the right vertical side of C_N we have

$$|\cot(z)| = \left|\cot\left(\frac{\pi}{2} + N\pi + iy\right)\right| = |\tan(iy)| = \left|\frac{e^y - e^{-y}}{e^y + e^{-y}}\right| < 1, \tag{2.12}$$

while on the upper horizontal side we have

$$|\cot(z)| = \left|\frac{e^{2iz} + 1}{e^{2iz} - 1}\right| = \left|\frac{1 + e^{-(2N+1)\pi}e^{2ix}}{1 - e^{-(2N+1)\pi}e^{2ix}}\right| \le \frac{1 + e^{-(2N+1)\pi}}{1 - e^{-(2N+1)\pi}}$$

Together with (2.12) and the fact that $|\cot(z)| = |\cot(-z)|$ this implies

$$|\cot(z)| \le \frac{1 + e^{-(2N+1)\pi}}{1 - e^{-(2N+1)\pi}} =: k_N, \quad z \in C_N.$$

For $z \in C_N$ have

$$\begin{aligned}\frac{|(z\cot(z) + b - ac)^2|}{|c^2(a^2 + b^2 + z^2)|^d} &\le \frac{(|z\cot(z)| + |b - ac|)^2}{|c^2(a^2 + b^2 + z^2)|^d} \\ &\le \left(\frac{k_N}{c} + \frac{|b - ac|}{|cz|}\right)^2 \left|\frac{z^2}{(a^2 + b^2 + z^2)^d}\right|.\end{aligned}$$

Since $\lim_{n\to\infty} k_n = 1$, the last expression tends to $(2-d)/c^2 < 1$ uniformly in z as $N \to \infty$, so for sufficiently large N we have

$$|(z\cot(z) + b - ac)^2| < |c^2(a^2 + b^2 + z^2)|^d, \quad z \in C_N.$$

Therefore, by Rouché's theorem the number of roots of (2.11) inside C_N is equal to the number of poles of $z \mapsto (z\cot(z) + b - ac)^2 - c^2(a^2 + b^2 + z^2)^d$ inside C_N plus the number of zeros of $z \mapsto c^2(a^2 + b^2 + z^2)^d$ inside C_N (considering

[2]A weaker version of this result (dealing with the case of real roots only) appears as Problem E1295 in *American Mathematical Monthly*, Vol. 65., No. 6, p. 450.

their multiplicities). For sufficiently large N those two numbers are $4N$ and $2d$ respectively, whence the equation (2.11) has $4N + 2d$ roots inside C_N.

Consider now $d = 1, 0 < c < 1$. Let D_N be the square vertices at $(\pm N\pi, \pm Ni\pi)$, and let D_N^ϵ denote D_N extended with semicircles of radius ϵ so that the poles of $\cot(z)$ at $\pm N\pi$ are inside D_N^ϵ, but the real zeros of (2.11) in $(N\pi, (N + 1/2)\pi)$ and $(-(N + 1/2)\pi, -N\pi)$ described in Lemma A.1 remain outside. For ease of exposition in what follows we make ϵ smaller if necessary, which can be done without invalidating previously established statements.

Similarly as before, on the right vertical side of D_N we have

$$|\tan(z)| = |\tan(N\pi + iy)| = |\tan(iy)| = \left|\frac{e^y - e^{-y}}{e^y + e^{-y}}\right| < 1, \tag{2.13}$$

while on the upper horizontal side we have

$$|\tan(z)| = \left|\frac{e^{2iz} - 1}{e^{2iz} + 1}\right| = \left|\frac{1 - e^{-2N\pi}e^{2ix}}{1 + e^{-2N\pi}e^{2ix}}\right| \le \frac{1 + e^{-2N\pi}}{1 - e^{-2N\pi}}. \tag{2.14}$$

Together with (2.13) and the fact that $|\cot(z)| = |\cot(-z)|$ this implies that for sufficiently small $\epsilon > 0$

$$|\cot(z)| \ge \frac{1 - e^{-2N\pi}}{1 + e^{-2N\pi}} =: k_n, \quad z \in D_N^\epsilon. \tag{2.15}$$

On D_N^ϵ we have

$$\frac{|c^2(a^2 + b^2 + z^2)|}{|(z\cot(z) + b - ac)^2|} \le \frac{c^2}{\left(|\cot(z)| - \left|\frac{b-ac}{z}\right|\right)^2} + \frac{|c^2(a^2 + b^2)|}{\left(|z||\cot(z)| - |b - ac|\right)^2},$$

so for N large enough

$$\frac{|c^2(a^2 + b^2 + z^2)|}{|(z\cot(z) + b - ac)^2|} \le \frac{c^2}{\left(k_N - \left|\frac{b-ac}{z}\right|\right)^2} + \frac{|c^2(a^2 + b^2)|}{\left(|z|\,k_N - |b - ac|\right)^2}.$$

Since $\lim_{n\to\infty} k_n = 1$, the last expression tends to $c^2 < 1$ uniformly in z as $N \to \infty$, so for sufficiently large N we have

$$|c^2(a^2 + b^2 + z^2)| < |(z\cot(z) + b - ac)^2|, \quad z \in D_N^\epsilon.$$

Therefore, by Rouché's theorem the number of roots of (2.11) inside D_N^ϵ is equal to the number of poles of $z \mapsto (z\cot(z) + b - ac)^2 - c^2(a^2 + b^2 + z^2)$ inside D_N^ϵ plus the number of zeros minus the number of poles of $z \mapsto (z\cot(z) + b - ac)^2$ inside D_N^ϵ (considering their multiplicities). The two mappings have common poles, so we are left with number of zeros of the second mapping, which for sufficiently small ϵ, according to Lemma A.6, is $4N$. Therefore, from Lemma A.5, and taking into account two real zeros in $(-(N+1/2)\pi, -N\pi) \cup (N\pi, (N+1/2)\pi)$ whose existence is established in Lemma A.1, we conclude that for $d = 1$, $0 < c < 1$ and sufficiently large N there are $4N+2$ zeros inside C_N.

Finally, consider the case $c = 1$, $d = 1$. Put $\alpha = b - ac$, $\beta^2 = a^2 + b^2$, so that we get

$$\frac{\cos(2z)}{\sin^2(z)}z^2 + \alpha z \frac{\sin(2z)}{\sin^2(z)} - (\beta^2 - \alpha^2) = 0, \tag{2.16}$$

or, equivalently,

$$2\cot(2z)\cot(z)z^2 + 2\alpha z\cot(z) - (\beta^2 - \alpha^2) = 0.$$

On D_N^ϵ we have

$$\frac{|2\alpha z\cot(z) - (\beta^2 - \alpha^2)|}{|2\cot(2z)\cot(z)z^2|} \leq \frac{\left|2\alpha - \frac{\beta^2 - \alpha^2}{z\cot(z)}\right|}{2|z||\cot(2z)|} \leq \frac{|2\alpha| + \frac{|\beta^2 - \alpha^2|}{|z|k_N}}{2|z|k_{2N}}.$$

Since $\lim_{n\to\infty} k_n = 1$, the last expression tends to zero uniformly in z as $N \to \infty$, so for sufficiently large N we obtain

$$|2\alpha z\cot(z) - (\beta^2 - \alpha^2)| < |2\cot(2z)\cot(z)z^2|, \quad z \in D_N^\epsilon,$$

that is,

$$\left|\frac{\cos(2z)}{\sin^2(z)}z^2\right| > \left|\alpha z\frac{\sin(2z)}{\sin^2(z)} - (\beta^2 - \alpha^2)\right|, \quad z \in D_N^\epsilon.$$

Thus, by Rouché's theorem this implies that the number of roots of (2.16) inside D_N^ϵ equals the number of zeros of $z \mapsto \frac{\cos(2z)}{\sin^2(z)}z^2$ inside D_N^ϵ, which is $4N$. Therefore, reasoning as in the previous part of the proof we conclude that for $c = 1$ we have $4N+2$ zeros of (2.11) inside C_N for N large enough. □

Acknowledgments I wish to thank Tomislav Šekara of University of Belgrade and the anonymous referee for their comments and suggestions.

Addendum

The first version of this paper appeared on SSRN in 2007 (following the author's investigation into applicability of the Talbot's numerical inversion method in transform analysis of option prices). Since then several publications have appeared using the main result of the present work, which we list below for completeness.

Based on Theorem 2.1, in Ferreiro-Castilla [2] and del Baño Rollin et al. [1] a smoothness result for the density of the log-spot in the Heston model is presented, together with an alternative proof of our main result. Theorem 2.1 was also used in Friz et al. [3] in the study of the asymptotic behaviour of the stock price density in the negatively correlated Heston model. Finally, Lemma 6.1 from Gulisashvili et al. [4], used in the study of the asymptotic behaviour of the mixing distribution density in the uncorrelated Heston model, is quite close in spirit to the results presented here.

Appendix

Lemma A.1 *For N sufficiently large, equation* (2.9) *has* $4N-2$ *real roots in* $(-(N+1/2)\pi, -\pi) \cup (\pi, (N+1/2)\pi)$.

Proof By Lemma A.5 for every $N > 1$ equation (2.9) as two real roots in each of the intervals $(-(k+1)\pi, -k\pi)$ and $(k\pi, (k+1)\pi)$, $k = 1, 2, \ldots N-1$.

Rewrite (2.9) as

$$z\cot(z) = -(b - ac) \pm c\sqrt{a^2 + b^2 + z^2}. \tag{A.1}$$

For $N > 0$

$$\lim_{z \to N\pi+} z\cot(N\pi) = +\infty, \quad z\cot(N\pi + \pi/2) = 0, \tag{A.2}$$

so we conclude that for sufficiently large N the equation with plus sign has one real root in $(N\pi, (N+1/2)\pi)$, hence by symmetry in $(-(N+1/2)\pi, -N\pi)$. □

Lemma A.2 *For N sufficiently large equation* (2.10) *has* $2N$ *real roots in* $(-(N+1/2)\pi, -\pi) \cup (\pi, (N+1/2)\pi)$ *if* $c > 0$, *and* $2N-2$ *real roots if* $c < 0$.

Proof By Lemma A.5 for every $N > 1$ equation (2.10) has one real root in each of the intervals $(-(k+1)\pi, -k\pi)$ and $(k\pi, (k+1)\pi)$, $k = 1, 2, \ldots N-1$.

Rewrite (2.10) as

$$z\cot(z) = -b + c(b^2 + z^2). \tag{A.3}$$

From (A.2) and (A.3) we conclude that if $c > 0$ for sufficiently large N equation (2.10) has one real root in $(N\pi, (N+1/2)\pi)$, hence by symmetry in $(-(N+1/2)\pi, -N\pi)$. □

Lemma A.3 *For real a, nonnegative b, and* $c > 0$ *equation* (2.9) *has either four real roots in* $(-\pi, \pi)$, *or two real roots in* $(-\pi, \pi)$ *and two imaginary roots.*

Proof The proof follows by simple geometrical considerations. For $z = 0$ the right-hand side of (A.1) assumes two values

$$\alpha_1 := -(b - ac) + c\sqrt{a^2 + b^2}, \quad \alpha_1 := -(b - ac) - c\sqrt{a^2 + b^2}.$$

Since $ac - c\sqrt{b^2 + a^2} \leq 0$ we have $\alpha_2 \leq 0$. On the other hand, the function $x \mapsto x\cot(x)$ is zero at the origin and strictly decreases on $[0, \pi)$, with a discontinuity of the second kind at π. Thus, (A.1) has one real root corresponding to the intersection of $x \mapsto x\cot(x)$ and $x \mapsto -(b - ac) - c\sqrt{a^2 + b^2 + z^2}$ on $(0, \pi)$.

If $\alpha_1 < 1$ following the same argument we conclude that there is another real root in $(0, \pi)$ corresponding to the intersection of $x \mapsto x\cot(x)$ and $x \mapsto -(b - ac) + c\sqrt{a^2 + b^2 + z^2}$. If $\alpha_1 = 1$ we have a double root at zero.

Thus, based on the above considerations and the symmetry around the origin it follows that in $(-\pi, \pi)$ equation (A.1) has four real roots if $\alpha_1 \leq 1$, and two real roots if $\alpha_1 > 1$. Therefore, to complete the proof we show that (A.1) has two imaginary roots if $\alpha_1 > 1$.

Put $z = iy$, $y \in \mathbb{R}$ in (A.1) to get

$$y\coth(y) = -(b - ac) \pm c\sqrt{a^2 + b^2 - y^2}. \tag{A.4}$$

On the left-hand side we have a continuous function equal to one at the origin that tends to infinity as y increases. Note that $\alpha_1 > 1$ implies $a^2 + b^2 > 0$ Thus, on the right-hand side we have a semi-circle starting at $(0, \alpha_1)$ on the ordinate, entering into the right half-plane, and ending at $(0, \alpha_2)$ on the ordinate, half-encircling the point $(0, 1)$ (as $\alpha_1 > 1$ and $\alpha_2 \leq 0$). Therefore, there must exist $y_0 > 0$ for which the equality holds in (A.4). Since $-y_0$ also solves (A.4), we have two imaginary solutions. □

Lemma A.4 *Assume $b \geq 0$. For $c > 0$ equation* (2.10) *has either two real roots in $(-\pi, \pi)$ or two imaginary roots. If $c < 0$ equation* (2.10) *has two real roots in $(-\pi, \pi)$ and two imaginary roots.*

Proof At $z = 0$ the right-hand side of (A.1) equals $-b + cb^2$. The function $x \mapsto x\cot(x)$ is zero at the origin and strictly decreases on $[0, \pi)$, with a discontinuity of the second kind at π. Thus, if $c < 0$ or $c > 0$ and $-b + cb^2 < 1$ there is one real root in $(0, \pi)$, hence by symmetry in $(-\pi, 0)$. If $-b + cb^2 = 1$ we have a double root at the origin. Next, with $z = iy$, $y \in \mathbb{R}$ equation (2.10) becomes

$$y\coth(y) = -b + c(b^2 - y^2). \tag{A.5}$$

Therefore, if $c > 0$ and $-b+cb^2 > 1$ the right-hand side dominates the left-hand side at the origin, while the opposite is true for sufficiently large y. From the continuity of the two functions it then follows that (A.5) has one positive root, hence by symmetry one negative root. Finally, if $c < 0$ the left-hand side dominates the right-hand side at the origin, while the opposite is true for sufficiently large y, giving a pair of imaginary roots. □

Lemma A.5 *For every positive integer k equation* (2.11) *has two real roots in each of the intervals $(-(k + 1)\pi, -k\pi)$ and $(k\pi, (k + 1)\pi)$.*

Proof The result follows from the fact that on each of those intervals the range of the map $x \mapsto x\cot(x)$ is the whole real line, while the maps $x \mapsto -b + ac \pm c(a^2 + b^2 + x^2)^{d/2}$ are bounded. □

Lemma A.6 *For $a \in \mathbb{R}$ the equation*

$$z\cot(z) = a, \quad z \in \mathbb{Z} \tag{A.6}$$

has $2N$ roots inside the square with vertices $(\pm N\pi, \pm Ni\pi)$. The roots are real or pure imaginary.

Proof For $a = 0$ the roots are the zeros of $\cos(z)$. If $a \neq 0$ from (2.13) and (2.14) we conclude that for sufficiently large N

$$|a\tan(z)| < |z|, \quad z \in D_N.$$

Thus, by Rouché's theorem

$$z = a\tan(z)$$

has $2N + 1$ roots inside the square with vertices $(\pm N\pi, \pm Ni\pi)$. If $k > 0$ it has two real roots in $(-(k+1)\pi/2, k\pi/2) \cup (k\pi/2, (k+1)\pi/2)$ if either $a > 0$ and k is even, or $a < 0$ and k is odd.

On the other hand, in $(-\pi, \pi)$ there are three roots (counting their multiplicities) if $a \geq 1$ and one root if $0 < a < 1$. In the latter case there are two imaginary roots (c.f. example on p. 255 of Hille [6]). Since (A.6) has one root less at the origin, the result follows. □

References

1. del Baño Rollin, S., Ferreiro-Castilla, A., Utzet, F.: On the density of log-spot in the Heston volatility model. Stoch. Process. Appl. **120**, 2037–2063 (2010)
2. Ferreiro-Castilla, A.: Stochastic Calculus and Analytic Characteristic Functions: Applications to Finance. Ph.D. thesis, Universitat Autònoma de Barcelona (2011)
3. Friz, P., Gerhold, S., Gulisashvili, A., Sturm, S.: On refined volatility smile expansion in the Heston model. Quant. Financ. **11**, 1151–1164 (2011)
4. Gulisashvili, A., Stein, E.M.: Asymptotic behavior of the stock price distribution density and implied volatility in stochastic volatility models. Appl. Math. Optim. **61**, 287–315 (2010)
5. Heston, S.L.: A closed-form solution for options with stochastic volatility with applications to bond and currency options. Rev. Financ. Stud. **6**(2), 327–343 (1993)
6. Hille, E.: Analytic Function Theory, vol. 1. Blaisdell, New York (1965)
7. Lewis, A.L.: Option Valuation Under Stochastic Volatility. Finance Press, Newport Beach (2000)
8. Lukacs, E.: Characteristic Functions. Charles Griffin & Co., London (1970)
9. Widder, D.V.: The Laplace Transform. Princeton University Press, Princeton (1946)

On the Probability Density Function of Baskets

Christian Bayer, Peter K. Friz and Peter Laurence

Abstract The state price density of a basket, even under uncorrelated Black–Scholes dynamics, does not allow for a closed form density. (This may be rephrased as statement on the sum of lognormals and is especially annoying for such are used most frequently in Financial and Actuarial Mathematics.) In this note we discuss short time and small volatility expansions, respectively. The method works for general multi-factor models with correlations and leads to the analysis of a system of ordinary (Hamiltonian) differential equations. Surprisingly perhaps, even in two asset Black–Scholes situation (with its flat geometry), the expansion can degenerate at a critical (basket) strike level; a phenomena which seems to have gone unnoticed in the literature to date. Explicit computations relate this to a phase transition from a unique to more than one "most-likely" paths (along which the diffusion, if suitably conditioned, concentrates in the afore-mentioned regimes). This also provides a (quantifiable) understanding of how precisely a presently out-of-money basket option may still end up in-the-money.

Keywords Sums of lognormals · Focality · Pricing of butterfly spreads on baskets

C. Bayer
Weierstrass Institute, Mohrenstrasse 39, 10117 Berlin, Germany
e-mail: christian.bayer@wias-berlin.de

P.K. Friz (✉)
Institut für Mathematik, Technische Universität Berlin, Berlin, Germany
e-mail: peter.friz@math.tu-berlin.de

P.K. Friz
Weierstraß-Institut für Angewandte Analysis und Stochastik, Berlin, Germany

P. Laurence
Dipartimento di Matematica, Università di Roma 1 Piazzale Aldo Moro 2, 00185 Rome, Italy

P. Laurence
Courant Institute of Mathematical Sciences, New York University, 251 Mercer Street, New York, NY 10012, USA

P.K. Friz et al. (eds.), *Large Deviations and Asymptotic Methods in Finance*, Springer Proceedings in Mathematics & Statistics 110,
DOI 10.1007/978-3-319-11605-1_16

1 Introduction

As is well known, the sum of independent log-normal variable does not admit a closed-form density. And yet, there are countless applications in Finance and Actuarial Mathematics where such sums play a crucial role, consider for instance the law of a Black–Scholes basket B at time T, i.e. the weighted average of d geometric Brownian motions.

As a consequence, there is a natural interest in approximations and expansions, see e.g. [9] and the references therein. This article contains a detailed investigation in small volatility and short time regimes. Forthcoming work of A. Gulisashvili and P. Tankov [12] deals with tail asymptotics. Our methods are not restricted to the geometric Brownian motion case: in principle, each Black–Scholes component could be replaced by the asset price in a stochastic volatility model, such as the the Stein–Stein model [16], with full correlation between all assets and their volatilities. In the end, explicit solutions only depend on the analytical tractability of a system of ordinary differential equations. If such tractability is not given, one can still proceed with numerical ODE solvers.

As a matter of fact, our aim here is not to push the generality in which our methods work: one can and should expect involved answers in complicated models. Rather, our main—and somewhat surprising—insight is that *unexpected phenomena* are already present in the *simplest possible setting*: to this end, our first focus will be on the case of $d = 2$ independent Black–Scholes assets, without drift and correlation, with unit spot and unit volatility). To be more specific, if C_B denotes the fair value of an (out-of-money) call option on the basket B struck at K, one naturally expects, for a small maturity T,

$$\frac{\partial^2}{\partial K^2} C_B(K,T) \sim (\text{const}) \exp\left(-\frac{\Lambda(K)}{T}\right)\frac{1}{\sqrt{T}}.$$

And yet, while true for *most* strikes, it fails for $K = K^*$; in fact,

$$\left\{\frac{\partial^2}{\partial K^2} C_B(K,T)\right\}_{K=K^*} \sim (\text{const}) \exp\left(-\frac{\Lambda(K^*)}{T}\right)\frac{1}{T^{3/4}}.$$

To the best of our knowledge, and despite the seeming triviality of the situation (two independent Black–Scholes assets!), the existence of a "special" strike level K^*, at which the value of a basket option (here: butterfly spread[1]) has a "special" decay behavior, as maturity approaches 0, seems to be new. There are different proofs of this fact; the most elementary argument—based on the analysis of a convolution integral—is given in Sect. 2. However, this approach—while telling us *what* happens—does not tell us *how* it happens.

The main contribution of this note is precisely a good understanding of the latter. In fact, there is clear picture that comes with K^*. For $K < K^*$ and conditional on the

[1] Extensions to spreads and vanilla options are possible and will be discussed elsewhere.

option to expire on the money, there is a unique "most likely" path around which the underlying asset price process will concentrate as maturity approaches 0. For $K > K^*$, however, this ceases to be true: there will be two distinct (here: equally likely) paths around which concentration occurs. What underlies this interpretation is that large deviation theory not only characterizes the probability of unlikely events (such as expiration in-the-money, if presently out-of-the-money, as time to maturity goes to zero) but also the mechanism via which these events can occur. Such understanding was already crucial in previous works on baskets aiming at quantification of basket (implied vol) skew relative to its components, starting with [1, 2]. As a matter of fact, the analysis in these papers relied on the statement that "generically there is a unique arrival point (of a unique energy minimizing path) on the (basket-strike) arrival manifold". The situation, however, even in the Black–Scholes model, is more involved. And indeed, we shall establish existence of a critical strike K^*, at which one sees the phase-transition from one to two energy minimizing, "most likely", paths.[2] And this information will have meaning to traders (as long as they believe in a diffusion model as maturity approaches 0, which may or may not be a good idea …) as it tells them the possible scenarios in which an out-of-the money basket option may still expire in the money.

Let us conclude this introduction with a few technical notes. We view the evolution of the basket price—even in the Black–Scholes model—as a stochastic volatility evolution model; by which we mean $dB_t/B_t = \sigma(t, \omega)dW_t$ (as opposed to a local vol evolution where $\sigma = \sigma(t, B_t)$). This should explain why the methods developed in Part I of [6, 7] for the analysis of stochastic volatility models (then used in Part II, [7], to solve the concrete smile problem (shape of the wings) for the correlated Stein–Stein model), are also adequate for the analysis of baskets.

2 Computations Based on Saddle-Point Method

In terms of a standard d-dimensional Wiener process $\left(W^1, \ldots, W^d\right)$,

$$B_T = \sum_{i=1}^{d} S_0^i \exp\left(\mu^i T + \sigma^i W_T^i\right).$$

Write $f = f_T(K)$ for the probability density function of B_T; i.e. for $\mathbb{P}\left[B_T \in [K, K + dK]\right]/dK$. Of course, it is given by some $(d-1)$-dimensional convolution integral, explicit asymptotic expansions are—in principle—possible with the

[2]It can be shown that, sufficiently close to the arrival manifold, there is in fact a unique energy minimizing paths. The (near-the-money) analysis of [1, 2] is then justified.

saddle point method. It will be enough for our purposes to illustrate the method in the afore-mentioned simplest possible setting:

$$d = 2,\ S_0^1 = S_0^2 = 1,\ \mu^1 = \mu^2 = 0,\ \sigma^1 = \sigma^2 = 1.$$

In other words, $B_T = \exp\left(W_T^1\right) + \exp\left(W_T^2\right)$. We claim that for some constant $c_0 = c_0(K) > 0$

$$f(K) = \begin{cases} \exp\left(-\frac{\Lambda(K)}{T}\right) \frac{1}{\sqrt{T}} \left(c_0 + O(T)\right), & \text{when } K \neq K^*, \quad \text{(1a)} \\ \exp\left(-\frac{\Lambda(K^*)}{T}\right) \frac{1}{T^{3/4}} \left(c_0 + O(T)\right), & \text{when } K = K^*, \quad \text{(1b)} \end{cases} \tag{1}$$

with

$$K^* = 2e \approx 5.43656$$

and

$$\Lambda(K) = \inf\{h_K(x) \mid x \in [0, K]\}$$

with

$$h_K(x) := (\log x)^2 + (\log(K - x))^2. \tag{2}$$

Note that for $K \leq K^*$ we can explicitly solve this minimization problem and obtain $\Lambda(K) = \log(K/2)^2$ with corresponding minimizer $x^* = K/2$, corresponding to the single local extremum of h_K. For $K > K^*$, we have two global minima, which cannot be given in closed form, and hence $\Lambda(K)$ can only be computed numerically (Fig. 1).

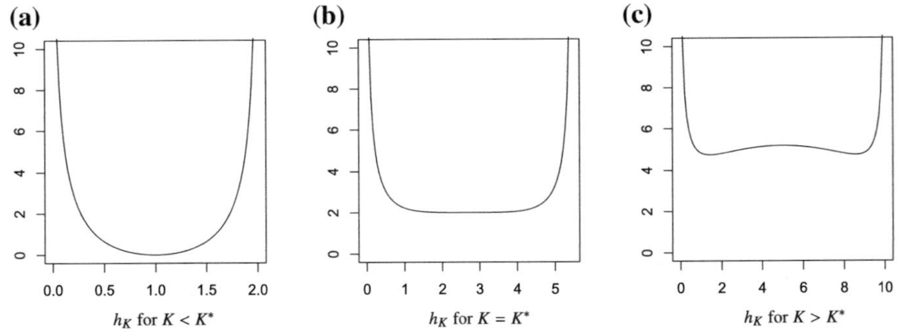

Fig. 1 Plot of h_K for different choices of K. **a** For $K < K^*$ there is a unique global minimum at $x^* = K/2$ which is non-degenerate in the sense that $h''(x^*) > 0$. **b** For $K = K^*$ there is a unique global minimum at $x = K/2$ which is degenerate in the sense that $h''(x^*) = 0$. **c** For $K > K^*$, $x = K/2$ gives a local maximum. There are two symmetric global minimizers, which are not given in closed form

The stock price S^i_T has a log-normal distribution with parameters $\mu^i = 0$ and $\xi^i = \sigma^i \sqrt{T} = \sqrt{T}$, where the density of the log-normal distribution is given by

$$f_{\mu,\xi}(x) = \frac{1}{\sqrt{2\pi}\xi x} \exp\left(-\frac{(\log x - \mu)^2}{2\xi^2} \right). \tag{3}$$

Obviously, the density of the sum of these two independent log-normal random variables satisfies

$$f(K) = \int_0^K f_{\mu^1,\xi^1}(K - x) f_{\mu^2,\xi^2}(x) dx. \tag{4}$$

Using our special parameters, the integrand is of the form

$$f_{\mu^1,\xi^1}(K - x) f_{\mu^2,\xi^2}(x) = \frac{1}{2\pi T x(K - x)} \exp\left(-\frac{h_K(x)}{2T} \right).$$

In order to apply the Laplace approximation to (4), we compute the minimizer for h_K, which is found by the first order condition

$$h'_K(x) = 0 \iff \frac{\log x}{x} - \frac{\log(K - x)}{K - x} = 0. \tag{5}$$

Clearly, this equation is solved by choosing $x^* = K/2$—which is the unique global minimizer iff $K \leq 2e$ and a local maximizer otherwise, in which case we have two global minima $x^*_1 < K/2 < x^*_2$. Assuming $K \leq 2e$, we can check degeneracy of that minimum directly by computing

$$h''_K(x^*) = h''_K(K/2) = 16 \frac{1 - \log(K/2)}{K^2}.$$

and

$$h''_K(x^*) = 0 \iff K = 2e. \tag{6}$$

With more work one can see that also the global minima x^*_1, x^*_2, in the case $K > 2e$, are non-degenerate. Hence, whenever $K \neq 2e$ a standard Laplace method leads to the expansion (1)a. In the remainder of this section, we consider the degenerate case and establish (1)b.

Choosing $K = 2e$ and, correspondingly, $x^* = e$, we obtain the Taylor expansion $h_K(x) = h_K(x^*) + \frac{h_K^{(4)}(x^*)}{24}(x - x^*)^4 + O((x - x^*)^5)$, with $h_K(x^*) = 2$ and $h_K^{(4)}(x^*) = 20e^{-4}$, we obtain the Laplace approximation

$$\begin{aligned} f(K) &= \int_0^K \frac{1}{2\pi T(K-x)x} \exp\left(-\frac{h_K(x)}{2T}\right) dx \\ &= \frac{1}{2\pi T e^2} \int_0^K \exp\left(-\frac{1}{T}\right) \exp\left(-\frac{5e^{-4}(x - K/2)^4}{12T}\right) dx\,(1 + O(T)) \\ &= \frac{3^{1/4}\Gamma(1/4)}{5^{1/4} 2\sqrt{2}\pi e} \exp\left(-\frac{1}{T}\right) \frac{1}{T^{3/4}}\,(1 + O(T)), \end{aligned}$$

where we used

$$\int_{-\infty}^{\infty} \exp(-\alpha x^4) dx = \frac{\Gamma(1/4)}{2\alpha^{1/4}}, \quad \alpha > 0.$$

Thus, we arrive at (1)b.

3 Large Deviations Approach

Our main tool here are novel marginal density expansions in small-noise regime [6]. This was used in order to compute the large-strike behavior of implied volatility in the correlated Stein–Stein model; [11, 16].[3]

In fact, the technical assumptions of [6] were satisfied in the analysis of the Stein–Stein model whereas in the (seemingly) trivial case of two IID Black-Scholes assets, the technical assumptions of [6] are indeed violated for a critical strike $K = K^*$. The necessity of this condition is then highlighted by the fact, as was seen in the previous section,

$$\left\{\frac{\partial^2}{\partial K^2} C_B(K, T)\right\}_{K=K^*} \not\sim \text{(const)} \exp\left(-\frac{\Lambda(K^*)}{T}\right) \frac{1}{T^{1/2}}.$$

The computation of K^* can be achieved either via a geometric construction borrowed from Riemannian geometry, which relies on the Weingarten map, or by some (fairly) elementary analysis of a system of Hamiltonian ODEs. In fact, the Hamiltonian point of view extends naturally when one introduces correlation, local and even stochastic volatility. Explicit answers then depend on the analytical tractability of these (boundary value) ODE problems. (Of course, the numerical solution of such problems is well-known.)

---

[3] Similar investigations have recently been conducted in the Heston model; [10, 13] and the references therein.

In the following, we review [6]. Consider a d-dimensional diffusion $\left(X_t^\varepsilon\right)_{t\geq 0}$ given by the stochastic differential equation

$$dX_t^\varepsilon = b\left(\varepsilon, X_t^\varepsilon\right) dt + \varepsilon\sigma\left(X_t^\varepsilon\right) dW_t, \quad \text{with } X_0^\varepsilon = x_0^\varepsilon \in \mathbb{R}^d, \tag{7}$$

and where $W = (W^1, \dots, W^m)$ is an m-dimensional Brownian motion. Unless otherwise stated, we assume $b : [0, 1) \times \mathbb{R}^d \to \mathbb{R}^d$, $\sigma = (\sigma_1, \dots, \sigma_m) : \mathbb{R}^d \to \mathrm{Lin}\left(\mathbb{R}^m, \mathbb{R}^d\right)$ and $x_0^{\cdot} : [0, 1) \to \mathbb{R}^d$ to be smooth, bounded with bounded derivatives of all orders. Set $\sigma_0 = b(0, \cdot)$ and assume that, for every multi-index α, the drift vector fields $b(\varepsilon, \cdot)$ converges to σ_0 in the sense[4]

$$\partial_x^\alpha b(\varepsilon, \cdot) \to \partial_x^\alpha b(0, \cdot) = \partial_x^\alpha \sigma_0(\cdot) \text{ uniformly on compacts as } \varepsilon \downarrow 0. \tag{8}$$

We shall also assume that

$$\partial_\varepsilon b(\varepsilon, \cdot) \to \partial_\varepsilon b(0, \cdot) \text{ uniformly on compacts as } \varepsilon \downarrow 0 \tag{9}$$

and

$$x_0^\varepsilon = x_0 + \varepsilon \hat{x}_0 + o(\varepsilon) \text{ as } \varepsilon \downarrow 0. \tag{10}$$

Theorem 1 (Small noise) *Let* (X^ε) *be the solution process to*

$$dX_t^\varepsilon = b\left(\varepsilon, X_t^\varepsilon\right) dt + \varepsilon\sigma\left(X_t^\varepsilon\right) dW_t, \quad \text{with } X_0^\varepsilon = x_0^\varepsilon \in \mathbb{R}^d.$$

Assume $b(\varepsilon, \cdot) \to \sigma_0(\cdot)$ *in the sense of (8), (9), and* $X_0^\varepsilon \equiv x_0^\varepsilon \to x_0$ *as* $\varepsilon \to 0$ *in the sense of (10). Assume non-degeneracy of* σ *in the sense that* $\sigma.\sigma^T$ *is strictly positive definite everywhere in space.*[5] *Fix* $y \in \mathbb{R}^l$, $N_y := (y, \cdot)$ *and let* $\mathcal{K}_y$ *be the the space of all* $h \in H$*, the Cameron-Martin space of absolutely continuous paths with derivatives in* $L^2([0, T], \mathbb{R}^m)$*, s.t. the solution to*

$$d\phi_t^h = \sigma_0\left(\phi_t^h\right) dt + \sum_{i=1}^m \sigma_i\left(\phi_t^h\right) dh_t^i, \ \phi_0^h = x_0 \in \mathbb{R}^d$$

satisfies $\phi_T^h \in N_y$*. In a neighborhood of* y*, assume smoothness of*[6]

$$\Lambda(y) = \inf\left\{\frac{1}{2}\|h\|_H^2 : h \in \mathcal{K}_y\right\}.$$

---

[4]If (7) is understood in Stratonovich sense, so that dW is replaced by $\circ dW$, the drift vector field $b(\varepsilon, \cdot)$ is changed to $\tilde{b}(\varepsilon, \cdot) = b(\varepsilon, \cdot) - \left(\varepsilon^2/2\right)\sum_{i=1}^m \sigma_i \cdot \partial\sigma_i$. In particular, σ_0 is also the limit of $\tilde{b}(\varepsilon, \cdot)$ in the sense of (8).

[5]This may be relaxed to a weak Hoermander condition with an explicit controllability condition.

[6]If $\#\mathcal{K}_y^{\min} = 1$ smoothness of the energy can be shown and need not be assumed; [6]. Note also that in our application to tail asymptotics, with θ-scaling, $\theta \in \{1, 2\}$, the energy must be linear resp. quadratic (by scaling) and hence smooth.

Assume also (i) there are only finitely many minimizers, i.e. $\mathcal{K}_{\mathrm{y}}^{\min} < \infty$ *where*

$$\mathcal{K}_{\mathrm{y}}^{\min} := \left\{ \mathrm{h}_0 \in \mathcal{K}_{\mathrm{y}} : \frac{1}{2} \|\mathrm{h}_0\|_H^2 = \Lambda(\mathrm{y}) \right\};$$

(ii) x_0 *is non-focal for* N_{y} *in the sense of* [6]. *(We shall review below how to check this.) Then there exists* $c_0 = c_0(\mathrm{x}_0, \mathrm{y}, T) > 0$ *such that*

$$\mathrm{Y}_T^{\varepsilon} = \Pi_l \mathrm{X}_T^{\varepsilon} = \left(X_T^{\varepsilon,1}, \ldots, X_T^{\varepsilon,l} \right), \quad 1 \leq l \leq d,$$

admits a density with expansion

$$f_{\varepsilon}(\mathrm{y}, T) = e^{-\frac{\Lambda(\mathrm{y})}{\varepsilon^2}} e^{\frac{\max\left\{\Lambda'(\mathrm{y}) \cdot \hat{\mathrm{Y}}_T(\mathrm{h}_0) : \mathrm{h}_0 \in \mathcal{K}_{\mathrm{y}}^{\min}\right\}}{\varepsilon}} \varepsilon^{-l} \left(c_0 + O(\varepsilon)\right) \text{ as } \varepsilon \downarrow 0,$$

where Λ' *denotes the gradient of* Λ.

Here $\hat{\mathrm{Y}} = \hat{\mathrm{Y}}(\mathrm{h}_0) = \left(\hat{Y}^1, \ldots, \hat{Y}^l\right)$ *is the projection,* $\hat{\mathrm{Y}} = \Pi_l \hat{\mathrm{X}}$, *of the solution to the following (ordinary) differential equation*

$$\begin{aligned} d\hat{\mathrm{X}}_t &= \left(\partial_x b\left(0, \phi_t^{\mathrm{h}_0}(\mathrm{x}_0)\right) + \partial_x \sigma(\phi_t^{\mathrm{h}_0}(\mathrm{x}_0)) \dot{\mathrm{h}}_0(t)\right) \hat{\mathrm{X}}_t dt + \partial_{\varepsilon} b\left(0, \phi_t^{\mathrm{h}_0}(\mathrm{x}_0)\right) dt, \qquad (11) \\ \hat{\mathrm{X}}_0 &= \hat{\mathrm{x}}_0. \end{aligned}$$

Remark 2 (Localization) The assumptions on the coefficients b, σ in Theorem 1 (smooth, bounded with bounded derivatives of all orders) are typical in this context (cf. Ben Arous [3, 4] for instance) but rarely met in practical examples from finance. This difficulty can be resolved by a suitable localization. For instance, as detailed in [6], an estimate of the form

$$\lim_{R \to \infty} \limsup_{\varepsilon \to 0} \varepsilon^2 \log \mathbb{P}[\tau_R \leq T] = -\infty. \qquad (12)$$

with $\tau_R := \inf\left\{t \in [0, T] : \sup_{s \in [0,t]} \left|\mathrm{X}_s^{\varepsilon}\right| \geq R\right\}$ will allow to bypass the boundedness assumptions.

3.1 Short Time Asymptotics

The reduction of *short time expansions* to small noise expansions by Brownian scaling is classical. In the present context, we have the following statement, taken from [6, Sect. 2.1].

Corollary 3 (Short time) *Consider* $dX_t = b(X_t)\,dt + \sigma(X_t)\,dW$, *started at* $X_0 = x_0 \in \mathbb{R}^d$, *with* C^∞*-bounded vector fields which are non-degenerate in the sense that* $\sigma.\sigma^T$ *is strictly positive definite everywhere in space. Fix* $y \in \mathbb{R}^l$, $N_y := (y, \cdot)$ *and assume (i), (ii) as in Theorem 1. Let* $f(t, \cdot) = f(t, y)$ *be the density of* $Y_t = (X_t^1, \ldots, X_t^l)$. *Then*

$$f(t, y) \sim (const)\frac{1}{t^{l/2}} \exp\left(-\frac{d^2(x_0, y)}{2t}\right) \text{ as } t \downarrow 0$$

where $d(x_0, y)$ *is the sub-Riemannian distance, based on* $(\sigma_1, \ldots, \sigma_m)$, *from the point* x_0 *to the affine subspace* N_y.

3.2 Computational Aspects

We present here the *mechanics* of the actual computations, in the spirit of the Pontryagin maximum principle (e.g. [15]). For details we refer to [6].

- **The Hamiltonian**. Based on the SDE (7), with diffusion vector fields $\sigma_1, \ldots, \sigma_m$ and drift vector field σ_0 (in the $\varepsilon \to 0$ limit) we define the *Hamiltonian*

$$\begin{aligned}\mathcal{H}(x, p) &:= \langle p, \sigma_0(x)\rangle + \frac{1}{2}\sum_{i=1}^{m} \langle p, \sigma_i(x)\rangle^2 \\ &= \langle p, \sigma_0(x)\rangle + \frac{1}{2}\left\langle p, \left(\sigma\sigma^T\right)(x)\, p\right\rangle.\end{aligned}$$

  Remark the driving Brownian motions $W^1, \ldots, W^m$ were assumed to be independent. Many stochastic models, notably in finance, are written in terms of correlated Brownian motions, i.e. with a non-trivial correlation matrix $\Omega = \left(\omega^{i,j} : 1 \le i, j \le m\right)$, where $d\left\langle W^i, W^j\right\rangle_t = \omega^{i,j}dt$. The Hamiltonian then becomes

$$\mathcal{H}(x, p) = \langle p, \sigma_0(x)\rangle + \frac{1}{2}\left\langle p, \left(\sigma\Omega\sigma^T\right)(x)\, p\right\rangle. \tag{13}$$

- **The Hamiltonian ODEs**. The following system of ordinary differential equations,

$$\begin{pmatrix} \dot{x}(t) \\ \dot{p}(t) \end{pmatrix} = \begin{pmatrix} \partial_p \mathcal{H}(x(t), p(t)) \\ -\partial_x \mathcal{H}(x(t), p(t)) \end{pmatrix}, \tag{14}$$

  gives rise to a solution flow, denoted by $H_{t\leftarrow 0}$, so that

$$H_{t\leftarrow 0}(x_0, p_0)$$

is the unique solution to the above ODE with initial data $(\mathrm{x}_0, \mathrm{p}_0)$. Our standing (regularity) assumption are more than enough to guarantee uniqueness and local ODE existence. As in [5, p. 37], the vector field $(\partial_{\mathrm{p}}\mathcal{H}, -\partial_{\mathrm{x}}\mathcal{H})$ is complete, i.e., one has global existence. It can be useful to start the flow backwards with time-T terminal data, say $(\mathrm{x}_T, \mathrm{p}_T)$; we then write

$$\mathrm{H}_{t \leftarrow T}(\mathrm{x}_T, \mathrm{p}_T)$$

for the unique solution to (14) with given time-T terminal data. Of course,

$$\mathrm{H}_{t \leftarrow T}(\mathrm{H}_{T \leftarrow 0}(\mathrm{x}_0, \mathrm{p}_0)) = \mathrm{H}_{t \leftarrow 0}(\mathrm{x}_0, \mathrm{p}_0).$$

- **Solving the Hamiltonian ODEs as boundary value problem**. Given the target manifold $N_{\mathrm{a}} = (\mathrm{a}, \cdot)$, the analysis in [6] requires solving the Hamiltonian ODEs (14) with mixed initial-, terminal—and transversality conditions,

$$\begin{aligned} \mathrm{x}(0) &= \mathrm{x}_0 \in \mathbb{R}^d, \\ \mathrm{x}(T) &= (\mathrm{y}, \cdot) \in \mathbb{R}^l \oplus \mathbb{R}^{d-l}, \\ \mathrm{p}(T) &= (\cdot, 0) \in \mathbb{R}^l \oplus \mathbb{R}^{d-l}. \end{aligned} \tag{15}$$

Note that this is a $2d$-dimensional system of ordinary differential equations, subject to $d + l + (d - l) = 2d$ conditions. In general, boundary problems for such ODEs may have more than one, exactly one or no solution. In the present setting, there will always be one or more than one solution. After all, we know by [6] that there exists at least one minimizing control h_0 and that can be reconstructed via the solution of the Hamiltonian ODEs, as explained in the following step.

- **Finding the minimizing controls**. The Hamiltonian ODEs, as boundary value problem, are effectively first order conditions (for minimality) and thus yield *candidates* for the minimizing control $\mathrm{h}_0 = \mathrm{h}_0(\cdot)$, given by

$$\dot{\mathrm{h}}_0 = \begin{pmatrix} \langle \sigma_1(\mathrm{x}(\cdot)), \mathrm{p}(\cdot)\rangle \\ \cdots \\ \langle \sigma_m(\mathrm{x}(\cdot)), \mathrm{p}(\cdot)\rangle \end{pmatrix}. \tag{16}$$

Each such candidate is indeed admissible in the sense $\mathrm{h}_0 \in \mathcal{K}_{\mathrm{a}}$ but may fail to be a minimizer. We thus compute the energy $\|\mathrm{h}_0\|_H^2 = \mathcal{H}(x_0, p_0)$ for each candidate and identify those ("$\mathrm{h}_0 \in \mathcal{K}_{\mathrm{a}}^{\min}$") with minimal energy. The procedure via Hamiltonian flows also yields a unique $\mathrm{p}_0 = \mathrm{p}_0(\mathrm{h}_0)$. If $\sigma_0 = 0$—as in our case—the energy is equal to $\mathcal{H}(x_0, p_0)$, otherwise the formula is slightly more complicated.

- **Checking non-focality**. By definition [6], x_0 is **non-focal** for $N = (\mathrm{y}, \cdot)$ along $\mathrm{h}_0 \in \mathcal{K}_{\mathrm{a}}^{\min}$ in the sense that, with $(\mathrm{x}_T, \mathrm{p}_T) := \mathrm{H}_{T \leftarrow 0}(\mathrm{x}_0, \mathrm{p}_0(\mathrm{h}_0)) \in \mathcal{T}^*\mathbb{R}^d$,

$$\partial_{(\mathfrak{z},\mathfrak{q})}|_{(\mathfrak{z},\mathfrak{q})=(0,0)}\pi \mathrm{H}_{0\leftarrow T}\left(\mathrm{x}_T + \begin{pmatrix} 0 \\ \mathfrak{z} \end{pmatrix}, \mathrm{p}_T + (\mathfrak{q}, 0)\right)$$

is non-degenerate (as $d \times d$ matrix; here we think of $(\mathfrak{z}, \mathfrak{q}) \in \mathbb{R}^{d-l} \times \mathbb{R}^l \cong \mathbb{R}^d$ and recall that π denotes the projection from $\mathcal{T}^*\mathbb{R}^d$ onto $\mathbb{R}^d$; in coordinates $\pi(\mathrm{x}, \mathrm{p}) = \mathrm{x}$). Note that in the point-point setting, $\mathrm{x}_T = \mathrm{y}$ is fixed and only perturbations of the arrival "velocity" p_T—without restrictions, i.e. without transversality condition—are considered. Non-degeneracy of the resulting map should then be called **non-conjugacy** (between two points; here: x_T and x_0). In the absence of the drift vector field σ_0, this is consistent with the usual meaning of non-conjugacy; after identifying tangent- and cotangent-space $\partial_{\mathfrak{q}}|_{\mathfrak{q}=0}\pi \mathrm{H}_{0\leftarrow T}$ is precisely the differential of the exponential map.

- **The explicit marginal density expansion**. We then have

$$f^{\varepsilon}(\mathrm{y}, T) = e^{-c_1/\varepsilon^2} e^{c_2/\varepsilon} \varepsilon^{-l} \left(c_0 + O(\varepsilon)\right) \text{ as } \varepsilon \downarrow 0.$$

with $c_1 = \Lambda(\mathrm{y})$. The second-order exponential constant c_2 then requires the solution of a finitely many ($\#\mathcal{K}_{\mathrm{a}}^{\min} < \infty$) auxiliary ODEs, cf. Theorem 1.

4 Analysis of the Black–Scholes Basket

For a general multi-dimensional Black-Scholes model, we have a Hamiltonian

$$\mathcal{H}(x, p) = \frac{1}{2}\left\langle p, (\sigma(x)\Omega\sigma(x)^T)p\right\rangle,$$

with $\sigma(x) = (\sigma^1 x^1, \ldots, \sigma^m x^m)$. While the corresponding Hamiltonian ODEs can be solved in closed form, the boundary conditions lead to systems of non-linear equations, which we cannot solve explicitly any more. While numerical solutions are, of course, possible, we restrict ourselves to the extremely simple setting of Sect. 2, in order to keep maximal tractability.

Consequently, we have the Hamiltonian $\mathcal{H}(x, p) = \frac{1}{2}\left((\sigma x^1 p^1)^2 + (\sigma x^2 p^2)^2\right)$. The solutions of the Hamiltonian ODEs started at (x_0, p_0) satisfy

$$H_{t\leftarrow 0}(x_0, p_0) = \begin{pmatrix} x_0^1 e^{\sigma^2 x_0^1 p_0^1 t} \\ x_0^2 e^{\sigma^2 x_0^2 p_0^2 t} \\ p_0^1 e^{-\sigma^2 x_0^1 p_0^1 t} \\ p_0^2 e^{-\sigma^2 x_0^2 p_0^2 t} \end{pmatrix}, \tag{17}$$

which can be easily seen from the observation that $\mathcal{H}$ is constant along solutions of the Hamiltonian ODEs together with symmetry between (x^1, p^1) and (x^2, p^2). This immediately implies that the inverse flow is given by

$$H_{0\leftarrow t}(x_t, p_t) = \begin{pmatrix} x_t^1 e^{-\sigma^2 x_t^1 p_t^1 t} \\ x_t^2 e^{-\sigma^2 x_t^2 p_t^2 t} \\ p_t^1 e^{\sigma^2 x_t^1 p_t^1 t} \\ p_t^2 e^{\sigma^2 x_t^2 p_t^2 t} \end{pmatrix}. \tag{18}$$

Now we introduce the boundary conditions. Note that, contrary to Theorem 1, we now project to the linear subspace $\{x : x^1 + x^2 = K\}$. Thus, the terminal condition on x translates into $x_T^1 + x_T^2 = K$—we need to end at the target manifold—, whereas the transversality condition translates to p_T being orthogonal to the target manifold. Evaluating these conditions at $T = 1$, we get

$$\begin{aligned} x_0^1 &= S_0^1 = 1, \\ x_0^2 &= S_0^2 = 1, \\ x_1^1 + x_1^2 &= K, \\ p_1^1 - p_1^2 &= 0. \end{aligned}$$

It is a pleasant exercise to check that solving for $x_1^1 =: x$ and $x_x^2 = K - x$ then leads exactly to the first order condition (5) encountered in Sect. 2. With identical arguments, assuming $K \le 2e$ from here on (and disregarding the case $K > 2e$ where closed form computations are not available), we find that the optimal configuration must satisfy $x_1^* = (K/2, K/2)$. Inserting this value into the first two components of (17), we obtain the equation

$$\frac{K}{2} = e^{\sigma^2 p_0^i} \iff p_0^i = \log\left(\frac{K}{2}\right)/\sigma^2, \quad i = 1, 2.$$

This implies that $p_1^* = \left(\frac{2}{\sigma^2 K}\log(K/2), \frac{2}{\sigma^2 K}\log(K/2)\right)$. Moreover, we see that the minimizing control satisfies

$$\dot{h}_0(t) = \begin{pmatrix} \sigma x^1(t) p^1(t) \\ \sigma x^2(t) p^2(t) \end{pmatrix} = \begin{pmatrix} \sigma p_0^1 \\ \sigma p_0^2 \end{pmatrix} = \begin{pmatrix} \frac{\log(K/2)}{\sigma} \\ \frac{\log(K/2)}{\sigma} \end{pmatrix}, \tag{19}$$

see (16), implying that the minimal energy is given by

$$\Lambda(K) = \frac{1}{2}\|h_0\|_H^2 = \frac{\log(K/2)^2}{\sigma^2} = \mathcal{H}(x_0, p_0). \tag{20}$$

Regarding focality, we have to check that the matrix:

$$M(x_1, p_1) := \begin{pmatrix} \frac{\partial}{\partial\epsilon}\Big|_{\epsilon=0} H^1_{0\leftarrow 1}(x_1 + \epsilon(1,-1), p_1) & \frac{\partial}{\partial\eta}\Big|_{\eta=0} H^1_{0\leftarrow 1}(x_1, p_1 + \eta(1,1)) \\ \frac{\partial}{\partial\epsilon}\Big|_{\epsilon=0} H^2_{0\leftarrow 1}(x_1 + \epsilon(1,-1), p_1) & \frac{\partial}{\partial\eta}\Big|_{\eta=0} H^2_{0\leftarrow 1}(x_1, p_1 + \eta(1,1)) \end{pmatrix} \tag{21}$$

is non-degenerate when evaluated at the optimal configuration (x_1^*, p_1^*). A simple calculation shows that

$$M(x_1, p_1) = \begin{pmatrix} e^{-\sigma^2 x_1^1 p_1^1} - x_1^1 p_1^1 \sigma^2 e^{-\sigma^2 x_1^1 p_1^1} & -\sigma^2 (x_1^1)^2 e^{-\sigma^2 x_1^1 p_1^1} \\ -e^{-\sigma^2 x_1^2 p_1^2} + x_1^2 p_1^2 \sigma^2 e^{-\sigma^2 x_1^2 p_1^2} & -\sigma^2 (x_1^2)^2 e^{-\sigma^2 x_1^2 p_1^2} \end{pmatrix},$$

implying that

$$M(x_1^*, p_1^*) = \begin{pmatrix} \frac{2}{K}(1 - \log(K/2)) & -\frac{\sigma^2 K}{2} \\ \frac{2}{K}(-1 + \log(K/2)) & -\frac{\sigma^2 K}{2} \end{pmatrix},$$

and we can conclude that

$$\det M(x_1^*, p_1^*) = 2\sigma^2 \left(\log(K/2) - 1\right),$$

which is zero if and only if $K = 2e$. We summarize the results of this calculation as follows:

- In the generic case $K \neq 2e$, the non-focality condition of Theorem 1 holds true, and we obtain (from Corollary 3) the following (short time) density expansion of $B_T = \exp(\sigma W_T^1) + \exp(\sigma W_T^2)$, expansion

$$K \mapsto \exp\left(-\frac{\Lambda(K)}{T}\right) \frac{1}{\sqrt{T}} \left(c_0 + O(T)\right)$$

  When specialized to unit volatility, we recover precisely (1)a.
- For $K = 2e$, the initial stock price is focal for the minimizing configuration, so the non-focality condition of Theorem 1 fails. And indeed, we *want* it to fail for the actual expansion in this case, namely (1)b, is not at all of the generic form predicted by our theorem.

Remark 4 It is immediate to use this analysis to deal also with the case of non-unit (but identical) spots $S_0^1 = S_0^2$ by scaling the Black-Scholes dynamics accordingly, i.e., by replacing K with K/S_0^1. Hence, in this case focality happens when $\log\left(\frac{K}{2S_0^1}\right) = 1$, i.e., when $K = 2S_0^1 e$.

Remark 5 The question arises if the critical ("focal") case $K = 2e$, with atypical algebraic factor $T^{-3/4}$ cf. (1)b, can also be recovered by a general theorem. Related results in [14] and also [17] suggests that this may indeed be the case but would require substantial additional work.

5 Extensions: Correlation, Local and Stochastic Vol

5.1 Analysis of the Black–Scholes Basket, Small Noise

In Sect. 4 we analyzed the density of a simple Black–Scholes basket with dynamics

$$dB_t = S_t^1 \sigma dW_t^1 + S_t^2 \sigma dW_t^2.$$

As explained in Sect. 3 the analysis is really based on a small noise (small vol) expansion of

$$dB_t^\epsilon = S_t^{1,\epsilon} \sigma \epsilon dW_t^1 + S_t^{2,\epsilon} \sigma \epsilon dW_t^2,$$

run til time $T = 1$. Consider now a situation with small rates, also of order ϵ. In other words,

$$dS_t^{i,\epsilon} = r S_t^{i,\epsilon} \epsilon dt + S_t^{i,\epsilon} \sigma \epsilon dW^i,$$

and then $B_t^\epsilon = S_t^{1,\epsilon} + S_t^{2,\epsilon}$ as before. We still assume $S_0^i = 1$. A look at Theorem 1 (now we cannot use Corollary 3) reveals that the entire leading order computation remains unchanged (at least at unit time and with trivial changes otherwise). The resulting (now: small noise) density expansion of $B_T^\epsilon|_{T=1}$ is more involved and takes the form

$$K \mapsto \exp\left(-\frac{\Lambda(K)}{\epsilon^2}\right) \exp\left(\frac{2r\log(K/2)}{\sigma^2 \log(2)\epsilon}\right) \frac{1}{\epsilon} (c_0 + O(\epsilon)). \tag{22}$$

Here $\Lambda(K)$ is given in closed form, cf. (20), so that $\Lambda'(K) = \frac{2\log(K/2)}{\sigma^2 K}$ is also explicitly known. Furthermore, under similar restrictions on K as before, h_0 is (still) given by (19), so that

$$\phi_t^{h_0} = \begin{pmatrix} (K/2)^t \\ (K/2)^t \end{pmatrix}.$$

Thus, the ODE for $\hat{X}$ (see Theorem 1) is given by

$$\frac{d\hat{X}_t}{dt} = \log(K/2)\hat{X}_t + r\begin{pmatrix} (K/2)^t \\ (K/2)^t \end{pmatrix}, \quad \hat{X}_0 = \hat{x}_0 = 0,$$

which has the solution

$$\hat{X}_t^i = r\left(1 - \left(\frac{1}{2}\right)^t\right)\frac{K^t}{\log 2},$$

implying that $\hat{Y}_1 = \hat{X}_1^1 + \hat{X}_1^2 = rK/\log(2)$. Thus, the second exponential term has the form given above.

5.2 Basket Analysis Under Local, Stochastic Vol etc.

One can immediately write down the Hamiltonian associated to, say two, or $d > 2$ assets, each of which is governed by local vol dynamics or stochastic vol, based on additional factors. In general, however, one will be stuck with the analysis of the resulting boundary value problem for the Hamiltonian ODEs; numerical (e.g. shooting) methods will have to be used. In some models, including the Stein–Stein model, we believe (due to the analysis carried out in [7]) that, in special cases, closed form answers are possible but we will not pursue this here. Instead, we continue with a few more computation in the Black–Scholes case for d assets.

5.3 Multi-variate Black–Scholes Models

In the multi-variate case $d > 2$ of a general, d-dimensional Black Scholes model with correlation matrix (ρ_{ij}), the Hamiltonian has the form

$$\mathcal{H}(x,p) = \frac{1}{2}\sum_{i,j=1}^{d} \rho_{ij}\sigma^i p^i x^i \sigma^j x^j p^j.$$

Thus, the Hamiltonian ODEs have the form

$$\dot{x}^l = \sigma^l x^l \sum_{i=1}^{d} \rho_{li}\sigma^i p^i x^i, \quad i = 1,\dots,d$$

$$\dot{p}^l = -\sigma^l p^l \sum_{i=1}^{d} \rho_{li}\sigma^i p^i x^i, \quad i = 1,\dots,d.$$

Consequently, it is again easy to see that $\frac{\partial}{\partial t}x^l(t)p^l(t) = 0$, implying that $x^l(t)p^l(t) = x_0^l p_0^l$. The Hamiltonian flow has the form

$$H_{t\leftarrow 0}(x_0, p_0) = \begin{pmatrix} \left(x_0^l \exp\left[\sigma^l \left(\sum_{i=1}^d \rho_{li}\sigma^i p_0^i x_0^i\right) t\right]\right)_{l=1}^d \\ \left(p_0^l \exp\left[-\sigma^l \left(\sum_{i=1}^d \rho_{li}\sigma^i p_0^i x_0^i\right) t\right]\right)_{l=1}^d \end{pmatrix}. \tag{23}$$

Using again that $p^l(t)x^l(t) = p^l(0)x^l(0)$ for any l, we obtain the inverse Hamiltonian flow

$$H_{0\leftarrow t}(x_t, p_t) = \begin{pmatrix} \left(x_t^l \exp\left[-\sigma^l \left(\sum_{i=1}^d \rho_{li}\sigma^i p_t^i x_t^i\right) t\right]\right)_{l=1}^d \\ \left(p_t^l \exp\left[\sigma^l \left(\sum_{i=1}^d \rho_{li}\sigma^i p_t^i x_t^i\right) t\right]\right)_{l=1}^d \end{pmatrix}. \tag{24}$$

The boundary conditions—at $T = 1$—are now given by

$$x_0 = S_0 \tag{25a}$$

$$\sum_{l=1}^d x^l(1) = K \tag{25b}$$

$$p^1(1) = p^2(1) = \cdots = p^d(1). \tag{25c}$$

Indeed, the transversality condition (25c) says that the final momentum $p(1)$ is orthogonal to the surface $\left\{ \sum_{l=1}^d y^l = K \right\}$, whose tangent space is spanned by the collection of vectors $\mathbf{e}_1 - \mathbf{e}_l$, $l = 2, \ldots, d$, with $\mathbf{e}_1, \ldots, \mathbf{e}_d$ the standard basis of $\mathbb{R}^d$. The equations (25) are certainly not difficult to solve numerically, but an explicit solution is not available, neither in the general case nor in the case of d uncorrelated assets.

Remark 6 The main point of this calculation is that while explicit solutions are no longer possible in a general Black-Scholes model, the phenomenon (1) potentially appears in all Black-Scholes models. Moreover, we stress that the non-focality conditions are easily checked numerically.

Remark 7 Note that the discretely monitored Asian option can be considered as a special case of a basket option on correlated assets. Indeed, let us consider an option on

$$\frac{1}{N}\sum_{i=1}^N S_{t_i}, \text{ with (for simplicity) } t_i = i\Delta t, \quad i = 1, \ldots, N.$$

For each individual $i \in \{1, \ldots, N\}$ we have, for fixed $\Delta t > 0$, the equality in law

$$S_{t_i} = S_0 e^{\sigma B_{i\Delta t} - \frac{1}{2}\sigma^2 i\Delta t} = S_0 e^{\sigma^i W^i_{\Delta t} - \frac{1}{2}(\sigma^i)^2\Delta t}$$

for $\sigma^i := \sqrt{i}\sigma$ and $W^i_{\Delta t} := B_{i\Delta t}/\sqrt{i}$. In law, the vector $\left(W^1_{\Delta t}, \dots, W^N_{\Delta t}\right)$ corresponds to the marginal distribution of an N-dimensional Brownian motion at time Δt with correlation $\rho_{ij} = \frac{\min(i,j)}{\sqrt{ij}}$, $1 \le i,\ j \le N$. Thus, the Asian option corresponds to an option on the basket with $S^i_0 \equiv S_0, \sigma^i$ as above and a correlation matrix ρ_{ij} with maturity Δt. Moreover, the asymptotic expansion of the price of the Asian option as $\Delta t \to 0$ corresponds to the short-time asymptotics of the basket.

Remark 8 A small-noise asymptotic expansion of the continuous Asian option on $\int_0^T S_t dt$ is also possible by the techniques of Sect. 3 (with ellipticity conditions replaced by weak Hörmander conditions). Essentially, this is equivalent to letting $N \to \infty$ in Remark 7—but more direct.

As in the two-dimensional case, the boundary conditions can be solved explicitly in the fully symmetric case, when $\sigma^l \equiv \sigma$ and, say, $S^l_0 \equiv 1$. For suitable K the optimal configuration is

$$x_0^* = (1, \dots, 1)^T, \quad x_1^* = (K/d, \dots, K/d)^T$$

$$p_0^* = \left(\frac{\log(K/d)}{\sigma^2}, \dots, \frac{\log(K/d)}{\sigma^2}\right)^T, \quad p_1^* = \left(\frac{d}{\sigma^2 K}\log(K/d), \dots, \frac{d}{\sigma^2 K}\log(K/d)\right)^T.$$

Introducing

$$\mathfrak{q} = \epsilon_1 \begin{pmatrix} 1 \\ 1 \\ \vdots \\ 1 \end{pmatrix}, \quad \mathfrak{z} = \begin{pmatrix} \epsilon_2 + \dots + \epsilon_d \\ -\epsilon_2 \\ \vdots \\ -\epsilon_d \end{pmatrix},$$

we obtain (for the case of d uncorrelated assets)

$$\begin{aligned} M(x_1, p_1) &:= \partial_{(\mathfrak{z},\mathfrak{q})}\big|_{(\mathfrak{z},\mathfrak{q})=0}\, \pi H_{0\leftarrow 1}(x_1 + \mathfrak{z}, p_1 + \mathfrak{q}) \\ &= \begin{pmatrix} a_1 & \mathbf{b} \\ \mathbf{a} & G \end{pmatrix}, \end{aligned}$$

where $\mathbf{a} = (a_2, \dots, a_d)^T \in \mathbb{R}^{(d-1)\times 1}$, $\mathbf{b} = b(1, \dots, 1) \in \mathbb{R}^{1\times(d-1)}$, $G = \operatorname{diag}(g_2, \dots, g_d) \in \mathbb{R}^{(d-1)\times(d-1)}$ with

$$a_l = -(\sigma^l)^2 (x_1^l)^2 e^{-(\sigma^l)^2 p_1^l x_1^l}, \quad l = 1, \dots d,$$

$$b = \left[1 - (\sigma^1)^2 x_1^1 p_1^1\right] e^{-(\sigma^1)^2 p_1^1 x_1^1},$$

$$g_l = -\left[1 - (\sigma^l)^2 x_1^l p_1^l\right] e^{-(\sigma^l)^2 p_1^l x_1^l}, \quad l = 2, \dots d.$$

In the symmetric case, we can evaluate M at the optimal configuration and obtain

$$M(x_1^*, p_1^*) = \begin{pmatrix} -\sigma^2 \frac{K}{d} & \left[1 - \log(K/d)\right] \frac{d}{K} & \cdots & \left[1 - \log(K/d)\right] \frac{d}{K} \\ -\sigma^2 \frac{K}{d} & -\left[1 - \log(K/d)\right] \frac{d}{K} & \cdots & 0 \\ \vdots & \vdots & \ddots & \vdots \\ -\sigma^2 \frac{K}{d} & 0 & \cdots & -\left[1 - \log(K/d)\right] \frac{d}{K} \end{pmatrix},$$

whose determinant can be seen to be

$$\det M(x_1^*, p_1^*) = (-1)^d \sigma^2 K \left[(1 - \log(K/d)) \frac{d}{K} \right]^{d-1}.$$

Thus, the non-focality condition fails if and only if $K = de$. Moreover, we obtain the energy

$$\Lambda(K) = \mathcal{H}(x_0^*, p_0^*) = \frac{d}{2} \frac{\log(K/d)^2}{\sigma^2}.$$

6 A Geometric Approach to Focality

In this final section we take a more geometrical look at the non-focality condition appearing in Sect. 3.2. Consider the Black Scholes model

$$dS_t^i = \sigma^i S_t^i dW_t^i, \quad \left\langle dW^i, dW^j \right\rangle_t = \rho_{i,j} dt.$$

We change parameters $\mathbf{S} \to \mathbf{y} \to \mathbf{x}$, by

$$y^i := \frac{\log\left(\frac{S^i}{S_0^i}\right)}{\sigma^i}, \quad x^i = L_{ip} y^p, \quad i = 1, \ldots, d,$$

where $\boldsymbol{\rho}$ denotes the correlation matrix of $\mathbf{W}$ and $\boldsymbol{\rho} = LL^T$ its Cholesky factorization. Obviously, $S^i = S_0^i e^{\sigma_i y^i}$. In terms of the $\mathbf{x}$-coordinates we have

$$x^i = x^i(\mathbf{F}) = L_{ip} \log\left(S^p / S_0^p\right) / \sigma^p,$$
$$S^i = S^i(\mathbf{x}) = S_0^i e^{\sigma_i L^{ip} x^p}.$$

The advantage of using the chart $\mathbf{x}$ is that the corresponding Riemannian metric tensor is the usual Euclidean metric tensor. Thus, we simply have

$$d(\mathbf{S}_0, \mathbf{S}) = |\mathbf{x}_0 - \mathbf{x}|$$

and the geodesics are straight lines as seen from the $\mathbf{x}$-chart. Note furthermore that $\mathbf{S} = \mathbf{S}_0$ is transformed to $\mathbf{x} = \mathbf{0}$.

The payoff function of the option is given by $\left(\sum w_i S_T^i - K\right)^+$. We normalize $w_i \equiv 1$ and $T \equiv 1$. The strike surface $F = \left\{\mathbf{S} \in \mathbb{R}_+^d \,\middle|\, \sum_{i=1}^d S^i = K\right\}$, which is (a sub-set of) a hyperplane in $\mathbf{S}$ coordinates is, however, transformed to a much more complicated submanifold in $\mathbf{x}$ coordinates. Re-phrasing the equation $\sum_i S^i = K$ in $\mathbf{y}$-coordinates and solving for y^d gives

$$y^d = \log\left[\left(K - \sum_{i=1}^{d-1} S_0^i e^{\sigma^i \sum_{p=1}^d L^{ip} x^p}\right) / S_0^d\right] / \sigma^d,$$

with $(L^{ij}) = (L_{ij})^{-1}$, which implies—using that L and L^{-1} are lower-triangular matrices –

$$L^{dd} x^d = \log\left[\left(K - \sum_{i=1}^{d-1} S_0^i e^{\sigma^i \sum_{p=1}^i L^{ip} x^p}\right) / S_0^d\right] / \sigma^d - \sum_{k=1}^{d-1} L^{dk} x^k.$$

For sake of clarity, let us introduce the notation $\mathbf{q} = (q^1, \ldots, q^{d-1}) := (x^1, \ldots, x^{d-1})$. A parametrization of the strike surface F is then given by the map $\varphi : U \subset \mathbb{R}^{d-1} \to \mathbb{R}^d$ with

$$U := \left\{\mathbf{q} \in \mathbb{R}^{d-1} \,\middle|\, \sum_{i=1}^{d-1} S_0^i e^{\sigma^i \sum_{p=1}^i L^{ip} q^p} < K\right\},$$

and

$$\varphi(\mathbf{q}) := \left(\mathbf{q}, \frac{1}{L^{dd}}\left\{\log\left[\left(K - \sum_{i=1}^{d-1} S_0^i e^{\sigma^i \sum_{p=1}^i L^{ip} q^p}\right) / S_0^d\right] / \sigma^d - \sum_{k=1}^{d-1} L^{dk} q^k\right\}\right).$$

Note that by the change of coordinates, we are implicitly assuming that $S^i > 0$ for all i. Moreover, the standard basis $\mathbf{e}_1(\mathbf{p}), \ldots, \mathbf{e}_{d-1}(\mathbf{p})$ of the tangent space $T_{\mathbf{p}}F$ to F at $\mathbf{p} = \varphi(\mathbf{q})$ is given by the columns of the Jacobi matrix of φ evaluated at $\mathbf{q}$, more precisely we have

$$\mathbf{e}_i(\mathbf{p}) = \left((\delta_i^j)_{j=1}^{d-1}, -\frac{1}{L^{dd}}\left[\frac{1}{\sigma^d} \frac{\sum_{j=i}^{d-1} \sigma^j L^{ji} S_0^j e^{\sigma^j \sum_{r=1}^j L^{jr} q^r}}{K - \sum_{j=1}^{d-1} S_0^j e^{\sigma^j \sum_{r=1}^j L^{jr} q^r}} + L^{di}\right]\right)$$

for $i = 1, \dots, d-1$ and $\mathbf{p} = \varphi(\mathbf{q})$. Consequently, the normal vector field N to S at $\mathbf{p} = \varphi(\mathbf{q})$ is given by

$$N(\mathbf{p}) = \alpha(\mathbf{p}) \left(\left(\frac{1}{L^{dd}} \left[\frac{1}{\sigma^d} \frac{\sum_{j=i}^{d-1} \sigma^j L^{ji} S_0^j e^{\sigma^j \sum_{r=1}^{j} L^{jr} q^r}}{K - \sum_{j=1}^{d-1} S_0^j e^{\sigma^j \sum_{r=1}^{j} L^{jr} q^r}} + L^{di} \right] \right)_{i=1}^{d-1}, 1 \right) = N \circ \varphi(\mathbf{q}),$$

where α is a normalization factor guaranteeing that $|N(\mathbf{p})| = 1$, i.e.,

$$\alpha(\mathbf{p}) = \left(1 + \sum_{i=1}^{d-1} \frac{1}{(L^{dd})^2} \left[\frac{1}{\sigma^d} \frac{\sum_{j=i}^{d-1} \sigma^j L^{ji} S_0^j e^{\sigma^j \sum_{r=1}^{j} L^{jr} q^r}}{K - \sum_{j=1}^{d-1} S_0^j e^{\sigma^j \sum_{r=1}^{j} L^{jr} q^r}} + L^{di} \right]^2 \right)^{-1/2}.$$

The *Weingarten map* or *shape operator* $L_{\mathbf{p}} : T_{\mathbf{p}}F \to T_{\mathbf{p}}F$ is defined by

$$L_{\mathbf{p}} \left(d\varphi_{\varphi^{-1}(\mathbf{p})}(\mathbf{v}) \right) = -d(N \circ \varphi)(\varphi^{-1}(\mathbf{p})) \cdot \mathbf{v},$$

$\mathbf{v} \in \mathbb{R}^{d-1} = T_{\varphi^{-1}(\mathbf{p})}U$, see [8]. In other words, for $\varphi(\mathbf{q}) = \mathbf{p}$, we interpret N as a map in $\mathbf{q}$ and $-L_{\mathbf{p}}$ is the directional derivative of that map. We study the Weingarten map since it gives us the curvature of the surface F. Indeed, the eigenvalues $k_1(\mathbf{p}), \dots, k_{d-1}(\mathbf{p})$ of the linear map $L_{\mathbf{p}} : T_{\mathbf{p}}F \to T_{\mathbf{p}}F$ are called *principal curvatures* of F. Then the *focal points* of F at $\mathbf{p}$ are given by

$$\{\mathbf{p} + \frac{1}{k_i(\mathbf{p})} N(\mathbf{p}) | 1 \le i \le d-1 \text{ such that } k_i(\mathbf{p}) \ne 0\}.$$

In order to compute the eigenvalues of the shape operator, we need to compute the representation of $L_{\mathbf{p}}$ in the standard basis $(\mathbf{e}_1(\mathbf{p}), \dots, \mathbf{e}_{d-1}(\mathbf{p}))$. Let us denote this matrix by $\overline{L}(\mathbf{p})$, then we obviously have

$$\overline{L}(\varphi(\mathbf{q}))_{ij} = -\langle \frac{\partial}{\partial q^j} (N \circ \varphi)(\mathbf{q}), \mathbf{e}_i(\varphi(\mathbf{q})) \rangle, \quad i, j = 1, \dots, d-1.$$

The principal curvatures $k_1(\mathbf{p}), \dots, k_{d-1}(\mathbf{p})$ are, thus, the eigenvalues of the $(d-1)$-dimensional matrix $\overline{L}(\mathbf{p})$.

Since the calculations become too complicated in the general case, we now again concentrate on the case of *two uncorrelated* assets, i.e., $d = 2$ and $\boldsymbol{\rho} = L = I_2$. In this case, we have

$$\mathbf{e}_1(\mathbf{p}) = \left(1, -\frac{\sigma^1}{\sigma^2} \frac{S_0^1 e^{\sigma^1 q^1}}{K - S_0^1 e^{\sigma^1 q^1}} \right),$$

$$N(\varphi(\mathbf{q})) = \frac{1}{\sqrt{(\sigma^1)^2 (S_0^1)^2 e^{2\sigma^1 q^1} + (\sigma^2)^2 \left(K - S_0^1 e^{\sigma^1 q^1} \right)^2}} \left(\sigma^1 S_0^1 e^{\sigma^1 q^1}, \sigma^2 \left(K - S_0^1 e^{\sigma^1 q^1} \right) \right).$$

Thus, the Weingarten map is given by

$$L_{\mathbf{p}}(v\mathbf{e}_1(\mathbf{p})) = v\kappa(\mathbf{p})\mathbf{e}_1(\mathbf{p}),$$

where for $\mathbf{q} = (q^1) \in \mathbb{R}$

$$\kappa(\varphi(\mathbf{q})) = k_1(\varphi(\mathbf{q})) = \frac{K(\sigma^1)^2(\sigma^2)^2 S_0^1 e^{\sigma^1 q^1}\left(S_0^1 e^{\sigma^1 q^1} - K\right)}{\left[(\sigma^1)^2(S_0^1)^2 e^{2\sigma^1 q^1} + (\sigma^2)^2\left(S_0^1 e^{\sigma^1 q^1} - K\right)^2\right]^{3/2}}$$

is the *curvature* of the curve F in $\mathbb{R}^2$. We see that $\kappa = 0$ if and only if $K = S_0^1 e^{\sigma^1 q^1}$, i.e., at the boundary of the surface F. Otherwise, κ is negative.

Here, both components of $N(\mathbf{p})$ are positive on F. Consequently, for any $\mathbf{p} = \varphi(\mathbf{q}) \in S$ there is precisely one focal point $\mathfrak{f} = \mathfrak{f}(\mathbf{p}) \in \mathbb{R}^2$, which is given by

$$\mathfrak{f}^1 = q^1 + \frac{S_0^1 e^{\sigma^1 q^1}\left[2(\sigma^2)^2 K - ((\sigma^1)^2 + (\sigma^2)^2) S_0^1 e^{\sigma^1 q^1}\right] - (\sigma^2)^2 K^2}{\sigma^1(\sigma^2)^2 K\left(K - S_0^1 e^{\sigma^1 q^1}\right)},$$

$$\mathfrak{f}^2 = \frac{1}{\sigma^2}\log\left(\frac{K - S_0^1 e^{\sigma^1 q^1}}{S_0^2}\right) + 2\frac{\sigma^2}{(\sigma^1)^2} - \frac{\sigma^2 K e^{-\sigma^1 q^1}}{(\sigma^1)^2 S_0^1} - \frac{((\sigma^1)^2 + (\sigma^2)^2) S_0^1 e^{\sigma^1 q^1}}{(\sigma^1)^2 \sigma^2 K}.$$

Denoting $\mathbf{p} = (x^1, x^2)$ and re-introducing the short-cut notation $S^i = S_0^i e^{\sigma^i x^i}$, $i = 1, 2$, (noting that $S^1 + S^2 = K$) we can express $\mathfrak{f}$ as

$$\mathfrak{f}^1 = x^1 + \frac{S^1\left[2(\sigma^2)^2 K - ((\sigma^1)^2 + (\sigma^2)^2) S^1\right] - (\sigma^2)^2 K^2}{\sigma^1(\sigma^2)^2 K S^2},$$

$$\mathfrak{f}^2 = x^2 + \frac{S^1\left[2(\sigma^2)^2 K - ((\sigma^1)^2 + (\sigma^2)^2) S^1\right] - (\sigma^2)^2 K^2}{(\sigma^1)^2 \sigma^2 K S^1}.$$

In the current setting, let $\mathbf{q}^*$ be the optimal configuration in $\mathbf{q}$-coordinates, i.e., the point on F with smallest Euclidean norm. Then the non-focality condition of Theorem 1 is satisfied, if 0 is not a focal point to $\varphi(\mathbf{q}^*)$, see the discussion in the proof of [6, Prop. 6].

Remark 9 As both components of the normal vector N are non-negative on F and the curvature κ is negative, $\mathbf{0}$ can only be a focal point if F has a non-empty intersection with the positive quadrant. Inserting into the parametrization of F, we see that this can only be the case if $K > S_0^1 + S_0^2$. In other words: if the option is in the money, then the non-focality condition is always satisfied (in the two-dimensional, uncorrelated case).

Let us again use the parameters of Sect. 2, i.e., $S_0^1 = S_0^2 = 1, \sigma^1 = \sigma^2 = \sigma$. Then we consider $\mathbf{S}^* = (K/2, K/2)$, which translates into $\mathbf{x}^* = \left(\frac{\log(K/2)}{\sigma}, \frac{\log(K/2)}{\sigma}\right)$. Inserting into the formulas for the focal points, we obtain

$$\mathfrak{f}^1(\mathbf{x}^*) = \mathfrak{f}^2(\mathbf{x}^*) = \frac{\log\left(\frac{K}{2}\right) - 1}{\sigma}.$$

So, 0 is focal to the optimal configuration, if and only if

$$K = 2e,$$

and we recover, once more, the results of Sects. 2 and 4—recall that $\mathbf{S}_0$ corresponds to 0 in $\mathbf{x}$-coordinates.

In Figs. 2 and 3 the focal points are visualized for two different configurations of two uncorrelated baskets. We plot the surface F as a submanifold of $\mathbb{R}^2$. We have seen above that for any $\mathbf{p} \in F$ there is precisely one focal point $\mathfrak{f}(\mathbf{p})$. Hence, we additionally plot the surface $\{\mathfrak{f}(\mathbf{p}) | \mathbf{p} \in F\}$—more precisely, part of this surface. In Fig. 2 we show the case constructed above where the non-focality condition is violated. In Fig. 3 the option is ITM. As explained above, in the ITM case the manifold F does not intersect the positive quadrant, implying that the non-focality condition is satisfied.

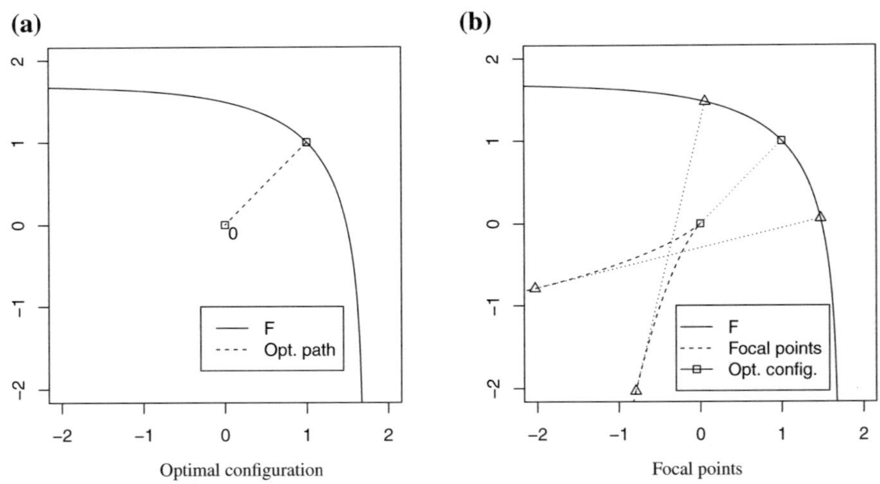

Fig. 2 Optimal configuration and focal points for two independent assets with $\sigma^1 = \sigma^2 = 1$, $\mathbf{S}_0 = (1, 1)$, $K = 2e$. **a** The *dashed line* depicts the optimal path between the spot price $\mathbf{S}_0$ (0 in the $\mathbf{q}$-chart) and the optimal configuration. **b** *Dotted lines* connect some selected points on the manifold F with the corresponding focal points. *Points* marked with a *triangle* visualize the construction of the focal points. We see that 0 is, indeed, focal to the optimal configuration

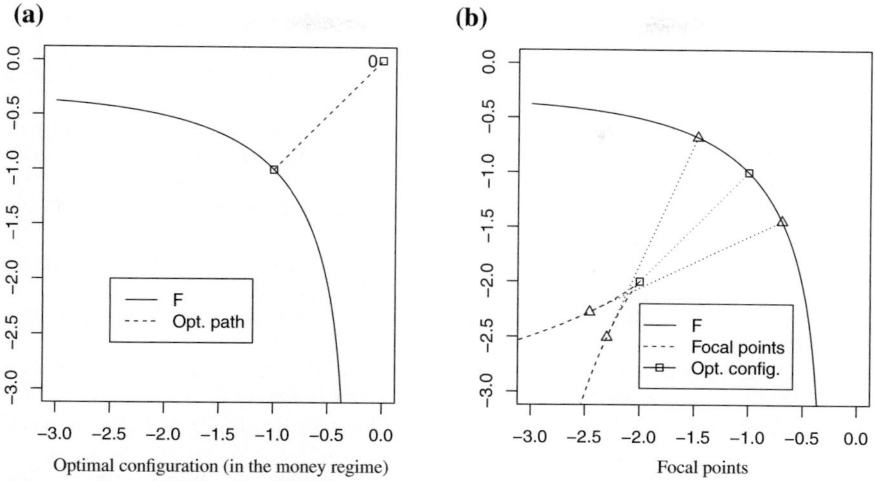

Fig. 3 Optimal configuration and focal points for two independent assets with $\sigma^1 = \sigma^2 = 1$, $\mathbf{S}_0 = (1, 1)$, $K = 2/e$. **a** The *dashed line* depicts the optimal path between the spot price $\mathbf{S}_0$ (0 in the **q**-chart) and the optimal configuration. **b** *Dotted lines* connect some selected points on the manifold F with the corresponding focal points. *Points* marked with a *triangle* visualize the construction of the focal points. This example illustrates the fact that the non-focality condition always holds when the basket option is in the money

Acknowledgments Martin Forde kindly informed us about some misleading formulations in a previous version. P.K.F. has received partial funding from the European Research Council under the European Union's Seventh Framework Program (FP7/2007-2013) / ERC grant agreement nr. 258237.

References

1. M. Avellaneda, Boyer-Olson, D., Busca, J., Friz, P.: Application of large deviation methods to the pricing of index options in finance, Comptes Rendus de l'Académie des Sciences—Series I—Mathematique (2003)
2. Avellaneda, M., Boyer-Olson, D., Busca, J., Friz, P.: Reconstructing volatility. RISK (2004)
3. Ben Arous, G.: Développement asymptotique du noyau de la chaleur hypoelliptique hors du cut-locus. Annales Scientifiques de l'Ecole Normale Supérieure **4**(21), 307–331 (1988)
4. Ben Arous, G.: Methods de Laplace et de la phase stationnaire sur l'espace de Wiener. Stochastics **25**, 125–153 (1988)
5. Bismut, J.M.: Malliavin Calculus and Large Deviations. Birkhauser, Boston (1984)
6. Deuschel, J.D., Friz, P.K., Jacquier, A., Violante, S.: Marginal density expansions for diffusions and stochastic volatility, part I: theoretical foundations. Commun. Pure Appl. Math. **67**(1), 40–82 (2014)
7. Deuschel, J.D., Friz, P.K., Jacquier, A., Violante, S.: Marginal density expansions for diffusions and stochastic volatility, part II: applications. Commun. Pure Appl. Math. **67**(2), 321–350 (2014)
8. do Carmo, M.P.: Differential Geometry of Curves and Surfaces. Prentice-Hall, Englewood Cliffs (1976)

9. Dufresne, D.: The log-normal approximation in financial and other computations. Adv. Appl. Probab. **36**, 747–773 (2004)
10. Gulisashvili, A.: Analytically tractable stochastic stock price models. Springer Finance. Springer, London (2012)
11. Gulisashvili, A., Stein, E.: Asymptotic behavior of the stock price distribution density and implied volatility in stochastic volatility models. Appl. Math. Optim. **61**(3), 287–315. doi:10.1007/s00245-009-9085-x
12. Gulisashvili, A., Tankov, P.: Tail behavior of sums and differences of log-normal random variables, Bernoulli, to appear
13. Heston, S.: A closed-form solution for options with stochastic volatility, with application to bond and currency options. Rev. Financ. Stud. **6**, 327–343 (1993)
14. Molchanov, S.A.: Diffusion processes and Riemannian geometry. Russ. Math. Surv. **30**(1), 1–63 (1975)
15. Seierstad, A., Sydsaeter, K.: Optimal Control Theory with Economic Applications. Advanced Textbooks in Economics, vol. 24. North- Holland, Amsterdam (1987)
16. Stein, E.M., Stein, J.C.: Stock price distributions with stochastic volatility: an analytic approach. Rev. Financ. Stud. **4**, 727–752 (1991)
17. Takanobu, S., Watanabe, S.: Asymptotic expansion formulas of the Schilder type for a class of conditional Wiener functional integration. In Asymptotics problems in probability theory: Wiener functionals and asymptotics. In: Elworthy, K.D., Ikeda, N. (eds.) Pitman Research Notes in Mathematics Series, vol. 284, pp. 194–241. (1993)

On Small-Noise Equations with Degenerate Limiting System Arising from Volatility Models

Giovanni Conforti, Stefano De Marco and Jean-Dominique Deuschel

Abstract The one-dimensional SDE with non Lipschitz diffusion coefficient

$$dX_t = b(X_t)dt + \sigma X_t^\gamma dB_t, \quad X_0 = x, \quad \gamma < 1 \tag{1}$$

is widely studied in mathematical finance. Several works have proposed asymptotic analysis of densities and implied volatilities in models involving instances of (1), based on a careful implementation of saddle-point methods and (essentially) the explicit knowledge of Fourier transforms. Recent research on tail asymptotics for heat kernels (Deuschel et al. Comm. in Pure and Applied Math., 67(1):40–82, 2014, [11]) suggests to work with the rescaled variable $X^\varepsilon := \varepsilon^{1/(1-\gamma)}X$: while allowing to turn a space asymptotic problem into a small-ε problem, the process X^ε satisfies a SDE in Wentzell–Freidlin form (i.e. with driving noise εdB). We prove a pathwise large deviation principle for the process X^ε as $\varepsilon \to 0$. As it will be seen, the limiting ODE governing the large deviations admits infinitely many solutions, a non-standard situation in the Wentzell–Freidlin theory. As for applications, the ε-scaling allows to derive leading order asymptotics for path functionals: while on the one hand the resulting formulae are confirmed by the CIR-CEV benchmarks, on the other hand the large deviation approach (i) applies to equations with a more general drift term and (ii) potentially opens the way to heat kernel analysis for higher-dimensional diffusions involving (1) as a component.

Keywords Pathwise large deviations · Square-root diffusions · Tail asymptotics · Freidlin-Wentzell · Large deviations · Degenerate diffusions · CIR process

---

G. Conforti
Universität Potsdam, Potsdam, Germany
e-mail: giovanniconfort@gmail.com

S. De Marco (✉)
Ecole Polytechnique, Route de Saclay, Palaiseau Cedex 91128, France
e-mail: demarco@cmap.polytechnique.fr

J.-D. Deuschel
Technische Universität Berlin, Berlin, Germany
e-mail: deuschel@mail.math.tuberlin.de

P.K. Friz et al. (eds.), *Large Deviations and Asymptotic Methods in Finance*, Springer Proceedings in Mathematics & Statistics 110,
DOI 10.1007/978-3-319-11605-1_17

1 Introduction

The Wentzell–Freidlin large deviation theory studies the asymptotic behavior of the distribution on path space of the solution to the equation $dX_t^\varepsilon = b(X^\varepsilon)dt + \varepsilon\sigma(X_t^\varepsilon)dB_t,\ X_0^\varepsilon = x$ as $\varepsilon \to 0$, where B is a Brownian motion. When the coefficients b and σ are, say, Lipschitz functions, it is easy to see (with an application of Gronwall's Lemma) that the trajectories of X^ε converge in law to the deterministic solution of the ordinary differential equation $d\varphi_t = b(\varphi_t)dt$, $\varphi_0 = x$. The theory of large deviations accounts for the rate of this convergence: denoting W the Wiener measure, the large deviation principle (LDP)

$$W(X^\varepsilon \in \Gamma) \approx e^{-\frac{1}{\varepsilon^2}\inf_{\phi\in\Gamma} I(\phi)}$$

holds for subsets Γ of the path space $C([0,T])$.[1] Denote $\varphi(h)$ the unique solution of the ODE $d\varphi_t = b(\varphi_t)dt + \sigma(\varphi_t)dh_t$, $\varphi_0 = x$, where the control h is an absolutely continuous path with square integrable derivative $\dot{h}$. The rate function I is given by $I(\phi) = \frac{1}{2}|\dot{h}|^2_{L^2}$, where h is the control steering the trajectory of the deterministic system along the given path ϕ, that is $\varphi(h) = \phi$. When the diffusion coefficient σ is invertible, the control h is identified by $\dot{h}_t = \sigma(\varphi_t)^{-1}(\dot{\varphi}_t - b(\varphi_t))$, yielding the typical form of the rate function

$$I(\phi) = \frac{1}{2}\int_0^T \frac{(\dot{\phi}_t - b(\phi_t))^2}{\sigma(\phi_t)^2}dt.$$

The intuition behind such a result is that we can write $X^\varepsilon(\omega) = X(\varepsilon\omega)$, where X is the 'pathwise' solution of $dX = b(X)dt + \sigma(X)dB$, $X_0 = x$. If we accept that such a map X exists and is regular enough, then the contraction principle in conjunction with Schilder's theorem for large deviations of Brownian paths [12, Chap. 1] provides the LDP and the rate function for X^ε. The standard assumptions under which such a program is carried are conditions of global Lipschitz continuity and ellipticity for the coefficients, see [10, 12]. Several works have aimed at weakening these assumptions and extending the class of equations for which the LDP holds. Dependence on ε in both the drift and the starting point can be introduced, and global Lipschitz continuity can be replaced with (essentially) local Lipschitz-continuity and conditions for the non explosion of the solution (building on the idea of Azencott [3] to exploit the quasi-continuity property of the Itô map, that only relies on local properties of the equation coefficients). We refer to [4] for a nice recent summary of sets of conditions under which the Wentzell–Freidlin estimate holds.

---

[1]The precise statement here is $-\inf_{\phi\in\mathring{\Gamma}} I(\phi) \le \liminf_{\varepsilon\to 0}\varepsilon^2 \log W(X^\varepsilon \in \Gamma) \le \limsup_{\varepsilon\to 0}\varepsilon^2 \log W(X^\varepsilon \in \Gamma) \le -\inf_{\phi\in\overline{\Gamma}} I(\phi)$.

Recent research on heat kernel asymptotics [11] focuses on the tail behavior for correlated stochastic volatility models. Exploiting the space-scaling properties of the log-price process Y_t in some parametric models (namely: there exists $\theta > 0$ such that the rescaled variable $Y_t^\varepsilon := \varepsilon^\theta Y_t$ has the same law as the log-price in a stochastic volatility model with driving noise εdB_t), the approach of [11] is to convert the asymptotic problem for the tail distribution, $W(Y_t > R)$ as $R \to \infty$, to the problem of small-noise probabilities, $W(Y_t^\varepsilon > 1)$ as $\varepsilon \to 0$. Then, a large deviation principle for the rescaled process serves as a building block to study the asymptotic behavior of the corresponding heat kernel (using the tools of Malliavin calculs and the Laplace method on path space, see [5, 7]). This approach can be fully justified, and explicit computations are possible, for the stochastic volatility model of Stein and Stein [25] (also known as Schöbel–Zhu [24] in the correlated case), where the stochastic volatility follows an Ornstein–Uhlenbeck process with constant diffusion coefficient, which is the main case-study of [11]. As pointed out in [11, Sect. 5.3], in the framework of models where the volatility has square-root diffusion coefficient (main example: Heston), or more generally a diffusion coefficient of the form x^γ, $\gamma < 1$ (as in [2, 21]), such a space-scaling approach leads to a situation where the same approach is not justified anymore (and a formal application of the resulting expansion even leads to a wrong conclusion). Quoting [11, Sect. 5.3], "curiously then even a large deviation principle for (the rescaled volatility process) as given above presently lacks justification".

To be more specific, consider the equation $dX_t = (\alpha + \beta X_t)dt + \sigma X_t^\gamma dB_t$ with positive initial condition $X_0 = x > 0$. Looking for a value of θ such that $\varepsilon^\theta X$ satisfies an equation with small-noise ε leads to define the rescaled process $X^\varepsilon := \varepsilon^{1/(1-\gamma)} X$, which indeed satisfies the equation

$$dX_t^\varepsilon = (\alpha^\varepsilon + \beta X_t^\varepsilon)dt + \varepsilon\sigma(X_t^\varepsilon)^\gamma dB_t, \qquad X_0^\varepsilon = x^\varepsilon \tag{1.1}$$

with

$$\alpha^\varepsilon := \varepsilon^{1/(1-\gamma)}\alpha \qquad x^\varepsilon := \varepsilon^{1/(1-\gamma)}x.$$

Of course, this change of variables allows to write $W(X_t > R) = W(X_t^\varepsilon > 1)$ using $\varepsilon = R^{-1/(1-\gamma)}$. As mentioned above, the question is whether a large deviation principle holds *at all* for $W(X_t^\varepsilon \in \cdot)$ as $\varepsilon \to 0$. Note that both the initial condition x_0^ε and the constant term α^ε in the drift coefficient tend to zero as $\varepsilon \to 0$. On the one hand, it is not difficult to see that $X^\varepsilon \to 0$ in law with respect to the uniform topology on $C([0, T])$. On the other hand, writing down formally the limiting ODE that should govern the large deviations, one gets

$$\dot{\varphi}_t = \beta\varphi_t + \sigma|\varphi_t|^\gamma \dot{h}_t, \qquad \varphi_0 = 0. \tag{1.2}$$

The Eq. (1.2) is known to admit infinitely many solutions. When $\dot{h}_t \geq 0$, the set of solutions contains the one-parameter family $\varphi_t^{(\theta)} = e^{\beta t}\Big(\sigma(1-\gamma)\int_\theta^t e^{-\beta(1-\gamma)s}\dot{h}_s ds\Big)^{1/(1-\gamma)} 1_{\{t\geq\theta\}}$, with $\theta \geq 0$.[2] Then, the definition itself of the map $h \mapsto \varphi(h)$ associating the control with the corresponding solution of the ODE is not anymore possible.

We will occasionally address this situation as "degenerate". Let us note straight away that large deviations for diffusions with non-Lipschitz coefficients have been studied in Baldi and Caramellino [4] Donati-Martin et al. [13], Klebaner and Lipster [19] and Robertson [23]. In [4, Theorem 1.2] a large deviation principle is derived for the family of equations $dX_t^\varepsilon = b(X_t^\varepsilon)dt + \varepsilon\sigma(X_t^\varepsilon)dB_t,\ X_0^\varepsilon = x > 0$ (note the strictly positive initial condition), where the function $\sigma(\cdot)$ roughly behaves like σx^γ (see [4, Assumption (A1.1)] for precise conditions) and $b : [0,\infty) \mapsto \mathbb{R}$ is a locally Lipschitz function with sub-linear growth and $b(0) > 0$. The conditions for both a drift term b and an initial datum independent of ε, such that $b(0) > 0$ and $x > 0$, are violated in the situation we consider here. In [13], $b(0) = 0$ and $x = 0$ are allowed, but the analysis is limited to the square-root case $\gamma = 1/2$, and b and x remain independent of ε. Note in this respect that setting $b(0) = x = 0$ implies $X^\varepsilon \equiv 0$ for *all* ε, and in this case a LDP trivially holds with the rate function $I(0) = 0$, $I(\phi) = \infty$ for $\phi \not\equiv 0$ (as stated in [13, Theorem 1.3]); in contrast with (1.1), where both $b^\varepsilon(0) = \alpha^\varepsilon$ and x^ε do tend to zero as $\varepsilon \to 0$, but coming from strictly positive values, so that the solution of the SDE is non trivial for every value of ε. In both these works, uniqueness for the limiting ODE is a key point (and appears as a part of [4, Assumption (A2.3)] and is exploited in [13, Sect. 5]). In order to study the asymptotic behavior of the ruin probability $W(\tau_0 \leq T)$ with $\tau_0 = \inf\{t : X_t = 0\}$ as the initial condition x tends to infinity, Klebaner and Lipster [19] exploit a similar space scaling by working with the 'normed' process $X_t^x = X_t/x$, and show that a LDP holds for the process X^x as $x \to \infty$. The major difference with our setting is that the initial condition $X_0^x = 1$ in [19] is fixed and does not tend to zero as x^ε in (1.1), which is one of the difficulties to encompass in our analysis. Robertson [23] derives LDP for a class of stochastic volatility models, including the Heston model with square-root volatility process. One of the assumptions used there is that the small noise problem for the volatility process has the same form as in Donati-Martin et al. [13], see [23, Assumption 2.1], and the work carried out is to transfer the LDP to the second component of the process (the log-price). Therefore, the work of [23] does not cover small-noise problems in the form of (1.1).

We establish a LDP for a generalized version of Eq. (1.1), allowing α to be a function of the process. That is, we start from Eq. (1) under the assumptions:

(H1) $\gamma \in [1/2, 1), \sigma > 0, x > 0$.
(H2) $b(y) = \alpha(y) + \beta y$, where α is a Lipschitz continuous and bounded function, and $\alpha(y) \geq 0$ in a neighbourhood of 0.

---

[2] When $\beta = 0, \gamma = 1/2$ and $\dot{h} \equiv 1$, one retrieves the textbook example of ODE for which uniqueness fails, $\dot{\varphi}_t = \sigma\sqrt{|\varphi_t|}$, whose solutions from $\varphi_0 = 0$ are given by the one-parameter family $\varphi_t^{(\theta)} = \frac{\sigma^2}{4}(t-\theta)^2 1_{\{t\geq\theta\}}$.

Under (H1)–(H2), (1) is known to admit a positive solution, which is pathwise unique by Yamada and Watanabe's uniqueness theorem.

Theorem 1.1 *Assume conditions (H1)–(H2), and let $(X_t)_{t\geq 0}$ be the unique strong solution to* (1). *Set $X^\varepsilon := \varepsilon^{1/(1-\gamma)}X$; then X^ε satisfies* (1.1) *with the constant α replaced by the function $\alpha(\cdot)$. Then, the family $\{X^\varepsilon\}_\varepsilon$ satisfies a large deviation principle on the path space $C([0,T],\mathbb{R}_+)$ with inverse speed ε^2 and rate function*

$$I_T(\varphi) = \frac{1}{2\sigma^2}\int_0^T \left(\frac{\dot{\varphi}_t - \beta\varphi_t}{\varphi_t^\gamma}\right)^2 1_{\{\varphi_t\neq 0\}}dt,$$

and $I_T(\varphi) = +\infty$ whenever $\varphi(0) \neq 0$ or φ is not absolutely continuous.

Let us note that in the definition of I_T above, the expression $\frac{1}{\varphi_t{}^\gamma}1_{\varphi_t\neq 0}$ is intended to be well defined for any $\varphi_t \in \mathbb{R}_+$, and it is equal to zero when $\varphi_t = 0$. It is easy to see that the unique zero of I_T is $\varphi \equiv 0$, consistently with the fact that $X^\varepsilon \overset{W}{\to} 0$ as $\varepsilon \to 0$. Roughly speaking, Theorem 1.1 allows to write $W(X^\varepsilon \in \Gamma) = \exp\left(-\frac{1}{\varepsilon^2}\left(\inf_{\phi\in\Gamma} I_T(\phi) + \psi(\varepsilon)\right)\right)$ for subsets Γ of $C(0,T)$ such that $\inf_{\phi\in\bar{\Gamma}} I_T(\phi) = \inf_{\phi\in\mathring{\Gamma}} I_T(\phi)$, where the function $\psi(\varepsilon)$ vanishes as $\varepsilon \to 0$; we refer to Theorem 2.1 in Sect. 2 for the precise statements.

According to our definition of X^ε, one has $W(X_t^\varepsilon \geq 0, \forall t \geq 0, \forall\varepsilon > 0) = W(X_t \geq 0, \forall t \geq 0) = 1$. A criterium for the strict positivity of the trajectories of X^ε, based on Feller's test for explosion, can also be given (see [9, Proposition 3.1]: when $\gamma > 1/2$, $a(0) > 0$ implies $W(X_t^\varepsilon > 0, t \geq 0) = 1$, while for $\gamma = 1/2$, the same conclusion is guaranteed by $2\alpha(y)/\sigma^2 \geq 1$ for y in a right neighborhood of zero—yielding the familiar Feller condition $2\alpha/\sigma^2 \geq 1$ when α is constant). Note that Theorem 1.1 does not assume any of these conditions for the non-attainability of zero; in particular for the CIR diffusion, we do not assume the Feller condition on the coefficients α and σ.

From Theorem 1.1, tail asymptotics for some functionals of the process X can be derived (which is exactly why the ε-scaling leading to X^ε was introduced!). The pathwise LDP allows to consider path functionals of the process, such as the running supremum, or the time average.

Theorem 1.2 *Let $(X_t)_{t\geq 0}$ be the unique strong solution to* (1) *under conditions (H1)–(H2), and let $T > 0$. Then, as $R \to \infty$*

$$W(X_T \geq R) = e^{-R^{2(1-\gamma)}(c_T + o(1))} \tag{1.3}$$

and

$$W\left(\sup_{t\in[0,T]} X_t \geq R\right) = e^{-R^{2(1-\gamma)}(c_T + o(1))} \tag{1.4}$$

and

$$W\left(\frac{1}{T}\int_0^T X_t dt \geq R\right) = e^{-R^{2(1-\gamma)}(\nu_T + o(1))}. \tag{1.5}$$

The constant c_T, resp. ν_T are explicitly known in terms of the model parameters, and are provided below in Proposition 2.5, resp. Proposition 3.14 for the case $\gamma = 1/2$.

The estimates in Theorem 1.2 can be compared with the explicit formulae available for cumulative distributions and critical exponents in the CIR and CEV models: these consistency checks are done in Sects. 2.1 and 3.4, showing that the estimates in Theorem 1.2 are correct on the log-scale. While in the one-dimensional setting the large deviation approach yield by Theorem 1.1 applies to equations with a more general drift term than a purely affine function, it also opens the way to heat kernel analysis for higher-dimensional diffusions involving (1) as a component, which is exactly the case left open in [11].

Let us finally note that, due to the non uniqueness of solutions for the limiting system, the problem we consider here appears to be related to the issue of regularization by noise of ODEs. Leaving further discussions to future work, let us just point out here a structural difference with that setting: in that context, one considers an SDE of the form $dX_t^\varepsilon = b(X_t^\varepsilon)dt + \varepsilon dB_t$, with unit dispersion coefficient, seen as a perturbation of the deterministic system $\dot{x}_t = b(x_t)$ with non-Lipschitz drift b (e.g. $b(x) = sign(x)|x|^\gamma$). Among the possible solutions of the deterministic system, one then looks at the (few) ones supporting the limiting law of X^ε, obtaining the so-called zero noise limits of the equation; see [27] and references therein. In our framework, the equation for X^ε already possesses a Lipschitz continuous drift $b(x) = \alpha_\varepsilon + \beta x$. Correspondingly, the limiting system $\dot{x}_t = \beta x$, $x_0 = 0$, *already has* a unique solution (here: the null path $x = 0$), which then gives the unique weak limit for X^ε (in contrast to [27, Corollary 1.2], where the limit is a probability distribution supported on two trajectories). As we pointed out, the difficulties in our setting come from the non-Lipschitz diffusion coefficient and appear at the level of the definition of the rate function via the control system (1.2).

In the remainder of the document, Sect. 2 is devoted to the proof of Theorem 1.1, while in Sect. 3.4 we prove the different statements of Theorem 1.2. We collect in Appendix A the proofs of some of the more technical material.

2 Main Theoretical Estimates

Let $\Omega := C([0, T], \mathbb{R})$, $\Omega_{\geq 0} := C([0, T], \mathbb{R}_+)$ denote the space of continuous (resp. continuous non negative) functions on $[0, T]$. $(\Omega, \mathcal{F}_t, \mathcal{F})$ denotes the canonical Wiener space, W the Wiener measure on $(\Omega, \mathcal{F}_t, \mathcal{F})$, and $\mathbb{E}$ the expectation under W. We denote $H = \{h \in AC([0, T], R) : \dot{h} \in L^2\}$ the space of absolutely continuous paths on $[0, T]$ with square-integrable derivative (usually referred to as Cameron-Martin space). For a set of coefficients $\alpha(\cdot), \beta, \gamma, \sigma$ satisfying conditions (H1)–(H2), we denote X the W almost-surely unique strong solution of (1). We define the rescaled process $X^\varepsilon := \varepsilon^{\frac{1}{1-\gamma}} X$; it is clear that X^ε solves Eq. (1.1) with coefficients identified by $\alpha^\varepsilon(x) = \varepsilon^{1/(1-\gamma)}\alpha(x)$ and $x^\varepsilon = \varepsilon^{1/(1-\gamma)}x$. Denote $b^\varepsilon(x) := \alpha^\varepsilon(x) + x$.

The following theorem gives the precise LDP announced in Theorem 1.1 in the Introduction. We recall that the expression $\frac{1}{y^\gamma}1_{y\neq 0}$ is well defined for any $y \in \mathbb{R}_+$, and it is equal to zero when $y = 0$.

Theorem 2.1 *Let X^ε be the unique strong solution to* (1.1). *Then,*

$$\begin{aligned} \limsup_{\varepsilon\to 0} \varepsilon^2 \log W(X^\varepsilon \in F) &\leq -\inf_F I_T(\varphi) \\ \liminf_{\varepsilon\to 0} \varepsilon^2 \log W(X^\varepsilon \in G) &\geq -\inf_G I_T(\varphi) \end{aligned} \tag{2.1}$$

for every closed set $F \subseteq \Omega_{\geq 0}$ and every open set $G \subseteq \Omega_{\geq 0}$, where the rate function $I_T(\varphi)$ is defined by

$$I_T(\varphi) := \frac{1}{2\sigma^2}\int_0^T \left(\frac{\dot\varphi_t - \beta\varphi_t}{\varphi_t^\gamma}\right)^2 1_{\{\varphi_t\neq 0\}}dt, \tag{2.2}$$

and $I_T(\varphi) = +\infty$ whenever $\varphi(0) \neq 0$ or φ is not absolutely continuous.

Remark 2.2 We could state the large deviation principle of Theorem 2.1 on $\Omega = C([0, T], \mathbb{R})$, setting the rate function $I_T(\varphi)$ to $+\infty$ whenever $\varphi \notin \Omega_{\geq 0}$. Since the process X^ε is known to be positive W-*a.s.* for every $\varepsilon > 0$, with such a definition of the rate function the LDP (2.1) holds for every closed subset F and every open subset G of Ω.

Remark 2.3 As pointed out in the Introduction, the rate function for a family $\{X^\varepsilon\}_\varepsilon$ satisfying $dX^\varepsilon = b(X^\varepsilon)dt + \varepsilon\sigma(X^\varepsilon)dB_t,\ X_0^\varepsilon = x$, can be written as

$$\overline{I}_T(\varphi) = \inf\left\{\frac{1}{2}|\dot h|_{L^2} : h \in H, \varphi(h) = \varphi\right\} \tag{2.3}$$

where $\varphi(h)$ is the solution to the limiting ODE controlled by h, $\dot\varphi = b(\varphi) + \sigma(\varphi)\dot h$ and $\varphi_0 = x$, provided this solution is unique. In our setting, consider $\varphi \in S(u)$, where now $S(u)$ denotes the set of positive solutions of the degenerate ODE (1.2) with control parameter $h = u \in H$: on the set $\{\varphi > 0\}$, u is uniquely determined by φ via $\dot u_t = \frac{\dot\varphi_t - \beta\varphi_t}{\varphi_t^\gamma}$; on the set $\{\varphi = 0\}$, the function φ is seen to satisfy Eq. (1.2) for *any* control parameter h. This means that the set of h such that $\varphi \in S(h)$ contains the infinitely many elements given by

$$\dot h_t = \frac{\dot\varphi_t - \beta\varphi_t}{\varphi_t^\gamma}1_{\{\varphi_t>0\}} + \frac{d\tilde h_t}{dt}1_{\{\varphi_t=0\}}, \qquad \tilde h \in H.$$

The control h_0 achieving the minimum norm is obtained setting $\tilde{h} \equiv 0$. This gives $\frac{1}{2}|\dot{h}_0|_{L^2} = \inf\{\frac{1}{2}|\dot{h}|_{L^2} : h \in H, \varphi \in S(h)\} = I_T(\varphi)$ for the rate function I_T defined in (2.2).

Remark 2.4 Assume that $b : [0,\infty) \to \mathbb{R}$ is a locally Lipschitz function with sublinear growth and $b(0) > 0$, and that $\overline{X}^\varepsilon$ satisfies $d\overline{X}^\varepsilon_t = b(\overline{X}^\varepsilon_t)dt + \varepsilon\sigma\big(\overline{X}^\varepsilon_t\big)^\gamma dB_t$ and $\overline{X}^\varepsilon_0 = x > 0$. Then it is known from [4, Theorem 2.1] or [8, Theorem 4.2] that $\overline{X}^\varepsilon$ satisfies a LDP with rate function

$$J_T(\varphi) := \frac{1}{2\sigma^2}\int_0^T \left(\frac{\dot{\varphi}_t - b(\varphi_t)}{\varphi_t^\gamma}\right)^2 dt,$$

and $J_T(\varphi) = \infty$ if φ is not absolutely continuous, where one classically agrees that $1/\varphi_t$ is equal to $+\infty$ if $\varphi_t = 0$. We stress that the latter rate function is radically different from I_T defined in (2.2): whenever $\varphi = 0$ on some non trivial interval $K \subset [0,1]$, then $J_T(\varphi) = \infty$, while in such a case the integrand in (2.2) gives zero contribution to I_T on K. In other words, while trajectories with a zero-set of positive measure require infinite energy to be followed by the process $\overline{X}^\varepsilon$ in the small-noise limit, they are favoured by the rate function of the process X^ε.

2.1 Tail Asymptotics

The space-scaling $X^\varepsilon = \varepsilon^{1/(1-\gamma)}X$ together with the large deviation principle (2.1) allow to work out tail asymptotics for functionals of the process X. The following proposition provides the precise constants appearing in Theorem 1.2 in the Introduction.

Proposition 2.5 *The asymptotic formulas* (1.3) *and* (1.4) *in Theorem 1.2 hold with the constant* c_T *given by*

$$c_T = \begin{cases} \frac{\beta e^{-2\beta(1-\gamma)T}}{\sigma^2(1-\gamma)(1-e^{-2\beta(1-\gamma)T})} & \text{if } \beta \neq 0 \\ \frac{1}{2\sigma^2(1-\gamma)^2 T} & \text{if } \beta = 0. \end{cases} \tag{2.4}$$

One can see that c_T does not depend on the function $\alpha(\cdot)$ in the drift of X, nor on the initial condition x.

Remark 2.6 Some comments are in order.

(i) **Comparison with explicit formulae for the CEV process**. The asymptotic behavior (1.3) can be compared with the explicit formulae available for the density of the CEV process. When $\alpha \equiv 0$ in (1), X can be obtained as a deterministic time-change of a power of a squared Bessel process (see [16, Sect. 6.4.3]). As a consequence, for every $T > 0$ the random variable X_T is known to admit a density with respect to the Lebesgue measure on the positive real line, given by

$$f_{X_T}(y) = \frac{(1-\gamma)}{d(T)} e^{\beta(-2(1-\gamma)+1/2)T} \exp\left(-\frac{1}{2d(T)} (x^{2(1-\gamma)} + y^{2(1-\gamma)} e^{-2\beta(1-\gamma)T} \right)$$
$$\times x^{1/2} y^{-2\gamma+1/2} I_{1/2(1-\gamma)} \left(\frac{1}{d(T)} x^{1-\gamma} y^{1-\gamma} e^{-\beta(1-\gamma)T} \right), \qquad y > 0, \tag{2.5}$$

where I_ν is the modified Bessel function of the first kind of index $\nu > 0$, and $d(T) = \frac{(1-\gamma)\sigma^2}{2\beta}(1 - e^{-2\beta(1-\gamma)T})$ (note *en passant* that one has $d(T) > 0$ for every choice of the sign of β).[3] The formula (2.5) is also valid for $\beta = 0$, when one replaces all the β-dependent constants with their limits as $\beta \to 0$, such as $d(T)|_{\beta=0} = (1-\gamma)^2\sigma^2 T$. Using the asymptotic behavior (see [1, Sect. 9.7.1]) of the modified Bessel function $I_\nu(z) \sim \frac{e^z}{\sqrt{2\pi z}}$ as $z \to \infty$ for fixed $\nu > 0$, one immediately obtains

$$\log f_{X_T}(y) =: g(y) \sim -\frac{e^{-2\beta(1-\gamma)T}}{2d(T)} y^{2(1-\gamma)} = -c_T y^{2(1-\gamma)}, \qquad x \to \infty,$$

with the constant c_T defined in (2.4). Using some standard tools of regular variation [6], one can then easily prove that $\log W(X_T > y) = \log \int_y^\infty e^{g(z)} dz \sim g(y) \sim -c_T y^{2(1-\gamma)}$ as $y \to \infty$, thus showing that estimate (1.3) is exact on the log-scale.

(ii) The asymptotic estimate $f_{X_T}(y) \le A_T\, e^{-a_T y^{2(1-\gamma)}}$, $y > 1$, for the density of X_T was proven in [9] for the solutions of a class of SDEs containing (1) under conditions (H1)–(H2) (namely, in [9] the coefficients β and γ are also allowed to depend smoothly on X), relying on techniques of Malliavin calculus and transformations for 1-dimensional SDEs. The constant a_T provided there is not optimal. While the estimates in [9] remain valid for more general equations, the large deviation principle in Theorem 2.1 allows to obtain a sharp estimate on the log-scale.

The asymptotic behavior $W\big(\frac{1}{T}\int_0^T X_t dt\big) = \exp\big(-R^{2(1-\gamma)}(\nu_T + o(1))\big)$ for the time average of the process can also be proven using Theorem 2.1: see Proposition 3.14 in Sect. 3.4, where an expression of the constant ν_T is provided in the case $\gamma = 1/2$.

---

[3] When $\gamma \in [1/2, 1)$, the law of X_T also possesses an atom at zero, $\mathbb{P}(X_T = 0) = m_T > 0$, and an explicit formula for the mass m_T is available (see again [16, Chap. 6]). From our point of view, this only means that the density f_{X_T} does not integrate to 1 on $(0, \infty)$, without affecting our analysis of the tail asymptotics at ∞.

3 Proof of the Main Estimates

We prove the large deviation principle in Theorem 2.1 by first showing the exponential tightness of the family $\{X^\varepsilon\}_\varepsilon$, namely for every $m < 0$ there exists a compact set $K_m \subset C([0,T])$ such that $\limsup_{\varepsilon\to 0} \varepsilon^2 \log W(X^\varepsilon \in K_m^c) \le m$. We then prove the weak upper bound

$$\limsup_{R\to 0}\ \limsup_{\varepsilon\to 0} \varepsilon^2 \log W(X^\varepsilon \in B(\varphi,R)) \le -I_T(\varphi) \qquad \forall \varphi \in \Omega_{\ge 0},$$

and the weak lower bound

$$\liminf_{R\to 0}\ \liminf_{\varepsilon\to 0} \varepsilon^2 \log W(X^\varepsilon \in B(\varphi,R)) \ge -I_T(\varphi) \qquad \forall \varphi \in \Omega_{\ge 0}$$

where $B(\varphi,R)$ denotes the closed ball in $C([0,T])$ of radius R, $B(\varphi,R) := \{\tilde\varphi : |\tilde\varphi - \varphi|_\infty \le R\}$. It is a general fact that exponential tightness combined with the weak upper bound yields the large deviation upper bound in (2.1) for any closed set after a covering argument (see [12, Chaps. 1 and 2]). On the other hand, the weak lower bound trivially provides the full lower bound in (2.1), observing that open sets are neighborhoods of their points.

3.1 Exponential Tightness

We prove the exponential tightness considering balls in the Hölder norm $\|\omega\|_\eta := \sup_{s,t\le T, s\ne t} \frac{|\omega_t - \omega_s|}{|t-s|^\eta}$ and a natural bound on the initial condition ω_0. More precisely, we define

$$K_R := \{\|\omega\|_\eta \le R\} \cap \{\omega_0 \in (0,x]\}. \tag{3.1}$$

It is classical that these sets are compact in $C([0,T])$.

Proposition 3.1 *The family of measures $W(X^\varepsilon \in \cdot)$ is exponentially tight in scale ε^2, i.e.*

$$\lim_{R\to +\infty}\ \limsup_{\varepsilon\to 0} \varepsilon^2 \log W\left(X^\varepsilon \in K_R^c\right) = -\infty$$

for every $0 < \eta < \frac{1}{2}$.

We follow [13] in the proof of Proposition 3.1. First, let us observe that for $\varepsilon \le 1$, $W(X_0^\varepsilon \in (0,x]) = 1$ so that we just need to estimate the Hölder norm of X^ε. To this end, we use a version of Garsia-Rodemich-Rumsey's Lemma, and the existence of exponential moments for a process bounding X^ε from above.

Lemma 3.2 *Consider* $(\tilde{X}_t, t \geq 0)$ *the strong solution to*

$$d\tilde{X}_t = (|\alpha|_\infty + |\beta|\tilde{X}_t)dt + \sigma(\tilde{X}_t)^\gamma dB_t, \qquad \tilde{X}_0 = x$$

and define $\tilde{X}^\varepsilon := \varepsilon^{1/(1-\gamma)}\tilde{X}$. *Then, there exist positive constants c and C such that:*

$$\mathbb{E}\left(\exp\left(c\varepsilon^{-2}(\tilde{X}_t^\varepsilon)^{2(1-\gamma)}\right)\right) \leq C, \quad \forall t \in [0, T], \quad \forall \varepsilon > 0. \tag{3.2}$$

Proof According to the definition of $\tilde{X}^\varepsilon$, one has $\varepsilon^{-2}(\tilde{X}_t^\varepsilon)^{2(1-\gamma)} = \tilde{X}_t^{2(1-\gamma)}$, so that (3.2) holds if and only if $\mathbb{E}\left[\exp\left(c\tilde{X}_t^{2(1-\gamma)}\right)\right] \leq C$ for all $t \in [0, T]$. When $\gamma = 1/2$, (3.2) follows from the asymptotic behavior of the density of the CIR process for large arguments (see e.g. [16, Sect. 6.3.2, p. 358]); for general γ and $\beta = 0$, from the asymptotic behavior of the density of the classical CEV process as stated for example in [16, Lemma 6.4.3.1, p. 368]. For general γ and β, we rely on a slight generalization of the proof of [9, Proposition 3.3]; we leave the details to Appendix A.

The next proposition is a direct consequence of Garsia-Rodemich-Rumsey's Lemma; see Appendix A for a statement of this lemma and a proof of Proposition 3.3.

Proposition 3.3 *Let* $\omega \in \Omega$. *Fix* ε, $R > 0$, $\eta \in (0, \frac{1}{2})$. *Assume that:*

$$\int_0^T \int_0^T \exp\left(\frac{|\omega_t - \omega_s|}{\varepsilon^2\sqrt{|t-s|}}\right)dsdt \leq K_{\varepsilon,\eta}(R) \tag{3.3}$$

with $K_{\varepsilon,\eta}(R) := \frac{1}{4}\exp\left(T^{\eta-1/2}\left(\frac{R}{8\varepsilon^2} - 4T^{1/2-\eta} - K_\eta\right)\right) - \frac{1}{4}T^2$ *and* $K_\eta := \sup_{u\in[0,T]} 2u^{1/2-\eta}\log(u^{-1}) < \infty$. *Then,*

$$\|\omega\|_\eta \leq R. \tag{3.4}$$

In the proof of Proposition 3.1, we exploit a localization procedure: for any $\varepsilon > 0$ and $n \in \mathbb{N}$, define the process $X^{\varepsilon,n}$ as the strong solution of the SDE with truncated coefficients:

$$dX_t^{\varepsilon,n} = b^\varepsilon(X_t^{\varepsilon,n} \wedge n)dt + \sigma\varepsilon\left(X_t^{\varepsilon,n} \wedge n\right)^\gamma dB_t, \quad X_0^{\varepsilon,n} = x^\varepsilon. \tag{3.5}$$

The paths of $X^{\varepsilon,n}$ can be decomposed in their martingale part and locally bounded variation part

$$dX_t^{\varepsilon,n} = dA_t^{\varepsilon,n} + dM_t^{\varepsilon,n}$$

with $dM_t^{\varepsilon,n} = \varepsilon\sigma(X_t^{\varepsilon,n} \wedge n)^\gamma dB_t$ and $dA_t^{\varepsilon,n} = b^\varepsilon(X_t^{\varepsilon,n} \wedge n)dt$. We shall also define for every n, ε the stopping time $T^{\varepsilon,n} := \inf\left\{t \geq 0 : X_t^\varepsilon \geq n\right\}$. By the pathwise uniqueness for Eq. (1) (equivalently, (3.5)), we have that up to time $T^{\varepsilon,n}$ the processes

$\left(X_t^\varepsilon\right)_{t\in[0,T]}$ and $\left(X_t^{\varepsilon,n}\right)_{t\in[0,T]}$ coincide almost surely. More precisely, $\forall n \in \mathbb{N}$ and $\varepsilon > 0$

$$W\left(X^\varepsilon_{t\wedge T^{\varepsilon,n}} = X^{\varepsilon,n}_{t\wedge T^{\varepsilon,n}}, \forall t \in [0,T]\right) = 1. \tag{3.6}$$

Proof of Proposition 3.1 Let us fix $\eta \in (0, \frac{1}{2})$. By (3.6),

$$\begin{aligned} W\left(\|X^\varepsilon\|_\eta \geq R\right) &\leq W\left(\|X^{\varepsilon,n}\|_\eta \geq R, T^{\varepsilon,n} \geq T\right) + W\left(T^{\varepsilon,n} \leq T\right) \\ &\leq W\left(\|X^{\varepsilon,n}\|_\eta \geq R\right) + W\left(T^{\varepsilon,n} \leq T\right). \end{aligned} \tag{3.7}$$

Let us estimate the first term in (3.7). Using Proposition 3.3 and Markov's inequality we have for every ε, n:

$$\begin{aligned} W(\|M^{\varepsilon,n}\|_\eta \geq R) &\leq W\left(\int_0^T\int_0^T \exp\left(\varepsilon^{-2}\frac{|M_t^{\varepsilon,n} - M_s^{\varepsilon,n}|}{\sqrt{|t-s|}}\right) dsdt \geq K_{\varepsilon,\eta}(R)\right) \\ &\leq \frac{1}{K_{\varepsilon,\eta}(R)}\int_0^T\int_0^T \mathbb{E}\left(\exp\left(\varepsilon^{-2}\frac{|M_t^{\varepsilon,n} - M_s^{\varepsilon,n}|}{\sqrt{|t-s|}}\right)\right) dsdt. \end{aligned}$$

Applying the exponential martingale inequality $\mathbb{E}\left(\exp(\lambda M_t)\right) \leq \sqrt{\mathbb{E}\left(\exp\left(2\lambda^2\langle M\rangle_t\right)\right)}$ [22, Chap. IV] with $\lambda = \frac{1}{\varepsilon^2\sqrt{|t-s|}}$, for $t > s$ one has

$$\begin{aligned} \mathbb{E}\left(\exp\left(\frac{|M_t^{\varepsilon,n} - M_s^{\varepsilon,n}|}{\varepsilon^2\sqrt{t-s}}\right)\right) &\leq 2\sqrt{\mathbb{E}\left(\exp\left(\frac{2\sigma^2}{\varepsilon^2(t-s)}\int_s^t \left(X_r^{\varepsilon,n}\wedge n\right)^{2\gamma} dr\right)\right)} \\ &\leq 2\exp\left(\sigma^2\varepsilon^{-2}n^{2\gamma}\right). \end{aligned}$$

Therefore, using the definition of the constant $K_{\varepsilon,\eta}(R)$ in Proposition 3.3

$$\limsup_{\varepsilon\to 0} \varepsilon^2 \log W\left(\|M^{\varepsilon,n}\|_\eta \geq R\right) \leq -T^{\eta-1/2}\frac{R}{8} + \sigma^2 n^{2\gamma}. \tag{3.8}$$

For the bounded variation part $A^{\varepsilon,n}$, we observe that

$$W\left(\|A^{\varepsilon,n}\|_\eta \geq R\right) \leq W\left(T^{1-\eta}\sup_{t\in[0,T]} b^\varepsilon\left(X_t^{\varepsilon,n}\wedge n\right) \geq R\right).$$

Under hypothesis (H), $b^\varepsilon(x) \leq |\alpha|_\infty + \beta x$ for every x. Therefore, for every ε, n

$$W\left(\|A^{\varepsilon,n}\|_\eta \geq R\right) \leq W\left(T^{1-\eta}(|\alpha|_\infty + \beta n) \geq R\right) = 0, \tag{3.9}$$

where the last identity holds as soon as $R > T^{1-\eta}(|\alpha|_\infty + \beta n)$.

We now deal with the second term in (3.7). It follows from the comparison theorem for one-dimensional SDEs [17, Proposition 5.2.18], that $X_t^\varepsilon \le \tilde{X}_t^\varepsilon, t \le T$, almost surely, where $\tilde{X}^\varepsilon$ is defined in Lemma 3.2. For every fixed γ and $a > 0$, it is a simple exercise to show that the function $y \mapsto \exp(a\varepsilon^{-2}(1+y)^{2(1-\gamma)}), y > 0$, is increasing and convex if ε is small enough.[4] For such values of ε, since $\tilde{X}_t^\varepsilon$ is a submartingale, so is $\exp\left(a\varepsilon^{-2}(1+\tilde{X}_t^\varepsilon)^{2(1-\gamma)}\right)$. Then, we can apply Markov's inequality and Doob's L^2-inequality, obtaining:

$$\begin{aligned} W\left(T^{n,\varepsilon} \le T\right) &= W\left(\sup_{t\in[0,T]} X_t^\varepsilon \ge n\right) \\ &\le W\left(\sup_{t\in[0,T]} \exp\left(a\varepsilon^{-2}\left(1+\tilde{X}_t^\varepsilon\right)^{2(1-\gamma)}\right)\right. \\ &\qquad \left.\ge \exp\left(a\varepsilon^{-2}(1+n)^{2(1-\gamma)}\right)\right) \\ &\le \exp\left(-a\varepsilon^{-2}(1+n)^{2(1-\gamma)}\right) \times 4\,\mathbb{E}\left(\exp\left(a\varepsilon^{-2}(1+\tilde{X}_T^\varepsilon)^{2(1-\gamma)}\right)\right). \end{aligned} \tag{3.10}$$

Using the elementary inequality $\exp(a(1+y)^{2(1-\gamma)}) \le \exp(a2^{2(1-\gamma)}) + \exp(a(2y)^{2(1-\gamma)})$, and choosing a such that $a \times 2^{2(1-\gamma)} = c$ where c is the constant in Lemma 3.2, it follows from this lemma and estimate (3.10) that

$$W\left(T^{n,\varepsilon} \le T\right) \le \exp\left(-a\varepsilon^{-2}n^{2(1-\gamma)}\right) \times 4\left[\exp(c\varepsilon^{-2}) + C\right], \tag{3.11}$$

where C is the second constant in Lemma 3.2. Now choosing $n := \lfloor\sqrt{R}\rfloor$, the condition under which (3.9) holds true is satisfied for R large enough. Passing to the limit as $\varepsilon \to 0$ in (3.7) and using (3.8), (3.9) and (3.11), we obtain

$$\limsup_{\varepsilon\to 0} \varepsilon^2 \log\left(W\left(\|X^\varepsilon\|_\eta \ge R\right)\right) \le \max\left\{-\frac{R}{8} + \sigma^2 R^\gamma, -aR^{(1-\gamma)} + c\right\}.$$

Letting $R \to \infty$, the conclusion follows. ■

3.2 Weak Upper Bound

This section is devoted to the proof of the following proposition.

---

[4]The second derivative reads $e^{a\varepsilon^{-2}(1+y)^{2(1-\gamma)}} \times 2a\varepsilon^{-2}(1-\gamma)(1+y)^{-2\gamma} \times [1 - 2\gamma + \frac{2a}{\varepsilon^2}(1-\gamma)(1+y)^{2(1-\gamma)}]$.

Proposition 3.4 $\forall \varphi \in \Omega_{\geq 0} \cap H$*:*

$$\limsup_{R\to 0}\limsup_{\varepsilon\to 0} \varepsilon^2 \log W\left(X^\varepsilon \in B(\varphi, R)\right) \leq -I_T(\varphi). \tag{3.12}$$

For every $h \in H$, $\varepsilon > 0$ and $\phi \in \Omega_{\geq 0}$, define

$$\begin{aligned} F^\varepsilon(\phi, h) := h_T\phi_T - h_0\phi_0 - h_T\int_0^T b^\varepsilon(\phi_s)ds - \int_0^T \Big(\phi_s - \int_0^s b^\varepsilon(\phi_r)\,dr\Big)\dot{h}_s ds \\ - \frac{\sigma^2}{2}\int_0^T h_s^2\phi_s^{2\gamma} ds. \end{aligned} \tag{3.13}$$

By setting $\varepsilon = 0$ in (3.13), we can define the functional $F^0(\phi, h)$. Note that $F^\varepsilon(\cdot, h)$ is continuous $\forall h \in H$ on the whole space $\Omega_{\geq 0}$ with respect to the sup-norm topology, and converges to $F^0(\cdot, h)$ uniformly on $\Omega_{\geq 0}$ as $\varepsilon \to 0$.

Remark 3.5 Applying the integration by parts formula to the product $h_t X_t^\varepsilon$, one has

$$\begin{aligned} \varepsilon\sigma\int_0^T h_t(X_t^\varepsilon)^\gamma dB_t &= h_T X_T^\varepsilon - h_0 x_0^\varepsilon - \int_0^T [\dot{h}_t X_t^\varepsilon + h_t b^\varepsilon(X_t^\varepsilon)]dt \\ &= h_T X_T^\varepsilon - h_0 x_0^\varepsilon \\ &\quad - h_T\int_0^T b^\varepsilon(X_t^\varepsilon)dt - \int_0^T \dot{h}_t\Big(X_t^\varepsilon - \int_0^t b^\varepsilon(X_s^\varepsilon)ds\Big)dt, \end{aligned}$$

hence

$$F^\varepsilon(X_\cdot^\varepsilon, h) = \varepsilon\sigma\int_0^T h_s(X_s^\varepsilon)^\gamma dB_s - \frac{\sigma^2}{2}\int_0^T h_s^2(X_s^\varepsilon)^{2\gamma} ds.$$

According to Remark 3.5, the random variable

$$M_T^{\varepsilon,h}(\omega) := \exp\left(\frac{1}{\varepsilon^2}F^\varepsilon\left(X^\varepsilon(\omega), h\right)\right) \tag{3.14}$$

is the value at time T of the local exponential martingale associated to $\frac{\sigma}{\varepsilon}\int_0^\cdot h_s(X_s^\varepsilon)^\gamma dB_s$. It should be stressed that, for any $h \in H$ and $\varepsilon > 0$, the functionals $F^\varepsilon(\phi, h)$ and $M_T^{\varepsilon,h}(\phi)$ are well defined for every $\phi \in \Omega_{\geq 0}$, and not only almost surely.

Proof of Proposition 3.4 Since any positive local martingale is a supermartingale, we have

$$\mathbb{E}\big[M_T^{\varepsilon,h}\big] \leq 1. \tag{3.15}$$

Fix now a trajectory $\varphi \in \Omega_{\geq 0}$. Using the remark above:

$$W(X^\varepsilon \in B(\varphi, R)) = \mathbb{E}\left[e^{-\frac{1}{\varepsilon^2}F^\varepsilon(X^\varepsilon,h)} M_T^{\varepsilon,h} 1_{\{X^\varepsilon \in B(\varphi,R)\}}\right]$$
$$\leq \sup_{\phi \in B(\varphi,R)} \exp\left(-\frac{1}{\varepsilon^2}F^\varepsilon(\phi, h)\right) \mathbb{E}\left(M_T^{\varepsilon,h}\right)$$
$$\leq \sup_{\phi \in B(\varphi,R)} \exp\left(-\frac{1}{\varepsilon^2}F^\varepsilon(\phi, h)\right).$$

Since $\sup_{\phi \in B(\varphi,R)} |F^\varepsilon(\phi, h) - F^0(\phi, h)| \to 0$, we have that

$$\limsup_{\varepsilon \to 0} \varepsilon^2 \log W(X^\varepsilon \in B(\varphi, R)) \leq \sup_{\phi \in B(\varphi,R)} (-F^0(\phi, h)).$$

Therefore, by the continuity of $\phi \mapsto F^0(\phi, h)$,

$$\limsup_{R \to 0} \limsup_{\varepsilon \to 0} \varepsilon^2 \log(W(X^\varepsilon \in B(\varphi, R) \leq -F^0(\varphi, h), \qquad \forall h \in H.$$

In the next proposition we prove that:

$$\sup_{h \in H} F^0(\varphi, h) = I_T(\varphi)$$

which concludes the proof of (3.12). ■

Proposition 3.6 $\forall\, \varphi \in \Omega_{\geq 0}$ *we have that:*

$$\sup_{h \in H} F^0(\varphi, h) = I_T(\varphi) \tag{3.16}$$

Proof Assume $\varphi \in \Omega_{\geq 0} \cap H$ is such that $I_T(\varphi) < \infty$. Then, the function u defined by $u_0 = 0$, $\dot{u}_s = \frac{\dot{\varphi}_s - b\varphi_s}{\sigma \varphi_s^\gamma} 1_{\varphi_s \neq 0}$ is by definition an element of H, and φ satisfies by construction the ODE (1.2) with control u. Repeating the computations in Remark 3.5, one can see that

$$F^0(\varphi, h) = \sigma \int_0^T h_s \varphi_s^\gamma \dot{u}_s ds - \frac{\sigma^2}{2} \int_0^T h_s^2 \varphi_s^{2\gamma} ds.$$

Note that $F^0(\varphi, h)$ is concave in h, hence if it has a critical point, this must be a maximum. The Fréchet differential $D^h F^0(\varphi, h)$ at h, applied to the generical element $k \in H$, reads

$$D^h F^0(\varphi, h)[k] = \sigma \int_0^T k_s \left[\varphi_s^\gamma \dot{u}_s - \sigma h_s \varphi_s^{2\gamma}\right] ds.$$

Therefore, $D^h F^0(\varphi, h)|_{h=h^*} = 0$ at any h^* such that $h_s^* = \frac{\dot{u}_s}{\sigma \varphi_s^\gamma}$ on $\{s : \varphi_s \neq 0\}$ (while h_s^* can take any arbitrary value on $\{s : \varphi_s = 0\}$). For such h^*, one has

$$F^0(\varphi, h^*) = \int_0^T (\dot{u}_s)^2 1_{\varphi_s \neq 0} ds - \frac{1}{2} \int_0^T (\dot{u}_s)^2 1_{\varphi_s \neq 0} ds = \frac{1}{2} \int_0^T (\dot{u}_s)^2 1_{\varphi_s \neq 0} ds = I_T(\varphi).$$

On the other hand, if φ is absolutely continuous and such that $I_T(\varphi) = +\infty$, one can approximate the function $\frac{\dot{\varphi}_s - \beta \varphi_s}{\varphi_s^{2\gamma}}$ with a sequence $h^n \in H$ such that $F^0(\varphi, h^n) \to +\infty$. ■

3.3 Weak Lower Bound

This section is devoted to the proof of

Proposition 3.7 *For all $\varphi \in \Omega_{\geq 0}$, we have*

$$\liminf_{R \to 0} \liminf_{\varepsilon \to 0} \varepsilon^2 \log W\left(X^\varepsilon \in B(\varphi, R)\right) \geq -I_T(\varphi). \tag{3.17}$$

In the spirit of Lamperti's transformation, we introduce the process $Y^\varepsilon := (X^\varepsilon)^{1-\gamma}$. Y^ε satisfies a SDE with constant diffusion coefficient and a drift coefficient that we will be able to control. We will prove a large deviation weak lower bound for Y^ε, and then transfer it to X^ε by means of the contraction principle.

Proposition 3.8 *Define*

$$\mathcal{I}_T(\psi) := \frac{1}{2\sigma^2(1-\gamma)^2} \int_0^T \left(\dot{\psi}_t - \beta(1-\gamma)\psi_t\right)^2 dt$$

for $\psi \in \Omega_{\geq 0}$, where $\mathcal{I}_T(\psi) = +\infty$ if $\psi(0) \neq 0$ or ψ is not absolutely continuous. Then, for all ψ such that $\mathcal{I}_T(\psi) < +\infty$, one has

$$\liminf_{R \to 0} \liminf_{\varepsilon \to 0} \varepsilon^2 \log W\left(Y^\varepsilon \in B(\psi, R)\right) \geq -\mathcal{I}_T(\psi). \tag{3.18}$$

In other words, the family Y^ε satisfies a large deviation weak lower bound on $C([0, T], \mathbb{R}_+)$, with rate function $\mathcal{I}_T(\psi)$.

Once we are provided with Proposition 3.8, it is straightforward to prove the weak lower bound for X^ε.

Proof of Proposition 3.7 Consider $\psi \in \Omega_{\geq 0}$ absolutely continuous. By Lemma 3.45 in [20], $\dot{\psi} = 0$ $a.s.$ on $\{\psi = 0\}$. Therefore, $\mathcal{I}_T$ defined in Proposition 3.8 can be rewritten as $\mathcal{I}_T(\psi) = \frac{1}{2\sigma^2(1-\gamma)^2} \int_0^T \left(\dot{\psi}_t - \beta(1-\gamma)\psi_t\right)^2 1_{\psi_t \neq 0} dt$. Using the definition of Y^ε and (3.18), since the map $\psi \mapsto \varphi = \psi^{\frac{1}{1-\gamma}}$ is continuous on $\Omega_{\geq 0}$, we can apply the contraction principle and obtain that $W(X^\varepsilon \in .)$ satisfies a large deviation weak lower bound with rate function $\bar{I}_T$. Let us describe $\bar{I}_T(\varphi)$ when φ is absolutely continuous and such that $I_T(\varphi) < \infty$ (where I_T was defined in (2.2)). Let $\psi_t = \varphi_t^{1-\gamma}$. On $\{\varphi = 0\}$, one has $\psi = 0$ as well, while for a point t in the open set $\{\varphi > 0\}$ such that $\dot{\varphi}_t$ exists, one has $\dot{\psi}_t = (1-\gamma)\frac{\dot{\varphi}_t}{\varphi_t^\gamma}$. Then, noting that $I_T(\varphi) < \infty$ implies that $\frac{\dot{\varphi}_t}{\varphi_t^\gamma}1_{\varphi_t>0}$ is integrable on $[0, T]$, ψ is also absolutely continuous on $[0, T]$ (see [20, Corollary 3.41]), with derivative $\dot{\psi}_t = (1-\gamma)\frac{\dot{\varphi}_t}{\varphi_t^\gamma}1_{\varphi>0}$ This yields

$$\bar{I}_T(\varphi) = \mathcal{I}_T(\psi(\varphi)) = \frac{1}{2\sigma^2(1-\gamma)^2}$$
$$\int_0^T \left((1-\gamma)\frac{\dot{\varphi}_t}{\varphi_t^\gamma} - \beta(1-\gamma)\varphi_t^{1-\gamma}\right)^2 1_{\varphi_t \neq 0} dt = I_T(\varphi) < \infty. \tag{3.19}$$

If $I(\varphi) = \infty$, there is nothing to prove in (3.17), and the claim follows.

3.3.1 Proof of Proposition 3.8

This section is devoted to the proof of the large deviation weak lower bound for the process Y^ε in (3.18). While postponing some of the most technical elements to Appendix A, we will make use here of the following notation: for every $h \in H$, $y \in \mathbb{R}$, we define $\mathcal{S}_y(h)$ to be the unique solution on $[0, T]$ of the ODE

$$\dot{\psi}_t = \beta(1-\gamma)\psi_t + \sigma(1-\gamma)\dot{h}_t, \quad \psi_0 = y. \tag{3.20}$$

We denote $W^{\varepsilon,h}$ the measure on Ω associated to the Girsanov shift $-\frac{1}{\varepsilon}\int_0^T \dot{h}_t dt$,

$$\frac{dW^{\varepsilon,h}}{dW}(\omega) = \exp\left(\frac{1}{\varepsilon}\int_0^T \dot{h}_t dB_t - \frac{1}{2\varepsilon^2}\int_0^T \dot{h}_t^2 dt\right). \tag{3.21}$$

An application of Girsanov's Theorem shows that $W\left(X^{\varepsilon,h} \in \cdot\right) \stackrel{d}{=} W^{\varepsilon,h}(X^\varepsilon \in \cdot)$, where $X^{\varepsilon,h}$ solves:

$$dX_t^{\varepsilon,h} = b^\varepsilon(X_t^{\varepsilon,h})dt + \sigma|X_t^{\varepsilon,h}|^\gamma \dot{h}_t dt + \varepsilon\sigma|X_t^{\varepsilon,h}|^\gamma dB_t, \quad X_0^{\varepsilon,h} = \varepsilon^{\frac{1}{1-\gamma}}x. \tag{3.22}$$

We also define the process $Y^{\varepsilon,h} := |X^{\varepsilon,h}|^{1-\gamma}$.

Remark 3.9 Note that for (3.22) there exists a weak solution, which we construct directly from a solution of (1.1) applying Girsanov's Theorem. Since pathwise uniqueness holds for the couple (b, σ), another application of the same theorem shows that pathwise uniqueness for (1.1) implies pathwise uniqueness for (3.22). Therefore we can always assume that $X^{\varepsilon,h}$ solves (3.22) with the Brownian motion B.

Two main ingredients enter in the proof of Proposition 3.8: the convergence in law (under some conditions on h) of the process $Y^{\varepsilon,h}$ to the deterministic limit $S_0(h)$ under the measure W (equivalently: the weak convergence of the measure $W^{\varepsilon,h}(Y^\varepsilon \in .)$ to $\delta_{S_0(h)}$), and a lower bound for the probability $W(Y^\varepsilon \in B(\psi, R))$ depending explicitly on the relative entropy between the two measures $W^{\varepsilon,h}$ and W. This is the content of the two following lemmas.

Lemma 3.10 (Convergence in law of $Y^{\varepsilon,h}_{\cdot}$) *Let $h \in H$ be such that*

$$(i)\ S_0(h)_t > 0, \quad \forall t \in (0, T]; \quad (ii)\ \dot{h}_t > k \text{ in a neighborhood of } 0, \text{ for some } k > 0. \tag{3.23}$$

Then, the process $Y^{\varepsilon,h}$ converges in law to $S_0(h)$ under W, as $\varepsilon \to 0$.

Lemma 3.11 (Relative entropy bound) *Let $(\Omega, \mathcal{F})$ be a probability space and P, Q two probability measures on $(\Omega, \mathcal{F})$ such that $dQ = F dP$. The relative entropy $H(Q|P)$ is defined as:*

$$H(Q|P) := \int_\Omega F \log(F) dP$$

Then, $\forall A \in \mathcal{F}$ we have:

$$\log\left(\frac{P(A)}{Q(A)}\right) \geq -\frac{e^{-1} + H(Q|P)}{Q(A)}. \tag{3.24}$$

Proof Applying Jensen's inequality, one has

$$\begin{aligned} \log\left(\frac{P(A)}{Q(A)}\right) &\geq \log\left(\int_A F^{-1} \frac{dQ}{Q(A)}\right) \\ &\geq -\frac{1}{Q(A)} \int_A \log(F) dQ \geq -\frac{1}{Q(A)} \int_A (\log(F)F)^+ dP. \end{aligned}$$

Using the elementary fact that $\inf_{x \geq 0} x \log(x) \geq -\frac{1}{e}$:

$$-\frac{1}{Q(A)} \int_A (\log(F)F)^+ dP \geq -\frac{e^{-1} + H(Q|P)}{Q(A)},$$

which proves (3.24). ∎

The relative entropy $H(W^{\varepsilon,h}|W)$ is easily computed using the martingale property of $F_t^{\varepsilon,h} = \exp\left(\frac{1}{\varepsilon}\int_0^t \dot{h}_s dB_s - \frac{1}{2\varepsilon^2}\int_0^t \dot{h}_s^2 ds\right)$ and Itô isometry:

$$\begin{aligned} H(W^{\varepsilon,h}|W) &= \mathbb{E}\left(F_T^{\varepsilon,h}\left(\frac{1}{\varepsilon}\int_0^T \dot{h}_t dB_t - \frac{1}{2\varepsilon^2}\int_0^T \dot{h}_t^2 dt\right)\right) \\ &= \mathbb{E}\left(\frac{1}{\varepsilon}\int_0^T F_t^{\varepsilon,h}\dot{h}_t dB_t \times \frac{1}{\varepsilon}\int_0^T \dot{h}_t dB_t\right) - \frac{1}{2\varepsilon^2}\int_0^T \dot{h}_t^2 dt \\ &= \frac{1}{\varepsilon^2}\int_0^T \dot{h}_t^2 dt - \frac{1}{2\varepsilon^2}\int_0^T \dot{h}_t^2 dt, \end{aligned}$$

therefore

$$H\left(W^{\varepsilon,h}|W\right) = \frac{1}{2\varepsilon^2}\int_0^T \dot{h}_t^2 dt. \tag{3.25}$$

The proof of Lemma 3.10 is postponed to Appendix A; using this lemma and Lemma 3.11, we can achieve here the proof of Proposition 3.8, completing the proof of the large deviation weak lower bound for the process X^ε.

Proof of Proposition 3.8 If $\mathcal{I}_T(\psi) = \infty$, (3.18) is trivially true. Then, consider $\psi \in \Omega_{\geq 0}$ such that $\mathcal{I}_T(\psi) < \infty$, and define $h \in H$ by setting $\dot{h}_t = \frac{\dot{\psi}_t - \beta(1-\gamma)\psi_t}{\sigma(1-\gamma)}$, so that $\mathcal{S}_0(h) = \psi$.

Step 1. Assume that h is such that (3.23) holds true. An application of the relative entropy bound (3.24) with $P = W$, $Q = W^{\varepsilon,h}$ yields

$$\begin{aligned} \varepsilon^2 \log\left(W\left(Y^\varepsilon \in B(\psi, R)\right)\right) \geq -\varepsilon^2 \frac{\left(e^{-1} + H(W^{\varepsilon,h}|W)\right)}{W^{\varepsilon,h}\left(Y^\varepsilon \in B(\psi, R)\right)} \\ + \varepsilon^2 \log W^{\varepsilon,h}(Y^\varepsilon \in B(\psi, R)). \end{aligned}$$

Using $W^{\varepsilon,h}(Y^\varepsilon \in B(\psi, R)) = W^\varepsilon(Y^{\varepsilon,h} \in B(\psi, R)) \to 1$ for every $R > 0$ by Proposition 3.10, and the expression of $H(W^{\varepsilon,h}|W)$ from (3.25), taking the limit as $\varepsilon \to 0$ we obtain (3.18).

Step 2. Assume now $\psi \in C^1([0, 1])$. Let h be defined as above, and define $h^n \in H$, $n \in \mathbb{N}$, by

$$\dot{h}_t^n := \dot{h}_t + 1/n. \tag{3.26}$$

We claim that $\forall n \in \mathbb{N}$, h^n satisfies (3.23). Let us first prove that condition (ii) in (3.23) holds. Observe that $\psi \geq 0$ and $\psi_0 = 0$ imply $\dot{\psi}_0 \geq 0$, hence $\dot{h}_0^n \geq 1/n$. By the continuity of $\dot{h}^n$, ensured by the fact that $\psi \in C^1([0, T])$, it follows that the condition (ii) in (3.23) holds with, say, $k = 1/(2n)$. In order to prove condition (i), we observe that the comparison principle for ODEs implies that $\forall t \in (0, T]$, $\mathcal{S}_0(h^n)_t > \mathcal{S}_0(h)_t = \psi_t \geq 0$; condition (i) is then proved. Furthermore, by the continuity of the solution to (3.20) with respect to the control parameter h, one has

$$\|\mathcal{S}_0(h^n) - \psi\|_\infty \to 0 \qquad \text{as } n \to \infty. \tag{3.27}$$

It follows from (3.27) that, for any $R > 0$

$$W\left(Y^{\varepsilon} \in B(\psi, R)\right) \geq W\left(Y^{\varepsilon} \in B(\mathcal{S}_0(h^n), R/2)\right) \tag{3.28}$$

if n is large enough. In the first part of the proof, we have shown that the weak lower bound holds for $W\left(Y^{\varepsilon} \in B(\mathcal{S}_0(h^n), R/2)\right)$; then, taking the limits as $\varepsilon \to 0$ and $R \to 0$ in (3.28), one has

$$\liminf_{R\to 0} \liminf_{\varepsilon\to 0} \varepsilon^2 \log W\left(Y^{\varepsilon} \in B(\psi, R)\right) \geq -\mathcal{I}_T(\mathcal{S}_0(h^n)) \quad \text{for every } n \in \mathbb{N}.$$

Since $\mathcal{I}_T(\mathcal{S}_0(h^n)) = \frac{1}{2}\int_0^T (\dot{h}^n)^2 dt \to \frac{1}{2}\int_0^T (\dot{h})^2 dt = \mathcal{I}_T(\psi)$, the bound (3.18) follows. Finally, a standard density argument of $C^1([0, 1])$ functions in $C([0, 1])$ allows to extend the claim to any $\psi \in \Omega_{\geq 0}$ such that $\mathcal{I}_T(\psi) < +\infty$. ■

Remark 3.12 In a classical situation, the claim would be the lower bound (3.17) for a process X^{ε} satisfying, say, $dX^{\varepsilon} = b_{\varepsilon}(X^{\varepsilon}) + \varepsilon\sigma(X^{\varepsilon})dB$ with Lipschitz coefficients σ and $b_{\varepsilon} \to b_0$, and $X_0^{\varepsilon} = x^{\varepsilon} \to x$. In this setting, fixing a control $h \in H$ and defining $X^{\varepsilon,h}$ from X^{ε} by shifting the Brownian motion B as in (3.22), it is straightforward (in fact: an application of Gronwall's Lemma) to show that $X^{\varepsilon,h}$ converges in law to the unique solution of the deterministic limit equation $d\varphi = b_0(\varphi)dt + \sigma(\varphi)dh$, $\varphi_0 = x$. In the present (degenerate) situation, the deterministic limit equation for the process $X^{\varepsilon,h}$ (obtained setting $\varepsilon = 0$ in (3.22)) coincides with the ODE (1.2) which admits infinitely many solutions. When circumventing this problem by passing through the transformed process $Y^{\varepsilon,h}$, we actually show that the convergence in law of $X^{\varepsilon,h}$ to a *particular* solution φ^* of the limiting equation is restored. Indeed, assume as in Proposition 3.10 that h is such that the unique solution ψ of the well-posed equation (3.20) with $y = 0$ is positive for every $t > 0$, and $Y^{\varepsilon,h}$ converges in law to ψ. The function ψ is easily computed, namely $\psi_t = \sigma(1-\gamma)e^{\beta(1-\gamma)t}\int_0^t e^{-\beta(1-\gamma)s}\dot{h}_s ds$. By definition, one has $X^{\varepsilon,h} = \left(Y^{\varepsilon,h}\right)^{\frac{1}{1-\gamma}} \xrightarrow{W} \psi^{\frac{1}{1-\gamma}} =: \varphi^*$. By direct computation, φ^* is absolutely continuous and such that $\varphi_0^* = 0$ and $\dot{\varphi}^* = \beta\varphi^* + \sigma(\varphi^*)^{\gamma}\dot{h}$, hence φ^* is *a* solution to (1.2); in particular,

$$\varphi_t^* := e^{\beta t}\left(\sigma(1-\gamma)\int_0^t e^{-\beta(1-\gamma)s}\dot{h}_s ds\right)^{\frac{1}{1-\gamma}}. \tag{3.29}$$

Therefore, in the small noise limit, the stochastic dynamics (3.22) performs a selection among the solutions of the limiting deterministic system (1.2), selecting the strictly positive one, φ^*. This looks reasonable in light of the fact that, though converging to zero, the drift parameter α^{ε} and the initial condition x^{ε} of the process remain strictly positive for all $\varepsilon > 0$.[5] Figure 1 shows the convergence of simulated

[5]By perturbing the initial condition and the drift in (1.2), one can retrieve the trajectory φ^* in (3.29) as the limit as $\rho \to 0$ of the solution of the equation $d\varphi_t = \rho + \beta\varphi_t dt + \sigma\varphi_t^{\gamma} dh$, $\varphi_0 = \rho$, for which existence and uniqueness hold.

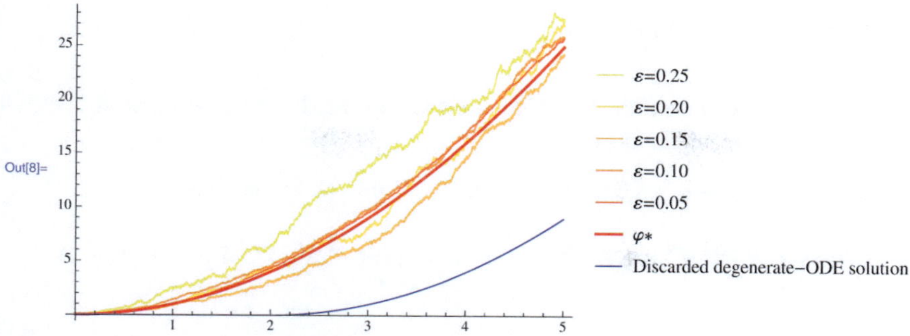

Fig. 1 An illustration of the convergence of the process $X^{\varepsilon,h}$ in (3.22) to a particular solution φ^* of the limiting deterministic sytem. Trajectories have been simulated for different values of the noise parameter ε and $\gamma = 1/2, \alpha(x) \equiv 1, \beta = 0, \sigma = 2, \dot{h} = 1, x = 0$

trajectories of the process $X^{\varepsilon,h}$ to φ^* in (3.29) as $\varepsilon \to 0$, for a given choice of the control parameter h.

Remark 3.13 **(Lower bound from the upper bound)** In general, the weak convergence of the controlled process $X^{\varepsilon,h}$ can be shown exploiting the large deviation upper bound. This goes as follows: in the notation of Remark 3.12, assume X^ε satisfies $dX^\varepsilon = b^\varepsilon(X^\varepsilon) + \varepsilon\sigma(X^\varepsilon)dB$ with Lipschitz coefficients, and define $X^{\varepsilon,h}$ from X^ε as in (3.22). Assume one has proven a large deviation upper bound analogous to (3.12) for the process $X^{\varepsilon,h}$, with a good rate function I^h depending on the control parameter h, $I^h(\psi) := \frac{1}{2}\int_0^T \left(\frac{\dot{\psi}_t - b_0(\psi_t) - \sigma(\psi_t)\dot{h}_t}{\sigma(\psi_t)}\right)^2 dt$. It is clear that I^h admits as a unique zero the solution $\varphi(h)$ of $\dot{\psi}_t = b_0(\psi_t) + \sigma(\psi_t)\dot{h}_t$. Using the compactness of the level sets of I^h and the large deviation upper bound, it is easy to conclude that

$$\lim_{\varepsilon \to 0} W\left(X^{\varepsilon,h} \notin B(\varphi(h), R)\right) = 0 \quad \forall R > 0,$$

hence $X^{\varepsilon,h} \to \varphi(h)$ in law. This provides a way of "bootstrapping" the large deviation lower bound from the upper bound (via weak convergence, together with the bound on relative entropy in Lemma 3.11). When the limit ODE has several solutions, this approach is not possible anymore: in the present case, the rate function $I^h(\psi) = \frac{1}{2}\int_0^T \left(\frac{\dot{\psi}_t - \beta\psi_t - \psi_t^\gamma \dot{h}_t}{\psi_t^\gamma}\right)^2 \mathbb{1}_{\{\psi_t > 0\}} dt$ has uncountably many zeroes, corresponding to the possible solutions of the degenerate ODE (1.2). While one is expecting that converging subsequences of the family of measures $\{W(X^{\varepsilon,h} \in \cdot)\}_\varepsilon$ converge to a probability distribution supported by the set of solutions, it is not obvious a priori how to restore a unique limit for $X^{\varepsilon,h}$ (which is why we pass through the transformed process $Y^{\varepsilon,h}$). When uniqueness for the limiting equation is granted, such an approach remains efficient, and applies outside the Markovian framework (see [8] for a treatment of delayed equations. In the setting of [8], uniqueness of solutions for the deterministic sytem is essential, and enters via their condition (H4)).

3.4 Proof of Tail Estimates

In this section, we prove the asymptotic estimates that have been stated in Sect. 2.1 and that follow from Theorem 2.1.

Proof of Proposition 2.5 Setting $\varepsilon := R^{-(1-\gamma)}$ into (2.1), one has

$$\limsup_{R\to+\infty} R^{-2(1-\gamma)} \log W\left(X_T \geq R\right) = \limsup_{\varepsilon\to 0} \varepsilon^2 \log W\left(X_T^{\varepsilon} \geq 1\right) \leq -P$$

where

$$\begin{aligned} P &= \inf\left\{I_T(\varphi) : \varphi_0 = 0, \varphi \geq 0, \varphi_T \geq 1\right\} \\ &= \inf_{y\geq 1} \inf\left\{I_T(\varphi) : \varphi_0 = 0, \varphi \geq 0, \varphi_T \geq y\right\} =: \inf_{y\geq 1} P(y). \end{aligned}$$

Fix $y \geq 1$ and a function φ in the admissible set of $P(y)$, such that $I_T(\varphi) < \infty$. Set $\psi_t = \varphi_t^{1-\gamma}$. On $\{\varphi = 0\}$, one has $\psi = 0$ as well, while for a point t in the open set $\{\varphi > 0\}$ such that $\dot{\varphi}_t$ exists, one has $\dot{\psi}_t = (1-\gamma)\frac{\dot{\varphi}_t}{\varphi_t^{\gamma}}$. Then, noting that $I_T(\varphi) < \infty$ implies that $\frac{\dot{\varphi}_t}{\varphi_t^{\gamma}} 1_{\varphi_t>0}$ is integrable on $[0, T]$, ψ is also absolutely continuous on $[0, T]$ (see [20, Corollary 3.41]). Moreover, $I_T(\varphi) = \frac{1}{2\sigma^2}\int_0^T \left(\frac{\dot{\varphi}_t - \beta\varphi_t}{\varphi_t^{\gamma}}\right)^2 1_{\varphi_t>0} dt = \frac{1}{2\sigma^2(1-\gamma)^2}\int_0^T (\dot{\psi}_t - \beta(1-\gamma)\psi_t)^2 1_{\psi_t>0} dt$. Noting that the inverse transformation $\varphi = \psi^{\frac{1}{(1-\gamma)}}$ also maps AC positive functions to AC positive functions (as $\frac{1}{(1-\gamma)} > 1$), one has

$$\begin{aligned} P(y) = \frac{1}{2\sigma^2(1-\gamma)^2} \inf\Bigg\{ \int_0^T \left(\dot{\psi}_t - \beta(1-\gamma)\psi_t\right)^2 1_{\psi_t>0} dt : \psi \text{ is abs. cont.}, \\ \psi_0 = 0, \psi \geq 0, \psi_T = y^{1-\gamma}\Bigg\}. \end{aligned}$$

When $\beta = 0$, the minimizer of this problem is $\psi_t^*(y) = y^{1-\gamma}t/T$. When $\beta \neq 0$, the solution of the Euler-Lagrange equation associated with the Lagrangian $(\dot{\psi} - \beta(1-\gamma)\psi)^2$ and the boundary conditions $\psi_0 = 0$, $\psi_T = y^{1-\gamma}$ yields the minimizer

$$\psi_t^*(y) = \frac{y^{1-\gamma}}{e^{\beta(1-\gamma)T} - e^{-\beta(1-\gamma)T}}\left(e^{\beta(1-\gamma)t} - e^{-\beta(1-\gamma)t}\right).$$

In both cases, $\psi_t^*(y) > 0$ for all $t \in (0, T]$, and the positivity constraint in $P(y)$ can be dropped. Using the monotonicity of ψ^* w.r.t. y, this yields $\inf_{y\geq 1} P(y) = P(1) = \frac{1}{2\sigma^2(1-\gamma)^2}\int_0^T \left(\dot{\psi}_t^*(1) - \beta(1-\gamma)\psi_t^*(1)\right)^2 dt$. An application of the large deviation lower bound (2.1) gives $\liminf_{R\to+\infty} R^{-2(1-\gamma)} \log W\left(X_T > R\right) = \liminf_{\varepsilon\to 0} \varepsilon^2 \log W\left(X_T^{\varepsilon} > 1\right) = -\inf_{y>1} P(y) = -P(1)$. Finally, the explicit evaluation of the integral in $P(1)$ over the function ψ^* yields the expression of the constant c_T in (2.4).

Let us consider the running maximum process. Another application of the large deviation principle (2.1) with $\varepsilon = R^{-(1-\gamma)}$ gives

$$\liminf_{R\to+\infty} R^{-2(1-\gamma)} \log W\Big(\sup_{t\in[0,T]} X_t > R\Big) \geq -\underline{c}_T$$

where $\underline{c}_T \inf\{I_T(\varphi) : \varphi_0 = 0, \varphi \geq 0, \sup_{t\in[0,T]} \varphi_t > 1\}$. Since $W\big(\sup_{t\in[0,T]} X_t > R\big) \geq W(X_t > R)$ for every $t \leq T$, one has $\underline{c}_T \leq \inf_{t\in[0,T]} c_t = c_T$, where the last identity holds for c_t is a decreasing function of t. On the other hand, $\limsup_{R\to+\infty} R^{-2(1-\gamma)} \log W\big(\sup_{t\in[0,T]} X_t \geq R\big) \leq -\bar{c}_T := -\inf\{I_T(\varphi) : \varphi_0 = 0, \varphi \geq 0, \sup_{t\in[0,T]} \varphi_t \geq 1\}$. Since

$$\begin{aligned}
\bar{c}_T &= \inf\Big\{I_T(\varphi) : \varphi_0 = 0, \varphi \geq 0, \sup_{t\in[0,T]} \varphi_t = 1, \varphi_t \geq 0\Big\} \\
&\geq \inf_{t\in[0,T]} \inf\{I_t(\phi) : \phi \text{ is abs. cont. on}[0,t], \phi_0 = 0, \phi \geq 0, \phi_t = 1\} \\
&= \inf_{t\in[0,T]} c_t = c_T
\end{aligned}$$

one has $\underline{c}_T = \bar{c}_T = c_T$, and the claim is proved. ■

As addressed in Sect. 2.1, Theorem 2.1 can also be used to obtain the leading-order asymptotics for the distribution of the time average of the process. Such a result can be used to derive the leading-order behavior of the implied volatility of Asian options $\mathbb{E}\big[\big(\frac{1}{T}\int_0^T X_t dt - K\big)^+\big]$ for large strike K.

Proposition 3.14 *Estimate* (1.5) *in Theorem 1.2 holds with* $\nu_T > 0$. *When* $\gamma = 1/2$, *the constant* ν_T *is given by*

$$\nu_T = \begin{cases} \frac{1}{2\sigma^2}\Big(T\beta^2 + \frac{4\omega^2}{T}\Big) & \textit{if } T\beta/2 < 1 \\ \frac{1}{2\sigma^2}\Big(T\beta^2 - \frac{4\omega^2}{T}\Big) & \textit{if } T\beta/2 \geq 1 \end{cases} \tag{3.30}$$

where

$$\omega = \begin{cases} \textit{the } \omega \in (0,\pi) \textit{ such that } \omega\cos\omega = T\beta/2\sin(\omega) & \textit{if } T\beta/2 < 1 \\ 0 & \textit{if } T\beta/2 = 1 \\ \textit{the } \omega \in (0,\infty) \textit{ such that } \omega\cosh(\omega) = T\beta/2\sinh(\omega) & \textit{if } T\beta(1-\gamma) \geq 1. \end{cases} \tag{3.31}$$

Remark 3.15 Following the lines of the proof of Proposition 3.14, one can prove the analogous asymptotic relation for a general time-average functional $\int_0^T X_t\mu(dt)$, where μ is a bounded signed measure on $[0, T]$. One gets

$$W\left(\int_0^T X_t\mu(dt) \geq R\right) = e^{-R^{2(1-\gamma)}(\mathcal{V}_T+\psi(R))} \qquad \text{as } R \to \infty,$$

where $\mathcal{V}_T$ is characterised by the variational formula $\mathcal{V}_T := \inf\Big\{I_T(\varphi) : \int_0^T \varphi_t\mu(dt) \geq 1, \varphi_t \geq 0, \forall t \in [0, T]\Big\}$.

Proof of Proposition 3.14 An application of th large deviation principle (2.1) with $\varepsilon := R^{-(1-\gamma)}$ yields $\limsup_{R\to+\infty} R^{-2(1-\gamma)} \log W\big(\frac{1}{T}\int_0^T X_t dt \geq R\big) = \limsup_{\varepsilon^2\to 0} \varepsilon^2 \log W\big(\frac{1}{T}\int_0^T X_t^\varepsilon dt \geq 1\big) \leq -\nu_T$, with $\nu_T = \inf\{I_T(\varphi) : \varphi_0 = 0, \varphi \geq 0, \frac{1}{T}\int_0^T \varphi_t dt \geq 1\}$. Proceeding as in the proof of Proposition 2.5, and in particular exploiting the endomorphism of $AC([0, T], \mathbb{R}_+)$ $\varphi \to \psi = \varphi^{1-\gamma}$ together with the chain rule $\dot{\psi} = \dot{\varphi}/\varphi^\gamma 1_{\varphi>0}$, one has

$$\begin{aligned}
\nu_T &= \inf\left\{I_T(\varphi) : \varphi_0 = 0, \varphi \geq 0, \frac{1}{T}\int_0^T \varphi_t dt \geq 1\right\} \\
&= \frac{T}{2\sigma^2(1-\gamma)^2} \inf\left\{\int_0^1 \left(\dot{\psi}_{Tt} - \beta(1-\gamma)\psi_{Tt}\right)^2 dt : \psi_0 = 0,\right. \\
&\qquad\qquad \left.\psi \geq 0, \int_0^1 \psi_{Tt}^{1/(1-\gamma)} dt \geq 1\right\} \\
&= \frac{1}{2T\sigma^2(1-\gamma)^2} \inf\left\{\int_0^1 \left(\frac{d}{dt}(\psi_{Tt}) - T\beta(1-\gamma)\psi_{Tt}\right)^2 dt : \psi_0 = 0,\right. \\
&\qquad\qquad \left.\psi \geq 0, \int_0^1 \psi_{Tt}^{1/(1-\gamma)} dt \geq 1\right\} \\
&= \inf_{\eta\geq 1} \frac{1}{2T\sigma^2(1-\gamma)^2} \inf\left\{\int_0^1 \left(\dot{\phi}_t - T\beta(1-\gamma)\phi_t\right)^2 dt : \phi_0 = 0,\right. \\
&\qquad\qquad \left.\phi \geq 0, \int_0^1 \phi_t^{1/(1-\gamma)} dt = \eta\right\} =: \inf_{\eta\geq 1} J(\eta).
\end{aligned}$$

When $\gamma = 1/2$, the latter variational problem was studied in [12, Exercise 2.1.13]. The explicit solution for J provides the expression of the constant $\nu_T = \inf_{\eta\geq 1} J(\eta) = J(1)$ given in (3.30). The large deviation lower bound yields $\liminf_{R\to+\infty} R^{-2(1-\gamma)} \log W\big(\frac{1}{T}\int_0^T X_t dt > R\big) = \liminf_{\varepsilon^2\to 0} \varepsilon^2 \log W\big(\frac{1}{T}\int_0^T X_t^\varepsilon dt > 1\big) \geq -J(1) = \eta_T$, and the claim is proved. ■

Consistency check with the explicit formulae for the integrated CIR process. Let us consider the case $\gamma = 1/2$, and compare Proposition 3.14 with the moment explosion of the integrated CIR process, corresponding to $\alpha(x) \equiv \alpha \geq 0$ in condition (H2). We focus on the (common) case of a mean-reverting drift, i.e. $\beta < 0$; computations for $\beta > 0$ are similar. Estimate (1.5) establishes that $\frac{1}{T}\int_0^T X_t dt$ has finite exponential moments up to order ν_T: more precisely,

$$u^* := \sup\{u > 0 : \mathbb{E}\Big[\exp\Big(\frac{u}{T}\int_0^T X_t dt\Big)\Big] < \infty\} = \sup\{\nu > 0 : \mathbb{P}\Big(\frac{1}{T}\int_0^T X_t dt > x\Big) = O(e^{-\nu x}) \text{ as } x \to \infty\} = \nu_T \tag{3.32}$$

(for the central identity, see for example [15, Sect. 4]); in other words, ν_T is the positive critical exponent of $\frac{1}{T}\int_0^T X_t dt$. Critical exponents for integrated CIR have been assessed by [2, 14, 18] relying (essentially) on the affine structure of the process. It is typical to obtain u^* by inverting an explicit explosion time: following [2, Corollary 3.3], $\mathbb{E}[\exp(\frac{u}{T}\int_0^T X_t dt)]$ is always finite if $u \le T\beta^2/(2\sigma^2)$, and if $u > T\beta^2/(2\sigma^2)$, the expectation is finite for $T < T^*(u)$ and infinite for $T > T^*(u)$, where T^* reads

$$T^*(u) = 2\frac{\pi + \arctan\left(\frac{\gamma(u)}{\beta}\right)}{\gamma(u)},$$

where $\gamma(u) = \sqrt{2\sigma^2\frac{u}{T} - \beta^2}$. Fixing T and using the monotonicity of T^*, this means that the expectation becomes infinite for $u > u^*$ with u^* the solution to

$$\pi + \arctan\left(\frac{\gamma(u)}{\beta}\right) = \frac{T}{2}\gamma(u) \tag{3.33}$$

As an equation in γ, it is easy to see that (3.33) has a unique root γ^* on $\mathbb{R}^+$ such that $\frac{T}{2}\gamma^* \in (\frac{\pi}{2}, \pi)$. From the definition of γ,

$$u^* = \frac{1}{2\sigma^2}(T\beta^2 + T(\gamma^*)^2) = \frac{1}{2\sigma^2}\left(T\beta^2 + \frac{4}{T}\left(\frac{T\gamma^*}{2}\right)^2\right) = \frac{1}{2\sigma^2}\left(T\beta^2 + \frac{4}{T}(\omega^*)^2\right)$$

setting $\omega^* = \frac{T\gamma^*}{2}$. From (3.33), ω^* is the unique solution to $\omega = \pi + \arctan\left(\frac{2\omega}{T\beta}\right)$, which is equivalent to $\tan(\omega) = \frac{2\omega}{T\beta}$ together with $\omega \in (\frac{\pi}{2}, \pi)$: one sees that this definition coincides with the one for ω in (3.31) (noticing we are in the first case when $\beta < 0$).

Acknowledgments We would like to thank an anonymous referee for the careful reading of the paper and for several valuable comments which helped to improve the presentation. We thank Peter Friz for stimulating discussions and Antoine Jacquier for useful references on integrated CIR processes. SDM (affiliated with TU-Berlin when this work was started) acknowledges partial financial support from Matheon. GC acknowledges financial support from Berlin Mathematical School. SDM and GC acknowledge financial support for travel expenses from the research program 'Chaire Risques Financiers' of the Fondation du Risque.

Appendix A

We complete the proof of Proposition 3.2 here.

Proof of Proposition 3.2 Let us define an auxiliary process $\overline{X}$ by

$$d\overline{X}_t = |\alpha|_\infty dt + \sigma \exp(-(1-\gamma)|\beta|t)\overline{X}_t^\gamma dB_t, \quad \overline{X}_0 = x;$$

after a simple application of the product rule, one has that the process $Z_t := \exp(|\beta|t)\overline{X}_t$ is a solution to

$$dZ_t = \big(|\alpha|_\infty \exp(|\beta|t) + |\beta|Z_t\big)dt + \sigma Z_t^\gamma dB_t, \quad Z_0 = x.$$

Since $|\alpha|_\infty \exp(|\beta|t) \geq |\alpha|_\infty$, an application of the comparison principle for SDE's [17, Proposition 5.2.18] yields $Z_t \geq \tilde{X}_t$, for all $t \geq 0$. Therefore, if $\overline{X}^{2(1-\gamma)}$ admits (some) exponential moments, so does $Z_t^{2(1-\gamma)}$ and by comparison $\tilde{X}_t^{2(1-\gamma)}$. In this sense, the process $\overline{X}$ is not covered by Proposition 3.3 in [9], since the latter deals with the case of a diffusion coefficient that does not depend on time (see [9, Eq. (3.1)]); nonetheless, the essential condition that [9, Proposition 3.3] relies on is the presence of a non-strictly positive slope coefficient, say b in the drift term $a + bX$ (cf. [9, Eq. (3.3)]). Since this is the case for the process $\overline{X}$ (which has zero slope coefficient b), it is straightforward to extend the proof to the present setting: in particular, in the spirit of Lamperti's change-of-variable argument, one still defines the function $\varphi(x) = \int_0^x \frac{1}{\sigma x^\gamma} = \frac{1}{\sigma(1-\gamma)} x^{1-\gamma}$ and studies the process $\tilde{\varphi}(X_t)$, where the function $\tilde{\varphi}$ is a modification of φ identically null around zero. Itô's formula shows that $\tilde{\varphi}(X_t)$ is an Itô process with bounded quadratic variation and a bounded drift term; the existence of quadratic exponential moments for $\tilde{\varphi}(X_t)$, then, is a consequence of Dubins–Schwarz time-change argument and Fernique's theorem. As a consequence, there exist $c', C > 0$ such that $\sup_{t \leq T} \mathbb{E}[\exp(c'\overline{X}_t^{2(1-\gamma)})] \leq C$; it follows $\sup_{t \leq T} \mathbb{E}[\exp(c\tilde{X}_t^{2(1-\gamma)})] \leq \sup_{t \leq T} \mathbb{E}[\exp(cZ_t^{2(1-\gamma)})] \leq C$ with $c := c' \exp(-2|\beta|(1-\gamma)T)$, and the claim is proved. ∎

We report the statement given in [26, Chap. 2, Theorem 2.13].

Lemma A.1 (Garsia-Rodemich-Rumsey's Lemma) *Let p and Ψ be continuous, strictly increasing functions on $[0, +\infty)$ such that $p(0) = \Psi(0) = 0$ and $\lim_{t \to +\infty} \Psi(t) = +\infty$. If $\omega \in \Omega$ is such that:*

$$\int_0^T \int_0^T \Psi\left(\frac{|\omega_t - \omega_s|}{p(|t-s|)}\right) dsdt \leq K, \tag{A.1}$$

then

$$|\omega_t - \omega_s| \leq 8 \int_0^{|t-s|} \Psi^{-1}\left(\frac{4K}{u^2}\right) dp(u). \tag{A.2}$$

Lemma A.1 allows us to prove Proposition 3.3:

Proof of Proposition 3.3 Assume that (3.3) holds true with the left hand side replaced by $K > 0$. Applying Lemma A.1 with the choice of functions $\Psi(y) = \exp(\varepsilon^{-2}y) - 1$, $p(y) = \sqrt{y}$, one has for all s, t

$$
\begin{aligned}
|\omega_t - \omega_s| \leq 8 \int_0^{|t-s|} \Psi^{-1}\left(\frac{4K}{u^2}\right) dp(u) &= 8\varepsilon^2 \int_0^{|t-s|} \log\left(\frac{4K}{u^2} + 1\right) dp(u) \\
&\leq 8\varepsilon^2 \left[\int_0^{|t-s|} \log\left(4K + T^2\right) dp(u)\right. \\
&\quad \left. + \int_0^{|t-s|} \log\left(u^{-2}\right) dp(u)\right] \\
&\leq 8\varepsilon^2 \left[\sqrt{|t-s|} \log\left(4K + T^2\right)\right. \\
&\quad \left. + \sqrt{|t-s|}\,(4 - 2\log(|t-s|))\right].
\end{aligned}
$$

Dividing on both sides by $(t-s)^\eta$ and taking suprema we obtain

$$
\|\omega\|_\eta \leq 8\varepsilon^2 \left(\log\left(4K + T^2\right) T^{1/2-\eta} + 4T^{1/2-\eta} + K_\eta\right).
$$

Since the right hand side in the last estimate is $K_{\varepsilon,\eta}^{-1}(K)$, (3.3) yields (3.4). ∎

Finally, we prove Lemma 3.10.

Proof of Lemma 3.10 Denote T^ε the stopping time

$$
T^\varepsilon(\omega) = \inf\left\{t \geq 0 : \omega_t \leq \frac{1}{2}\varepsilon x^{1-\gamma}\right\}. \tag{A.3}
$$

We can apply Itô formula to the function $f(x) = x^{1-\gamma}$ up to time $T^\varepsilon(Y^{\varepsilon,h})$, and obtain

$$
Y_t^{\varepsilon,h} - \varepsilon x^{1-\gamma} = \int_0^t \tilde{b}^\varepsilon(Y_s^{\varepsilon,h})ds + \sigma(1-\gamma)h_t + \varepsilon\sigma(1-\gamma)B_t, \quad \forall t \leq T^\varepsilon(Y^{\varepsilon,h}), \quad a.s. \tag{A.4}
$$

where $\tilde{b}^\varepsilon$ is given by

$$
\tilde{b}_\varepsilon(y) := (1-\gamma)\varepsilon^{\frac{1}{1-\gamma}} \alpha(\varepsilon^{-\frac{1}{(1-\gamma)}} y^{\frac{1}{(1-\gamma)}}) \frac{1}{y^{\frac{\gamma}{1-\gamma}}} - \frac{\sigma^2\gamma(1-\gamma)}{2}\varepsilon^2 \frac{1}{y} + \beta(1-\gamma)y \tag{A.5}
$$

We need to prove

$$\lim_{\varepsilon\to 0} W\left(\sup_{t\in[0,T]} |Y_t^{\varepsilon,h} - \mathcal{S}_0(h)_t| \leq R\right) = 1 \qquad \forall R > 0. \tag{A.6}$$

In order to simplify the notation, there is no ambiguity in writing Y instead of $Y^{\varepsilon,h}$ inside this proof.

Step 1. We first prove (A.6) under the assumption

$$k := \inf_{t\in[0,T]} \dot{h}_t > 0 \tag{A.7}$$

Let us fist show that

$$\lim_{\varepsilon\to 0} W\left(T^{\varepsilon}\left(Y^{\varepsilon,h}\right) \leq T\right) = 0 \tag{A.8}$$

A direct computation shows that there exist a constant $c > 0$ depending on x, σ, $\alpha(\cdot)$ such that:

$$\inf_{y\geq \frac{1}{2}\varepsilon x^{1-\gamma}} \left\{\tilde{b}^{\varepsilon}(y) - \beta(1-\gamma)y\right\} \geq -c\varepsilon. \tag{A.9}$$

Define $(Z_t)_{t\in[0,T]}$ by

$$Z_t = \varepsilon x^{1-\gamma} + (-c\varepsilon + \sigma(1-\gamma)k)\,t + \beta(1-\gamma)\int_0^t Z_s ds + \varepsilon\sigma(1-\gamma)B_t \tag{A.10}$$

Using (A.9), it follows from the comparison principle for SDEs that

$$Y_t \geq Z_t \quad \forall\, t \leq T^{\varepsilon}(Y), \quad a.s. \tag{A.11}$$

We claim that

$$W\left(T^{\varepsilon}(Z) \leq T\right) \to 0 \tag{A.12}$$

holds true. Since $W\left(T^{\varepsilon}(Y) \leq T\right) \leq W\left(T^{\varepsilon}(Z) \leq T\right)$ by (A.11), then (A.8) holds. We prove (A.12) later on. Now, it follows from the definition of $\mathcal{S}_0(h)_t$ and an application of Gronwall's Lemma that

$$|Y_t - \mathcal{S}_0(h)_t| \leq \varepsilon\left(c + \sigma(1-\gamma)\sup_{t\in[0,T]}|B_t|\right)e^{|\beta|(1-\gamma)T} =: \Theta_T \qquad \forall t \leq T^{\varepsilon}(Y),$$

therefore, for any $R > 0$ and ε small enough

$$\begin{aligned} W\left(\sup_{t\in[0,T^{\varepsilon}]}|Y_t - \mathcal{S}_0(h)_t| \leq R\right) &\geq W\left(\left\{\sup_{t\in[0,T^{\varepsilon}(Y)]}|Y_t - \mathcal{S}_0(h)_t| \leq \Theta_T\right\} \cap \left\{\Theta_T^{\varepsilon} \leq R\right\}\right) \\ &\geq W\left(\left\{T^{\varepsilon}(Y) \geq T\right\} \cap \left\{\Theta_T^{\varepsilon} \leq R\right\}\right). \end{aligned}$$

Since both the events in the right hand side of the last inequality have probability converging to 1, (A.6) follows, and Lemma 3.10 is proved under condition (A.7).

Step 2. We assume that (A.7) holds only on the time interval $[0, \rho]$, that is $\dot{h}_t \geq k$ for every $t \leq \rho$, for some $k, \rho > 0$. Repeating the argument of Step 1 with $T = \rho$, we have

$$\lim_{\varepsilon \to 0} W\left(\sup_{t\in[0,\rho]} |Y_t - \mathcal{S}_0(h)_t| \leq R^{'} \right) = 1, \quad \forall R^{'} > 0 \qquad \text{(A.13)}$$

We apply estimate (A.13) together with a localization argument. Define a time-shift operator $\tau_\rho \omega$, for every $\omega \in \Omega$, by $(\tau_\rho \omega)_t = \omega_{\rho+t}$ for all $t \in [0, T - \rho]$. For any fixed $y > 0$, denote $X^{y,\rho}$ the strong solution of the SDE:

$$X_t^{y,\rho} = y^{\frac{1}{(1-\gamma)}} + \int_0^t b^\varepsilon(X_s^{y,\rho}) + \sigma |X_s^{y,\rho}|^\gamma \dot{h}_{\rho+s} ds + \varepsilon\sigma \int_0^t |X_s^{y,\rho}|^\gamma dB_s$$

and set

$$Y^{y,\rho} := (X^{y,\rho})^{1-\gamma}.$$

Note that $Y^{y,\rho}$ is well defined since $X^{y,\rho} \geq 0$ for all $t \in [0, T]$, W-almost surely. If $h = 0$ the non negativity of the trajectories of $X^{y,\rho}$ follows from an application Proposition 3.1 in [9] and extends to $h \in H$ by an application of the Girsanov theorem. By definition of Y and $Y^{y,\rho}$, the Markov property yields

$$\mathbb{E}(f(\tau_\rho Y)|\mathcal{F}_\rho) = \mathbb{E}(f(Y^{Y_\rho,\rho}))$$

By the continuity of the map $(h, y) \mapsto \mathcal{S}_y(h)$ we can choose $R' > 0$ such that

$$\sup_{y\in B(\mathcal{S}_0(h)_\rho, R')} \sup_{t\in[0,T-\rho]} |\mathcal{S}_y(\tau_\rho h)_t - \mathcal{S}_{\mathcal{S}_0(h)_\rho}(\tau_\rho h)_t| \leq \frac{R}{2} \qquad \text{(A.14)}$$

Therefore, using (A.14) the following inclusion of events holds (assume w.lo.g $R' \leq \frac{R}{2}$):

$$\left\{ \sup_{t\in[0,T]} |Y_t - \mathcal{S}_0(h)_t| \leq R \right\} \supseteq \left\{ \sup_{[0,\rho]} |Y_t - \mathcal{S}_0(h)_t| \leq R' \right\}$$
$$\cap \left\{ \sup_{t\in[0,T-\rho]} |\tau_\rho(Y)_t - \mathcal{S}_{Y_\rho}(\tau_\rho h)_t| \leq \frac{R}{2} \right\}$$

Applying the Markov property

$$W\left(\sup_{t\in[0,T]}|Y_t-\mathcal{S}_0(h)_t|\leq R\right) \tag{A.15}$$
$$\geq \mathbb{E}\left(\mathbb{1}_{\{\sup_{t\in[0,\rho]}|Y_t-\mathcal{S}_0(h)_t|\leq R'\}}W\left(\sup_{t\in[0,T-\rho]}|Y_t^{Y_\rho,\rho}-S_{Y_\rho}(\tau_\rho h)_t|\leq\frac{R}{2}\right)\right)$$
$$\geq W\left(\sup_{t\in[0,\rho]}|Y_t-\mathcal{S}_0(h)_t|\leq R'\right)$$
$$\times\inf_{y\in B(\mathcal{S}_0(h)_\rho,R')}W\left(\sup_{t\in[0,T-\rho]}|Y_t^{y,\rho}-\mathcal{S}_y(\tau_\rho h)_t|\leq\frac{R}{2}\right)$$

We want to show that

$$\lim_{\varepsilon\to 0}\inf_{y\in B(\mathcal{S}_0(h)_\rho,R')}W\left(\sup_{t\in[0,T-\rho]}|Y_t^{y,\rho}-\mathcal{S}_y(\tau_\rho h)_t|\leq\frac{R}{2}\right)=1 \tag{A.16}$$

It follows from the hypothesis $\mathcal{S}_0(h)_t>0\ \forall t>0$ and the continuity of the map $(y,h)\mapsto\mathcal{S}_y(h)$ that, if R', R are small enough

$$y^*:=\inf_{y\in B(\mathcal{S}_0(h)_\rho,R')}\inf_{t\in[0,T-\rho]}\mathcal{S}_y(\tau_\rho h)_t-\frac{R}{2}>0. \tag{A.17}$$

Define $U^{y,\rho}$ as the unique strong solution of the SDE:

$$U_t^{y,\rho}=y+\int_0^t\left(\tilde{b}_u^\varepsilon(U_s^{y,\rho})+\sigma(1-\gamma)\dot{h}_{s+\rho}\right)ds+\varepsilon\sigma(1-\gamma)B_t,$$

where

$$\tilde{b}_u^\varepsilon(y)=\begin{cases}\tilde{b}^\varepsilon(y) & \text{if } y\geq y^*\\ \beta(1-\gamma)y+(1-\gamma)\varepsilon^{\frac{1}{1-\gamma}}\alpha(\varepsilon^{-\frac{1}{(1-\gamma)}}(y^*)^{\frac{1}{(1-\gamma)}})\frac{1}{(y^*)^{\frac{\gamma}{1-\gamma}}}-\frac{\sigma^2\gamma(1-\gamma)}{2}\varepsilon^2\frac{1}{y^*} & \text{if } y<y^*.\end{cases}$$

Then one has

$$W\left(\sup_{t\in[0,T-\rho]}|Y_t^{y,\rho}-S_y(\tau_\rho h)_t|\leq\frac{R}{2}\right)=W\left(\sup_{t\in[0,T-\rho]}|U_t^{y,\rho}-S_y(\tau_\rho h)_t|\leq\frac{R}{2}\right). \tag{A.18}$$

Now observing that $\tilde{b}^u_\varepsilon$ is globally Lipschitz continuous $\forall \varepsilon > 0$ and $C^\varepsilon := \sup_{y\in\mathbb{R}} |\tilde{b}^\varepsilon_u(y) - \beta(1-\gamma)y| \to 0$, an application of Gronwall's lemma gives

$$\mathbb{E}\left(\sup_{t\in[0,T-\rho]} \left|U^{y,\rho}_t - \mathcal{S}_y(\tau_\rho h)_t\right|\right) \le (C^\varepsilon T + 2\varepsilon\sigma(1-\gamma)\sqrt{T})\exp(|\beta(1-\gamma)|T). \tag{A.19}$$

By letting $\varepsilon \to 0$ and applying the Markov inequality, observing that the right hand side of (A.19) does not depend on y, we have proven (A.16). By letting $\varepsilon \to 0$ in (A.15) and applying (A.13) and (A.16), the proof of Lemma 3.10 is complete. ■

Proof of (A.12). Observe that $\tilde{Z} := \frac{1}{\varepsilon} Z$ is an Ornstein-Uhlenbeck process,

$$\tilde{Z}_t = x^{1-\gamma} + \mu_\varepsilon t + \beta(1-\gamma)\int_0^t \tilde{Z}_s ds + \sigma(1-\gamma)B_t \tag{A.20}$$

where $\mu_\varepsilon := \frac{1}{\varepsilon}(-c\varepsilon + \sigma(1-\gamma)k) = -c + \frac{\sigma(1-\gamma)k}{\varepsilon}$. It is immediate by the definition of $\tilde{Z}$ that $W(T^\varepsilon(Z) \le T) = W\left(\inf_{t\in[0,T]} \tilde{Z} \le \frac{x^{1-\gamma}}{2}\right)$. The explicit representation of $\tilde{Z}$ reads

$$\tilde{Z}_t := x^{1-\gamma}e^{\beta(1-\gamma)t} + f_\varepsilon(t) + \sigma(1-\gamma)\exp(\beta(1-\gamma)t)\int_0^t \exp(-\beta(1-\gamma)s)dB_s \tag{A.21}$$

with $f_\varepsilon(t) = -\frac{\mu_\varepsilon(1-\exp(\beta(1-\gamma)t))}{\beta(1-\gamma)}$. Consider a deterministic time τ_ε with $\tau_\varepsilon \to 0$ as $\varepsilon \to 0$, to be chosen precisely later on. Noting that f_ε is a decreasing function, for $\tau_\varepsilon \le t \le T$ one has

$$\tilde{Z}_t \ge f_\varepsilon(\tau_\varepsilon) - \sigma(1-\gamma)\left|\int_0^t \exp(-\beta(1-\gamma)s)dB_s\right|; \tag{A.22}$$

hence, using Markov's inequality and Doob's inequality

$$\begin{aligned} W\left(\inf_{t\in[\tau_\varepsilon,T]} \tilde{Z}_t \le x^{1-\gamma}/2\right) &\le W\left(\sup_{t\in[\tau_\varepsilon,T]} \sigma(1-\gamma)\left|\int_0^t \exp(-\beta(1-\gamma)s)dB_s\right|\right. \\ &\qquad \left.\ge f_\varepsilon(\tau_\varepsilon) - x^{1-\gamma}/2\right) \\ &\le C\sigma(1-\gamma)\left(f_\varepsilon(\tau_\varepsilon) - x^{1-\gamma}/2\right)^{-1} \\ &\qquad \times \left(\int_0^T \exp(-2\beta(1-\gamma)s)ds\right)^{\frac{1}{2}}. \end{aligned}$$

Now, the choice $\tau_\varepsilon = \sqrt{\varepsilon}$ gives $f_\varepsilon(\tau_\varepsilon) \sim \mu_\varepsilon \tau_\varepsilon \to \infty$ as $\varepsilon \to 0$, so that $\left(f_\varepsilon(\tau_\varepsilon) - x^{1-\gamma}/2\right)^{-1} \to 0$. On the other hand, $\inf_{t\in[0,\tau_\varepsilon]} \tilde{Z}_t \to x^{1-\gamma}$ a.s. as $\varepsilon \to 0$, hence $W\left(\inf_{t\in[0,\tau_\varepsilon]} \tilde{Z}_t \leq x/2\right) \to 0$ as $\varepsilon \to 0$, and the claim is proven.

References

1. Abramowitz, M., Stegun, I.A.: Handbook of Mathematical Functions with Formulas, Graphs, and Mathematical Tables, 10th edn. Dover, New York (1972)
2. Andersen, L., Piterbarg, V.: Moment explosions in stochastic volatility models. Financ. Stoch. **11**, 29–50 (2007)
3. Azencott, R.: Grandes déviations et applications. Ecole d'été de Probabilités de Saint-Flour VIII-1978. Lecture Notes in Mathematics, vol. 774, pp. 1–176. Springer, Berlin (1980)
4. Baldi, P., Caramellino, L.: General Freidlin-Wentzell large deviations and positive diffusions. Stat. Probab. Lett. **81**, 1218–1229 (2011)
5. Ben Arous, G.: Développement asymptotique du noyau de la chaleur hypoelliptique hors du cut-locus. Annales scientifiques de l'Ecole Normale Supérieure **4**(21), 307–331 (1988)
6. Bingham, N.H., Goldie, C.M., Teugels, J.L.: Regular Variation. Cambridge University Press, Cambridge (1987)
7. Bismut, J.-M.: Large Deviations and the Malliavin Calculus. Birkhäuser, Boston (1984)
8. Chiarini, A., Fischer, M.: On large deviations for small noise Itô processes. Adv. Appl. Probab. **46**(4), 1126–1147 (2014)
9. De Marco, S.: Smoothness and asymptotic estimates of densities for SDEs with locally smooth coefficients and applications to square root-type diffusions. Ann. Appl. Probab. **4**(21), 1282–1321 (2011)
10. Dembo, A., Zeitouni, O.: Large Deviations Techniques and Applications, 2nd edn. Applications of Mathematics, Springer, New York (1998)
11. Deuschel, J.-D., Friz, P., Jacquier, A., Violante, S.: Marginal density expansions for diffusions and stochastic volatility, II: Applications. Comm. in Pure and Applied Math. **67**(1), 40–82 (2014)
12. Deuschel, J-D., Stroock., W.: Large Deviations. Pure and Applied Mathematics. American Mathematical Society, New York, London. Revised edition of: An introduction to the theory of large deviations/D.W. Stroock. cop.1984 (2000)
13. Donati-Martin, C., Rouault, A., Yor, M., Zani, M.: Large deviations for squares of Bessel and Ornstein-Uhlenbeck processes. Probab. Theory Relat. Fields **129**, 261–289 (2004)
14. Dufresne, D.: The integrated square-root process. Research Paper no. 90, Centre for Actuarial Studies, University of Melbourne (2001)
15. Gulisashvili, A.: Asymptotic formulas with error estimates for call pricing functions and the implied volatility at extreme strikes. SIAM J. Financ. Math. **1**(1), 609–641 (2010)
16. Jeanblanc, M., Yor, M., Chesney, M.: Mathematical Methods for Financial Markets. Springer finance. Springer, Dordrecht, Heidelberg, London (2009)
17. Karatzas, I., Shreve, S.: Brownian Motion and Stochastic Calculus, 2nd edn. Springer, New York (1991)
18. Keller-Ressel, M.: Moment explosions and long-term behavior of affine stochastic volatility models. Math. Financ. **21**, 73–98 (2011)
19. Klebaner, F., Liptser, R.: Asymptotic analysis of ruin in the constant elasticity of variance model. Theory Probab. Appl. **55**(2), 291–297 (2011)
20. Leoni, G.: A First Course in Sobolev Spaces. Graduate studies in mathematics. American Mathematical Society, Cambridge (2009)
21. Lions, P.-L., Musiela, M.: Correlations and bounds for stochastic volatility models. Annales de l'Institut H. Poincaré **24**, 1–16 (2007)

22. Revuz, D., Yor, M.: Continuous Martingales and Brownian Motion, 3rd edn. Springer, New York (1999)
23. Robertson, S.: Sample path large deviations and optimal importance sampling for stochastic volatility models. Stoch. Process. Appl. **120**(1), 66–83 (2010)
24. Schöbel, R., Zhu, J.: Stochastic volatility with an Ornstein-Uhlenbeck process: an extension. Eur. Financ. Rev. **3**(1), 23–46 (1999)
25. Stein, E.M., Stein, J.C.: Stock price distribution with stochastic volatility: an analytic approach. Rev. Financ. Stud. **4**, 727–752 (1991)
26. Stroock, D., Varadhan, S.R.S.: Multidimensional Diffusion Processes. Grundlehren der mathematischen Wissenschaften. Fundamental Principles of Mathematical Sciences, vol. 233. Springer, Berlin (1979). Reprinted in 2006
27. Trevisan, D.: Zero noise limits using local times. Electron. Commun. Probab. **18**(31), 1–7 (2013)

Long Time Asymptotics for Optimal Investment

Huyên Pham

Abstract This survey reviews portfolio selection problem for long-term horizon. We consider two objectives: (i) maximize the probability for outperforming a target growth rate of wealth process (ii) minimize the probability of falling below a target growth rate. We study the asymptotic behavior of these criteria formulated as large deviations control problems, that we solve by duality method leading to ergodic risk-sensitive portfolio optimization problems. Special emphasis is placed on linear factor models where explicit solutions are obtained.

Keywords Long-term investment · Large deviations · Risk-sensitive control · Ergodic HJB equation · Risk-sensitive control problems · Hamilton-Jacobi-Bellman equations · Large-time asymptotic · Large deviations

MSC Classification (2000) 60F10 · 91G10 · 93E20

1 Introduction

Dynamic portfolio selection looks for strategies maximizing some performance criterion. It is a main topic in mathematical finance, first solved in continuous time in the seminal paper [13], and extended in various directions by taking into account stochastic investment opportunities, market imperfections and/or transaction costs.

Contribution to Springer Proceedings in Asymptotic Methods in Finance (Editors Friz-Gatheral-Gulisashvili-Jacquier-Teichmann), in memory of Peter Laurence.

H. Pham (✉)
Laboratoire de Probabilités et Modèles Aléatoires CNRS, UMR 7599,
Université Paris Diderot, Paris, France
e-mail: pham.xuanhuyen@gmail.com
URL: http://www.math.univ-paris-diderot.fr

H. Pham
CREST-ENSAE, Malakoff, France

P.K. Friz et al. (eds.), *Large Deviations and Asymptotic Methods in Finance*, Springer Proceedings in Mathematics & Statistics 110,
DOI 10.1007/978-3-319-11605-1_18

We refer for instance to the textbooks [10, 11] or [19], and the recent survey paper [12] for developments on this subject.

Classical criterion for investment decision is the expected utility maximization from terminal wealth, which requires to specify on one hand the utility function representing the investor's preference, and subjective by nature, and on the other hand the finite horizon. We consider in this paper an alternative behavioral foundation, with an objective criterion over long term. More precisely, we are concerned with the performance of a portfolio relative to a given target, and are interested in maximizing (resp. minimizing) the probability to outperform (resp. to fall below) a target growth rate when time horizon goes to infinity. Such criterion, formulated as a large deviations portfolio optimization problem, has been proposed by [22] in a static framework, studied in a continuous-time framework for the maximization of upside chance probability by [17], and then by [9], see also [21] in discrete-time models and [18] for a survey paper. The asymptotics of minimizing the downside risk probability is studied in [8, 15].

Large deviations portfolio optimization is a nonstandard stochastic control problem, and is tackled by duality approach. The dual control problem is an ergodic risk-sensitive portfolio optimization problem studied in [6] by dynamic programming PDE methods in a Markovian setting, see also [7], and leads to particularly tractable results with time-homogenous policies. A nice feature of the duality approach is also to relate the target level in the objective probability of upside chance maximization or downside risk minimization to the subjective degree of risk aversion, hence to make endogenous the utility function of the investor.

The rest of this paper is organized as follows. Section 2 formulates the large deviations criterion. In Sect. 3, we state the general duality relation for the large deviations optimization problem, both for the upside chance probability maximization and downside risk minimization. We illustrate in Sect. 4 our results in the Black-Scholes toy model with constant proportion portfolio. Finally, we consider in Sect. 5 a factor model for assets price, and characterize the optimal strategy of the large deviations optimization problem via the resolution of an ergodic Hamilton-Jacobi-Bellman equation from the risk-sensitive dual control. Explicit solutions are provided in the linear Gaussian factor model.

2 Large Deviations Criterion

We study a portfolio choice criterion, which is preferences-free, i.e. objective, and horizon-free, i.e. over long term investment. This is formulated as a large deviations criterion that we now describe in an abstract set-up. On a filtered probability space $(\Omega, \mathcal{F}, \mathbb{F} = (\mathcal{F}_t)_{t\geq 0}, \mathbb{P})$ supporting all the random quantities appearing in the sequel, we consider a frictionless financial market with d assets of positive price process $S = (S^1, \ldots, S^d)$. There is an agent investing at any time t a fraction π_t of her wealth in the assets based on the available information $\mathcal{F}_t$. We denote by $\mathcal{A}$ the set of admissible control strategies $\pi = (\pi_t)_{t\geq 0}$, and X^π the associated positive wealth process of dynamics:

$$dX_t^\pi = X_t^\pi \pi_t' \text{diag}(S_t)^{-1} dS_t, \quad t \geq 0, \tag{2.1}$$

where $\text{diag}(S_t)^{-1}$ denotes the diagonal $d \times d$ matrix of i-th diagonal term $1/S_t^i$. We then define the so-called *growth rate portfolio*, i.e. the logarithm of the wealth process X^π:

$$L_t^\pi := \ln X_t^\pi, \quad t \geq 0.$$

We set by $\bar{L}^\pi$ the average growth rate portfolio over time:

$$\bar{L}_t^\pi := \frac{L_t^\pi}{t}, \quad t > 0.$$

We shall then consider two problems on the long time asymptotics for the average growth rate:

(i) **Upside chance probability**: given a target growth rate ℓ, the agent wants to maximize over portfolio strategies $\pi \in \mathcal{A}$

$$\mathbb{P}\big[\bar{L}_T^\pi \geq \ell\big] \quad \text{when} \quad T \to \infty.$$

(ii) **Downside risk probability**: given a target growth rate ℓ, the agent wants to minimize over portfolio strategies $\pi \in \mathcal{A}$

$$\mathbb{P}\big[\bar{L}_T^\pi \leq \ell\big] \quad \text{when} \quad T \to \infty.$$

Actually, when horizon time T goes to infinity, the probabilities of upside chance or downside risk have typically an exponential decay in time, and we are led to the following mathematical formulations of large deviations criterion:

$$v_+(\ell) := \sup_{\pi \in \mathcal{A}} \limsup_{T \to \infty} \frac{1}{T} \ln P\big[\bar{L}_T^\pi \geq \ell\big], \tag{2.2}$$

$$v_-(\ell) := \inf_{\pi \in \mathcal{A}} \liminf_{T \to \infty} \frac{1}{T} \ln P\big[\bar{L}_T^\pi \leq \ell\big]. \tag{2.3}$$

This criterion depends on the objective probability $\mathbb{P}$, and the target growth rate ℓ, but there is no exogenous utility function, and finite horizon. Large deviations control problem (2.2) and (2.3) are nonstandard in the literature on stochastic control, and we shall study these problems by a duality approach.

3 Duality

We derive in this section the dual formulation of the large deviations criterion introduced in (2.2) and (2.3). Given $\pi \in \mathcal{A}$, if the average growth rate portfolio $\bar{L}_T^\pi$ satisfies a large deviations principle, then large deviations theory states that its rate function $I(., \pi)$ should be related to its limiting log-Laplace transform $\Gamma(., \pi)$ by duality via the Gärtner-Ellis theorem:

$$I(\ell, \pi) = \sup_\theta \big[\theta\ell - \Gamma(\theta, \pi)\big], \tag{3.1}$$

where $I(., \pi)$ is the rate function associated to the LDP of $\bar{L}_T^\pi$:

$$\limsup_{T\to\infty} \frac{1}{T} \ln P\big[\bar{L}_T^\pi \geq \ell\big] = -\inf_{\ell' \geq \ell} I(\ell', \pi) \;=\; I(\ell, \pi), \quad \ell \geq \lim_{T\to\infty} \bar{L}_T^\pi, \tag{3.2}$$

and $\Gamma(., \pi)$ is the limiting log-Laplace transform of $\bar{L}_T^\pi$:

$$\Gamma(\theta, \pi) := \limsup_{T\to\infty} \frac{1}{T} \ln \mathbb{E}\big[e^{\theta T \bar{L}_T^\pi}\big], \quad \theta \in \mathbb{R},$$

The issue is now to extend this duality relation (3.1) when optimizing over control π. To fix the ideas, let us formally derive from (3.1) and (3.2) the maximization of upside chance probability.

$$\begin{aligned}
\sup_\pi \limsup_{T\to\infty} \frac{1}{T} \ln P\big[\bar{L}_T^\pi \geq \ell\big] &= \sup_\pi \big[-I(\ell, \pi)\big] \\
&= \sup_\pi \Big[-\sup_\theta \big[\theta\ell - \Gamma(\theta, \pi)\big]\Big] \\
&= \sup_\pi \inf_\theta \big[\Gamma(\theta, \pi) - \theta\ell\big] \\
\text{(if we can invert sup and inf)} &= \inf_\theta \big[\sup_\pi \Gamma(\theta, \pi) - \theta\ell\big].
\end{aligned}$$

We thus expect that

$$v_+(\ell) = \inf_\theta \big[\Lambda_+(\theta) - \theta\ell\big], \tag{3.3}$$

where Λ_+ is defined by

$$\Lambda_+(\theta) = \sup_\pi \Gamma(\theta, \pi).$$

In other words, we should have a duality relation between the value function v_+ of the large deviations control problem, and the value function Λ_+, which is known in the mathematical finance literature, as an ergodic risk-sensitive portfolio optimization problem.

Let us now state rigorously the duality relation in an abstract (model-free) setting. We first consider the upside chance large deviations probability, and define the corresponding dual control problem:

$$\Lambda_+(\theta) := \sup_{\pi\in\mathcal{A}} \limsup_{T\to\infty} \frac{1}{T} \ln \mathbb{E}\big[e^{\theta T \bar{L}_T^{\pi}}\big], \quad \theta \geq 0. \tag{3.4}$$

We easily see from Hölder inequality that Λ_+ is convex on $\mathbb{R}_+$. The following result is due to [17].

Theorem 3.1 *Suppose that Λ_+ is finite and differentiable on $(0, \bar{\theta})$ for some $\bar{\theta} \in (0, \infty]$, and there exists $\hat{\pi}(\theta) \in \mathcal{A}$ solution to $\Lambda_+(\theta)$ for any $\theta \in (0, \bar{\theta})$. Then, for all $\ell < \Lambda'_+(\bar{\theta})$, we have:*

$$v_+(\ell) = \inf_{\theta\in[0,\bar{\theta})} \big[\Lambda_+(\theta) - \theta\ell\big].$$

Moreover, an optimal control for $v_+(\ell)$, when $\ell \in (\Lambda'_+(0), \Lambda'_+(\bar{\theta}))$, is

$$\pi^{+,\ell} = \hat{\pi}(\theta(\ell)), \quad \textit{with} \;\; \Lambda'_+(\theta(\ell)) \; = \; \ell,$$

while a nearly-optimal control for $v_+(\ell) = 0$, when $\ell \leq \Lambda'_+(0)$, is:

$$\pi^{+(n)} = \hat{\pi}(\theta_n), \quad \textit{with} \;\; \theta_n = \theta\left(\Lambda'_+(0) + \frac{1}{n}\right) \stackrel{n\to\infty}{\longrightarrow} 0,$$

in the sense that

$$\lim_{n\to\infty} \limsup_{T\to\infty} \frac{1}{T} \ln \mathbb{P}\big[\bar{L}_T^{\pi^{+(n)}} \geq \ell\big] = v_+(\ell).$$

Proof Step 1: Let us consider the Fenchel-Legendre transform of the convex function Λ_+ on $[0, \bar{\theta})$:

$$\Lambda_+^*(\ell) = \sup_{\theta\in[0,\bar{\theta})} [\theta\ell - \Lambda_+(\theta)], \quad \ell \in \mathbb{R}. \tag{3.5}$$

Since Λ_+ is C^1 on $(0, \bar{\theta})$, it is well-known (see e.g. Lemma 2.3.9 in [4]) that the function Λ_+^* is convex, nondecreasing and satisfies:

$$\Lambda_+^*(\ell) = \begin{cases} \theta(\ell)\ell - \Lambda_+(\theta(\ell)), & \text{if } \; \Lambda'_+(0) < \ell < \Lambda'_+(\bar{\theta}) \\ 0, & \text{if } \; \ell \leq \Lambda'_+(0), \end{cases} \tag{3.6}$$

$$\theta(\ell)\ell - \Lambda_+^*(\ell) > \theta(\ell)\ell' - \Lambda_+^*(\ell'), \quad \forall \Lambda_+'(0) < \ell < \Lambda_+'(\bar{\theta}), \ \forall \ell' \neq \ell, \tag{3.7}$$

where $\theta(\ell) \in (0, \bar{\theta})$ is s.t. $\Lambda_+'(\theta(\ell)) = \ell \in (\Lambda_+'(0), \Lambda_+'(\bar{\theta}))$. Moreover, Λ_+^* is continuous on $(-\infty, \Lambda_+'(\bar{\theta}))$.

Step 2: Upper bound. For all $\ell \in \mathbb{R}$, $\pi \in \mathcal{A}$, an application of Chebycheff's inequality yields:

$$\mathbb{P}[\bar{L}_T^\pi \geq \ell] \leq \exp(-\theta\ell T)\mathbb{E}[\exp(\theta T \bar{L}_T^\pi)], \quad \forall \, \theta \in [0, \bar{\theta}),$$

and so

$$\limsup_{T\to\infty} \frac{1}{T} \ln \mathbb{P}[\bar{L}_T^\pi \geq \ell] \leq -\theta\ell + \limsup_{T\to\infty} \frac{1}{T} \ln \mathbb{E}[\exp(\theta T \bar{L}_T^\pi)], \quad \forall \, \theta \in [0, \bar{\theta}).$$

By definitions of Λ_+ and Λ_+^*, we deduce:

$$\sup_{\pi\in\mathcal{A}} \limsup_{T\to\infty} \frac{1}{T} \ln \mathbb{P}[\bar{L}_T^\pi \geq \ell] \leq -\Lambda_+^*(\ell). \tag{3.8}$$

Step 3: Lower bound. Consider first the case $\ell \in (\Lambda_+'(0), \Lambda_+'(\bar{\theta}))$, and let us define the probability measure $\mathbb{Q}_T$ on $(\Omega, \mathcal{F}_T)$ via:

$$\frac{d\mathbb{Q}_T}{d\mathbb{P}} = \exp\Big[\theta(\ell) L_T^{\pi^{+,\ell}} - \Gamma_T(\theta(\ell), \pi^{+,\ell})\Big], \tag{3.9}$$

where

$$\Gamma_T(\theta, \pi) = \ln \mathbb{E}[\exp(\theta T \bar{L}_T^\pi)], \quad \theta \in [0, \bar{\theta}), \ \pi \in \mathcal{A}.$$

For any $\varepsilon > 0$, we have:

$$\begin{aligned}
\frac{1}{T} \ln \mathbb{P}[\ell - \varepsilon < \bar{L}_T^{\pi^{+,\ell}} < \ell + \varepsilon] &= \frac{1}{T} \ln\left(\int \frac{d\mathbb{P}}{d\mathbb{Q}_T} 1_{\left\{\ell-\varepsilon<\bar{L}_T^{\pi^{+,\ell}}<\ell+\varepsilon\right\}} d\mathbb{Q}_T\right) \\
&\geq -\theta(\ell)\big(\ell+\varepsilon\big) + \frac{1}{T}\Gamma_T(\theta(\ell), \pi^{+,\ell}) \\
&\quad + \frac{1}{T} \ln \mathbb{Q}_T\big[\ell - \varepsilon < \bar{L}_T^{\pi^{+,\ell}} < \ell + \varepsilon\big],
\end{aligned}$$

where we use (3.9) in the last inequality. By definition of the dual problem, this yields:

$$\begin{aligned}
\liminf_{T\to\infty} \quad & \frac{1}{T} \ln \mathbb{P}[\ell - \varepsilon < \bar{L}_T^{\pi^{+,\ell}} < \ell + \varepsilon] \\
&\geq -\theta(\ell)\big(\ell+\varepsilon\big) + \Lambda_+(\theta(\ell))
\end{aligned}$$

$$
\begin{aligned}
&+\liminf_{T\to\infty}\frac{1}{T}\ln\mathbb{Q}_T\left[\ell-\varepsilon<\bar{L}_T^{\pi^{+,\ell}}<\ell+\varepsilon\right]\\
&\geq -\Lambda_+^*(\ell)-\theta(\ell)\varepsilon\\
&+\liminf_{T\to\infty}\frac{1}{T}\ln\mathbb{Q}_T\left[\ell-\varepsilon<\bar{L}_T^{\pi^{+,\ell}}<\ell+\varepsilon\right],
\end{aligned} \tag{3.10}
$$

where the second inequality follows by the definition of Λ_+^* (and actually holds with equality due to (3.6)). We now show that:

$$
\liminf_{T\to\infty}\frac{1}{T}\ln\mathbb{Q}_T\left[\ell-\varepsilon<\bar{L}_T^{\pi^{+,\ell}}<\ell+\varepsilon\right]=0. \tag{3.11}
$$

Denote by $\tilde{\Gamma}_T$ the c.g.f. under $\mathbb{Q}_T$ of $L_T^{\pi^{+,\ell}}$. For all $\zeta\in\mathbb{R}$, we have by (3.9):

$$
\begin{aligned}
\tilde{\Gamma}_T(\zeta) &:= \ln\mathbb{E}^{\mathbb{Q}_T}[\exp(\zeta L_T^{\pi^{+,\ell}})]\\
&= \Gamma_T(\theta(\ell)+\zeta,\pi^{+,\ell})-\Gamma_T(\theta(\ell),\pi^{+,\ell}).
\end{aligned}
$$

Therefore, by definition of the dual control problem (3.4), we have for all $\zeta\in[-\theta(\ell),\bar{\theta}-\theta(\ell))$:

$$
\limsup_{T\to\infty}\frac{1}{T}\tilde{\Gamma}_T(\zeta)\leq\Lambda_+(\theta(\ell)+\zeta)-\Lambda_+(\theta(\ell)). \tag{3.12}
$$

As in part (1) of this proof, by Chebycheff's inequality, we have for all $\zeta\in[0,\bar{\theta}-\theta(\ell))$:

$$
\begin{aligned}
\limsup_{T\to\infty}\frac{1}{T}\ln\mathbb{Q}_T\left[\bar{L}_T^{\pi^{+,\ell}}\geq\ell+\varepsilon\right] &\leq -\zeta(\ell+\varepsilon)+\limsup_{T\to\infty}\frac{1}{T}\tilde{\Gamma}_T(\zeta)\\
&\leq -\zeta\,(\ell+\varepsilon)+\Lambda_+(\zeta+\theta(\ell))-\Lambda_+(\theta(\ell)),
\end{aligned}
$$

where the second inequality follows from (3.12). We deduce

$$
\begin{aligned}
\limsup_{T\to\infty}\frac{1}{T}\ln\mathbb{Q}_T\left[\bar{L}_T^{\pi^{+,\ell}}\geq\ell+\varepsilon\right] &\leq -\sup\{\zeta\,(\ell+\varepsilon)-\Lambda_+(\zeta):\zeta\in[\theta(\ell),\bar{\theta})\}\\
&\quad -\Lambda_+(\theta(\ell))+\theta(\ell)\,(\ell+\varepsilon)\\
&\leq -\Lambda_+^*\,(\ell+\varepsilon)-\Lambda_+(\theta(\ell))+\theta(\ell)\,(\ell+\varepsilon)\,,\\
&= -\Lambda_+^*\,(\ell+\varepsilon)+\Lambda_+^*(\ell)+\varepsilon\theta(\ell),
\end{aligned} \tag{3.13}
$$

where the second inequality and the last equality follow from (3.6). Similarly, we have for all $\zeta \in [-\theta(\ell), 0]$:

$$\limsup_{T\to\infty} \frac{1}{T} \ln \mathbb{Q}_T \left[\bar{L}_T^{\pi^{+,\ell}} \le \ell - \varepsilon \right] \le -\zeta(\ell-\varepsilon) + \limsup_{T\to\infty} \frac{1}{T} \tilde{\Gamma}_T(\zeta)$$
$$\le -\zeta(\ell-\varepsilon) + \Lambda_+(\theta(\ell)+\zeta) - \Lambda_+(\theta(\ell)),$$

and so:

$$\limsup_{T\to\infty} \frac{1}{T} \ln \mathbb{Q}_T \left[\bar{L}_T^{\pi^{+,\ell}} \le \ell - \varepsilon \right] \le -\sup\{\zeta(\ell-\varepsilon) - \Lambda_+(\zeta) : \zeta \in [0, \theta(\ell)]\}$$
$$-\Lambda_+(\theta(\ell)) + \theta(\ell)(\ell-\varepsilon)$$
$$\le -\Lambda_+^*(\ell-\varepsilon) + \Lambda_+^*(\theta(\ell)) - \varepsilon\theta(\ell). \quad (3.14)$$

By (3.13) and (3.14), we then get:

$$\limsup_{T\to\infty} \frac{1}{T} \ln \mathbb{Q}_T \left[\left\{ \bar{L}_T^{\pi^{+,\ell}} \le \ell - \varepsilon \right\} \cup \left\{ \bar{L}_T^{\pi^{+,\ell}} \ge \ell + \varepsilon \right\} \right]$$
$$\le \max\left\{ \limsup_{T\to\infty} \frac{1}{T} \ln \mathbb{Q}_T \left[\bar{L}_T^{\pi^{+,\ell}} \ge \ell + \varepsilon \right]; \limsup_{T\to\infty} \frac{1}{T} \ln \mathbb{Q}_T \left[\bar{L}_T^{\pi^{+,\ell}} \le \ell - \varepsilon \right] \right\}$$
$$\le \max\left\{ -\Lambda_+^*(\ell+\varepsilon) + \Lambda_+^*(\ell) + \varepsilon\theta(\ell); -\Lambda_+^*(\ell-\varepsilon) + \Lambda_+^*(\theta(\ell)) - \varepsilon\theta(\ell) \right\}$$
$$< 0,$$

where the strict inequality follows from (3.7). This implies that $\mathbb{Q}_T[\{\bar{L}_T^{\pi^{+,\ell}} \le \ell - \varepsilon\} \cup \{\bar{L}_T^{\pi^{+,\ell}} \ge \ell + \varepsilon\}] \to 0$ and hence $\mathbb{Q}_T[\ell - \varepsilon < \bar{L}_T^{\pi^{+,\ell}} < \ell + \varepsilon] \to 1$ as T goes to infinity. In particular (3.11) is satisfied, and by sending ε to zero in (3.10), we get for any $\ell' < \ell < \Lambda'_+(\bar{\theta})$:

$$\liminf_{T\to\infty} \frac{1}{T} \ln \mathbb{P}[\bar{L}_T^{\pi^{+,\ell}} > \ell'] \ge \lim_{\varepsilon\to 0} \liminf_{T\to\infty} \frac{1}{T} \ln \mathbb{P}[\ell - \varepsilon < \bar{L}_T^{\pi^{+,\ell}} < \ell + \varepsilon]$$
$$\ge -\Lambda_+^*(\ell).$$

By continuity of Λ_+^* on $(-\infty, \Lambda'_+(\bar{\theta}))$, we obtain

$$\liminf_{T\to\infty} \frac{1}{T} \ln \mathbb{P}[\bar{L}_T^{\pi^{+,\ell}} \ge \ell] \ge -\Lambda_+^*(\ell).$$

This last inequality combined with (3.8) proves the assertion for $v_+(\ell)$ when $\ell \in (\Lambda'_+(0), \Lambda'_+(\bar{\theta}))$.

Now, consider the case $\ell \le \Lambda'_+(0)$, and define $\ell_n = \Lambda'_+(0) + \frac{1}{n}$, $\pi^{+(n)} = \hat{\pi}(\theta(\ell_n))$. Then, by the same arguments as in (3.10) with $\ell_n \in (\Lambda'_+(0), \Lambda'_+(\bar{\theta}))$, we have

$$\liminf_{T\to\infty} \frac{1}{T} \ln \mathbb{P}[\bar{L}_T^{\pi^{+(n)}} \geq \ell] \geq \lim_{\varepsilon\to 0} \liminf_{T\to\infty} \frac{1}{T} \ln \mathbb{P}[\ell_n - \varepsilon < \bar{L}_T^{\pi^{+(n)}} < \ell_n + \varepsilon]$$
$$\geq -\Lambda_+^*(\ell_n).$$

By sending n to infinity, together with the continuity of Λ_+^*, we get

$$\liminf_{n\to\infty} \liminf_{T\to\infty} \frac{1}{T} \ln \mathbb{P}[\bar{L}_T^{\pi^{+(n)}} \geq \ell] \geq -\Lambda_+^*(\Lambda_+'(0)) = 0,$$

which combined with (3.8), ends the proof. □

Remark 3.1 Theorem 3.1 shows that the upside chance large deviations control problem can be solved via the resolution of the dual control problem. When the target growth rate level ℓ is smaller than $\Lambda_+'(0)$, then one can achieve almost surely over long term an average growth term above ℓ, in the sense that $v_+(\ell) = 0$, with a nearly optimal portfolio strategy which does not depend on this level. When the target level ℓ lies between $\Lambda_+'(0)$ and $\Lambda_+'(\bar{\theta})$, the optimal strategy depends on this level and is obtained from the optimal strategy for the dual control problem $\Lambda_+(\theta)$ at point $\theta = \theta(\ell)$. When $\Lambda_+'(\bar{\theta}) = \infty$, i.e. Λ_+ is steep, we have a complete resolution of the large deviations control problem for all values of ℓ. Otherwise, the problem remains open for $\ell > \Lambda_+'(\bar{\theta})$. □

Let us next consider the downside risk probability, and define the corresponding dual control problem:

$$\Lambda_-(\theta) := \inf_{\pi\in\mathcal{A}} \liminf_{T\to\infty} \frac{1}{T} \ln \mathbb{E}\big[e^{\theta T \bar{L}_T^\pi}\big], \quad \theta \leq 0. \tag{3.15}$$

Convexity of Λ_- is not so straightforward as for Λ_+, and requires the additional condition that the set of admissible controls $\mathcal{A}$ is convex. Indeed, under this condition, we observe from the dynamics (2.1) that a convex combination of wealth process is a wealth process. Thus, for any $\theta_1, \theta_2 \in (-\infty, 0)$, $\lambda \in (0, 1)$, $\pi^1, \pi^2 \in \mathcal{A}$, there exists $\pi \in \mathcal{A}$ such that:

$$\frac{\lambda\theta_1}{\lambda\theta_1 + (1-\lambda)\theta_2} X_T^{\pi^1} + \frac{(1-\lambda)\theta_2}{\lambda\theta_1 + (1-\lambda)\theta_2} X_T^{\pi^2} = X_T^\pi.$$

By concavity of the logarithm function, we then obtain

$$\ln X_T^\pi \geq \frac{\lambda\theta_1}{(\lambda\theta_1 + (1-\lambda)\theta_2)} \ln X_T^{\pi^1} + \frac{(1-\lambda)\theta_2}{(\lambda\theta_1 + (1-\lambda)\theta_2)} \ln X_T^{\pi^2},$$

and so, by setting $\theta = \lambda\theta_1 + (1-\lambda)\theta_2 < 0$:

$$\theta T \bar{L}_T^\pi \leq \lambda\theta_1 T \bar{L}_T^{\pi^1} + (1-\lambda)\theta_2 T \bar{L}_T^{\pi^2}.$$

Taking exponential and expectation on both sides of this relation, and using Hölder inequality, we get:

$$\mathbb{E}\big[e^{\theta T \bar{L}_T^{\pi}}\big] \leq \Big(\mathbb{E}\big[e^{\theta_1 T \bar{L}_T^{\pi^1}}\big]\Big)^{\lambda} \Big(\mathbb{E}\big[e^{\theta_2 T \bar{L}_T^{\pi^2}}\big]\Big)^{1-\lambda}.$$

Taking logarithm, dividing by T, sending T to infinity, and since π^1, π^2 are arbitrary in $\mathcal{A}$, we obtain by definition of Λ_-:

$$\Lambda_-(\theta) \leq \lambda\Lambda_-(\theta_1) + (1-\lambda)\Lambda_-(\theta_2),$$

i.e. the convexity of Λ_- on $\mathbb{R}_-$. Since $\Lambda_-(0) = 0$, the convex function Λ_- is either infinite on $(-\infty, 0)$ or finite on $\mathbb{R}_-$. We now state the duality relation for downside risk large deviations probability, whose proof can be found in [15].

Theorem 3.2 *Suppose that Λ_- is differentiable on $(-\infty, 0)$, and there exists $\hat{\pi}(\theta) \in \mathcal{A}$ solution to $\Lambda_-(\theta)$ for any $\theta < 0$. Then, for all $\ell < \Lambda'_-(0)$, we have:*

$$v_-(\ell) = \inf_{\theta \leq 0} \big[\Lambda_-(\theta) - \theta\ell\big],$$

and an optimal control for $v_-(\ell)$, when $\ell \in (\Lambda'_-(-\infty), \Lambda'_-(0))$ is:

$$\pi^{-,\ell} = \hat{\pi}(\theta(\ell)), \quad \text{with } \Lambda'_-(\theta(\ell)) = \ell,$$

while $v_-(\ell) = -\infty$ when $\ell < \Lambda'_-(-\infty)$.

Remark 3.2 Theorem 3.2 shows that the downside risk large deviations control problem can be solved via the resolution of the dual control problem. When the target growth rate level ℓ is smaller than $\Lambda'_-(-\infty)$, then one can find a portfolio strategy so that the average growth term almost never fall below ℓ over the long term, in the sense that $v_-(\ell) = -\infty$. When the target level ℓ lies between $\Lambda'_-(-\infty)$ and $\Lambda'_-(0)$, the optimal strategy depends on this level and is obtained from the optimal strategy for the dual control problem $\Lambda_-(\theta)$ at point $\theta = \theta(\ell)$. □

Interpretation of the dual problem
For $\theta \neq 0$, the dual problem can be written as

$$\frac{1}{\theta}\Lambda_\pm(\theta) = \sup_{\pi \in \mathcal{A}} \limsup_{T \to \infty} J_T(\theta, \pi),$$

with

$$J_T(\theta, \pi) := \frac{1}{\theta T} \ln \mathbb{E}\big[e^{\theta T \bar{L}_T^{\pi}}\big],$$

and is known in the literature as a risk-sensitive control problem. A Taylor expansion around $\theta = 0$ puts in evidence the role played by the risk sensitivity parameter θ:

$$J_T(\theta, \pi) \simeq \mathbb{E}\big[\bar{L}_T^\pi\big] + \theta T \mathrm{Var}(\bar{L}_T^\pi) + O(\theta^2).$$

This relation shows that risk-sensitive control amounts to making dynamic the Markowitz problem: one maximizes the expected average growth rate subject to a constraint on its variance. Risk-sensitive portfolio criterion on finite horizon T has been studied in [2, 3], and in the ergodic case $T \to \infty$, by [6, 16].

Endogenous utility function

Recalling that growth rate is the logarithm of wealth process, the duality relation for the upside large deviations probability means formally that for large horizon T:

$$\begin{aligned}\mathbb{P}\big[\bar{L}_T^{\pi^{+,\ell}} \geq \ell\big] &\simeq \exp\big(v_+(\ell)T\big)\\ &= \exp\big(\Lambda_+(\theta(\ell))T - \theta(\ell)\ell T\big)\\ &\simeq \mathbb{E}\Big[\big(X_T^{\pi^{+,\ell}}\big)^{\theta(\ell)}\Big]e^{-\theta(\ell)T}, \quad \text{with } \theta(\ell) > 0.\end{aligned}$$

Similarly, we have for the downside risk probability:

$$\mathbb{P}\big[\bar{L}_T^{\pi^{-,\ell}} \leq \ell\big] \simeq \mathbb{E}\Big[\big(X_T^{\pi^{-,\ell}}\big)^{\theta(\ell)}\Big]e^{-\theta(\ell)T}, \quad \text{with } \theta(\ell) < 0.$$

In other words, the target growth rate level ℓ determines endogenously the risk aversion parameter $1-\theta(\ell)$ of an agent with Constant Relative Risk Aversion (CRRA) utility function and large investment horizon. Moreover, the optimal strategy $\pi^{\pm,\ell}$ for $v_\pm(\ell)$ is expected to provide a good approximation for the solution to the CRRA utility maximization problem

$$\sup_{\pi \in \mathcal{A}} \mathbb{E}\big[(X_T^\pi)^{\theta(\ell)}\big],$$

with a large but finite time horizon.

4 A Toy Model: The Black Scholes Case

We illustrate the results of the previous section in a toy example, namely the Black-Scholes model, with one stock of price process

$$dS_t = S_t\big(bd + \sigma dW_t\big), \quad t \geq 0.$$

We also consider an agent with constant proportion portfolio strategies. In other words, the set of admissible controls $\mathcal{A}$ is equal to $\mathbb{R}$. Given a constant proportion π

$\in \mathbb{R}$ invested in the stock, and starting w.l.o.g. with unit capital, the average growth rate portfolio of the agent is equal to

$$\bar{L}_T^\pi = \frac{L_T^\pi}{T} = \left(b\pi - \frac{\sigma^2\pi^2}{2}\right) + \sigma\pi\frac{W_T}{T}.$$

It follows that $\bar{L}_T^\pi$ is distributed according to a Gaussian law:

$$\bar{L}_T^\pi \rightsquigarrow \mathcal{N}\Big(b\pi - \frac{\sigma^2\pi^2}{2}, \frac{\sigma^2\pi^2}{T}\Big),$$

and its (limiting) Log-Laplace function is equal to

$$\Gamma(\theta,\pi) := (\lim_{T\to\infty})\frac{1}{T}\ln \mathbb{E}\Big[e^{\theta T \bar{L}_T^\pi}\Big] = \theta\Big[b\pi - (1-\theta)\frac{\sigma^2\pi^2}{2}\Big]$$

• **Upside chance probability**.

The dual control problem in the upside case is then given by

$$\Lambda_+(\theta) = \sup_{\pi\in\mathbb{R}} \Gamma(\theta,\pi) = \begin{cases} \infty, & \text{if } \theta \geq 1, \\ \Gamma(\theta,\hat{\pi}(\theta)) = \frac{b^2}{2\sigma^2}\frac{\theta}{1-\theta}, & \text{if } 0 \leq \theta < 1, \end{cases}$$

with

$$\hat{\pi}(\theta) = \frac{b}{\sigma^2(1-\theta)}.$$

Hence, Λ_+ differentiable on $[0,1)$ with: $\Lambda_+'(0) = \frac{b^2}{2\sigma^2}$, and $\Lambda_+'(1) = \infty$, i.e. Λ_+ is steep. From Theorem 3.1, the value function of the upside large deviations probability is explicitly computed as:

$$\begin{aligned} v_+(\ell) &:= \sup_{\pi\in\mathbb{R}} \limsup_{T\to\infty} \frac{1}{T}\ln \mathbb{P}\big[\bar{L}_T^\pi \geq \ell\big] \\ &= \inf_{0\leq\theta<1}\big[\Lambda_+(\theta) - \theta\ell\big] \\ &= \begin{cases} 0, & \text{if } \ell \leq \Lambda_+'(0) = \frac{b^2}{2\sigma^2} \\ -\big(\sqrt{\Lambda_+'(0)} - \sqrt{\ell}\big)^2, & \text{if } \ell > \Lambda_+'(0) \end{cases} \end{aligned}$$

with an optimal strategy:

$$\pi^{+,\ell} = \begin{cases} \frac{b}{\sigma^2}, & \text{if } \ell \leq \Lambda_+'(0) \\ \sqrt{\frac{2\ell}{\sigma^2}}, & \text{if } \ell > \Lambda_+'(0). \end{cases}$$

Notice that, when $\ell \leq \Lambda'_+(0)$, we have not only a nearly optimal control as stated in Theorem 3.1, but an optimal control given by $\pi^+ = b/\sigma^2$, which is precisely the optimal portfolio for the classical Merton problem with logarithm utility function. Indeed, in this model, we have by the law of large numbers: $\bar{L}_T^{\pi^+} \to \frac{b^2}{2\sigma^2} = \Lambda'_+(0)$, as T goes to infinity, and so $\lim_{T\to\infty} \frac{1}{T} \ln \mathbb{P}[\bar{L}_T^{\pi^+} \geq \ell] = 0 = v_+(\ell)$. Otherwise, when $\ell > \Lambda'_+(0)$, the optimal strategy depends on ℓ, and the larger the target growth rate level, the more one has to invest in the stock.

• **Downside risk probability**.

The dual control problem in the downside case is then given by

$$\Lambda_-(\theta) \;=\; \inf_{\pi\in\mathbb{R}} \Gamma(\theta,\pi) = \Gamma(\theta,\hat{\pi}(\theta)) \;=\; \frac{b^2}{2\sigma^2}\frac{\theta}{1-\theta}, \qquad \theta \leq 0,$$

with

$$\hat{\pi}(\theta) = \frac{b}{\sigma^2(1-\theta)}.$$

Hence, Λ_- is differentiable on $\mathbb{R}_-$ with: $\Lambda'_-(-\infty) = 0$, and $\Lambda'_-(0) = \frac{b^2}{2\sigma^2}$. From Theorem 3.1, the value function of the downside large deviations probability is explicitly computed as:

$$\begin{aligned} v_-(\ell) &:= \inf_{\pi\in\mathbb{R}} \liminf_{T\to\infty} \frac{1}{T} \ln \mathbb{P}\big[\bar{L}_T^\pi \leq \ell\big] \\ &= \inf_{\theta\leq 0} \big[\Lambda_-(\theta) - \theta\ell\big] \\ &= \begin{cases} -\infty, & \text{if } \ell < 0 \\ -\big(\sqrt{\Lambda'_-(0)} - \sqrt{\ell}\big)^2, & \text{if } 0 \leq \ell \leq \Lambda'_-(0) \;=\; \frac{b^2}{2\sigma^2} \end{cases} \end{aligned}$$

with an optimal strategy:

$$\pi^{-,\ell} = \sqrt{\frac{2\ell}{\sigma^2}}, \quad \text{if } 0 \leq \ell \leq \Lambda'_-(0).$$

Moreover, when $\ell < 0$, and by choosing $\pi^- = 0$, we have $\bar{L}_T^{\pi^-} = 0$, so that $\mathbb{P}[\bar{L}_T^{\pi^-} \leq \ell] = 0$, and thus $v_-(\ell) = -\infty$. In other words, when the target growth rate $\ell < 0$, by doing nothing, we have an optimal strategy for $v_-(\ell)$.

Remark 4.1 The above direct calculations rely on the fact that we restrict portfolio π to be constant in proportion. Actually, the explicit forms of the value function and optimal strategy remain the same if we allow a priori portfolio strategies $\pi \in \mathcal{A}$ to change over time based on the available information, i.e. to be $\mathbb{F}$-predictable. This requires more advanced tools from stochastic control and PDEs to be presented in the sequel in a more general framework. □

5 Factor Model

We consider a market model with one riskless asset price $S^0 = 1$, and d stocks of price process S governed by

$$dS_t = \mathrm{diag}(S_t)\big(b(Y_t)dt + \sigma(Y_t)dW_t\big)$$
$$dY_t = \eta(Y_t)dt + \gamma(Y_t)dW_t,$$

where Y is a factor process valued in $\mathbb{R}^m$, and W is a $d+m$ dimensional standard Brownian motion. The coefficients b, σ, η, γ are assumed to satisfy regular conditions ensuring existence of a unique strong solution to the above stochastic differential equation, and σ is also of full rank, i.e. the $d \times d$-matrix $\sigma\sigma'$ is invertible.

A portfolio strategy π is an $\mathbb{R}^d$-valued adapted process, representing the fraction of wealth invested in the d stocks. The admissibility condition for π in $\mathcal{A}$ will be precised later, but for the moment π is required to satisfy the integrability conditions:

$$\int_0^T |\pi_t' b(Y_t)|dt + \int_0^T |\pi_t'\sigma(Y_t)|^2 dt < \infty, \quad a.s. \text{ for all } T > 0.$$

The growth rate portfolio is then given by:

$$L_T^\pi = \int_0^T \Big(\pi_t' b(Y_t) - \frac{\pi_t'\sigma\sigma'(Y_t)\pi_t}{2}\Big)dt + \int_0^T \pi_t'\sigma(Y_t)dW_t.$$

For any $\theta \in \mathbb{R}$, and π, we compute the Log-Laplace function of the growth rate portfolio:

$$\Gamma_T(\theta, \pi) := \ln \mathbb{E}\big[e^{\theta L_T^\pi}\big]$$
$$= \ln \mathbb{E}\Big[\mathcal{E}\Big(\int_0^T \theta\pi_t'\sigma(Y_t)dW_t\Big)e^{\theta\int_0^T f(\theta, Y_t, \pi_t)dt}\Big],$$

where $\mathcal{E}(.)$ denotes the Doléans-Dade exponential, and f is the function:

$$f(\theta, y, \pi) = \pi' b(y) - \frac{1-\theta}{2}\pi'\sigma\sigma'(y)\pi.$$

We now impose the admissibility condition that π lies in $\mathcal{A}$ if the Doléans-Dade local martingale $\mathcal{E}\Big(\int_0^. \theta\pi_t'\sigma(Y_t)dW_t\Big)_{0\le t\le T}$ is a true martingale for any $T > 0$, which is ensured, for instance, by the Novikov condition. In this case, this Doléans-Dade exponential defines a probability measure $\mathbb{Q}_\pi$ equivalent to $\mathbb{P}$ on $(\Omega, \mathcal{F}_T)$, and we have:

$$\Gamma_T(\theta, \pi) = \ln \mathbb{E}^{\mathbb{Q}_\pi}\Big[\exp\Big(\theta \int_0^T f(\theta, Y_t, \pi_t)dt\Big)\Big],$$

where Y is governed under $\mathbb{Q}_\pi$ by

$$dY_t = \big(\eta(Y_t) + \theta\gamma(Y_t)\sigma'(Y_t)\pi_t\big)dt + \gamma(Y_t)dW_t^\pi.$$

with W^π a $\mathbb{Q}_\pi$-Brownian motion from Girsanov's theorem.

We then consider the dual control problems:

- **Upside chance**: for $\theta \geq 0$,

$$\Lambda_+(\theta) = \sup_{\pi\in\mathcal{A}} \limsup_{T\to\infty} \frac{1}{T} \ln \mathbb{E}^{\mathbb{Q}_\pi}\Big[\exp\Big(\theta \int_0^T f(\theta, Y_t, \pi_t)dt\Big)\Big].$$

- **Downside risk**: for $\theta \leq 0$,

$$\Lambda_-(\theta) = \inf_{\pi\in\mathcal{A}} \liminf_{T\to\infty} \frac{1}{T} \ln \mathbb{E}^{\mathbb{Q}_\pi}\Big[\exp\Big(\theta \int_0^T f(\theta, Y_t, \pi_t)dt\Big)\Big].$$

These problems are known in the literature as ergodic risk-sensitive control problems, and studied by dynamic programming methods in [1, 5, 14]. Let us now formally derive the ergodic equations associated to these risk-sensitive control problems. We consider the finite horizon risk-sensitive stochastic control problems:

$$u_+(T, y; \theta) = \sup_{\pi\in\mathcal{A}} \mathbb{E}^{\mathbb{Q}_\pi}\Big[\exp\Big(\theta \int_0^T f(\theta, Y_t, \pi_t)dt\Big)\Big|Y_0 = y\Big], \ \theta \geq 0$$

$$u_-(T, y; \theta) = \inf_{\pi\in\mathcal{A}} \mathbb{E}^{\mathbb{Q}_\pi}\Big[\exp\Big(\theta \int_0^T f(\theta, Y_t, \pi_t)dt\Big)\Big|Y_0 = y\Big], \ \theta \leq 0,$$

and by using the formal substitution:

$$\ln u_\pm(T, y; \theta) \simeq \Lambda_\pm(\theta)T + \varphi_\pm(y; \theta), \quad \text{for large } T,$$

in the corresponding Hamilton-Jacobi-Bellman (HJB) equations for $u_\pm$:

$$\frac{\partial u_\pm}{\partial T} = \sup_{\pi\in\mathbb{R}^d} \big[\theta f(\theta, y, \pi)u_\pm + (\eta(y) + \theta\gamma(y)\sigma'(y)\pi)' D_y u_\pm + \frac{1}{2}\mathrm{tr}(\gamma\gamma'(y)D_y^2 u_\pm)\big],$$

we obtain the ergodic HJB equation for the pair $(\Lambda_\pm(\theta), \varphi_\pm(., \theta))$ as:

$$\Gamma(\theta) = \eta(y)' D_y\varphi + \frac{1}{2}\mathrm{tr}(\gamma\gamma'(y)D_y^2\varphi) + \frac{1}{2}|\gamma'(y)D_y\varphi|^2$$
$$+ \theta \sup_{\pi\in\mathbb{R}^d} \Big[\pi'(b(y) + \sigma(y)\gamma'(y)D_y\varphi) - \frac{1-\theta}{2}\pi'\sigma\sigma'(y)\pi\Big],$$

which is well-defined for $\theta < 1$. In the above equation $\Gamma(\theta)$ is a candidate for $\Lambda_\pm(\theta)$ while φ is a candidate solution for $\varphi_\pm$. This can be rewritten as a semi-linear ergodic PDE with quadratic growth in the gradient:

$$\begin{aligned}\Gamma(\theta) &= \big(\eta(y) + \frac{\theta}{1-\theta}\gamma\sigma'(\sigma\sigma')^{-1}b(y)\big).D_y\varphi + \frac{1}{2}\mathrm{tr}(\gamma\gamma'(y)D_y^2\varphi)\\ &+ \frac{1}{2}D_y\varphi'\gamma(y)\big[I_{d+m} + \frac{\theta}{1-\theta}\sigma'(\sigma\sigma')^{-1}\sigma(y)\big]\gamma'(y)D_y\varphi\\ &+ \frac{\theta}{2(1-\theta)}b'(\sigma\sigma')^{-1}b(y),\end{aligned} \tag{5.1}$$

and a candidate for optimal feedback control of the dual problem:

$$\hat{\pi}(y;\theta) = \frac{1}{1-\theta}(\sigma\sigma')^{-1}(y)\big[b(y) + \sigma\gamma'(y)D_y\varphi(y;\theta)\big]. \tag{5.2}$$

We now face the questions:

- Existence of a pair solution $(\Gamma(\theta), \varphi(.,\theta))$ to the ergodic PDE (5.1)?
- Do we have $\Gamma(\theta) = \Lambda_\pm(\theta)$, and what is the domain of Γ?

We give some assumptions, which allows us to answer the above issues.

(H1) b, σ, η and γ are smooth C^2 and globally Lipschitz.
(H2) $\sigma\sigma'(y)$ and $\gamma\gamma'(y)$ are uniformly elliptic: there exist $\delta_1, \delta_2 > 0$ s.t.

$$\delta_1|\xi|^2 \le \xi'\sigma\sigma'(y)\xi \le \delta_2|\xi|^2, \quad \forall\xi, y \in \mathbb{R}^m,$$
$$\delta_1|\xi|^2 \le \xi'\gamma\gamma'(y)\xi \le \delta_2|\xi|^2, \quad \forall\xi, y \in \mathbb{R}^m.$$

(H3) There exist $c_1 > 0$ and $c_2 \ge 0$ s.t.

$$b(\sigma\sigma')^{-1}b(y) \ge c_1|y|^2 - c_2, \quad \forall y \in \mathbb{R}^m.$$

(H4) Stability condition: there exist $c_3 > 0$ and $c_4 \ge 0$ s.t.

$$\big(\eta(y) - \gamma\sigma'(\sigma\sigma')^{-1}b(y)\big).y \le -c_3|y|^2 + c_4$$

According to [1] (see also [15, 20]), the next result states the existence of a smooth solution to the ergodic equation.

Proposition 5.1 *Under* **(H1)–(H4)**, *there exists for any* $\theta < 1$, *a solution* $(\Gamma(\theta), \varphi(.;\theta))$ *with* $\varphi(.;\theta)$ C^2, *to the ergodic HJB equation s.t:*

- *For $\theta < 0$, $\varphi(.;\theta)$ is upper-bounded*

$$\varphi(y;\theta) \longrightarrow -\infty, \quad \textit{as } |y| \to \infty,$$

- *For $\theta \in (0,1)$, $\varphi(.;\theta)$ is lower-bounded*

$$\varphi(y;\theta) \longrightarrow \infty, \quad \textit{as } |y| \to \infty,$$

and

$$\big|D_y\varphi(y;\theta)\big| \leq C_\theta(1+|y|).$$

We now relate a solution to the ergodic equation to the dual risk-sensitive control problem. In other words, this means the convergence of the finite horizon risk-sensitive stochastic control to the component Γ of the ergodic equation. We distinguish the downside and upside cases.

• **Downside risk**: In this case, it is shown in [15] that for all $\theta < 0$, the solution $(\Gamma(\theta), \varphi(.;\theta)$ to (5.1), with $\varphi(.,\theta)$ C^2 and upper bounded, is unique (up to an additive constant for $\varphi(.;\theta)$), and we have:

$$\Gamma(\theta) = \Lambda_-(\theta), \quad \theta < 0.$$

Moreover, there is an admissible optimal feedback control $\hat{\pi}(.,\theta)$ for $\Lambda_-(\theta)$ given by (5.2), and for which the factor process Y is ergodic under $\mathbb{Q}_{\hat{\pi}}$. It is also proved in [15] that $\Gamma = \Lambda_-$ is differentiable on $(-\infty, 0)$. Therefore, from Theorem 3.2, the solution to the downside risk large deviations probability is given by:

$$v_-(\ell) = \inf_{\theta \leq 0}\big[\Gamma(\theta) - \theta\ell\big], \quad \ell < \Gamma'(0),$$

with an optimal control:

$$\pi_t^{-,\ell} = \hat{\pi}(Y_t;\theta(\ell)), \quad \Gamma'(\theta(\ell)) = \ell, \quad \forall \ell \in (\Gamma'(-\infty), \Gamma'(0)),$$

while $v_-(\ell) = -\infty$ for $\ell < \Gamma'(-\infty)$.

• **Upside chance**: In this case, $0 < \theta < 1$, there is no unique solution $(\Gamma(\theta), \varphi(.;\theta))$ to the ergodic equation, with $\varphi(.;\theta)$ C^2 lower-bounded, even up to an additive constant, as pointed out in [6]. In general, we only have a verification type result, which states that if the process Y is ergodic under $\mathbb{Q}_{\hat{\pi}}$, then

$$\Gamma(\theta) = \Lambda_+(\theta),$$

and $\hat{\pi}(.,\theta)$ is an optimal feedback control for $\Lambda_+(\theta)$.

In the next paragraph, we consider a linear factor model for which explicit calculations can be derived.

5.1 Linear Gaussian Factor Model

We consider the linear factor model:

$$
\begin{aligned}
dS_t &= \mathrm{diag}(S_t)\big((B_1 Y_t + B_0)dt + \sigma dW_t\big) \quad \text{in } \mathbb{R}^d, \\
dY_t &= K Y_t dt + \gamma dW_t, \quad \text{in } \mathbb{R}^m,
\end{aligned}
$$

with K a stable matrix in $\mathbb{R}^m$, B_1 a constant $d \times m$ matrix, B_0 a non-zero vector in $\mathbb{R}^d$, σ a $d \times (d+m)$-matrix of rank d, and γ a nonzero $m \times (d+m)$ matrix. We are searching for a candidate solution to the ergodic equation (5.1) in the quadratic form:

$$
\varphi(y;\theta) = \frac{1}{2} C(\theta) y.y + D(\theta) y, \quad y \in \mathbb{R}^m,
$$

for some $m \times m$ matrices $C(\theta)$ and $D(\theta)$. Plugging this form of φ into (5.1), we find that $C(\theta)$ must solve the algebraic Riccati equation:

$$
\begin{aligned}
&\frac{1}{2} C(\theta)'\gamma\big(I_{d+m} + \frac{\theta}{1-\theta}\sigma'(\sigma\sigma')^{-1}\sigma\big)\gamma' C(\theta) \\
&+ \big(K + \frac{\theta}{1-\theta}\gamma\sigma'(\sigma\sigma')^{-1}B_1\big)' C(\theta) + \frac{1}{2}\frac{\theta}{1-\theta} B_1'(\sigma\sigma')^{-1} B_1 = 0, \qquad (5.3)
\end{aligned}
$$

while $B(\theta)$ is determined by

$$
\begin{aligned}
&\Big(K + \frac{\theta}{1-\theta}\gamma\sigma'(\sigma\sigma')^{-1}B_1 + \gamma\big(I_{d+m} + \frac{\theta}{1-\theta}\sigma'(\sigma\sigma')^{-1}\sigma\big)\gamma' C(\theta)\Big)' D(\theta) \\
&\qquad + \frac{\theta}{1-\theta}(\sigma\gamma' C(\theta) + B_1)'(\sigma\sigma')^{-1} B_0 = 0.
\end{aligned}
$$

Then, $\Gamma(\theta)$ is given by:

$$
\begin{aligned}
\Gamma(\theta) &= \frac{1}{2}\mathrm{tr}(\gamma\gamma' C(\theta)) + \frac{1}{2} D(\theta)'\gamma(I_{d+m} + \frac{\theta}{1-\theta}\sigma'(\sigma\sigma')^{-1}\sigma)\gamma' D(\theta) \\
&\quad + \frac{\theta}{1-\theta} B_0'(\sigma\sigma')^{-1}\sigma\gamma' D(\theta) + \frac{1}{2}\frac{\theta}{1-\theta} B_0'(\sigma\sigma')^{-1} B_0,
\end{aligned}
$$

and a candidate for the optimal feedback control is:

$$
\hat{\pi}(y;\theta) = \frac{1}{1-\theta}(\sigma\sigma')^{-1}\big[(B_1 + \sigma\gamma' C(\theta))y + B_0 + \sigma\gamma' D(\theta)\big].
$$

In [6], it is shown that there exists some positive $\bar{\theta}$ small enough, s.t. for $\theta < \bar{\theta}$, there exists a solution $C(\theta)$ to the Riccati equation (5.3) s.t. Y is ergodic under $\mathbb{Q}_{\hat{\pi}}$, and so by verification theorem, $\Gamma(\theta) = \Lambda_{\pm}(\theta)$. In the one-dimensional asset and factor model, as studied in [17], we obtain more precise results. Indeed, in this case: $d = m = 1$, the Riccati equation is a second-order polynomial equation in $C(\theta)$, which admits two explicits roots given by:

$$C_\pm(\theta) = -\frac{K}{|\gamma|^2}\left[\frac{1-\theta\big(1-\rho\frac{|\gamma|B_1}{K|\sigma|}\big) \pm \sqrt{(1-\theta)(1-\theta\beta)}}{1-\theta(1-\rho^2)}\right],$$

for all $\theta \le \bar{\theta}$, with

$$\bar{\theta} = \frac{1}{\beta}\wedge 1, \;\; \beta = 1-\rho^2 + \Big(\rho - \frac{|\gamma|B_1}{K|\sigma|}\Big)^2 > 0,$$

where $|\gamma|$ (resp. $|\sigma|$) is the Euclidian norm of γ (resp. σ), and $\rho \in [-1,1]$ is the correlation between S and Y, i.e. $\rho = \frac{\gamma\sigma'}{|\gamma||\sigma|}$. Actually, only the solution $C(\theta) = C_-(\theta)$ is relevant in the sense that for this root, Y is ergodic under $Q_{\hat{\pi}}$, and thus by verification theorem:

$$\begin{aligned}\Lambda_\pm(\theta) &= \Gamma(\theta) = \frac{1}{2}|\gamma|^2 C_-(\theta) + \frac{1}{2}|\gamma|^2 D(\theta)^2\big(1+\frac{\theta}{1-\theta}\rho^2\big)\\ &\quad + \frac{\theta}{1-\theta}\frac{B_0}{|\sigma|}\rho|\gamma|D(\theta) + \frac{1}{2}\frac{\theta}{1-\theta}\frac{B_0^2}{|\sigma|^2}, \quad \theta < \bar{\theta},\end{aligned}$$

where

$$D(\theta) = -\frac{B_0}{K|\sigma|}\frac{\theta\big(\rho|\gamma|C_-(\theta) + \frac{B_1}{|\sigma|}\big)}{\sqrt{(1-\theta)(1-\theta\beta)}},$$

and with optimal control for $\Lambda_\pm(\theta)$ given by:

$$\hat{\pi}(y;\theta) = \frac{1}{(1-\theta)|\sigma|}\Big[\Big(\frac{B_1}{|\sigma|} + \rho|\gamma|C_-(\theta)\Big)y + \frac{B_0}{|\sigma|} + \rho|\gamma|D(\theta)\Big].$$

Moreover, it is also proved in [17], that

$$\Gamma'(0) = \frac{B_0^2}{2|\sigma|^2} - \frac{B_1^2|\gamma|}{4|\sigma|^2 K} > 0,$$

(recall that $K < 0$) and the function Γ is steep, i.e.

$$\lim_{\theta\uparrow\bar{\theta}} \Gamma'(\theta) = \infty.$$

From Theorems 3.1 and 3.2, the solutions to the upside chance and downside risk large deviations probability are given by:

$$v_+(\ell) = \inf_{0 \le \theta < \bar{\theta}} \big[\Gamma(\theta) - \theta\ell\big], \quad \ell \in \mathbb{R},$$
$$v_-(\ell) = \inf_{\theta \le 0} \big[\Gamma(\theta) - \theta\ell\big], \quad \ell < \Gamma'(0),$$

with optimal control and nearly optimal control for $v_+(\ell)$:

$$\pi_t^{+,\ell} = \hat{\pi}(Y_t; \theta(\ell)), \quad \Gamma'(\theta(\ell)) \; = \; \ell, \quad \text{when } \ell > \Gamma'(0),$$
$$\pi_t^{+(n)} = \hat{\pi}(Y_t; \theta_n), \quad \text{with } \theta_n = \theta(\Gamma'(0) + \frac{1}{n}) \overset{n\to\infty}{\longrightarrow} 0, \quad \text{when } \ell \le \Gamma'(0),$$

and optimal control for $v_-(\ell)$:

$$\pi_t^{-,\ell} = \hat{\pi}(Y_t; \theta(\ell)), \quad \Gamma'(\theta(\ell)) \; = \; \ell, \quad \forall \ell \in (\Gamma'(-\infty), \Gamma'(0)).$$

5.2 *Examples*

- **Black-Scholes model**. This corresponds to the case where $B_1 = 0$. Then, $\beta = \bar{\theta} = 1$, $C_-(\theta) = D(\theta) = 0$, and so

$$\Lambda_\pm(\theta) \; = \; \Gamma(\theta) = \frac{1}{2}\frac{\theta}{1-\theta}\frac{B_0^2}{|\sigma|^2}, \quad \forall \theta < 1.$$

We thus obtain the same optimal strategy as described in Sect. 4.

- **Platen-Rebolledo model**. In this model, the logarithm of the stock price S is governed by an Ornstein-Uhlenbeck process Y, and this corresponds to the case where $B_1 = K < 0$, $B_0 = \frac{1}{2}|\gamma|^2 > 0$, $\gamma = \sigma$, and thus $\rho = 1$. Then, $\beta = 0$, $\bar{\theta} = 1$,

$$C_-(\theta) \; = \; \frac{|K|}{|\sigma|^2}\big[1 - \sqrt{1-\theta}\big], \; D(\theta) \; = \; -\frac{1}{2}\theta,$$

and so

$$\Gamma(\theta) = \frac{|K|}{2}\big[1 - \sqrt{1-\theta}\big] + \theta\frac{|\sigma|^2}{8}, \quad \theta < 1,$$
$$\Gamma'(0) = \bar{\ell} \; := \; \frac{|K|}{4} + \frac{|\sigma|^2}{8}, \qquad \Gamma'(-\infty) \; = \; \underline{\ell} \; := \; \frac{|\sigma|^2}{8},$$

$$\theta(\ell) = 1 - \left(\frac{\bar{\ell} - \underline{\ell}}{\ell - \underline{\ell}}\right)^2, \quad \forall \ell > \underline{\ell}.$$

The solution to the upside chance large deviations probability is then given by:

$$v_+(\ell) = \begin{cases} -\frac{(\ell-\bar{\ell})^2}{\ell-\bar{\ell}+\frac{|K|}{4}}, & \text{if } \ell > \bar{\ell} \\ 0, & \text{if } \ell \leq \bar{\ell}. \end{cases}$$

with optimal (resp. nearly optimal) portfolio strategy:

$$\pi_t^{+,\ell} = \frac{K - 4(\ell - \bar{\ell})}{|\sigma|^2} Y_t + \frac{1}{2}, \quad \text{if } \ell > \bar{\ell}$$
$$\pi_t^{+(n)} = \frac{K - 1/n}{|\sigma|^2} Y_t + \frac{1}{2}, \quad \text{if } \ell \leq \bar{\ell}.$$

The solution to the downside risk large deviations probability is given by:

$$v_-(\ell) = \begin{cases} -\frac{(\ell-\bar{\ell})^2}{\ell-\underline{\ell}}, & \text{if } \underline{\ell} < \ell \leq \bar{\ell} \\ -\infty, & \text{if } \ell \leq \underline{\ell}, \end{cases}$$

with optimal portolio strategy:

$$\pi_t^{-,\ell} = -\frac{4(\ell - \underline{\ell})}{|\sigma|^2} Y_t + \frac{1}{2}, \quad \text{if } \underline{\ell} < \ell \leq \bar{\ell}$$

References

1. Bensoussan, A., Frehse, J.: On Bellman equations of ergodic control in $\mathbb{R}^N$. J. Reine Angew. Math. **429**, 125–160 (1992)
2. Bielecki, T.R., Pliska, S.R.: Risk-sensitive dynamic asset management. Appl. Math. Optim. **39**, 337–360 (1999)
3. Davis, M., Lleo, S.: Risk-sensitive benchmarked asset management. Quant. Financ. **8**, 415–426 (2008)
4. Dembo, A., Zeitouni, O.: Large Deviations Techniques and Applications, 2nd edn. Springer, New York (1998)
5. Fleming, W., McEneaney, W.: Risk sensitive control on an infinite horizon. SIAM J. Control Optim. **33**, 1881–1915 (1995)
6. Fleming, W., Sheu, S.J.: Risk sensitive control and an optimal investment model. Math. Financ. **10**, 197–213 (2000)
7. Guasoni, P., Robertson, S.: Portfolios and risk premia for the long run. Ann. Appl. Probab. **22**, 239–284 (2012)
8. Hata, H., Nagai, H., Sheu, S.J.: Asymptotics of the probability minimizing a down-side risk. Ann. Appl. Probab. **20**, 52–89 (2010)

9. Hata, H., Sekine, J.: Solving long term investment problems with Cox-Ingersoll-Ross interest rates. Adv. Math. Econ. **8**, 231–255 (2005)
10. Karatzas, I., Shreve, S.: Methods of Mathematical Finance. Springer, New York (1998)
11. Korn, R.: Optimal Portfolios: Stochastic Models for Optimal Investment and Risk Management in Continuous-time. World Scientific, Singapore (1997)
12. Liu, R., Muhle-Karbe, J.: Portfolio choice with stochastic investment opportunities: a user's guide, Preprint (2013)
13. Merton, R.: Optimum consumption and portfolio rules in a continuous-time model. J. Econ. Theory **3**, 373–413 (1971)
14. Nagai, H.: Bellman equations of risk sensitive control. SIAM J. Control Optim. **34**, 74–101 (1996)
15. Nagai, H.: Downside risk minimization via a large deviation approach. Ann. Appl. Probab. **22**, 608–669 (2012)
16. Nagai, H., Peng, S.: Risk-sensitive dynamic portfolio optimization with partial information on infinite time horizon. Ann. Appl. Probab **12**, 173–195 (2002)
17. Pham, H.: A large deviations approach to optimal long term investment. Financ. Stoch. **7**, 169–195 (2003)
18. Pham, H.: Some applications and methods of large deviations in finance and insurance. Paris-Princeton Lectures on Mathematical Finance, Lecture Notes in Mathematics, vol. 1919 (2007)
19. Pham, H.: Continuous Time Stochastic Control and Optimization with Financial Applications. SMAP. Springer, New York (2009)
20. Robertson, S., Xing, H.: Large time behavior of solutions to semi-linear equations with quadratic growth in the gradient. SIAM J. Control Optim. **53**(1), 185–212 (2015)
21. Stettner, L.: Duality and risk-sensitive portfolio optimization. In: Yin, G., Zhang, Q. (eds.) Mathematics of Finance, Contemporary Mathematics, vol. 351, pp. 333–347 (2004)
22. Stutzer, M.: Portfolio choice with endogenous utility: a large deviations approach. J. Econom. **116**, 365–386 (2003)

Systemic Risk and Default Clustering for Large Financial Systems

Konstantinos Spiliopoulos

Abstract As it is known in the finance risk and macroeconomics literature, risk-sharing in large portfolios may increase the probability of creation of default clusters and of systemic risk. We review recent developments on mathematical and computational tools for the quantification of such phenomena. Limiting analysis such as law of large numbers and central limit theorems allow to approximate the distribution in large systems and study quantities such as the loss distribution in large portfolios. Large deviations analysis allow us to study the tail of the loss distribution and to identify pathways to default clustering. Sensitivity analysis allows to understand the most likely ways in which different effects, such as contagion and systematic risks, combine to lead to large default rates. Such results could give useful insights into how to optimally safeguard against such events.

Keywords Systemic risk · Default clustering · Large portfolios · Loss distribution · Asymptotic methods · Rare events

1 Introduction

The past several years have made clear the need to better understand the behaviour in large interconnected financial systems. Almost all areas of modern life are touched by a financial crisis. The recent financial crisis of 2007–2009 brought into focus the networked structure of the financial world. It challenged the mathematical finance community to understand connectedness in financial systems. The understanding of systemic risk, i.e., the risk that a large numbers of components of an interconnected financial system fails within a short time leading to the failure of the system itself, becomes an important issue to investigate.

Interconnections often make a system robust, but they can also act as conduits for risk. Even things that may seemingly be unrelated, may become related as risk

K. Spiliopoulos (✉)
Department of Mathematics & Statistics, Boston University, Boston, MA 02215, USA
e-mail: kspiliop@bu.edu

P.K. Friz et al. (eds.), *Large Deviations and Asymptotic Methods in Finance*, Springer Proceedings in Mathematics & Statistics 110,
DOI 10.1007/978-3-319-11605-1_19

restrictions, may for example, force a sale of one type of a well-performing asset to compensate for the poor behavior of another asset. Thus, appropriate mathematical models need to be developed, in order to help in the understanding of how risk can propagate between financial objects.

It is possible that initial shocks could trigger contagion effects (e.g., [1]). Examples of such shocks include: changes in interest rate values, in currencies values, changes of commodities prices, or reduction in global economic growth. Then, there may be a transmission mechanism which causes other institutions in the system to be affected by the initial shock. An example of such a mechanism is financial linkages among economies. Another reason could simply be investor irrationality. In either case, systemic risk causes the perceived risk-return trade-off in the economy to change. Uncertainty becomes an issue and market participants fear subsequent losses in asset prices with a large dispersion in regards to the magnitude of the crisis. Reduce-form point process models of correlated default are many times used (a): to assess portfolio credit risk and (b): to value securities exposed to correlated default risk. The workhorses of these models are counting processes. In this work we focus on using dynamic portfolio credit risk models to study large portfolio asymptotics and default clustering.

Large portfolio asymptotic were first studied in [2]. The model in [2] is a static model of a homogeneous pool and firms default independently of one another conditional on a normally distributed random variable representing a systematic risk factor. Alternative distributions of the systematic factor were examined in [3, 4] and the case of heterogeneous portfolios was studied in [5]. In [6], the authors extend the model of [2] dynamically and the systematic risk factor follows a Brownian motion. In [6], the authors study a structural model for distance to default process in a pool of names. A firm defaults when the default process hits zero. Exploiting conditional independence of defaults, [7, 8] have studied the tail of the loss distribution in the static case. Large deviations arguments were also used in [9] to study stochastic recovery effects on large static pools of credit assets.

Reduced-form models of correlated default timing have appeared in the finance literature under different forms. Giesecke and Weber [10] take the intensity of a name as a function of the state of the names in a specified neighborhood of that name. The authors in [11, 12] take the intensity to be a function of the portfolio loss and each name can be either in a good or in a distressed financial state. These papers prove law of large numbers for the portfolio loss distribution and develop Gaussian approximations to the portfolio loss distribution based on central limit theorems. Cvitanić et al. [13] consider the typical behavior of a mean field system with permanent default impact.

Sircar and Zariphopoulou [14] study large portfolio asymptotics for utility indifference valuation of securities exposed to the losses in the pool. In [15], the authors study systematic risk via a mean field model of interacting agents. Using a model of a two well potential, agents can move freely from a healthy state to a failed state. The authors study probabilities of transition from the healthy to the failed state using large deviations ideas. In [16] the authors propose and study a model for inter-bank lending and study its stochastic stability.

The authors in [17] employ jump-diffusion models driven by Hawkes processes to empirically study default clustering and the time dimension of systemic risk. Duan [18] proposes a hierarchical model with individual shocks and group specific shocks. The work of [19] reviews intensity models that are governed by exogenous and endogenous Markov Chains. In [20], the authors proposed a dynamic point process model of correlated default timing in a portfolio of firms ("names"). The model incorporates different sources of default clustering identified in recent empirical research, including idiosyncratic risks, exposure to systematic risk factors and contagion in financial markets, see [21, 22]. Based on the weak convergence ideas of [20], the authors in [23] obtain and study formulas for the bilateral counterparty valuation adjustment of a credit default swaps portfolio referencing an asymptotically large number of entities.

The model in [20] can be naturally understood as an interacting particle system that is influenced by an exogenous source of randomness. There is a central source of interconnections and failure of any of the components stresses the central 'bus', which in turn can cause the failure of other components (a contagion effect). Computing the distribution of the loss from default in such models tends to be a difficult task and while Monte-Carlo simulation methods are broadly applicable, they can be slow for large portfolios or large time horizons as it is commonly the interest in practice. Mathematical and computational tools for the approximation to the distribution of the loss from default in large heterogeneous portfolios were then developed in [24], Gaussian correction theory was developed in [25] and analysis of tail events and most likely paths to failure via the lens of large deviations theory was then developed in [26]. We remark here that to a large extend systemic risk refers to the tail of the distribution. The authors in [27] combine the large pool asymptotic results of [1, 3, 4, 9, 10, 24–34] with maximum likelihood ideas to construct tractable statistical inference procedures for parameter estimation in large financial systems.

Such mathematical results lead to new computational tools for the measurement and prediction of risk in high-dimensional financial networks. These tools mainly include approximations of the distribution of losses from defaults and of portfolio risk measures, and efficient computational tools for the analysis of extreme default events. The mathematical results also yield important insights into the behavior of systemic risk as a function of the characteristics of the names in the system, and in particular their interaction.

Financial institutions (banks, pension funds, etc.) often hold large portfolios in order to diversify away a number of idiosyncratic effects of individual assets. Deposit insurance premia depend upon meaningful models and assessment of the macroeconomic effect of the various phenomena that drive defaults. Development of related mathematical and computational tools can help inform the design of regulatory policy, improve the pricing of federal deposit insurance, and lead to more accurate risk measurement at financial institutions.

In this paper, we focus on dynamic default timing models for large financial systems that fall into the category of intensity models in portfolio credit risk. Based on the default timing model developed in [20], we address several of the issues just mentioned and that are typically of interest. The mathematical and computational tools developed allow to reach to financial related conclusions for the behavior of such large financial systems.

Although the primary interest of this work is risk in financial systems, models of the type discussed in this paper are generic enough to allow for modifications that make them relevant in other domains, including systems reliability, insurance and epidemiology. In reliability, a large system of interacting components might have a central connection, and be influenced by an external environment (temperature, for example). The failure of an individual component (which could be governed by an intensity model appropriate for the particular application) increases the stress on the central connection and thus the other components, making the entire system more likely to fail. In insurance, the system could represent a pool of insurance policies. The effect of wildfires might, in that example, be modelled by a contagion term. Systematic risk in the form of environmental conditions has an impact on the whole pool.

The rest of the article is structured as follows. In Sect. 2 we describe the correlated default timing proposed in [20]. Section 3 studies the typical behavior of the loss distribution in such portfolios as the number of names (agents) in the pool grow to infinity. Section 4 focuses on developing the Gaussian correction theory. As we shall see there, Gaussian corrections are very useful because they make the approximations accurate even for portfolios of relatively small sizes. In Sect. 5, we study the tail of the loss distribution using arguments from the large deviations theory. We also study the most likely path to systemic failure and to the development of default clusters. An understanding of the preferred paths to large default rates and the most likely path to the creation of default clusters can give useful insights into how to optimally safeguard against such events. Importance sampling techniques can then be used to construct asymptotically efficient estimators for tail event probabilities, see Sect. 6. Conclusions are in Sect. 7. A large part of the material presented in this work, but not all, is related to recent work of the author described in [20, 24–26].

2 A Dynamic Correlated Default Timing Model

One of the issues of fundamental importance in financial markets is systemic risk, which may be understood as the likelihood of failure of a substantial fraction of firms in the economy. There are a number of ways of interpreting this, but our focus will be the behavior of actual *defaults*. Defaults are discrete events, so one can frame the interest within the language of point processes. Empirically, defaults tend to happen in groups; feedback and exposure to market forces (along the lines of "regimes") tend to produce correlation among defaults.

Let us fix a probability space $(\Omega, \mathscr{F}, \mathbb{P})$ where all random variables will be defined. Denote by τ^n the stopping time at which the nth component (or particle) in our system fails. Then, as $\delta \searrow 0$, a failure time τ^n has intensity process λ^n, which satisfies

$$\mathbb{P}\{\tau^n \in (t, t+\delta] | \mathscr{F}_t,\ \tau^n > t\} \approx \lambda^n_t \delta, \tag{1}$$

where $\mathscr{F}_t$ is the sigma-algebra generated by the entire system up to time t. Hence, we essentially have that the process defined by $1_{\{\tau^n \le t\}} - \int_0^t \lambda_s^n 1_{\{\tau^n > s\}} ds$ is a martingale.

Motivated by the empirical studies in [21, 22], we may model the intensity λ^n in such a way that it depends on three factors: a mean reverting idiosyncratic source of risk, the portfolio loss rate and a systematic risk factor. Heterogeneity can be addressed by allowing the intensity parameters of each name to be different. The mean reverting character of the idiosyncratic source of risk is there to guarantee that the effect of a default in the pool has a transient effect on the default intensities of the surviving names. The dependence on the portfolio loss rate, denoted by $L_\cdot^N$ is the term that is responsible for the contagious effects, whereas the systematic risk factor, denoted by $X_\cdot$ is an exogenous source of risk. To be precise, the default intensities, λ^n's, are governed by the following interacting system of stochastic differential equations (SDEs)

$$d\lambda_t^n = -\alpha_n(\lambda_t^n - \bar{\lambda}_n)dt + \sigma_n\sqrt{\lambda_t^n}dW_t^n + \beta_n^C dL_t^N + \varepsilon\beta_n^S\lambda_t^n dX_t, \quad \lambda_0^n = \lambda_\circ^n. \quad (2)$$

where, $\{W^n\}_{n\in\mathbb{N}}$ be a countable collection of independent standard Brownian motions.

The process L_t^N represents the empirical failure rate in the system, i.e.,

$$L_t^N = \frac{1}{N}\sum_{n=1}^{N} 1_{\{\tau^n \le t\}}, \quad (3)$$

where by letting $\{\mathfrak{e}_n\}_{n\in\mathbb{N}}$ to be an i.i.d. collection of standard exponential random variables we have

$$\tau^n = \inf\left\{t \ge 0 : \int_{s=0}^{t} \lambda_s^n ds \ge \mathfrak{e}_n\right\}. \quad (4)$$

The process X_t represents the systematic risk, which can be modeled to be the solution to some SDE

$$dX_t = b_0(X_t)dt + \sigma_0(X_t)dV_t, \quad X_0 = x_\circ. \quad (5)$$

where V is a standard Brownian motion which is independent of the W^n's and $\mathfrak{e}_n$'s. Plausible models for X_t could be an Ornstein-Uhlenbeck process or a Cox-Ingersoll-Ross (CIR) process.

In the case $\beta_n^C = \beta_n^S = 0$ for all $n \in \{1, \ldots, N\}$, one recovers the classical CIR process model in credit risk, e.g., [35]. Namely, the intensity SDE (1) extends the widely-used CIR process by including two additional terms that generate correlation between failure times. The term $\varepsilon\beta_n^S\lambda_t^n dX_t$ induces correlated diffusive movements of the component intensities; the process X represents the state of the macro-economy, which affects all assets in the pool. The term $\beta_n^C dL_t^N$ introduces a feedback (contagion) effect. The standard term $-\alpha_n(\lambda_t^n - \bar{\lambda}_n)dt$ is a mean reverting term allowing the component to "heal" after a shock (i.e., a failure). This parsimonious formulation allows us to take advantage of the wealth of knowledge about CIR-type processes. The parameter $\varepsilon > 0$ allows us to later on focus on rare events.

The process L^N of (3), which simply gives us the fraction of components which have already failed by time t, affects each of the remaining components in a natural way. Each failure corresponds to a Dirac function in the measure dL^N; the term $\beta_n^C dL_t^N$ thus leads to upward impulses in λ^n's, which leads (via (4)) to sooner failure of the remaining functioning components. We might think of a central "bus" in a system of components. Each of the components depends on this bus, which in turn sensitive to failures in the various components. In the financial application that was considered in [20], this feedback mechanism is empirically observed to be an important channel for the clustering of defaults in the U.S. (see [21]).

In order to allow for heterogeneity, the parameters in (2) depend on the index n. Define the "type"

$$\mathsf{p}_t^n = (\lambda_t^n, \alpha_n, \bar{\lambda}_n, \sigma_n, \beta_n^C, \beta_n^S) \tag{6}$$

for each $n \in \mathbb{N}$ and $t \geq 0$. The p_t^n's take value in $\mathcal{P} = \mathbb{R}_+^3 \times \mathbb{R} \times \mathbb{R}_+ \times \mathbb{R} \subset \mathbb{R}^6$. The parameters $(\lambda_0^n, \alpha_n, \bar{\lambda}_n, \sigma_n, \beta_n^C, \beta_n^S)$ are assumed to be bounded uniformly in $n \in \mathbb{N}$.

We can capture the heterogeneity of the system by defining $U_N = \frac{1}{N}\sum_{n=1}^N \delta_{\mathsf{p}_n}$ and assuming that this empirical type frequency has a (weak) limit. In particular we make the following assumption

Assumption 2.1 We assume that $U = \lim_{N\to\infty} U_N$ exists (in $\mathscr{P}(\mathcal{P})$).

Proposition 3.3 in [20] guarantees that under the assumption of an existence of a unique strong solution for the SDE for X. process, the system (2)–(5) has a unique strong solution such that $\lambda_t^n \geq 0$ for every $N \in \mathbb{N}$, $n \in \{1, \ldots, N\}$ and $t \geq 0$. The model (2)–(5) is a mean-field type model; the feedback occurs through the empirical average of the pool of names. It is somewhat similar to certain genetic models (most notably the Fleming-Viot process; see [36], [37, Chap. 10], and [38]). However, as it is also demonstrated in [20, 24], the structure of the system (2)–(5) presents several difficulties that bring the analysis of such systems outside the scope of the standard setup.

3 Typical Behavior: Law of Large Numbers

The system (2)–(5) can naturally be understood as an interacting particle system. This suggests how to understand its large-scale behavior. The structure of the feedback (the empirical average L^N) is of mean-field type (roughly within the class of McKean-Vlasov models; see [31, 39]). An understanding of "typical" behavior of a system as $N \to \infty$ is fundamental in identifying "atypical" or "rare" events.

To formulate the law of large numbers result, we define the empirical distribution of the p^n's corresponding to the names that have survived up to time t, as follows:

$$\mu_t^N = \frac{1}{N}\sum_{n=1}^{N}\delta_{\mathsf{p}_t^N}1_{\{\tau^n>t\}}.$$

This captures the entire dynamics of the model (including the effect of the heterogeneities). We can directly calculate the failure rate from the μ^N's:

$$L_t^N = 1-\mu_t^N(\mathcal{P}), \qquad t\geq 0. \tag{7}$$

Let us then identify the limit of $\mu_t^N(\mathcal{P})$ as $N\to\infty$. This is a law of large numbers (LLN) result and it identifies the baseline "typical" behavior of the system. For $f\in C^2(\mathcal{P})$, let

$$\begin{aligned}
(\mathcal{L}_1 f)(\mathsf{p}) &= \frac{1}{2}\sigma^2\lambda\frac{\partial^2 f}{\partial\lambda^2}(\mathsf{p})-\alpha(\lambda-\bar{\lambda})\frac{\partial f}{\partial\lambda}(\mathsf{p})-\lambda f(\mathsf{p})\\
(\mathcal{L}_2 f)(\mathsf{p}) &= \beta^C\frac{\partial f}{\partial\lambda}(\mathsf{p})\\
(\mathcal{L}_3^x f)(\mathsf{p}) &= \varepsilon\beta^S\lambda b_0(x)\frac{\partial f}{\partial\lambda}(\mathsf{p})+\frac{\varepsilon^2}{2}(\beta^S)^2\lambda^2\sigma_0^2(x)\frac{\partial^2 f}{\partial\lambda^2}(\mathsf{p})\\
(\mathcal{L}_4^x f)(\mathsf{p}) &= \varepsilon\beta^S\lambda\sigma_0(x)\frac{\partial f}{\partial\lambda}(\mathsf{p}) \qquad \text{and} \qquad \mathcal{Q}(\mathsf{p})=\lambda
\end{aligned} \tag{8}$$

for $\mathsf{p}=(\lambda,\alpha,\bar{\lambda},\sigma,\beta^C,\beta^S)$. The generator $\mathcal{L}_1$ corresponds to the diffusive part of the intensity with killing rate λ, and $\mathcal{L}_2$ is the macroscopic effect of contagion on the surviving intensities at any given time. The operators $\mathcal{L}_3^x$ and $\mathcal{L}_4^x$ capture the dynamics due to the exogenous systematic risk X. Then μ^N tends in distribution (in the natural topology of subprobability measures on $\mathcal{P}$) to a measure-valued process $\bar{\mu}$. Letting

$$\langle f,\mu\rangle = \int_{\mathsf{p}\in\mathcal{P}} f(\mathsf{p})\mu(d\mathsf{p})$$

for all $f\in C^2(\mathcal{P})$, the limit $\bar{\mu}$ satisfies the stochastic evolution equation

$$d\langle f,\bar{\mu}_t\rangle = \left\{\langle\mathcal{L}_1 f,\bar{\mu}_t\rangle+\langle\mathcal{Q},\bar{\mu}_t\rangle\langle\mathcal{L}_2 f,\bar{\mu}_t\rangle+\left\langle\mathcal{L}_3^{X_t}f,\bar{\mu}_t\right\rangle\right\}dt+\left\langle\mathcal{L}_4^{X_t}f,\bar{\mu}_t\right\rangle dV_t \quad \text{a.s.} \tag{9}$$

With sufficient regularity, this is equivalent to the stochastic integro-partial differential equation (SIPDE)

$$dv = \mathcal{L}_1^* v dt+\left(\int\mathcal{Q}v\right)\mathcal{L}_2^* v dt+\mathcal{L}_3^{X_t,*}v dt+\varepsilon\mathcal{L}_4^{X_t,*}v dV_t \quad \text{a.s.} \tag{10}$$

where * denotes adjoint in the appropriate sense (for notational simplicity, we have written (10) to include the types as one of the coordinates; in a heterogeneous collection in practice we would often use only λ in solving (10)). We recall the rigorous statement in Theorem 3.1.

The SIPDE (10) gives us a "large system approximation" of the failure rate:

$$L_t^N \approx 1 - \bar{\mu}_t(\mathcal{P}) = 1 - \int_{\mathcal{P}} \upsilon(t, \mathsf{p}) d\mathsf{p}. \tag{11}$$

The computation of the first-order approximation (11) suggested by the LLN requires solving the SIPDE (10) governing the density of the limiting measure. In [24] a numerical method for this purpose is proposed. The method is based on an infinite system of SDE's for certain moments of the limiting measure. These SDEs are driven by the systematic risk process X and a truncated system can be solved using a discretization or random ODE scheme. The solution to the SDE system leads to the solution to the SIPDE via an inverse moment problem.

The approximation (11) has significant computational advantages over a naive Monte Carlo simulation of the high-dimensional original stochastic system (2)–(5) and its accuracy is demonstrated in the left of Fig. 1 for a specific choice of parameters. It also provides information about catastrophic failure.

The tail represents extreme default scenarios, and these are at the center of risk measurement and management applications in practice. The analysis of the limiting distribution generates important insights into the behavior of the tails as a function of the characteristics of the system (2)–(5). For example, we see that the tail is heavily influenced by the sensitivity of a name to the variations of the systematic risk X. The bigger the sensitivity the fatter the tail, and the larger the likelihood of large losses in the system (see the right of Fig. 1). Insights of this type can help understand the

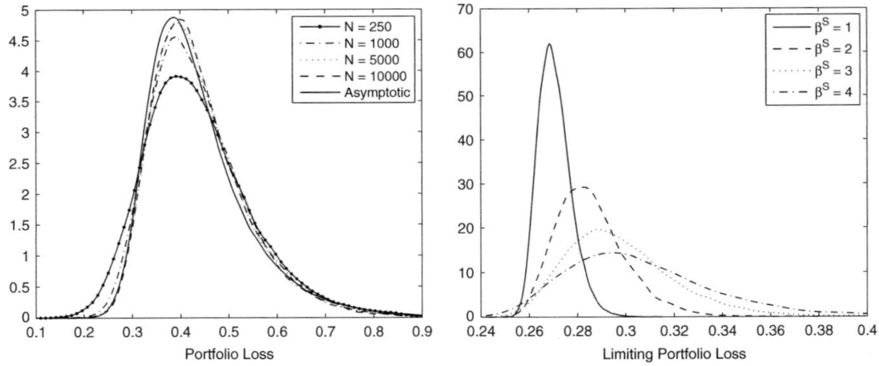

Fig. 1 On the *left* Comparison of distributions of failure rate L_t^N for different N at $t = 1$. Parameter choices: $(\sigma, \alpha, \bar{\lambda}, \lambda_0, \beta^C, \beta^S) = (0.9, 4, 0.2, 0.2, 4, 8)$. On the *right* Comparison of distribution of limiting failure rate $1 - \bar{\mu}_t(\mathcal{P})$ for different values of the systematic risk sensitivity β^S at $t = 1$. Parameter choices: $(\sigma, \alpha, \bar{\lambda}, \lambda_0, \beta^C) = (0.9, 4, 0.2, 0.2, 2)$

role of contagion and systematic risk, and how they interact to produce atypically large failure rates. This, in turn, leads to ways to minimize or "manage" catastrophic failures.

Let us next present the statement of the mathematical result. We denote by E the collection of sub-probability measures (i.e., defective probability measures) on $\mathcal{P}$; i.e., E consists of those Borel measures ν on $\mathcal{P}$ such that $\nu(\mathcal{P}) \leq 1$.

Theorem 3.1 (Theorem 3.1 in [24]) *We have that $\mu^N_\cdot$ converges in distribution to $\bar{\mu}_\cdot$ in $D_E[0,T]$. The evolution of $\bar{\mu}_\cdot$ is given by the measure evolution equation*

$$d\langle f, \bar{\mu}_t\rangle_E = \left\{\langle \mathcal{L}_1 f, \bar{\mu}_t\rangle_E + \langle \mathcal{Q}, \bar{\mu}_t\rangle_E \langle \mathcal{L}_2 f, \bar{\mu}_t\rangle_E + \left\langle \mathcal{L}_3^{X_t} f, \bar{\mu}_t\right\rangle_E\right\} dt$$
$$+ \left\langle \mathcal{L}_4^{X_t} f, \bar{\mu}_t\right\rangle_E dV_t, \quad \forall f \in C^\infty(\mathcal{P}) \text{ a.s.}$$

Suppose there is a solution of the nonlinear SPDE

$$dv(t,\mathsf{p}) = \left\{\mathcal{L}_1^* v(t,\mathsf{p}) + \mathcal{L}_3^{*,X_t} v(t,\mathsf{p}) + \left(\int_{\mathsf{p}'\in\mathcal{P}} \mathcal{Q}(\mathsf{p}')v(t,\mathsf{p}')d\mathsf{p}'\right)\mathcal{L}_2^* v(t,\mathsf{p})\right\} dt \tag{12}$$
$$+ \mathcal{L}_4^{*,X_t} v(t,\mathsf{p})dV_t, \quad t>0, \quad \mathsf{p}\in\mathcal{P}$$

where $\mathcal{L}_i^*$ denote adjoint operators, with initial condition

$$\lim_{t\searrow 0} v(t,\mathsf{p})d\mathsf{p} = U(d\mathsf{p}).$$

Then

$$\bar{\mu}_t = v(t,\mathsf{p})d\mathsf{p}.$$

We close this section, by briefly describing the method of moments that leads to the numerical computation of the loss from default. We focus our discussion on the homogeneous case and we refer the reader to [24] for the general case.

Firstly, we remark that the SPDE (12) can be supplied with appropriate boundary conditions, which as it is mentioned in [24], are

$$v(t,\lambda=0) = v(t,\lambda=\infty) = 0.$$

Secondly, it turns out that for $k \in \mathbb{N}$, the moments $u_k(t) = \int_0^\infty \lambda^k v(t,\lambda)d\lambda$ exist almost surely. By (11) is clear that we want to compute $u_0(t)$. In particular, note that the limiting loss $L_t = 1 - u_0(t)$.

By an integration by parts and using the boundary conditions at $\lambda = 0$ and at $\lambda = \infty$, we can prove that they follow the following system of stochastic differential equations

$$\begin{aligned} du_k(t) &= \big\{u_k(t)\big(-\alpha k + \beta^S b_0(X_t)k + 0.5(\beta^S)^2\sigma_0^2(X_t)k(k-1)\big) \\ &\quad + u_{k-1}(t)\big(0.5\sigma^2 k(k-1) + \alpha\bar{\lambda}k + \beta^C k u_1(t)\big) - u_{k+1}(t)\big\}dt \\ &\quad + \beta^S \sigma_0(X_t) k u_k(t) dV_t, \\ u_k(0) &= \int_0^\infty \lambda^k \Lambda_\circ(\lambda) d\lambda, \end{aligned} \tag{13}$$

where $\Lambda_\circ(\lambda) = \lim_{N\to\infty} \frac{1}{N}\sum_{n=1}^N \delta_{\lambda_0^n}(\lambda)$.

The system (13) is a non-closed system since to determine $u_k(t)$, one needs to know $u_{k+1}(t)$. So, in practice one must perform a truncation at some level $k = K$ where we let $u_{K+1} = u_K$ (that is, we use the first $K+1$ moments). As it is shown in [24] one needs relatively small numbers of moments in order to compute the zero-th moment $u_0(t)$ with good accuracy. Then, by solving backwards, one computes $u_0(t)$ and from this one gets the limiting loss distribution

$$L_t = 1 - u_0(t).$$

4 Central Limit Theorem Correction

The asymptotics of (10) give via (11) the limiting behavior of the system as the number of components becomes large. Starting with that result, the results in [25] develop Gaussian fluctuation theory analogous to the central limit theory (see for example [11, 12, 32, 40] for some related literature). This result provides the leading order asymptotics correction to the law of large numbers approximation developed in Sect. 3. In practical terms, the usefulness of such of a result is twofold: (a) the approximation is accurate even for portfolios of moderate size, see [25], and (b) one can make use of the approximation to develop tractable statistical inference procedures for the statistical calibration of such models, see [27].

To be more precise, let us define the signed measure

$$\Xi_t^N = \sqrt{N}\left\{\mu_t^N - \bar{\mu}_t\right\};$$

as $N \to \infty$. Conditional on the exogenous systematic risk process X, a central limit theorem applies and $\bar{\Xi} = \lim_{N\to\infty} \Xi^N$ exists in an appropriate space of distributions and is Gaussian. Unconditionally, it may not be Gaussian but is of mean zero (since we have removed the bias $\bar{\mu}$ from μ^N).

The usefulness of the fluctuation analysis is that it leads to a second-order approximation to the distribution of the portfolio loss L^N in large pools. The fluctuations analysis yields an approximation which improves the first-order approximation (11) suggested by the LLN, especially for smaller system sizes N.

In particular, Theorem 4.1 implies that

$$\mathbb{P}(\sqrt{N}(L_t^N - L_t) \geq \ell) \approx \mathbb{P}(\bar{\Xi}_t(\mathcal{P}) \leq -\ell)$$

for large N. This motivates the approximation

$$\mu_t^N = \frac{1}{\sqrt{N}}\Xi_t^N + \bar{\mu}_t \overset{d}{\approx} \frac{1}{\sqrt{N}}\bar{\Xi}_t + \bar{\mu}_t,$$

which then implies the following second-order approximation for the portfolio loss.

$$L_t^N \overset{d}{\approx} L_t - \frac{1}{\sqrt{N}}\bar{\Xi}_t(\mathcal{P}). \tag{14}$$

The numerical computation of the second-order approximation (14) suggested by the fluctuation analysis is amenable to a moment method similar to that used for computing the first-order approximation (11). In addition to solving the LLN SIPDE, we would also need to solve for the fluctuation limit. This limit is governed by a stochastic evolution equation, which gives rise to an additional system of "fluctuation moments." This system is driven by the exogenous systematic risk process X and the martingale $\bar{\mathcal{M}}_t$ in Theorem 4.1 that is conditionally Gaussian given X.

Left of Fig. 2 compares the approximate loss distribution with the actual loss distribution for specific parameter choices. It is evident from the numerical comparisons that the second-order approximation has increased accuracy, especially for smaller portfolios and in the tail of the distribution. The right of Fig. 2 compares for the 95 and 99 percent value at risk (VaR) between the actual loss, LLN approximation (11), and approximation (14) for a pool of $N = 1{,}000$ names. It is also evident from the figure that the approximation for the VaR based on (14) is much more accurate than the law of large numbers approximation.

Let us close this section, with a few words on the actual mathematical result. It turns out that the convergence $\bar{\Xi} = \lim_{N\to\infty} \Xi^N$ happens in an appropriate weighted Hilbert space, which we denote by $W_0^J(w, \rho)$, with w and ρ the appropriate weight functions, $J \in \mathbb{N}$ and $W_0^{-J}(w, \rho)$ will be its dual. Such weighted Sobolev spaces were introduced in [33] and further generalized in [41] to study stochastic partial differential equations with unbounded coefficients. These weighted spaces turn out to be convenient for the present situation, see [25].

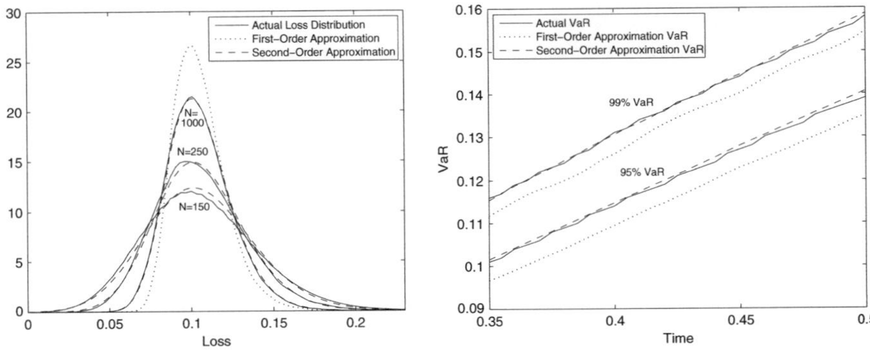

Fig. 2 On the *left* Comparison of approximate and actual loss distributions of failure rate L_t^N for different N at $t = 0.5$. Parameter choices: $(\sigma, \alpha, \bar{\lambda}, \lambda_0, \beta^C, \beta^S) = (0.9, 4, 0.2, 0.2, 1, 1)$. On the *right* Comparison of approximate and actual VaR. Parameter choices: $(\sigma, \alpha, \bar{\lambda}, \lambda_0, \beta^C, \beta^S) = (0.9, 4, 0.2, 0.2, 1, 1)$. In both cases, X is an OU process with reversion speed 2, volatility 1, initial value 1 and mean 1

In order to state the convergence result, we introduce some operators. Let $\mathsf{p} \in \mathcal{P} \subset \mathbb{R}^6$ and for $f \in C_b^2(\mathcal{P})$, define

$$
\begin{aligned}
(\mathcal{G}_{x,\mu} f)(\mathsf{p}) &= (\mathcal{L}_1 f)(\mathsf{p}) + (\mathcal{L}_3^x f)(\mathsf{p}) + \langle \mathcal{Q}, \mu \rangle (\mathcal{L}_2 f)(\mathsf{p}) + \langle \mathcal{L}_2 f, \mu \rangle \mathcal{Q}(\mathsf{p}) \\
(\mathcal{L}_5(f,g))(\mathsf{p}) &= \sigma^2 \frac{\partial f}{\partial \lambda}(\mathsf{p}) \frac{\partial g}{\partial \lambda}(\mathsf{p}) \lambda \\
(\mathcal{L}_6(f,g))(\mathsf{p}) &= f(\mathsf{p}) g(\mathsf{p}) \lambda \\
(\mathcal{L}_7 f)(\mathsf{p}) &= f(\mathsf{p}) \lambda
\end{aligned}
$$

Then, we have the following theorem related to the fluctuations analysis.

Theorem 4.1 (Theorem 4.1 in [25]) *For $J > 0$ large enough and for appropriate weight functions (w, ρ), the sequence $\{\Xi_t^N, t \in [0, T]\}_{N \in \mathbb{N}}$ is relatively compact in $D_{W_0^{-J}(w,\rho)}[0, T]$. For any $f \in W_0^J(w, \rho)$, the limit accumulation point of Ξ^N, denoted by $\bar{\Xi}$, is unique in $W_0^{-J}(w, \rho)$ and satisfies the stochastic evolution equation*

$$
\langle f, \bar{\Xi}_t \rangle = \langle f, \bar{\Xi}_0 \rangle + \int_0^t \langle \mathcal{G}_{X_s, \bar{\mu}_s} f, \bar{\Xi}_s \rangle ds + \int_0^t \left\langle \mathcal{L}_4^{X_s} f, \bar{\Xi}_s \right\rangle dV_s + \langle f, \bar{\mathcal{M}}_t \rangle, \text{ a.s.} \tag{15}
$$

for any $f \in W_0^J(w, \rho)$, where $\bar{\mathcal{M}}$ is a distribution-valued martingale with predictable variation process

$$
\begin{aligned}
[\langle f, \bar{\mathcal{M}} \rangle]_t = \int_0^t \Big[&\langle \mathcal{L}_5(f, f), \bar{\mu}_s \rangle + \langle \mathcal{L}_6(f, f), \bar{\mu}_s \rangle + \langle \mathcal{L}_2 f, \bar{\mu}_s \rangle^2 \langle \mathcal{Q}, \bar{\mu}_s \rangle \\
&- 2 \langle \mathcal{L}_7 f, \bar{\mu}_s \rangle \langle \mathcal{L}_2 f, \bar{\mu}_s \rangle \Big] ds.
\end{aligned}
$$

Conditional on the σ-algebra $\mathcal{V}_t$ that is generated by the V−Brownian motion, $\bar{\mathcal{M}}_t$ is centered Gaussian with covariance function, for $f, g \in W_0^J(w, \rho)$, given by

$$\mathrm{Cov}\left[\langle f, \bar{\mathcal{M}}_{t_1}\rangle, \langle g, \bar{\mathcal{M}}_{t_2}\rangle \,\Big|\, \mathcal{V}_{t_1\vee t_2}\right] = \mathbb{E}\Bigg[\int_0^{t_1\wedge t_2} [\langle \mathcal{L}_5(f,g), \bar{\mu}_s\rangle + \langle \mathcal{L}_6(f,g), \bar{\mu}_s\rangle + \langle \mathcal{L}_2 f, \bar{\mu}_s\rangle \langle \mathcal{L}_2 g, \bar{\mu}_s\rangle \langle \mathcal{Q}, \bar{\mu}_s\rangle - \langle \mathcal{L}_7 g, \bar{\mu}_s\rangle \langle \mathcal{L}_2 f, \bar{\mu}_s\rangle - \langle \mathcal{L}_7 f, \bar{\mu}_s\rangle \langle \mathcal{L}_2 g, \bar{\mu}_s\rangle] ds \,\Big|\, \mathcal{V}_{t_1\vee t_2}\Bigg]. \tag{16}$$

It is clear that if $\beta_n^S = 0$ for all n, then the limiting distribution-valued martingale $\bar{\mathcal{M}}$ is centered Gaussian with covariance operator given by the (now deterministic) term within the expectation in (16).

The main idea for the derivation of (15) comes from the proof of the convergence to the solution of (9). Define

$$(\mathcal{L}_1^\circ f)(\mathsf{p}) = \frac{1}{2}\sigma^2\lambda \frac{\partial^2 f}{\partial \lambda^2}(\mathsf{p}) - \alpha(\lambda - \bar{\lambda})\frac{\partial f}{\partial \lambda}(\mathsf{p})$$

for $\mathsf{p} = (\lambda, \alpha, \bar{\lambda}, \sigma, \beta^C, \beta^S)$. Let's also assume for the moment that $\beta_n^S = 0$ for every $n \in \mathbb{N}$, i.e., let's neglect exposure to the exogenous risk X and focus on contagion. Then we can write the evolution of $\langle f, \mu_t^N\rangle$ as

$$\begin{aligned} d\left\langle f, \mu_t^N\right\rangle &= \frac{1}{N}\sum_{n=1}^N \mathcal{L}^\circ f(\mathsf{p}_t^N)1_{\{t<\tau_n\}}dt - \frac{1}{N}\sum_{n=1}^N f(\mathsf{p}_t^n)\lambda_t^n 1_{\{t<\tau^n\}}dt \\ &\quad + \frac{1}{N}\sum_{n=1}^N\sum_{m=1}^N \left\{ f\left(\mathsf{p}^n + \frac{\beta_n^C}{N}e_1\right) - f(\mathsf{p}_t^n)\right\} 1_{\{\tau_n<t\}}\lambda_t^m 1_{\{\tau_m\le t\}}dt + dM_t \\ &\approx \left\langle \mathcal{L}_1 f, \mu_t^N\right\rangle dt + \left\langle \mathcal{L}_2 f, \mu_t^N\right\rangle\left\langle \mathcal{Q}, \mu_t^N\right\rangle dt + dM_t \end{aligned}$$

where M is a martingale which may change from line to line. This leads to (9), when $\beta_n^S = 0$ for every $n \in \mathbb{N}$, see [20].

To get the Gaussian correction, we see that

$$d\left\langle f, \Xi_t^N\right\rangle \approx \left\langle \mathcal{L}_1 f, \Xi_t^N\right\rangle + \left\{\left\langle \mathcal{L}_2 f, \Xi_t^N\right\rangle\left\langle \mathcal{Q}, \mu_t^N\right\rangle + \langle \mathcal{L}_2 f, \bar{\mu}_t\rangle\left\langle \mathcal{Q}, \Xi_t^N\right\rangle\right\} dt + dM_t$$

where M is a martingale. For large N, M should be Gaussian, in which case Ξ^N is indeed a Gaussian process. Putting the systematic risk process X back into (2)–(5), one recovers the result of Theorem 4.1.

5 Analysis of Tail Events: Large Deviations

Once we have identified what is typical, we can study the structure of atypically large failure rates. Large deviations outlines a circle of ideas and calculations for understanding the origination and transformation of rare events (see [42, 43]). Large deviation arguments allow us to identify the "dominant" way that rare events will occur in complex systems. This is the feature that is being exploited in [26], i.e., how different sources of stochasticity can lead to system collapse.

By the discussion in Sect. 3, we have that the pool has a default rate $L_T = 1 - \bar{\mu}_T(\mathcal{P})$ at time T. Let's fix $\ell > L_T$. Then $\lim_{N\to\infty} \mathbb{P}\{L_T^N \geq \ell\} = 0$; it is a *rare event* that the default rate in the pool exceeds ℓ. We want to understand as much as possible about $\{L_T^N \geq \ell\}$.

Using, the theory of large deviations, we can understand both how rare this event is, and what the "most likely" way is for this rare event to occur. Events far from equilibrium crucially depend on how rare events propagate through the system. Large deviations gives rigorous ways to understand these effects, and we want to use this machinery to understand the structure of atypically large default clusters in the portfolio. A reference for large deviations is [44].

If we have that

$$\mathbb{P}\{L^N \approx \varphi\} \approx \exp[-NI(\varphi)], \text{ as } N \to \infty$$

for some appropriate functional I, then by the contraction principle we should have that

$$\mathbb{P}\{L_T^N \approx \ell\} \approx \exp\left[-NI'(\ell)\right], \text{ as } N \to \infty \tag{17}$$

where

$$I'(\ell) = \inf\{I(\varphi) : \varphi(T) = \ell\} \tag{18}$$

(in other words, I' is the large deviations rate function for L_T^N). This gives us the rate at which the tail of the default rate L_T^N decays as the diversification parameter grows. More importantly, though, the variational problem (18) gives us the *preferred* way which atypically large default rates occur. Namely, if there is a $\varphi_\ell^* : [0, T] \to [0, 1]$ such that

$$I'(\ell) = I(\varphi_\ell^*)$$

then for any $\delta > 0$, the Gibbs conditioning principle suggests that

$$\lim_{N\to\infty} \mathbb{P}\{\|L_T^N - \varphi_\ell^*\| \geq \delta | L_T^N \geq \ell\} = 0.$$

Insights into large deviations of (2)–(5) have been developed in [26] when $\varepsilon \downarrow 0$ and when $\varepsilon = O(1)$ as $N \nearrow \infty$. We note here that in the case $\varepsilon = O(1)$, the large deviations principle is conditional on the systematic risk X. Such results allow us to study the comparative effect of the systematic risk process X and of the contagion feedback on the tails of the loss distribution.

Before presenting the result, let us first investigate numerically a test case, which is indicative of the kind of results that large deviations theory can give us. Apart from approximating the tail of the distribution, large deviations can give quantitative insights into the most likely path to failure of a system.

For presentation purposes and for the rest of this section, we assume that $\varepsilon = \varepsilon_N \downarrow 0$ as $N \uparrow \infty$. Consider a heterogeneous test portfolio composed initially of $N = 200$ names. Let us assume that we can separate the names in the portfolio into three types: Type A is 16.67 % of the names, Type B is 33.33 % of the names and Type C is 50 % of the names. For presentation purposes, we assume that all parameters but the contagion parameter are the same among the different types. In particular, we have the following choice of parameters.

It is instructive to compare the different cases, based on whether there are contagion effects in the default intensities or not. In particular, we compare two different cases, (a) Systematic risk only: $\beta^S \neq 0$, $\beta^C = 0$, and (b) Systematic risk and contagion: $\beta^S \neq 0$, $\beta^C \neq 0$. In each case, the time horizon is $T = 1$.

Using the methods of Sect. 3, one can compute that the typical loss in such a pool at time $T = 1$. If contagion effects are not present, i.e., if $\beta^C_A = \beta^C_B = \beta^C_C = 0$, then the typical loss in such a portfolio at time $T = 1$ is $L_T = 42.5\,\%$. If on the other hand, contagion (feedback) effects are present and the β^C parameters take the values of Table 1, then the typical loss in such a portfolio at time $T = 1$ has been increased to $L_T = 72.1\,\%$. In Fig. 3, we plot the large deviations rate functions for each of the two different cases. As we saw in the beginning of this section, the rate function governs the asymptotics of the tail of the loss distribution. Notice that in every case, the rate function is convex and it becomes zero at the corresponding law of large numbers.

Moreover, since the contagion parameter of Type A is higher than the contagion parameter for Type B or C, one expects that names of Type A will be more prompt to the contagious impact of defaults. Indeed, after computing the rate function and the associated extremals, as defined by large deviations theory, one gets the most likely paths to failure as seen in Figs. 4 and 5. The $\varphi(t)$ trajectories correspond to the contagion extremals for each of the three types, whereas the $\psi(t)$ corresponds to the systematic risk extremal.

One can make two conclusions out of Figs. 4 and 5. The first conclusion is related to the φ extremals (Fig. 4). We notice that at any given time t, the extremal for Type

Table 1 Parameter values for a test portfolio composed of three types of assets

	α	$\bar{\lambda}$	σ	λ_0	γ	β^S	β^C
Type A	0.5	2	0.5	0.2	1	1	10
Type B	0.5	2	0.5	0.2	1	1	3
Type C	0.5	2	0.5	0.2	1	1	1

We take $\varepsilon_N = \frac{1}{\sqrt{N}}$

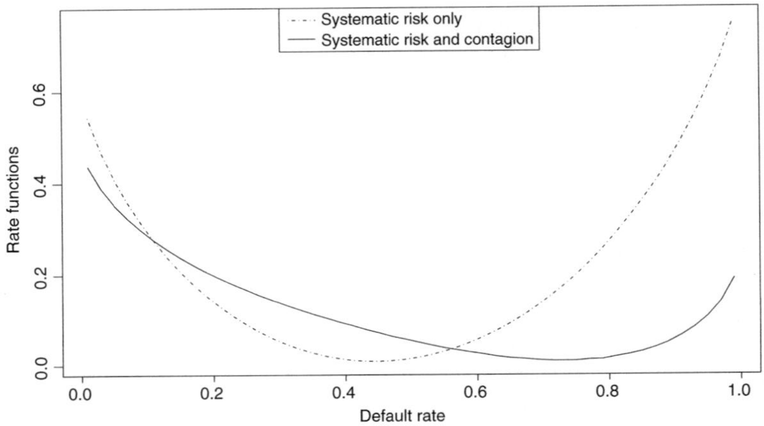

Fig. 3 Rate function governing the log-asymtptotics of the tail of the loss distribution

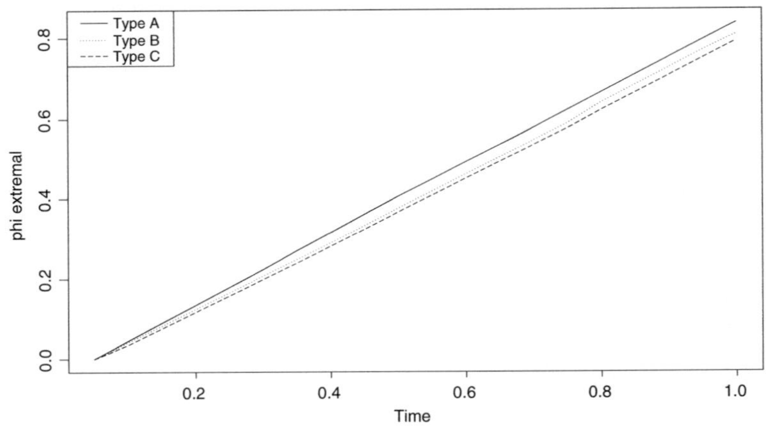

Fig. 4 Optimal $\varphi(t)$ trajectories for the three different types in the pool for $t \in [0, 1]$ and $\ell = 0.81$

A is bigger than the extremal for Type B, which in turn is bigger than the extremal of Type C. This implies that unlikely large losses for components of Type A are more likely than unlikely large losses for components of Type B, which are more likely than large losses for components of Type C. Thus, components of Type A affect the pool more than components of Type B, which in turn affect the pool more than components of Type C even though Type A composes 16.67 % of the pool, whereas Type B, composes 33.33 % of the pool and Type C composes 50 % of the pool. The second conclusion is related to the ψ extremals (Fig. 5). We notice that the effect of the systematic risk is most profound in the beginning but then its significance decreases.

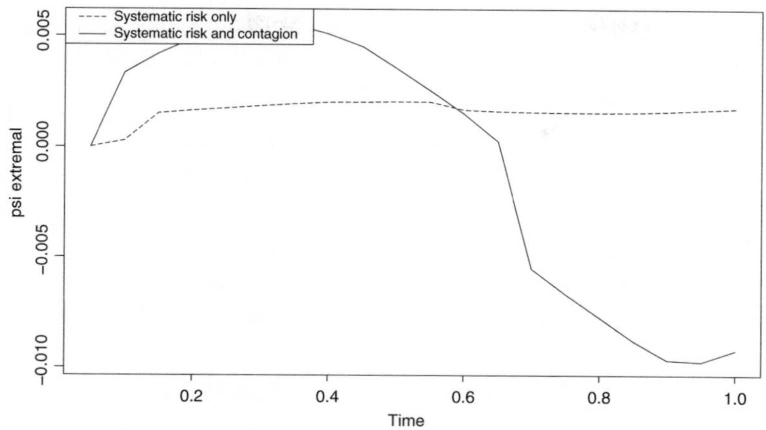

Fig. 5 Comparing optimal $\psi(t)$ trajectories in the case of absence and presence of the contagion effects for $t \in [0, 1]$ and $\ell = 0.81$

Namely, if a large cluster were to occur, the systematic risk factor is likely to play an important role in the beginning, but then the contagion effects become more important. Assets of Type A are likely to contribute to the default clustering effect more, followed by assets of Type B and the ones that will contribute the least to the default cluster are assets of Type C.

As it is also seen in the numerical experiments done in [26], the large deviations analysis help quantify the effect that the contagion and the systematic risk factor have on the behavior of the extremals (the most likely path to failure). An understanding of the role of the preferred paths to large default rates and the most likely ways in which contagion and systematic risk combine to lead to large default rates would give useful insights into how to optimally hedge against such events.

Let us next proceed by motivating the development of the large deviations principle for the default timing model (2)–(5) that is considered in this paper.

We denote scenarios, i.e., defaults, that are not in $[0, T]$ by an abstract point $\star$ not in $[0, T]$ and define the Polish space

$$\mathcal{T} = [0, T] \cup \{\star\}$$

To motivate things, let's first assume for simplicity that $\beta^C = \beta^S = 0$ and that the system is homogeneous, i.e., that $\mathsf{p}^n = \mathsf{p}$ for all n. Define

$$d\lambda_t = -\alpha(\lambda_t - \bar{\lambda})dt + \sigma\sqrt{\lambda_t}dW_t \qquad t > 0$$

with $\lambda_0 = \lambda_\circ$. This Feller diffusion will represent the conditional intensity of a "randomly-selected" component of our (homogeneous and independent) system. Define the measure $\mu_0 \in \mathscr{P}(\mathbb{R}_+)$ by setting

$$\mu_0[0,t] = 1 - \mathbb{E}\left[\exp\left[-\int_0^t \lambda_s ds\right]\right]$$

for all $t > 0$; μ_0 is the common law of the default times τ_n's.

In the independent case, i.e., when $\beta^C = 0$, standard Sanov's theorem [44], implies that $\{dL^N\}_{N\in\mathbb{N}}$ has a large deviations principle with rate function

$$H(\nu,\mu_0) = \int_{t\in\mathcal{T}} \ln \frac{d\nu}{d\mu_0}(t)\nu(dt)$$

if $\nu \ll \mu_0$ and $H(\nu,\mu_0) = \infty$ if $\nu \not\ll \mu_0$ (i.e., $H(\nu,\mu_0)$ is the relative entropy of ν with respect to μ_0). By the contraction principle, the rate function for L_T^N is

$$I^{\mathrm{ind},\prime}(\ell) = \inf\left\{H(\nu,\mu_0) : \nu \in \mathscr{P}(\mathbb{R}_+),\, \nu[0,t] = \varphi(t) \text{ for all } t \in [0,T] \text{ and } \nu[0,T] = \ell\right\}$$

In the independent case, we can actually compute both the extremal φ that achieves the infimum and the corresponding rate function $I^{\mathrm{ind},\prime}(\ell)$ in closed form.

Assume that $\mu_0[0,T] \in (0,1)$ and $\ell \in (0,1)$. Fix $\nu \in \mathscr{P}(\mathcal{T})$ such that $\nu[0,T] = \ell$. Define

$$\mu_{0,-}(A) = \frac{\mu_0(A\cap[0,T])}{\mu_0[0,T]} \quad \text{and} \quad \nu_-(A) = \frac{\nu(A\cap[0,T])}{\ell}$$

for all $A \in \mathscr{B}[0,T]$. Then μ_- and ν_- are in $\mathscr{P}[0,T]$. We can write that

$$H(\nu,\mu_0) = \ell\left\{\hbar(\nu_-,\mu_{0,-}) + \ln\frac{\ell}{\mu_0[0,T]}\right\} + \ln\frac{\nu\{\star\}}{\mu_0\{\star\}}\nu\{\star\} \tag{19}$$

where $\hbar$ is entropy on $\mathscr{P}[0,T]$. We can minimize the $\hbar$ term by setting $\nu_- = \mu_{0,-}$, and we get that

$$\begin{aligned} I^{\mathrm{ind},\prime}(\ell) &= \ell\ln\frac{\ell}{\mu_0[0,T]} + (1-\ell)\ln\frac{1-\ell}{\mu_0\{\star\}} \\ &= \ell\ln\frac{\ell}{\mu_0[0,T]} + (1-\ell)\ln\frac{1-\ell}{1-\mu_0[0,T]}. \end{aligned} \tag{20}$$

This is in fact obvious; $L_T^N = \frac{1}{N}\sum_{n=1}^N 1_{\{\tau_n \le T\}}$, and in this case the $1_{\{\tau_n \le T\}}$'s are i.i.d. Bernoulli random variables with common bias $\mu_0[0, T]$. The rate function $I^{\mathrm{ind},\prime}(\ell)$ of (20) is the entropy of Bernoulli coin flips. Of more interest, however, is the optimal path. In setting $\nu_- = \mu_-$ in (19), we essentially identify the optimal path

$$\varphi(t) = \ell \frac{\mu_0[0,t]}{\mu_0[0,T]},$$

where the last relation holds since we also require $\varphi(T) = \ell$.

It turns out that one can extend this result to give a generalized Sanov's theorem for the case $\beta^C > 0$, where dL^N feeds back into the dynamics of the λ^n's. The case $\beta^S > 0$ can be treated using a conditioning argument and the well developed theory of large deviations for small noise diffusions. For the heterogeneous case, one needs an additional variational step which minimizes over all the possible ways that losses are distributed among systems of different types. Even though an explicit closed form expression for the extremals and for the corresponding rate function is no longer possible, one can still rely on numerically computing them. Let us make this discussion precise.

To fix the discussion, let us assume (see [26] for the general case) that the exogenous risk X is of Ornstein-Uhlenbeck type, i.e.,

$$\begin{aligned} dX_t &= -\gamma X_t dt + dV_t \\ X_0 &= x_\circ \end{aligned}$$

Let W^* be a reference Brownian motion. Fix a name in the pool $\mathsf{p} = (\lambda_\circ, \alpha, \bar{\lambda}, \sigma, \beta^C, \beta^S) \in \mathcal{P}$ and time horizon $T > 0$.

The Freidlin-Wentzell theory of large deviations for SDE's gives us a natural starting point. In the Freidlin-Wentzell analysis, a dominant ODE is subjected to a small diffusive perturbation; informally, the Freidlin-Wentzell theory tells us that if we want to find the probability that the randomly-perturbed path is close to a reference trajectory, we should use that reference trajectory in the dynamics. This leads to the correct LDP rate function for the original SDE. If we want to find the asymptotics of the probability that $\left(dL^N \approx d\varphi, \varepsilon_N dX \approx d\psi\right)$ for some absolutely continuous functions φ and ψ, i.e., $\varphi, \psi \in AC([0,T], \mathbb{R})$, we should consider the stochastic hazard functions

$$\begin{aligned} d\lambda_t^{\varphi,\psi} &= -\alpha(\lambda_t^{\varphi,\psi} - \bar{\lambda})dt + \sigma\sqrt{\lambda_t^{\varphi,\psi}}\, dW_t^* + \beta^C d\varphi(t) + \beta^S \lambda_t^{\varphi,\psi} d\psi(t) \qquad t \in [0,T] \\ \lambda_0 &= \lambda_\circ. \end{aligned}$$

This will represent the conditional intensity of a "randomly-selected" name in our pool. Define next

$$f^{\mathsf{p}}_{\varphi,\psi}(t) = \mathbb{E}\left[\lambda_t^{\varphi,\psi} \exp\left[-\int_{s=0}^t \lambda_s^{\varphi,\psi} ds\right]\right],$$

where, we have used the superscript p to denote the dependence on the particular type. Then for every $t \in [0, T]$ we have that

$$\int_{s=0}^{t} f^{\mathsf{p}}_{\varphi,\psi}(s)ds = 1 - \mathbb{E}\left[\exp\left[-\int_{s=0}^{t} \lambda^{\varphi,\psi}_s ds\right]\right] = \mathbb{P}\left\{\int_{s=0}^{t} \lambda^{\varphi,\psi}_s ds > \mathfrak{e}\right\}$$

where $\mathfrak{e}$ is an exponential(1) random variable which is independent of W^*. In other words, $f^{\mathsf{p}}_{\varphi,\psi}$ is the density (up to time T) of a default time whose conditional intensity is $\lambda^{\varphi,\psi}$. In fact, due to the affine structure of the model, we have an explicit expression for $f^{\mathsf{p}}_{\varphi,\psi}$ (see Lemma 4.1 in [26]).

For given trajectories φ and ψ in $AC([0, T]; \mathbb{R})$, define $\mu^{\mathsf{p}}_{\varphi,\psi} \in \mathscr{P}(\mathcal{T})$ as

$$\mu^{\mathsf{p}}_{\varphi,\psi}(A) = \int_{t \in A \cap [0,T]} f^{\mathsf{p}}_{\varphi,\psi}(t)dt + \delta_\star(A)\left\{1 - \int_0^T f^{\mathsf{p}}_{\varphi,\psi}(t)dt\right\}$$

for all $A \in \mathscr{B}(\mathcal{T})$.

At a heuristic level one can derive the large deviations principle as follows. Let us assume that we can establish that

$$\mathbb{P}\{L^N \approx \varphi | X^N \approx \psi\} \approx \exp\left[-N I^\circ(\varphi, \psi)\right]$$

and that $\left\{X^N_\cdot = \varepsilon_N X_\cdot, N < \infty\right\}$ also has large deviations principle in $C([0, T]; \mathbb{R})$ with action functional J_X; i.e.,

$$\mathbb{P}\left\{X^N \approx \psi\right\} \approx \exp\left[-\frac{1}{\varepsilon_N^2} J_X(\psi)\right]$$

as $N \nearrow \infty$. Then, we should have that

$$\mathbb{P}\{L^N \approx \varphi,\ X^N \approx \psi\} \approx \exp\left[-N I^\circ(\varphi, \psi) - \frac{1}{\varepsilon_N^2} J_X(\psi)\right].$$

In fact, the previous heuristics can be carried out rigorously and in the end one derives the following rigorous large deviations result.

Theorem 5.1 (Theorem 3.8 in [26]) *Consider the system defined in* (2)–(5) *with* $\lim_{N\to\infty} \varepsilon_N = 0$ *such that* $\lim_{N\to\infty} N\varepsilon_N^2 = c \in (0, \infty)$ *and let* $T < \infty$. *Under the appropriate assumptions the family* $\{L^N_T, N \in \mathbb{N}\}$ *satisfies the large deviation principle, with rate function*

$$I'(\ell) = \inf\left\{I(\varphi, \psi) : \varphi \in C(\mathcal{P} \times [0, T]), \psi \in C([0, T]), \psi(0) = \varphi(\mathsf{p}, 0) = 0,\right.$$
$$\left.\bar{\varphi}(s) = \int_{\mathcal{P}} \varphi(\mathsf{p}, s) U(d\mathsf{p}), \bar{\varphi}(T) = \ell\right\}$$

where if $\varphi \in AC(\mathcal{P} \times [0,T])$, $\psi \in AC([0,T])$, $\psi(0) = \varphi(\mathsf{p}, 0) = 0$, *then*

$$I(\varphi, \psi) = \int_{\mathcal{P}} H\left(\varphi(\mathsf{p}), \mu^{\mathsf{p}}_{\bar{\varphi},\psi}\right) U(d\mathsf{p}) + \frac{1}{c} J_X(\psi)$$

and $I(\varphi, \psi) = \infty$ *otherwise. Here,* $J_X(\psi)$ *is the rate function for the process* $\{\varepsilon_N X^N, N < \infty\}$. *Namely, for* $\psi \in AC([0,T]; \mathbb{R})$ *with* $\psi(0) = 0$ *we have*

$$J_X(\psi) = \frac{1}{2} \int_0^T \left|\dot{\psi}(s) + \gamma \psi(s)\right|^2 ds$$

and $J_X(\psi) = \infty$ *otherwise.* $I'(\ell)$ *has compact level sets.*

If the heterogeneous portfolio is composed by K different types of assets with homogeneity within each type, then Theorem 5.1 simplifies to the following expression.

For $\xi, \varphi, \psi \in AC([0,T])$ let us define the functional

$$g^{\mathsf{p}}(\xi, \varphi, \psi) = \int_0^T \ln\left(\frac{\dot{\xi}(t)}{f^{\mathsf{p}}_{\varphi,\psi}(t)}\right) \dot{\xi}(t) dt + \ln\left(\frac{1 - \xi(T)}{1 - \int_0^T f^{\mathsf{p}}_{\varphi,\psi}(t) dt}\right)(1 - \xi(T))$$

Due to the affine structure of the model, we have an explicit expression for $f^{\mathsf{p}}_{\varphi,\psi}$ (see Lemma 4.1 in [26]).

Assume that κ_i % of the names are of type A_i with $i = 1, \ldots, K$ and $\sum_{i=1}^K \kappa_i = 100$. Setting $\varphi(\mathsf{p}, s) = \sum_{i=1}^K \frac{\kappa_i}{100} \varphi_{A_i}(s) \chi_{\{\mathsf{p}_{A_i}\}}(\mathsf{p})$, we get the following simplified expression for the rate function

$$I'(\ell) = \inf \left\{ \sum_{i=1}^K \frac{\kappa_i}{100} g^{\mathsf{p}_{A_i}}(\varphi_{A_i}, \varphi, \psi) + \frac{1}{c} J_X(\psi) : \varphi(t) = \sum_{i=1}^K \frac{\kappa_i}{100} \varphi_{A_i}(t) \text{ for every } t \in [0,T] \right.$$
$$\left. \varphi(T) = \ell, \varphi_{A_i}(0) = \psi(0) = 0, \varphi_{A_i}, \psi \in AC([0,T]) \text{ for every } i = 1, \ldots, K \right\}.$$

An optimization algorithm can then be employed to solve the minimization problem associated with $I'(\ell)$ and compute the extremals φ_{A_i} for $i = 1, \ldots, K$ and ψ. This is the formula that the numerical example presented in Figs. 4 and 5 was based on. In the numerical example that was considered there we had three types, i.e., $K = 3$.

The large deviations results have a number of important applications. Firstly, they lead to an analytical approximation of the tail of the distribution of the failure rate L^N for large systems. These approximations complement the first- and second- order approximations suggested by the law of large numbers and fluctuations analysis of Sects. 3 and 4 respectively and facilitates the estimation of the likelihood of systemic collapse. Secondly, the large deviations results provide an understanding of the "preferred" ways of collapse, which can also be used to design "stress tests" for the system. In particular, this understanding can guide the selection of meaningful stress scenarios to be analyzed. Thirdly, they can motivate the design of asymptotically efficient importance sampling schemes for the tail of the portfolio loss. We discuss some of the related issues in Sect. 6.

6 Monte Carlo Methods for Estimation of Tail Events: Importance Sampling

Suppose we want to computationally simulate $\mathbb{P}\{L_T^N \geq \ell\}$, where $\lim_{N\to\infty} \mathbb{P}\{L_T^N \geq \ell\} = 0$ again holds. Accurate estimates of such rare-event probabilities are important in many applications areas of our system (2)–(5), including credit risk management, insurance, communications and reliability. Monte Carlo methods are widely used to obtain such estimates in large complex systems such as ours; see, for example, [29, 30, 45–51].

Standard Monte Carlo sampling techniques perform very poorly in estimating rare events (for which, by definition, most samples can be discarded). Importance sampling, which involves a change of measure, can be used to address this issue. In general, large deviations theory provides an optimal way to 'tilt' measures. The variational problems identified by large deviations usually lead to measure transformations under which pre-specified rare events become much more likely, but which give unbiased estimates of probabilities of interest; see for example [28, 34, 52–56].

Let Γ^N be any unbiased estimator of $\mathbb{P}\{L_T^N \geq \ell\}$ that is defined on some probability space with probability measure $\mathbb{Q}$. In other words, Γ^N is a random variable such that $\mathbb{E}^{\mathbb{Q}}\Gamma^N = \mathbb{P}\{L_T^N \geq \ell\}$, where $\mathbb{E}^{\mathbb{Q}}$ is the expectation operator associated with $\mathbb{E}$. In our setting, it takes the form

$$\Gamma^N = 1_{\{L_T^N > \ell\}} \frac{d\mathbb{P}}{d\mathbb{Q}},$$

where $\frac{d\mathbb{P}}{d\mathbb{Q}}$ is the associated Radon-Nikodym derivative.

Importance sampling involves the generation of independent copies of Γ^N under $\mathbb{Q}$; the estimate is the sample mean. The specific number of samples required depends on the desired accuracy, which is measured by the variance of the sample mean. However, since the samples are independent it suffices to consider the variance of a single sample. Because of unbiasedness, minimizing the variance is equivalent to minimizing the second moment. An application of Jensen's inequality, shows that if

$$\liminf_{N\to\infty} \frac{1}{N} \ln \mathbb{E}^{\mathbb{Q}}(\Gamma^N)^2 = -2I'(\ell),$$

then Γ^N achieves this best decay rate, and is said to be *asymptotically optimal*. One wants to choose $\mathbb{Q}$ such that asymptotic optimality is attained.

To motivates things let us assume for the moment that $\beta^C = \beta^S = 0$ and that the system is homogeneous, i.e., that $\mathsf{p}^n = \mathsf{p}$ for all n. In the independent and homogeneous case, $\Xi_n = 1_{\{\tau_n \le T\}}$ are i.i.d. random variables such that for every $t \in [0, T]$

$$\mathbb{P}\{\tau_n \le t\} = \mathbb{P}\left\{\int_0^t \lambda_s^{0,0} ds > \mathfrak{e}\right\} = 1 - \mathbb{E}\left[\exp\left[-\int_0^t \lambda_s^{0,0} ds\right]\right] = \int_0^t f_{0,0}(s) ds$$

For notational convenience, we shall define

$$p = \int_0^T f_{0,0}(s) ds$$

It is easy to see that,

$$NL_T^N \sim \text{Binomial}(N, p)$$

To minimize the variance, we need to increase the probability of defaults. Define

$$\Lambda^N(\theta; t) = \ln \mathbb{E}\left[e^{\theta L_t^N}\right]$$

A simple computation shows that

$$\bar{\Lambda}(\theta; t) = \lim_{N \to \infty} \frac{1}{N} \Lambda^N(N\theta; t) = \ln\left(p\left(e^\theta - 1\right) + 1\right)$$

Define

$$p_\theta = \frac{pe^\theta}{1 + p(e^\theta - 1)}$$

Clearly $p_0 = p$. Notice that the density of a Binomial(N, p) with respect to a Binomial(N, p_θ) is

$$\begin{aligned} \mathcal{Z}_\theta &= \prod_{n=1}^N \left(\frac{p}{p_\theta}\right)^{\Xi_n} \left(\frac{1-p}{1-p_\theta}\right)^{1-\Xi_n} = \prod_{n=1}^N \left[\left(1 + p(e^\theta - 1)\right) e^{-\theta \Xi_n}\right] \\ &= e^{N\left(-\theta L_T^N + \bar{\Lambda}(\theta; T)\right)} \end{aligned}$$

Therefore, for θ fixed, the suggestion is to simulate under a new change of measure, under which $NL_T^N \sim$ Binomial(N, p_θ) and to return the estimator

$$\Gamma = \frac{1}{M} \sum_{i=1}^M 1_{\{L_T^{N,i} > \ell\}} e^{N\left(-\theta L_T^{N,i} + \bar{\Lambda}(\theta; T)\right)}$$

It is clear that this estimator is unbiased. We want to choose θ that minimizes the variance, or equivalently the second moment. For this purpose, we define the second moment

$$Q(\ell,\theta)=\mathbb{E}_{\theta}\Gamma^{2}=\mathbb{E}_{\theta}\left[1_{\{L_T>\ell\}}e^{2N\left(-\theta L_T^N+\bar{\Lambda}(\theta;T)\right)}\right]$$

Notice that

$$-\frac{1}{N}\ln Q(\ell,\theta)\geq -2\frac{1}{N}N\left(-\theta\ell+\bar{\Lambda}(\theta;T)\right)=2(\theta\ell-\bar{\Lambda}(\theta;T))$$

Due to convexity of $\bar{\Lambda}(\theta;T)$, we have that the maximizer over $\theta\in[0,\infty)$ of the lower bound is at θ^* such that $\frac{\partial\bar{\Lambda}(\theta^*;T)}{\partial\theta}=\ell$. In particular, (recall that $\frac{\partial\bar{\Lambda}(0;T)}{\partial\theta}=p$) we have

$$\theta^*=\begin{cases}\ln\frac{\ell(1-p)}{p(1-\ell)}, & \text{if } \ell>p\\ 0, & \text{if } \ell<p\end{cases}$$

This construction means that under the new measure, we have

$$\mathbb{P}_{\theta^*}\{\tau_n\leq T\}=p_{\theta^*}=\ell.$$

In fact, we have the following theorem.

Theorem 6.1 *Let $\theta^*>0$ such that $\frac{\partial\bar{\Lambda}(\theta^*;T)}{\partial\theta}=\ell$. Then asymptotic optimality holds, in the sense that*

$$\lim_{N\to\infty}-\frac{1}{N}\ln Q(\ell,\theta^*)=2I^{ind,\prime}(\ell)$$

where $I^{ind,\prime}(\ell)$ is defined in (20).

Proof By Jensen's inequality we clearly have the upper bound. Namely, for every $\theta\in[0,\infty)$

$$\limsup_{N\to\infty}-\frac{1}{N}\ln Q(\ell,\theta)\leq 2I^{\text{ind}}(\ell) \tag{21}$$

Now, we need to prove that the lower bound is achieved for $\theta=\theta^*$, i.e., that

$$\liminf_{N\to\infty}-\frac{1}{N}\ln Q(\ell,\theta^*)\geq 2I^{\text{ind}}(\ell) \tag{22}$$

Recalling that $\theta^* = \ln \frac{\ell(1-p)}{p(1-\ell)}$ and $p = \int_{s=0}^{T} f_{0,0}(s)ds$, we easily see that

$$\begin{aligned}\liminf_{N\to\infty} -\frac{1}{N} \ln Q(\ell,\theta^*) &\geq 2\left(\theta^*\ell - \bar{\Lambda}(\theta^*;T)\right)\\ &= 2\left(\theta^*\ell - \ln\left(p(e^{\theta^*}-1)+1\right)\right)\\ &= 2\left(\ell \ln\frac{\ell}{p} + (1-\ell)\ln\frac{1-\ell}{1-p}\right)\\ &= 2I^{\text{ind}}(\ell)\end{aligned}$$

This concludes the proof of the theorem. □

In the heterogeneous case, i.e., if p^n can be different for each $n \in \mathbb{N}$, then $NL_T^N = \sum_{n=1}^{N} 1_{\{\tau_n \leq T\}}$ is no longer Binomial, but it is a sum of independent (but not identically distributed) Bernoulli random variables with success probability

$$p_n = \int_0^T f_{0,0}^{\mathsf{p}^n}(s)ds$$

indexed by n. Due to independence, similar methods as the one described above can be used to construct asymptotically efficient importance sampling schemes in the heterogeneous case.

The scheme just presented essentially amounts to a twist in the intensity of the defaults. However, in contrast to the independent case, i.e., when $\beta^C = \beta^S = 0$, the situation in the general dependent case $\beta^C, \beta^S \neq 0$ is more complicated. Notice also if at least one of the β_n^C's is not zero, then the model (2)–(5) does not fall into the category of the doubly-stochastic models, so techniques as the ones used in [45] do not apply. Also, implementation of interacting particle schemes for Markov Chain models as the ones developed in [29, 47] do not readily apply for such intensity models. The re-sampling schemes of [48] could apply in this setting, but one would need to construct an appropriate mimicking Markov Chain, something which is not clear how to do in the current setting.

We briefly present here an importance sampling scheme for the case that there exists at least one $\beta_n^C \neq 0$ and also applies independently of whether the systematic effects are present in the model or not. The suggested measure change essentially mimics the principal idea behind the measure change for the independent case. To be more precise, one directly twists the intensity of $NL_T^N = \sum_{n=1}^{N} 1_{\{\tau_n \leq T\}}$.

Let $\{S_k\}$ be the arrival times of NL_T^N and notice that $\left\{L_T^N \geq \ell\right\} = \left\{S_{\lceil \ell N \rceil} \leq T\right\}$. Let $M_s^n = 1_{\{\tau^n > s\}}$ and $\theta_s^N \geq 1$ be some progressively measurable twisting process. Then, define the measure $\mathbb{Q}$ via the Radon-Nicodym derivative

$$Z_N = e^{-\int_0^{S_{\lceil \ell N \rceil}} \log(\theta_{s-}^N) d(NL_s^N) - \int_0^{S_{\lceil \ell N \rceil}} (1-\theta_s^N) \sum_{n=1}^{N} \lambda_s^n M_s^n ds}.$$

It is known that if $\mathbb{E}\left[e^{-\sum_{k=1}^{\lceil \ell N \rceil}\log\left(\theta^N_{S_k-}\right)}\right] < \infty$, then $\mathbb{Q}$ defined by $\frac{d\mathbb{P}}{d\mathbb{Q}} = Z_N$ is a probability measure and it can be shown that NL^N_s admits $\mathbb{Q}-$intensity $\theta^N_s \sum_{n=1}^N \lambda^n_s M^n_s$ on the interval $[0, S_{\lceil \ell N \rceil})$.

This construction gives us some freedom into choosing appropriately the twisting process θ^N_s. Different choices of the twisting process θ^N_s are of course possible. For tractability purposes we restrict attention to a one-parameter family and set

$$\theta^N_s = \frac{\beta N}{\sum_{n=1}^N \lambda^n_s M^n_s} + 1.$$

For any $\beta \geq 0$ and under the measure induced by Z_N, i.e., under $\mathbb{Q}_\beta$, the process NL^N_s has intensity $\sum_{n=1}^N \lambda^n_s M^n_s + \beta N$ on $[0, S_{\lceil \ell N \rceil})$, i.e., it amounts to an additive shift of the intensity. Thus, β is a superimposed default rate and its role is to increase the default rate in the whole portfolio.

The purpose then is to optimize the limit as $N \to \infty$ of the upper bound of the second moment of the resulting estimator over β. This is the measure change that is investigated in [57], and it is shown there that there is a choice of $\beta = \beta^*$ for which asymptotic optimality can be established. Namely, there is a choice of $\beta = \beta^*$ that minimizes the second moment of the estimator in the limit as $N \to \infty$. We refer the interested reader to [57] for implementation details on this change of measure for related intensity models and for corresponding simulation results.

7 Conclusions

We presented an empirically motivated model of correlated default timing for large portfolios. Large portfolio analysis allows to approximate the distribution of the loss from default, whereas Gaussian corrections make the approximation valid even for portfolios of moderate size. The results can be used to compute the loss distribution and to approximate portfolio risk measures such as Value-at-Risk or Expected Shortfall. Then, large deviations analysis can help understand the tail of the loss distribution and find the most-likely paths to systemic failure and to the creation of default clusters. Such results give useful insights into the behavior of systemic risk as a function of the characteristics of the names in the portfolio and can be also potentially used to determine how to optimally safeguard against rare large losses. Importance sampling techniques can be used to construct asymptotically efficient estimators for tail event probabilities.

Acknowledgments The author was partially supported by the National Science Foundation (DMS 1312124).

References

1. Meinerding, C.: Asset allocation and asset pricing in the face of systemic risk: a literature overview and assessment. Int. J. Theor. Appl. Financ. (IJTAF) **15**(03), 1250023-1–1250023-27 (2012)
2. Vasicek, O.: Limiting loan loss probability distribution. Technical Report, KMV Corporation (1991)
3. Lucas, A., Klaassen, P., Spreij, P., Straetmans, S.: An analytic approach to credit risk of large corporate bond and loan portfolios. J. Bank. Financ. **25**, 1635–1664 (2001)
4. Schloegl, L., O'Kane, D.: A note on the large homogeneous portfolio approximation with the student-t copula. Financ. Stoch. **9**(4), 577–584 (2005)
5. Gordy, M.B.: A risk-factor model foundation for ratings-based bank capital rules. J. Financ. Intermed. **12**, 199–232 (2003)
6. Bush, N., Hambly, B., Haworth, H., Jin, L., Reisinger, C.: Stochastic evolution equations in portfolio credit modelling. SIAM J. Financ. Math. **2**, 627–664 (2011)
7. Dembo, A., Deuschel, J.-D., Duffie, D.: Large portfolio losses. Financ. Stoch. **8**, 3–16 (2004)
8. Glasserman, P., Kang, W., Shahabuddin, P.: Large deviations in multifactor portfolio credit risk. Math. Financ. **17**(3), 345–379 (2007)
9. Spiliopoulos, K., Sowers, R.: Recovery rates in investment-grade pools of credit assets: a large deviations analysis. Stoch. Process. Appl. **121**(12), 2861–2898 (2011)
10. Giesecke, K., Weber, S.: Credit contagion and aggregate losses. J. Econ. Dyn. Control **30**, 741–767 (2006)
11. Dai Pra, P., Runggaldier, W., Sartori, E., Tolotti, M.: Large portfolio losses: a dynamic contagion model. Ann. Appl. Probab. **19**, 347–394 (2009)
12. Dai Pra, P., Tolotti, M.: Heterogeneous credit portfolios and the dynamics of the aggregate losses. Stoch. Process. Appl. **119**, 2913–2944 (2009)
13. Cvitanić, J., Ma, J., Zhang, J.: The law of large numbers for self-exciting correlated defaults. Stoch. Process. Appl. **122**(8), 2781–2810 (2012)
14. Sircar, R., Zariphopoulou, T.: Utility valuation of credit derivatives and application to CDOs. Quant. Financ. **10**(2), 195–208 (2010)
15. Garnier, J., Papanicolaou, G., Yang, T.-W.: Large deviations for a mean field model of systemic risk. SIAM J. Financ. Math. **4**, 151–184 (2012)
16. Fouque, J.-P., Ichiba, T.: Stability in a model of inter-bank lending. SIAM J. Financ. Math. **4**, 784–803 (2013)
17. Ait-Sahalia, Y., Cacho-Diaz, J., Laeven, R.: Modeling financial contagion using mutually exciting jump processes. To appear in Journal of Financial Economics
18. Duan, J.-C.: Maximum likelihood estimation using price data of the derivative contract. Math. Financ. **4**, 155–167 (1994)
19. Bielecki, T., Crépey, S., Herbertsson, A.: Markov chain models of portfolio credit risk. Oxford Handbook of Credit Derivatives. Oxford University Press, New York (2011)
20. Giesecke, K., Spiliopoulos, K., Sowers, R.: Default clustering in large portfolios: typical events. Ann. Appl. Probab. **23**(1), 348–385 (2013)
21. Azizpour, S., Giesecke, K., Schwenkler, G.: Exploring the sources of default clustering. Working paper, Stanford University (2010)
22. Duffie, D., Saita, L., Wang, K.: Multi-period corporate default prediction with stochastic covariates. J. Financ. Econ. **83**(3), 635–665 (2006)
23. Bo, L., Capponi, A.: Bilateral credit valuation adjustment for large credit derivatives portfolios. Financ. Stoch. **18**(2), 431–482 (2014)
24. Giesecke, K., Spiliopoulos, K., Sowers, R., Sirignano, J.A.: Large portfolio asymptotics for loss from default. Math. Financ. **25**(1), 77–114 (2015)
25. Spiliopoulos, K., Sirignano, J.A., Giesecke, K.: Fluctuation analysis for the loss from default. Stoch. Process. Appl. **124**(7), 2322–2362 (2014)
26. Spiliopoulos, K., Sowers, R.: Default clustering in large pools: large deviations. SIAM J. Financ. Math. **6**, 86–116 (2015)

27. Sirignano, J.A., Schwenkler, G., Giesecke, K.: Likelihood estimation for large financial systems. Working paper, Stanford University (2013)
28. Glasserman, P., Wang, Y.: Counterexamples in importance sampling for large deviations probabilities. Ann. Appl. Probab. **7**, 731–746 (1997)
29. Carmona, R., Fouque, J.-P., Douglas, V.: Interacting particle systems for the computation of rare credit portfolio losses. Financ. Stoch. **13**(4), 613–633 (2009)
30. Deng, S., Giesecke, K., Lai, T.L.: Sequential importance sampling and resampling for dynamic portfolio credit risk. Oper. Res. **60**(1), 78–91 (2012)
31. Kotelenez, P.M., Kurtz, T.G.: Macroscopic limits for stochastic partial differential equations of Mckeanvlasov type. Probab. Theory Relat. Fields **146**(1–2), 189–222 (2010)
32. Kurtz, T.G., Xiong, J.: A stochastic evolution equation arising from the fluctuations of a class of interacting particle systems. Commun. Math. Sci. **2**(3), 325–358 (2004)
33. Purtukhia, O.G.: On the equations of filtering of multi-dimensional diffusion processes (unbounded coefficients). Thesis, Moscow, Lomonosov University 1984 (in Russian) (1984)
34. Sadowsky, S.J.: On Monte Carlo estimation of large deviations probabilities. Ann. Appl. Probab. **6**, 399–722 (1996)
35. Duffie, D., Pan, J., Singleton, K.: Transform analysis and asset pricing for affine jump-diffusions. Econometrica **68**, 1343–1376 (2000)
36. Dawson, D.A., Hochberg, K.J.: Wandering random measures in the Fleming-Viot model. Ann. Probab. **10**(3), 554–580 (1982)
37. Ethier, S.N., Kurtz, T.G.: Markov Processes: Characterization and Convergence. Wiley, New York (1986)
38. Fleming, W.H., Viot, M.: Some measure-valued Markov processes in population genetics theory. Indiana Univ. Math. J. **28**(5), 817–843 (1979)
39. Gartner, J.: On the Mckean-Vlasov limit for interacting diffusions. Mathematische Nachrichten **137**(1), 197–248 (1988)
40. Fernandez, B., Mélèard, S.: A Hilbertian approach for fluctuations on the Mckean-Vlasov model. Stoch. Process. Appl. **71**, 33–53 (1997)
41. Gyöngy, I., Krylov, N.: Stochastic partial differential equations with unbounded coefficients and applications, I. Stoch. Stoch. Rep. **32**, 53–91 (1990)
42. Freidlin, M., Wentzell, A.: Random Perturbations of Dynamical Systems, 2nd edn. Springer, New York (1984)
43. Varadhan, S.R.S.: Large Deviations and Applications. CBMS-NSF Regional Conference Series in Applied Mathematics, vol. 46. Society for Industrial and Applied Mathematics (SIAM), Philadelphia (1984)
44. Dembo, A., Zeitouni, O.: Large Deviations Techniques and Applications, 2nd edn. Springer, New York (1988)
45. Bassamboo, A., Jain, S.: Efficient importance sampling for reduced form models in credit risk. In: Proceedings of the 2006 Winter Simulation Conference, pp. 741–748 (2006)
46. Juneja, S., Bassamboo, A., Zeevi, A.: Portfolio credit risk with extremal dependence: asymptotic analysis and efficient simulation. Oper. Res. **56**(3), 593–606 (2008)
47. Carmona, R., Crépey, S.: Particle methods for the estimation of Markovian credit portfolio loss distributions. Int. J. Theor. Appl. Financ. **13**(4), 577–602 (2010)
48. Giesecke, K., Kakavand, H., Mousavi, M., Takada, H.: Exact and efficient simulation of correlated defaults. SIAM J. Financ. Math. **1**, 868–896 (2010)
49. Glasserman, P., Li, J.: Importance sampling for portfolio credit risk. Manag. Sci **51**(11), 1643–1656 (2005)
50. Glasserman, P.: Tail approximations for portfolio credit risk. J. Deriv. **12**(2), 24–42 (2004)
51. Zhang, X., Blanchet, J., Giesecke, K., Glynn, P.: Affine point processes: asymptotic analysis and efficient rare-event simulation. Working paper, Stanford University (2011)
52. Asmussen, S., Peter Glynn, W.: Stochastic Simulation: Algorithms and Analysis. Grundlehren der Mathematischen Wissenschaften. Springer, New York (2007)
53. Bucklew, J.: Introduction to Rare Event Simulation. Grundlehren der Mathematischen Wissenschaften. Springer, New York (2004)

54. Dupuis, P., Spiliopoulos, K., Wang, H.: Importance sampling for multiscale diffusions. Multiscale Model. Simul. **12**, 1–27 (2012)
55. Dupuis, P., Wang, H.: Importance sampling, large deviations and differential games. Stoch. Stoch. Rep. **76**, 481–508 (2004)
56. Dupuis, P., Wang, H.: Subsolutions of an Isaacs equation and efficient schemes for importance sampling. Math. Oper. Res. **32**, 723–757 (2007)
57. Giesecke, K., Shkolnik, A.: Asymptotic optimal importance sampling of default times. Working paper, Stanford University (2011)

Estimation of Volatility Functionals: The Case of a $\sqrt{n}$ Window

Jean Jacod and Mathieu Rosenbaum

Abstract We consider a multidimensional Itô semimartingale regularly sampled on $[0, t]$ at high frequency $1/\Delta_n$, with Δ_n going to zero. The goal of this paper is to provide an estimator for the integral over $[0, t]$ of a given function of the volatility matrix, with the optimal rate $1/\sqrt{\Delta_n}$ and minimal asymptotic variance. To achieve this, we use spot volatility estimators based on observations within time intervals of length $k_n\Delta_n$. In [5], this was done with $k_n \to \infty$ and $k_n\sqrt{\Delta_n} \to 0$, and a central limit theorem was given after suitable de-biasing. Here we do the same with the choice $k_n \asymp 1/\sqrt{\Delta_n}$. This results in a smaller bias, although more difficult to eliminate.

Keywords Semimartingale · High frequency data · Volatility estimation · Central limit theorem · Efficient estimation · Estimation of volatility functionals · Asymptotic aspects

MSC2010 60F05 · 60G44 · 62F12

1 Introduction

Consider an Itô semimartingale X_t, whose squared volatility c_t (a $d \times d$ matrices-valued process if X is d-dimensional) is itself another Itô semimartingale. The process X is observed at discrete times $i\Delta_n$ for $i = 0, 1, \ldots$, the time lag Δ_n being small (high-frequency setting) and eventually going to 0. The aim is to estimate integrated functionals of the volatility, that is $\int_0^t g(c_s)\,ds$ for arbitrary (smooth enough) functions g, on the basis of the observations at stage n and within the time interval $[0, t]$.

J. Jacod
Institut de Mathématiques de Jussieu, CNRS – UMR 7586 and Université Pierre et Marie Curie, 4 Place Jussieu, 75252 Paris Cedex 05, France
e-mail: jean.jacod@upmc.fr

M. Rosenbaum (✉)
Laboratoire de Probabilités et Modèles Aléatoires, CNRS – UMR 7599 and Université Pierre et Marie Curie, 4 Place Jussieu, 75252 Paris Cedex 05, France
e-mail: mathieu.rosenbaum@upmc.fr

P.K. Friz et al. (eds.), *Large Deviations and Asymptotic Methods in Finance*, Springer Proceedings in Mathematics & Statistics 110,
DOI 10.1007/978-3-319-11605-1_20

This is of course a quite well understood problem when g is the identity function. In particular, when X is one-dimensional and continuous, $\int_0^t g(c_s)\,ds$ corresponds then to the integrated (squared) volatility, which can be efficiently estimated using the so-called realized volatility, that is the sum of the squared increments of X. However, many other functions g are of interest. For example, in the case of the realized volatility mentioned above, the quantity $\int_0^t c_s^2\,ds$, called quarticity and corresponding to $g(x) = x^2$, appears in the asymptotic variance of the estimator. Therefore, estimating the quarticity becomes necessary if one wants to build confidence intervals for the integrated volatility. Actually, in the context of volatility estimation, for most statistical procedures, the asymptotic variance is a combination of terms of the form $\int_0^t g(c_s)\,ds$, see [4]. Hence the statistician needs to be able to estimate such quantities. Note that the functions g involved in limiting variances are often polynomial. Nevertheless, more complicated expressions may also be found, in particular in the multi-dimensional setting in the presence of jumps. We refer to [5] for more details on the motivation for estimating general integrated functionals of the volatility process.

In [5], we have exhibited estimators which are consistent and asymptotically optimal, in the sense that they asymptotically achieve the best rate $1/\sqrt{\Delta_n}$, and also the minimal asymptotic variance in the cases where optimality is well-defined (namely, when X is continuous and has a Markov type structure, in the sense of [2]). These estimators have this rate and minimal asymptotic variance as soon as the jumps of X are summable, plus some mild technical conditions.

The aim of this paper is to complement [5] with another estimator, of the same type, but using spot volatility estimators based on a different window size. In this introduction, we explain the differences between the estimator in [5] and the one presented here.

For the sake of simplicity, we consider the case when X is continuous and one-dimensional (the discontinuous and multi-dimensional case is considered later), that is of the form

$$X_t = X_0 + \int_0^t b_s\,ds + \int_0^t \sigma_s\,dW_s$$

and $c_t = \sigma_t^2$ is the squared volatility. Natural estimators for $V(g)_t = \int_0^t g(c_s)\,ds$ are

$$V(g)_t^n = \Delta_n \sum_{i=1}^{[t/\Delta_n]-k_n+1} g(\widehat{c}_i^n), \quad \text{where } \widehat{c}_i^n = \frac{1}{k_n\Delta_n} \sum_{j=0}^{k_n-1} (X_{(i+j)\Delta_n} - X_{(i+j-1)\Delta_n})^2 \tag{1.1}$$

for an arbitrary sequence of integers such that $k_n \to \infty$ and $k_n\Delta_n \to 0$. One knows that $V(g)_t^n \xrightarrow{\mathbb{P}} V(g)_t$ (when g is continuous and of polynomial growth).

The variables $\widehat{c}_i^n$ are spot volatility estimators, and according to [4] we know that $\widehat{c}_{[t/\Delta_n]}^n$ estimates c_t, with a rate depending on the "window size" k_n. The optimal rate $1/\Delta_n^{1/4}$ is achieved by taking $k_n \asymp 1/\sqrt{\Delta_n}$.[1] When k_n is smaller, the rate is $\sqrt{k_n}$

[1] By $k_n \asymp 1/\sqrt{\Delta_n}$, we mean $a_1/\sqrt{\Delta_n} \le k_n \le a_2/\sqrt{\Delta_n}$, for some $a_1 > 0$ and $a_2 > 0$.

and the estimation error is a purely "statistical error"; when k_n is bigger, the rate is $1/\sqrt{k_n\Delta_n}$ and the estimation error is due to the variability of the volatility process c_t itself (its volatility and its jumps). With the optimal choice $k_n \asymp 1/\sqrt{\Delta_n}$, the estimation error is a mixture of the statistical error and the error due to the variability of c_t.

In [5], we have used a "small" window, that is $k_n \ll 1/\sqrt{\Delta_n}$. Somewhat surprisingly, this allows for optimality in the estimation of $\int_0^t g(c_s)\,ds$ (rate $1/\sqrt{\Delta_n}$ and minimal asymptotic variance). However, the price to pay is the need of a de-biasing term to be subtracted from $V(g)^n$, without which the rate is smaller and no Central Limit Theorem is available.

Here, we consider the window size $k_n \asymp 1/\sqrt{\Delta_n}$. This leads to a convergence rate $1/\sqrt{\Delta_n}$ for $V(g)^n$ itself, and the limit is again conditionally Gaussian with the "minimal" asymptotic variance, but with a bias that depends on the volatility of the volatility c_t, and on its jumps. It is however possible to subtract from $V(g)^n$ a de-biasing term again, so that the limit becomes (conditionally) centered.

Section 2 is devoted to presenting assumptions and results, and all proofs are gathered in Sect. 3. The reader is referred to [5] for motivation and various comments and a detailed discussion of optimality. However, in order to make this paper readable, we basically give the full proofs, even though a number of partial results have already been proved in the above-mentioned paper, and with the exception of a few well designated lemmas.

2 The Results

2.1 Setting and Assumptions

The underlying process X is d-dimensional, and observed at the times $i\Delta_n$ for $i = 0, 1, \ldots$, within a fixed interval of interest $[0, t]$. For any process we write $\Delta_i^n Y = Y_{i\Delta_n} - Y_{(i-1)\Delta_n}$ for the increment over the ith observation interval. We assume that the sequence Δ_n goes to 0. The precise assumptions on X are as follows.

First, X is an Itô semimartingale on a filtered space $(\Omega, \mathcal{F}, (\mathcal{F}_t)_{t\geq 0}, \mathbb{P})$. It can be written in its Grigelionis form, as follows, using a d-dimensional Brownian motion W and a Poisson random measure μ on $\mathbb{R}_+ \times E$, where E is an auxiliary Polish space and with the (non-random) intensity measure $\nu(dt, dz) = dt \otimes \lambda(dz)$ for some σ-finite measure λ on E:

$$X_t = X_0 + \int_0^t b_s\,ds + \int_0^t \sigma_s\,dW_s + \int_0^t \int_E \delta(s,z)\,1_{\{\|\delta(s,z)\|\leq 1\}}\,(\mu-\nu)(ds,dz) \\ + \int_0^t \int_E \delta(s,z)\,1_{\{\|\delta(s,z)\|>1\}}\,\mu(ds,dz). \tag{2.1}$$

This is a vector-type notation: the process b_t is $\mathbb{R}^d$-valued optional, the process σ_t is $\mathbb{R}^d \otimes \mathbb{R}^d$-valued optional, $\delta = \delta(\omega, t, z)$ is a predictable $\mathbb{R}^d$-valued function on $\Omega \times \mathbb{R}_+ \times E$ and $\|\cdot\|$ is the euclidean norm on $\mathbb{R}^d$.

The spot volatility process $c_t = \sigma_t \sigma_t^*$ (* denotes transpose) takes its values in the set $\mathcal{M}_d^+$ of all nonnegative symmetric $d \times d$ matrices. We suppose that c_t is again an Itô semimartingale, which can be written as

$$c_t = c_0 + \int_0^t \widetilde{b}_s\, ds + \int_0^t \widetilde{\sigma}_s\, dW_s + \int_0^t \int_E \widetilde{\delta}(s,z)\, 1_{\{\|\widetilde{\delta}(s,z)\| \le 1\}}\, (\mu - \nu)(ds,dz) \\ + \int_0^t \int_E \widetilde{\delta}(s,z)\, 1_{\{\|\widetilde{\delta}(s,z)\| > 1\}}\, \mu(ds,dz), \tag{2.2}$$

with the same W and μ as in (2.1). This is indeed *not a restriction*: if X and c are two Itô semimartingales, we have a representation as above for the pair (X, c) and, if the dimension of W exceeds the dimension of X, one can always add fictitious component to X, arbitrarily set to 0, so that the dimensions of X and W agree.

In (2.2), $\widetilde{b}$ and $\widetilde{\sigma}$ are optional and $\widetilde{\delta}$ is as δ; moreover $\widetilde{b}$ and $\widetilde{\delta}$ are $\mathbb{R}^{d^2}$-valued. Finally, we need the spot volatility of the volatility and "spot covariation" of the continuous martingale parts of X and c, which are

$$\widetilde{c}_t^{ij,kl} = \sum_{m=1}^d \widetilde{\sigma}_t^{ij,m} \widetilde{\sigma}_t^{kl,m}, \qquad \widetilde{c}_t^{\prime i,jk} = \sum_{l=1}^d \sigma_t^{il} \widetilde{\sigma}_t^{jk,l}.$$

The precise assumptions on the coefficients are as follows, with r a real in $[0, 1)$.

Assumption (A'-r): There are a sequence (J_n) of nonnegative bounded λ-integrable functions on E and a sequence (τ_n) of stopping times increasing to ∞, such that

$$t \le \tau_n(\omega) \implies \|\delta(\omega,t,z)\|^r \wedge 1 + \|\widetilde{\delta}(\omega,t,z)\|^2 \wedge 1 \le J_n(z).$$

Moreover, the processes $b'_t = b_t - \int \delta(t,z)\, 1_{\{\|\delta(t,z)\| \le 1\}}\, \lambda(dz)$ (which is well defined), $\widetilde{c}_t$ and $\widetilde{c}'_t$ are càdlàg or càglàd, and the maps $t \mapsto \widetilde{\delta}(\omega,t,z)$ are càglàd (recall that $\widetilde{\delta}$ should be predictable), as well as the processes $\widetilde{b}_t + \int \widetilde{\delta}(t,z)\big(\kappa(\|\widetilde{\delta}(t,z)\|) - 1_{\{\|\widetilde{\delta}(t,z)\| \le 1\}}\big)\, \lambda(dz)$ for one (hence for all) continuous function κ on $\mathbb{R}_+$ with compact support and equal to 1 on a neighborhood of 0.

The bigger r, the weaker Assumption (A'-r), and when (A'-0) holds the process X has finitely many jumps on each finite interval. The part of (A'-r) concerning the jumps of X implies that $\sum_{s \le t} \|\Delta X_s\|^r < \infty$ a.s. for all $t < \infty$, and it is in fact "almost" implied by this property. Since $r < 1$, this implies $\sum_{s \le t} \|\Delta X_s\| < \infty$ a.s.

Remark 2.1 (A'-r) above is basically the same as Assumption (A-r) in [5], albeit (slightly) stronger (hence its name): some degree of regularity in time seems to be needed for $\widetilde{b}, \widetilde{c}, \widetilde{c}', \widetilde{\delta}$ in the present case.

2.2 A First Central Limit Theorem

For defining the estimators of the spot volatility, we first choose a sequence k_n of integers which satisfies, as $n \to \infty$:

$$k_n \sim \frac{\theta}{\sqrt{\Delta_n}}, \qquad \theta \in (0, \infty), \tag{2.3}$$

and a sequence u_n in $(0, \infty]$. The $\mathcal{M}_d^+$-valued variables $\widehat{c}_i^n$ are defined, componentwise, as

$$\widehat{c}_i^{n,lm} = \frac{1}{k_n \Delta_n} \sum_{j=0}^{k_n-1} \Delta_{i+j}^n X^l \, \Delta_{i+j}^n X^m \, 1_{\{\|\Delta_{i+j}^n X\| \le u_n\}}, \tag{2.4}$$

and they implicitly depend on Δ_n, k_n, u_n.

One knows that $\widehat{c}_{[t/\Delta_n]}^n \stackrel{\mathbb{P}}{\longrightarrow} c_t$ for any t, and there is an associated Central Limit Theorem under (A'-2), with rate $1/\Delta_n^{1/4}$: the choice (2.3) is optimal, in the sense that it allows us to have the fastest possible rate by a balance between the involved "statistical error" which is of order $1/\sqrt{k_n}$, and the variation of c_t over the interval $[t, t + k_n\Delta_n]$, which is of order $\sqrt{k_n\Delta_n}$ because c_t is an Itô semimartingale (and even when it jumps), see [1, 4].

By Theorem 9.4.1 of [4], one also knows that under (A'-r) and if $u_n \asymp \Delta_n^{\varpi}$ for some $\varpi \in \left[\frac{p-1}{2p-r}, \frac{1}{2}\right)$ we have

$$V(g)_t^n := \Delta_n \sum_{i=1}^{[t/\Delta_n]-k_n+1} g(\widehat{c}_i^n) \stackrel{\text{u.c.p.}}{\Longrightarrow} V(g)_t := \int_0^t g(c_s)\, ds \tag{2.5}$$

(convergence in probability, uniformly over each compact interval; by convention $\sum_{i=a}^b v_i = 0$ if $b < a$), as soon as the function g on $\mathcal{M}_d^+$ is continuous with $|g(x)| \le K(1 + \|x\|^p)$ for some constants K, p. Actually, for this to hold we need much weaker assumptions on X, but we do not need this below. Note also that when X is continuous, the truncation in (2.4) is useless: one may use (2.4) with $u_n \equiv \infty$, which reduces to (1.1) in the one-dimensional case.

Now, we want to determine at which rate the convergence (2.5) takes place. This amounts to proving an associated Central Limit Theorem. For an appropriate choice of the truncation levels, such a CLT is available for $V(g)^n$, with the rate $1/\sqrt{\Delta_n}$, but the limit exhibits a bias term. Below, g is a smooth function on $\mathcal{M}_d^+$, and the two first partial derivatives are denoted as $\partial_{jk} g$ and $\partial^2_{jk,lm} g$, since any $x \in \mathcal{M}_d^+$ has d^2 components x^{jk}. The family of all partial derivatives of order j is simply denoted as $\partial^j g$.

Theorem 2.2 *Assume (A'-r) for some $r < 1$. Let g be a C^3 function on $\mathcal{M}_d^+$ such that*

$$\|\partial^j g(x)\| \le K(1 + \|x\|^{p-j}), \qquad j = 0, 1, 2, 3 \tag{2.6}$$

for some constants $K > 0$, $p \ge 3$. Either suppose that X is continuous and $u_n/\Delta_n^{\varepsilon} \to \infty$ for some $\varepsilon < 1/2$ (for example, $u_n \equiv \infty$, so there is no truncation at all), or suppose that

$$u_n \asymp \Delta_n^{\varpi}, \qquad \frac{2p-1}{2(2p-r)} \le \varpi < \frac{1}{2}. \tag{2.7}$$

Then we have the finite-dimensional (in time) stable convergence in law

$$\frac{1}{\sqrt{\Delta_n}}\,(V(g)^n_t - V(g)_t) \stackrel{\mathcal{L}_{f-s}}{\longrightarrow} A^1_t + A^2_t + A^3_t + A^4_t + Z_t,$$

where Z is a process defined on an extension $(\widetilde{\Omega}, \widetilde{\mathcal{F}}, (\widetilde{\mathcal{F}}_t)_{t\geq 0}, \widetilde{\mathbb{P}})$ of $(\Omega, \mathcal{F}, (\mathcal{F}_t)_{t\geq 0}, \mathbb{P})$, which conditionally on $\mathcal{F}$ is a continuous centered Gaussian martingale with variance

$$\widetilde{\mathbb{E}}\big((Z_t)^2 \mid \mathcal{F}\big) = \sum_{j,k,l,m=1}^{d} \int_0^t \partial_{jk} g(c_s)\, \partial_{lm} g(c_s)\, \big(c^{jl}_s c^{km}_s + c^{jm}_s c^{kl}_s\big)\, ds, \qquad (2.8)$$

and where, with the notation

$$G(x,y) = \int_0^1 \big(g(x+wy) - wg(x+y) - (1-w)g(x)\big)\, dw, \qquad (2.9)$$

we have

$$\begin{aligned} A^1_t &= -\tfrac{\theta}{2}\,\big(g(c_0) + g(c_t)\big) \\ A^2_t &= \tfrac{1}{2\theta} \sum_{j,k,l,m=1}^{d} \int_0^t \partial^2_{jk,lm} g(c_s)\, \big(c^{jl}_s c^{km}_s + c^{jm}_s c^{kl}_s\big)\, ds \\ A^3_t &= -\tfrac{\theta}{12} \sum_{j,k,l,m=1}^{d} \int_0^t \partial^2_{jk,lm} g(c_s)\, \widetilde{c}^{jk,lm}_s\, ds \\ A^4_t &= \theta \sum_{s\leq t} G(c_{s-}, \Delta c_s). \end{aligned}$$

Note that $|G(x,y)| \leq K(1+\|x\|)^p\, \|y\|^2$, so the sum defining A^4_t is absolutely convergent, and vanishes when c_t is continuous.

Remark 2.3 The bias has four parts:

(1) The first one is due to a border effect: indeed, the formula giving $V(g)^n_t$ contains $[t/\Delta_n] - k_n + 1$ summands only, whereas the natural (unfeasible) approximation $\Delta_n \sum_{i=1}^{[t/\Delta_n]} g(c_{(i-1)\Delta_n})$ contains $[t/\Delta_n]$ summands. The sum of the lacking $k_n - 1$ summands is of order of magnitude $(k_n - 1)\Delta_n$, which goes to 0 and thus does not impair consistency, but it creates an obvious bias after normalization by $1/\sqrt{\Delta_n}$. Removing this source of bias is straightforward: since $g(c_s)$ is "under-represented" when s is close to 0 or to t, we add to $V(g)^n_t$ the variable

$$\frac{(k_n-1)\Delta_n}{2}\,\big(g(\widehat{c}^n_1) + g(\widehat{c}^n_{[t/\Delta_n]-k_n+1})\big).$$

Of course, other weighted averages of $g(\widehat{c}^n_i)$ for i close to 0 or to $[t/\Delta_n] - k_n + 1$ would be possible.

(2) The second part A^2 is continuous in time and is present even in the toy model given by $X_t = \sqrt{c}\, W_t$ with c a constant and $\Delta_n = \frac{1}{n}$ and $T = 1$. In this simple case, the interpretation is as follows: instead of taking the "optimal" $g(\widehat{c}_n)$ for estimating $g(c)$, with $\widehat{c}_n = \sum_{i=1}^n (\Delta_i^n X)^2$, one takes $\frac{1}{n}\sum_{i=1}^n g(\widehat{c}_i^n)$ with $\widehat{c}_i^n$ a "local" estimator of c. This adds a statistical error which results in a bias. Note that, even in the general case, this bias would disappear, were we taking in (2.3) the (forbidden) value $\theta = \infty$ (with still $k_n \Delta_n \to 0$), at the expense of a slower rate of convergence.

(3) The third and fourth parts A^3 and A^4 are respectively continuous and purely discontinuous, due to the continuous part and to the jumps of the volatility process c_t itself. These two biases disappear if we take $\theta = 0$ in (2.3) (with still $k_n \to \infty$), again a forbidden value, and again at the expense of a slower rate of convergence.

The only test function g for which the last three biases disappear is the identity $g(x) = x$. This is because, in this case, and up to the border terms, $V(g)_t^n$ is nothing but the realized quadratic variation itself and the spot estimators $\widehat{c}_i^n$ actually merge together and disappear as such.

Remark 2.4 Observe that (2.7) implies $r < 1$. This restriction is not a surprise, since one needs $r \le 1$ in order to estimate the integrated volatility by the (truncated) realized volatility, with a rate of convergence $1/\sqrt{\Delta_n}$. When $r = 1$, it is likely that the CLT still holds for an appropriate choice of the sequence u_n, and with another additional bias, see e.g. [6] for a slightly different context. Here we let this borderline case aside.

2.3 Estimation of the Bias

Now we proceed to "remove" the bias, which means subtracting consistent estimators for the bias from $V(g)_t^n$. As written before, we have

$$A_t^{n,1} = -\frac{k_n\sqrt{\Delta_n}}{2}\left(g(\widehat{c}_1^n) + g(\widehat{c}_{[t/\Delta_n]-k_n+1}^n)\right) \xrightarrow{\mathbb{P}} A_t^1 \tag{2.10}$$

(this comes from $\widehat{c}_1^n \xrightarrow{\mathbb{P}} c_0$ and $\widehat{c}_{[t/\Delta_n]-k_n+1}^n \xrightarrow{\mathbb{P}} c_{t-}$, plus $c_{t-} = c_t$ a.s.). Next, observe that $A^2 = \frac{1}{\theta} V(h)$ for the test function h defined on $\mathcal{M}_d^+$ by

$$h(x) = \frac{1}{2}\sum_{j,k,l,m=1}^d \partial_{jk,lm}^2 g(x)\left(x^{jl}x^{km} + x^{jm}x^{kl}\right).$$

Therefore

$$A_t^{n,2} = \frac{1}{k_n\sqrt{\Delta_n}} V(h)_t^n \xrightarrow{\mathbb{P}} A_t^2. \tag{2.11}$$

The term A_t^3 involves the volatility of the volatility, for which estimators have been provided in the one-dimensional case by Vetter [7]; namely, if $d = 1$ and under suitable technical assumptions (slightly stronger than here), *plus* the continuity of X_t and c_t, he proves that

$$\frac{3}{2k_n} \sum_{i=1}^{[t/\Delta_n]-2k_n+1} (\widehat{c}^n_{i+k_n} - \widehat{c}^n_i)^2$$

converges to $\int_0^t \big(\widetilde{c}_s + \frac{6}{\theta^2}\,(c_s)^2\big)\,ds$. Of course, we need to modify this estimator here, in order to include the function $\partial^2 g$ in the limit and account for the possibilities of having $d \geq 2$ and having jumps in X. We propose to take

$$A_t^{n,3} = -\frac{\sqrt{\Delta_n}}{8} \sum_{i=1}^{[t/\Delta_n]-2k_n+1} \sum_{j,k,l,m=1}^{d} \partial^2_{jk,lm} g(\widehat{c}^n_i)\,(\widehat{c}^{n,jk}_{i+k_n} - \widehat{c}^{n,jk}_i)\,(\widehat{c}^{n,lm}_{i+k_n} - \widehat{c}^{n,lm}_i). \tag{2.12}$$

When X and c are continuous, one may expect the convergence to $A_t^3 - \frac{1}{2}\,A_t^2$ (observe that $\frac{\sqrt{\Delta_n}}{8} \sim \frac{3}{2k_n}\,\frac{\theta}{12}$), and one may expect the same when X jumps and c is still continuous, because in (2.4) the truncation basically eliminates the jumps of X. In contrast, when c jumps, the limit should rather be related to the "full" quadratic variation of c. Indeed we have the following theorem.

Theorem 2.5 *Under the assumptions of Theorem 2.2, for all $t \geq 0$ we have*

$$A_t^{n,3} \xrightarrow{\mathbb{P}} -\frac{1}{2}\,A_t^2 + A_t^3 + A_t'^4,$$

where

$$A_t'^4 = \theta \sum_{s \leq t} G'(c_{s-}, \Delta c_s)$$

and

$$G'(x,y) = -\frac{1}{8} \sum_{j,k,l,m} \int_0^1 \big(\partial^2_{jk,lm}\, g(x) + \partial^2_{jk,lm}\, g(x + (1-w)y)\big)\, w^2\, y^{jk}\, y^{lm}\, dw. \tag{2.13}$$

At this stage, it remains to find consistent estimators for $A_t^4 - A_t'^4$, which has the form

$$A_t^4 - A_t'^4 = \theta \sum_{s \leq t} G''(c_{s-}, \Delta c_s), \quad \text{where } G'' = G - G'.$$

More generally, we aim at estimating

$$\mathcal{V}(F)_t = \sum_{s \le t} F(c_{s-}, \Delta c_s),$$

at least when the function F on $\mathcal{M}_d^+ \times \mathcal{M}_d$, where $\mathcal{M}_d$ is the set of all $d \times d$ matrices, is C^1 and $|F(x, y)| \le K\|y\|^2$ uniformly in x within any compact set, as is the function G'' above.

The solution to this problem is not as simple as it might appear at first glance. We first truncate from below, taking any sequence u'_n of truncation levels satisfying

$$u'_n \to 0, \qquad \frac{u'_n}{\Delta_n^{\varpi'}} \to \infty \quad \text{for some } \varpi' \in \Big(0, \frac{1}{8}\Big). \tag{2.14}$$

Second, we resort on the following trick. Since $\widehat{c}_i^n$ is "close" to the average of c_t over the interval $(i\Delta_n, (i+k_n)\Delta_n]$, we (somehow wrongly) pretend that, for all j:

$$\begin{gathered}\exists s \in ((j-1)k_n\Delta_n, jk_n\Delta_n] \text{ with } \|\Delta c_s\| > u'_n \\ \Leftrightarrow \ \|\widehat{c}^n_{jk_n} - \widehat{c}^n_{(j-2)k_n}\| > u'_n \quad \Delta c_s \sim \widehat{c}^n_{jk_n} - \widehat{c}^n_{(j-2)k_n}, \\ \|\widehat{c}^n_{(j-1)k_n} - \widehat{c}^n_{(j-3)k_n}\| \bigvee \|\widehat{c}^n_{(j+1)k_n} - \widehat{c}^n_{(j-1)k_n}\| < \|\widehat{c}^n_{jk_n} - \widehat{c}^n_{(j-2)k_n}\|.\end{gathered}$$

The condition (2.14) implies that for n large enough there is at most one jump of size bigger than u'_n in each interval $((i-1)\Delta_n, (i-1+k_n)\Delta_n]$ within $[0, t]$, and no two consecutive intervals of this form contain such jumps. Despite this, the statement above is of course not true, the main reason being that $\widehat{c}_i^n$ and c_i^n do not exactly agree. However it is "true enough" to allow for the next estimators to be consistent for $\mathcal{V}(F)_t$:

$$\begin{gathered}\mathcal{V}(F)^n_t = \sum_{j=3}^{[t/k_n\Delta_n]-3} F(\widehat{c}^n_{(j-3)k_n+1}, \delta^n_j\widehat{c})\, 1_{\{\|\delta^n_{j-1}\widehat{c}\| \vee \|\delta^n_{j+1}\widehat{c}\| \vee u'_n < \|\delta^n_j\widehat{c}\|\}}, \\ \text{where } \delta^n_j\widehat{c} = \widehat{c}^n_{jk_n+1} - \widehat{c}^n_{(j-2)k_n+1}.\end{gathered} \tag{2.15}$$

Since this is a sum of approximately $[t/k_n\Delta_n]$ terms, the rate of convergence of $\mathcal{V}(F)^n_t$ towards $\mathcal{V}(F)_t$ in law is probably $1/\Delta_n^{1/4}$ only. However, here we are looking for consistent estimators, and the rate is not of concern to us. Note that, again, the upper limit in the sum above is chosen in such a way that $\mathcal{V}(F)^n_t$ is computable on the basis of the observations within the interval $[0, t]$.

Theorem 2.6 *Assume all hypotheses of Theorem 2.2, and let F be a continuous function on $\mathbb{R}_+ \times \mathbb{R}$ satisfying, with the same $p \ge 3$ as in (2.7),*

$$|F(x, y)| \le K(1 + \|x\| + \|y\|)^{p-2}\, \|y\|^2. \tag{2.16}$$

Then for all $t \ge 0$ we have

$$\mathcal{V}(F)^n_t \xrightarrow{\mathbb{P}} \mathcal{V}(F)_t. \tag{2.17}$$

2.4 An Unbiased Central Limit Theorem

At this stage, we can set, with the notation (2.11), (2.12) and (2.15), and also (2.9) and (2.13) for G and G':

$$\begin{aligned}\overline{V}(g)^n_t &= V(g)^n_t + \frac{k_n\Delta_n}{2}\,\big(g(\widehat{c}^n_1) + g(\widehat{c}^n_{[t/\Delta_n]-k_n+1}\big) \\ &\quad - \sqrt{\Delta_n}\,\Big(\frac{3}{2}\,A^{n,2}_t + A^{n,3}_t\Big) - k_n\Delta_n\,\mathcal{V}(G-G')^n_t.\end{aligned}$$

We then have the following theorem, which is a straightforward consequence of the three previous theorems and of $k_n\sqrt{\Delta_n} \to \theta$, plus (2.10) and (2.11) and the fact that the function $G - G'$ satisfies (2.16) when g satisfies (2.6).

Theorem 2.7 *Under the assumptions of Theorem 2.2, and with Z as in this theorem, for all $t \geq 0$ we have the finite-dimensional stable convergence in law*

$$\frac{1}{\sqrt{\Delta_n}}\,(\overline{V}(g)^n_t - V(g)_t) \stackrel{\mathcal{L}_{f-s}}{\longrightarrow} Z_t.$$

Note that θ no longer explicitly appears in this statement, so one can replace (2.3) by the weaker statement

$$k_n \asymp \frac{1}{\sqrt{\Delta_n}}$$

(this is easily seen by taking subsequences n_l such that $k_{n_l}\sqrt{\Delta_{n_l}}$ converge to an arbitrary limit in $(0,\infty)$).

It is simple to make this CLT "feasible", that is, usable in practice for determining a confidence interval for $V(g)_t$ at any time $t > 0$. Indeed, we can define the following function on $\mathcal{M}^+_d$:

$$\overline{h}(x) = \sum_{j,k,l,m=1}^{d} \partial_{jk}\,g(x)\,\partial_{lm}\,g(x)\,\big(x^{jl}x^{km} + x^{jm}x^{kl}\big).$$

We then have $V(\overline{h})^n \stackrel{\text{u.c.p.}}{\Longrightarrow} V(\overline{h})$, where $V(\overline{h})_t$ is the right hand side of (2.8). Then we readily deduce:

Corollary 2.8 *Under the assumptions of the previous theorem, for any $t > 0$ we have the following stable convergence in law, where Y is an $\mathcal{N}(0,1)$ variable:*

$$\frac{\overline{V}(g)^n_t - V(g)_t}{\sqrt{\Delta_n\,V(\overline{h})^n_t}} \stackrel{\mathcal{L}-s}{\longrightarrow} Y, \quad \textit{in restriction to the set } \{V(\overline{h})_t > 0\}.$$

Finally, let us mention that the estimators $\overline{V}(g)_t^n$ enjoy exactly the same asymptotic efficiency properties as the estimators in [5], and we refer to this paper for a discussion of this topic.

Example 2.9 (*Quarticity*) Suppose $d = 1$ and take $g(x) = x^2$, so we want tho estimate the quarticity $\int_0^t c_s^2\,ds$. In this case we have

$$h(x) = 2x^2, \qquad G(x,y) - G'(x,y) = 0.$$

Then the "optimal" estimator for the quarticity is

$$\Delta_n\big(1 - \frac{3}{k_n}\big) \sum_{i=1}^{[t/\Delta_n]-k_n+1} (\widehat{c}_i^n)^2 + \frac{\Delta_n}{4} \sum_{i=1}^{[t/\Delta_n]-2k_n+1} (\widehat{c}_{i+k_n}^n - \widehat{c}_i^n)^2$$
$$+ \frac{(k_n-1)\Delta_n}{2} \big((\widehat{c}_1^n)^2 + (\widehat{c}_{[t/\Delta_n]-k_n+1}^n)^2\big).$$

The asymptotic variance is $8\int_0^t c_s^4\,ds$, to be compared with the asymptotic variance of the more usual estimator $\frac{1}{3\Delta_n}\sum_{i=1}^{[t/\Delta_n]}(\Delta_i^n X)^4$, which is $\frac{32}{3}\int_0^t c_s^4\,ds$.

3 Proofs

3.1 Preliminaries

According to the localization Lemma 4.4.9 of [4] (for the assumption (K) in that lemma), it is enough to show all four Theorems 2.2, 2.5–2.7 under the following stronger assumption.

Assumption (SA'-r): We have (A'-r). Moreover, we have for a λ-integrable function J on E and a constant A:

$$\|b\|,\ \|b'\|,\ \|\widetilde{b}\|,\ \|c\|,\ \|\widetilde{c}\|,\ \|\widetilde{c}'\|,\ J \le A,$$
$$\|\delta(\omega,t,z)\|^r \le J(z), \quad \|\widetilde{\delta}(\omega,t,z)\|^2 \le J(z). \tag{3.1}$$

In the sequel, we thus suppose that X satisfies (SA'-r), and also that (2.3) holds: these assumptions are typically not recalled. Below, all constants are denoted by K, and they vary from line to line. They may implicitly depend on the process X (usually through A in (3.1)). When they depend on an additional parameter p, we write K_p.

We will usually replace the discontinuous process X by the continuous process

$$X'_t = \int_0^t b'_s\,ds + \int_0^t \sigma_s\,dW_s, \tag{3.2}$$

connected with X by $X_t = X_0 + X'_t + \sum_{s \le t} \Delta X_s$. Note that b' is bounded, and without loss of generality we will use below its càdlàg version. Note also that, since the jumps of c are bounded, one can rewrite (2.2) as

$$c_t = c_0 + \int_0^t \widetilde{b}_s\, ds + \int_0^t \widetilde{\sigma}_s\, dW_s + \int_0^t \int_E \widetilde{\delta}(s,z)\, (\mu - \nu)(ds, dz).$$

This amounts to replacing $\widetilde{b}$ in (2.2) by $\widetilde{b}_{t+} + \int_E \delta(t+,z)(\kappa(\|\widetilde{\delta}(t+,z)\|) - 1_{\{\|\widetilde{\delta}(t+,z)\| \le 1\}})\, \lambda(dz)$, where κ is a continuous function with compact support, equal to 1 on the set $[0, A]$. Note that the new process $\widetilde{b}$ is bounded càdlàg.

With any process Z we associate the variables

$$\eta(Z)_{t,s} = \sqrt{\mathbb{E}\big(\sup_{v \in (t,t+s]} \|Z_{t+v} - Z_t\|^2 \mid \mathcal{F}_t\big)}, \tag{3.3}$$

and we recall Lemma 4.2 of [5]:

Lemma 3.1 *For all $t > 0$, all bounded càdlàg processes Z, and all sequences $v_n \ge 0$ of real numbers tending to 0, we have $\Delta_n \mathbb{E}\big(\sum_{i=1}^{[t/\Delta_n]} \eta(Z)_{(i-1)\Delta_n, v_n}\big) \to 0$, and for all $0 \le v \le s$ we have $\mathbb{E}(\eta(Z)^n_{t+v,s} \mid \mathcal{F}_t) \le \eta(Z)_{t,s}$.*

3.2 An Auxiliary Result on Itô Semimartingales

In this subsection we give some simple estimates for a d-dimensional semimartingale

$$Y_t = \int_0^t b^Y_s\, ds + \int_0^t \sigma^Y_s\, dW_s + \int_0^t \int_E \delta^Y(s,z)\, (\mu - \nu)(ds, dz)$$

on some space $(\Omega, \mathcal{F}, (\mathcal{F}_t)_{t \ge 0}, \mathbb{P})$, which may be different from the one on which X is defined, as well as W and μ, but we still suppose that the intensity measure ν is the same. Note that $Y_0 = 0$ here. We assume that for some constant A and function J^Y we have, with $c^Y = \sigma^Y \sigma^{Y,*}$:

$$\|b^Y\| \le A, \ \|c^Y\| \le A^2, \quad \|\delta^Y(\omega, t, z)\|^2 \le J^Y(z) \le A^2, \quad \int_E J^Y(z)\, \lambda(dz) \le A^2. \tag{3.4}$$

The compensator of the quadratic variation of Y is of the form $\int_0^t \overline{c}^Y_s\, ds$, where $\overline{c}^Y_t = c^Y_t + \int_E \delta^Y(t,z)\, \delta^Y(t,z)^*\, \lambda(dz)$. Moreover, if the process c^Y is itself an Itô semimartingale, the quadratic covariation of the continuous martingale parts of Y and c^Y is also of the form $\int_0^t \widehat{c}'^Y_s\, ds$ for some process $\widehat{c}'^Y$, necessarily bounded if both Y and c^Y satisfy (3.4) (and, if $Y = X$, we have $c^Y = c$ and $\widehat{c}'^Y = \widehat{c}'$).

Lemma 3.2 *Below we assume (3.4), and the constant K only depends on A.*
(a) We have for $t \in [0, 1]$:

$$\begin{aligned} &\big\|\mathbb{E}(Y_t \mid \mathcal{F}_0) - t b_0^Y\big\| \le t\,\eta(b^Y)_{0,t} \le Kt \\ &\big|\mathbb{E}(Y_t^j Y_t^m \mid \mathcal{F}_0) - t\overline{c}_0^{Y,jm}\big| \le Kt(t + \sqrt{t}\,\eta(b^Y)_{0,t} + \eta(\overline{c}^Y)_{0,t}) \le Kt, \end{aligned} \tag{3.5}$$

and if further $\|\mathbb{E}(\overline{c}_t^Y - \overline{c}_0^Y \mid \mathcal{F}_0)\| \le A^2 t$ for all t, we also have

$$\big|\mathbb{E}(Y_t^j Y_t^m \mid \mathcal{F}_0) - t\overline{c}_0^{Y,jm}\big| \le 2t^{3/2}(2A^2\sqrt{t} + A\eta(b^Y)_{0,t}) \le Kt^{3/2}. \tag{3.6}$$

(b) When Y is continuous, and if $\mathbb{E}(\|\overline{c}_t^Y - \overline{c}_0^Y\|^2 \mid \mathcal{F}_0) \le A^4 t$ for all t, we have

$$\big|\mathbb{E}\big(Y_t^j Y_t^k Y_t^l Y_t^m \mid \mathcal{F}_0\big) - t^2(c_0^{Y,jk} c_0^{Y,lm} + c_0^{Y,jl} c_0^{Y,km} + c_0^{Y,jm} c_0^{Y,kl})\big| \le Kt^{5/2}. \tag{3.7}$$

(c) When c^Y is a (possibly discontinuous) semimartingale satisfying the same conditions (3.4) as Y, and if Y itself is continuous, we have

$$\big|\mathbb{E}\big((Y_t^j Y_t^k - tc_0^{Y,jk})(c_t^{Y,lm} - c_0^{Y,lm}) \mid \mathcal{F}_0\big)\big| \le Kt^{3/2}(\sqrt{t} + \eta(\widetilde{c}'^Y)_{0,t}). \tag{3.8}$$

Proof The first part of (3.5) follows by taking the $\mathcal{F}_0$-conditional expectation in the decomposition $Y_t = M_t + tb_0^Y + \int_0^t (b_s^Y - b_0^Y)\,ds$, where M is a d-dimensional martingale with $M_0 = 0$. For the second part, we deduce from Itô's formula that $Y^j Y^m$ is the sum of a martingale vanishing at 0 and of

$$\begin{aligned} &b_0^j \int_0^t Y_s^m\,ds + b_0^m \int_0^t Y_s^j\,ds + \int_0^t Y_s^m(b_s^j - b_0^j)\,ds + \int_0^t Y_s^j(b_s^m - b_0^m)\,ds \\ &\quad + \overline{c}_0^{Y,jm} t + \int_0^t (\overline{c}_s^{Y,jm} - \overline{c}_0^{Y,jm})\,ds. \end{aligned}$$

Since $\mathbb{E}(\|Y_t\| \mid \mathcal{F}_0) \le KA\sqrt{t}$, as in (3.9), we deduce the second part of (3.5) and also (3.6) by taking again the conditional expectation and by using the Cauchy-Schwarz inequality and the first part.

Equation (3.7) is a part of Lemma 4.1 of [5]. For (3.8), we first observe that $Y_t^j Y_t^k - tc_0^{Y,jk} = B_t + M_t$ and $c_t^{Y,lm} - c_0^{Y,lm} = B_t' + M_t'$, with M and M' martingales (M is continuous). The processes B, B', $\langle M, M\rangle$, $\langle M', M'\rangle$ and $\langle M, M'\rangle$ are absolutely continuous, with densities $\overline{b}_s$, $\overline{b}_s'$, h_s, h_s' and h_s'' satisfying, by (3.4) for Y and c^Y:

$$|\overline{b}_s| \le 2\|Y_s\|\,\|b_s^Y\| + \|c_s^Y - c_0^Y\|, \quad |\overline{b}_s'| \le K, \quad |h_s| \le K\|Y_s\|^2, \quad |h_s'| \le K,$$

where $h_s'' = Y_s^j \widetilde{c}'^{Y,k,lm} + Y_s^k \widetilde{c}'^{Y,j;lm}$. Again as in (3.9) below, $\mathbb{E}(\|Y_t\|^q \mid \mathcal{F}_0) \le K_q t^{q/2}$ for all q, and $\mathbb{E}(\|c_t^Y - c_0^Y\|^2 \mid \mathcal{F}_0) \le Kt$. This yields $\mathbb{E}(B_t^2 \mid \mathcal{F}_0) \le Kt^3$ and $\mathbb{E}(M_t^2 \mid \mathcal{F}_0) \le Kt^2$. Since $|B_t'| \le Kt$ and $\mathbb{E}(M_t'^2 \mid \mathcal{F}_0) \le Kt$, we deduce that the $\mathcal{F}_0$- conditional expectations of $B_t B_t'$, $B_t M_t'$ and $M_t B_t'$ are smaller than Kt^2.

Finally $\mathbb{E}(M_t M'_t \mid \mathcal{F}_0) = \mathbb{E}(\langle M, M'\rangle_t \mid \mathcal{F}_0)$, and $\langle M, M'\rangle_t$ is the sum of $\widehat{c}_0^{\prime Y,k,lm} \int_0^t Y_s^j\,ds + \int_0^t Y_s^j (\widehat{c}_s^{\prime Y,k,lm} - \widehat{c}_0^{\prime Y,k,lm})\,ds$ and a similar term with k and j exchanged. Then using again $\mathbb{E}(\|Y_t\|^2 \mid \mathcal{F}_0) \le Kt$, plus $\|\mathbb{E}(Y_t \mid \mathcal{F}_0)\| \le Kt$ and Cauchy-Schwarz inequality, we obtain that the above conditional expectation is smaller than $K(t^2 + t^{3/2}\eta(\widehat{c}^{\prime Y})_t)$. This completes the proof of (3.8). □

3.3 Some Estimates

(1) We begin with well known estimates for X' and c, under (3.1) and for $s, t \ge 0$ and $q \ge 0$, see [4] for details:

$$\begin{array}{ll} \mathbb{E}\big(\sup_{w\in[0,s]} \|X'_{t+w} - X'_t\|^q \mid \mathcal{F}_t\big) \le K_q\, s^{q/2}, & \|\,\mathbb{E}(X'_{t+s} - X'_t \mid \mathcal{F}_s)\| \le Ks \\ \mathbb{E}\big(\sup_{w\in[0,s]} \|c_{t+w} - c_t\|^q \mid \mathcal{F}_t\big) \le K_q\, s^{1\wedge(q/2)}, & \|\,\mathbb{E}(c_{t+s} - c_t \mid \mathcal{F}_s)\| \le Ks. \end{array} \tag{3.9}$$

Next, it is much easier (although unfeasible in practice) to replace $\widehat{c}_i^n$ in (2.5) by the estimators based on the process X' given by (3.2). Namely, we will replace $\widehat{c}_i^n$ by the following:

$$\widehat{c}_i^{\prime n} = \frac{1}{k_n \Delta_n} \sum_{j=0}^{k_n-1} \Delta_{i+j}^n X'\, \Delta_{i+j}^n X'^{*}.$$

The difference between $\widehat{c}_i^n$ and $\widehat{c}_i^{\prime n}$ is estimated by the following inequality, valid when $u_n \asymp \Delta_n^{\varpi}$ and $q \ge 1$, and where a_n denotes a sequence of numbers (depending on u_n), going to 0 as $n \to \infty$ (this is Eq. 4.8 of [5]):

$$\mathbb{E}\big(\|\widehat{c}_i^n - \widehat{c}_i^{\prime n}\|^q\big) \le K_q a_n\, \Delta_n^{(2q-r)\varpi+1-q}. \tag{3.10}$$

(2) The jumps of c also potentially cause troubles. So we will eliminate the "big" jumps as follows. For any $\rho > 0$ we consider the subset $E_\rho = \{z : J(z) > \rho\}$, which satisfies $\lambda(E_\rho) < \infty$, and we denote by $\mathcal{G}^\rho$ the σ-field generated by the variables $\mu([0,t] \times A)$, where $t \ge 0$ and A runs through all Borel subsets of E_ρ. The process

$$N_t^\rho = \mu((0,t] \times E_\rho) \tag{3.11}$$

is a Poisson process and we let $S_1^\rho, S_2^\rho, \ldots$ be its successive jump times, and $\Omega_{n,t,\rho}$ be the set on which $S_j^\rho \notin \{i\Delta_n : i \ge 1\}$ for all $j \ge 1$ such that $S_j^\rho < t$, and $S_{j+1}^\rho > t \wedge S_j^\rho + (6k_n + 1)\Delta_n$ for all $j \ge 0$ (with the convention $S_0^\rho = 0$; taking $6k_n$ here instead of the more natural k_n will be needed in the proof of Theorem 2.6, and makes no difference here). All these objects are $\mathcal{G}^\rho$-measurable, and $\mathbb{P}(\Omega_{n,t,\rho}) \to 1$ as $n \to \infty$, for all $t, \rho > 0$.

We define the processes

$$\widetilde{b}(\rho)_t = \widetilde{b}_t - \int_{E_\rho} \widetilde{\delta}(t+,z)\,\lambda(dz), \qquad \overline{c}(\rho)_t = \widetilde{\sigma}_t\,\widetilde{\sigma}_t^* + \int_{(E_\rho)^c} \widetilde{\delta}(t+,z)\,\widetilde{\delta}(t+,z)^*\,\lambda(dz)$$

$$\begin{array}{l} c(\rho)_t = c_t - \int_0^t \int_{E_\rho} \widetilde{\delta}(s,z)\,\mu(ds,dz) \;=\; c^{(1)}(\rho)_t + c^{(2)}(\rho)_t, \quad \text{where} \\ \quad c^{(1)}(\rho)_t = c_0 + \int_0^t \widetilde{b}(\rho)_s\,ds + \int_0^t \widetilde{\sigma}_s\,dW_s \\ \quad c^{(2)}(\rho)_t = \int_0^t \int_{(E_\rho)^c} \widetilde{\delta}(t-,z)\,(\mu-\nu)(ds,dz), \end{array} \tag{3.12}$$

so $\overline{c}(\rho)$, which is $\mathbb{R}^{d^2}\otimes\mathbb{R}^{d^2}$-valued, is the càdlàg version of the density of the predictable quadratic variation of $c(\rho)$. Moreover $\mathcal{G}^\rho = \{\emptyset,\Omega\}$ and $(\widetilde{b}(\rho), c(\rho)) = (\widetilde{b}, c)$ when ρ exceeds the bound of the function J. Note also that $\widetilde{b}(\rho)$ and $\overline{c}(\rho)$ are càdlàg.

By Lemma 2.1.5 and Proposition 2.1.10 in [4] applied to each components of X' and $c^{(2)}(\rho)$, plus the property $\|\widetilde{b}(\rho)\| \leq K/\rho$, for all $t \geq 0$, $s \in [0,1]$, $\rho \in (0,1]$, $q \geq 2$, we have

$$\begin{array}{l} \mathbb{E}\big(\sup_{w\in[0,s]} \|X'_{t+w} - X'_t\|^q \mid \mathcal{F}_t \vee \mathcal{G}^\rho\big) \leq K_q\,s^{q/2} \\ \|\,\mathbb{E}(X'_{t+s} - X'_t \mid \mathcal{F}_s \vee \mathcal{G}^\rho)\| + \|\,\mathbb{E}(c(\rho)_{t+s} - c(\rho)_t \mid \mathcal{F}_s \vee \mathcal{G}^\rho)\| \leq Ks \\ \mathbb{E}\big(\sup_{w\in[0,s]} \|c^{(2)}(\rho)_{t+w} - c^{(2)}(\rho)_t\|^q \mid \mathcal{F}_t \vee \mathcal{G}^\rho\big) \leq K_q\,\phi_\rho\,(s + s^{q/2}) \\ \mathbb{E}\big(\sup_{w\in[0,s]} \|c(\rho)_{t+w} - c(\rho)_t\|^q \mid \mathcal{F}_t \vee \mathcal{G}^\rho\big) \leq K_q\big(\phi_\rho\,s + s^{q/2} + \frac{s^q}{\rho^q}\big) \leq K_{q,\rho}\,s. \end{array} \tag{3.13}$$

where $\phi_\rho = \int_{(E_\rho)^c} J(z)\,\lambda(dz) \to 0$ as $\rho \to 0$. Note also that $\|\widetilde{b}(\rho)_t\| \leq K/\rho$.

(3) For convenience, we put

$$\begin{array}{lll} b_i^n = b_{(i-1)\Delta_n}, & c_i^n = c_{(i-1)\Delta_n}, & \\ \widetilde{b}(\rho)_i^n = \widetilde{b}(\rho)_{(i-1)\Delta_n}, & \overline{c}(\rho)_i^n = \overline{c}(\rho)_{(i-1)\Delta_n}, & c(\rho)_i^n = c(\rho)_{(i-1)\Delta_n}, \\ \mathcal{F}_i^n = \mathcal{F}_{(i-1)\Delta_n}, & \mathcal{F}_i^{n,\rho} = \mathcal{F}_i^n \vee \mathcal{G}^\rho. & \end{array} \tag{3.14}$$

All the above variables are $\mathcal{F}_i^{n,\rho}$-measurable. Recalling (3.3), and writing $\eta(Z,(\mathcal{H}_t))_{t,s}$ if we use the filtration $(\mathcal{H}_t)$ instead of $(\mathcal{F}_t)$, we also set

$$\begin{array}{l} \eta(\rho)_{i,j}^n = \max(\eta(Y, (\mathcal{G}^\rho \bigvee \mathcal{F}_t))_{(i-1)\Delta_n, j\Delta_n} \,:\, Y = b', \widetilde{b}(\rho), c, \overline{c}(\rho), \widehat{c}'), \\ \qquad \eta(\rho)_i^n = \eta(\rho)_{i,i+2k_n}^n. \end{array}$$

Therefore, Lemma 3.1 yields for all $t, \rho > 0$ and j, k such that $j + k \leq 2k_n$:

$$\Delta_n \mathbb{E}\Big(\sum_{i=1}^{[t/\Delta_n]} \eta(\rho)_i^n\Big) \to 0, \qquad \mathbb{E}(\eta(\rho)_{i+j,k}^n \mid \mathcal{F}_i^{n,\rho}) \leq \eta(\rho)_i^n. \tag{3.15}$$

We still need some additional notation. First, define $\mathcal{G}^\rho$-measurable (random) set of integers:

$$L(n,\rho) = \{i = 1, 2, \ldots : N^\rho_{(i+2k_n)\Delta_n} - N^\rho_{(i-1)\Delta_n} = 0\} \tag{3.16}$$

(taking above $2k_n$ instead of k_n is necessary for the proof of Theorem 2.5). Observe that

$$i \in L(n,\rho),\ 0 \le j \le 2k_n + 1 \Rightarrow c^n_{i+j} - c^n_i = c(\rho)^n_{i+j} - c(\rho)^n_i. \tag{3.17}$$

Second, we define the following $\mathbb{R}^d \otimes \mathbb{R}^d$-valued variables

$$\begin{aligned}
\alpha^n_i &= \Delta^n_i X' \, \Delta^n_i X'^* - c^n_i \, \Delta_n \\
\beta^n_i &= \widehat{c}^n_i - c^n_i = \tfrac{1}{k_n \Delta_n} \textstyle\sum_{j=0}^{k_n-1} \big(\alpha^n_{i+j} + (c^n_{i+j} - c^n_i)\Delta_n\big) \\
\gamma^n_i &= \widehat{c}^n_{i+k_n} - \widehat{c}^n_i = \beta^n_{i+k_n} - \beta^n_i + c^n_{i+k_n} - c^n_i.
\end{aligned} \tag{3.18}$$

(4) Now we proceed with estimates. (3.13) yields, for all $q \ge 0$:

$$\begin{array}{ll}
\mathbb{E}(\|\alpha^n_i\|^q \mid \mathcal{F}^{n,\rho}_i) \le K_q \Delta^q_n, & \|\mathbb{E}(\alpha^n_i \mid \mathcal{F}^{n,\rho}_i)\| \le K\, \Delta_n^{3/2}, \\
\mathbb{E}\big(\| \sum_{j=0}^{k_n-1} \alpha^n_{i+j} \|^q \mid \mathcal{F}^{n,\rho}_i\big) \le K_q \Delta_n^{3q/4}, & \mathbb{E}(\|\widehat{c}^n_i\|^q \mid \mathcal{F}^{n,\rho}_i) \le K_q,
\end{array} \tag{3.19}$$

the third inequality following from the first two ones, plus Burkholder-Gundy and Hölder inequalities, and the last inequality from the third one and the boundedness of c_t. Moreover, since the set $\{i \in L(n,\rho)\}$ is $\mathcal{G}^\rho$-measurable, the last part of (3.13), (3.17), and Hölder's inequality, readily yield

$$q \ge 2,\ i \in L(n,\rho) \ \Rightarrow\ \mathbb{E}\big(\|\beta^n_i\|^q \mid \mathcal{F}^{n,\rho}_i)\big| \le K_q \Big(\sqrt{\Delta_n}\, \phi_\rho + \Delta_n^{q/4} + \frac{\Delta_n^{q/2}}{\rho^q}\Big). \tag{3.20}$$

(5) The previous estimates are not enough for us. We will apply the estimates of Lemma 3.2 with $Y_t = X'_{(i-1)\Delta_n+t} - X'_{(i-1)\Delta_n}$ for any given pair n, i, and with the filtration $(\mathcal{F}_{(i-1)\Delta_n+t} \vee \mathcal{G}^\rho)_{t\ge 0}$. We observe that on the set $A(\rho,n,i) = \{\exists j \le 2k_n : i - j \in L(n,\rho)\}$, which is $\mathcal{G}^\rho$-measurable, and because of (3.17), the process c^Y coincides with $c(\rho)_{(i-1)\Delta_n+t} - c(\rho)_{(i-1)\Delta_n}$ if $t \in [0,\Delta_n]$. Then in restriction to this set, by (3.6) and (3.7) and by the definition of $\eta(\rho)^n_{i,1}$, we have

$$\begin{aligned}
&\big|\mathbb{E}(\Delta^n_i X'^j \, \Delta^n_i X'^m \mid \mathcal{F}^{n,\rho}_i) - c^{n,jm}_i \Delta_n\big| \le K_\rho \Delta_n^{3/2} (\sqrt{\Delta_n} + \eta(\rho)^n_{i,1}) \\
&\big|\mathbb{E}\big(\Delta^n_i X'^j \, \Delta^n_i X'^k \, \Delta^n_i X'^l \, \Delta^n_i X'^m \mid \mathcal{F}^{n,\rho}_i\big) \\
&- (c^{n,jk}_i c^{n,lm}_i + c^{n,jl}_i c^{n,km}_i + c^{n,jm}_i c^{n,kl}_i) \Delta_n^2\big| \le K_\rho \Delta_n^{5/2}
\end{aligned}$$

(the constant above depends on ρ, through the bound K/ρ for the drift of $c(\rho)$). Then a simple calculation gives us

$$\left.\begin{array}{l}\| \mathbb{E}(\alpha_i^n \mid \mathcal{F}_i^{n,\rho})\| \leq K_\rho \Delta_n^{3/2}(\sqrt{\Delta_n} + \eta(\rho)_{i,1}^n) \\ \left|\mathbb{E}\big(\alpha_i^{n,jk}\alpha_i^{n,lm} \mid \mathcal{F}_i^{n,\rho}\big) - (c_i^{n,jl}c_i^{n,km} + c_i^{n,jm}c_i^{n,kl})\Delta_n^2\right| \leq K_\rho \Delta_n^{5/2}\end{array}\right\} \quad \text{on } A(\rho,n,i). \tag{3.21}$$

Next, we apply Lemma 3.2 to the process $Y_t = c(\rho)_{(i-1)\Delta_n+t} - c(\rho)_{(i-1)\Delta_n}$ for any given pair n, i, and with the filtration $(\mathcal{F}_{(i-1)\Delta_n+t} \vee \mathcal{G}^\rho)_{t\geq 0}$. We then deduce from (3.5), plus again (3.17), that

$$\begin{array}{l} i \in L(n,\rho),\ 0 \leq t \leq k_n\Delta_n \Rightarrow \\ \left|\mathbb{E}((c^{jk}_{(i-1)\Delta_n+t} - c^{jk}_{(i-1)\Delta_n})(c^{lm}_{(i-1)\Delta_n+t} - c^{lm}_{(i-1)\Delta_n}) \mid \mathcal{F}_i^{n,\rho}) - t\overline{c}(\rho)_i^{n,jklm}\right| \\ \quad \leq K_\rho t\, \eta(\rho)^n_{i,k_n} \\ \left\|\mathbb{E}(c_{(i-1)\Delta_n+t} - c_{(i-1)\Delta_n} \mid \mathcal{F}_i^{n,\rho}) - t\widetilde{b}(\rho)_i^n\right\| \leq K_\rho t\,\eta(\rho)^n_{i,k_n} \leq K_p\, t. \end{array} \tag{3.22}$$

Moreover, the Cauchy-Schwarz inequality and (3.19) on the one hand, and (3.8) applied with the process $Y_t = X'_{(i-1)\Delta_n+t} - X'_{(i-1)\Delta_n}$ on the other hand, give us

$$i \in L(n,\rho) \Rightarrow \left\{\begin{array}{l} \left|\mathbb{E}\big(\alpha_i^{n,kl}\ \Delta_i^n\widetilde{b}(\rho)^{ms} \mid \mathcal{F}_i^{n,\rho}\big)\right| \leq K\Delta_n\eta(\rho)^n_{i,1} \\ \left|\mathbb{E}\big(\alpha_i^{n,kl}\ \Delta_i^n c^{ms} \mid \mathcal{F}_i^{n,\rho}\big)\right| \leq K_\rho\Delta_n^{3/2}(\sqrt{\Delta_n} + \eta(\rho)^n_{i,1}). \end{array}\right. \tag{3.23}$$

(6) We now proceed to estimates on β_i^n.

Lemma 3.3 *We have on the set where i belongs to $L(n,\rho)$:*

$$\begin{array}{r} \left|\mathbb{E}(\beta_i^{n,jk}\,\beta_i^{n,lm} \mid \mathcal{F}_i^{n,\rho}) - \frac{1}{k_n}(c_i^{n,jl}c_i^{n,km} + c_i^{n,jm}c_i^{n,kl}) - \frac{k_n\Delta_n}{3}\,\overline{c}(\rho)_i^{n,jklm}\right| \\ \leq K_\rho\sqrt{\Delta_n}\,(\Delta_n^{1/4} + \eta(\rho)_i^n) \\ \left|\mathbb{E}(\beta_i^{n,jk}(c_{i+k_n}^{n,lm} - c_i^{n,lm}) \mid \mathcal{F}_i^{n,\rho}) - \frac{k_n\Delta_n}{2}\,\overline{c}(\rho)_i^{n,jklm}\right| \leq K_\rho\sqrt{\Delta_n}\,(\sqrt{\Delta_n} + \eta(\rho)_i^n). \end{array}$$

Proof We set $\zeta_{i,j}^n = \alpha_{i+j}^n + (c_{i+j}^n - c_i^n)\Delta_n$ and write $\beta_i^{n,jk}\beta_i^{n,lm}$ as

$$\frac{1}{k_n^2\Delta_n^2}\sum_{u=0}^{k_n-1}\zeta_{i,u}^{n,jk}\zeta_{i,u}^{n,lm} + \frac{1}{k_n^2\Delta_n^2}\sum_{u=0}^{k_n-2}\sum_{v=u+1}^{k_n-1}\zeta_{i,u}^{n,jk}\zeta_{i,v}^{n,lm} + \frac{1}{k_n^2\Delta_n^2}\sum_{u=0}^{k_n-2}\sum_{v=u+1}^{k_n-1}\zeta_{i,u}^{n,lm}\zeta_{i,v}^{n,jk}. \tag{3.24}$$

For the estimates below, we implicitly assume $i \in L(n,\rho)$ and $u, v \in \{0,\dots,k_n-1\}$.

First, we deduce from (3.21) and (3.22), plus (3.23) and successive conditioning, that

$$\big|\mathbb{E}(\zeta_{i,u}^{n,jk}\zeta_{i,u}^{n,lm}\mid\mathcal{F}_i^{n,\rho})-(c_i^{n,jl}c_i^{n,km}+c_i^{n,jm}c_i^{n,kl})\Delta_n^2\big|\leq K\Delta_n^{5/2}. \tag{3.25}$$

Second, if $u<v$, the same type of arguments and the boundedness of $\widetilde{b}(\rho)_t$ and c_t yield

$$\begin{aligned}
&|\mathbb{E}(\zeta_{i,v}^{n,jk}\mid\mathcal{F}_{i+u+1}^{n,\rho})-(c_{i+u+1}^{n,jk}-c_i^{n,jk})\Delta_n-\widetilde{b}(\rho)_{i+u+1}^{n,jk}\Delta_n^2(v-u-1)|\\
&\qquad\qquad\qquad\leq K\Delta_n^{3/2}(k_n\sqrt{\Delta_n}+\eta(\rho)_{i+v,1}^n)\\
&|\mathbb{E}(\alpha_{i+u}^{n,lm}(c_{i+u+1}^{n,jk}-c_{i+u}^{n,jk})\mid\mathcal{F}_{i+u}^{n,\rho})|\leq K_\rho\Delta_n^{3/2}(\sqrt{\Delta_n}+\eta(\rho)_{i+u,1}^n)\\
&|\mathbb{E}(\alpha_{i+u}^{n,lm}(c_{i+u}^{n,jk}-c_i^{n,jk})\mid\mathcal{F}_{i+u}^{n,\rho})|\leq K\Delta_n^{3/2}(\sqrt{\Delta_n}+\eta_{i+u,1}^n)\\
&|\mathbb{E}(\alpha_{i+u}^{n,lm}(\widetilde{b}(\rho)_{i+u+1}^{n,jk}-\widetilde{b}(\rho)_{i+u}^{n,jk})\mid\mathcal{F}_{i+u}^{n,\rho})|\leq K_\rho\Delta_n^{3/2}(\sqrt{\Delta_n}+\eta(\rho)_{i+u,1}^n)\\
&|\mathbb{E}(\alpha_{i+u}^{n,lm}\widetilde{b}(\rho)_{i+u}^{n,jk}\mid\mathcal{F}_{i+u}^{n,\rho})|\leq K_\rho\Delta_n^{3/2}(\sqrt{\Delta_n}+\eta(\rho)_{i+u,1}^n)\\
&|\mathbb{E}((c_{i+u}^{n,lm}-c_i^{n,lm})(c_{i+u+1}^{n,jk}-c_i^{n,jk})\mid\mathcal{F}_i^{n,\rho})-\overline{c}(\rho)_i^{n,jklm}\Delta_n u|\leq K_\rho\Delta_n\,\eta(\rho)_i^n\\
&|\mathbb{E}((c_{i+u}^{n,lm}-c_i^{n,lm})\,\widetilde{b}(\rho)_{i+u+1}^{n,jk}\mid\mathcal{F}_i^{n,\rho})|\leq K_\rho\Delta_n^{1/4}.
\end{aligned}$$

Since $\sum_{u=0}^{k_n-2}\sum_{v=u+1}^{k_n-1}u=k_n^3/6+\mathrm{O}(k_n^2)$, we easily deduce that the $\mathcal{F}_i^{n,\rho}$-conditional expectation of the last term in (3.24) is $\frac{1}{6}\,\overline{c}(\rho)_i^{n,jklm}k_n\Delta_n$, up to a remainder term which is $\mathrm{O}(\sqrt{\Delta_n}\,(\Delta_n^{1/4}+\eta(\rho)_i^n))$, and the same is obviously true for the second term. The first claim of the lemma readily follows from this and (3.24) and (3.25).

The proof of the second claim is similar. Indeed, we have

$$\beta_i^{n,jk}(c_{i+k_n}^{n,lm}-c_i^{n,lm})=\frac{1}{k_n\Delta_n}\sum_{u=0}^{k_n-1}\big(\alpha_{i,u}^{n,jk}+(c_{i+u}^{n,jk}-c_i^{n,jk})\Delta_n\big)\big(c_{i+k_n}^{n,lm}-c_i^{n,lm}\big)$$

and

$$\begin{aligned}
&\big|\mathbb{E}(c_{i+k_n}^{n,lm}-c_i^{n,lm}\mid\mathcal{F}_{i+u+1}^{n,\rho})-c_{i+u+1}^{n,lm}-c_i^{n,lm}-\widetilde{b}(\rho)_{i+u+1}^{n,lm}\Delta_n(k_n-u-1)\big|\\
&\quad\leq K\Delta_n\eta(\rho)_{i+u+1,k_n-u}^n.
\end{aligned}$$

Using the previous estimates, we conclude as for the first claim. □

Finally, we deduce the following two estimates on the variables γ_i^n of (3.18), for any $q\geq 2$:

$$i\in L(n,\rho)\ \Rightarrow\ \begin{cases}\big|\mathbb{E}\big(\gamma_i^{n,jk}\gamma_i^{n,lm}\mid\mathcal{F}_i^{n,\rho}\big)-\frac{2}{k_n}\,(c_i^{n,jl}c_i^{n,km}+c_i^{n,jm}c_i^{n,kl})\\
\qquad\qquad -\frac{2k_n\Delta_n}{3}\,\overline{c}(\rho)_i^{n,jklm}\big|\leq K_\rho\sqrt{\Delta_n}\,\big(\Delta_n^{1/8}+\eta(\rho)_i^n\big)\\
\mathbb{E}(\|\gamma_i^n\|^q\mid\mathcal{F}_i^{n,\rho})\leq K_q\big(\sqrt{\Delta_n}\,\phi_\rho+\Delta_n^{q/4}+\frac{\Delta_n^{q/2}}{\rho^q}\big).\end{cases} \tag{3.26}$$

To see that the first claim holds, one expands the product $\gamma_i^{n,jk}\gamma_i^{n,lm}$ and uses successive conditioning, the Cauchy-Schwarz inequality and (3.13), (3.17) and (3.22), and Lemma 3.3; the contributing terms are

$$\begin{aligned}&\beta_i^{n,jk}\,\beta_i^{n,lm}+\beta_{i+k_n}^{n,jk}\,\beta_{i+k_n}^{n,lm}+(c_{i+k_n}^{n,jk}-c_i^{n,jk})(c_{i+k_n}^{n,lm}-c_i^{n,lm})\\&\quad-\beta_i^{n,jk}(c_{i+k_n}^{n,lm}-c_i^{n,lm})-\beta_i^{n,lm}(c_{i+k_n}^{n,jk}-c_i^{n,jk}).\end{aligned}$$

For the second claim we use (3.13), (3.17) and (3.20), and it holds for all $q \geq 2$.

3.4 The Behavior of Some Functionals of $c(\rho)$

For $\rho > 0$, we set

$$\begin{aligned}U(\rho)_t^n&=\sum_{j=3}^{[t/k_n\Delta_n]-3}\|\mu(\rho)_j^n\|^2\,1_{\{\|\mu(\rho)_j^n\|>u_n'/4\}},\quad\text{where}\\\mu(\rho)_j^n&=\frac{1}{k_n}\sum_{w=0}^{k_n-1}(c(\rho)_{jk_n+w}^n-c(\rho)_{(j-2)k_n+w}^n).\end{aligned}\tag{3.27}$$

The aim of this subsection is to prove the following lemma.

Lemma 3.4 *Under (SA'-r) and (2.14) we have*

$$\lim_{\rho\to 0}\ \limsup_{n\to\infty}\ \mathbb{E}\big(U(\rho)_t^n\big)\ =\ 0.$$

Assumption (SA'-r) is of course not fully used. One only needs the assumptions concerning the process c_t.

Proof With the notation (3.12), and for $l = 1, 2$ we define $\mu^{(l)}(\rho)_j^n$ and $U^{(l)}(\rho)_t^n$ as above, upon substituting $c(\rho)$ and $u_n'/4$ with $c^{(l)}(\rho)$ and $u_n'/8$. Since $U(\rho)_t^n \leq 4U^{(1)}(\rho)_t^n + 4U^{(2)}(\rho)_t^n$, it suffices to prove the result for each $U^{(l)}(\rho)_t^n$.

First, $\|\mu^{(1)}(\rho)_j^n\|^2\,1_{\{\|\mu^{(1)}(\rho)_i^n\|>u_n'/8\}}$ is smaller than $K\|\mu^{(1)}(\rho)_j^n\|^4/u_n'^2$, whereas (recalling $\|\widetilde{b}(\rho)\| \leq K/\rho$) classical estimates yield $\mathbb{E}\big(\|\mu^{(1)}(\rho)_j^n\|^4\big) \leq K\Delta_n(1+\Delta_n/\rho)$. Thus the expectation of $U^{(1)}(\rho)_t^n$ is less than $K\Delta_n^{1/2-2\varpi'}(1+\Delta_n/\rho)$, yielding the result for $U^{(1)}(\rho)_t^n$.

Secondly, we have $U^{(2)}(\rho)_t^n \leq \sum_{j=3}^{[t/k_n\Delta_n]}\|\mu^{(2)}(\rho)_i^n\|^2$ and the first part of (3.13) yields $\mathbb{E}\big(\|\mu^{(2)}(\rho)_i^n\|^2\big) \leq K\phi_\rho\sqrt{\Delta_n}$. Since $\phi_\rho \to 0$ as $\rho \to 0$, the result for $U^{(1)}(\rho)_t^n$ follows. □

3.5 A Basic Decomposition

We start the proof of Theorem 2.2 by giving a decomposition of $V(g)^n - V(g)$, with quite a few terms. It is based on the key property $\widehat{c}_i^n = c_i^n + \beta_i^n$ and on the definition

(3.18) of α_i^n and β_i^n. A simple calculation shows that $\frac{1}{\sqrt{\Delta_n}}\,(V(g)_t^n - V(g)_t) = \sum_{j=1}^5 V_t^{n,j}$, as soon as $t > k_n\Delta_n$, where (the sums on components below always extend from 1 to d):

$$
\begin{aligned}
V_t^{n,1} &= \sqrt{\Delta_n}\,\sum_{i=1}^{[t/\Delta_n]-k_n+1} \big(g(\widehat{c}_i^n) - g(\widehat{c}_i'^n)\big)\\
V_t^{n,2} &= \frac{1}{\sqrt{\Delta_n}} \sum_{i=1}^{[t/\Delta_n]-k_n+1} \int_{(i-1)\Delta_n}^{i\Delta_n} (g(c_i^n) - g(c_s))\,ds\\
V_t^{n,3} &= \frac{1}{k_n\sqrt{\Delta_n}} \sum_{i=1}^{[t/\Delta_n]-k_n+1} \sum_{l,m} \partial_{lm} g(c_i^n) \sum_{u=0}^{k_n-1} \alpha_{i+u}^{n,lm}\\
V_t^{n,4} &= \frac{\sqrt{\Delta_n}}{k_n} \sum_{i=1}^{[t/\Delta_n]-k_n+1} \sum_{l,m} \partial_{lm} g(c_i^n) \sum_{u=1}^{k_n-1} (c_{i+u}^{n,lm} - c_i^{n,lm})\\
&\quad - \frac{1}{\sqrt{\Delta_n}} \int_{\Delta_n([t/\Delta_n]-k_n+1}^{t} g(c_s)\,ds\\
V_t^{n,5} &= \sqrt{\Delta_n} \sum_{i=1}^{[t/\Delta_n]-k_n+1} \Big(g(c_i^n + \beta_i^n) - g(c_i^n) - \sum_{l,m} \partial_{lm} g(c_i^n)\,\beta_i^{n,lm}\Big).
\end{aligned}
$$

The leading term is $V^{n,3}$, the bias comes from the terms $V^{n,4}$ and $V^{n,5}$, and the first two terms are negligible, in the sense that they satisfy

$$
j = 1,2 \;\Rightarrow\; V_t^{n,j} \xrightarrow{\mathbb{P}} 0 \qquad \text{for all } t > 0. \tag{3.28}
$$

We end this subsection with the proof of (3.28).

The case $j = 1$: (2.6) implies

$$
\begin{aligned}
|g(\widehat{c}_i^n) - g(\widehat{c}_i'^n)| &\le K(1 + \|\widehat{c}_i^n\| + \|\widehat{c}_i'^n\|)^{p-1}\,\|\widehat{c}_i^n - \widehat{c}_i'^n\|\\
&\le K(1 + \|\widehat{c}_i'^n\|)^{p-1}\,\|\widehat{c}_i^n - \widehat{c}_i'^n\| + K\|\widehat{c}_i^n - \widehat{c}_i'^n\|^p.
\end{aligned}
$$

Recalling the last part of (3.19), we deduce from (3.10), together with the fact that $1 - r\varpi - p(1-2\varpi) < \frac{(2-r)\varpi}{2q}$ for all $q > 1$ small enough and Hölder's inequality that $\mathbb{E}(|g(\widehat{c}_i^n) - g(\widehat{c}_i'^n)|) \le K a_n \Delta_n^{(2p-r)\varpi+1-p}$. Therefore

$$
\mathbb{E}\Big(\sup_{s\le t} |V_s^{n,1}|\Big) \le K t a_n \Delta_n^{(2p-r)\varpi+1/2-p}
$$

and (3.28) for $j = 1$ follows.

The case $j = 2$: Since g is C^2 and c_t is an Itô semimartingale with bounded characteristics, the convergence $V^{n,2} \stackrel{\text{u.c.p.}}{\Longrightarrow} 0$ is well known: see for example the proof of (5.3.24) in [4], in which one replaces $\rho_{c_s}(f)$ by $g(c_s)$.

3.6 The Leading Term $V^{n,3}$

Our aim here is to prove that

$$V^{n,3} \stackrel{\mathcal{L}-s}{\Longrightarrow} Z \tag{3.29}$$

(functional stable convergence in law), where Z is the process defined in Theorem 2.2.

A change of order of summation allows us to rewrite $V^{n,3}$ as

$$V_t^{n,3} = \frac{1}{\sqrt{\Delta_n}} \sum_{i=1}^{[t/\Delta_n]} \sum_{l,m} w_i^{n,lm}\, \alpha_i^{n,lm}, \quad \text{where } w_i^{n,lm} = \frac{1}{k_n} \sum_{j=(i-[t/\Delta_n]+k_n-1)^+}^{(i-1)\wedge(k_n-1)} \partial_{lm} g(c_{i-j}^n).$$

Observe that w_i^n and α_i^n are measurable with respect to $\mathcal{F}_i^n$ and $\mathcal{F}_{i+1}^n$, respectively, so by Theorem IX.7.28 of [3] (with $G = 0$ and $Z = 0$ in the notation of that theorem) it suffices to prove the following four convergences in probability, for all $t > 0$ and all component indices:

$$\frac{1}{\sqrt{\Delta_n}} \sum_{i=1}^{[t/\Delta_n]-k_n+1} w_i^{n,lm}\, \mathbb{E}(\alpha_i^{n,lm} \mid \mathcal{F}_i^n) \stackrel{\mathbb{P}}{\longrightarrow} 0 \tag{3.30}$$

$$\frac{1}{\Delta_n} \sum_{i=1}^{[t/\Delta_n]-k_n+1} w_i^{n,jk}\, w_i^{n,lm}\, \mathbb{E}(\alpha_i^{n,jk}\, \alpha_i^{n,lm} \mid \mathcal{F}_i^n) \stackrel{\mathbb{P}}{\longrightarrow} \int_0^t \partial_{jk} g(c_s)\, \partial_{lm} g(c_s)\, \big(c_s^{jl} c_s^{km} + c_s^{jm} c_s^{kl}\big)\, ds \tag{3.31}$$

$$\frac{1}{\Delta_n^2} \sum_{i=1}^{[t/\Delta_n]-k_n+1} \|w_i^n\|^4\, \mathbb{E}(\|\alpha_i^n\|^4 \mid \mathcal{F}_i^n) \stackrel{\mathbb{P}}{\longrightarrow} 0 \tag{3.32}$$

$$\frac{1}{\sqrt{\Delta_n}} \sum_{i=1}^{[t/\Delta_n]-k_n+1} w_i^{n,lm}\, \mathbb{E}(\alpha_i^{n,lm}\, \Delta_i^n N \mid \mathcal{F}_i^n) \stackrel{\mathbb{P}}{\longrightarrow} 0, \tag{3.33}$$

where $N = W^j$ for some j, or is an arbitrary bounded martingale, orthogonal to W.

For proving these properties, we pick a ρ bigger than the upper bound of the function J, so $\mathcal{G}^\rho$ becomes the trivial σ-field and $\mathcal{F}_i^n = \mathcal{F}_i^{n,\rho}$ and $L(n,\rho) = \mathbb{N}$. In such a way, we can apply all estimates of the previous subsections with the conditioning σ-fields $\mathcal{F}_i^n$. Therefore (3.19) and the property $\|w_i^n\| \leq K$ readily imply (3.30) and (3.32). In view of the form of α_i^n, a usual argument (see e.g. [4]) shows that in fact $\mathbb{E}(\alpha_i^{n,lm}\,\Delta_i^n N \mid \mathcal{F}_i^n) = 0$ for all N as above, hence (3.33) holds.

For (3.31), by (3.21) it suffices to prove that

$$\Delta_n \sum_{i=1}^{[t/\Delta_n]-k_n+1} w_i^{n,jk}\, w_i^{n,lm}\,(c_i^{n,jl} c_i^{n,km} + c_i^{n,jm} c_i^{n,kl}) \\ \stackrel{\mathbb{P}}{\longrightarrow} \int_0^t \partial_{jk} g(c_s)\,\partial_{lm} g(c_s)\,\big(c_s^{jl} c_s^{km} + c_s^{jm} c_s^{kl}\big)\,ds.$$

In view of the definition of w_i^n, for each t we have $w_{i(n,t)}^{n,jk} \to \partial_{jk} g(c_t)$ and $c_{i(n,t)}^{n,jk} \to c_t^{jk}$ almost surely if $|i(n,t)\Delta_n - t| \leq k_n \Delta_n$ (recall that c is almost surely continuous at t, for any fixed t), and the above convergence follows by the dominated convergence theorem, thus ending the proof of (3.29).

3.7 The Term $V^{n,4}$

In this subsection we prove that, for all t,

$$V_t^{n,4} \stackrel{\mathbb{P}}{\longrightarrow} \frac{\theta}{2} \sum_{l,m} \int_0^t \partial_{lm} g(c_{s-})\,dc_s^{lm} - \theta\, g(c_t). \tag{3.34}$$

We call $V_t^{\prime n,4}$ and $V_t^{\prime\prime n,4}$, respectively, the first sum, and the last integral, in the definition of $V_t^{n,4}$. Since $k_n\sqrt{\Delta_n} \to \theta$ and c is a.s. continuous at t, it is obvious that $V_t^{\prime\prime n,4}$ converges almost surely to $-\theta\, g(c_t)$, and it remains to prove the convergence of $V_t^{\prime n,4}$ to the first term in the right side of (3.34).

We first observe that $c_{i+u}^n - c_i^n = \sum_{v=0}^{u-1} \Delta_{i+v}^n c$. Then, upon changing the order of summation, we can rewrite $V_t^{\prime n,4}$ as

$$V_t^{\prime n,4} = \sum_{i=1}^{[t/\Delta_n]-1} \sum_{l,m} w_i^{n,lm}\,\Delta_i^n c^{lm},$$
$$w_i^{n,lm} = \frac{\sqrt{\Delta_n}}{k_n} \sum_{u=0\vee(i+k_n-1-[t/\Delta_n])}^{(i-1)\wedge(k_n-2)} (k_n - 1 - u)\partial_{lm} g(c_{i-u}^n).$$

In other words, recalling $k_n\sqrt{\Delta_n} \leq K$ and $\|\partial g(c_s)\| \leq K$, we see that

$$V_t^{\prime n,4} = \sum_{l,m} \int_0^t H(n,t)_s^{lm}\, dc_s^{lm},$$

where $H(n,t)_s$ is a $d \times d$-dimensional predictable process, bounded uniformly (in n, s, ω) and given on the set $[k_n\Delta_n, t - k_n\Delta_n]$ by

$$(i-1)\Delta_n < s \leq i\Delta_n \quad \Rightarrow \quad H(n,t)_s^{lm} = \frac{\sqrt{\Delta_n}}{k_n} \sum_{u=0}^{k_n-2} (k_n - 1 - u)\partial_{lm} g(c_{i-u}^n)$$

(its expression on $[0, k_n\Delta_n)$ and on $(t-k_n\Delta_n, t]$ is more complicated, but not needed, apart from the fact that it is uniformly bounded). Now, since $\sum_{u=0}^{k_n-2}(k_n - 1 - u) = k_n^2/2 + \mathrm{O}(k_n)$ as $n \to \infty$, we observe that $H(n,t)_s^{lm}$ converges to $\frac{\theta}{2}\,\partial_{lm} g(c_{s-})$ for all $s \in (0,t)$. Since c is a.s. continuous at t, we deduce from the dominated convergence theorem for stochastic integrals that $V_t^{\prime n,4}$ indeed converges in probability to the first term in the right side of (3.34).

3.8 The Term $V^{n,5}$

The aim of this subsection is to prove the convergence

$$V_t^{n,5} \stackrel{\mathbb{P}}{\longrightarrow} A_t^2 - 2A_t^3 + \theta \sum_{s \leq t} \int_0^1 \Big(g(c_{s-} + w\Delta c_s) - g(c_{s-}) - w \sum_{l,m} \partial_{lm} g(c_{s-})\, \Delta c_s^{lm}\Big)\, dw. \tag{3.35}$$

We have $V_t^{n,5} = \sum_{i=1}^{[t/\Delta_n]-k_n+1} v_i^n$, where

$$v_i^n = \sqrt{\Delta_n}\,\Big(g(c_i^n + \beta_i^n) - g(c_i^n) - \sum_{l,m} \partial_{lm} g(c_i^n)\, \beta_i^{n,lm}\Big).$$

We also set

$$\begin{gathered}\overline{\alpha}_i^n = \tfrac{1}{k_n\Delta_n} \textstyle\sum_{u=0}^{k_n-1} \alpha_{i+u}^n, \qquad \overline{\beta}_i^n = \beta_i^n - \overline{\alpha}_i^n = \tfrac{1}{k_n} \textstyle\sum_{u=1}^{k_n-1} (c_{i+u}^n - c_i^n), \\ v_i^{\prime n} = \sqrt{\Delta_n}\,\Big(g(c_i^n + \overline{\beta}_i^n) - g(c_i^n) - \sum_{l,m} \partial_{lm} g(c_i^n)\, \overline{\beta}_i^{n,lm}\Big), \qquad v_i^{\prime\prime n} = v_i^n - v_i^{\prime n}.\end{gathered} \tag{3.36}$$

We take $\rho \in (0,1]$, and will eventually let it go to 0. With the sets $L(n,\rho)$ of (3.16), we associate

$$\begin{aligned} L(n,\rho,t) &= \{1, \ldots, [t/\Delta_n] - k_n + 1\} \cap L(n,\rho) \\ \overline{L}(n,\rho,t) &= \{1, \ldots, [t/\Delta_n] - k_n + 1\} \backslash L(n,\rho). \end{aligned}$$

We split the sum giving $V_t^{n,5}$ into three terms:

$$U_t^{n,\rho} = \sum_{i\in L(n,\rho,t)} v_i^n, \qquad U_t^{\prime n,\rho} = \sum_{i\in \overline{L}(n,\rho,t)} v_i^{\prime n}, \qquad U_t^{\prime\prime n,\rho} = \sum_{i\in \overline{L}(n,\rho,t)} v_i^{\prime\prime n}. \quad (3.37)$$

(A) The processes $U^{n,\rho}$. A Taylor expansion and (2.6) give us

$$v_i^n = v(1)_i^n + v(2)_i^n + v(3)_i^n, \text{ where } \begin{cases} v(1)_i^n = \frac{\sqrt{\Delta_n}}{2} \sum_{j,k,l,m} \partial^2_{jk,lm} g(c_i^n)\, \mathbb{E}(\beta_i^{n,jk}\, \beta_i^{n,lm} \mid \mathcal{F}_i^{n,\rho}) \\ v(2)_i^n = \frac{\sqrt{\Delta_n}}{2} \sum_{j,k,l,m} \partial^2_{jk,lm} g(c_i^n)\, \beta_i^{n,jk}\, \beta_i^{n,lm} - v(1)_i^n \\ |v(3)_i^n| \leq K\sqrt{\Delta_n}\,(1 + \|\beta_i^n\|)^{p-3}\, \|\beta_i^n\|^3. \end{cases}$$

Therefore

$$U^{n,\rho} = \sum_{j=1}^{3} U(j)^{n,\rho}, \quad \text{where } \ U(j)_t^n = \sum_{i\in L(n,\rho,t)} v(j)_i^n. \qquad (3.38)$$

On the one hand, letting

$$w(\rho)_i^n = \sum_{j,k,l,m} \partial^2_{jk,lm} g(c_i^n) \Big(\frac{1}{2k_n\sqrt{\Delta_n}}\, (c_i^{n,jl} c_i^{n,km} + c_i^{n,jm} c_i^{n,kl}) + \frac{k_n\sqrt{\Delta_n}}{6}\, \overline{c}(\rho)_i^{n,jklm} \Big),$$

the càdlàg property of c and $\overline{c}(\rho)$ and $k_n\sqrt{\Delta_n} \to \theta$ imply

$$W(\rho)_t^n := \Delta_n \sum_{i=1}^{[t/\Delta_n]-k_n+1} w(\rho)_i^n \stackrel{\mathbb{P}}{\longrightarrow} U(1)_t^\rho := A_t^2 + \frac{\theta}{6} \sum_{j,k,l,m} \int_0^t \partial^2_{jk,lm} g(c_s)\, \overline{c}(\rho)_s^{jklm}\, ds.$$

On the other hand, Lemma 3.3 yields $|v(1)_i^n - \Delta_n w(\rho)_i^n| \leq K_\rho \Delta_n\, (\Delta_n^{1/4} + \eta(\rho)_i^n)$ when $i \in L(n,\rho)$, whereas $|w(\rho)_i^n| \leq K$ always. Therefore

$$\mathbb{E}\big(|U(1)_t^{n,\rho} - W(\rho)_t^n|\big) \leq K_\rho \Delta_n\, \mathbb{E}\Big(\sum_{i=1}^{[t/\Delta_n]} (\sqrt{\Delta_n} + \eta(\rho)_i^n) \Big) + K\Delta_n \mathbb{E}\big(\#(\overline{L}(n,\rho,t))\big).$$

Now, $\#(\overline{L}(n,\rho,t))$ is not bigger than $(2k_n+1)N_t^\rho$, implying that $\Delta_n\mathbb{E}(\#(L'(n,\rho,t))) \leq K_\rho\sqrt{\Delta_n}$. Taking advantage of (3.15), we deduce that the above expectation goes to 0 as $n \to \infty$, and thus

$$U(1)_t^{n,\rho} \stackrel{\mathbb{P}}{\longrightarrow} U(1)_t^\rho. \qquad (3.39)$$

Next, $v(2)_i^n$ is $\mathcal{F}_{i+k_n}^{n,\rho}$-measurable, with vanishing $\mathcal{F}_i^{n,\rho}$-conditional expectation, and each set $\{i \in L(n,\rho)\}$ is $\mathcal{F}_0^{n,\rho}$-measurable. It follows that

$$\mathbb{E}\big((U(2)_t^{n,\rho})^2\big) \leq 2k_n\, E\Big(\sum_{i\in L(n,\rho,t)} \mathbb{E}\big(|v(2)_i^n|^2 \mid \mathcal{F}_i^{n,\rho}\big)\Big)$$
$$\leq K k_n \Delta_n\, E\Big(\sum_{i\in L(n,\rho,t)} \mathbb{E}\big(|\beta_i^n|^4 \mid \mathcal{F}_i^{n,\rho}\big)\Big) \;\leq\; Kt\phi_\rho + K_\rho t\sqrt{\Delta_n},$$

where we have applied (3.20) for the last inequality. Another application of the same estimate gives us

$$\mathbb{E}\big(|U(3)_t^n|\big) \;\leq\; Kt\phi_\rho + K_\rho t\Delta_n^{1/4}.$$

These two results and the property $\phi_\rho \to 0$ as $\rho \to 0$ clearly imply

$$\lim_{\rho\to 0}\ \limsup_{n\to\infty}\ \mathbb{E}(|U(2)_t^{n,\rho}| + |U(3)_t^{n,\rho}|) \;=\; 0. \tag{3.40}$$

(B) The processes $U'^{n,\rho}$. We will use here the jump times $S_1^\rho, S_2^\rho, \ldots$ of the Poisson process N^ρ, and will restrict our attention to the set $\Omega_{n,t,\rho}$ defined before (3.12), whose probability goes to 1 as $n \to \infty$. On this set, $\overline{L}(n,\rho,t)$ is the collection of all integers i which are between $[S_q^\rho/\Delta_n] - 2k_n + 2$ and $[S_q^\rho/\Delta_n] + 1$, for some q between 1 and N_t^ρ. Thus

$$U_t'^{n,\rho} = \sum_{q=1}^{N_t^\rho} H(n,\rho,q), \quad \text{where } H(n,\rho,q) = \sum_{i=[S_q^\rho/\Delta_n]-2k_n+1}^{[S_q^\rho/\Delta_n]+1} v_i'^n. \tag{3.41}$$

The behavior of each $H(n,\rho,q)$ is a pathwise question. We fix q and set $S = S_q^\rho$ and $a_n = [S/\Delta_n]$, so $S > a_n\Delta_n$ because S is not a multiple of Δ_n. For further reference we consider a case slightly more general than strictly needed here. We have $c_i^n \to c_{S-}$ when $a_n - 6k_n + 1 \leq i \leq a_n + 1$ and $c_i^n \to c_S$ when $a_n + 2 \leq i \leq a_n + 6k_n$, uniformly in i (for each given outcome ω). Hence

$$\overline{\beta}_i^n - \frac{(k_n - a_n + i - 2)^+ \wedge (k_n - 1)}{k_n}\,\Delta c_S \to 0$$
$$\text{uniformly in } i \in \{a_n - 6k_n + 2, \ldots, a_n + 5k_n\}. \tag{3.42}$$

Thus, the following convergence holds, uniform in $i \in \{a_n - 2k_n + 1, \ldots, a_n + 1\}$:

$$\frac{1}{\sqrt{\Delta_n}}\, v_i'^n - \Big(g\big(c_{S-} + \tfrac{k_n - a_n + i - 2}{k_n}\,\Delta c_S\big) - g(c_{S-})$$
$$- \sum_{l,m} \partial_{lm} g(c_{S_-})\,\big(c_{S-}^{jk} + \tfrac{k_n - a_n + i - 2}{k_n}\,\Delta c_S^{lm}\big)\Big) \to 0,$$

which implies

$$H(n,\rho,q) - \sqrt{\Delta_n}\sum_{u=1}^{k_n-3}\Big(g\big(c_{S_q-} + \frac{u}{k_n}\,\Delta c_{S_q}\big) - g(c_{S_q-}) - \sum_{l,m}\partial_{lm} g(c_{S_q-})\,\frac{u}{k_n}\,\Delta c_{S_q}^{lm}\Big) \to 0$$

and by Riemann integration this yields

$$H(n,\rho,q) \to \theta \int_0^1 \Big(g(c_{S_q-} + w\Delta c_{S_q}) - g(c_{S_q-}) - w \sum_{l,m} \partial_{lm} g(c_{S_q-})\, \Delta c_S^{lm}\Big)\, dw.$$

Henceforth, we have

$$U_t^{\prime n,\rho} \stackrel{\mathbb{P}}{\longrightarrow} U_t^{\prime \rho} := \theta \sum_{q=1}^{N_t^\rho} \int_0^1 \Big(g(c_{S_q-} + w\Delta c_{S_q}) - g(c_{S_q-}) - w \sum_{l,m} \partial_{lm} g(c_{S_q-})\, \Delta c_{S_q}^{lm}\Big)\, dw. \tag{3.43}$$

(C) The processes $U^{\prime\prime n,\rho}$. Since $|\overline{\beta}_i^n| \leq K$ we deduce from (2.6) that $|v_i^{\prime\prime n}| \leq K\sqrt{\Delta_n}\,(\|\overline{\alpha}_i^n\| + \|\overline{\alpha}_i^n\|^p)$. (3.19) yields $\mathbb{E}(\|\overline{\alpha}_i^n\|^q \mid \mathcal{F}_i^{n,\rho}) \leq K_q \Delta_n^{q/4}$ for all $q > 0$. Therefore

$$\mathbb{E}\big(|U_t^{\prime\prime n,\rho}|\big) \;\leq\; K\Delta_n^{3/4}\, \mathbb{E}(\#(\overline{L}(,n,\rho,t))) \;\leq\; K_\rho \Delta_n^{1/4},$$

by virtue of what precedes (3.39). We then deduce

$$U_t^{\prime\prime n,\rho} \stackrel{\mathbb{P}}{\longrightarrow} 0. \tag{3.44}$$

(D) Proof of (3.35). On the one hand, $V^{n,5} = U(1)^{n,\rho} + U(2)^{n,\rho} + U(3)^{n,\rho} + U^{\prime n,\rho} + U^{\prime\prime n,\rho}$; on the other hand, the dominated convergence theorem (observe that $\overline{c}(\rho)_t \to \widetilde{\sigma}_t^2$ for all t) yields that $U(1)_t^\rho \stackrel{\mathbb{P}}{\longrightarrow} A^2 - \frac{1}{2}\, A_t^3$ and

$$U_t^{\prime\rho} \stackrel{\mathbb{P}}{\longrightarrow} \theta \sum_{s\leq t} \int_0^1 \Big(g(c_{s-} + w\Delta c_s) - g(c_{s-}) - w \sum_{l,m} \partial_{lm} g(c_{s-})\, \Delta c_s^{lm}\Big)\, dw$$

as $\rho \to 0$ (for the latter convergence, note that $|g(x+y) - g(x) - \sum_{l,m} \partial_{lm} g(x) y^{lm}| \leq K\|y\|^2$ when x, y stay in a compact set). Then the property (3.35) follows from (3.39), (3.40), (3.43) and (3.44).

(E) Proof of Theorem 2.2. We are now ready to prove Theorem 2.2. Recall that $\frac{1}{\sqrt{\Delta_n}}\,(V(g)n_t - V(g)) = \sum_{j=1}^5 V^{n,j}$. By virtue of (3.28), (3.29), (3.34), (3.35), it is enough to check that

$$\begin{aligned} A_t^1 + A_t^3 + A_t^4 + A_t^5 &= \tfrac{\theta}{2} \textstyle\sum_{l,m} \int_0^t \partial_{lm} g(c_{s-})\, dc_s^{lm} - \theta\, g(c_t) \\ &\quad - 2A_t^3 + \theta \textstyle\sum_{s\leq t} \int_0^1 \big(g(c_{s-} + w\Delta c_s) - g(c_{s-}) - w \sum_{l,m} \partial_{lm} g(c_{s-})\, \Delta c_s^{lm}\big)\, dw. \end{aligned}$$

To this aim, we observe that Itô's formula gives us

$$\begin{aligned} g(c_t) = g(c_0) + \sum_{l,m} \int_0^t \partial_{lm} g(c_{s-})\, dc_s^{lm} - \frac{6}{\theta} A_t^3 + \sum_{s\leq t} \Big(g(c_{s-} + \Delta c_s) - g(c_{s-}) \\ - \sum_{l,m} \partial_{lm} g(c_{s-})\, \Delta c_s^{lm}\Big), \end{aligned}$$

so the desired equality is immediate (use also $\int_0^1 w\,dw = \frac{1}{2}$), and the proof of Theorem 2.2 is complete.

3.9 Proof of Theorem 2.5

The proof of Theorem 2.5 follows the same line as in Sect. 3.8, and we begin with an auxiliary step.

Step (1) Replacing $\widehat{c}_i^n$ by $\widehat{c}_i^{\prime n}$. The summands in the definition (2.12) of $A_t^{n,3}$ are $R(\widehat{c}_i^n, \widehat{c}_{i+k_n}^n)$, where $R(x,y) = \sum_{j,k,l,m} \partial^2_{jk,lm} g(x)(y^{jk} - x^{jk})(y^{lm} - x^{lm})$, and we set

$$A_t^{\prime n,3} = -\frac{\sqrt{\Delta_n}}{8} \sum_{i=1}^{[t/\Delta_n]-2k_n+1} R(\widehat{c}_i^{\prime n}, \widehat{c}_{i+k_n}^{\prime n}).$$

We prove here that

$$A_t^{n,3} - A_t^{\prime n,3} \xrightarrow{\mathbb{P}} 0 \tag{3.45}$$

for all t, and this is done as in the step $j=1$ in Sect. 3.5. The function R is C^1 on $\mathbb{R}_+^2$ with $\|\partial^j R(x,y)\| \leq K(1+\|x\|+\|y\|)^{p-j}$ for $j=0,1$, by (2.6). Thus

$$\begin{aligned}
&|R(\widehat{c}_i^n, \widehat{c}_{i+k_n}^n) - R(\widehat{c}_i^{\prime n}, \widehat{c}_{i+k_n}^{\prime n})| \\
&\quad \leq K(1+\|\widehat{c}_i^{\prime n}\| + \|\widehat{c}_{i+k_n}^{\prime n})\|)^{p-1}(\|\widehat{c}_i^n - \widehat{c}_i^{\prime n}\| + \|\widehat{c}_{i+k_n}^n - \widehat{c}_{i+k_n}^{\prime n}\|) \\
&\qquad + K\|\widehat{c}_i^n - \widehat{c}_i^{\prime n}\|^p + K\|\widehat{c}_{i+k_n}^n - \widehat{c}_{i+k_n}^{\prime n}\|^p.
\end{aligned}$$

Then, exactly as in the case afore-mentioned, we conclude (3.45), and it remains to prove that, for all t, we have

$$A_t^{\prime n,3} \xrightarrow{\mathbb{P}} -\frac{1}{2} A_t^2 + A_t^3 + A_t^{\prime 4}.$$

Step (2) From now on we use the same notation as in Sect. 3.8, although they denote different variables or processes. For any $\rho \in (0,1]$ we have $A^{\prime n,3} = U^{n,\rho} + U^{\prime n,\rho} + U^{\prime\prime n,\rho}$, as defined in (3.37), but with

$$\begin{aligned}
v_i^n &= -\tfrac{\sqrt{\Delta_n}}{8} R(c_i^n + \beta_i^n, c_{i+k_n}^n + \beta_{i+k_n}^n) \\
v_i^{\prime n} &= -\tfrac{\sqrt{\Delta_n}}{8} R(c_i^n + \overline{\beta}_i^n, c_{i+k_n}^n + \overline{\beta}_{i+k_n}^n), \quad v_i^{\prime\prime n} = v_i^n - v_i^{\prime n}.
\end{aligned}$$

Recalling γ_i^n in (3.18), the decomposition (3.38) holds with

$$v(1)_i^n = -\tfrac{\sqrt{\Delta_n}}{8} \sum_{j,l,k,m} \partial^2_{jl,km}\, g(c_i^n)\, \mathbb{E}\big(\gamma_i^{n,jk}\, \gamma_i^{n,lm} \mid \mathcal{F}_i^{n,\rho}\big)$$
$$v(2)_i^n = -\tfrac{\sqrt{\Delta_n}}{8} \sum_{j,l,k,m} \partial^2_{jl,km}\, g(c_i^n)\, \gamma_i^{n,jk}\, \gamma_i^{n,lm} - v(1)_i^n$$
$$v(3)_i^n = v_i^n - v(1)_i^n - v(2)_i^n.$$

Use $\widehat{c}_i^{\prime n} - c_i^n = \beta_i^n$ and (2.6) and a Taylor expansion to check that

$$|v(3)_i^n| \le K\sqrt{\Delta_n}\, \|\gamma_i^n\|^2\, \|\beta_i^n\|\, (1 + \|\beta_i^n\|)^{p-3}.$$

We also have $|v(2)_i^n| \le K\sqrt{\Delta_n}\, \|\gamma_i^n\|^2$, hence (3.20) and (3.26) yield

$$\mathbb{E}(|v(3)_i^n| \mid \mathcal{G}^\rho) + \mathbb{E}(|v(2)_i^n|^2 \mid \mathcal{G}^\rho) \le K\Delta_n \Big(\phi_\rho + \Delta_n^{1/4} + \frac{\Delta_n}{\rho^p}\Big),$$

and thus (3.40) holds here as well, by the same argument. Moreover, (3.26) again yields (3.39), with now

$$U_t^\rho = -\sum_{j,k,l,m} \int_0^t \partial^2_{jk,lm}\, g(c_s)\, \Big(\frac{\theta}{12}\, \overline{c}(\rho)_s^{jklm} + \frac{1}{4\theta}\, (c_s^{jl} c_s^{km} + c_s^{jm} c_s^{kl})\Big)\, ds.$$

This goes to $A_t^3 - \frac{1}{2} A_t^2$ as $\rho \to 0$.

Another application of (2.6) gives us

$$|v_i^{\prime\prime n}| \le K\sqrt{\Delta_n}\, \big(1 + \|\gamma_i^n\|^2\big)\big(\|\overline{\alpha}_i^n\| + \|\overline{\alpha}_{i+k_n}^n\| + \|\overline{\alpha}_i^n\|^p + \|\overline{\alpha}_{i+k_n}^n\|^p\big).$$

Then another application of (3.19), (3.20) and (3.26) yields $\mathbb{E}(|v_i^{\prime\prime n}| \mid \mathcal{G}^\rho) \le K\Delta_n^{3/4}$ and we conclude (3.44) as previously. We are thus left to prove that

$$\rho > 0 \;\Rightarrow\; U_t^{\prime n,\rho} \stackrel{\mathbb{P}}{\longrightarrow} U_t^{\prime\rho}, \quad \text{with, as } \rho \to 0,\; U_t^{\prime\rho} \stackrel{\mathbb{P}}{\longrightarrow} A_t^{\prime 4}. \tag{3.46}$$

Step (3) On the set $\Omega_{n,t,\rho}$ we have (3.41) and we study $H(n,\rho,q)$, in the same way as before, on the set $\Omega_{n,t,\rho}$. We fix q and set $S = S_q$ and $a_n = [S/\Delta_n]$. We then apply (3.42) and also $c_i^n \to c_{S-}$ or $c_i^n \to c_S$, according to whether $a_n - 2k_n + 1 \le i \le a_n + 1$ or $a_n + 2 \le i \le a_n + k_n$, to obtain $v_i^{\prime n} - \overline{v}_i^{\prime n} \to 0$, uniformly in i between $a_n - 2k_n + 1$ and $a_n + 1$, where

$$\overline{v}_i^{\prime n} = \begin{cases} 0 & \text{if } a_n - 2k_n + 1 \le i \le a_n - 2k_n + 2 \\ -\frac{(2k_n - a_n + i - 2)^2 \sqrt{\Delta_n}}{8k_n^2} \sum_{j,k,l,m} \partial^2_{jk,lm}\, g(c_{S-})\, \Delta c_S^{jk}\, \Delta c_S^{lm} & \\ & \text{if } a_n - 2k_n + 3 \le i \le a_n - k_n + 1 \\ \frac{(a_n - i + 2)^2 \sqrt{\Delta_n}}{8k_n^2} \sum_{j,k,l,m} \partial^2_{jk,lm}\, g\Big(c_{S-} + \frac{k_n - a_n + i + 2}{k_n}\, \Delta c_S\Big)\, \Delta c_S^{jk}\, \Delta c_S^{lm} & \\ & \text{if } a_n - k_n + 2 \le i \le a_n + 1. \end{cases}$$

We then deduce, by Riemann integration, that

$$H(n,\rho,q) \to -\frac{\theta}{8}\sum_{j,k,l,m}\int_0^1 \big(\partial^2_{jk,lm}\, g(c_{S_q-}) + \partial^2_{jk,lm}\, g(c_{S_q-}+(1-w)\Delta c_{S_q})\big)\, w^2\, \Delta c^{jk}_{S_q}\, \Delta c^{lm}_{S_q}\, dw,$$

which is $\theta G'(c_{S_q-}, \Delta c_{S_q})$, hence the first part of (3.46), with $U'^{\rho}_t = \theta \sum_{q=1}^{N^\rho_t} G'(c_{S^\rho_q-}, \Delta c_{S^\rho_q})$. The second part of (3.46) follows from the dominated convergence theorem, and the proof of Theorem 2.5 is complete.

3.10 Proof of Theorem 2.6

The proof is once more somewhat similar to the proof of Sect. 3.8, although the way we replace $\widehat{c}^n_i$ by $\widehat{c}'^n_i$ and further by $\overline{\alpha}^n_i + \beta^n_i$ is different.

(A) Preliminaries. The jth summand in (2.15) involves several estimators $\widehat{c}^n_i$, spanning the time interval $((j-3)k_n\Delta_n, (j+2)k_n\Delta_n]$. It is thus convenient to replace the sets $L(n,\rho)$, $L(n,\rho,t)$ and $\overline{L}(n,\rho,t)$, for $\rho, t > 0$, by the following ones:

$$\begin{aligned}
L'(n,\rho) &= \{j = 3,4,\ldots :\ N^\rho_{(j+2)k_n\Delta_n} - N^\rho_{(j-3)k_n\Delta_n} = 0\}\\
L'(n,\rho,t) &= \{3,\ldots,[t/k_n\Delta_n]-3\} \cap L'(n,\rho)\\
\overline{L}'(n,\rho,t) &= \{3,\ldots,[t/k_n\Delta_n]-3\} \cap (\mathbb{N}\backslash L'(n,\rho)).
\end{aligned}$$

For any $\rho \in (0,1]$ we write $\mathcal{V}(F)^n_t = \mathcal{V}^{n,\rho}_t + \overline{\mathcal{V}}^{n,\rho}_t$, where

$$\begin{aligned}
v^n_j &= F(\widehat{c}^n_{(j-3)k_n+1}, \delta^n_j\widehat{c})\, 1_{\{\|\delta^n_{j-1}\widehat{c}\| \vee \|\delta^n_{j+1}\widehat{c}\| \vee u'_n < \|\delta^n_j\widehat{c}\|\}}\\
\mathcal{V}^{n,\rho}_t &= \sum\nolimits_{j\in L'(n,\rho,t)} v^n_j, \qquad \overline{\mathcal{V}}^{n,\rho}_t = \sum\nolimits_{j\in \overline{L}'(n,\rho,t)} v^n_j.
\end{aligned}$$

We also set

$$\begin{aligned}
&\delta^n_j\widehat{c}' = \widehat{c}'^n_{jk_n+1} - \widehat{c}'^n_{(j-2)k_n+1}, \qquad\qquad \delta^n_j\beta = \beta^n_{jk_n+1} - \beta^n_{(j-2)k_n+1},\\
&w^n_j = \sum\nolimits_{m=-3}^{2} \|\widehat{c}^n_{(j+m)k_n+1} - \widehat{c}'^n_{(j+m)k_n+1}\|,\ w'^n_j = (1+\|\widehat{c}'^n_{(j-3)k_n+1}\|)^{p-1}\,(1+\|\delta^n_j\widehat{c}\|)^2.
\end{aligned}$$

Equation (3.10) and the last part of (3.19) yield

$$q \geq 1 \ \Rightarrow\ \mathbb{E}\big((w^n_j)^q\big) \leq K_q\, \Delta_n^{(2q-r)\varpi+1-q}, \qquad \mathbb{E}((w'^n_i)^q) \leq K_q. \tag{3.47}$$

Observe that $\delta^n_j\widehat{c}'$ is analogous to γ^n_i, with a doubled time lag, so it satisfies a version of (3.26) and, for $q \geq 2$, we have

$$i \in L'(n,\rho) \ \Rightarrow\ \mathbb{E}\big(\|\delta^n_j\widehat{c}'\|^q \mid \mathcal{F}^{n,\rho}_{(j-2)k_n+1}\big) \leq K_q\big(\sqrt{\Delta_n}\,\phi_\rho + \Delta_n^{q/4} + \frac{\Delta_n^{q/2}}{\rho^q}\big). \tag{3.48}$$

(B) The processes $\mathcal{V}^{n,\rho}$. (2.16) yields

$$|v_j^n| \leq K(1+\|\widehat{c}^n_{(j-3)k_n+1}\|)^{p-2}\,\|\delta_j^n\widehat{c}\|^2\,1_{\{\|\delta_j^n\widehat{c}\|>u'_n\}} + K\|\delta_j^n\widehat{c}\|^p.$$

Thus a (tedious) computation shows that, with the notation

$$a_j^n = (1+\|\widehat{c}'^n_{(j-3)k_n+1}\|)^{p-2}\,\|\delta_j^n\widehat{c}'\|^2\,1_{\{\|\delta_j^n\widehat{c}'\|>u'_n/2\}},$$
$$a_j'^n = w_j'^n\Big(w_i^n + (w_i^n)^p + \frac{(w_i^n)^v}{u_n'^v}\Big),$$

with $v>0$ arbitrary, we have $|v_j^n| \leq K(a_j^n + \|\delta_j^n\widehat{c}'\|^p + a_j'^n)$ (with K depending on v). Therefore we have $|\mathcal{V}_t^{n,\rho}| \leq K(B_t^{n,\rho} + C_t^{n,\rho} + D_t^n)$, where

$$B_t^{n,\rho} = \sum_{j\in L'(n,\rho,t)} a_j^n, \qquad C_t^{n,\rho} = \sum_{j\in L'(n,\rho,t)} \|\delta_j^n\widehat{c}'\|^p, \qquad D_t^n = \sum_{j=3}^{[t/k_n\Delta_n]} a_j'^n.$$

First, (3.47) and Hölder's inequality give us $\mathbb{E}(a_j'^n) \leq K_{q,v}\Delta_n^{l(q,v)}$ for any $q>1$ and $v>0$, where (recalling (2.7) and (2.14) for ϖ and ϖ') we have set $l(q,v) = \frac{1-r\varpi}{q} - \big(p(1-2\varpi)\vee v(1-2\varpi+\varpi')\big)$. Upon choosing v small enough and q close enough to 1, and in view of (2.7), we see that $l(q,v)>\frac{1}{2}$, thus implying

$$\mathbb{E}(D_t^n) \to 0. \tag{3.49}$$

Next, we deduce from (3.48) that

$$\mathbb{E}\big(C_t^{n,\rho}\big) \leq K\mathbb{E}\Big(\mathbb{E}\Big(\sum_{i\in L'(n,\rho,t)} \|\delta_j^n\widehat{c}'\|^p \mid \mathcal{G}^\rho\Big)\Big) \leq Kt\Big(\phi_\rho + \Delta_n^{p/4} + \frac{\Delta_n^{p/2}}{\rho^p}\Big),$$

and thus, since $p\geq 3$,

$$\lim_{\rho\to 0}\limsup_{n\to\infty} \mathbb{E}(|C_t^{n,\rho}|) = 0. \tag{3.50}$$

The analysis of $B_t^{n,\rho}$ is more complicated. We have $\delta_j^n\widehat{c}' = z_j^n + z_j'^n$, where

$$z_j^n = \overline{\alpha}^n_{jk_n+1} - \overline{\alpha}^n_{(j-2)k_n+1}, \quad z_j'^n = \frac{1}{k_n}\sum_{m=1}^{k_n}(c^n_{jk_n+m} - c^n_{(j-2)k_n+m})$$

(recall (3.36) for $\overline{\alpha}_i^n$), hence

$$a_j^n \leq 4(1+\|\widehat{c}'^n_{(j-3)k_n+1}\|)^{p-2}\Big(\|z_j^n\|^2\,1_{\{\|z_i^n\|>u'_n/4\}} + \|z_j'^n\|^2\,1_{\{\|z_i'^n\|>u'_n/4\}}\Big).$$

It easily follows that for all $A > 1$,

$$B_t^{n,\rho} \leq 16\, B_t^{n,\rho,1} + 4A^{p-2}\, B_t^{n,\rho,2} + \frac{2^p}{A}\, B_t^{n,\rho,3}, \tag{3.51}$$

where

$$\begin{array}{c} B_t^{n,\rho,m} = \sum_{j\in L'(n,\rho,t)} a(m)_j^n, \qquad a(1)_j^n = (1+\|\widehat{c}'^n_{(j-3)k_n+1}\|)^{p-2}\, \frac{\|z_j^n\|^3}{u'_n}, \\ a(2)_j^n = \|z_j'^n\|^2\, 1_{\{\|z_i'^n\|>u'_n/4\}}, \quad a(3)_j^n = \|\widehat{c}'^n_{(j-3)k_n+1}\|^{p-1}\, \|z_j'^n\|^2. \end{array}$$

On the one hand, (3.19) and Hölder's inequality yield $\mathbb{E}(a(1)_j^n \mid \mathcal{G}^\rho) \leq K\,\Delta_n^{3/4-\varpi'}$ and, since $\varpi' < \frac{1}{4}$, we deduce

$$\mathbb{E}\big(B_t^{n,\rho,1}\big) \;\to\; 0. \tag{3.52}$$

On the other hand, observe that $z_j'^n = \mu(\rho)_j^n$, with the notation (3.27), and as soon as $j \in L'(n,\rho)$, so Lemma 3.4 gives us

$$\lim_{\rho\to 0}\; \limsup_{n\to\infty}\; \mathbb{E}\big(B_t^{n,\rho,2}\big) \;=\; 0. \tag{3.53}$$

Finally, (3.13) shows that $\mathbb{E}(\|z_j'^n\|^q \mid \mathcal{F}^{n,\rho}_{(j-2)k_n+1}) \leq K_{q,\rho}\sqrt{\Delta_n}$ for all $q \geq 2$ and $j \in L'(n,\rho)$, whereas $\widehat{c}'^n_{(j-3)k_n+1}$ is $\mathcal{F}^n_{(j-2)k_n+1}$-measurable, so (3.13), (3.19) and successive conditioning yield $\mathbb{E}(a(3)_j^n \mid \mathcal{G}^\rho) \leq K_{q,\rho}\sqrt{\Delta_n}$. Then, again as for (3.52), one obtains

$$\mathbb{E}\big(B_t^{n,\rho,3}\big) \;\leq\; K_\rho\, t. \tag{3.54}$$

At this stage, we gather (3.49)–(3.54) and obtain, by letting first $n \to \infty$, then $\rho \to 0$, then $A \to \infty$, that

$$\lim_{\rho\to 0}\; \limsup_{n\to\infty}\; \mathbb{E}\big(|\mathcal{V}_t^{n,\rho}|\big) \;=\; 0. \tag{3.55}$$

(C) The processes $\overline{\mathcal{V}}^{n,\rho}$. With the previous notation S_j^ρ and N_t^ρ, and on the set $\Omega_{n,\rho,t}$, we have

$$\overline{\mathcal{V}}_t^{n,\rho} = \sum_{m=1}^{N_t^\rho} \sum_{j=-2}^{2} v^n_{[S_m^\rho/k_n\Delta_n]+j}\,. \tag{3.56}$$

This is a finite sum (bounded in n for each ω). Letting $S = S_m^\rho$ for m and ρ fixed and $w_n = \frac{S}{k_n\Delta_n} - \big[\frac{S}{k_n\Delta_n}\big]$, we know that for any given $j \in \mathbb{Z}$ the variable $\widehat{c}^n_{([S/k_n\Delta_n]+j)k_n+1}$ converge in probability to c_{S-} if $j < 0$ and to c_S if $j > 0$, whereas for $j = 0$ we have $\widehat{c}^n_{[S/k_n\Delta_n]k_n+1} - w_n c_S - (1-w_n)c_S \stackrel{\mathbb{P}}{\longrightarrow} 0$. This in turn implies

$$j < 0 \text{ or } j > 2 \Rightarrow \delta^n_{[S/k_n\Delta_n]+j}\widehat{c} \stackrel{\mathbb{P}}{\longrightarrow} 0,$$
$$\delta^n_{[S/k_n\Delta_n]}\widehat{c} - (1-w_n)\Delta c_S \stackrel{\mathbb{P}}{\longrightarrow} 0, \quad \delta^n_{[S/k_n\Delta_n]+1}\widehat{c} \stackrel{\mathbb{P}}{\longrightarrow} \Delta c_S, \quad \delta^n_{[S/k_n\Delta_n]+2}\widehat{c} - w_n\Delta c_S \stackrel{\mathbb{P}}{\longrightarrow} 0.$$

By virtue of the definition of v^n_j, and since $u'_n \to 0$ and also since w_n is almost surely in (0, 1) and F is continuous and $F(x,0)=0$, one readily deduces that

$$v^n_{[S/k_n\Delta_n]+j} \stackrel{\mathbb{P}}{\longrightarrow} \begin{cases} F(c_{S-}, \Delta c_S) & \text{if } j = 1 \\ 0 & \text{if } j \neq 1. \end{cases}$$

Coming back to (3.56), we deduce that

$$\overline{\mathcal{V}}^{n,\rho}_t \stackrel{\mathbb{P}}{\longrightarrow} \overline{\mathcal{V}}^{\rho}_t := \sum_{m=1}^{N^\rho_t} F(c_{S^\rho_m-}, \Delta c_{S^\rho_m}). \tag{3.57}$$

In view of (2.16), an application of the dominated convergence theorem gives $\overline{\mathcal{V}}^{\rho}_t \to \mathcal{V}(F)_t$. Then (2.17) follows from $\mathcal{V}(F)^n_t = \mathcal{V}^{n,\rho}_t + \overline{\mathcal{V}}^{n,\rho}_t$ and (3.55) and (3.57). The proof of Theorem 2.6 is complete.

Acknowledgments We are grateful to the referee for his/her very careful reading of the paper.

References

1. Alvarez, A., Panloup, P., Pontier, M., Savy, N.: Estimation of the instantaneous volatility. Stat. Inference Stoch. Process. **15**, 27–59 (2010)
2. Clément, E., Delattre, S., Gloter, A.: An infinite dimensional convolution theorem with applications to the efficient estimation of the integrated volatility. Stoch. Process. Appl. **123**, 2500–2521 (2013)
3. Jacod, J., Shiryaev, A.N.: Limit Theorems for Stochastic Processes, 2nd edn. Springer, Berlin (2003)
4. Jacod, J., Protter, P.: Discretization of Processes. Springer, Berlin (2012)
5. Jacod, J., Rosenbaum, M.: Quarticity and other functionals of volatility: efficient estimation. Ann. Stat. **41**, 1462–1484 (2013)
6. Vetter, M.: Limit theorems for bipower variation of semimartingales. Stoch. Process. Appl. **120**, 22–38 (2010)
7. Dette, H., Podolskij, M., Vetter, M.: Estimation of integrated volatility in continuous-time financial models with applications to goodness-of-fit testing. Scand. J. Stat. **33**(2), 259–278 (2006)

MIX
Papier aus verantwortungsvollen Quellen
Paper from responsible sources
FSC® C105338

If you have any concerns about our products,
you can contact us on
ProductSafety@springernature.com

In case Publisher is established outside the EU,
the EU authorized representative is:
Springer Nature Customer Service Center GmbH
Europaplatz 3, 69115 Heidelberg, Germany

Printed by Libri Plureos GmbH
in Hamburg, Germany